AF595321

Phospholipases: From Structure to Biological Function

Phospholipases: From Structure to Biological Function

Editors

María A. Balboa
Jesús Balsinde

MDPI • Basel • Beijing • Wuhan • Barcelona • Belgrade • Manchester • Tokyo • Cluj • Tianjin

Editors

María A. Balboa
Instituto de Biología y
Genética Molecular (IBGM)
Consejo Superior de
Investigaciones Científicas
(CSIC)
Valladolid
Spain

Jesús Balsinde
Instituto de Biología y
Genética Molecular (IBGM)
Consejo Superior de
Investigaciones Científicas
(CSIC)
Valladolid
Spain

Editorial Office
MDPI
St. Alban-Anlage 66
4052 Basel, Switzerland

This is a reprint of articles from the Special Issue published online in the open access journal *Biomolecules* (ISSN 2218-273X) (available at: www.mdpi.com/journal/biomolecules/special_issues/Phospholipases).

For citation purposes, cite each article independently as indicated on the article page online and as indicated below:

LastName, A.A.; LastName, B.B.; LastName, C.C. Article Title. *Journal Name* **Year**, *Volume Number*, Page Range.

ISBN 978-3-0365-4064-1 (Hbk)
ISBN 978-3-0365-4063-4 (PDF)

© 2022 by the authors. Articles in this book are Open Access and distributed under the Creative Commons Attribution (CC BY) license, which allows users to download, copy and build upon published articles, as long as the author and publisher are properly credited, which ensures maximum dissemination and a wider impact of our publications.
The book as a whole is distributed by MDPI under the terms and conditions of the Creative Commons license CC BY-NC-ND.

Contents

About the Editors

María A. Balboa

María A. Balboa holds a doctorate in Immunology from the Autónoma University of Madrid, Spain. To specialize in lipid signaling, she moved to the U.S. to work in the Department of Pharmacology of the University of California at San Diego, under the supervision of Prof. Paul Insel. Later, she moved to the Department of Chemistry and Biochemistry of the same university to work under the supervision of Prof. Edward Dennis. She returned to Spain as a Spanish National Research Council's Assistant Professor in the University of Valladolid School of Medicine. She is currently a Professor in this institution. Her work focuses on how innate immune cells respond to stimuli by changing their lipid composition. Her recent work has revealed a relationship between the regulation of lipid metabolism and the intensity of the immune response.

Jesús Balsinde

Jesús Balsinde is a Spanish National Research Council (CSIC) Distinguished Professor of Biology and Biomedicine, and an Investigator of the Spanish National Network on Diabetes and Related Metabolic Disorders (CIBERDEM). He received an MS in Chemistry and Biochemistry and a PhD in Biochemistry and Molecular Biology from Complutense University of Madrid, and received postdoctoral training at the University of California, San Diego. He joined the faculty of the University of Valladolid School of Medicine in 2000.

His research focuses on applying mass spectrometry-based lipidomics approaches to the study of the innate immune response in humans and murine animal systems. Particular emphasis is placed on the search for stimulus-specific lipid activation markers whose metabolic pathways of synthesis can provide targets for pharmacological intervention. The phospholipase A2 family of enzymes and the routes for phospholipid fatty incorporation and remodeling are also long standing interests. His work has been continuously supported by state and regional founding agencies, charities, private foundations and the industry, and he is past and present member of various editorial and scientific boards and grant review panels.

Preface to "Phospholipases: From Structure to Biological Function"

The phospholipases regulate much of lipid metabolism and generate numerous lipid products that control cellular signaling. Thus, in a way, lipids could be regarded as the most important biomolecules because, when deregulated, lipid signaling contributes to a wide range of diseases, including cardiovascular and degenerative diseases, and cancer. This book brings together international experts in phospholipase research to provide a comprehensive view of the recent advances in this rapidly moving field. This book is addressed to all the people interested in lipid science, as it covers mechanistic and functional studies at molecular and cellular levels, as well as animal models of disease.

The editors are indebted to all the contributors who have made this book possible and helped to prove again that high-caliber lipid science is alive and well.

María A. Balboa and Jesús Balsinde
Editors

ditorial

’hospholipases: From Structure to Biological Function

1aría A. Balboa [1,2,*] and Jesús Balsinde [1,2,*]

1 Instituto de Biología y Genética Molecular, Consejo Superior de Investigaciones Científicas (CSIC), 47003 Valladolid, Spain

2 Centro de Investigación Biomédica en Red de Diabetes y Enfermedades Metabólicas Asociadas (CIBERDEM), 28029 Madrid, Spain

* Correspondence: mbalboa@ibgm.uva.es (M.A.B.); jbalsinde@ibgm.uva.es (J.B.); Tel.: +34-983-184-833 (M.A.B.); +34-983-423-062 (J.B.)

itation: Balboa, M.A.; Balsinde, J. hospholipases: From Structure to ological Function. *Biomolecules* **2021**, !, 428. https://doi.org/10.3390/ om11030428

eceived: 7 January 2021
ccepted: 26 February 2021
ublished: 15 March 2021

ublisher's Note: MDPI stays neutral ith regard to jurisdictional claims in ublished maps and institutional affil- tions.

opyright: © 2021 by the authors. icensee MDPI, Basel, Switzerland. his article is an open access article stributed under the terms and onditions of the Creative Commons ttribution (CC BY) license (https:// eativecommons.org/licenses/by/ 0/).

Phospholipases are enzymes that cleave ester bonds within phospholipids. As a consequence of these hydrolysis reactions, a variety of lipid products are generated, which control much of cellular signaling. Phospholipase A_2s are key enzymes in this regard due to their role as primary generators of free polyunsaturated fatty acids, which are precursors of various families of compounds playing multiple roles in inflammation [1–3]. Phospholipase A_2s hydrolyze the ester bond at the sn-2 position of the glycerol backbone of phospholipids. Based on their dependence on Ca^{2+} and cellular localization, these enzymes are generally classified into several families [4–6]. Three of these families are the most studied in terms of cellular signaling and lipid mediator production: the Ca^{2+}-dependent cytosolic phospholipase A_2s, the Ca^{2+}-dependent secreted phospholipase A_2s, and the Ca^{2+}-independent phospholipase A_2s. In this respect, the contribution by Murakami et al. [7] provides a timely overview of the functioning of members of these three major phospholipase A_2 families in several pathophysiological conditions, ranging from host defense and metabolism to cancer and skin barrier function.

The Ca^{2+}-dependent group IVA phospholipase A_2, also known as cytosolic phospholipase $A_2\alpha$ (cPL$A_2\alpha$), is a key enzyme in inflammation because of its ability to initiate the production of eicosanoids by preferentially releasing arachidonic acid in response to cell activation. Due to this property, cPL$A_2\alpha$ has been the subject of continuing pharmaceutical interest for the development of anti-inflammatory drugs. However, no phospholipase A_2 inhibitor has yet reached the market. Targeting and inhibiting the phospholipase A_2 reaction has proved problematic since numerous enzymes with phospholipase A_2 activity co-exist in cells with overlapping activation properties, and the lipid metabolic pathways are complex, often redundant, and highly interconnected. Thus, the search for improved formulations continues. In this context, the work by Koutoulogenis et al. [8] shows that modifying a 2-oxoester inhibitor of cPL$A_2\alpha$ by inserting a methyl group on the α-carbon atom of the oxoester results in molecules with enhanced metabolic stability that retain considerable inhibitor potency against cPL$A_2\alpha$. This increases the potential for pharmaceutical development of this class of inhibitors. In another study, Ashcroft et al. [9] show that the anti-psoriatic effects of inhibiting cPL$A_2\alpha$ with the selective inhibitor AVX001, an n-3 polyunsaturated fatty acid derivative, results from a combination of anti-inflammatory and anti-proliferative effects. Thus, the therapeutic mode of action of AVX001 in psoriasis could depend both on reducing inflammatory eicosanoid production and inhibition of the hyperproliferative state of keratinocytes.

The group V secreted phospholipase A_2 is known to participate in a number of functions of macrophages exposed to interleukin 4 (IL-4), which polarizes the cells to an anti-inflammatory phenotype. Koganesawa et al. [10] performed a mass spectrometry analysis of phospholipase A_2 products and substrates using macrophages deficient or not in the group V enzyme, and polarized to either anti-inflammatory (IL-4-treated) or pro-inflammatory (bacterial lipopolysaccharide plus interferon-γ) phenotypes. The authors identify selective changes between the different activation regimes, suggesting the existence of novel lipid pathways and functions critical for inflammation, which may rely on the group V enzyme.

Secreted phospholipase A_2s often act in concert with cPLA$_2\alpha$ to elicit certain biological responses [11,12]. However, the molecular mechanisms involved in this cross-talk are still poorly defined. Using mass-spectrometry-based lipidomics, Rodríguez et al. [13 characterized the human monocyte response to a group IIA secreted phospholipase A from the venom of the Central American snake *Bothrops asper*. The data reveal significant interactions between the secreted phospholipase A_2 and the cPLA$_2\alpha$ of the monocytes in terms of lipid droplet biogenesis, eicosanoid responses, and biochemical pathways, which contribute to initiating the inflammatory response. Using the same secreted phospholipase A_2 as a prototypical group IIA enzyme, Leiguez et al. [14] show that the enzyme can directly activate adipocytes to release prostaglandin E_2 via a pathway involving cPLA$_2\alpha$ activation, which in turn mediates the production of various adipokines. Thus, these findings increase our understanding of the role of group IIA secreted phospholipase A_2 in obesity and associated metabolic disorders.

Crotoxin is a secreted phospholipase A_2 abundantly present in the venom of the South American rattlesnake *Crotalus durissus terrificus*. Crotoxin is responsible for many of the pathological conditions associated with snakebite, including neurotoxicity, myotoxicity and immune alterations. Sartim et al. [15] describe the effects of crotoxin on respiratory failure, and introduce the use of nicotinic blockers and prostanoid biosynthesis inhibitors as possible therapeutic agents to counteract rattlesnake envenomation.

Determining the involvement of the Ca^{2+}-independent group VIA phospholipase A (commonly referred to as Ca^{2+}-independent phospholipase $A_2\beta$ or iPLA$_2\beta$) in physiology and pathophysiology has been shown to be very complex. Thanks to the generation of knock-out mice for this enzyme [16], its involvement in a variety of pathophysiological conditions could be established. In this regard, Chamulitrat et al. [17] discuss recent data with iPLA$_2\beta$-deficient mice, implicating the enzyme in the development of hepatic steatosis and inflammation in obese and non-obese murine models. Likewise, Turk et al. [18 describe the production of mice with selective deficiency of iPLA$_2\beta$ in macrophages or insulin-secreting β cells, and the contrasting metabolic phenotype exhibited by these two kinds of cells. In studies with cultured cells, Monge et al. [19] show that in addition to cPLA$_2\alpha$, iPLA$_2\beta$ is instrumental for activated cells to mobilize adrenic acid, the two-carbon chain elongation product of arachidonic acid. As with arachidonic acid, free adrenic acid can be metabolized to a number of oxygenated bioactive metabolites. Thus, adrenic acid mobilization shares regulatory features with arachidonic acid mobilization regarding cPLA$_2\alpha$ involvement, but seems to be a more complex process, as it involves participation of a second enzyme that is not involved in arachidonate release, i.e., iPLA$_2\beta$.

Phospholipid phosphatases have recently received considerable attention because of their involvement in a number of pathophysiological states. Although not immediately recognized by some, phospholipid phosphatases are also phospholipases in their own right as they hydrolyze the ester bond between the glycerol/sphingosine and phosphate moieties of certain phospholipids (i.e., a phospholipase C-type cleavage). Some have a very strict substrate specificity, such as the lipins, which hydrolyze phosphatidic acid only, whereas others are less specific, being able to use multiple substrates. The latter are known as lipid phosphate phosphatases, of which there are three different groups. Tang and Brindley provide a comprehensive view of what is currently known about the role of lipid phosphate phosphatases in cancer [20]. The authors discuss the structure and cellular localization of lipid phosphate phosphatases, as well as their expression levels and effects on different cancers. The authors also review the impact of two major lipid phosphate phosphatase substrates, i.e., lysophosphatidic acid and sphingosine 1-phosphate, on different aspects of tumor development.

Phosphatidic acid, the substrate of lipins, is a fundamental intermediate for the synthesis of triacylglycerol and all glycerophospholipids, and plays key roles in intracellular and intercellular signaling. The contribution of Lutkewitte and Finck [21] focuses on how the lipin-mediated control of phosphatidic acid concentrations regulates metabolism and signaling in the heart, skeletal muscle, liver, and adipose tissue. Schilke et al. [22] describe the contribution of both lipid phosphate phosphatases and lipin-1 to generate pro-inflammatory and/or pro-resolving lipid products during atherosclerosis that may influence disease progression or regression. This work is not limited to discussing the roles of phospholipid phosphatases; it also includes information on the role of other phospholipases, i.e., various members of the phospholipase A_2 family, phospholipase C, and phospholipase D.

Majeed syndrome is a rare autoinflammatory disease that develops in patients who carry loss of function mutations in the *LPIN2* gene, which encodes for lipin-2. Ferguson and El-Shanti [23] present a comprehensive and up-to-date review of this disorder, emphasizing clinical features, genetics, pathogenesis, and cellular and molecular mechanisms. The latter include exacerbated activation of the NLPR3 inflamasome in macrophages, as well as increased production of osteoclastogenic mediators by M2-polarized macrophages.

Sphingomyelinases are a special group of phospholipase C-type enzymes that specifically hydrolyze sphingomyelin to generate ceramide, a potent inductor of apoptosis. Cells can evade apoptosis by converting ceramide to sphingosine-1-phosphate. In this regard, Hawkins et al. [24] discuss the role of sphingomyelinases and sphingosine 1-phosphate in glioblastoma and metastatic brain tumors. There are two main types of sphingomyelinases: neutral and acid. Both have been well-characterized for their contribution to signaling pathways and their roles in diverse pathologies, including liver diseases. Insausti-Urkia et al. [25] summarize the physiological functions of neutral and acid sphingomyelinases and their role in chronic and metabolic liver diseases.

Collectively, the papers included in this Special Issue provide forefront information related to the role of phospholipases in a number of physiological and pathophysiological settings. It is expected that these novel and exciting findings will help stimulate further interest in the highly competitive and rapidly moving field of lipid signaling.

Funding: Work in the authors' laboratories was funded by the Spanish Ministry of Science and Innovation, grant numbers SAF2016-80883-R and PID2019-105989RB-I00. CIBERDEM is an initiative of Instituto de Salud Carlos III.

Conflicts of Interest: The authors declare no conflict of interest.

References

1. Astudillo, A.M.; Balboa, M.A.; Balsinde, J. Selectivity of phospholipid hydrolysis by phospholipase A_2 enzymes in activated cells leading to polyunsaturated fatty acid mobilization. *Biochim. Biophys. Acta* **2019**, *1864*, 772–783. [CrossRef]
2. Astudillo, A.M.; Balgoma, D.; Balboa, M.A.; Balsinde, J. Dynamics of arachidonic acid mobilization by inflammatory cells. *Biochim. Biophys. Acta* **2012**, *1821*, 249–256. [CrossRef] [PubMed]
3. Serhan, C.N. Pro-resolving lipid mediators are leads for resolution physiology. *Nature* **2014**, *510*, 92–101. [CrossRef] [PubMed]
4. Mouchlis, V.D.; Dennis, E.A. Phospholipase A_2 catalysis and lipid mediator lipidomics. *Biochim. Biophys. Acta* **2019**, *1864*, 766–771. [CrossRef] [PubMed]
5. Guijas, C.; Rodríguez, J.P.; Rubio, J.M.; Balboa, M.A.; Balsinde, J. Phospholipase A_2 regulation of lipid droplet formation. *Biochim. Biophys. Acta* **2014**, *1841*, 1661–1671. [CrossRef]
6. Murakami, M. Novel functions of phospholipase A_2s. *Biochim. Biophys. Acta* **2019**, *1864*, 763–765. [CrossRef] [PubMed]
7. Murakami, M.; Sato, H.; Taketomi, Y. Updating phospholipase A_2 biology. *Biomolecules* **2020**, *10*, 1457. [CrossRef] [PubMed]
8. Koutoulogenis, G.S.; Kokotou, M.G.; Hayashi, D.; Mouchlis, V.D.; Dennis, E.A.; Kokotos, G. 2-Oxoester phospholipase A_2 inhibitors with enhanced metabolic stability. *Biomolecules* **2020**, *10*, 491. [CrossRef]
9. Ashcroft, F.J.; Mahammad, N.; Midtun Flatekvål, H.; Jullumstrø Feuerherm, A.; Johansen, B. cPLA$_2\alpha$ enzyme inhibition attenuates inflammation and keratinocyte proliferation. *Biomolecules* **2020**, *10*, 1402. [CrossRef]
10. Koganesawa, M.; Yamaguchi, M.; Samuchiwal, S.K.; Balestrieri, B. Lipid profile of activated macrophages and contribution of group V phospholipase A_2. *Biomolecules* **2021**, *11*, 25. [CrossRef]
11. Balboa, M.A.; Pérez, R.; Balsinde, J. Amplification mechanisms of inflammation: Paracrine stimulation of arachidonic acid mobilization by secreted phospholipase A_2 is regulated by cytosolic phospholipase A_2-derived hydroperoxyeicosatetraenoic acid. *J. Immunol.* **2003**, *171*, 989–994. [CrossRef]

12. Ruipérez, V.; Astudillo, M.A.; Balboa, M.A.; Balsinde, J. Coordinate regulation of TLR-mediated arachidonic acid mobilization in macrophages by group IVA and group V phospholipase A_2s. *J. Immunol.* **2009**, *182*, 3877–3883. [CrossRef]
13. Rodríguez, J.P.; Leiguez, E.; Guijas, C.; Lomonte, B.; Gutiérrez, J.M.; Teixeira, C.; Balboa, M.A.; Balsinde, J. A lipidomic perspective of the action of group IIA secreted phospholipase A_2 on human monocytes: Lipid droplet biogenesis and activation of cytosolic phospholipase $A_2\alpha$. *Biomolecules* **2020**, *10*, 891. [CrossRef]
14. Leiguez, E.; Motta, P.; Marques, R.M.; Lomonte, B.; Sampaio, S.V.; Teixeira, C. A representative group IIA phospholipase A_2 activates preadipocytes to produce inflammatory mediators implicated in obesity development. *Biomolecules* **2020**, *10*, 1593. [CrossRef] [PubMed]
15. Sartim, M.A.; Souza, C.O.S.; Diniz, C.R.A.F.; da Fonseca, V.M.B.; Sousa, L.O.; Peti, A.P.F.; Costa, T.R.; Lourenço, A.G.; Borges, M.C.; Sorgi, C.A.; et al. Crotoxin-induced mice lung impairment: Role of nicotinic acetylcholine receptors and COX-derived prostanoids. *Biomolecules* **2020**, *10*, 794. [CrossRef] [PubMed]
16. Turk, J.; White, T.D.; Nelson, A.J.; Lei, X.; Ramanadham, S. iPLA$_2\beta$ and its role in male fertility, neurological disorders, metabolic disorders, and inflammation. *Biochim. Biophys. Acta* **2019**, *1864*, 846–860. [CrossRef] [PubMed]
17. Chamulitrat, W.; Jansakun, C.; Li, H.; Liebisch, G. Rescue of hepatic phospholipid remodeling defect in iPLA$_2\beta$-null mice attenuates obese but not non-obese fatty liver. *Biomolecules* **2020**, *10*, 1332. [CrossRef] [PubMed]
18. Turk, J.; Song, H.; Wohltmann, M.; Frankfater, C.; Lei, X.; Ramanadham, S. Metabolic effects of selective deletion of group VIA phospholipase A_2 from macrophages or pancreatic islet beta-cells. *Biomolecules* **2020**, *10*, 1455. [CrossRef] [PubMed]
19. Monge, P.; Garrido, A.; Rubio, J.M.; Magrioti, V.; Kokotos, G.; Balboa, M.A.; Balsinde, J. The contribution of cytosolic group IVA and calcium-Independent group VIA phospholipase A_2s to adrenic acid mobilization in murine macrophages. *Biomolecules* **2020**, *10*, 542. [CrossRef] [PubMed]
20. Tang, X.; Brindley, D.N. Lipid phosphate phosphatases and cancer. *Biomolecules* **2020**, *10*, 1263. [CrossRef] [PubMed]
21. Lutkewitte, A.J.; Finck, B.N. Regulation of signaling and metabolism by lipin-mediated phosphatidic acid phosphohydrolase activity. *Biomolecules* **2020**, *10*, 1386. [CrossRef] [PubMed]
22. Schilke, R.M.; Blackburn, C.M.R.; Bamgbose, T.T.; Woolard, M.D. Interface of phospholipase activity, immune cell function, and atherosclerosis. *Biomolecules* **2020**, *10*, 1449. [CrossRef] [PubMed]
23. Ferguson, P.J.; El-Shanti, H. Majeed Syndrome: A Review of the Clinical, Genetic and Immunologic Features. *Biomolecules* **2021**, *11*, 367. [CrossRef] [PubMed]
24. Hawkins, C.C.; Ali, T.; Ramanadham, S.; Hjelmeland, A.B. Sphingolipid metabolism in glioblastoma and metastatic brain tumors: A review of sphingomyelinases and sphingosine-1-phosphate. *Biomolecules* **2020**, *10*, 1357. [CrossRef]
25. Insausti-Urkia, N.; Solsona-Vilarrasa, E.; García-Ruiz, C.; Fernández-Checa, J.C. Sphingomyelinases and liver diseases. *Biomolecules* **2020**, *10*, 1497. [CrossRef] [PubMed]

Review

Updating Phospholipase A_2 Biology

Makoto Murakami *, Hiroyasu Sato and Yoshitaka Taketomi

Laboratory of Microenvironmental and Metabolic Health Sciences, Center for Disease Biology and Integrative Medicine, Graduate School of Medicine, The University of Tokyo, Bunkyo-ku, Tokyo 113-8655, Japan; sato-hr@m.u-tokyo.ac.jp (H.S.); taketomiys@m.u-tokyo.ac.jp (Y.T.)

* Correspondence: makmurak@m.u-tokyo.ac.jp; Tel.: +81-3-5841-1431

Received: 24 September 2020; Accepted: 15 October 2020; Published: 19 October 2020

Abstract: The phospholipase A_2 (PLA_2) superfamily contains more than 50 enzymes in mammals that are subdivided into several distinct families on a structural and biochemical basis. In principle, PLA_2 has the capacity to hydrolyze the *sn*-2 position of glycerophospholipids to release fatty acids and lysophospholipids, yet several enzymes in this superfamily catalyze other reactions rather than or in addition to the PLA_2 reaction. PLA_2 enzymes play crucial roles in not only the production of lipid mediators, but also membrane remodeling, bioenergetics, and body surface barrier, thereby participating in a number of biological events. Accordingly, disturbance of PLA_2-regulated lipid metabolism is often associated with various diseases. This review updates the current state of understanding of the classification, enzymatic properties, and biological functions of various enzymes belonging to the PLA_2 superfamily, focusing particularly on the novel roles of PLA_2s in vivo.

Keywords: fatty acid; knockout mouse; lipid mediator; lipidomics; lysophospholipid; membrane; phospholipase A_2; phospholipid

1. Introduction

Based on their structural relationships, the PLA_2 superfamily is classified into several families, including the secreted PLA_2 ($sPLA_2$), cytosolic PLA_2 ($cPLA_2$), Ca^{2+}-independent PLA_2 ($iPLA_2$, also called patatin-like phospholipase (PNPLA)), platelet-activating factor acetylhydrolase (PAF-AH), lysosomal PLA_2 ($LPLA_2$), PLA/acyltransferase (PLAAT), α/β hydrolase (ABHD), and glycosylphosphatidylinositol (GPI)-specific PLA_2 families. PLA_2s trigger the production of lipid mediators by releasing polyunsaturated fatty acids (PUFAs) and lysophospholipids from membrane phospholipids, and also participate in membrane homeostasis by altering phospholipid composition, in energy production by supplying fatty acids for β-oxidation, in generation of barrier lipids, or in fine-tuning of the microenvironmental balance between saturated and unsaturated fatty acids, among others. Many of the PLA_2 enzymes recognize the differences in the fatty acyl and/or head group moieties of their substrate phospholipids, and several enzymes catalyze even non-PLA_2 reactions such as phospholipase A_1 (PLA_1), lysophospholipase, neutral lipid lipase, and transacylase reactions. The in vivo functions of individual PLA_2s rely on their enzymatic, biochemical, and cell biological properties, their tissue and cellular distributions, lipid composition in target membranes, the spatiotemporal availability of downstream lipid-metabolizing enzymes, or the presence of cofactor(s) that can modulate the enzymatic function, in various pathophysiological settings.

During the last two decades, the functions of various PLA_2s have been clarified by studies based on not only gene-manipulated (knockout and transgenic) mice but also human diseases caused by mutations of these enzymes. Here, we provide an overview of the biological roles of various PLA_2s and their underlying lipid pathways, focusing mainly on new findings in the last five years. Readers interested in older views as a starting point for further readings should refer to our current reviews describing the classification of the PLA_2 superfamily [1], those covering PLA_2s and lipid mediators broadly [2,3], and those focusing on $sPLA_2$s [4–7].

2. The $sPLA_2$ Family

2.1. General Features

The $sPLA_2$ family comprises low-molecular-mass, Ca^{2+}-requiring enzymes with a conserved His-Asp catalytic dyad. There are 11 mammalian $sPLA_2$s (IB, IIA, IIC, IID, IIE, IIF, III, V, X, XIIA, and XIIB), which are structurally subdivided into group I/II/V/X, group III, and group XII branches [8]. Individual $sPLA_2$s exhibit unique tissue or cellular distributions and enzymatic properties, suggesting their distinct biological roles. With regard to the substrate specificity of $sPLA_2$s as assessed by an assay using tissue-extracted natural membranes, $sPLA_2$-IB, -IIA and -IIE do not discriminate *sn*-2 fatty acid species, $sPLA_2$-V tends to prefer those with a lower degree of unsaturation such as oleic acid (OA; C18:1) and linoleic acid (LA; C18:2), and $sPLA_2$-IID, -IIF, -III and -X tend to prefer PUFAs such as ω6 arachidonic acid (AA; C20:4) and ω3 docosahexaenoic acid (DHA; C22:6). With regard to the polar head groups, $sPLA_2$-III, -V and -X efficiently hydrolyze phosphatidylcholine (PC), while $sPLA_2$s in the group II subfamily hydrolyze phosphatidylethanolamine (PE) much better than PC. Individual $sPLA_2$s exert their specific functions by producing lipid mediators, by altering membrane phospholipid composition, by degrading foreign phospholipids in microorganisms or diets, or by modifying extracellular non-cellular lipid components such as lipoproteins, pulmonary surfactant or microvesicles in response to given microenvironmental cues. In certain cases, the $sPLA_2$-binding protein PLA2R1 modulates the functions of $sPLA_2$s in either a positive or negative way. The pathophysiological roles of individual $sPLA_2$s, as revealed by studies using $sPLA_2$ knockout or transgenic mice in combination with comprehensive lipidomics, have been detailed in several recent reviews [3,6,7,9].

2.2. $sPLA_2$-IB in Digestion and Immunity

$sPLA_2$-IB (encoded by *PLA2G1B* in human) is synthesized as an inactive zymogen in the pancreas, and its *N*-terminal propeptide is cleaved by trypsin to yield an active enzyme in the duodenum [10]. The main role of $sPLA_2$-IB, a "pancreatic $sPLA_2$", is to digest dietary and biliary phospholipids in the intestinal lumen. Perturbation of this process by gene disruption or pharmacological inhibition of $sPLA_2$-IB leads to resistance to diet-induced obesity, insulin resistance, and atherosclerosis due to decreased phospholipid digestion and absorption in the gastrointestinal tract [11–14]. Indeed, the human *PLA2G1B* gene maps to an obesity susceptibility locus [15]. These functions of $sPLA_2$-IB have been summarized in a recent review [16].

Beyond the well-established role of $sPLA_2$-IB as a "digestive $sPLA_2$" as outlined above, two recent studies have uncovered novel immunological functions of this $sPLA_2$. Entwistle et al. [17] showed that $sPLA_2$-IB is induced in a population of intestinal epithelial cells during helminth infection and is responsible for killing tissue-embedded larvae. *Pla2g1b*$^{-/-}$ mice fail to expel the intestinal helminths *Heligmosomoides polygyrus* and *Nippostrongylus brasiliensis*. Treatment of the parasite with $sPLA_2$-IB hydrolyzes worm phospholipids (e.g., PE) and impairs development to the adult stage, suggesting that exposure to this enzyme in the intestine is an important mechanism of host-mediated defense against such parasites.

Pothlichet et al. [18] reported that $sPLA_2$-IB is involved in $CD4^+$ T cell lymphopenia in patients infected with human immunodeficiency virus (HIV). $sPLA_2$-IB, in synergy with the HIV gp41 envelope protein, induces $CD4^+$ T cell anergy, inhibiting the responses to IL-2, IL-4, and IL-7 as well as activation, proliferation, and survival of $CD4^+$ T cells. Other $sPLA_2$s fail to display a similar function, implying a specific action of $sPLA_2$-IB. Importantly, the effects of HIV on $CD4^+$ T cell anergy can be blocked by a $sPLA_2$-IB-specific neutralizing antibody in vivo. Thus, the $sPLA_2$-IB/gp41 pair constitutes a new mechanism of immune dysfunction, although the cellular source of plasma $sPLA_2$-IB in this context remains to be determined.

2.3. $sPLA_2$-IIA in Host Defense, Sterile Inflammation, and Colon Cancer

The best-known physiological function of $sPLA_2$-IIA (encoded by *PLA2G2A*) is the degradation of bacterial membranes, particularly those in Gram-positive bacteria, thereby providing the first line of antimicrobial defense as a "bactericidal $sPLA_2$" [19,20]. The ability of $sPLA_2$-IIA to hydrolyze PE and phosphatidylglycerol, which are abundant in bacterial membranes, appears to fit with its anti-bacterial function. In the lungs of patients with cystic fibrosis, $sPLA_2$-IIA-resistant Gram-negative *Pseudomonas aeruginosa* upregulates the expression of $sPLA_2$-IIA, which then eradicates $sPLA_2$-IIA-sensitive Gram-positive *Staphylococcus aureus*, allowing the former bacterium to become dominant within the niche [21]. Thus, $sPLA_2$-IIA-mediated regulation of the bacterial community in the lung microenvironment crucially affects the pathology of cystic fibrosis.

$sPLA_2$-IIA is often referred to as an "inflammatory $sPLA_2$", as its expression is induced by pro-inflammatory cytokines and lipopolysaccharide (LPS) [22]. Besides its action on bacterial membranes as noted above, $sPLA_2$-IIA targets phospholipids in extracellular microvesicles (EVs), particularly those in extracellular mitochondria (organelles that have evolved from bacteria), which are released from activated platelets or leukocytes at sites of inflammation [23]. Hydrolysis of EV phospholipids by $sPLA_2$-IIA results in the production of lipid mediators as well as the release of mitochondrial DNA as a danger-associated molecular pattern (DAMP) that contributes to amplification of inflammation. Specifically, the AA released from platelet-derived EVs by $sPLA_2$-IIA is metabolized by platelet-type 12-lipoxygenase to 12-hydroxyeicosatetraenoic acid (HETE), which then acts on the BLT2 receptor on neutrophils facilitating the uptake of the EVs [24]. Thus, $sPLA_2$-IIA is primarily involved in host defense by killing bacteria and triggering innate immunity, while over-amplification of the response leads to exacerbation of inflammation by hydrolyzing EVs (Figure 1). These dual functions of $sPLA_2$-IIA were summarized in a recent review [25].

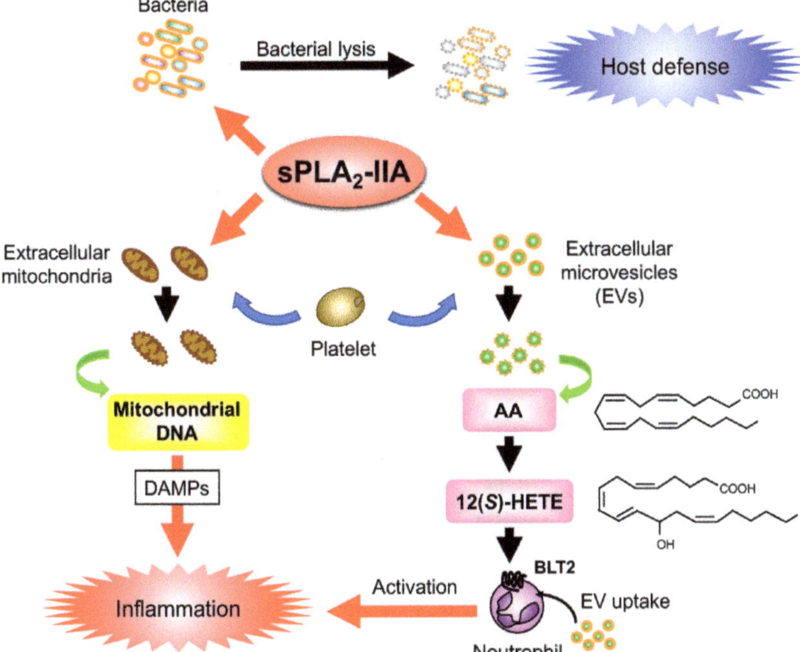

Figure 1. The roles of $sPLA_2$-IIA in anti-bacterial defense by degrading bacterial membrane and in sterile inflammation by releasing pro-inflammatory eicosanoids from extracellular microvesicles (EVs) derived from inflammatory cells.

$sPLA_2$-IIA also appears to play a role in host defense against the malaria pathogen *Plasmodium falciparum*. Several $sPLA_2$s, including $sPLA_2$-IIF, -V and -X, which efficiently hydrolyze plasma lipoproteins to release free fatty acids, have the capacity to inhibit parasite growth in vitro, yet these $sPLA_2$s are undetectable in human plasma. $sPLA_2$-IIA, though hardly hydrolyzing normal lipoproteins, is increased in the plasma of malaria patients and hydrolyzes "oxidized" lipoproteins to block *Plasmodium* growth [26]. Injection of recombinant $sPLA_2$-IIA into *Plasmodium*-infected mice reduces the peak of parasitemia when the level of plasma peroxidation is increased during infection. Thus, malaria-induced oxidation of lipoproteins converts them into a preferential substrate for $sPLA_2$-IIA, thus promoting its parasite-killing effect.

PLA2G2A-transgenic mice display notable skin abnormalities with hair loss and epidermal hyperplasia [27]. Using *K14*-driven, skin-specific *PLA2G2A*-transgenic mice, Chovatiya et al. [28] recently demonstrated that $sPLA_2$-IIA promotes hair follicle stem cell proliferation through JNK signaling and that $sPLA_2$-IIA-knockdown skin cancers xenografted into NOD-SCID mice show a concomitant reduction of tumor volume and decreased JNK signaling. Kuefner et al. [29] showed that *PLA2G2A*-transgenic mice are protected from diet-induced obesity and become more prone to adipocyte browning with increased expression of thermogenic markers. However, since C57BL/6 mice do not express $sPLA_2$-IIA endogenously due to a frameshift mutation in the *Pla2g2a* gene [30], the physiological relevance of these results obtained from transgenic overexpression of $sPLA_2$-IIA in this mouse strain should be interpreted with caution. The increased energy expenditure in *PLA2G2A*-transgenic mice might be simply due to their lack of fur, which results in increased heat dissipation from the body surface. Indeed, transgenic mice overexpressing $sPLA_2$-IIF or $sPLA_2$-X are also furless and lean [31]. Studies using a series of $sPLA_2$-knockout mice have suggested that $sPLA_2$-IIF and $sPLA_2$-IID are the main endogenous regulators of skin homeostasis and adipocyte browning, respectively, as described below.

Although $sPLA_2$-IIA has been identified as a genetic modifier of mouse intestinal tumorigenesis [30], the underlying mechanism has long remained unclear. Schewe et al. [32] recently provided a potential mechanism by which $sPLA_2$-IIA, expressed by Paneth cells in the small intestine, suppresses colon cancer. In a normal state, $sPLA_2$-IIA inhibits Wnt signaling through intracellular activation of Yap1. Upon inflammation, $sPLA_2$-IIA is secreted into the intestinal lumen, where it promotes inflammation through prostaglandin (PG) E_2 synthesis via the PLA2R1-$cPLA_2\alpha$ pathway and Wnt signaling. Transgenic overexpression of $sPLA_2$-IIA delays recovery from colonic inflammation but decreases colon cancer susceptibility due to perturbation of its homeostatic Wnt-inhibitory function. Thus, this trade-off effect could provide a mechanism whereby $sPLA_2$-IIA acts as a genetic modifier of colonic inflammation and cancer. However, the inflammatory $sPLA_2$-IIA-PLA2R1-$cPLA_2\alpha$-PGE_2 signaling axis proposed in this study may require further clarification, since contrary to this hypothesis, there is ample evidence that $cPLA_2\alpha$ and its product PGE_2 contribute to attenuation of colitis and promotion of colon cancer [33]. Other factors, such as intestinal dysbiosis due to loss of $sPLA_2$-IIA, might primarily underlie the increased susceptibility to gastrointestinal cancer—a possibility that awaits future study.

Apart from its potential signaling role, PLA2R1 acts as a clearance receptor for $sPLA_2$s [8,34]. In a model of experimental autoimmune myocarditis induced by immunizing BALB/c mice, a strain possessing a normal *Pla2g2a* gene, with the murine α-myosin heavy chain [30], PLA2R1 deficiency markedly increases $sPLA_2$-IIA and -IB proteins (but not mRNAs) in the myocardium, probably as a result of their impaired clearance [35]. In the affected myocardium, PLA2R1 and these two $sPLA_2$s are localized in α-SMA$^+$ myofibroblasts and infiltrating neutrophils, respectively. The *Pla2r1*$^{-/-}$ myocardium shows increased areas of inflammatory cell infiltration, accompanied by an increase in PGE_2, which promotes IL-23-induced expansion of Th17 cells. Thus, it appears that this increase of $sPLA_2$-IIA and -IB proteins due to their defective clearance contributes to exacerbation of autoimmune myocarditis, although it remains unclear whether these $sPLA_2$s act through PGE_2 synthesis or through other mechanisms, whether some other $sPLA_2$s that are expressed in the myocardium and have the capacity to bind to PLA2R1 are also involved in this process, or whether the effect of PLA2R1 ablation is $sPLA_2$-independent.

2.4. $sPLA_2$-IID in Immunosuppression, Host Defense, and Adaptive Thermogenesis

$sPLA_2$-IID (encoded by *PLA2G2D*), constitutively expressed in dendritic cells (DCs) in lymphoid organs, is a "resolving $sPLA_2$" that attenuates DC-mediated adaptive immunity by hydrolyzing PE to mobilize anti-inflammatory ω3 PUFAs and their metabolites such as DHA-derived resolvin D1 [36]. *Pla2g2d*$^{-/-}$ mice exhibit more severe contact hypersensitivity (Th1 response) and psoriasis (Th17 response), accompanied by marked reductions of ω3 PUFAs and their metabolites in the draining lymph nodes and spleen. On the other hand, *Pla2g2d*$^{-/-}$ mice are protected against skin cancer likely because of the enhanced anti-tumor immunity in association with increased IFN-γ^{+}CD8^{+} T cells [37]. The immunosuppressive role of $sPLA_2$-IID in the Th1, Th2, and Th17 responses was detailed in a recent review [9].

Chronic low-grade inflammation is associated with age-related immune dysfunction in the lung, and is countered by enhanced expression of pro-resolving/anti-inflammatory factors to maintain tissue homeostasis. Vijay et al. [38] reported that $sPLA_2$-IID with an anti-inflammatory property contributes to worse outcomes in mice infected with severe acute respiratory syndrome coronavirus (SARS-CoV) or influenza A virus. Strikingly, *Pla2g2d*$^{-/-}$ mice are highly protected against SARS-CoV infection, with enhanced migration of DCs to the lymph nodes, augmented anti-viral T cell responses, reduced lung damage, and increased survival. In the context of the current worldwide SARS-CoV2 pandemic, inhibition of $sPLA_2$-IID in the lungs of older patients with severe respiratory infections could be a potentially attractive therapeutic intervention for restoration of immune function.

ω3 PUFAs confer health benefits by preventing inflammation and obesity and by increasing thermogenesis in brown and beige adipocytes. $sPLA_2$-IID is constitutively expressed in M2 macrophages in white adipose tissue (WAT) and downregulated during obesity [39]. Global or macrophage-specific $sPLA_2$-IID deficiency decreases energy expenditure and thermogenesis by preventing adipocyte browning, thus exacerbating diet-induced obesity, insulin resistance, and WAT inflammation [39]. In WAT, PLA2G2D constitutively supplies a pool of ω3 PUFAs, which acts on the PUFA receptor GPR120 and thereby promotes the thermogenic program as a "thermogenic $sPLA_2$". Importantly, dietary supplementation with ω3 PUFAs normalizes the metabolic derangement in *Pla2g2d*$^{-/-}$ mice. These findings highlight the contribution of the macrophage-driven PLA2G2D-ω3 PUFA axis to metabolic health (Figure 2). Possibly in relation to this, a polymorphism in the human *PLA2G2D* gene is linked to body weight changes in patients with chronic lung disease [40].

2.5. $sPLA_2$-IIE and $sPLA_2$-IIF in the Skin

$sPLA_2$-IIE (encoded by *PLA2G2E*) is hardly detected in human tissues, whereas $sPLA_2$-IIE instead of $sPLA_2$-IIA is upregulated in several mouse tissues under inflammatory or other conditions. For instance, $sPLA_2$-IIE is highly upregulated in adipocytes during diet-induced obesity [41], and is expressed in hair follicles in correlation with the growth phase of the hair cycle [42]. $sPLA_2$-IIE hydrolyzes PE without apparent fatty acid selectivity in lipoproteins and hair follicles, and *Pla2g2e*$^{-/-}$ mice display modest metabolic and hair follicle abnormalities.

$sPLA_2$-IIF (encoded by *PLA2G2F*) is an "epidermal $sPLA_2$" expressed in the suprabasal epidermis and upregulated by the Th17 cytokines IL-17A and IL-22 in psoriatic skin [31]. $sPLA_2$-IIF preferentially hydrolyzes PUFA-containing plasmalogen-type PE to produce lysoplasmalogen (plasmalogen-type lysophosphatidylethanolamine; P-LPE), which in turn promotes epidermal hyperplasia. Accordingly, *Pla2g2f*$^{--}$ mice are protected against psoriasis and skin cancer, while *Pla2g2f*-transgenic mice spontaneously develop psoriasis-like skin and are more susceptible to skin cancer [31]. Overall, two skin $sPLA_2$s, $sPLA_2$-IIE in the outer root sheath of hair follicles and $sPLA_2$-IIF in epidermal keratinocytes, play non-redundant roles in distinct compartments of mouse skin, underscoring the functional diversity of multiple $sPLA_2$s in the coordinated regulation of skin homeostasis and diseases. The roles of $sPLA_2$-IIE and -IIF were detailed in a recent review [9].

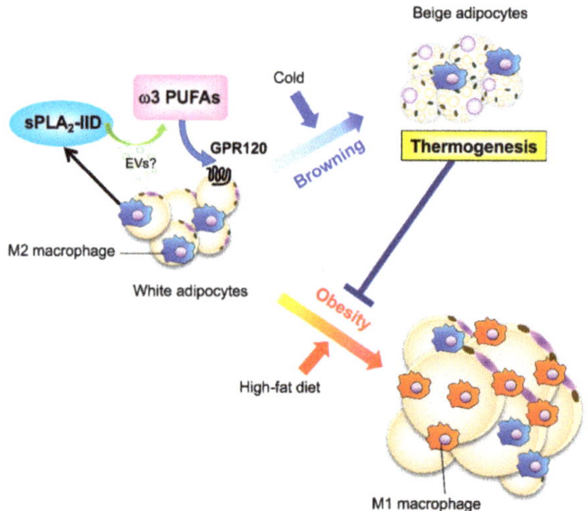

Figure 2. The role of $sPLA_2$-IID expressed in M2 macrophages in adipocyte browning and adaptive thermogenesis. $sPLA_2$-IID releases ω3 PUFAs, which then act on GPR120 to drive the thermogenic and anti-inflammatory programs toward metabolic health. Impairment of this $sPLA_2$-IID-driven lipid pathway leads to impaired thermogenesis and exacerbated diet-induced obesity.

2.6. $sPLA_2$-III in Male Reproduction, Anaphylaxis, and Colonic Diseases

$sPLA_2$-III (encoded by *PLA2G3*) consists of three domains, in which the central $sPLA_2$ domain similar to bee venom group III $sPLA_2$ is flanked by large and unique *N*- and *C*-terminal domains [43]. The enzyme is processed to the $sPLA_2$ domain-only form that retains full enzymatic activity [44]. $sPLA_2$-III is expressed in the epididymal epithelium and acts on immature sperm cells passing through the epididymal duct in a paracrine manner to allow sperm membrane phospholipid remodeling, a process that is prerequisite for sperm motility [45]. Homozygous and even heterozygous *Pla2g3*-deficient sperm have impaired motility and thereby fail to fertilize oocytes, leading to hypofertility. In the context of allergy, $sPLA_2$-III secreted from immature mast cells is functionally coupled with lipocalin-type PGD_2 synthase (L-PGDS) in neighboring fibroblasts to supply a microenvironmental pool of PGD_2, which in turn acts on the PGD_2 receptor DP1 on mast cells to promote their proper maturation [46]. Accordingly, mice lacking $sPLA_2$-III, as well as those lacking L-PGDS or DP1, have immature mast cells and display reduced local and systemic anaphylactic responses.

Several lines of evidence suggest a potential link between $sPLA_2$-III and the development of colon cancer. For instance, $sPLA_2$-III-transfected colon cancer cells xenografted into nude mice show increased growth [47], higher expression of $sPLA_2$-III in human colorectal cancer is positively correlated with a higher rate of lymph node metastasis and shorter survival [48], and polymorphisms in the human *PLA2G3* gene are significantly associated with a higher risk of colorectal cancer [49]. Importantly, $Pla2g3^{-/-}$ mice are resistant to several models of colon cancer [50]. Furthermore, $Pla2g3^{-/-}$ mice are less susceptible to colitis, with lower expression of pro-inflammatory and pathogenic Th17 cytokines and higher expression of epithelial barrier genes [50], implying that amelioration of colonic inflammation by $sPLA_2$-III ablation underlies the protective effect against colon cancer. The $Pla2g3^{-/-}$ colon displays significant reduction of several lysophospholipids including lysophophatidic acid (LPA) and lysophosphatidylinositol (LPI) [50], which may promote colon inflammation or cancer through their receptors LPA_2 and GPR55, respectively [51,52]. Overall, these results establish a role for $sPLA_2$-III in the aggravation of colonic inflammation and cancer and point to $sPLA_2$-III as a novel druggable target for colorectal diseases. The biological roles of $sPLA_2$-III were summarized in a recent review [9].

2.7. $sPLA_2$-V in Obesity, Type-2 Immunity, and Aortic Protection

Although $sPLA_2$-V (encoded by *PLA2G5*) was previously thought to be a regulator of AA metabolism [53,54], it has become obvious that this $sPLA_2$ has a preference for phospholipids bearing *sn*-2 fatty acids with a lower degree of unsaturation. Transgenic overexpression of $sPLA_2$-V results in neonatal death due to a respiratory defect attributable to the ability of $sPLA_2$-V to potently hydrolyze dipalmitoyl-PC, a major component of lung surfactant [55]. Mice that are transgenic for other $sPLA_2$s do not exhibit such a phenotype, implying the particular ability of $sPLA_2$-V to hydrolyze PC with *sn*-2 palmitic acid (C16:0) in the lung microenvironment. $sPLA_2$-V is markedly induced in adipocytes during obesity as a "metabolic $sPLA_2$" and hydrolyzes PC in hyperlipidemic LDL to release OA and LA, which counteract adipose tissue inflammation and thereby ameliorate metabolic disorders [41]. Impairment of this process in *Pla2g5*$^{-/-}$ mice leads to exacerbation of diet-induced obesity and insulin intolerance, accompanied by elevated phospholipid levels in plasma LDL. This phenotype is reminiscent of clinical evidence that a *PLA2G5* polymorphism is associated with plasma LDL levels in patients with type 2 diabetes [56] and that the levels of *PLA2G5* mRNA expression in WAT are inversely correlated with plasma LDL levels in obese subjects [41].

$sPLA_2$-V is a "Th2-prone $sPLA_2$" induced in M2 macrophages by the Th2 cytokines IL-4 and IL-13 and promotes Th2-driven pathologies such as asthma. Gene ablation of $sPLA_2$-V perturbs proper polarization and function of M2 macrophages in association with decreased Th2 immunity [57]. *Pla2g5*$^{-/-}$ mice show reduced activation of type 2 innate lymphoid cells (ILC2) and infiltration of eosinophils in the lung following repetitive inhalation of the fungal allergen *Alternaria Alternata* [58]. Adoptive transfer experiments have revealed the contribution of $sPLA_2$-V expressed in both macrophages and non-hematopoietic cells (probably bronchial epithelial cells) to the pathology. Lipidomics analysis has demonstrated reduction of OA and LA in the lung and macrophages in *Pla2g5*$^{-/-}$ mice. Exogenous administration of these unsaturated fatty acids to *Pla2g5*$^{-/-}$ mice restores IL-33-induced inflammation and ILC2 expansion, implying that macrophage-associated $sPLA_2$-V contributes to type 2 immunity by promoting ILC2 activation though the release of OA and LA. The biological roles of $sPLA_2$-V in asthma were summarized in a recent review [59]. Probably because of the alteration in the macrophage phenotype, *Pla2g5*$^{-/-}$ macrophages have a reduced ability to phagocytose extracellular materials, thereby being more susceptible to fungal infection and arthritis due to defective clearance of hazardous fungi and immune complexes, respectively [60,61]. Likewise, *Pla2g5*$^{-/-}$ mice suffer from more severe lung inflammation caused by bacterial infection [62], which could also be explained by poor clearance of these microbes by alveolar macrophages. Additionally, local generation of LPE in the plasma membrane by $sPLA_2$-V may also contribute to macrophage phagocytosis [63].

Aortic dissection is a life-threatening aortopathy involving separation of the aortic wall. Since aortic dissection occurs suddenly without preceding clinical signs and current treatment strategies are limited mainly to antihypertensive agents and emergency surgery, biomarkers that can predict fragility and/or therapeutic targets for stabilization of the aortic wall are needed in order to improve patient outcomes. Behind its proposed role in atherosclerosis development [64], $sPLA_2$-V is a primary "endothelial $sPLA_2$" that protects against aortic dissection by endogenously mobilizing vasoprotective fatty acids [65]. Global and endothelial cell-specific deletion of $sPLA_2$-V leads to dissection of the thoracic ascending aorta shortly after infusion of angiotensin II (AT-II). In the AT-II-treated aorta, endothelial $sPLA_2$-V mobilizes OA and LA, which attenuate endoplasmic reticulum (ER) stress and increase the expression of lysyl oxidase, an enzyme that crosslinks extracellular matrix (ECM) proteins, thereby stabilizing the ECM in the aorta. Of note, dietary supplementation with OA and LA reverses the increased susceptibility of *Pla2g5*$^{-/-}$ mice to aortic dissection. These findings reveal an unexplored functional link between $sPLA_2$-driven phospholipid metabolism and aortic stability (Figure 3), possibly contributing to the development of improved diagnostic and/or therapeutic strategies for preventing aortic dissection. Importantly, this work provides in vivo relevance for the actions of this $sPLA_2$ that had been proposed by several in vitro studies: (i) it releases OA and LA in preference to PUFAs, (ii) it preferentially acts on membranes of agonist-stimulated rather than quiescent cells, and (iii) it is retained on the cell surface through binding to heparan sulfate proteoglycan. Furthermore, this avenue of cardiovascular research has revealed a potential mechanism that could underlie the benefits of the olive oil-rich (i.e., OA-rich) Mediterranean diet in terms of cardiovascular health.

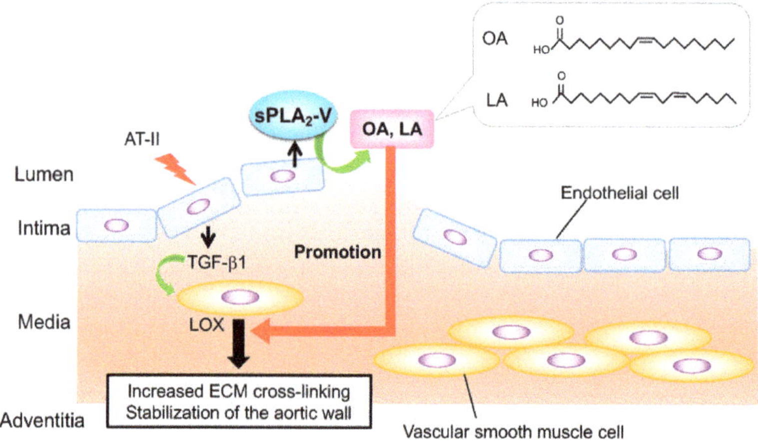

Figure 3. The role of endothelial $sPLA_2$-V in aortic stability. $sPLA_2$-V is a major $sPLA_2$ isoform expressed in aortic endothelial cells (ECs) and is largely retained on the luminal surface of the aortic endothelium likely through binding to heparin sulfate proteoglycans. Endothelial $sPLA_2$-V acts on membrane phospholipids of AT-II-activated ECs to mobilize oleic acid (OA) and linoleic acid (LA), which in turn promote AT-II-induced upregulation of lysyl oxidase (LOX) that facilitates ECM crosslinking, thereby stabilizing the aortic wall. Impairment of this $sPLA_2$-V-driven lipid pathway leads to increased susceptibility to aortic dissection.

2.8. $sPLA_2$-X in Sperm Activation, Colitis, and Asthma

Among the mammalian $sPLA_2$s, $sPLA_2$-X (encoded by *PLA2G10*) has the highest activity on PC leading to release of fatty acids, particularly PUFAs, and is activated by cleavage of the *N*-terminal propeptide by furin-type convertases [66]. In mice, $sPLA_2$-X is expressed abundantly in the testis and gastrointestinal tract and to a much lesser extent in the lung, whereas its expression in other tissues is very low. In the process of reproduction, $sPLA_2$-X is secreted from the acrosomes of activated spermatozoa and hydrolyzes sperm membrane phospholipids to release DHA, docosapentaenoic acid (DPA, C22:5), and LPC, which facilitate in vitro fertilization with oocytes [33,67]. *Pla2g10*$^{-/-}$ spermatozoa also show impairment of the late phase of the progesterone-induced acrosome reaction ex vivo [68]. Thus, the two particular $sPLA_2$s expressed in male reproductive organs, i.e., $sPLA_2$-III secreted from the epididymal epithelium (see above) and $sPLA_2$-X secreted from sperm acrosomes, act as "reproductive $sPLA_2$s" to coordinately regulate male reproduction.

$sPLA_2$-X is expressed abundantly in colorectal epithelial and goblet cells and plays a protective role against colitis by mobilizing anti-inflammatory ω3 PUFAs such as EPA and DHA [33]. Accordingly, *Pla2g10*$^{-/-}$ mice display more severe epithelial damage and inflammation with reduction of colonic ω3 PUFAs rather than ω6 AA in a colitis model, while *PLA2G10*-transgenic mice exhibit global anti-inflammatory phenotypes in association with elevation of systemic levels of ω3 PUFAs and their metabolites [33]. Supplementation with exogenous EPA restores the colitis phenotype in *Pla2g10*$^{-/-}$ mice. Furthermore, *Pla2g10*$^{-/-}$ mice have lower fecal LA levels [69], suggesting that gastrointestinal $sPLA_2$-X may have a role in the digestion of dietary and biliary phospholipids (as in the case of $sPLA_2$-IB) or, alternatively, contribute to shaping of the gut microbiota. The latter possibility may help to explain the fact that *Pla2g10*$^{-/-}$ mice display certain inflammatory, cardiovascular, and metabolic phenotypes that are not necessarily consistent among different laboratories [69–72].

$sPLA_2$-X is expressed constitutively in the airway epithelium and increased after antigen challenge in mice, and also in asthma patients. *Pla2g10*$^{-/-}$ mice are protected from antigen-induced asthma, with marked reductions of airway hyperresponsiveness, eosinophil and T cell trafficking to the airways, airway occlusion, secretion of type-2 cytokines, generation of antigen-specific IgE, and synthesis of pulmonary eicosanoids including cysteinyl leukotrienes [73]. Further, *Pla2g10*$^{-/-}$ mice have reduced IL-33 levels and fewer ILC2 cells in the lung, lower IL-33-induced IL-13 expression in mast cells, and a marked reduction in both the number of newly recruited macrophages and the M2 polarization of these macrophages in the lung [74]. These results indicate that $sPLA_2$-X serves as a key regulator of both innate and adaptive immune responses to allergens. Interestingly, as in the case of bee venom group III $sPLA_2$ that elicits strong type-2 immune responses, exogenous administration of $sPLA_2$-X serves as an adjuvant, leading to augmented type-2 immune responses with increased airway hypersensitivity and antigen-specific type-2 inflammation following peripheral sensitization and subsequent airway challenge with the antigen [75]. The biological roles of $sPLA_2$-X in asthma were detailed in a recent review [76].

3. The $sPLA_2$ Family

3.1. General Features

The cytosolic PLA_2 ($cPLA_2$) family comprises 6 isoforms (α–ζ), among which $cPLA_2\beta$, δ, ε, and ζ map to the same chromosomal locus [77]. There is structural similarity between the $cPLA_2$ and $iPLA_2$ families in that their catalytic domain is characterized by a three-layer $\alpha/\beta/\alpha$ architecture employing a conserved Ser/Asp catalytic dyad [78,79]. It appears that these two families were evolved from a common ancestral gene, with the $cPLA_2$ family emerging from the $iPLA_2$ family at the branching point of vertebrates in correlation with the development of the lipid mediator signaling pathways. Enzymes belonging to the $cPLA_2$ family are characterized by the presence of a C2 domain at their *N*-terminal region, with the exception of $cPLA_2\gamma$ in which this domain is absent. The C2 domain is responsible for the Ca^{2+}-dependent association with membranes. Herein, we will overview several recent topics for $cPLA_2\alpha$ as well as potential functions of other $cPLA_2$ isoforms.

3.2. New Insights into $cPLA_2\alpha$

$cPLA_2\alpha$ (group IVA PLA_2; encoded by *PLA2G4A*) is the best known PLA_2 that plays a central role in stimulus-coupled AA metabolism. $cPLA_2\alpha$ is the only PLA_2 that shows a striking substrate specificity for phospholipids containing AA (and also those containing EPA, if this ω3 PUFA is present in cell membranes at substantial levels). Upon cell activation, $cPLA_2\alpha$ translocates from the cytosol to the perinuclear (particularly Golgi) membranes in response to an increase in the μM range of cytosolic Ca^{2+} concentration, and is maximally activated by phosphorylation through mitogen-activated protein kinases and other kinases [80,81]. The phosphoinositide PIP_2 modulates the subcellular localization and activation of $cPLA_2\alpha$ [82]. The regulatory roles of $cPLA_2\alpha$ in eicosanoid generation in various pathophysiological events, as revealed by biochemical analyses as well as by studies using *Pla2g4a*$^{-/-}$ mice, have been well summarized in several elegant reviews [83,84].

By means of comprehensive lipidomics, Slatter et al. [85] showed that human platelets acutely increase mitochondrial energy generation following stimulation and that the substrates for this, including multiple fatty acids and oxidized species that support energy generation via β-oxidation, are exclusively provided by $cPLA_2\alpha$. This implies that $cPLA_2\alpha$ is a central regulator of both lipid mediator generation and energy flux in human platelets and that acute phospholipid membrane remodeling is required to support energy demands during platelet activation.

Ceramide-1-phosphate (C1P), a sphingolipid-derived bioactive lipid, directly binds to and activates $cPLA_2\alpha$ to stimulate the production of eicosanoids in vitro [86], but in vivo evidence for this event has been lacking. Recently, MacKnight et al. [87] addressed this issue by generating knockin mice in which endogenous $cPLA_2\alpha$ is replaced with a mutant form having an ablated C1P-interaction site.

In a skin wound healing model, wound maturation, rather than wound closure, is enhanced in the mutant cPLA$_2\alpha$-knockin mice compared to control mice. Primary dermal fibroblasts from the knockin mice show substantially increased collagen deposition and migration. The knockin mice also show an altered eicosanoid profile, with a reduction of PGE_2 and TXB_2 (a stable end-metabolite of TXA_2) as well as an increase of HETE species, which enhances the migration and collagen deposition of dermal fibroblasts. This gain-of-function role for the mutant cPLA$_2\alpha$ is associated with its relocalization to the cytoplasm and cytoplasmic vesicles. These findings clarify the key mechanisms by which wound healing is regulated by cPLA$_2\alpha$-C1P interaction in vivo and provide insight into the roles of cPLA$_2\alpha$ and eicosanoids in orchestrating wound repair.

Chao et al. [88] reported that the C2 domain in cPLA$_2\alpha$ interacts with the CARD domain in mitochondrial antiviral signaling protein (MAVS), boosting NF-κB-driven transcriptional programs that promote experimental autoimmune encephalomyelitis, a model of multiple sclerosis. cPLA$_2\alpha$ recruitment to MAVS also disrupts MAVS-hexokinase 2 interactions, decreasing hexokinase activity and the production of lactate involved in the metabolic support of neurons. These findings define a novel role of cPLA$_2\alpha$ in driving pro-inflammatory astrocyte activities in cooperation with MAVS through protein–protein interaction in the context of neuroinflammation. It remains unknown whether some cPLA$_2\alpha$-driven lipid mediators are involved in this process.

Oncogenic *PIK3CA* (encoding a PI3K isoform) results in an increase of AA and eicosanoids, thus promoting cell proliferation to beyond a cell-autonomous degree. Mechanistically, mutant PIK3CA drives a multimodal signaling network involving mTORC2-PKCζ-mediated activation of cPLA$_2\alpha$ [89]. Notably, inhibition of cPLA$_2\alpha$ acts synergistically with a fatty acid-free diet to restore immunogenicity and selectively reduce mutant PIK3CA-induced tumorigenicity. This reveals an important role for activated PI3K signaling in regulation of AA metabolism, highlighting a targetable metabolic vulnerability that depends largely on dietary fat restriction.

3.3. *cPLA$_2\beta$ Fusion Protein in Carcinoma Cell Proliferation and Survival*

cPLA$_2\beta$ (group IVB PLA$_2$; encoded by *PLA2G4B*) displays PLA$_1$, PLA$_2$ and more potent lysophospholipase activities in vitro. A kinetic study has demonstrated that cPLA$_2\beta$ associates with a membrane surface that is rich in phosphoinositides when intracellular Ca^{2+} is low, whereas it moves to a cardiolipin-rich membrane such as the mitochondrial membrane when intracellular Ca^{2+} rises. Among three splice variants termed cPLA$_2\beta$1, β2 and β3, only the β3 form is identified as an endogenous protein and is constitutively associated with mitochondrial and early endosomal membranes [90].

Cheng et al. [91] reported that *JMJD7-PLA2G4B*, a read-through fusion gene formed by splicing of the neighboring *JMJD7* (jumonji domain containing 7) and *PLA2G4B* genes, is expressed in human squamous cell carcinoma (HNSCC), as well as several other cancers. Ablation of *JMJD7-PLA2G4B*, but not *JMJD7* or *PLA2G4B* alone, significantly inhibits proliferation of SCC cells by promoting G1 arrest and increases starvation-induced cell death. These findings provide a novel insight into the oncogenic control of JMJD7-PLA2G4B in HNSCC cell proliferation and survival and suggest that this fusion protein may serve as an important therapeutic target and prognostic marker for HNSCC development and progression. cPLA$_2\beta$ has also been implicated in age-related changes in phospholipids and decreased energy metabolism in monocytes [92]. However, it remains unknown whether certain lipid metabolites generated by cPLA$_2\beta$ would be involved in these events.

3.4. *cPLA$_2\gamma$ in Lipid Droplet Formation*

Human cPLA$_2\gamma$ (group IVC PLA$_2$; encoded by *PLA2G4C*), lacking the C2 domain characteristic of the cPLA$_2$ family, is C-terminally farnesylated, is tightly associated with membranes, and possesses lysophospholipase and transacylase activities in addition to PLA$_2$ activity [93]. Several lines of evidence suggest that cPLA$_2\gamma$ acts mainly as a CoA-independent transacylase, transferring a fatty acid from one phospholipid to the other phospholipid, in cells [94,95]. cPLA$_2\gamma$ is widely expressed in human tissues with a tendency for higher expression in the heart and skeletal muscle, whereas its expression in most mouse tissues is very low, making it difficult to address the in vivo roles of $cPLA_2\gamma$ using a knockout strategy. Exceptionally, mouse $cPLA_2\gamma$ is highly expressed in oocytes during the stage of germinal vesicle breakdown, when it dynamically relocates from the cortex to the nuclear envelope, suggesting its possible role in nuclear membrane remodeling in developing oocytes [96].

Lipid droplet (LD) accumulation in hepatocytes is a typical characteristic of steatosis. Hepatitis C virus (HCV) infection, one of the risk factors related to hepatic steatosis, induces LD accumulation in human hepatocytes. $cPLA_2\gamma$ has been identified as a host factor upregulated by HCV infection and involved in HCV replication, where it promotes LD biogenesis and HCV assembly [97,98]. $cPLA_2\gamma$, through the domain around the amino acid residues 260–292, is tightly associated normally with ER membranes and relocated into LDs. Importantly, *PLA2G4C* knockdown hampers LD formation upon HCV stimulation, while *PLA2G4C* overexpression leads to LD formation in hepatocytes and enhances LD accumulation in the liver of mice fed a high-fat diet, suggesting its potential role in fatty liver disease.

3.5. $cPLA_2\delta$ in Psoriasis

$cPLA_2\delta$ (group IVD PLA_2; encoded by *PLA2G4D*) was first identified as a keratinocyte-specific $cPLA_2$ isoform that is induced during psoriasis and releases LA selectively [99]. However, subsequent studies showed that the PLA_2 activity of $cPLA_2\delta$ is much weaker than that of $cPLA_2\alpha$ [77] and that its PLA_1 activity is superior to its PLA_2 activity [90]. Cheung et al. [100] recently demonstrated the expression of $cPLA_2\delta$ in psoriatic mast cells, and found unexpectedly that its activity is extracellular. IFN-α-stimulated mast cells release exosomes, which transfer cytoplasmic $cPLA_2\delta$ to neighboring Langerhans cells expressing CD1a, which present lipid antigens to T cells. Thus, the exosome-mediated transfer of $cPLA_2\delta$ from mast cells to Langerhans cells leads to the generation of neolipid antigens and subsequent recognition by lipid-specific CD1a-reactive T cells, inducing the production of IL-22 and IL-17A. These data offer an alternative model of psoriasis pathogenesis in which lipid-specific CD1a-reactive T cells contribute to psoriatic inflammation, suggesting that $cPLA_2\delta$ inhibition or CD1a blockade may have potential for treatment of psoriasis. However, given that CD1a is present in humans but not in mice, and that $cPLA_2\delta$ is located mainly in epidermal keratinocytes, the regulatory roles of $cPLA_2\delta$ in psoriatic skin require further exploration.

3.6. $cPLA_2\varepsilon$ as an N-Acyltransferase for N-Acylethanolamine Biosynthesis

N-acylethanolamines (NAEs) represent a group of endocannabinoid lipid mediators, including arachidonoylethanolamine (also known as anandamide; AEA), which acts on the endocannabinoid receptors CB1 or CB2, and palmitoyletanolamine (PEA) and oleoylethanolamine (OEA), which act on the nuclear receptor $PPAR\alpha$ or through other mechanisms. The biosynthesis of NAEs, particularly PEA and OEA, occurs in two steps; transfer of *sn*-1 saturated or monounsaturated fatty acid of PC to the amino group of PE by *N*-acyltransferases to generate *N*-acyl-PE (NAPE), followed by hydrolysis mainly by NAPE-specific phospholipase D (NAPE-PLD) to give rise to NAEs [101]. There are two types of *N*-acyltransferase, i.e., Ca^{2+}-dependent and -independent enzymes. Recently, it has been shown that $cPLA_2\varepsilon$ (group IVE PLA_2; encoded by *PLA2G4E*) functions as a Ca^{2+}-dependent *N*-acyltransferase [102]. In response to Ca^{2+} ionophore stimulation, HEK293 cells overexpressing $cPLA_2\varepsilon$ produce various NAPE and NAE species, accompanied by concomitant decreases in PE and PC and increases in LPE and LPC, indicating that $cPLA_2\varepsilon$ produces NAPEs by utilizing the diacyl- and plasmalogen types of PE as acyl acceptors and the diacyl types of PC and PE as acyl donors [103] (Figure 4A). The activity of $cPLA_2\varepsilon$ is markedly enhanced by the presence of phosphatidylserine (PS) or other anionic phospholipids, and $cPLA_2\varepsilon$ largely co-localizes with PS in the plasma membrane and organelles involved in the endocytic pathway [104,105]. This localization might be related to the observation that $cPLA_2\varepsilon$ drives recycling through the clathrin-independent endocytic route [106].

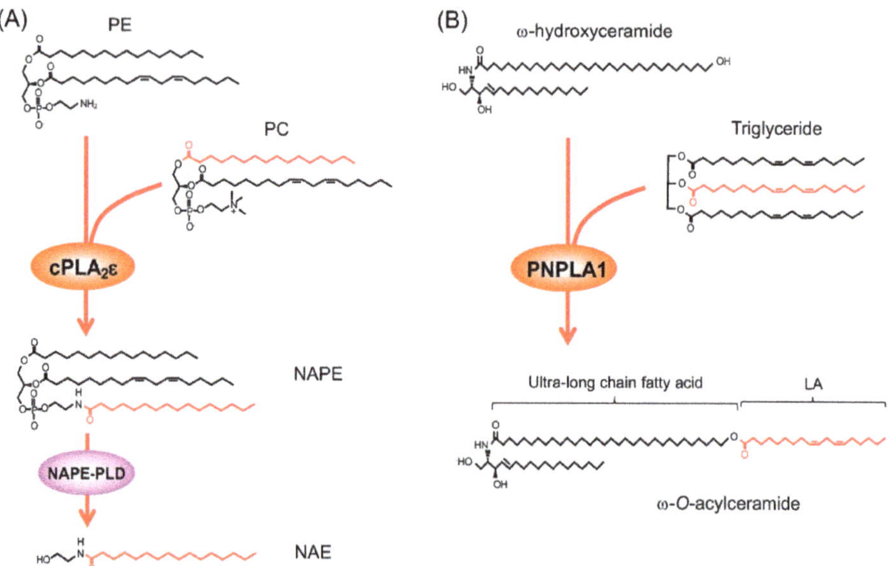

Figure 4. Examples of PLA_2 enzymes that have unique enzymatic activity. (**A**) $cPLA_2\varepsilon$ catalyzes an *N*-acyltransferase reaction, transferring the *sn*-1 fatty acid of PC to the amino group of PE to produce NAPE, which is then converted to NAE by NAPE-PLD. (**B**) PNPLA1 acts as a unique transacylase, transferring LA from triglyceride to ω-hydroxyceramide to give rise to ω-*O*-acylceramide, which is essential for skin barrier function.

However, it still remains to be determined whether $cPLA_2\varepsilon$ indeed contributes to NAPE and NAE biosynthesis under certain in vivo conditions. Since genetic deletion or pharmacological inhibition of NAPE-PLD disturbs lipid metabolism in the liver, intestine, and adipose tissue [107], as well as emotional behavior [108], it is tempting to speculate that $cPLA_2\varepsilon$, which lies upstream of NAPE-PLD for NAE biosynthesis, may also participate in these events to improve metabolism and neuronal functions. The potential neuroprotective role of $cPLA_2\varepsilon$ is supported by the association between reduced expression of $cPLA_2\varepsilon$ and dementia, where adenoviral overexpression of $cPLA_2\varepsilon$ in hippocampal neurons completely restores cognitive deficits in the elderly APP/PS1 mouse, a model of Alzheimer's disease [109].

3.7. $cPLA_2\zeta$ in Myocardial Mitochondria

The observation that multiple fatty acids are non-selectively released from Ca^{2+}-stimulated *Pla2g4a*$^{-/-}$ lung fibroblasts, an event that is suppressed by $cPLA_2\alpha$ inhibitors, suggests that other $cPLA_2$ isoform(s) might contribute to this event. This fatty acid release and PGE_2 production by *Pla2g4a*$^{-/-}$ fibroblasts depend on $cPLA_2\zeta$ (group IVF PLA_2, encoded by *PLA2G4F*) [110]. In response to ionomycin, $cPLA_2\zeta$ translocates to ruffles and dynamic vesicular structures, while $cPLA_2\alpha$ translocates to the Golgi and ER, suggesting distinct mechanisms of regulation for the two enzymes.

Moon et al. [111] reported that mitochondria isolated from human heart contain at least two PLA_2s—$cPLA_2\zeta$ and $iPLA_2\gamma$—of which $cPLA_2\zeta$ mediates Ca^{2+}-activated release of AA from mitochondria in normal heart. The AA pool mobilized by $cPLA_2\zeta$ is preferentially channeled into cytochrome P450 epoxygenases for the synthesis of epoxyeicosatrienoic acids (EETs), which have a protective effect against heart failure. In contrast, in the failing heart, $iPLA_2\gamma$ mainly mobilizes mitochondrial AA, which preferentially couples with lipoxygenases for the synthesis of toxic HETEs that open mitochondrial permeability transition pores, leading to further progression of heart failure.

These results reveal an unexplored biological role of cPLA$_2\zeta$ as well as iPLA$_2\gamma$ (see below) in the myocardium, although confirmation using *Pla2g4f*-null mice will be required.

4. The iPLA$_2$/PNPLA Family

4.1. General Features

The human genome encodes 9 Ca^{2+}-independent PLA$_2$ (iPLA$_2$) enzymes. These enzymes are now more generally known as patatin-like phospholipase domain-containing lipases (PNPLA1-9), since all members in this family share a patatin domain, which was initially discovered in patatin (iPLA$_2\alpha$), a potato protein [2,112]. Unlike the cPLA$_2$ family, which is present only in vertebrates, the iPLA$_2$/PNPLA family is widely expressed in many eukaryotes including yeast, ameba, nematode, fly and vertebrates. iPLA$_2$/PNPLA isoforms display lipid hydrolase or transacylase/acyltransferase activities with specificities for diverse lipids such as phospholipids, neutral lipids, sphingolipids, and retinol esters. In principle, enzymes with a large and unique *N*-terminal region (PNPLA6~9) act mainly on phospholipids as PLA$_1$, PLA$_2$ or lysophospholipases, whereas those lacking the *N*-terminal domain (PNPLA1~5) act on neutral lipids as lipases or transacylases. Analysis of gene disruption or mutation of the iPLA$_2$/PNPLA enzymes in mice and humans have provided valuable insights into their physiological roles in homeostatic lipid metabolism that are fundamental for life. Since there have been many excellent reviews on this enzyme family [2,3,112,113], we will herein highlight several recent topics that shed further light on the pathophysiological roles of this enzyme family.

4.2. iPLA$_2\beta$ in Neurodegeneration and Hepatic Steatosis

iPLA$_2\beta$ (also known as group VIA PLA$_2$; encoded by *PLA2G6*) is the only iPLA$_2$ isoform that acts primarily as a PLA$_2$ with poor fatty acid selectivity [114,115]. Different splice variants of iPLA$_2\beta$ are associated with the plasma membrane, mitochondria, ER, and the nuclear envelope. Although iPLA$_2\beta$ lacks a transmembrane domain, it has putative protein-interaction motifs such as ankyrin repeats, which are capable of interacting with multiple cognate receptor proteins, and a calmodulin-binding site, which interacts with the inhibitory calmodulin. The crystal structure of iPLA$_2\beta$ reveals a dimer formation of the catalytic domains, which are surrounded by ankyrin repeats that adopt an outwardly flared orientation, poised to interact with membrane proteins [116]. The closely integrated active sites are positioned for cooperative activation and internal transacylation. The structure also suggests allosteric inhibition by calmodulin, where a single calmodulin molecule interacts with two catalytic domains, altering the conformation of the dimer interface and active sites.

The roles of iPLA$_2\beta$ in male fertility, neuronal disorders, metabolic diseases, and inflammation, among others, have been studied in a number of *Pla2g6* knockout, knockdown, and overexpression studies and are well summarized in recent reviews [113,117]. In particular, the roles of iPLA$_2\beta$ in neuronal function have received extensive attention, since numerous mutations of the *PLA2G6* gene have been discovered in patients with neurodegenerative disorders such as infantile neuroaxonal dystrophy (INAD) and Parkinson's disease [118]. In fact, iPLA$_2\beta$ is also referred to as the parkinsonism-associated protein PARK14, mutations of which are associated with impaired Ca^{2+} signaling in dopaminergic neurons [119].

Three recent studies using mutant flies have provided novel insights into the regulatory roles of iPLA$_2\beta$ in the brain in the context of Parkinson's disease with α-synucleinopathy. Kinghorn et al. [120] reported that knockout of the *PLA2G6* gene in *Drosophila* results in reduced survival, locomotor deficits, and organismal hypersensitivity to oxidative stress, accompanied by mitochondrial abnormalities and increased lipid peroxidation levels. Inhibition of lipid peroxidation partially rescues the locomotor abnormalities and mitochondrial membrane potential caused by iPLA$_2\beta$ deficiency. Lin et al. [121] showed that the loss of iPLA$_2\beta$ causes an increase of brain ceramide, leading to lysosomal stress and neurodegeneration. iPLA$_2\beta$ binds to the retromer subunits Vps35 and Vps26 and enhances their function to promote protein and lipid recycling. Loss of iPLA$_2\beta$ impairs retromer function resulting in a progressive increase of ceramide, thus inducing a positive feedback loop that affects membrane fluidity and impairs retromer function and neuronal function. Mori et al. [122] showed that iPLA$_2\beta$ deficiency in *Drosophila* results in defective neurotransmission during the early developmental stages and progressive cell loss throughout the brain, including degeneration of the dopaminergic neurons. In the brain, iPLA$_2\beta$ loss results in shortening of the acyl-chain length of phospholipids, resulting in membrane lipid disequilibrium and thereby ER stress. Introduction of the mitochondria-ER contact site-resident protein C19orf12, another causal gene for Parkinson's disease, in *PLA2G6*-deficient flies rescues the phenotypes associated with altered lipid composition, ER stress, and neurodegeneration. Moreover, the acceleration of α-synuclein aggregation by iPLA$_2\beta$ deficiency is suppressed by administration of LA, which corrects the brain phospholipid composition. Thus, membrane remodeling by iPLA$_2\beta$ is required for the survival of dopaminergic neurons and α-synuclein stability.

The roles of iPLA$_2\beta$ in phospholipid remodeling in the context of hepatic lipid metabolism have recently been studied using *Pla2g6*$^{-/-}$ mice. iPLA$_2\beta$ deficiency attenuates obesity and hepatic steatosis in *ob/ob* mice through hepatic fatty-acyl phospholipid remodeling [123]. Aging sensitized by iPLA$_2\beta$ deficiency induces liver fibrosis and intestinal atrophy involving suppression of homeostatic genes and alteration of intestinal lipids and bile acids [124]. *Pla2g6*$^{-/-}$ mice fed a high-fat diet show attenuation of hepatic steatosis through correction of defective phospholipid remodeling [125]. Lastly, *Pla2g6*$^{-/-}$ mice fed a methionine/choline-deficient diet do not show correction of this defect, but the hepatocellular injury is attenuated via inhibition of lipid uptake genes [126].

4.3. *iPLA$_2\gamma$ in Bioenergetics and Signaling*

iPLA$_2\gamma$ (also known as group VIB PLA$_2$, encoded by *PNPLA8*) displays PLA$_2$ activity, but acts as a PLA$_1$ toward phospholipids bearing *sn*-2 PUFA [127,128]. Accordingly, hydrolysis of PUFA-bearing phospholipids by iPLA$_2\gamma$ typically gives rise to 2-lysophospholipids (having a PUFA at the *sn*-2 position) rather than 1-lysophospholipids (having a saturated or monounsaturated acid at the *sn*-1 position). *Pnpla8*$^{-/-}$ mice display multiple bioenergetic and neuronal dysfunctions, including growth retardation, kyphosis, muscle weakness with atrophy of myofilaments, cold intolerance, reduced exercise endurance, increased mortality due to cardiac stress, resistance to diet-induced obesity and insulin resistance, performance deficits in spatial learning and memory, and abnormal mitochondrial function with a dramatic decrease in fatty acid β-oxidation and oxygen consumption [129]. The alterations in myocardial and hippocampal cardiolipin content and composition indicate that iPLA$_2\gamma$ is involved in cardiolipin remodeling. These features of *Pnpla8*$^{-/-}$ mice are reminiscent of those seen in patients with Barth syndrome, a disease caused by mutations in the human *TAZ* gene, which encodes tafazzin, a mitochondrial transacylase required for cardiolipin remodeling [130]. Moreover, loss-of-function variants of the human *PNPLA8* gene recapitulate the mitochondriopathy observed in *Pnpla8*$^{-/-}$ mice [131].

iPLA$_2\gamma$ also participates in the generation of lipid mediators under certain conditions. *Pnpla8*$^{-/-}$ mice display reduced ADP- or collagen-stimulated TXA$_2$ generation by platelets ex vivo and show prolonged bleeding time and reduced pulmonary thromboembolism in vivo [132]. Cardiomyocyte-specific deletion of iPLA$_2\gamma$ decreases infarct size upon ischemia/reperfusion, with reduction of oxygenated metabolites of ω3 and ω6 PUFAs including PGs, HETEs, and hydroxy-DHAs [133]. iPLA$_2\gamma$ releases 9-hydroxyoctadecenoic acid (9-HODE), an oxygenated metabolite of LA, from cardiolipin, integrating mitochondrial bioenergetics and signaling [134]. Heart failure-induced activation of iPLA$_2\gamma$ leads to mitochondrial generation of HETEs that open the mitochondrial permeability transition pore, thus further amplifying myocardial damage [111]. Phospholipids bearing *sn*-2 AA are cleaved by the PLA$_1$ activity of iPLA$_2\gamma$ to give rise to 2-arachidonoyl-lysophospholipids, which are then oxygenized directly by cyclooxygenase-2 or 12-lipoxygenase for conversion to eicosanoid-esterified lysophospholipids [134,135]. The generation of eicosanoid-esterified lysophospholipids is attenuated by the absence of iPLA$_2\gamma$, underscoring an iPLA$_2\gamma$-initiated pathway generating new classes of lipid metabolites with potential signaling functions.

4.4. PNPNA6 and PNPLA7 as Lysophospholipases

PNPLA6 (iPLA$_2\delta$) and its closest paralog PNPLA7 (iPLA$_2\theta$) have a lysophospholipase activity that cleaves LPC to yield fatty acid and glycerophosphocholine (GPC) [136,137]. Counterparts of these enzymes in yeast and fly act as a phospholipase B, which converts PC to GPC by liberating both *sn*-1 and *sn*-2 fatty acids. PNPLA6, also referred to as neuropathy target esterase (NTE), was originally identified as a target enzyme for the poisonous effect of organophosphates, which cause a severe neurological disorder characterized by degeneration of long axons in the spinal cord and peripheral nerves, leading to paralysis of the lower limbs [138]. In cultured renal cells, the production of GPC, an osmoprotective metabolite, is enhanced by PNPLA6 overexpression and is diminished by its siRNA knockdown or its inhibitor organophosphate [139]. Global *Pnpla6*$^{-/-}$ mice die in utero due to placental defects, while neuron-specific *Pnpla6*$^{-/-}$ mice exhibit progressive neuronal degeneration, leading to prominent neuronal pathology in the hippocampus and thalamus and also defects in the cerebellum [140,141]. Neuronal absence of PNPLA6 results in disruption of the ER, vacuolation of nerve cell bodies and abnormal reticular aggregates, and sustained elevation of PC over many months is accompanied by progressive degeneration and massive swelling of axons in the sensory and motor spinal tracts and worsening hindlimb dysfunction [142]. In humans, *PNPLA6* mutations near the catalytic site cause a severe motor neuron disease characterized by progressive spastic paraplegia and distal muscle wasting [143] as well as childhood blindness with retinal degeneration, including Leber congenital amaurosis, Oliver McFarlane syndrome, and Boucher-Neuhäuser syndrome [144]. Although the in vivo role of PNPLA7, also known as NTE-related esterase (NRE), is still unknown, it is downregulated by insulin in WAT [136] and interacts with LDs through its catalytic domain [145], suggesting its metabolic role.

4.5. PNPLA2 and PNPLA3 in Triglyceride Metabolism

Although PNPLA2 and PNPLA3 correspond to iPLA$_2\zeta$ and iPLA$_2\varepsilon$, respectively, according to the PLA$_2$ classification, the latter names are not used here because these enzymes act essentially as neutral lipid lipases, but not as phospholipases. PNPLA2 is upregulated, while PNPLA3 is downregulated, upon starvation, and vice versa upon feeding, indicating that these two closely related lipases are nutritionally regulated in a reciprocal way [146]. PNPLA2, more generally known as adipose triglyceride lipase (ATGL), is a major lipase that hydrolyzes triglycerides in LDs to release fatty acids for β-oxidation-coupled energy production, a process known as lipolysis [147]. Genetic deletion or mutation of PNPLA2 leads to accumulation of triglycerides in multiple tissues leading to heart failure, while conferring protection from fatty liver and glucose intolerance, likely because these mice are able to utilize glucose but not free fatty acids as a fuel [148,149]. PNPLA2 deficiency also protects against cancer-associated cachexia by preventing fat loss [150]. The fatty acids released from LDs by PNPLA2 act as endogenous ligands for the nuclear receptor PPARα or PPARδ, which drives energy consumption [151,152]. In addition, the AA released from triglycerides by PNPLA2-driven lipolysis is utilized for eicosanoid generation in certain situations [153]. The activity of PNPLA2 is regulated positively by ABHD5 (also known as CGI-58) and negatively by perilipin, G0S2 and mysterin, which modulate the accessibility of PNPLA2 to LDs [152,154]. Adipocyte-specific *G0S2*-transgenic mice show attenuated lipolysis and adipocyte hypertrophy, accompanied by a reduced hepatic triglyceride level and increased insulin sensitivity [155], thus recapitulating the phenotypes observed in *Pnpla2*$^{-/-}$ mice [149]. Mutations in the human *PNPLA2* gene cause Chanarin–Dorfman syndrome, a condition in which triglycerides are stored abnormally in the body [156]. *ABHD5* mutations also cause a similar neutral lipid storage disease but also additionally cause ichthyosis [157], likely because ABHD5 acts as a cofactor for not only PNPLA2-mediated lipolysis, but also PNPLA1-driven ω-*O*-acylceramide synthesis in the skin (see below). The regulatory mechanisms and metabolic roles of PNPLA2 have been detailed in other elegant reviews [158,159].

Mutations in the human *PNPLA3* gene are highly associated with non-alcoholic fatty liver disease (NAFLD) and steatohepatitis (NASH) [160]. However, *Pnpla3*$^{-/-}$ mice do not display a fatty liver phenotype [161], whereas PNPLA3^{I158M} knockin mice develop hepatic steatosis [162], illustrating the complexity of the regulatory roles of PNPLA3 in hepatic lipid metabolism. PNPLA3 reportedly acts as an acyltransferase that alters the fatty acid composition of triglycerides [163], as a transacylase that promotes the transfer of PUFAs from triglycerides to phospholipids in hepatic LDs [164], or as a retinyl-palmitate lipase in hepatic stellate cells to fine-tune the plasma levels of retinoids, which might influence the differentiation of stellate cells to myofibroblasts [165]. It has now become recognized that PNPLA3 functions primarily as a triglyceride lipase and that I158M mutation leads to loss of function. Importantly, PNPLA3, and PNPLA3^{I158M} to an even greater extent, strongly interact with ABHD5, a cofactor for PNPLA2, thereby interfering with the lipolytic activity of PNPLA2 [166,167]. Overexpression of PNPLA3^{I158M} greatly suppresses PNPLA2-dependent lipolysis, leading to massive triglyceride accumulation in hepatocytes and brown adipocytes. Moreover, transgenic overexpression of PNPLA3^{I158M} increases hepatic triglyceride levels in WT mice, but not in *Abhd5*$^{-/-}$ mice, confirming that the pro-steatotic effects of PNPLA3 require the presence of ABHD5. Thus, the increased abundance of PNPLA3^{I148M} results in sequestration of ABHD5 on LDs, thereby limiting the availability of this cofactor for activation of PNPLA2. The question remains of how PNPLA2, PNPLA3, and ABHD5 each find their way to hepatocyte LDs in the correct stoichiometric proportions and function to regulate LD assembly and turnover under conditions of fasting or nutrient excess.

4.6. PNPLA1 in Acylceramide Synthesis for Skin Barrier Function

ω-*O*-acylceramide, a unique sphingolipid present specifically in the *stratum corneum* of the epidermis, is essential for skin barrier formation, and impairment of its biosynthesis leads to ichthyosis or atopic dermatitis. Mutations in the human *PNPLA1* gene cause autosomal recessive congenital ichthyosis [168]. Unlike most PNPLA isoforms that are ubiquitously expressed in many tissues, PNPLA1 is localized predominantly in the upper layer of the epidermis. PNPLA1 acts as a unique transacylase, catalyzing the transfer of LA in triglyceride to the ω-hydroxy group of ultra-long-chain fatty acid in ceramide to give rise to ω-*O*-acylceramide [169–171] (Figure 4B). Global or keratinocyte-specific deletion of PNPLA1 hampers epidermal ω-*O*-acylceramide formation, thereby severely impairing skin barrier function leading to neonatal death due to excessive dehydration. The enzymatic activity of PNPLA1 is enhanced by ABHD5, which may present triglycerides to PNPLA1 to facilitate substrate recognition [172,173], providing a mechanism whereby *ABHD5* mutations cause Chanarin-Dorfman syndrome accompanied by ichthyosis with impaired ω-*O*-acylceramide formation. The role of PNPLA1 in ω-*O*-acylceramide biosynthesis has been described in detail in a recent review [174].

5. Other PLA_2 Family

5.1. The PAF-AH Family

The PAF-acetylhydrolase (PAF-AH) family comprises one extracellular and three intracellular enzymes that were originally found to have the capacity to deacetylate and thereby inactivate the lysophospholipid-derived lipid mediator PAF [175,176]. Plasma-type PAF-AH (group VIIA PLA_2; encoded by *PLA2G7*) is a secreted protein produced by macrophages, mast cells or other sources, and is now more generally referred to as lipoprotein-associated PLA_2 (Lp-PLA_2), existing as a low-density lipoprotein (LDL)-bound form in human plasma [177]. Although PAF is a potent mediator of allergic responses, Lp-PLA_2 deficiency fails to augment airway inflammation or hyperresponsiveness after PAF+LPS treatment or passive or active allergic sensitization and challenge [178]. A series of clinical studies have revealed a correlation of Lp-PLA_2 with atherosclerosis, likely because this enzyme hydrolyzes oxidized phospholipids (i.e., phospholipids having an oxidized fatty acid at the *sn*-2 position) in modified LDL with pro-atherogenic potential [179,180]. Although darapladib, a potent Lp-PLA_2 inhibitor, failed to meet the primary endpoints of two large phase III trials for treatment of atherosclerosis [181], recent clinical and preclinical studies have revealed that Lp-PLA_2 inhibition may have therapeutic effects in diabetic macular edema and Alzheimer's disease [182,183]. Lp-PLA_2

deficiency in $Apc^{Min/+}$ mice leads to decreased intestinal polyposis and tumorigenesis, suggesting a role of PAF or some oxidized lipids in cancer development [184].

Type-I PAF-AH is a heterotrimer composed of two catalytic α1 and α2 subunits (group XIIIA and XIIIB PLA_2s, encoded by *PAFAH1B2* and *PAFAH1B3*, respectively), and a regulatory β subunit that is identical to LIS-1, a causative gene for a type of Miller-Dieker syndrome [185]. Loss of both the α1 and α2 catalytic subunits leads to male infertility [186], reduction of amyloid-β generation by promoting the degradation of amyloid precursor protein C-terminal fragments [187], and an increase in the size of the ganglionic eminences resulting from increased proliferation of GABAergic neurons through perturbation of the Wnt signaling pathway [188].

Type-II PAF-AH (PAF-AH2 or group VIIB PLA_2; encoded by *PAFAH2*) shows significant homology with Lp-PLA_2 and preferentially hydrolyzes oxidized phospholipids in cells. In a CCl_4-induced liver injury model, *Pafah2*$^{-/-}$ mice show a delay in hepatic injury recovery with unusual accumulation of 8-isoprostaglandin $F_{2\alpha}$ (8-iso-$PGF_{2\alpha}$) in membrane phospholipids, indicating that PAF-AH2 removes toxic oxidized lipids from cell membranes and thereby protects the tissue from oxidative stress-induced injury [189]. Mast cells spontaneously produce ω3 PUFA-derived epoxides (ω3 epoxides), such as EPA-derived 17,18-epoxyeicosatetraenoic acid (17,18-EpETE) and DHA-derived 19,20-epoxydocosapentaenoic acid (19,20-EpDPE), whose production depends on PAF-AH2-driven cleavage of ω3 epoxide-esterified phospholipids in mast cell membranes [190]. Genetic or pharmacological inactivation of PAF-AH2 reduces the steady-state production of ω3 epoxides, leading to attenuated mast cell activation and anaphylaxis following FcεRI crosslinking (Figure 5). Mechanistically, the ω3 epoxides promote IgE-mediated activation of mast cells by down-regulating Srcin1, a Src-inhibitory protein that counteracts FcεRI signaling, through a pathway involving PPARγ. Thus, the PAF-AH2–ω3 epoxide pathway ensures optimal mast cell activation and presents new potential drug targets for allergic diseases. The properties and functions of the PAF-AH family have been summarized in a recent review [191].

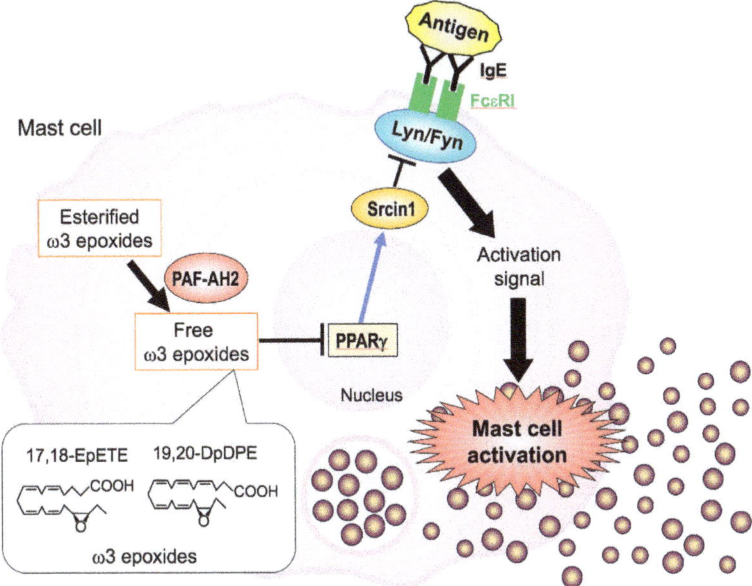

Figure 5. The role of PAF-AH2 in mast cell activation by producing ω3 epoxides. PAF-AH2 constitutively hydrolyzes ω3 epoxide-esterified phospholipids in cell membranes to liberate ω3 epoxides. These unique ω3 PUFA metabolites attenuate PPARγ signaling and downregulate Srcin1, which blocks activation of the Src family kinases Fyn and Lyn, thereby augmenting FcεRI signaling.

5.2. Lysosomal PLA$_2$s

Lysosomal PLA$_2$ (LPLA$_2$ or group XV PLA$_2$; encoded by *PLA2G15*) is catalytically active under mildly acidic conditions and structurally homologous with lecithin:cholesterol acyltransferase (LCAT), an enzyme that transfers the *sn*-2 fatty acid of PC to cholesterol to produce cholesteryl ester in high-density lipoprotein (HDL) [192]. LPLA$_2$ hydrolyzes both *sn*-1 and *sn*-2 fatty acids in phospholipids and contributes to phospholipid degradation in lysosomes. Genetic deletion of LPLA$_2$ results in unusual accumulation of non-degraded lung surfactant phospholipids in lysosomes of alveolar macrophages leading to phospholipidosis [193], reduced presentation of lysophospholipid antigens to CD1d by invariant natural killer T (iNKT) cells [194], and impairment of adaptive T cell immunity against mycobacterium [195].

Peroxiredoxin 6 (Prdx6) is another lysosomal PLA$_2$ that is also called acidic Ca^{2+}-independent PLA$_2$ (aiPLA$_2$). Prdx6 is a multifunctional enzyme since it also possesses glutathione peroxidase and lysophospholipid acyltransferase activities. The aiPLA$_2$ activity of Prdx6 has important physiological roles in the turnover (synthesis and degradation) of lung surfactant phospholipids [196], protection against LPS-induced lung injury [197], repair of peroxidized cell membranes [198], and sperm fertilizing competence [199]. The properties and functions of LPLA$_2$ and Prdx6/aiPLA$_2$ have been summarized in recent reviews [200,201].

5.3. The PLAAT Family

The PLAAT family, which comprises 3 enzymes in mice and 5 enzymes in humans, is structurally similar to lecithin:retinol acyltransferase (LRAT). Because of the ability of some members in this family to suppress *H-ras* signaling, it is also referred to as the HRASLS (for *H-ras*-like suppressor) family. Members of this family, including PLA2G16 (group XVI PLA$_2$, also known as PLAAT3 or HRASLS3), display PLA$_1$ and PLA$_2$ activities, as well as Ca^{2+}-independent *N*-acyltransferase activity that synthesizes NAPE, to various degrees [202]. PLA2G16 is highly expressed in adipocytes, and *Pla2g16*$^{-/-}$ mice are resistant to diet-induced obesity [203]. Adipocyte-derived LPC produced by PLA2G16 activates NLRP3 inflammasomes in adipocytes and adipose tissue macrophages mediating homocysteine-induced insulin resistance [204]. PLA2G16 and its paralogs in this family have also been implicated in tumor invasion and metastasis [205], vitamin A metabolism [206], peroxisome biogenesis [207], and cellular entry and clearance of Picornaviruses [208]. The properties and functions of the PLAAT/PLA2G16 family have been summarized in recent reviews [209,210].

5.4. The ABHD Family

The ABHD family is a newly recognized group of lipolytic enzymes, comprising at least 19 enzymes in humans [211]. Enzymes in this family typically possess both hydrolase and acyltransferase motifs. Although the functions of many of the ABHD isoforms still remain uncertain, some of them have been demonstrated to act on neutral lipids or phospholipids as lipid hydrolases. ABHD3 selectively hydrolyzes phospholipids with medium-chain fatty acids [212]. ABHD4 releases fatty acids from multiple classes of *N*-acyl-phospholipids to produce *N*-acyl-lysophospholipids [213]. ABHD6 acts as lysophospholipase or monoacylglycerol lipase, the latter being possibly related to signaling of 2-arachidonoyl glycerol, an endocannabinoid lipid mediator that plays a role in retrograde neurotransmission [214]. ABHD12 hydrolyzes lysophosphatidylserine (LysoPS), and is therefore referred to as LysoPS lipase [215]. Genetic or pharmacological blockade of ABHD12 stimulates immune responses in vivo, pointing to a key role for this enzyme in regulating immunostimulatory lipid pathways. Mutations in the human *ABHD12* gene result in accumulation of LysoPS in the brain and cause a disease known as PHARC, which is characterized by polyneuropathy, hearing loss, ataxia, retinitis pigmentosa, and cataract [216]. ABHD16A acts as a PS-selective PLA$_2$ (referred to as PS lipase), lying upstream of ABHD12 in the PS-catabolic pathway [217]. Disruption of ABHD12 and ABHD16A in macrophages respectively increases and decreases LysoPS levels and cytokine production. Although

ABHD5 does not have catalytic activity because of the absence of a serine residue in the catalytic center, it interacts with and modifies the functions of several PNPLA members including PNPLA1-3, as described above.

5.5. GPI-Specific PLA_2s

GPI is a complex glycolipid covalently linked to the C terminus of proteins on the plasma membrane, particularly in the raft microdomain. The biosynthesis of GPI and its attachment to proteins occur in the ER. GPI-anchoring proteins (GPI-APs) are subjected to fatty acid remodeling, which replaces an unsaturated fatty acid at the *sn*-2 position of the PI moiety with a saturated fatty acid. PGAP3, which resides in the Golgi, and PGAP6, which is localized mainly on the cell surface, are GPI-specific PLA_2s (GPI-PLA_2s) involved in fatty acid remodeling of GPI-APs [218]. PGAP3-dependent fatty acid remodeling of GPI-APs has a significant role in the control of autoimmunity, possibly by the regulation of apoptotic cell clearance and Th1/Th2 balance [219]. CRIPTO, a GPI-AP that plays critical roles in early embryonic development by acting as a Nodal co-receptor, is a highly sensitive substrate of PGAP6. CRIPTO is released by PGAP6 as a glycosyllysophosphatidylinositol-bound form and acts as a co-receptor in Nodal signaling. *Pgap6*$^{-/-}$ mice show defects in early embryonic development, particularly in the formation of the anterior-posterior axis, as do *Cripto*$^{-/-}$ embryos, suggesting that PGAP6 plays a critical role in Nodal signaling modulation through CRIPTO shedding [220].

6. Conclusions

In this review, we have provided an overview of the biological functions of a nearly full set of PLA_2s identified to date, particularly over the past five years during which considerable advances in this research field have been made. In the interests of brevity, we have referenced previous reviews whenever possible and apologize to the authors of the numerous original papers that were not explicitly cited. By applying lipidomics approaches to knockout or transgenic mice for various PLA_2s, it has become evident that individual enzymes regulate specific forms of lipid metabolism, perturbation of which can be eventually linked to distinct pathophysiological outcomes. Knowledge of individual PLA_2s acquired from studies using animal models is now being translated to humans. Note that the designation "PLA_2" is used more broadly for enzymes that have significant homology with prototypic PLA_2 subtypes, even if they do not necessarily exert PLA_2 activity. These examples include, for instance, *N*-acyltransferase (c$PLA_2\varepsilon$), ω-*O*-acylceramide synthase (PNPLA1), triglyceride lipase (PNPLA2 and PNPLA3), and lysophospholipase (PNPLA6 and PNPLA7). Several PLA_2s, such as c$PLA_2\delta$, i$PLA_2\gamma$, and PLA2G16, possess PLA_1 activity that is even superior to PLA_2 activity, although biological importance of the *sn*-1 cleavage by these PLA_2s is largely unclear. There are also several enzymes that possess PLA_2 activity but are not designated as PLA_2, as exemplified by the ABHD and GPI-specific PLA_2 families. Thus, the term "PLA_2" is somehow confusing and should be interpreted with caution if researchers encounter the genetic symbol "*PLA2*" on genome-wide transcriptome and proteome analyses.

Lipid metabolic pathways are complex, often redundant and highly interconnected, in some cases being counter regulatory. As recently reviewed by Kokotos and coworkers [221], various PLA_2 subtypes are being targeted pharmacologically to alleviate the symptoms of various disease models, but none of the PLA_2 inhibitors currently developed has reached the market yet. If the modulation of one PLA_2 pathway does not suffice therapeutically, targeting two or more pathways could be effective. On the other hand, blunting of related PLA_2 subtypes simultaneously may not be therapeutically efficient in some cases, since this strategy would block both positive and negative pathways regulated by distinct isoforms. For instance, pan-sPLA_2 inhibitors (e.g., varespladib), which inhibit group I/II/V/X sPLA_2s altogether, have been reported to exert a poor therapeutic effect on atherosclerosis, arthritis, and allergy, probably because they blunt both offensive and defensive sPLA_2s. This points to potential prophylactic or therapeutic use of an agent that could specifically inhibit a particular PLA_2 subtype, even though such a strategy will be a challenge since individual PLA_2s appear to play highly selective and often opposite roles in specific organs and disease states.

At present, the tools that are available for following the dynamics of individual PLA_2s and associated lipid metabolites and for monitoring their precise modifications and spatiotemporal localizations are still technically limited. The actions of lipids are frequently masked by the large steady-state mass of structural lipids in membranes, making it difficult to detect spatiotemporal lipid dynamics and functions. Further advances in this field and their integration for therapeutic use are likely to benefit from improved, time- and space-resolved lipidomics technology for monitoring individual PLA_2s and associated lipid metabolisms within tissue microenvironments. It seems that more work will be necessary to dissect each of the many regulated pathways of bioactive lipids and to define the mechanisms responsible for the regulatory actions of individual PLA_2s, as well as the roles of the pathways in specific responses at the cell, tissue, and organism level. Hopefully, the next decade will yield a more integrated view of the overall map of the PLA_2 network in biology, allowing the therapeutic application of inhibitors or the lipid products of some enzymes to human diseases.

Author Contributions: M.M. wrote the manuscript. Figures were prepared by H.S. and Y.T. All authors have read and agreed to the published version of the manuscript.

Funding: This work was supported by Grants-in-Aid for Scientific Research JP16H02613 and JP20H05691 from Japan Society for the Promotion of Science and AMED-CREST 20gm1210013 and FORCE 19gm4010005from the Japan Agency for Medical Research and Development (to M.M.).

Acknowledgments: We thank all past and present members of the PLA_2 group for their contributions to the scientific work emanating from our laboratory. This article summarizes work from a large number of published papers; however, due to space limitations we have been unable to include all relevant publications in our discussion. We apologize to those authors whose relevant work has not been included in this review article.

Conflicts of Interest: The authors declare no conflict of interest.

References

1. Murakami, M. Novel functions of phospholipase A_2s: Overview. *Biochim. Biophys. Acta Mol. Cell Biol. Lipids* **2019**, *1864*, 763–765. [CrossRef] [PubMed]
2. Murakami, M.; Taketomi, Y.; Miki, Y.; Sato, H.; Hirabayashi, T.; Yamamoto, K. Recent progress in phospholipase A_2 research: From cells to animals to humans. *Prog. Lipid Res.* **2011**, *50*, 152–192. [CrossRef] [PubMed]
3. Murakami, M. Lipoquality control by phospholipase A_2 enzymes. *Proc. Jpn. Acad. Ser. B* **2017**, *93*, 677–702. [CrossRef] [PubMed]
4. Murakami, M.; Taketomi, Y.; Girard, C.; Yamamoto, K.; Lambeau, G. Emerging roles of secreted phospholipase A_2 enzymes: Lessons from transgenic and knockout mice. *Biochimie* **2010**, *92*, 561–582. [CrossRef]
5. Murakami, M.; Taketomi, Y.; Sato, H.; Yamamoto, K. Secreted phospholipase A_2 revisited. *J. Biochem.* **2011**, *150*, 233–255. [CrossRef] [PubMed]
6. Murakami, M.; Sato, H.; Miki, Y.; Yamamoto, K.; Taketomi, Y. A new era of secreted phospholipase A_2. *J. Lipid Res.* **2015**, *56*, 1248–1261. [CrossRef]
7. Murakami, M.; Yamamoto, K.; Miki, Y.; Murase, R.; Sato, H.; Taketomi, Y. The roles of the secreted phospholipase A_2 gene family in immunology. *Adv. Immunol.* **2016**, *132*, 91–134. [CrossRef]
8. Lambeau, G.; Gelb, M.H. Biochemistry and physiology of mammalian secreted phospholipases A_2. *Annu. Rev. Biochem.* **2008**, *77*, 495–520. [CrossRef]
9. Murakami, M.; Miki, Y.; Sato, H.; Murase, R.; Taketomi, Y.; Yamamoto, K. Group IID, IIE, IIF and III secreted phospholipase A_2s. *Biochim. Biophys. Acta Mol. Cell Biol. Lipids* **2019**, *1864*, 803–818. [CrossRef]
10. Seilhamer, J.J.; Randall, T.L.; Yamanaka, M.; Johnson, L.K. Pancreatic phospholipase A_2: Isolation of the human gene and cDNAs from porcine pancreas and human lung. *DNA* **1986**, *5*, 519–527. [CrossRef]
11. Hollie, N.I.; Konaniah, E.S.; Goodin, C.; Hui, D.Y. Group 1B phospholipase A_2 inactivation suppresses atherosclerosis and metabolic diseases in LDL receptor-deficient mice. *Atherosclerosis.* **2014**, *234*, 377–380. [CrossRef] [PubMed]

12. Huggins, K.W.; Boileau, A.C.; Hui, D.Y. Protection against diet-induced obesity and obesity- related insulin resistance in group 1B PLA_2-deficient mice. *Am. J. Physiol. Metab.* **2002**, *283*, E994–E1001. [CrossRef] [PubMed]
13. Hui, D.Y.; Cope, M.J.; Labonté, E.D.; Chang, H.-T.; Shao, J.; Goka, E.; Abousalham, A.; Charmot, D.; Buysse, J. The phospholipase A_2 inhibitor methyl indoxam suppresses diet-induced obesity and glucose intolerance in mice. *Br. J. Pharmacol.* **2009**, *157*, 1263–1269. [CrossRef] [PubMed]
14. Labonté, E.D.; Kirby, R.J.; Schildmeyer, N.M.; Cannon, A.M.; Huggins, K.W.; Hui, D.Y. Phospholipase A_2-mediated lysophospholipid absorption directly contributes to postprandial hyperglycemia. *Diabetes* **2006**, *55*, 935–941. [CrossRef]
15. Wilson, S.G.; Adam, G.; Langdown, M.; Reneland, R.; Braun, A.; Andrew, T.; Surdulescu, G.L.; Norberg, M.; Dudbridge, F.; Reed, P.W.; et al. Linkage and potential association of obesity-related phenotypes with two genes on chromosome 12q24 in a female dizygous twin cohort. *Eur. J. Hum. Genet.* **2006**, *14*, 340–348. [CrossRef]
16. Hui, D.Y. Group 1B phospholipase A_2 in metabolic and inflammatory disease modulation. *Biochim. Biophys. Acta Mol. Cell Biol. Lipids* **2019**, *1864*, 784–788. [CrossRef]
17. Entwistle, L.J.; Pelly, V.S.; Coomes, S.M.; Kannan, Y.; Perez-Lloret, J.; Czieso, S.; Dos Santos, M.S.; Macrae, J.I.; Collinson, L.M.; Sesay, A.; et al. Epithelial-cell-derived phospholipase A_2 group 1B is an endogenous anthelmintic. *Cell Host Microbe* **2017**, *22*, 484–493.e5. [CrossRef]
18. Pothlichet, J.; Rose, T.; Bugault, F.; Jeammet, L.; Meola, A.; Haouz, A.; Saul, F.; Geny, D.; Alcami, J.; Ruiz-Mateos, E.; et al. PLA2G1B is involved in CD4 anergy and CD4 lymphopenia in HIV-infected patients. *J. Clin. Investig.* **2020**, *130*, 2872–2887. [CrossRef]
19. Laine, V.J.; Grass, D.S.; Nevalainen, T.J. Protection by group II phospholipase A_2 against *Staphylococcus aureus*. *J. Immunol.* **1999**, *162*, 7402–7408.
20. Weinrauch, Y.; Abad, C.; Liang, N.S.; Lowry, S.F.; Weiss, J. Mobilization of potent plasma bactericidal activity during systemic bacterial challenge. Role of group IIA phospholipase A_2. *J. Clin. Investig.* **1998**, *102*, 633–638. [CrossRef]
21. Pernet, E.; Guillemot, L.; Burgel, P.-R.; Martin, C.; Lambeau, G.; Sermet-Gaudelus, I.; Sands, D.; LeDuc, D.; Morand, P.C.; Jeammet, L.; et al. *Pseudomonas aeruginosa* eradicates *Staphylococcus aureus* by manipulating the host immunity. *Nat. Commun.* **2014**, *5*, 5105. [CrossRef]
22. Hamaguchi, K.; Kuwata, H.; Yoshihara, K.; Masuda, S.; Shimbara, S.; Oh-Ishi, S.; Murakami, M.; Kudo, I. Induction of distinct sets of secretory phospholipase A_2 in rodents during inflammation. *Biochim. Biophys. Acta Bioenerg.* **2003**, *1635*, 37–47. [CrossRef] [PubMed]
23. Boudreau, L.H.; Duchez, A.-C.; Cloutier, N.; Soulet, D.; Martin, N.; Bollinger, J.; Paré, A.; Rousseau, M.; Naika, G.S.; Lévesque, T.; et al. Platelets release mitochondria serving as substrate for bactericidal group IIA-secreted phospholipase A_2 to promote inflammation. *Blood* **2014**, *124*, 2173–2183. [CrossRef] [PubMed]
24. Duchez, A.-C.; Boudreau, L.H.; Naika, G.S.; Bollinger, J.; Belleannée, C.; Cloutier, N.; Laffont, B.; Mendoza-Villarroel, R.E.; Lévesque, T.; Rollet-Labelle, E.; et al. Platelet microparticles are internalized in neutrophils via the concerted activity of 12-lipoxygenase and secreted phospholipase A_2-IIA. *Proc. Natl. Acad. Sci. USA* **2015**, *112*, E3564–E3573. [CrossRef]
25. Dore, E.; Boilard, E. Roles of secreted phospholipase A_2 group IIA in inflammation and host defense. *Biochim. Biophys. Acta Mol. Cell Biol. Lipids* **2019**, *1864*, 789–802. [CrossRef] [PubMed]
26. Dacheux, M.; Sinou, V.; Payré, C.; Jeammet, L.; Parzy, D.; Grellier, P.; Deregnaucourt, C.; Lambeau, G. Antimalarial activity of human group IIA secreted phospholipase A_2 in relation to enzymatic hydrolysis of oxidized lipoproteins. *Infect. Immun.* **2019**, *87*, e00556-19. [CrossRef] [PubMed]
27. Grass, D.S.; Felkner, R.H.; Chiang, M.Y.; Wallace, R.E.; Nevalainen, T.J.; Bennett, C.F.; Swanson, M.E. Expression of human group II PLA_2 in transgenic mice results in epidermal hyperplasia in the absence of inflammatory infiltrate. *J. Clin. Investig.* **1996**, *97*, 2233–2241. [CrossRef]
28. Chovatiya, G.L.; Sunkara, R.R.; Roy, S.; Godbole, S.R.; Waghmare, S.K. Context-dependent effect of $sPLA_2$-IIA induced proliferation on murine hair follicle stem cells and human epithelial cancer. *EBioMedicine* **2019**, *48*, 364–376. [CrossRef] [PubMed]
29. Kuefner, M.S.; Deng, X.; Stephenson, E.J.; Pham, K.; Park, E.A. Secretory phospholipase A_2 group IIA enhances the metabolic rate and increases glucose utilization in response to thyroid hormone. *FASEB J.* **2018**, *33*, 738–749. [CrossRef]

30. Macphee, M.; Chepenik, K.P.; Liddell, R.A.; Nelson, K.K.; Siracusa, L.D.; Buchberg, A.M. The secretory phospholipase A_2 gene is a candidate for the *Mom1* locus, a major modifier of Apc^{Min}-induced intestinal neoplasia. *Cell* **1995**, *81*, 957–966. [CrossRef]
31. Yamamoto, K.; Miki, Y.; Sato, M.; Taketomi, Y.; Nishito, Y.; Taya, C.; Muramatsu, K.; Ikeda, K.; Nakanishi, H.; Taguchi, R.; et al. The role of group IIF-secreted phospholipase A_2 in epidermal homeostasis and hyperplasia. *J. Exp. Med.* **2015**, *212*, 1901–1919. [CrossRef] [PubMed]
32. Schewe, M.; Franken, P.F.; Sacchetti, A.; Schmitt, M.; Joosten, R.; Böttcher, R.; Van Royen, M.E.; Jeammet, L.; Payre, C.; Scott, P.M.; et al. Secreted phospholipases A_2 are intestinal stem cell niche factors with distinct roles in homeostasis, inflammation, and cancer. *Cell Stem Cell* **2016**, *19*, 38–51. [CrossRef] [PubMed]
33. Murase, R.; Sato, H.; Yamamoto, K.; Ushida, A.; Nishito, Y.; Ikeda, K.; Kobayashi, T.; Yamamoto, T.; Taketomi, Y.; Murakami, M. Group X secreted phospholipase A_2 releases ω3 polyunsaturated fatty acids, suppresses colitis, and promotes sperm fertility. *J. Biol. Chem.* **2016**, *291*, 6895–6911. [CrossRef] [PubMed]
34. Rouault, M.; Le Calvez, C.; Boilard, E.; Surrel, F.; Singer, A.; Ghomashchi, F.; Bezzine, S.; Scarzello, S.; Bollinger, J.; Gelb, M.H.; et al. Recombinant production and properties of binding of the full set of mouse secreted phospholipases A_2 to the mouse M-type receptor. *Biochemistry* **2007**, *46*, 1647–1662. [CrossRef]
35. Kishi, H.; Yamaguchi, K.; Watanabe, K.; Nakamura, K.; Fujioka, D.; Kugiyama, K. Deficiency of phospholipase A_2 receptor exacerbates autoimmune myocarditis in mice. *Inflammation* **2020**, *43*, 1097–1109. [CrossRef]
36. Miki, Y.; Yamamoto, K.; Taketomi, Y.; Sato, H.; Shimo, K.; Kobayashi, T.; Ishikawa, Y.; Ishii, T.; Nakanishi, H.; Ikeda, K.; et al. Lymphoid tissue phospholipase A_2 group IID resolves contact hypersensitivity by driving antiinflammatory lipid mediators. *J. Exp. Med.* **2013**, *210*, 1217–1234. [CrossRef]
37. Miki, Y.; Kidoguchi, Y.; Sato, M.; Taketomi, Y.; Taya, C.; Muramatsu, K.; Gelb, M.H.; Yamamoto, K.; Murakami, M. Dual roles of group IID phospholipase A_2 in inflammation and cancer. *J. Biol. Chem.* **2016**, *291*, 15588–15601. [CrossRef]
38. Vijay, R.; Hua, X.; Meyerholz, D.K.; Miki, Y.; Yamamoto, K.; Gelb, M.; Murakami, M.; Perlman, S. Critical role of phospholipase A_2 group IID in age-related susceptibility to severe acute respiratory syndrome–CoV infection. *J. Exp. Med.* **2015**, *212*, 1851–1868. [CrossRef]
39. Sato, H.; Taketomi, Y.; Miki, Y.; Murase, R.; Yamamoto, K.; Murakami, M. Secreted phospholipase PLA2G2D contributes to metabolic health by mobilizing ω3 polyunsaturated fatty acids in WAT. *Cell Rep.* **2020**, *31*, 107579. [CrossRef]
40. Takabatake, N.; Sata, M.; Inoue, S.; Shibata, Y.; Abe, S.; Wada, T.; Machiya, J.-I.; Ji, G.; Matsuura, T.; Takeishi, Y.; et al. A novel polymorphism in secretory phospholipase A_2-IID is associated with body weight loss in chronic obstructive pulmonary disease. *Am. J. Respir. Crit. Care Med.* **2005**, *172*, 1097–1104. [CrossRef]
41. Sato, H.; Taketomi, Y.; Ushida, A.; Isogai, Y.; Kojima, T.; Hirabayashi, T.; Miki, Y.; Yamamoto, K.; Nishito, Y.; Kobayashi, T.; et al. The adipocyte-inducible secreted phospholipases PLA2G5 and PLA2G2E play distinct roles in obesity. *Cell Metab.* **2014**, *20*, 119–132. [CrossRef] [PubMed]
42. Yamamoto, K.; Miki, Y.; Sato, H.; Nishito, Y.; Gelb, M.H.; Taketomi, Y.; Murakami, M. Expression and function of group IIE phospholipase A_2 in mouse skin. *J. Biol. Chem.* **2016**, *291*, 15602–15613. [CrossRef]
43. Valentin, E. Novel secreted phospholipase A_2 with homology to the group III bee venom enzyme. *J. Biol. Chem.* **2000**, *275*, 7492–7496. [CrossRef]
44. Murakami, M.; Masuda, S.; Shimbara, S.; Bezzine, S.; Lazdunski, M.; Lambeau, G.; Gelb, M.H.; Matsukura, S.; Kokubu, F.; Adachi, M.; et al. Cellular arachidonate-releasing function of novel classes of secretory phospholipase A_2s (groups III and XII). *J. Biol. Chem.* **2003**, *278*, 10657–10667. [CrossRef]
45. Sato, H.; Taketomi, Y.; Isogai, Y.; Miki, Y.; Yamamoto, K.; Masuda, S.; Hosono, T.; Arata, S.; Ishikawa, Y.; Ishii, T.; et al. Group III secreted phospholipase A_2 regulates epididymal sperm maturation and fertility in mice. *J. Clin. Investig.* **2010**, *120*, 1400–1414. [CrossRef] [PubMed]
46. Taketomi, Y.; Ueno, N.; Kojima, T.; Sato, H.; Murase, R.; Yamamoto, K.; Tanaka, S.; Sakanaka, M.; Nakamura, M.; Nishito, Y.; et al. Mast cell maturation is driven via a group III phospholipase A_2-prostaglandin D_2–DP1 receptor paracrine axis. *Nat. Immunol.* **2013**, *14*, 554–563. [CrossRef]
47. Murakami, M.; Masuda, S.; Shimbara, S.; Ishikawa, Y.; Ishii, T.; Kudo, I. Cellular distribution, post-translational modification, and tumorigenic potential of human group III secreted phospholipase A_2. *J. Biol. Chem.* **2005**, *280*, 24987–24998. [CrossRef] [PubMed]

48. Kazama, S.; Kitayama, J.; Hiyoshi, M.; Taketomi, Y.; Murakami, M.; Nishikawa, T.; Tanaka, T.; Tanaka, J.; Kiyomatsu, T.; Kawai, K.; et al. Phospholipase A_2 group III and group X have opposing associations with prognosis in colorectal cancer. *Anticancer Res.* **2015**, *35*, 2983–2990. [PubMed]
49. Hoeft, B.; Linseisen, J.; Beckmann, L.; Müller-Decker, K.; Canzian, F.; Hüsing, A.; Kaaks, R.; Vogel, U.; Jakobsen, M.U.; Overvad, K.; et al. Polymorphisms in fatty acid metabolism-related genes are associated with colorectal cancer risk. *Carcinog.* **2009**, *31*, 466–472. [CrossRef]
50. Murase, R.; Taketomi, Y.; Miki, Y.; Nishito, Y.; Saito, M.; Fukami, K.; Yamamoto, K.; Murakami, M. Group III phospholipase A_2 promotes colitis and colorectal cancer. *Sci. Rep.* **2017**, *7*, 12261. [CrossRef] [PubMed]
51. Lin, S.; Wang, D.; Iyer, S.; Ghaleb, A.M.; Shim, H.; Yang, V.W.; Chun, J.; Yun, C.C. The absence of LPA_2 attenuates tumor formation in an experimental model of colitis-associated cancer. *Gastroenterol.* **2009**, *136*, 1711–1720. [CrossRef] [PubMed]
52. Stančić, A.; Jandl, K.; Hasenöhrl, C.; Reichmann, F.; Marsche, G.; Schuligoi, R.; Heinemann, A.; Storr, M.A.; Schicho, R. The GPR55 antagonist CID16020046 protects against intestinal inflammation. *Neurogastroenterol. Motil.* **2015**, *27*, 1432–1445. [CrossRef] [PubMed]
53. Murakami, M.; Koduri, R.S.; Enomoto, A.; Shimbara, S.; Seki, M.; Yoshihara, K.; Singer, A.; Valentin, E.; Ghomashchi, F.; Lambeau, G.; et al. Distinct arachidonate-releasing functions of mammalian secreted phospholipase A_2s in human embryonic kidney 293 and rat mastocytoma RBL-2H3 cells through heparan sulfate shuttling and external plasma membrane mechanisms. *J. Biol. Chem.* **2000**, *276*, 10083–10096. [CrossRef] [PubMed]
54. Murakami, M.; Shimbara, S.; Kambe, T.; Kuwata, H.; Winstead, M.V.; Tischfield, J.A.; Kudo, I. The functions of five distinct mammalian phospholipase A_2s in regulating arachidonic acid release. Type IIa and type V secretory phospholipase A_2s are functionally redundant and act in concert with cytosolic phospholipase A_2. *J. Biol. Chem.* **1998**, *273*, 14411–14423. [CrossRef]
55. Ohtsuki, M.; Taketomi, Y.; Arata, S.; Masuda, S.; Ishikawa, Y.; Ishii, T.; Takanezawa, Y.; Aoki, J.; Arai, H.; Yamamoto, K.; et al. Transgenic expression of group V, but not group X, secreted phospholipase A_2 in mice leads to neonatal lethality because of lung dysfunction. *J. Biol. Chem.* **2006**, *281*, 36420–36433. [CrossRef]
56. Wootton, P.T.; Arora, N.L.; Drenos, F.; Thompson, S.R.; Cooper, J.A.; Stephens, J.W.; Hurel, S.J.; Hurt-Camejo, E.; Wiklund, O.; Humphries, S.E.; et al. Tagging SNP haplotype analysis of the secretory PLA_2-V gene, *PLA2G5*, shows strong association with LDL and oxLDL levels, suggesting functional distinction from $sPLA_2$-IIA: Results from the UDACS study. *Hum. Mol. Genet.* **2007**, *16*, 1437–1444. [CrossRef]
57. Ohta, S.; Imamura, M.; Xing, W.; Boyce, J.A.; Balestrieri, B. Group V secretory phospholipase A_2 is involved in macrophage activation and is sufficient for macrophage effector functions in allergic pulmonary inflammation. *J. Immunol.* **2013**, *190*, 5927–5938. [CrossRef]
58. Yamaguchi, M.; Samuchiwal, S.K.; Quehenberger, O.; Boyce, J.A.; Balestrieri, B. Macrophages regulate lung ILC2 activation via Pla2g5-dependent mechanisms. *Mucosal Immunol.* **2017**, *11*, 615–626. [CrossRef]
59. Samuchiwal, S.K.; Balestrieri, B. Harmful and protective roles of group V phospholipase A_2: Current perspectives and future directions. *Biochim. Biophys. Acta Mol. Cell Biol. Lipids* **2019**, *1864*, 819–826. [CrossRef]
60. Balestrieri, B.; Maekawa, A.; Xing, W.; Gelb, M.H.; Katz, H.R.; Arm, J.P. Group V secretory phospholipase A_2 modulates phagosome maturation and regulates the innate immune response against *Candida albicans*. *J. Immunol.* **2009**, *182*, 4891–4898. [CrossRef]
61. Boilard, E.; Lai, Y.; Larabee, K.; Balestrieri, B.; Ghomashchi, F.; Fujioka, D.; Gobezie, R.; Coblyn, J.S.; Weinblatt, M.E.; Massarotti, E.M.; et al. A novel anti-inflammatory role for secretory phospholipase A_2 in immune complex-mediated arthritis. *EMBO Mol. Med.* **2010**, *2*, 172–187. [CrossRef]
62. Degousee, N.; Kelvin, D.J.; Geisslinger, G.; Hwang, D.M.; Stefanski, E.; Wang, X.-H.; Danesh, A.; Angioni, C.; Schmidt, H.; Lindsay, T.F.; et al. Group V phospholipase A_2 in bone marrow-derived myeloid cells and bronchial epithelial cells promotes bacterial clearance after *Escherichia coli* pneumonia. *J. Biol. Chem.* **2011**, *286*, 35650–35662. [CrossRef] [PubMed]
63. Rubio, J.M.; Rodríguez, J.P.; Gil-De-Gómez, L.; Guijas, C.; Balboa, M.A.; Balsinde, J. Group V secreted phospholipase A_2 is upregulated by IL-4 in human macrophages and mediates phagocytosis via hydrolysis of ethanolamine phospholipids. *J. Immunol.* **2015**, *194*, 3327–3339. [CrossRef]
64. Nienaber, C.A.; Clough, R.E.; Sakalihasan, N.; Suzuki, T.; Gibbs, R.; Mussa, F.; Jenkins, M.P.; Thompson, M.M.; Evangelista, A.; Yeh, J.S.M.; et al. Aortic dissection. *Nat. Rev. Dis. Prim.* **2016**, *2*, 16054. [CrossRef] [PubMed]

65. Watanabe, K.; Taketomi, Y.; Miki, Y.; Kugiyama, K.; Murakami, M. Group V secreted phospholipase A_2 plays a protective role against aortic dissection. *J. Biol. Chem.* **2020**, *295*, 10092–10111. [CrossRef] [PubMed]
66. Jemel, I.; Ii, H.; Oslund, R.C.; Payré, C.; Dabert-Gay, A.-S.; Douguet, D.; Chargui, K.; Scarzello, S.; Gelb, M.H.; Lambeau, G. Group X secreted phospholipase A_2 proenzyme is matured by a furin-like proprotein convertase and releases arachidonic acid inside of human HEK293 cells. *J. Biol. Chem.* **2011**, *286*, 36509–36521. [CrossRef]
67. Escoffier, J.; Jemel, I.; Tanemoto, A.; Taketomi, Y.; Payré, C.; Coatrieux, C.; Sato, H.; Yamamoto, K.; Masuda, S.; Pernet-Gallay, K.; et al. Group X phospholipase A_2 is released during sperm acrosome reaction and controls fertility outcome in mice. *J. Clin. Investig.* **2010**, *120*, 1415–1428. [CrossRef] [PubMed]
68. Nahed, R.A.; Martinez, G.; Escoffier, J.; Yassine, S.; Karaouzene, T.; Hograindleur, J.-P.; Turk, J.; Kokotos, G.; Ray, P.F.; Bottari, S.P.; et al. Progesterone-induced acrosome exocytosis requires sequential involvement of calcium-independent phospholipase $A_2\beta$ (i$PLA_2\beta$) and group X secreted phospholipase A_2 ($sPLA_2$). *J. Biol. Chem.* **2015**, *291*, 3076–3089. [CrossRef] [PubMed]
69. Sato, H.; Isogai, Y.; Masuda, S.; Taketomi, Y.; Miki, Y.; Kamei, D.; Hara, S.; Kobayashi, T.; Ishikawa, Y.; Ishii, T.; et al. Physiological roles of group X-secreted phospholipase A_2 in reproduction, gastrointestinal phospholipid digestion, and neuronal function. *J. Biol. Chem.* **2011**, *286*, 11632–11648. [CrossRef]
70. Ait-Oufella, H.; Herbin, O.; Lahoute, C.; Coatrieux, C.; Loyer, X.; Joffre, J.; Laurans, L.; Ramkhelawon, B.; Blanc-Brude, O.; Karabina, S.; et al. Group X secreted phospholipase A_2 limits the development of atherosclerosis in LDL receptor–null mice. *Arter. Thromb. Vasc. Biol.* **2013**, *33*, 466–473. [CrossRef]
71. Li, X.; Shridas, P.; Forrest, K.; Bailey, W.; Webb, N.R. Group X secretory phospholipase A_2 negatively regulates adipogenesis in murine models. *FASEB J.* **2010**, *24*, 4313–4324. [CrossRef]
72. Zack, M.; Boyanovsky, B.B.; Shridas, P.; Bailey, W.; Forrest, K.; Howatt, D.A.; Gelb, M.H.; De Beer, F.C.; Daugherty, A.; Webb, N.R. Group X secretory phospholipase A_2 augments angiotensin II-induced inflammatory responses and abdominal aortic aneurysm formation in apoE-deficient mice. *Atheroscler.* **2011**, *214*, 58–64. [CrossRef] [PubMed]
73. Henderson, W.R.; Chi, E.Y.; Bollinger, J.G.; Tien, Y.-T.; Ye, X.; Castelli, L.; Rubtsov, Y.; Singer, A.G.; Chiang, G.K.; Nevalainen, T.; et al. Importance of group X–secreted phospholipase A_2 in allergen-induced airway inflammation and remodeling in a mouse asthma model. *J. Exp. Med.* **2007**, *204*, 865–877. [CrossRef]
74. Nolin, J.D.; Lai, Y.; Ogden, H.L.; Manicone, A.M.; Murphy, R.C.; An, D.; Frevert, C.W.; Ghomashchi, F.; Naika, G.S.; Gelb, M.H.; et al. Secreted PLA_2 group X orchestrates innate and adaptive immune responses to inhaled allergen. *JCI Insight* **2017**, *2*, e94929. [CrossRef] [PubMed]
75. Ogden, H.L.; Lai, Y.; Nolin, J.D.; An, D.; Frevert, C.W.; Gelb, M.H.; Altemeier, W.A.; Hallstrand, T.S. Secreted phospholipase A_2 group X acts as an adjuvant for type 2 inflammation, leading to an allergen-specific immune response in the lung. *J. Immunol.* **2020**, *204*, 3097–3107. [CrossRef]
76. Nolin, J.D.; Murphy, R.C.; Gelb, M.H.; Altemeier, W.A.; Henderson, W.R.; Hallstrand, T.S. Function of secreted phospholipase A_2 group-X in asthma and allergic disease. *Biochim. Biophys. Acta Mol. Cell Biol. Lipids* **2019**, *1864*, 827–837. [CrossRef] [PubMed]
77. Ohto, T.; Uozumi, N.; Hirabayashi, T.; Shimizu, T. Identification of novel cytosolic phospholipase A_2s, murine $cPLA_2\delta$, ϵ, and ζ, which form a gene cluster with $cPLA_2\beta$. *J. Biol. Chem.* **2005**, *280*, 24576–24583. [CrossRef]
78. Ghosh, M.; Loper, R.; Gelb, M.H.; Leslie, C.C. Identification of the expressed form of human cytosolic phospholipase $A_2\beta$ ($cPLA_2\beta$): $cPLA_2\beta3$ is a novel variant localized to mitochondria and early endosomes. *J. Biol. Chem.* **2006**, *281*, 16615–16624. [CrossRef]
79. Ghosh, M.; Loper, R.; Ghomashchi, F.; Tucker, D.E.; Bonventre, J.V.; Gelb, M.H.; Leslie, C.C. Function, activity, and membrane targeting of cytosolic phospholipase $A_2\zeta$ in mouse lung fibroblasts. *J. Biol. Chem.* **2007**, *282*, 11676–11686. [CrossRef] [PubMed]
80. Clark, J.D.; Lin, L.-L.; Kriz, R.W.; Ramesha, C.S.; Sultzman, L.A.; Lin, A.Y.; Milona, N.; Knopf, J.L. A novel arachidonic acid-selective cytosolic PLA_2 contains a Ca^{2+}-dependent translocation domain with homology to PKC and GAP. *Cell* **1991**, *65*, 1043–1051. [CrossRef]
81. Lin, L.-L.; Wartmann, M.; Lin, A.Y.; Knopf, J.L.; Seth, A.; Davis, R.J. $cPLA_2$ is phosphorylated and activated by MAP kinase. *Cell* **1993**, *72*, 269–278. [CrossRef]
82. Casas, J.; Gijón, M.A.; Vigo, A.G.; Crespo, M.S.; Balsinde, J.; Balboa, M.A. Phosphatidylinositol 4,5-bisphosphate anchors cytosolic group IVA phospholipase A_2 to perinuclear membranes and decreases its calcium requirement for translocation in live cells. *Mol. Biol. Cell* **2006**, *17*, 155–162. [CrossRef] [PubMed]

83. Leslie, C.C. Cytosolic phospholipase A_2: Physiological function and role in disease. *J. Lipid Res.* **2015**, *56*, 1386–1402. [CrossRef] [PubMed]
84. Shimizu, T. Lipid mediators in health and disease: Enzymes and receptors as therapeutic targets for the regulation of immunity and inflammation. *Annu. Rev. Pharmacol. Toxicol.* **2009**, *49*, 123–150. [CrossRef]
85. Slatter, D.A.; Aldrovandi, M.; O'Connor, A.; Allen, S.M.; Brasher, C.J.; Murphy, R.C.; Mecklemann, S.; Ravi, S.; Darley-Usmar, V.; O'Donnell, V.B. Mapping the human platelet lipidome reveals cytosolic phospholipase A_2 as a regulator of mitochondrial bioenergetics during activation. *Cell Metab.* **2016**, *23*, 930–944. [CrossRef]
86. Stahelin, R.V.; Subramanian, P.; Vora, M.; Cho, W.; Chalfant, C.E. Ceramide-1-phosphate binds group IVA cytosolic phospholipase A_2 via a novel site in the C2 domain. *J. Biol. Chem.* **2007**, *282*, 20467–20474. [CrossRef]
87. Macknight, H.P.; Stephenson, D.J.; Hoeferlin, L.A.; Benusa, S.D.; DeLigio, J.T.; Maus, K.D.; Ali, A.N.; Wayne, J.S.; Park, M.A.; Hinchcliffe, E.H.; et al. The interaction of ceramide 1-phosphate with group IVA cytosolic phospholipase A_2 coordinates acute wound healing and repair. *Sci. Signal.* **2019**, *12*, eaav5918. [CrossRef]
88. Chao, C.-C.; Gutiérrez-Vázquez, C.; Rothhammer, V.; Mayo, L.; Wheeler, M.A.; Tjon, E.C.; Zandee, S.E.; Blain, M.; De Lima, K.A.; Takenaka, M.C.; et al. Metabolic control of astrocyte pathogenic activity via $cPLA_2$-MAVS. *Cell* **2019**, *179*, 1483–1498.e22. [CrossRef]
89. Koundouros, N.; Karali, E.; Tripp, A.; Valle, A.; Inglese, P.; Perry, N.J.; Magee, D.J.; Virmouni, S.A.; Elder, G.A.; Tyson, A.L.; et al. Metabolic fingerprinting links oncogenic PIK3CA with enhanced arachidonic acid-derived eicosanoids. *Cell* **2020**, *181*, 1596–1611.e27. [CrossRef]
90. Ghomashchi, F.; Naika, G.S.; Bollinger, J.G.; Aloulou, A.; Lehr, M.; Leslie, C.C.; Gelb, M.H. Interfacial Kinetic and Binding Properties of mammalian group IVB phospholipase A_2 ($cPLA_2\beta$) and comparison with the other $cPLA_2$ isoforms. *J. Biol. Chem.* **2010**, *285*, 36100–36111. [CrossRef]
91. Cheng, Y.; Wang, Y.; Li, J.; Chang, I.; Wang, C.-Y. A novel read-through transcript JMJD7-PLA2G4B regulates head and neck squamous cell carcinoma cell proliferation and survival. *Oncotarget* **2016**, *8*, 1972–1982. [CrossRef]
92. Saare, M.; Tserel, L.; Haljasmägi, L.; Taalberg, E.; Peet, N.; Eimre, M.; Vetik, R.; Kingo, K.; Saks, K.; Tamm, R.; et al. Monocytes present age-related changes in phospholipid concentration and decreased energy metabolism. *Aging Cell* **2020**, *19*, e13127. [CrossRef] [PubMed]
93. Underwood, K.W.; Song, C.; Kriz, R.W.; Chang, X.J.; Knopf, J.L.; Lin, L.-L. A novel calcium-independent phospholipase A_2, $cPLA_2\gamma$, that is prenylated and contains homology to $cPLA_2$. *J. Biol. Chem.* **1998**, *273*, 21926–21932. [CrossRef] [PubMed]
94. Asai, K.; Hirabayashi, T.; Houjou, T.; Uozumi, N.; Taguchi, R.; Shimizu, T. Human group IVC phospholipase A_2 ($cPLA_2\gamma$). Roles in the membrane remodeling and activation induced by oxidative stress. *J. Biol. Chem.* **2003**, *278*, 8809–8814. [CrossRef] [PubMed]
95. Lebrero, P.; Astudillo, A.M.; Rubio, J.M.; Fernández-Caballero, L.; Kokotos, G.; Balboa, M.A.; Balsinde, J. Cellular plasmalogen content does not influence arachidonic acid levels or distribution in macrophages: A role for cytosolic phospholipase $A_2\gamma$ in phospholipid remodeling. *Cells* **2019**, *8*, 799. [CrossRef]
96. Vitale, A.; Perlin, J.; Leonelli, L.; Herr, J.; Wright, P.; Digilio, L.; Coonrod, S. Mouse $cPLA_2\gamma$, a novel oocyte and early embryo-abundant phospholipase $A_2\gamma$-like protein, is targeted to the nuclear envelope during germinal vesicle breakdown. *Dev. Biol.* **2005**, *282*, 374–384. [CrossRef] [PubMed]
97. Su, X.; Liu, S.; Zhang, X.; Lam, S.M.; Hu, X.; Zhou, Y.; Chen, J.; Wang, Y.; Wu, C.; Shui, G.; et al. Requirement of cytosolic phospholipase $A_2\gamma$ in lipid droplet formation. *Biochim. Biophys. Acta Mol. Cell Biol. Lipids.* **2017**, *1862*, 692–705. [CrossRef]
98. Xu, S.; Pei, R.; Guo, M.; Han, Q.; Lai, J.; Wang, Y.; Wu, C.; Zhou, Y.; Lu, M.; Chen, X. Cytosolic phospholipase $A_2\gamma$ is involved in hepatitis C virus replication and assembly. *J. Virol.* **2012**, *86*, 13025–13037. [CrossRef]
99. Chiba, H.; Michibata, H.; Wakimoto, K.; Seishima, M.; Kawasaki, S.; Okubo, K.; Mitsui, H.; Torii, H.; Imai, Y. Cloning of a gene for a novel epithelium-specific cytosolic phospholipase A_2, $cPLA_2\delta$, induced in psoriatic skin. *J. Biol. Chem.* **2004**, *279*, 12890–12897. [CrossRef]
100. Cheung, K.L.; Jarrett, R.; Subramaniam, S.; Salimi, M.; Gutowska-Owsiak, D.; Chen, Y.-L.; Hardman, C.; Xue, L.; Cerundolo, V.; Ogg, G. Psoriatic T cells recognize neolipid antigens generated by mast cell phospholipase delivered by exosomes and presented by CD1a. *J. Exp. Med.* **2016**, *213*, 2399–2412. [CrossRef]
101. Tsuboi, K.; Uyama, T.; Okamoto, Y.; Ueda, N. Endocannabinoids and related *N*-acylethanolamines: Biological activities and metabolism. *Inflamm. Regen.* **2018**, *38*, 28. [CrossRef] [PubMed]

102. Ogura, Y.; Parsons, W.H.; Kamat, S.S.; Cravatt, B.F. A calcium-dependent acyltransferase that produces *N*-acyl phosphatidylethanolamines. *Nat. Chem. Biol.* **2016**, *12*, 669–671. [CrossRef] [PubMed]
103. Mustafiz, S.S.B.; Uyama, T.; Morito, K.; Takahashi, N.; Kawai, K.; Hussain, Z.; Tsuboi, K.; Araki, N.; Yamamoto, K.; Tanaka, T.; et al. Intracellular Ca^{2+}-dependent formation of *N*-acyl-phosphatidylethanolamines by human cytosolic phospholipase $A_2\varepsilon$. *Biochim. Biophys. Acta Mol. Cell Biol. Lipids* **2019**, *1864*, 158515. [CrossRef] [PubMed]
104. Binte Mustafiz, S.S.; Uyama, T.; Hussain, Z.; Kawai, K.; Tsuboi, K.; Araki, N.; Ueda, N. The role of intracellular anionic phospholipids in the production of *N*-acyl-phosphatidylethanolamines by cytosolic phospholipase $A_2\varepsilon$. *J. Biochem.* **2019**, *165*, 343–352. [CrossRef]
105. Hussain, Z.; Uyama, T.; Kawai, K.; Mustafiz, S.S.B.; Tsuboi, K.; Earaki, N.; Ueda, N. Phosphatidylserine-stimulated production of *N*-acyl-phosphatidylethanolamines by Ca^{2+}-dependent *N*-acyltransferase. *Biochim. Biophys. Acta Mol. Cell Biol. Lipids* **2018**, *1863*, 493–502. [CrossRef]
106. Capestrano, M.; Mariggio, S.; Perinetti, G.; Egorova, A.V.; Iacobacci, S.; Santoro, M.; Di Pentima, A.; Iurisci, C.; Egorov, M.V.; Di Tullio, G.; et al. Cytosolic phospholipase $A_2\varepsilon$ drives recycling through the clathrin-independent endocytic route. *J. Cell Sci.* **2014**, *127*, 977–993. [CrossRef]
107. Everard, A.; Plovier, H.; Rastelli, M.; Van Hul, M.; D'Oplinter, A.D.W.; Geurts, L.; Druart, C.; Robine, S.; Delzenne, N.M.; Muccioli, G.G.; et al. Intestinal epithelial *N*-acylphosphatidylethanolamine phospholipase D links dietary fat to metabolic adaptations in obesity and steatosis. *Nat. Commun.* **2019**, *10*, 457. [CrossRef]
108. Mock, E.D.; Mustafa, M.; Gunduz-Cinar, O.; Cinar, R.; Petrie, G.N.; Kantae, V.; Di, X.; Ogasawara, D.; Varga, Z.V.; Paloczi, J.; et al. Discovery of a NAPE-PLD inhibitor that modulates emotional behavior in mice. *Nat. Chem. Biol.* **2020**, *16*, 667–675. [CrossRef]
109. Perez-Gonzalez, M.; Mendioroz, M.; Badesso, S.; Sucunza, D.; Roldan, M.; Espelosín, M.; Ursua, S.; Lujan, R.; Cuadrado-Tejedor, M.; García-Osta, A. PLA2G4E, a candidate gene for resilience in Alzheimer´s disease and a new target for dementia treatment. *Prog. Neurobiol.* **2020**, *191*, 101818. [CrossRef]
110. Ghosh, M.; Tucker, D.E.; Burchett, S.A.; Leslie, C.C. Properties of the group IV phospholipase A_2 family. *Prog. Lipid Res.* **2006**, *45*, 487–510. [CrossRef]
111. Moon, S.H.; Liu, X.; Cedars, A.M.; Yang, K.; Kiebish, M.A.; Joseph, S.M.; Kelley, J.; Jenkins, C.M.; Gross, R.W. Heart failure–induced activation of phospholipase $iPLA_2\gamma$ generates hydroxyeicosatetraenoic acids opening the mitochondrial permeability transition pore. *J. Biol. Chem.* **2017**, *293*, 115–129. [CrossRef]
112. Kienesberger, P.C.; Oberer, M.; Lass, A.; Zechner, R. Mammalian patatin domain containing proteins: A family with diverse lipolytic activities involved in multiple biological functions. *J. Lipid Res.* **2008**, *50*, S63–S68. [CrossRef]
113. Ramanadham, S.; Ali, T.; Ashley, J.W.; Bone, R.N.; Hancock, W.D.; Lei, X. Calcium-independent phospholipases A_2 and their roles in biological processes and diseases. *J. Lipid Res.* **2015**, *56*, 1643–1668. [CrossRef] [PubMed]
114. Larsson, P.K.A.; Claesson, H.-E.; Kennedy, B.P. Multiple splice variants of the human calcium-independent phospholipase A_2 and their effect on enzyme activity. *J. Biol. Chem.* **1998**, *273*, 207–214. [CrossRef] [PubMed]
115. Tang, J.; Kriz, R.W.; Wolfman, N.; Shaffer, M.; Seehra, J.; Jones, S.S. A novel cytosolic calcium-independent phospholipase A_2 contains eight ankyrin motifs. *J. Biol. Chem.* **1997**, *272*, 8567–8575. [CrossRef] [PubMed]
116. Malley, K.R.; Koroleva, O.; Miller, I.; Sanishvili, R.; Jenkins, C.M.; Gross, R.W.; Korolev, S. The structure of $iPLA_2\beta$ reveals dimeric active sites and suggests mechanisms of regulation and localization. *Nat. Commun.* **2018**, *9*, 765. [CrossRef]
117. Turk, J.; White, T.D.; Nelson, A.J.; Lei, X.; Ramanadham, S. $iPLA_2\beta$ and its role in male fertility, neurological disorders, metabolic disorders, and inflammation. *Biochim. Biophys. Acta Mol. Cell Biol. Lipids* **2019**, *1864*, 846–860. [CrossRef]
118. Morgan, N.V.; Westaway, S.K.; Morton, J.E.V.; Gregory, A.; Gissen, P.; Sonek, S.; Cangul, H.; Coryell, J.; Canham, N.; Nardocci, N.; et al. *PLA2G6*, encoding a phospholipase A_2, is mutated in neurodegenerative disorders with high brain iron. *Nat. Genet.* **2006**, *38*, 752–754. [CrossRef]
119. Zhou, Q.; Yen, A.; Rymarczyk, G.; Asai, H.; Trengrove, C.; Aziz, N.; Kirber, M.T.; Mostoslavsky, G.; Ikezu, T.; Wolozin, B.; et al. Impairment of PARK14-dependent Ca^{2+} signalling is a novel determinant of Parkinson's disease. *Nat. Commun.* **2016**, *7*, 10332. [CrossRef] [PubMed]

120. Kinghorn, K.J.; Castillo-Quan, J.I.; Bartolome, F.; Angelova, P.R.; Li, L.; Pope, S.; Cochemé, H.M.; Khan, S.; Asghari, S.; Bhatia, K.P.; et al. Loss of *PLA2G6* leads to elevated mitochondrial lipid peroxidation and mitochondrial dysfunction. *Brain* **2015**, *138*, 1801–1816. [CrossRef]
121. Lin, G.; Lee, P.-T.; Chen, K.; Mao, D.; Tan, K.L.; Zuo, Z.; Lin, W.-W.; Wang, L.; Bellen, H.J. Phospholipase PLA2G6, a parkinsonism-associated gene, affects Vps26 and Vps35, retromer function, and ceramide levels, similar to α-synuclein gain. *Cell Metab.* **2018**, *28*, 605–618.e6. [CrossRef]
122. Mori, A.; Hatano, T.; Inoshita, T.; Shiba-Fukushima, K.; Koinuma, T.; Meng, H.; Kubo, S.-I.; Spratt, S.; Cui, C.; Yamashita, C.; et al. Parkinson's disease-associated *iPLA$_2$-VIA*/PLA2G6 regulates neuronal functions and α-synuclein stability through membrane remodeling. *Proc. Natl. Acad. Sci. USA* **2019**, *116*, 20689–20699. [CrossRef] [PubMed]
123. Deng, X.; Wang, J.; Jiao, L.; Utaipan, T.; Tuma-Kellner, S.; Schmitz, G.; Liebisch, G.; Stremmel, W.; Chamulitrat, W. iPLA$_2$β deficiency attenuates obesity and hepatic steatosis in *ob/ob* mice through hepatic fatty-acyl phospholipid remodeling. *Biochim. Biophys. Acta Mol. Cell Biol. Lipids* **2016**, *1861*, 449–461. [CrossRef]
124. Jiao, L.; Gan-Schreier, H.; Zhu, X.; Wei, W.; Tuma-Kellner, S.; Liebisch, G.; Stremmel, W.; Chamulitrat, W. Ageing sensitized by iPLA$_2$β deficiency induces liver fibrosis and intestinal atrophy involving suppression of homeostatic genes and alteration of intestinal lipids and bile acids. *Biochim. Biophys. Acta Mol. Cell Biol. Lipids* **2017**, *1862*, 1520–1533. [CrossRef]
125. Otto, A.-C.; Gan-Schreier, H.; Zhu, X.; Tuma-Kellner, S.; Staffer, S.; Ganzha, A.; Liebisch, G.; Chamulitrat, W. Group VIA phospholipase A$_2$ deficiency in mice chronically fed with high-fat-diet attenuates hepatic steatosis by correcting a defect of phospholipid remodeling. *Biochim. Biophys. Acta Mol. Cell Biol. Lipids* **2019**, *1864*, 662–676. [CrossRef] [PubMed]
126. Zhu, X.; Gan-Schreier, H.; Otto, A.-C.; Cheng, Y.; Staffer, S.; Tuma-Kellner, S.; Ganzha, A.; Liebisch, G.; Chamulitrat, W. iPLA$_2$β deficiency in mice fed with MCD diet does not correct the defect of phospholipid remodeling but attenuates hepatocellular injury via an inhibition of lipid uptake genes. *Biochim. Biophys. Acta Mol. Cell Biol. Lipids* **2019**, *1864*, 677–687. [CrossRef]
127. Liu, X.; Moon, S.H.; Jenkins, C.M.; Sims, H.F.; Gross, R.W. Cyclooxygenase-2 mediated oxidation of 2-arachidonoyl-lysophospholipids identifies unknown lipid signaling pathways. *Cell Chem. Biol.* **2016**, *23*, 1217–1227. [CrossRef]
128. Moon, S.H.; Jenkins, C.M.; Liu, X.; Guan, S.; Mancuso, D.J.; Gross, R.W. Activation of mitochondrial calcium-independent phospholipase A$_2$γ (iPLA$_2$γ) by divalent cations mediating arachidonate release and production of downstream eicosanoids. *J. Biol. Chem.* **2012**, *287*, 14880–14895. [CrossRef] [PubMed]
129. Mancuso, D.J.; Sims, H.F.; Han, X.; Jenkins, C.M.; Guan, S.P.; Yang, K.; Moon, S.H.; Pietka, T.; Abumrad, N.A.; Schlesinger, P.H.; et al. Genetic ablation of calcium-independent phospholipase A$_2$γ leads to alterations in mitochondrial lipid metabolism and function resulting in a deficient mitochondrial bioenergetic phenotype. *J. Biol. Chem.* **2007**, *282*, 34611–34622. [CrossRef] [PubMed]
130. Bione, S.; D'Adamo, P.; Maestrini, E.; Gedeon, A.K.; Bolhuis, P.A.; Toniolo, D. A novel X-linked gene, G4.5. is responsible for Barth syndrome. *Nat. Genet.* **1996**, *12*, 385–389. [CrossRef] [PubMed]
131. Saunders, C.J.; Moon, S.H.; Liu, X.; Thiffault, I.; Coffman, K.; LePichon, J.B.; Taboada, E.; Smith, L.D.; Farrow, E.G.; Miller, N.; et al. Loss of function variants in human *PNPLA8* encoding calcium-independent phospholipase A$_2$γ recapitulate the mitochondriopathy of the homologous null mouse. *Hum. Mutat.* **2015**, *36*, 301–306. [CrossRef] [PubMed]
132. Yoda, E.; Rai, K.; Ogawa, M.; Takakura, Y.; Kuwata, H.; Suzuki, H.; Nakatani, Y.; Murakami, M.; Hara, S. Group VIB Calcium-independent phospholipase A$_2$ (iPLA$_2$γ) regulates platelet activation, hemostasis and thrombosis in mice. *PLoS ONE* **2014**, *9*, e109409. [CrossRef]
133. Moon, S.H.; Mancuso, D.J.; Sims, H.F.; Liu, X.; Nguyen, A.L.; Yang, K.; Guan, S.; Dilthey, B.G.; Jenkins, C.M.; Weinheimer, C.J.; et al. Cardiac myocyte-specific knock-out of calcium-independent phospholipase A$_2$γ (iPLA$_2$γ) decreases oxidized fatty acids during ischemia/reperfusion and reduces infarct size. *J. Biol. Chem.* **2016**, *291*, 19687–19700. [CrossRef] [PubMed]
134. Liu, G.-Y.; Moon, S.H.; Jenkins, C.M.; Li, M.; Sims, H.F.; Guan, S.; Gross, R.W. The phospholipase iPLA$_2$γ is a major mediator releasing oxidized aliphatic chains from cardiolipin, integrating mitochondrial bioenergetics and signaling. *J. Biol. Chem.* **2017**, *292*, 10672–10684. [CrossRef] [PubMed]

135. Liu, X.; Sims, H.F.; Jenkins, C.M.; Guan, S.; Dilthey, B.G.; Gross, R.W. 12-LOX catalyzes the oxidation of 2-arachidonoyl-lysolipids in platelets generating eicosanoid-lysolipids that are attenuated by iPLA$_2\gamma$ knockout. *J. Biol. Chem.* **2020**, *295*, 5307–5320. [CrossRef] [PubMed]
136. Kienesberger, P.C.; Lass, A.; Preiss-Landl, K.; Wolinski, H.; Kohlwein, S.D.; Zimmermann, R.; Zechner, R. Identification of an insulin-regulated lysophospholipase with homology to neuropathy target esterase. *J. Biol. Chem.* **2007**, *283*, 5908–5917. [CrossRef] [PubMed]
137. Quistad, G.B.; Barlow, C.; Winrow, C.J.; Sparks, S.E.; Casida, J.E. Evidence that mouse brain neuropathy target esterase is a lysophospholipase. *Proc. Natl. Acad. Sci. USA* **2003**, *100*, 7983–7987. [CrossRef]
138. Winrow, C.J.; Hemming, M.L.; Allen, D.M.; Quistad, G.B.; Casida, J.E.; Barlow, C. Loss of neuropathy target esterase in mice links organophosphate exposure to hyperactivity. *Nat. Genet.* **2003**, *33*, 477–485. [CrossRef]
139. Gallazzini, M.; Ferraris, J.D.; Kunin, M.; Morris, R.G.; Burg, M.B. Neuropathy target esterase catalyzes osmoprotective renal synthesis of glycerophosphocholine in response to high NaCl. *Proc. Natl. Acad. Sci. USA* **2006**, *103*, 15260–15265. [CrossRef]
140. Akassoglou, K.; Malester, B.; Xu, J.; Tessarollo, L.; Rosenbluth, J.; Chao, M.V. Brain-specific deletion of neuropathy target esterase/swisscheese results in neurodegeneration. *Proc. Natl. Acad. Sci. USA* **2004**, *101*, 5075–5080. [CrossRef]
141. Moser, M.; Li, Y.; Vaupel, K.; Kretzschmar, D.; Kluge, R.; Glynn, P.; Buettner, R. Placental failure and impaired vasculogenesis result in embryonic lethality for neuropathy target esterase-deficient mice. *Mol. Cell. Biol.* **2004**, *24*, 1667–1679. [CrossRef]
142. Read, D.J.; Li, Y.; Chao, M.V.; Cavanagh, J.B.; Glynn, P. Neuropathy target esterase is required for adult vertebrate axon maintenance. *J. Neurosci.* **2009**, *29*, 11594–11600. [CrossRef] [PubMed]
143. Rainier, S.; Bui, M.; Mark, E.; Thomas, D.; Tokarz, D.; Ming, L.; Delaney, C.E.; Richardson, R.J.; Albers, J.W.; Matsunami, N.; et al. Neuropathy target esterase gene mutations cause motor neuron disease. *Am. J. Hum. Genet.* **2008**, *82*, 780–785. [CrossRef]
144. Kmoch, S.; Majewski, J.; Ramamurthy, V.; Cao, S.; Fahiminiya, S.; Ren, H.; MacDonald, I.M.; Lopez, I.; Sun, V.; Keser, V.; et al. Mutations in *PNPLA6* are linked to photoreceptor degeneration and various forms of childhood blindness. *Nat. Commun.* **2015**, *6*, 5614. [CrossRef] [PubMed]
145. Heier, C.; Kien, B.; Huang, F.; Eichmann, T.O.; Xie, H.; Zechner, R.; Chang, P.-A. The phospholipase PNPLA7 functions as a lysophosphatidylcholine hydrolase and interacts with lipid droplets through its catalytic domain. *J. Biol. Chem.* **2017**, *292*, 19087–19098. [CrossRef] [PubMed]
146. Huang, Y.; He, S.; Li, J.Z.; Seo, Y.-K.; Osborne, T.F.; Cohen, J.C.; Hobbs, H.H. A feed-forward loop amplifies nutritional regulation of PNPLA3. *Proc. Natl. Acad. Sci. USA* **2010**, *107*, 7892–7897. [CrossRef]
147. Zimmermann, R.; Strauss, J.G.; Haemmerle, G.; Schoiswohl, G.; Birner-Gruenberger, R.; Riederer, M.; Lass, A.; Neuberger, G.; Eisenhaber, F.; Hermetter, A.; et al. Fat Mobilization in adipose tissue Is promoted by adipose triglyceride lipase. *Science* **2004**, *306*, 1383–1386. [CrossRef]
148. Girousse, A.; Tavernier, G.; Valle, C.; Moro, C.; Mejhert, N.; Dinel, A.-L.; Houssier, M.; Roussel, B.; Besse-Patin, A.; Combes, M.; et al. Partial inhibition of adipose tissue lipolysis improves glucose metabolism and insulin sensitivity without alteration of fat mass. *PLoS Biol.* **2013**, *11*, e1001485. [CrossRef]
149. Haemmerle, G.; Busemann, H.; Young, A.F.; Alexander, C.M.O.; Hoppe, P.; Mukhopadhyay, S.; Nittler, L.R. Lipolysis and Altered energy metabolism in mice lacking adipose triglyceride lipase. *Science* **2006**, *312*, 734–737. [CrossRef]
150. Das, S.K.; Eder, S.; Schauer, S.; Diwoky, C.; Temmel, H.; Guertl, B.; Gorkiewicz, G.; Tamilarasan, K.P.; Kumari, P.; Trauner, M.; et al. Adipose triglyceride lipase contributes to cancer-associated cachexia. *Science* **2011**, *333*, 233–238. [CrossRef]
151. Haemmerle, G.; Moustafa, T.; Woelkart, G.; Büttner, S.; Schmidt, A.; Van De Weijer, T.; Hesselink, M.; Jaeger, D.; Kienesberger, P.C.; Zierler, K.; et al. ATGL-mediated fat catabolism regulates cardiac mitochondrial function via PPAR-α and PGC-1. *Nat. Med.* **2011**, *17*, 1076–1085. [CrossRef]
152. Yang, X.; Lu, X.; Lombes, M.; Rha, G.B.; Chi, Y.I.; Guerin, T.M.; Smart, E.J.; Liu, J. The G0/G1 switch gene 2 regulates adipose lipolysis through association with adipose triglyceride lipase. *Cell Metab.* **2010**, *11*, 194–205. [CrossRef] [PubMed]
153. Schlager, S.; Goeritzer, M.; Jandl, K.; Frei, R.; Vujic, N.; Kolb, D.; Strohmaier, H.; Dorow, J.; Eichmann, T.O.; Rosenberger, A.; et al. Adipose triglyceride lipase acts on neutrophil lipid droplets to regulate substrate availability for lipid mediator synthesis. *J. Leukoc. Biol.* **2015**, *98*, 837–850. [CrossRef]

154. Sugihara, M.; Morito, D.; Ainuki, S.; Hirano, Y.; Ogino, K.; Kitamura, A.; Hirata, H.; Nagata, K. The AAA+ ATPase/ubiquitin ligase mysterin stabilizes cytoplasmic lipid droplets. *J. Cell Biol.* **2019**, *218*, 949–960. [CrossRef]
155. Heckmann, B.L.; Zhang, X.; Xie, X.; Saarinen, A.; Lu, X.; Yang, X.; Liu, J. Defective adipose lipolysis and altered global energy metabolism in mice with adipose overexpression of the lipolytic inhibitor G0/G1 switch gene 2 (G0S2). *J. Biol. Chem.* **2014**, *289*, 1905–1916. [CrossRef] [PubMed]
156. Lass, A.; Zimmermann, R.; Haemmerle, G.; Riederer, M.; Schoiswohl, G.; Schweiger, M.; Kienesberger, P.; Strauss, J.G.; Gorkiewicz, G.; Zechner, R. Adipose triglyceride lipase-mediated lipolysis of cellular fat stores is activated by CGI-58 and defective in Chanarin-Dorfman Syndrome. *Cell Metab.* **2006**, *3*, 309–319. [CrossRef]
157. Demerjian, M.; Crumrine, D.A.; Milstone, L.M.; Williams, M.L.; Elias, P.M. Barrier dysfunction and pathogenesis of neutral lipid storage disease with ichthyosis (Chanarin–Dorfman Syndrome). *J. Investig. Dermatol.* **2006**, *126*, 2032–2038. [CrossRef] [PubMed]
158. Lass, A.; Zimmermann, R.; Oberer, M.; Zechner, R. Lipolysis—A highly regulated multi-enzyme complex mediates the catabolism of cellular fat stores. *Prog. Lipid Res.* **2011**, *50*, 14–27. [CrossRef]
159. Young, S.G.; Zechner, R. Biochemistry and pathophysiology of intravascular and intracellular lipolysis. *Genes Dev.* **2013**, *27*, 459–484. [CrossRef]
160. Romeo, S.; Kozlitina, J.; Xing, C.; Pertsemlidis, A.; Cox, D.; Pennacchio, L.A.; Boerwinkle, E.; Cohen, J.C.; Hobbs, H.H. Genetic variation in *PNPLA3* confers susceptibility to nonalcoholic fatty liver disease. *Nat. Genet.* **2008**, *40*, 1461–1465. [CrossRef]
161. Basantani, M.K.; Sitnick, M.T.; Cai, L.; Brenner, D.S.; Gardner, N.P.; Li, J.Z.; Schoiswohl, G.; Yang, K.; Kumari, M.; Gross, R.W.; et al. Pnpla3/Adiponutrin deficiency in mice does not contribute to fatty liver disease or metabolic syndrome. *J. Lipid Res.* **2011**, *52*, 318–329. [CrossRef]
162. Smagris, E.; Basuray, S.; Li, J.; Huang, Y.; Lai, K.-M.V.; Gromada, J.; Cohen, J.C.; Hobbs, H.H. *Pnpla3I148M* knockin mice accumulate PNPLA3 on lipid droplets and develop hepatic steatosis. *Hepatology* **2014**, *61*, 108–118. [CrossRef] [PubMed]
163. Li, J.Z.; Huang, Y.; Karaman, R.; Ivanova, P.T.; Brown, H.A.; Roddy, T.; Castro-Perez, J.; Cohen, J.C.; Hobbs, H.H. Chronic overexpression of $PNPLA3^{I148M}$ in mouse liver causes hepatic steatosis. *J. Clin. Investig.* **2012**, *122*, 4130–4144. [CrossRef]
164. Mitsche, M.A.; Hobbs, H.H.; Cohen, J.C. Patatin-like phospholipase domain–containing protein 3 promotes transfers of essential fatty acids from triglycerides to phospholipids in hepatic lipid droplets. *J. Biol. Chem.* **2018**, *293*, 6958–6968. [CrossRef] [PubMed]
165. Pirazzi, C.; Valenti, L.; Motta, B.M.; Pingitore, P.; Hedfalk, K.; Mancina, R.M.; Burza, M.A.; Indiveri, C.; Ferro, Y.; Montalcini, T.; et al. PNPLA3 has retinyl-palmitate lipase activity in human hepatic stellate cells. *Hum. Mol. Genet.* **2014**, *23*, 4077–4085. [CrossRef] [PubMed]
166. Wang, Y.; Kory, N.; Basuray, S.; Cohen, J.C.; Hobbs, H.H. PNPLA3, CGI-58, and inhibition of hepatic triglyceride hydrolysis in mice. *Hepatology* **2019**, *69*, 2427–2441. [CrossRef] [PubMed]
167. Yang, A.; Mottillo, E.P.; Mladenovic-Lucas, L.; Zhou, L.; Granneman, J.G. Dynamic interactions of ABHD5 with PNPLA3 regulate triacylglycerol metabolism in brown adipocytes. *Nat. Metab.* **2019**, *1*, 560–569. [CrossRef] [PubMed]
168. Grall, A.; Guaguere, E.; Planchais, S.; Grond, S.; Bourrat, E.; Hausser, I.; Hitte, C.; Le Gallo, M.; Derbois, C.; Kim, G.-J.; et al. *PNPLA1* mutations cause autosomal recessive congenital ichthyosis in golden retriever dogs and humans. *Nat. Genet.* **2012**, *44*, 140–147. [CrossRef]
169. Grond, S.; Eichmann, T.O.; Dubrac, S.; Kolb, D.; Schmuth, M.; Fischer, J.; Crumrine, D.; Elias, P.M.; Haemmerle, G.; Zechner, R.; et al. PNPLA1 deficiency in mice and humans leads to a defect in the synthesis of omega-*O*-acylceramides. *J. Investig. Dermatol.* **2017**, *137*, 394–402. [CrossRef] [PubMed]
170. Hirabayashi, T.; Anjo, T.; Kaneko, A.; Senoo, Y.; Shibata, A.; Takama, H.; Yokoyama, K.; Nishito, Y.; Ono, T.; Taya, C.; et al. PNPLA1 has a crucial role in skin barrier function by directing acylceramide biosynthesis. *Nat. Commun.* **2017**, *8*, 14609. [CrossRef]
171. Ohno, Y.; Kamiyama, N.; Nakamichi, S.; Kihara, A. PNPLA1 is a transacylase essential for the generation of the skin barrier lipid ω-*O*-acylceramide. *Nat. Commun.* **2017**, *8*, 14610. [CrossRef] [PubMed]
172. Kien, B.; Grond, S.; Haemmerle, G.; Lass, A.; Eichmann, T.O.; Radner, F.P.W. ABHD5 stimulates PNPLA1-mediated ω-*O*-acylceramide biosynthesis essential for a functional skin permeability barrier. *J. Lipid Res.* **2018**, *59*, 2360–2367. [CrossRef]

173. Ohno, Y.; Nara, A.; Nakamichi, S.; Kihara, A. Molecular mechanism of the ichthyosis pathology of Chanarin-Dorfman syndrome: Stimulation of PNPLA1-catalyzed ω-O-acylceramide production by ABHD5. *J. Dermatol. Sci.* **2018**, *92*, 245–253. [CrossRef] [PubMed]
174. Hirabayashi, T.; Murakami, M.; Kihara, A. The role of PNPLA1 in ω-O-acylceramide synthesis and skin barrier function. *Biochim. Biophys. Acta Mol. Cell Biol. Lipids* **2019**, *1864*, 869–879. [CrossRef]
175. Hattori, M.; Arai, H. Intracellular PAF-acetylhydrolase type I. *Enzymes* **2015**, *38*, 23–36. [CrossRef]
176. Kono, N.; Arai, H. Intracellular Platelet-activating factor acetylhydrolase, type II. *Enzymes* **2015**, *38*, 43–54. [CrossRef] [PubMed]
177. Tjoelker, L.W.; Wilder, C.; Eberhardt, C.; Stafforinit, D.M.; Dietsch, G.; Schimpf, B.; Hooper, S.; Le Trong, H.; Cousens, L.S.; Zimmerman, G.A.; et al. Anti-inflammatory properties of a platelet-activating factor acetylhydrolase. *Nat. Cell Biol.* **1995**, *374*, 549–553. [CrossRef]
178. Jiang, Z.; Fehrenbach, M.L.; Ravaioli, G.; Kokalari, B.; Redai, I.G.; Sheardown, S.A.; Wilson, S.; Macphee, C.H.; Haczku, A. The effect of lipoprotein-associated phospholipase A_2 deficiency on pulmonary allergic responses in aspergillus fumigatus sensitized mice. *Respir. Res.* **2012**, *13*, 100. [CrossRef]
179. Corson, M.A. Phospholipase A_2 inhibitors in atherosclerosis: The race is on. *Lancet* **2009**, *373*, 608–610. [CrossRef]
180. Wilensky, R.L.; Shi, Y.; Mohler, E.R.; Hamamdzic, D.; Burgert, M.E.; Li, J.; Postle, A.; Fenning, R.S.; Bollinger, J.G.; Hoffman, B.E.; et al. Inhibition of lipoprotein-associated phospholipase A_2 reduces complex coronary atherosclerotic plaque development. *Nat. Med.* **2008**, *14*, 1059–1066. [CrossRef]
181. O'Donoghue, M.L.; Braunwald, E.; White, H.D.; Steen, D.P.; Lukas, M.A.; Tarka, E.; Steg, P.G.; Hochman, J.S.; Bode, C.; Maggioni, A.P.; et al. Effect of darapladib on major coronary events after an acute coronary syndrome. *JAMA* **2014**, *312*, 1006–1015. [CrossRef] [PubMed]
182. Acharya, N.K.; Levin, E.; Clifford, P.M.; Han, M.; Tourtellotte, R.; Chamberlain, D.; Pollaro, M.; Coretti, N.J.; Kosciuk, M.C.; Nagele, E.P.; et al. Diabetes and hypercholesterolemia increase blood-brain barrier permeability and brain amyloid deposition: Beneficial effects of the LpPLA_2 inhibitor darapladib. *J. Alzheimer's Dis.* **2013**, *35*, 179–198. [CrossRef]
183. Canning, P.; Kenny, B.-A.; Prise, V.; Glenn, J.; Sarker, M.H.; Hudson, N.; Brandt, M.; Lopez, F.J.; Gale, D.; Luthert, P.J.; et al. Lipoprotein-associated phospholipase A_2 (Lp-PLA_2) as a therapeutic target to prevent retinal vasopermeability during diabetes. *Proc. Natl. Acad. Sci. USA* **2016**, *113*, 7213–7218. [CrossRef] [PubMed]
184. Xu, C.; Reichert, E.C.; Nakano, T.; Lohse, M.; Gardner, A.A.; Revelo, M.P.; Topham, M.K.; Stafforini, D.M. Deficiency of phospholipase A_2 group 7 decreases intestinal polyposis and colon tumorigenesis in $Apc^{Min/+}$ mice. *Cancer Res.* **2013**, *73*, 2806–2816. [CrossRef]
185. Hattori, M.; Adachi, H.; Tsujimoto, M.; Arai, H.; Inoue, K. Miller-Dieker lissencephaly gene encodes a subunit of brain platelet-activating factor. *Nat. Cell Biol.* **1994**, *370*, 216–218. [CrossRef]
186. Koizumi, H.; Yamaguchi, N.; Hattori, M.; Ishikawa, T.-O.; Aoki, J.; Taketo, M.M.; Inoue, K.; Arai, H. Targeted disruption of intracellular type I platelet activating factor-acetylhydrolase catalytic subunits causes severe impairment in spermatogenesis. *J. Biol. Chem.* **2003**, *278*, 12489–12494. [CrossRef]
187. Page, R.M.; Münch, A.; Horn, T.; Kuhn, P.-H.; Colombo, A.; Reiner, O.; Boutros, M.; Steiner, H.; Lichtenthaler, S.F.; Haass, C. Loss of PAFAH1B2 reduces amyloid-β generation by promoting the degradation of amyloid precursor protein C-terminal fragments. *J. Neurosci.* **2012**, *32*, 18204–18214. [CrossRef] [PubMed]
188. Livnat, I.; Finkelshtein, D.; Ghosh, I.; Arai, H.; Reiner, O. PAF-AH catalytic subunits modulate the Wnt pathway in developing GABAergic neurons. *Front. Cell. Neurosci.* **2010**, *4*, 19. [CrossRef] [PubMed]
189. Kono, N.; Inoue, T.; Yoshida, Y.; Sato, H.; Matsusue, T.; Itabe, H.; Niki, E.; Aoki, J.; Arai, H. Protection against oxidative stress-induced hepatic injury by intracellular type II platelet-activating factor acetylhydrolase by metabolism of oxidized phospholipids in vivo. *J. Biol. Chem.* **2007**, *283*, 1628–1636. [CrossRef] [PubMed]
190. Shimanaka, Y.; Kono, N.; Taketomi, Y.; Arita, M.; Okayama, Y.; Tanaka, Y.; Nishito, Y.; Mochizuki, T.; Kusuhara, H.; Adibekian, A.; et al. Omega-3 fatty acid epoxides are autocrine mediators that control the magnitude of IgE-mediated mast cell activation. *Nat. Med.* **2017**, *23*, 1287–1297. [CrossRef]
191. Kono, N.; Arai, H. Platelet-activating factor acetylhydrolases: An overview and update. *Biochim. Biophys. Acta Mol. Cell Biol. Lipids* **2019**, *1864*, 922–931. [CrossRef] [PubMed]
192. Abe, A.; Hiraoka, M.; Wild, S.; Wilcoxen, S.E.; Paine, R.; Shayman, J.A. Lysosomal phospholipase A_2 is selectively expressed in alveolar macrophages. *J. Biol. Chem.* **2004**, *279*, 42605–42611. [CrossRef]

193. Hiraoka, M.; Abe, A.; Lu, Y.; Yang, K.; Han, X.; Gross, R.W.; Shayman, J.A. Lysosomal phospholipase A_2 and phospholipidosis. *Mol. Cell. Biol.* **2006**, *26*, 6139–6148. [CrossRef]
194. Paduraru, C.; Bezbradica, J.S.; Kunte, A.; Kelly, R.; Shayman, J.A.; Veerapen, N.; Cox, L.R.; Besra, G.S.; Cresswell, P. Role for lysosomal phospholipase A_2 in iNKT cell-mediated CD1d recognition. *Proc. Natl. Acad. Sci. USA* **2013**, *110*, 5097–5102. [CrossRef] [PubMed]
195. Schneider, B.; Behrends, J.; Hagens, K.; Harmel, N.; Shayman, J.A.; Schaible, U.E. Lysosomal phospholipase A_2: A novel player in host immunity to Mycobacterium tuberculosis. *Eur. J. Immunol.* **2014**, *44*, 2394–2404. [CrossRef] [PubMed]
196. Fisher, A.B.; Dodia, C.; Feinstein, S.I.; Ho, Y.-S. Altered lung phospholipid metabolism in mice with targeted deletion of lysosomal-type phospholipase A_2. *J. Lipid Res.* **2005**, *46*, 1248–1256. [CrossRef] [PubMed]
197. Vázquez-Medina, J.P.; Tao, J.-Q.; Patel, P.; Bannitz-Fernandes, R.; Dodia, C.; Sorokina, E.M.; Feinstein, S.I.; Chatterjee, S.; Fisher, A.B. Genetic inactivation of the phospholipase A_2 activity of peroxiredoxin 6 in mice protects against LPS-induced acute lung injury. *Am. J. Physiol. Cell. Mol. Physiol.* **2019**, *316*, L656–L668. [CrossRef]
198. Fisher, A.B.; Vasquez-Medina, J.P.; Dodia, C.; Sorokina, E.M.; Tao, J.-Q.; Feinstein, S.I. Peroxiredoxin 6 phospholipid hydroperoxidase activity in the repair of peroxidized cell membranes. *Redox Biol.* **2017**, *14*, 41–46. [CrossRef]
199. Moawad, A.R.; Fernandez, M.C.; Scarlata, E.; Dodia, C.; Feinstein, S.I.; Fisher, A.B.; O'Flaherty, C. Deficiency of peroxiredoxin 6 or inhibition of its phospholipase A_2 activity impair the in vitro sperm fertilizing competence in mice. *Sci. Rep.* **2017**, *7*, 12994. [CrossRef]
200. Fisher, A.B. The phospholipase A_2 activity of peroxiredoxin 6. *J. Lipid Res.* **2018**, *59*, 1132–1147. [CrossRef]
201. Shayman, J.A.; Tesmer, J.J.G. Lysosomal phospholipase A_2. *Biochim. Biophys. Acta Mol. Cell Biol. Lipids* **2019**, *1864*, 932–940. [CrossRef] [PubMed]
202. Uyama, T.; Ikematsu, N.; Inoue, M.; Shinohara, N.; Jin, X.-H.; Tsuboi, K.; Tonai, T.; Tokumura, A.; Ueda, N. Generation of *N*-acylphosphatidylethanolamine by members of the phospholipase A/acyltransferase (PLA/AT) Family. *J. Biol. Chem.* **2012**, *287*, 31905–31919. [CrossRef] [PubMed]
203. Jaworski, K.; Ahmadian, M.; Duncan, R.E.; Sarkadi-Nagy, E.; Varady, K.A.; Hellerstein, M.K.; Lee, H.-Y.; Samuel, V.T.; Shulman, G.I.; Kim, K.-H.; et al. AdPLA ablation increases lipolysis and prevents obesity induced by high-fat feeding or leptin deficiency. *Nat. Med.* **2009**, *15*, 159–168. [CrossRef]
204. Zhang, S.-Y.; Dong, Y.-Q.; Wang, P.; Zhang, X.; Yan, Y.; Sun, L.; Liu, B.; Zhang, D.; Zhang, H.; Liu, H.; et al. Adipocyte-derived lysophosphatidylcholine activates adipocyte and adipose tissue macrophage Nod-like receptor protein 3 inflammasomes mediating homocysteine-induced insulin resistance. *EBioMedicine* **2018**, *31*, 202–216. [CrossRef]
205. Xiong, S.; Tu, H.; Kollareddy, M.; Pant, V.; Li, Q.; Zhang, Y.; Jackson, J.G.; Suh, Y.-A.; Elizondo-Fraire, A.C.; Yang, P.; et al. Pla2g16 phospholipase mediates gain-of-function activities of mutant p53. *Proc. Natl. Acad. Sci. USA* **2014**, *111*, 11145–11150. [CrossRef]
206. Golczak, M.; Sears, A.E.; Kiser, P.D.; Palczewski, K. LRAT-specific domain facilitates vitamin A metabolism by domain swapping in HRASLS3. *Nat. Chem. Biol.* **2014**, *11*, 26–32. [CrossRef]
207. Uyama, T.; Kawai, K.; Kono, N.; Watanabe, M.; Tsuboi, K.; Inoue, T.; Araki, N.; Arai, H.; Ueda, N. Interaction of phospholipase A/acyltransferase-3 with Pex19p: A possible involvement in the down-regulation of peroxisomes. *J. Biol. Chem.* **2015**, *290*, 17520–17534. [CrossRef] [PubMed]
208. Staring, J.; Von Castelmur, E.; Blomen, V.A.; Hengel, L.G.V.D.; Brockmann, M.; Baggen, J.; Thibaut, H.J.; Nieuwenhuis, J.; Janssen, H.; Van Kuppeveld, F.J.M.; et al. PLA2G16 represents a switch between entry and clearance of *Picornaviridae*. *Nat. Cell Biol.* **2017**, *541*, 412–416. [CrossRef]
209. Hussain, Z.; Uyama, T.; Tsuboi, K.; Ueda, N. Mammalian enzymes responsible for the biosynthesis of *N*-acylethanolamines. *Biochim. Biophys. Acta Mol. Cell Biol. Lipids* **2017**, *1862*, 1546–1561. [CrossRef]
210. Mardian, E.B.; Bradley, R.M.; Duncan, R.E. The HRASLS (PLA/AT) subfamily of enzymes. *J. Biomed. Sci.* **2015**, *22*, 99. [CrossRef]
211. Thomas, G.; Brown, A.L.; Brown, J.M. In vivo metabolite profiling as a means to identify uncharacterized lipase function: Recent success stories within the alpha beta hydrolase domain (ABHD) enzyme family. *Biochim. Biophys. Acta Mol. Cell Biol. Lipids* **2014**, *1841*, 1097–1101. [CrossRef] [PubMed]

212. Long, J.Z.; Cisar, J.S.; Milliken, D.; Niessen, S.; Wang, C.; Trauger, S.A.; Siuzdak, G.; Cravatt, B.F. Metabolomics annotates ABHD3 as a physiologic regulator of medium-chain phospholipids. *Nat. Chem. Biol.* **2011**, *7*, 763–765. [CrossRef]
213. Lee, H.-C.; Simon, G.M.; Cravatt, B.F. ABHD4 regulates multiple classes of *N*-acyl phospholipids in the mammalian central nervous system. *Biochemistry* **2015**, *54*, 2539–2549. [CrossRef]
214. Marrs, W.R.; Blankman, J.L.; Horne, E.A.; Thomazeau, A.; Lin, Y.H.; Coy, J.; Bodor, A.L.; Muccioli, G.G.; Hu, S.S.-J.; Woodruff, G.; et al. The serine hydrolase ABHD6 controls the accumulation and efficacy of 2-AG at cannabinoid receptors. *Nat. Neurosci.* **2010**, *13*, 951–957. [CrossRef] [PubMed]
215. Blankman, J.L.; Long, J.Z.; Trauger, S.A.; Siuzdak, G.; Cravatt, B.F. ABHD12 controls brain lysophosphatidylserine pathways that are deregulated in a murine model of the neurodegenerative disease PHARC. *Proc. Natl. Acad. Sci. USA* **2013**, *110*, 1500–1505. [CrossRef] [PubMed]
216. Fiskerstrand, T.; Brahim, D.H.-B.; Johansson, S.; M'Zahem, A.; Haukanes, B.I.; Drouot, N.; Zimmermann, J.; Cole, A.J.; Vedeler, C.; Bredrup, C.; et al. Mutations in ABHD12 cause the neurodegenerative disease PHARC: An inborn error of endocannabinoid metabolism. *Am. J. Hum. Genet.* **2010**, *87*, 410–417. [CrossRef]
217. Kamat, S.S.; Camara, K.; Parsons, W.H.; Chen, D.-H.; Dix, M.M.; Bird, T.D.; Howell, A.R.; Cravatt, B.F. Immunomodulatory lysophosphatidylserines are regulated by ABHD16A and ABHD12 interplay. *Nat. Chem. Biol.* **2015**, *11*, 164–171. [CrossRef]
218. Maeda, Y.; Tashima, Y.; Houjou, T.; Fujita, M.; Yoko-o, T.; Jigami, Y.; Taguchi, R.; Kinoshita, T. Fatty acid remodeling of GPI-anchored proteins is required for their raft association. *Mol. Biol. Cell* **2007**, *18*, 1497–1506. [CrossRef]
219. Wang, Y.; Murakami, Y.; Yasui, T.; Wakana, S.; Kikutani, H.; Kinoshita, T.; Maeda, Y. Significance of glycosylphosphatidylinositol-anchored protein enrichment in lipid rafts for the control of autoimmunity. *J. Biol. Chem.* **2013**, *288*, 25490–25499. [CrossRef]
220. Lee, G.-H.; Fujita, M.; Takaoka, K.; Murakami, Y.; Fujihara, Y.; Kanzawa, N.; Murakami, K.-I.; Kajikawa, E.; Takada, Y.; Saito, K.; et al. A GPI processing phospholipase A_2, PGAP6, modulates Nodal signaling in embryos by shedding CRIPTO. *J. Cell Biol.* **2016**, *215*, 705–718. [CrossRef]
221. Nikolaou, A.; Kokotou, M.G.; Vasilakaki, S.; Kokotos, G. Small-molecule inhibitors as potential therapeutics and as tools to understand the role of phospholipases A_2. *Biochim. Biophys. Acta Mol. Cell Biol. Lipids* **2019**, *1864*, 941–956. [CrossRef]

Publisher's Note: MDPI stays neutral with regard to jurisdictional claims in published maps and institutional affiliations.

© 2020 by the authors. Licensee MDPI, Basel, Switzerland. This article is an open access article distributed under the terms and conditions of the Creative Commons Attribution (CC BY) license (http://creativecommons.org/licenses/by/4.0/).

Article

2-Oxoester Phospholipase A_2 Inhibitors with Enhanced Metabolic Stability

Giorgos S. Koutoulogenis [1], Maroula G. Kokotou [1], Daiki Hayashi [2], Varnavas D. Mouchlis [2], Edward A. Dennis [2,*] and George Kokotos [1,*]

[1] Department of Chemistry, National and Kapodistrian University of Athens, Athens 15771, Greece; gkoutoul@chem.uoa.gr (G.S.K.); mkokotou@chem.uoa.gr (M.G.K.)

[2] Department of Chemistry and Biochemistry and Department of Pharmacology, School of Medicine, University of California, San Diego, La Jolla, CA 92093-0601, USA; dhayashi@health.ucsd.edu (D.H.); vmouchlis@ucsd.edu (V.D.M.)

* Correspondence: edennis@ucsd.edu (E.A.D.); gkokotos@chem.uoa.gr (G.K.); Tel.: +1-858-534-3055 (E.A.D.); +30-210-7274462 (G.K.)

Received: 1 March 2020; Accepted: 20 March 2020; Published: 24 March 2020

Abstract: 2-Oxoesters constitute an important class of potent and selective inhibitors of human cytosolic phospholipase A_2 (GIVA $cPLA_2$) combining an aromatic scaffold or a long aliphatic chain with a short aliphatic chain containing a free carboxylic acid. Although highly potent 2-oxoester inhibitors of GIVA $cPLA_2$ have been developed, their rapid degradation in human plasma limits their pharmaceutical utility. In an effort to address this problem, we designed and synthesized two new 2-oxoesters introducing a methyl group either on the α-carbon to the oxoester functionality or on the carbon carrying the ester oxygen. We studied the in vitro plasma stability of both derivatives and their in vitro inhibitory activity on GIVA $cPLA_2$. Both derivatives exhibited higher plasma stability in comparison with the unsubstituted compound and both derivatives inhibited GIVA $cPLA_2$, however to different degrees. The 2-oxoester containing a methyl group on the α-carbon atom to the oxoester functionality exhibits enhancement of the metabolic stability and retains considerable inhibitory potency.

Keywords: inhibitor; metabolic stability; α-methylation; oxoesters; phospholipase A_2

---

1. Introduction

Phospholipase A_2 (PLA_2) enzymes are involved in a variety of inflammatory diseases, which has stimulated the interest of the scientific community in exploring the pathophysiological role of each type of PLA_2 and developing strategies for the modulation of their activity [1–3]. Among the various PLA_2s present in mammals, the cytosolic calcium-dependent PLA_2s ($cPLA_2$) is characterized by a marked preference for the hydrolysis of arachidonic acid from the phospholipid substrates initiating the eicosanoid cascade [4]. The most well studied enzyme of this group is Group IVA (GIVA) $cPLA_2$ and it is responsible for the biosynthesis of many diverse lipid signaling molecules contributing to inflammation [5]. Fundamental characteristics of the GIVA $cPLA_2$, such as its role and function, have been summarized in recent review articles [6,7].

Early studies using gene targeted mice that lack GIVA $cPLA_2$ have shown that these mice are much less prone to inflammatory pathological responses to disease, stresses and physical injuries, indicating the involvement of the enzyme in cellular and systemic damage [8,9]. The mechanisms regulating lipid peroxidation of arachidonic acid and docosahexaenoic acid in the central nervous system as well as the role of GIVA $cPLA_2$ in oxidative and inflammatory signaling pathways in the central nervous system have been recently reviewed [10–12], highlighting the metabolic events linking this enzyme to activation in neurons, astrocytes, microglial cells, and cerebrovascular cells.

A variety of small-molecule inhibitors have been developed as tools to understand the role of each type of PLA_2 enzyme and as new medicinal agents [13,14]. Our groups have developed several classes of small-molecule PLA_2 inhibitors [15–20] and have studied their in vitro inhibitory activities as well as their in vivo properties [21,22]. In particular, we have designed and synthesized a series of 2-oxoesters as a novel class of potent and selective inhibitors of human GIVA $cPLA_2$ [23], having the ability to modulate the production of eicosanoids. The structural characteristics of these inhibitors are either a long fatty chain or an aromatic scaffold on the left, a 2-oxoester functionality and a short fatty chain with a terminal carboxylic group on the right (Figure 1). More potent 2-oxoester inhibitors were reported by our group [24], maintaining the same structural characteristics, but changing the substitution to an aromatic scaffold or altering the number of carbon atoms either to the left or to the right of oxoester group. Structure-activity relationship studies of these inhibitors allowed us to determine the necessary building blocks for optimizing inhibitory activity against GIVA $cPLA_2$.

Among the many synthesized 2-oxoester GIVA $cPLA_2$ inhibitors, four of them stood out as showing potent inhibition; their structures are shown in Figure 1. On the oxoester functionality (left side) of these inhibitors, there is an aliphatic chain containing two to four carbons bearing an aromatic scaffold, which is either a biphenyl system or a benzene ring with an aliphatic alkoxy group and on the ester functionality (right side) there is again a short aliphatic chain containing two to four carbons bearing a carboxyl group. As was shown, the four carbon atom chains on the left and on the right of the oxoester functionality are essential for the optimum inhibition of GIVA $cPLA_2$, because when the number of carbon atoms on either side is decreased, the $X_I(50)$, defined elsewhere [25], increases, showing less potency [24].

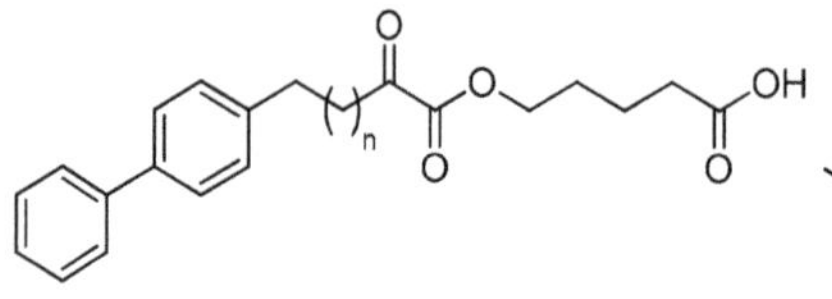

O O OH O O O OH O O m n O

1a, n=3, **GK452** $X_I(50) = 0.000078$
1b, n=1, **GK482** $X_I(50) = 0.00041$

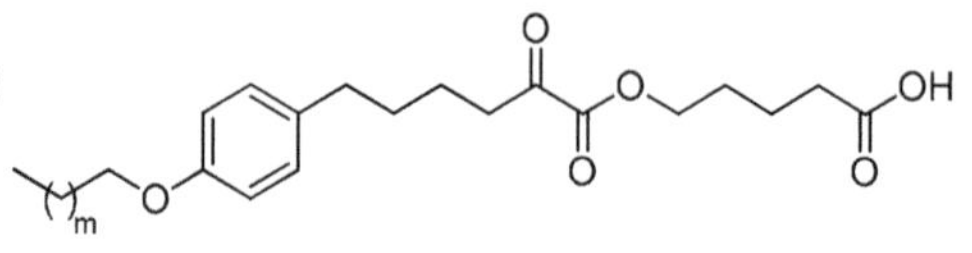

2a, n=2, **GK504** $X_I(50) = 0.000066$
2b, n=4, **GK484** $X_I(50) = 0.000019$

Figure 1. Known 2-oxoester inhibitors of GIVA $cPLA_2$ [23,24].

Although the developed 2-oxoester inhibitors so far exhibited a potent in vitro inhibition of GIVA $cPLA_2$, their rapid degradation in human plasma limits their pharmaceutical utility [24]. The aim of our work was to chemically modify the 2-oxoester inhibitors to increase their metabolic stability. We present herein the synthesis of methyl substituted 2-oxoesters, a study of their metabolic stability in plasma, and a study of their in vitro activity and specificity for three different PLA_2s.

2. Materials and Methods

2.1. General Chemistry Methods

Chromatographic purification of products was accomplished using forced-flow chromatography on Merck® (Merck, Darmstadt, Germany) Kieselgel 60 F_{254} 230–400 mesh. Thin-layer chromatography (TLC) was performed on aluminum backed silica plates (0.2 mm, 60 F_{254}). Visualization of the developed chromatogram was performed by fluorescence quenching using phosphomolybdic acid, ninhydrin or potassium permanganate stains. Melting points were determined on a Buchi® 530 (Buchi, Flawil, Switzerland) hot stage apparatus and are uncorrected. Mass spectra (ESI) were recorded on a Finningan® Surveyor MSQ LC-MS spectrometer (Thermo, Darmstadt, Germany). High resolution mass spectrometry (HRMS) spectra were recorded on a Bruker® Maxis Impact QTOF (Bruker Daltonics, Bremen, Germany) spectrometer. ^{1}H and ^{13}C NMR spectra were recorded on a Varian® Mercury (Varian, Palo Alto, CA, USA) (200 MHz and 50 MHz, respectively), and are internally referenced to residual solvent signals. Data for ^{1}H NMR are reported as follows: chemical shift (δ ppm), integration, multiplicity (s = singlet, d = doublet, t = triplet, m = multiplet, br s = broad signal), coupling constant and assignment. Data for ^{13}C NMR are reported in terms of chemical shift (δ ppm).

5-([1,1′-Biphenyl]-4-yl)-2-methylpentanoic acid (**4**): In a 25 mL round bottom flask containing a solution of diisopropylamine (1372 mg, 13.56 mmol) in extra dry tetrahydrofuran (THF) (6.0 mL) was added *n*-BuLi (8.47 mL, 1.6M, 13.56 mmol) under Ar, at 0 °C and the reaction mixture was left under stirring for 30 min. Next, a solution of compound **3** (1150 mg, 4.52 mmol) in extra dry THF (5 mL) was added and the resulting mixture was left stirring at 0 °C for 30 min. Hexamethylphosphoramide (HMPA) (2.20 mL) in extra dry THF (0.5 mL) was added and the resulting mixture continued stirring at 0 °C for 1 h. Finally, a solution of CH_3I (905 mg, 5.96 mmol) in extra dry THF (1.5 mL) was added and the reaction mixture was left stirring at room temperature for 16 h. After completion of the reaction, a saturated aqueous solution of NH_4Cl (15 mL) and Et_2O (40 mL) were added. Then, hot water (3 × 15 mL) was added in the resulting mixture in order to dissolve the remaining HMPA. The aqueous layer was extracted by Et_2O (3 × 15 mL) and the combined organic layers were dried over $MgSO_4$. The solvent was removed under reduced pressure and the product was purified by column chromatography (petroleum ether 40–60 °C:EtOAc) (8:2). Yield 78%; Yellowish oil; ^{1}H NMR (200 MHz, $CDCl_3$): δ = 11.46 (br s, 1H), 7.67–7.29 (m, 9H), 2.73 (t, *J* = 7.1 Hz, 2H), 2.65–2.49 (m, 1H), 1.95–1.51 (m, 4H), 1.26 (d, *J* = 6.9 Hz, 3H); ^{13}C NMR (50 MHz, $CDCl_3$): δ = 183.8, 141.5, 141.4, 139.1, 129.1, 129.00, 127.4, 127.3, 39.6, 35.7, 33.4, 29.2, 17.2; HRMS [M-H$^-$]: 267.1387; (calculated for $[C_{18}H_{19}O_2]^-$: 267.1391).

5-([1,1′-Biphenyl]-4-yl)-2-methylpentan-1-ol (**5**). To a solution of compound **4** (960 mg, 3.57 mmol) in dry THF (20 mL) was added ethyl chloroformate (0.51 mL, 5.35 mmol) followed by Et_3N (0.75 mL, 5.35 mmol) at −10 °C and the reaction mixture was left stirring for 40 min. Next, $NaBH_4$ (1080 mg, 28.56 mmol) was added and a subsequent dropwise addition of MeOH (25 mL) at 0 °C took place. The resulting reaction mixture was left stirring at room temperature for 16 h. The reaction mixture was evaporated under reduced pressure and was treated by HCl 1N until pH<7. The aqueous layer was extracted by EtOAc (3 × 20 mL) and the combined organic layers were washed by a 10% aqueous solution of $NaHCO_3$ (20 mL). The organic layer was dried over $MgSO_4$ and the solvent was removed under reduced pressure. The product was purified by column chromatography (petroleum ether 40–60 °C:EtOAc) (8:2). Yield 67%; Colourless oil; ^{1}H NMR (200 MHz, $CDCl_3$): δ = 7.68–7.29 (m, 9H), 3.62–3.42 (m, 2H), 2.71 (t, *J* = 7.5 Hz, 2H), 2.35 (br s, 1H), 1.85–1.52 (m, 4H), 1.35-1.19 (m, 1H), 1.00 (d, *J* = 6.6 Hz, 3H); ^{13}C NMR (50 MHz, $CDCl_3$): δ = 142.0, 141.4, 138.9, 129.1, 129.0, 127.3, 127.2, 68.4, 36.2, 35.9, 33.2, 29.12, 16.9; HRMS [M+Na]$^+$: 277.1562; (calculated for $[C_{18}H_{22}NaO]^+$ 277.1563).

5-([1,1′-Biphenyl]-4-yl)-2-methylpentanal (**6**). To a solution of compound **5** (600 mg, 2.36 mmol) in dry CH_2Cl_2 (25 mL), TEMPO (38.0 mg, 0.24 mmol) and iodobenzene diacetate (988 mg, 3.07 mmol) were added and the reaction mixture was left stirring at room temperature for 3 h. Next, a 10% aqueous solution of $Na_2S_2O_3$ (10 mL) was added and the resulting reaction mixture was left stirring for 5 min. The organic layers were washed by H_2O (20 mL) and dried over $MgSO_4$. The solvent was removed under reduced pressure, the product was purified by column chromatography (petroleum ether 40–60 °C:EtOAc) (9:1), and was used directly for the next experiment. Yield 85%; Colourless oil.

6-([1,1′-Biphenyl]-4-yl)-2-hydroxy-3-methylhexanenitrile (**7**). To a solution of compound **6** (506 mg, 2.00 mmol) in CH_2Cl_2 (15 mL), a saturated aqueous solution of $NaHSO_3$ (1.5 mL) was added and the reaction mixture was left stirring vigorously at room temperature for 30 min. The solvent was removed under reduced pressure and THF (15 mL) and H_2O (10 mL) were added, followed by dropwise addition of a 4N aqueous solution of KCN (1.5 mL, 6.00 mmol). Then, the reaction mixture was left stirring at room temperature for 16 h. The reaction mixture was evaporated under reduced pressure and Et_2O (30 mL) was added. The organic layer was washed with H_2O (2 x 10 mL) and dried over $MgSO_4$. The solvent was removed under reduced pressure and the product was purified by column chromatography (petroleum ether 40–60 °C:EtOAc) (8:2). Mixture of diastereomers; Yield 94%; Colourless oil; ^{1}H NMR (200 MHz, $CDCl_3$): δ = 7.65–7.20 (m, 9H), 4.36 (d, *J* = 5.5 Hz, 1H), 2.68 (t, *J* = 7.5 Hz, 2H), 2.61–2.43 (br s, 1H), 2.02–1.85 (m, 1H), 1.83–1.52 (m, 3*H*), 1.46–1.26 (m, 1H), 1.16–1.06 (m, 3H); ^{13}C NMR (50 MHz, $CDCl_3$): δ = 141.5, 141.4, 141.3, 139.1, 129.1, 129.0, 127.4, 127.3, 119.8, 119.5, 66.4, 66.0, 38.0, 35.8, 32.0, 31.5, 29.0, 28.9, 15.1, 14.9; HRMS $[M+Na]^+$: 302.1511; (calculated for $[C_{19}H_{21}NNaO]^+$ 302.1515).

Methyl 6-([1,1′-biphenyl]-4-yl)-2-hydroxy-3-methylhexanoate (**8**). A solution of compound **7** (525 mg, 1.88 mmol) in MeOH (6 mL) was treated with a freshly prepared 6N HCl in MeOH (6 mL) under stirring at room temperature for 16 h. The solvent was removed under reduced pressure and Et_2O (30 mL) was added. The organic layer was washed with H_2O (2 × 10 mL) and dried over $MgSO_4$. The solvent was removed under reduced pressure and the product was purified by column chromatography (petroleum ether 40–60 °C:EtOAc) (8:2). Mixture of diastereomers; Yield 36%; White oil; ^{1}H NMR (200 MHz, $CDCl_3$): δ = 7.75–7.15 (m, 9H), 4.28–4.08 (m, 1H), 3.98–3.67 (m, 3H), 2.95 (br s, 1H), 2.82–2.55 (m, 2H), 2.18–1.90 (m, 1H), 1.81–1.58 (m, 2H), 1.49-1.21 (m, 2H), 1.18–0.77 (m, 3H); ^{13}C NMR (50 MHz, $CDCl_3$): δ = 176.0, 175.6, 141.9, 141.8, 141.3, 138.9, 129.1, 129.0, 127.3, 75.2, 73.5, 52.7, 52.6, 37.4, 37.0, 35.9, 35.8, 33.1, 30.7, 29.5, 29.2, 16.1, 13.8; HRMS $[M+Na]^+$: 335.1613; (calculated for $[C_{20}H_{24}NaO_3]^+$ 335.1618).

6-([1,1′-Biphenyl]-4-yl)-2-hydroxy-3-methylhexanoic acid (**9**). To a solution of compound **8** (200 mg, 0.64 mmol) in MeOH (6 mL), a 1N aqueous solution of NaOH (5 mL) was added and the reaction mixture was left stirring at room temperature for 2 days. After acidification with 1N HCl (pH 1), the reaction mixture was extracted with Et_2O (3 × 20 mL). The combined organic layers were washed with H_2O (2 × 10 mL) and dried over $MgSO_4$. The solvent was removed under reduced pressure and the product was purified by column chromatography (petroleum ether 40–60 °C:EtOAc) (8:2). Mixture of diastereomers; Yield 64%; White solid; mp: 147–151 °C; ^{1}H NMR (200 MHz, CD_3OD): δ = 7.61–7.14 (m, 9H), 4.12–3.97 (m, 1H), 2.70–2.53 (m, 2H), 2.02–1.85 (m, 1H), 1.73–1.44 (m, 4H), 1.02–0.82 (m, 3H); ^{13}C NMR (50 MHz, CD_3OD): δ = 176.6, 176.2, 141.8, 141.2, 138.7, 128.8, 128.6, 126.8, 126.7, 126.6, 126.5, 74.5, 73.0, 37.0, 35.5, 35.4, 29.2, 15.4, 13.1; HRMS $[M-H]^-$: 297.1492; (calculated for $[C_{19}H_{21}O_3]^-$ 297.1496).

tert-Butyl 5-bromohexanoate (**12**). In a 25 mL round bottom flask, δ-hexalactone (342 mg, 3.00 mmol) and 33% HBr in AcOH (5 mL) were added and the reaction mixture was left stirring at 75 °C for 16 h. After completion of the reaction, the reaction mixture was extracted with CH_2Cl_2 (3 × 15 mL). The organic layers were collected, washed with H_2O (3 × 15 mL) and dried over $MgSO_4$. The solvent was removed under reduced pressure and the formed bromo carboxylic acid was used directly to the next step. To a solution of the bromo carboxylic acid (544 mg, 2.82 mmol) in dry CH_2Cl_2 (20 mL) in a pressure vessel, concentrated H_2SO_4 (0.45 mL) was added dropwise. The reaction mixture was cooled at −196 °C using liquid nitrogen and isobutylene (20 mL) was added to the solution. The reaction mixture was left stirring at room temperature for 5 d. The reaction mixture was poured to a 100 mL beaker and was left stirring for 30 min. The reaction mixture was washed with H_2O (3 × 15 mL) and the organic phase was dried over $MgSO_4$. The solvent was removed under reduced pressure and the product was purified by column chromatography (petroleum ether 40–60 °C:EtOAc) (9:1). Yield 83%; Yellowish oil; ^{1}H NMR (200 MHz, $CDCl_3$): δ = 4.12–4.02 (m, 1H), 2.23-2.14 (m, 2H), 1.79–1.62 (m, 7H), 1.39 (s, 9H); ^{13}C NMR (50 MHz, $CDCl_3$): δ = 172.7, 80.4, 51.1, 40.4, 34.8, 28.3, 26.6, 23.4; HRMS $[M+Na]^+$: 273.0467; (calculated for $[C_{10}H_{19}BrNaO_2]^+$ 273.0461).

General procedure for the synthesis of 2-hydroxy esters **15a,b**.

To a stirred solution of compound **9** or **13** (0.38 mmol) in THF (2.4 mL), a 20% aqueous solution of Cs_2CO_3 (124 mg, 0.38 mmol) was added (pH 9) and the reaction mixture was left stirring for 20 min at room temperature. The organic solvent was evaporated under reduced pressure and the residue was dissolved in DMF (6 mL). *tert*-Butyl ester **14** or **12** (0.46 mmol) was added and the reaction mixture was left stirring under reflux for 72 h to 168 h. Then, H_2O (10 mL) was added and the reaction mixture was extracted with EtOAc (3 × 10 mL). The combined organic layers were dried over $MgSO_4$ and the solvent was removed under reduced pressure. The product was purified by column chromatography (petroleum ether 40–60 °C:EtOAc) (8:2).

5-(tert-Butoxy)-5-oxopentyl 6-([1,1′-biphenyl]-4-yl)-2-hydroxy-3-methylhexanoate (**15a**). Mixture of diastereomers; Yield 35%; Colourless oil; ^{1}H NMR (200 MHz, $CDCl_3$): δ = 7.64–7.17 (m, 9H), 4.25–4.05 (m, 3H), 2.72-2.57 (m, 2H), 2.31–2.16 (m, 2H), 2.05-1.52 (m, 8H), 1.44 (s, 9H), 1.01 (d, *J* = 6.9 Hz, 1.5H), 0.84 (d, *J* = 6.9 Hz, 1.5H); ^{13}C NMR (50 MHz, $CDCl_3$): δ = 175.5, 175.1, 172.8, 141.9, 141.8, 141.3, 138.9, 134.1, 129.1, 129.0, 128.9, 127.2, 123.4, 80.6, 75.1, 73.4, 65.5, 65.4, 37.5, 36.9, 35.9, 35.1, 35.0, 33.1, 30.8, 29.6, 29.3, 28.3, 28.1, 21.7, 16.1, 13.7; HRMS $[M+Na]^+$: 477.2608; (calculated for $[C_{28}H_{38}NaO_5]^+$ 477.2611).

6-(tert-Butoxy)-6-oxohexan-2-yl 6-([1,1′-biphenyl]-4-yl)-2-hydroxyhexanoate (**15b**). Mixture of diastereomers; Yield 37%; Colourless oil; ^{1}H NMR (200 MHz, $CDCl_3$): δ = 7.64–7.17 (m, 9H), 5.02–4.90 (m, 1H), 4.18–4.07 (m, 1H), 2.98–2.54 (m, 4H), 2.29–2.10 (m, 2H), 1.90–1.50 (m, 8H), 1.43 (s, 9H), 1.28-1.19 (m, 3H); ^{13}C NMR (50 MHz, $CDCl_3$): δ = 175.2, 172.8, 172.7, 141.8, 141.7, 141.3, 138.9, 129.0, 128.9, 127.3, 127.2, 80.5, 72.6, 70.7, 70.5, 35.7, 35.6, 35.2, 34.6, 34.5, 31.5, 31.4, 28.3, 24.8, 24.6, 21.0, 20.1, 20.0; HRMS $[M+Na]^+$: 477.2614; (calculated for $[C_{28}H_{38}NaO_5]^+$ 477.2611).

General procedure for the synthesis of 2-oxoesters **16a,b**.

To a solution of compound **15** (0.13 mmol) in dry CH_2Cl_2 (2.0 mL) was added Dess-Martin periodinane (72 mg, 0.17 mmol) and the reaction mixture was left stirring at room temperature for 2 to 4 h. Then, a 10% aqueous solution of $Na_2S_2O_3$ (5 mL) was added and the reaction mixture was extracted with CH_2Cl_2 (3 × 5 mL). The combined organic layers were mixed and dried over $MgSO_4$ and the solvent was removed under reduced pressure. The product was purified by column chromatography (petroleum ether 40–60 °C:EtOAc) (8:2).

5-(tert-Butoxy)-5-oxopentyl 6-([1,1′-biphenyl]-4-yl)-3-methyl-2-oxohexanoate (**16a**). Yield 75%; Colourless oil; ^{1}H NMR (200 MHz, $CDCl_3$): δ = 7.63–7.19 (m, 9H) 4.24 (t, *J* = 6.3 Hz, 2H), 3.31–3.15 (m, 1H), 2.65 (t, *J* = 7.2 Hz, 2H), 2.25 (t, *J* = 6.9 Hz, 2H), 1.89-1.56 (m, 8H), 1.44 (s, 9H), 1.15 (d, *J* = 7.0 Hz, 3H); ^{13}C NMR (50 MHz, $CDCl_3$): δ = 198.1, 172.7, 162.1, 141.2, 139.0, 129.0, 128.9, 127.3, 127.2, 127.1, 80.6, 66.0, 42.3, 35.6, 35.1, 31.7, 29.0, 28.3, 28.0, 21.6, 15.4; HRMS $[M+Na]^+$: 475.2434; (calculated for $[C_{28}H_{36}NaO_5]^+$ 475.2455).

6-(tert-Butoxy)-6-oxohexan-2-yl 6-([1,1′-biphenyl]-4-yl)-2-oxohexanoate (**16b**). Yield 77%; Yellowish oil; ^{1}H NMR (200 MHz, $CDCl_3$): δ = 7.63–7.17 (m, 9H), 5.09–4.96 (m, 1H), 2.92–2.78 (m, 2H), 2.75–2.59 (m, 2H) 2.29–2.12 (m, 2H), 1.83–1.52 (m, 8H), 1.48–1.22 (m,12H); ^{13}C NMR (50 MHz, $CDCl_3$): δ = 194.9, 172.7, 161.1, 141.3, 139.0, 129.0, 128.9, 127.3, 127.2, 80.5, 73.7, 39.4, 35.4, 35.2, 35.1, 30.9, 28.3, 22.8, 21.0, 19.9; HRMS $[M+Na]^+$: 475.2456; (calculated for $[C_{28}H_{36}NaO_5]^+$ 475.2455).

General procedure for the synthesis of 2-oxoesters **17a,b**.

A solution of *tert*-butyl ester **16** (0.13 mmol) in dry CH_2Cl_2 (3.0 mL) and TFA (3.0 mL) was stirred at room temperature for 2 h. The solvent was removed under reduced pressure and then CH_2Cl_2 (5 mL) was added and re-evaporated twice. The product was purified by column chromatography (petroleum ether 40–60 °C:EtOAc) (8:2).

5-((6-([1,1′-Biphenyl]-4-yl)-3-methyl-2-oxohexanoyl)oxy)pentanoic acid (**17a**) (GK587). Yield 77%; Colourless oil; ^{1}H NMR (200 MHz, $CDCl_3$): δ = 7.63–7.16 (m, 9H), 4.25 (t, *J* = 6.3 Hz, 2H), 3.34–3.12 (m, 1H), 2.65 (t, *J* = 7.2 Hz, 2H), 2.39 (t, *J* = 6.8 Hz, 2H),1.91–1.54 (m, 8H), 1.15 (d, *J* = 7.0 Hz, 3H); ^{13}C NMR (50 MHz, $CDCl_3$): δ = 198.0, 179.4, 162.0, 141.2, 139.0, 129.0, 128.9, 127.3, 127.2, 65.9, 42.3, 35.6, 33.5, 31.6, 29.0, 27.9, 21.2, 15.4; HRMS $[M-H]^-$: 395.1854; (calculated for $[C_{24}H_{27}O_5]^-$ 395.1864).

5-((6-([1,1′-Biphenyl]-4-yl)-2-oxohexanoyl)oxy)hexanoic acid (**17b**) (GK639). Yield 87%; Colourless oil; ^{1}H NMR (200 MHz, $CDCl_3$): δ = 7.62-7.17 (m, 9H), 5.10–4.94 (m, 1H), 2.94–2.77 (m, 2H), 2.73–2.57 (m, 2H), 2.44-2.28 (m, 2H),1.85–1.52 (m, 8H), 1.30 (d, *J* = 7.0 Hz, 3H); ^{13}C NMR (50 MHz, $CDCl_3$): δ = 194.9, 179.5, 161.1, 141.3, 139.0, 129.0, 128.9, 127.3, 127.3, 127.2, 73.6, 39.4, 35.4, 35.0, 33.7, 30.9, 22.8, 20.6, 19.9; HRMS [M-H]$^-$: 395.1857; (calculated for $[C_{24}H_{27}O_5]^-$ 395.1864).

2.2. Plasma Stability Studies

For plasma stability, LC-MS studies were performed with an ABSciex Triple TOF 4600 (ABSciex, Darmstadt, Germany) combined with a micro-LC Eksigent (Eksigent, Darmstadt, Germany) and an autosampler set at 5 °C and a thermostated column compartment. Electrospray ionization (ESI) in negative mode was used for the MS experiments. The data acquisition method consisted of a TOF-MS full scan *m/z* 50–850 Da and several IDA-TOF-MS/MS (Information Dependent Acquisition) product ion scans using 40 V Collision Energy (CE) with 15 V (Collision Energy Spread) CES used for each candidate ion in each data acquisition cycle (1091). Halo C18 2.7 µm, 90 Å, 0.5 × 50 mm^2 from Eksigent was used as a column and the mobile phase consisted of a gradient (A: acetonitrile/0.01% formic acid/isopropanol 80/20 v/v; B: H_2O/0.01% formic acid). The elution gradient adopted started with 5% of phase B for 0.5 min, gradually increasing to 98% in the next 7.5 min. These conditions were kept constant for 0.5 min, and then the initial conditions (95% solvent B, 5% solvent A) were restored within 0.1 min to re-equilibrate the column for 1.5 min for the next injection (flow rate 55 µL/min). The data acquisition was carried out with MultiQuant 3.0.2 and PeakView 2.1 from AB SCIEX (ABSciex, Darmstadt, Germany).

2.3. In vitro PLA_2 Activity Assay

Group specific PLA_2 assays were employed to determine the inhibitory activity using a lipidomics-based mixed micelle assay as previously described [4,25]. The substrate for each enzyme consisted of 100 µM PAPC (except for GIVA $cPLA_2$ as noted), 400 µM of C12E8 surfactant, and 2.5 µM of 17:0 LPC internal standard. For GIVA $cPLA_2$, the total phospholipid concentration (100 µM) consisted of 97 µM PAPC and 3 µM of $PI(4,5)P_2$ which enhances the activity of the enzyme. A specific buffer was prepared to achieve optimum activity for each enzyme. The buffer for GIVA $cPLA_2$ contained 100 mM HEPES pH 7.5, 90 µM $CaCl_2$, and 2 mM DTT. For GVIA $iPLA_2$, the buffer consisted of 100 mM HEPES pH 7.5, 2 mM ATP, and 4 mM DTT. Finally, the buffer for GV $sPLA_2$ contained 50 mM Tris-HCl pH 8.0 and 5 mM $CaCl_2$. The enzymatic reaction was performed in a 96 well-plate using a Benchmark Scientific H5000-H MultiTherm heating shaker for 30 min at 40 °C. Each reaction was quenched with 120 µL of methanol/acetonitrile (80/20, v/v), and the samples were analyzed using a HPLC-MS system. A Shimadzu SCL-10A system controller with two LC-10AD liquid pumps (Shimadzu Corp.,Kyoto Kyoto, Japan) connected to a column controller instrument (Analytical Sales & Products, Inc, Flanders NJ, USA), and a CTC Analytics PAL autosampler platform (Leap Technologies, NC, USA) were used for HPLC analysis. An AB Sciex 4000 QTRAP triple quadrupole/linear ion trap hybrid mass spectrometer (AB Sciex LLC, Framingham MA, USA) was used for MS analysis. A blank experiment, which did not contain enzyme, was also included for each substrate to determine the non-enzymatic hydrolysis product and to detect any changes in the intensity of the 17:0 LPC internal standard.

3. Results and Discussion

3.1. Design and Synthesis of Inhibitors

We assumed that in human plasma, it is likely that the oxoester functionality is hydrolyzed by an esterase. We hypothesized that if a small organic functionality, such as a methyl group, was added, either on the α-carbon to the oxoester functionality or on the α-carbon to the ester oxygen, it would decrease the rate of destruction of these compounds. The methyl group is the smallest alkyl group and has played a beneficial role in drug design by often constituting the problem-solving key to lead optimization [26]. Mainly, the PharmacoKinetic (PK) and PharmacoDynamic (PD) properties of a compound can be modified by the addition of the methyl group, where in many cases an increase in the selectivity and potency of the pharmaceutical agent has resulted.

GK452 (**1a**, Figure 1) was one of the most potent 2-oxoester inhibitors of GIVA $cPLA_2$. Thus, we envisaged that the two methylated derivatives of this compound shown in Figure 2 should result in an increased metabolic stability.

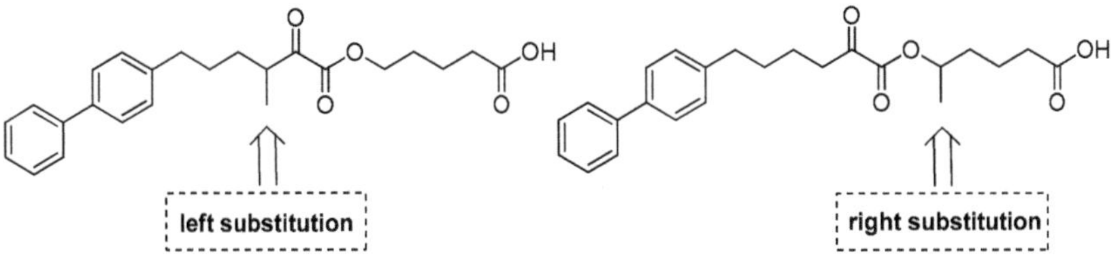

left substitution

right substitution

Figure 2. Design of α-methyl derivatives of **GK452** inhibitor of GIVA $cPLA_2$.

The synthetic strategy for how to insert the methyl group in the desired positions (left or right of the oxoester) was, firstly, with a lithium diisopropylamide (LDA) treatment followed by methylation for the left addition, and by a coupling reaction of an intermediate α-hydroxy carboxylic acid to a methylated secondary bromo *tert*-butyl ester for the right addition.

The synthesis of the needed α-hydroxy β-methyl carboxylic acid **9**, bearing a biphenyl scaffold on the left, is described in Scheme 1. The first reaction was an α-substitution using an LDA method, hexamethylphosphoramide (HMPA) to help stabilize the formed enolate and CH_3I as the electrophile [27], giving the α-methyl carboxylic acid **4** in a high yield. Then, the carboxylic moiety was reduced to an alcohol by converting it into the corresponding mixed anhydride and finally reducing it with $NaBH_4$ in the presence of MeOH [28]. Alcohol **5**, which was obtained in a high yield, was next oxidized using PCC to aldehyde **6**, which after reaction with KCN gave cyanohydrin **7** in an excellent yield. Acidic methanolysis by treatment with freshly prepared 6N HCl in MeOH led to the corresponding hydroxy methyl ester **8** in a good yield, which finally, after saponification, afforded the desired α-hydroxy carboxylic acid **9**.

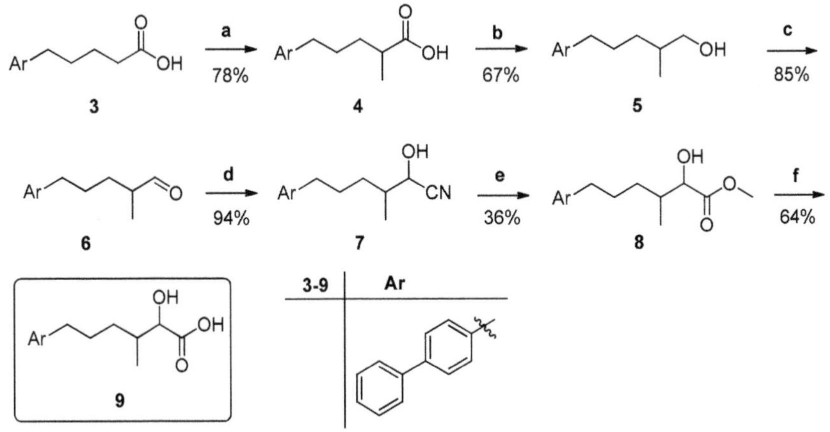

Ar OH 3 —a, 78%→ Ar OH 4 —b, 67%→ Ar OH 5 —c, 85%→

Ar O 6 —d, 94%→ Ar OH CN 7 —e, 36%→ Ar OH O O 8 —f, 64%→

Ar OH OH O 9

3-9	Ar

Scheme 1. Synthesis of 2-hydroxy acid **9**. (**a**) (i) LDA, dry THF (ii) HMPA, (iii) CH_3I; (**b**) (i) CH_3CH_2OCOCl, Et_3N, dry THF, (ii) $NaBH_4$, MeOH; (**c**) iodobenzene diacetate, TEMPO (10 mol%), dry CH_2Cl_2; (**d**) aq. sol. $NaHSO_3$, CH_2Cl_2, (ii) KCN, H_2O; (**e**) 6N HCl/MeOH; (**f**) 1N NaOH, MeOH.

For the synthesis of *tert*-butyl ester **12**, acidic bromination of commercially available delta-hexalactone **10**, using 33% HBr in AcOH [29], produced the intermediate bromo carboxylic acid **11** in almost quantitative yield. Next step was the protection of the carboxyl group using isobutylene in the presence of c. H_2SO_4, leading to *tert*-butyl ester **12** in a high yield (Scheme 2).

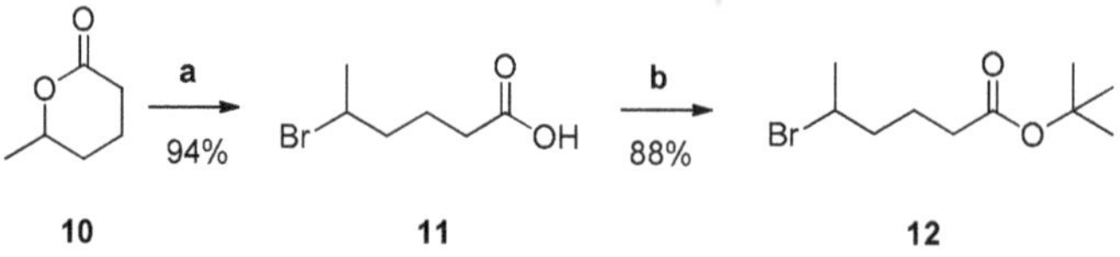

Scheme 2. Synthesis of bromo *tert*-butyl ester **12**. (**a**) 33% HBr in AcOH; (**b**) isobutylene, c.H_2SO_4, dry CH_2Cl_2.

After having synthesized the needed acid **9** and the appropriate hydroxy acid **13**, as previously described [23], a coupling reaction was performed by treatment with Cs_2CO_3 and the corresponding bromo *tert*-butyl esters (methylated or not) (Scheme 3). The coupling products **15a** and **15b** were produced in moderate yields 35–37%. For this step, it was observed that, when *tert*-butyl ester **12** was used, an increased reaction time was needed. Oxidation using Dess-Martin periodinane in dry CH_2Cl_2 gave oxoesters **16a** and **16b** in high yields. Final deprotection of the *tert*-butyl ester by using 50% TFA in dry CH_2Cl_2 led to the desired products **17a** (GK587) and **17b** (GK639) methylated derivatives.

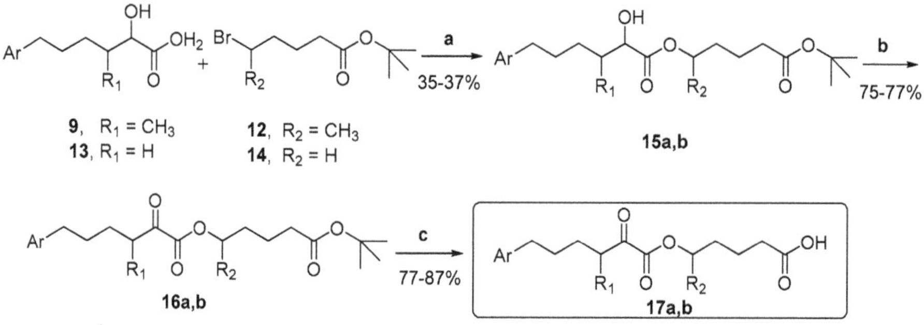

15-17	R_1	R_2
a	CH_3	H
b	H	CH_3

Scheme 3. Synthesis of 2-oxoesters. (**a**) (i) Cs_2CO_3, THF, H_2O (pH 9); (ii) DMF, reflux, 24–72 h; (**b**) Dess Martin, dry CH_2Cl_2, r.t., 3 h; (**c**) 50% TFA, dry CH_2Cl_2, r.t., 2 h.

3.2. *Plasma Stability Studies*

The in vitro stability of 2-oxoesters **17a** and **17b**, as well as inhibitor GK452 for comparison, was studied in human plasma. The reactions were initiated by the addition of test compound to a preheated plasma solution to yield a final concentration of 1 mg/L [24]. Samples (50 μL) were obtained and after the appropriate treatment were analyzed by LC-HRMS/MS (ABSciex, Darmstadt, Germany) and the results are shown in Figure 3 (see, also Supplementary Material).

An approximately 85% increase on the metabolic stability was observed at 15 min for both compounds; the parent compound remaining was increased from 25% to 45–46%. At 30 min, the effect is smaller, but still significant (70% enhancement, from 17% to 29%). Thus, these methyl substitutions at the left or the right of the oxoester functionality show that when steric hindrance around the oxoester functionality is increased, the stability of the oxoester in the plasma is increased as well.

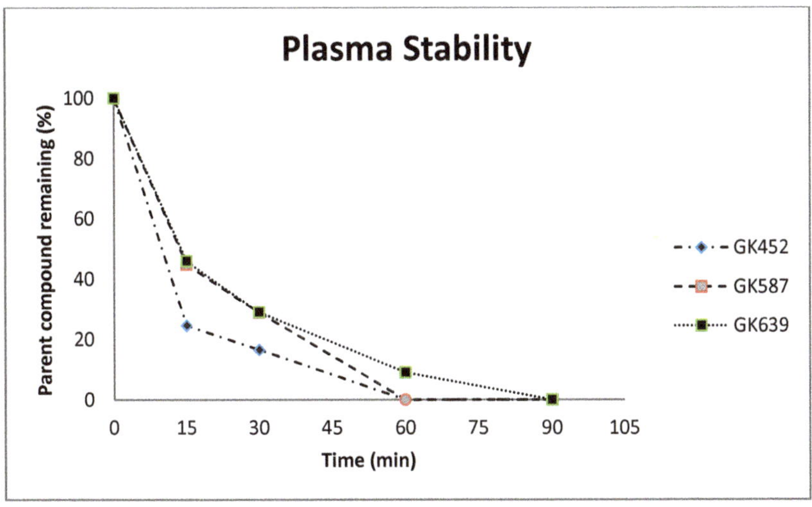

Figure 3. Plasma stability of GK587 and GK639 in comparison with GK452.

3.3. In Vitro Inhibitory Potency and Selectivity of Compounds GK587 and GK639

Both of the synthesized methyl substituted 2-oxoesters were tested for their in vitro inhibitory activity on recombinant human GIVA $cPLA_2$ using a LC/MS lipidomics based mixed micelle assay [25]. In addition, their selectivity over the intracellular human calcium-independent PLA_2 (GVIA $iPLA_2$) as well one secreted PLA_2 (GV $sPLA_2$) was also studied using similar group-specific mixed micelle assays. The inhibition results presented in Table 1 are expressed as both percent inhibition or as $X_I(50)$ values. First, the percent of inhibition for each PLA_2 enzyme at 0.091 mol fraction of each inhibitor was determined. Then, the $X_I(50)$ values were measured for compounds that displayed greater than 95% inhibition of GIVA $cPLA_2$. The $X_I(50)$ is the mole fraction of the inhibitor in the total substrate interface required to inhibit the enzyme activity by 50%. The dose-response inhibition curves for GIVA $cPLA_2$ inhibitors GK587 and GK639 are shown in Figure 4.

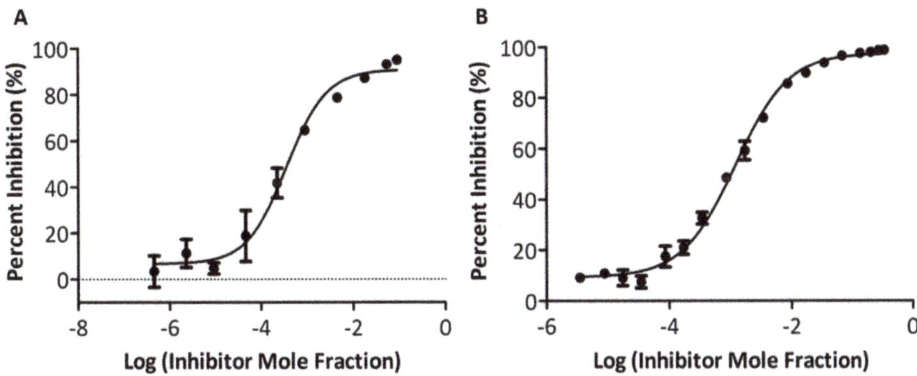

Figure 4. Dose-response inhibition curves for GIVA $cPLA_2$ inhibitors GK587 (**A**) and GK639 (**B**). The curves were generated using GraphPad Prism version 5.04 (GraphPad Software Inc., San Diego CA, USA) with a nonlinear regression targeted at symmetrical sigmoidal curves based on plots of % inhibition vs. log(inhibitor mole fraction). The reported $X_I(50)$ values were calculated from the resultant plots.

Both compounds GK587 and GK639 were found to inhibit GIVA $cPLA_2$ with $X_I(50)$ values of 0.00036 and 0.0012, respectively (Table 1). It is apparent that the introduction of a methyl group resulted in reduction of the inhibitory potency, when compared with the parent inhibitor GK452, which exhibits a $X_I(50)$ value of 0.000078 [23]. However, GK587 was approximately five times less potent than GK452, while GK639 was approximately fifteen times less potent. These findings indicate that the methyl substitution at either position leads to reduced inhibitory activity, however the substitution on the carbon carrying the ester oxygen causes more potent suppression of the potency. That means the methyl substitution on the α-carbon atom to the oxoester functionality is preferable. Both compounds GK587 and GK639 are selective inhibitors of GIVA $cPLA_2$, because both of them did not inhibit GVIA $iPLA_2$ and GV $sPLA_2$ (Table 1).

The results shown below lead to the conclusion that the methyl substitution on the α-carbon atom to the oxoester functionality is a successful strategy to increase the plasma stability of the oxoester inhibitors, ensuring that the inhibitor retains considerable inhibitory potency.

Table 1. In vitro Potency and Selectivity of Methyl Substituted 2-Oxoesters.

Compound	Structure	GIVA $cPLA_2$ % Inhibition[a]	GIVA $cPLA_2$ $X_I(50)$	GIVA $iPLA_2$ % Inhibition[a]	GIVA $sPLA_2$ % Inhibition[a]
GK452	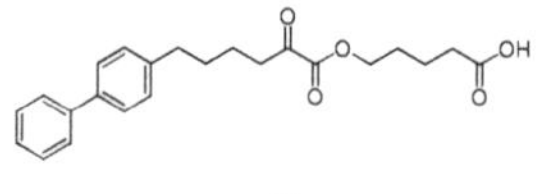	>95%	0.000078 ± 0.00001[b]	N.D.[b,c]	N.D.[b,c]
GK587	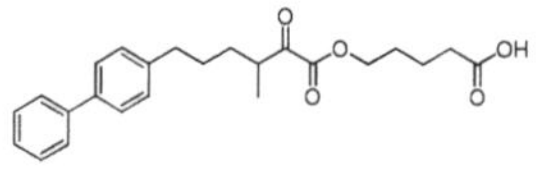	>95%	0.00036 ± 0.00007	N.D.[c]	N.D.[c]
GK639	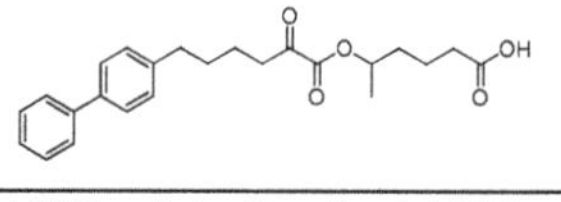	>95%	0.0012 ± 0.00008	N.D.[c]	N.D.[c]

[a] % Inhibition at 0.091 mole fraction of each inhibitor; [b] data taken from ref. [23]. [c] N.D. signifies compounds with less than 25% inhibition (or no detectable inhibition).

Synthetic GIVA $cPLA_2$ inhibitors are potential novel anti-inflammatory agents. For example, inhibitor GK470, previously developed by our group, was found to suppress the release of arachidonic acid in vitro and to exhibit an anti-inflammatory effect comparable to the reference drug methotrexate, in a prophylactic collagen-induced arthritis model, whereas in a therapeutic model, it showed results comparable to those of the reference drug Enbrel [20]. In addition, synthetic inhibitors may help in clarifying the biological role of GIVA $cPLA_2$ and the inter-connection with other enzymes. As an example, the involvement of both PLA_2 and phospholipase D (PLD) in the signaling through phosphatidic acid seems important for cancer cell survival [30,31].

4. Conclusions

In conclusion, in the present work we designed and synthesized two new 2-oxoester compounds in an effort to increase their metabolic stability. Determining their in vitro plasma stability and their in vitro inhibitory activity on GIVA $cPLA_2$, led us to conclude that inhibitor GK587, in which a methyl group was introduced on the α-carbon atom to the oxoester functionality, exhibits increased metabolic stability retaining at the same time considerable inhibitory potency. Thus, including an α-methyl substitution is a promising way of improving the pharmacological properties of 2-oxoester inhibitors and suggests that further chemical modification of this inhibitor class has the potential for pharmaceutical development of potent, metabolically stable, and selective inhibitors of phospholipase A_2.

Supplementary Materials: HRMS and plasma stability studies as well as copies of the ^{1}H and ^{13}C NMR spectra are available online at http://www.mdpi.com/2218-273X/10/3/491/s1.

Author Contributions: Synthesis and writing—original draft preparation, G.S.K.; plasma stability and LC-HRMS studies, M.G.K.; inhibitory activity studies, D.H. and V.D.M.; supervision and writing—review and editing, G.K. and E.A.D. All authors have read and agreed to the published version of the manuscript

Funding: This research was carried out within the framework of a Stavros Niarchos Foundation grant to the National and Kapodistrian University of Athens (GK) and NIH grant RO1 GM20501 to the University of California, San Diego (EAD).

Acknowledgments: G.S.K. would like to thank Stavros Niarchos Foundation (SNF) for a scholarship.

Conflicts of Interest: The authors have declared no conflict of interest.

References

1. Dennis, E.A.; Cao, J.; Hsu, Y.H.; Magrioti, V.; Kokotos, G. Phospholipase A_2 enzymes: Physical structure, biological function, disease implication, chemical inhibition, and therapeutic intervention. *Chem. Rev.* **2011**, *111*, 6130–6185. [CrossRef]
2. Murakami, M.; Nakatani, Y.; Atsumi, G.; Inoue, K.; Kudo, I. Regulatory functions of phospholipase A_2. *Crit. Rev. Immunol.* **2017**, *37*, 121–179. [CrossRef]
3. Vasquez, A.M.; Mouchlis, D.V.; Dennis, E.A. Review of four major distinct types of human phospholipase A_2. *Adv. Biol. Regul.* **2018**, *67*, 212–218. [CrossRef]
4. Mouchlis, V.D.; Chen, Y.; McCammon, J.A.; Dennis, E.A. Membrane allostery and unique hydrophobic sites promote enzyme substrate specificity. *J. Am. Chem. Soc.* **2018**, *140*, 3285–3291. [CrossRef] [PubMed]
5. Dennis, E.A.; Norris, P.C. Eicosanoid storm in infection and inflammation. *Nat. Rev. Immunol.* **2015**, *15*, 511–523. [CrossRef] [PubMed]
6. Leslie, C.C. Cytosolic phospholipase A_2: Physiological function and role in disease. *J. Lipid Res.* **2015**, *56*, 1386–1402. [CrossRef] [PubMed]
7. Kita, Y.; Shindou, H.; Shimizu, T. Cytosolic phospholipase A_2 and lysophospholipid acyltransferases. *BBA Mol. Cell Biol. Lipids* **2019**, *1864*, 838–845. [CrossRef] [PubMed]
8. Bonventre, J.V.; Huang, Z.; Taheri, M.R.; O'Leary, E.; Li, E.; Moskowitz, M.A.; Sapirstein, A. Reduced fertility and postischaemic brain injury in mice deficient in cytosolic phospholipase A_2. *Nature* **1997**, *390*, 622–625. [CrossRef] [PubMed]
9. Uozumi, N.; Kume, K.; Nagase, T.; Nakatani, N.; Ishii, S.; Tashiro, F.; Komagata, Y.; Maki, K.; Ikuta, K.; Ouchi, Y.; et al. Role of cytosolic phospholipase A_2 in allergic response and parturition. *Nature* **1997**, *390*, 618–622. [CrossRef] [PubMed]
10. Sun, G.Y.; Chuang, D.Y.; Zong, Y.; Jiang, J.; Lee, J.C.M.; Gu, Z.; Simonyi, A. Role of cytosolic phospholipase A_2 in oxidative and inflammatory signaling pathways in different cell types in the central nervous system. *Mol. Neurobiol.* **2014**, *50*, 6–14. [CrossRef]
11. Yang, B.; Li, R.; Greenlief, C.M.; Fritsche, K.L.; Gu, Z.; Cui, J.; Lee, J.C.; Beversdorf, D.Q.; Sun, G.Y. Unveiling anti-oxidative and anti-inflammatory effects of docosahexaenoic acid and its lipid peroxidation product on lipopolysaccharide-stimulated BV-2 microglial cells. *J. Neuroinfl.* **2018**, *15*, 202. [CrossRef] [PubMed]
12. Yang, B.; Fritsche, K.L.; Beversdorf, D.Q.; Gu, Z.; Lee, J.C.; Folk, W.R.; Greenlief, C.M.; Sun, G.Y. Yin-Yang mechanisms regulating lipid peroxidation of docosahexaenoic acid and arachidonic acid in the central nervous system. *Front. Neurol.* **2019**, *10*, 642. [CrossRef] [PubMed]
13. Kokotou, M.G.; Limnios, D.; Nikolaou, A.; Psarra, A.; Kokotos, G. Inhibitors of phospholipase A_2 and their therapeutic potential: An update on patents (2012–2016). *Expert Opin. Ther. Pat.* **2017**, *27*, 217–225. [CrossRef] [PubMed]
14. Nikolaou, A.; Kokotou, M.G.; Vasilakaki, S.; Kokotos, G. Small-molecule inhibitors as potential therapeutics and as tools to understand the role of phospholipases A_2. *BBA Mol. Cell Biol. Lipids* **2019**, *1864*, 941–956. [CrossRef] [PubMed]
15. Kokotos, G.; Six, D.A.; Loukas, V.; Smith, T.; Constantinou-Kokotou, V.; Hadjipavlou-Litina, D.; Kotsovolou, S.; Chiou, A.; Beltzner, C.C.; Dennis, E.A. Inhibition of group IVA cytosolic phospholipase A_2 by novel 2-oxoamides in vitro, in cells and in vivo. *J. Med. Chem.* **2004**, *47*, 3615–3628. [CrossRef] [PubMed]

16. Stephens, D.; Barbayanni, E.; Constantinou-Kokotou, V.; Peristeraki, A.; Six, D.A.; Cooper, J.; Harkewicz, R.; Deems, R.A.; Dennis, E.A.; Kokotos, G. Differential inhibition of group IVA and group VIA phospholipases A_2 by 2-oxoamides. *J. Med. Chem.* **2006**, *49*, 2821–2828. [CrossRef]
17. Six, D.A.; Barbayanni, E.; Loukas, V.; Constantinou-Kokotou, V.; Hadjipavlou-Litina, D.; Stephens, D.; Wong, A.C.; Magrioti, V.; Moutevelis-Minakakis, P.; Baker, S.F.; et al. Structure-activity relationship of 2-oxoamide inhibition of group IVA cytosolic phospholipase A_2 and group V secreted phospholipase A_2. *J. Med. Chem.* **2007**, *50*, 4222–4235. [CrossRef]
18. Baskakis, C.; Magrioti, V.; Cotton, N.; Stephens, D.; Constantinou-Kokotou, V.; Dennis, E.A.; Kokotos, G. Synthesis of polyfluoro ketones for selective inhibition of human phospholipase A_2 enzymes. *J. Med. Chem.* **2008**, *51*, 8027–8037. [CrossRef]
19. Kokotos, G.; Hsu, Y.-H.; Burke, J.E.; Baskakis, C.; Kokotos, C.G.; Magrioti, V.; Dennis, E.A. Potent and selective fluoroketone inhibitors of group VIA calcium-independent phospholipase A_2. *J. Med. Chem.* **2010**, *53*, 3602–3610. [CrossRef]
20. Kokotos, G.; Feuerherm, A.J.; Barbayianni, E.; Shah, I.; Sæther, M.; Magrioti, V.; Nguyen, T.; Constantinou-Kokotou, V.; Dennis, E.A.; Johansen, B. Inhibition of group IVA cytosolic phospholipase A_2 by thiazolyl ketones in vitro, ex vivo, and in vivo. *J. Med. Chem.* **2014**, *57*, 7523–7535. [CrossRef]
21. Kalyvas, A.; Baskakis, C.; Magrioti, V.; Constantinou-Kokotou, V.; Stephens, D.; Lopez-Vales, R.; Lu, J.Q.; Yong, V.W.; Dennis, E.A.; Kokotos, G.; et al. Differing roles for members of the phospholipase A_2 superfamily in experimental autoimmune encephalomyelitis. *Brain* **2009**, *132*, 1221–1235. [CrossRef] [PubMed]
22. Bone, R.N.; Gai, Y.; Magrioti, V.; Kokotou, M.G.; Ali, T.; Lei, X.; Tse, H.M.; Kokotos, G.; Ramanadham, S. Inhibition of Ca^{2+}-independent phospholipase $A_2\beta$ (iPL$A_2\beta$) ameliorates islet infiltration and incidence of diabetes in NOD mice. *Diabetes* **2015**, *64*, 541–554. [CrossRef] [PubMed]
23. Kokotou, M.G.; Galiatsatou, G.; Magrioti, V.; Koutoulogenis, G.; Barbayianni, E.; Limnios, D.; Mouchlis, V.D.; Satpathy, B.; Navratil, A.; Dennis, E.A.; et al. 2-Oxoesters: A novel class of potent and selective inhibitors of cytosolic group IVA phospholipase A_2. *Sci. Rep.* **2017**, *7*, 7025. [CrossRef] [PubMed]
24. Psarra, A.; Kokotou, M.G.; Galiatsatou, G.; Mouchlis, V.D.; Dennis, E.A.; Kokotos, G. Highly potent 2-oxoester inhibitors of cytosolic phospholipase A_2 (GIVA cPLA_2). *ACS Omega* **2018**, *3*, 8843–8853. [CrossRef] [PubMed]
25. Mouchlis, V.D.; Armando, A.M.; Dennis, E.A. Substrate specific inhibition constants for phospholipase A_2 acting on unique phospholipid substrates in mixed micelles and membranes using lipidomics. *J. Med. Chem.* **2019**, *62*, 1999–2007. [CrossRef]
26. Sun, S.; Fu, J. Methyl-containing pharmaceuticals: Methylation in drug design. *Bioorg. Med. Chem. Lett.* **2018**, *28*, 3283–3289. [CrossRef] [PubMed]
27. Li, X.-H.; Wan, S.-L.; Chen, D.; Liu, Q.R.; Ding, C.-H.; Fang, P.; Hou, X.-L. Enantioselective construction of quaternary carbon stereocenter via palladium-catalyzed asymmetric allylic alkylation of lactones. *Synthesis* **2016**, *48*, 1568–1572. [CrossRef]
28. Kokotos, G. A convenient one-pot conversion of N-protected amino acids and peptides into alcohols. *Synthesis* **1990**, *1990*, 299–301. [CrossRef]
29. Haruki, S.; Akemi, N.; Mamoru, K. A new one-pot synthetic method for selenium containing medium-sized α,β-unsaturated cyclic ketones. *Synthesis* **2008**, *2008*, 3229–3236.
30. Foster, D.A. Phosphatidic acid signaling to mTOR: Signals for the survival of human cancer cells. *Biochim. Biophys. Acta* **2009**, *1791*, 949–955. [CrossRef]
31. Mukhopadhyay, S.; Saqcena, M.; Chatterjee, A.; Garcia, A.; Frias, M.A.; Foster, D.A. Reciprocal regulation of AMP-activated protein kinase and phospholipase D. *J. Biol. Chem.* **2015**, *290*, 6986–6993. [CrossRef] [PubMed]

© 2020 by the authors. Licensee MDPI, Basel, Switzerland. This article is an open access article distributed under the terms and conditions of the Creative Commons Attribution (CC BY) license (http://creativecommons.org/licenses/by/4.0/).

Article

$cPLA_2\alpha$ Enzyme Inhibition Attenuates Inflammation and Keratinocyte Proliferation

Felicity J. Ashcroft [1], Nur Mahammad [1], Helene Midtun Flatekvål [1], Astrid J. Feuerherm [1,2] and Berit Johansen [1,*]

1 Department of Biology, Norwegian University of Science and Technology, Realfagbygget, 7491 Trondheim, Norway; felicity.ashcroft@ntnu.no (F.J.A.); nurm@ntnu.no (N.M.); helene.flatekval@lyse.net (H.M.F.); astfe@tkmidt.no (A.J.F.)

2 Center for Oral Health Services and Research (TkMidt), 7030 Trondheim, Norway

* Correspondence: berit.johansen@ntnu.no

Received: 7 September 2020; Accepted: 29 September 2020; Published: 2 October 2020

Abstract: As a regulator of cellular inflammation and proliferation, cytosolic phospholipase A_2 α ($cPLA_2\alpha$) is a promising therapeutic target for psoriasis; indeed, the $cPLA_2\alpha$ inhibitor AVX001 has shown efficacy against plaque psoriasis in a phase I/IIa clinical trial. To improve our understanding of the anti-psoriatic properties of AVX001, we sought to determine how the compound modulates inflammation and keratinocyte hyperproliferation, key characteristics of the psoriatic epidermis. We measured eicosanoid release from human peripheral blood mononuclear cells (PBMC) and immortalized keratinocytes (HaCaT) and studied proliferation in HaCaT grown as monolayers and stratified cultures. We demonstrated that inhibition of $cPLA_2\alpha$ using AVX001 produced a balanced reduction of prostaglandins and leukotrienes; significantly limited prostaglandin E_2 (PGE_2) release from both PBMC and HaCaT in response to pro-inflammatory stimuli; attenuated growth factor-induced arachidonic acid and PGE_2 release from HaCaT; and inhibited keratinocyte proliferation in the absence and presence of exogenous growth factors, as well as in stratified cultures. These data suggest that the anti-psoriatic properties of AVX001 could result from a combination of anti-inflammatory and anti-proliferative effects, probably due to reduced local eicosanoid availability.

Keywords: $cPLA_2\alpha$; psoriasis; proliferation; anti-inflammatory

---

1. Introduction

The phospholipase A_2 (PLA_2) superfamily of enzymes cleave phospholipids at the sn-2 position to release free fatty acids and lysophospholipids. These are the precursors to a multitude of lipid signaling molecules including the eicosanoids, which are metabolites of arachidonic acid (AA) and have important roles in inflammation and inflammatory diseases. Cytosolic phospholipase A_2 α ($cPLA_2\alpha$) is the only PLA_2 enzyme with high specificity for phospholipids carrying AA at the sn-2 position, placing it as an important upstream regulator of eicosanoid production [1]. When activated by extracellular stimuli, $cPLA_2\alpha$ undergoes Ca^{++}-dependent translocation from the cytoplasm to intracellular membranes and becomes predominantly localized to the peri-nuclear region of the cell [2–4]. This is where metabolism of AA by the cyclo-oxygenase (COX) and lipo-oxygenase (LOX) pathways typically occurs, producing prostaglandins and thromboxane A_2 (TxA_2), or leukotrienes, hydroxyeicosatetraenoic acids (HETEs), and hydroperoxyeicosatetraenoic acids (HPETEs), respectively. The importance of cPLA2α for stimulus-induced eicosanoid production and the pathogenesis of inflammation has been demonstrated by gene silencing both in vitro [5,6] and in animal models [7–11], and from the use of specific inhibitors of $cPLA_2\alpha$ in preclinical models of inflammatory diseases, as was recently reviewed by Nikolaou et al. [12]. Examples include the use of the indole-derivative

ZPL-5212372 in asthma and atopic dermatitis [13], the pyrrolidine-based compound RSC-3388 in a *Streptococcus pneumonia* infection model [14], and the ω-3 polyunsaturated fatty acid (PUFA) derivatives AVX001 and AVX002 in collagen-induced arthritis [15].

Plaque psoriasis (psoriasis vulgaris) is a disease with a chronic inflammatory phenotype that drives the hyperproliferation and aberrant differentiation of the epidermis [16]. Chronic inflammation in psoriasis is associated with higher expression of PLA_2 enzymes [17–19] and increased levels of eicosanoids [20–23]. Evidence for the involvement of eicosanoids in psoriasis is supported by mouse models of the disease—the leukotriene B_4 (LTB_4) receptor 1 and TxA_2 receptor have critical roles in imiquimod-induced skin inflammation [24–26], and prostaglandin E_2 (PGE_2) acting at prostaglandin receptors EP2 and EP4 is important for Th17-dependent inflammation in interleukin 23 (IL-23)-induced psoriasis [27]. Suppression of eicosanoid production is therefore an interesting prospect for treating psoriasis.

Non-steroidal anti-inflammatory drugs (NSAIDs) that inhibit AA metabolism via the COX pathway are commonly used for their analgesic, anti-inflammatory, and antithrombotic actions; however, their use is associated with many adverse gastrointestinal and cardiovascular effects (as reviewed in [28]) and they can induce or exacerbate psoriasis. The latter effect at least is postulated to be attributable to a skewed eicosanoid profile and the accumulation of leukotrienes (reviewed in [29,30]). Thus, it has been hypothesized that creating a more balanced suppression of eicosanoids using either dual COX-LOX inhibitors or by suppression of AA production using PLA_2 inhibitors (reviewed in [12,31]) would provide a better and safer therapeutic option.

The $cPLA_2\alpha$ inhibitor AVX001 is a ω-3 PUFA-derivative developed by Avexxin (now Coegin Pharma) that was demonstrated to be highly selective and to inhibit the in vitro activity of $cPLA_2\alpha$ with an IC_{50} of 120nM, being more potent than either docosahexaenoic acid (DHA) or the ω-6 PUFA derivative arachidonyl trifluoromethyl ketone ($AACOCF_3$, ATK) [15,32]. A topical application of AVX001 was trialed in a randomized, double-blind, placebo-controlled, dose-escalation first-in-man study to assess its safety and efficacy in patients with mild to moderate plaque psoriasis [33]. AVX001 showed significant efficacy and was well tolerated up to the maximum dose tested of 5%, supporting the targeting of $cPLA_2\alpha$ as a safe therapeutic strategy. The specificity and potency of $cPLA_2\alpha$ inhibition by AVX001 has been demonstrated [15,32], however, its mode of action in psoriasis remains to be determined. Given the documented role of $cPLA_2\alpha$ in mediating inflammatory signals in monocytes and keratinocytes [34–37] and the more recent interest in $cPLA_2\alpha$ as a driver of cellular proliferation [38], we sought to study potential modes of action of AVX001 in psoriatic skin by investigating its effects on inflammation and proliferation using human peripheral blood mononuclear cells (PBMC) and keratinocytes.

2. Materials and Methods

2.1. Materials

Cell culture media and chemicals were purchased from Sigma-Aldrich (St. Louis, MO, US) unless stated otherwise. A23178, naproxen, celecoxib calcipotriol hydrate, and lipopolysaccharide (LPS) were purchased from Sigma-Aldrich. Nordihydroguaiaretic acid (NDGA) was from Cayman chemicals (Ann Arbor, MI, USA) Recombinant human epidermal growth factor (EGF) and tumour necrosis factor (TNF)-α were from R&D systems (Abdingdon, UK). The fluoroketone AVX001 was synthesized and characterized according to Holmeide and Skattebol [39], and provided by Dr. Inger Reidun Aukrust and Dr. Marcel Sandberg (Synthetica AS, Oslo, Norway). AVX001 was stored at −80 °C as a 20 mM stock solution in dimethyl sulphoxide (DMSO) under argon gas to minimize oxidation.

2.2. PBMC Isolation and Treatment

Blood was recruited from healthy donors at St. Olavs Hospital HF, the Bloodbank (project approved by Regional Ethical Committee of Mid-Norway; #2016/553). Peripheral blood mononuclear cells (PBMC) were isolated using SepMate separation tubes with LymphoPrep density gradient medium from STEMCELL Technologies (Cambridge, UK), according to the manufacturer's recommendations. For experiments, 1 x 10^6 cells per well were plated in 1 mL Roswell Park Memorial Insitute (RPMI) medium supplemented with 5% fetal bovine serum (FBS), 0.3 mg/mL glutamine, and 0.1 mg/mL gentamicin. Inhibitors were added 2 h prior to the addition of the Ca^{++} ionophore A23178 (30 μM, 15 min) to activate cPLA$_2\alpha$ or lipopolysaccharide (LPS) (10 ng/mL, 72 h) as a potent inducer of inflammation. Following treatment, the cell suspensions were centrifuged to isolate the supernatant from the cell fraction. Samples were stored at −80 °C until analysis.

2.3. *Enzyme-Linked Immunoassay Detection of Eicosanoids*

Cell supernatant samples were analyzed by enzyme-linked immunosorbent assay (ELISA) for PGE$_2$ (Cayman #514435), LTB$_4$ (Cayman #10009292), TxB$_2$ (Cayman #501020), or 12S-HETE (Enzo Lifesciences #ADI-900-050) according to the manufacturers' protocols. Cell supernatants were assayed at dilutions of 1:100 for PGE$_2$, except supernatants from non-LPS-treated PBMC that were assayed undiluted in all assays. Supernatants were hybridized overnight, and the enzymatic conversion of the substrate was read at OD420 nm. Data were processed using a 4-parameter logistic fit model.

2.4. *Culture of HaCaT Keratinocytes*

2.4.1. Maintenance

The spontaneously immortalized skin keratinocyte cell line HaCaT [40] was kindly provided by Prof. N. Fusenig (Heidelberg, Deutsches Krebsforschungszentrum, Germany). These cells are commonly used to study proliferative and inflammatory responses in psoriasis research [41–46], as they express epidermal growth factor receptor (EGFR) and can proliferate both independently of, as well as in response to, stimulation with growth factors [47]. HaCaT were maintained in Dulbecco's modified Eagle Medium (DMEM) supplemented with 5% (*v*/*v*) FBS, 0.3 mg/mL glutamine, and 0.1 mg/mL gentamicin (DMEM-5) at 37 °C with 5% CO_2 in a humidified atmosphere at sub-confluency to prevent differentiation. Treatments were carried out in DMEM supplemented with 0.5% (*v*/*v*) FBS and 0.3 mg/mL glutamine (DMEM-0.5)

2.4.2. Eicosanoid Release

For analysis of eicosanoid release, we plated HaCaT in 12-well plates at 5 × 10^4 cells per well in DMEM-5 and cultured them for 3 days until reaching approximately 50% confluency, when the media was replaced with DMEM-0.5. The following day, the cells were stimulated with tumour necrosis factor (TNF)-α (30 ng/mL, 72 h), EGF (30 ng/mL, 24 h), or calcipotriol (10 nM, 72 h).

2.5. *[3H]-Arachidonic Acid Release Assay*

At 2 days post-confluency, we labelled HaCaT for 18 h with ^{3}H-AA (0.4 μCi/mL) in DMEM-0.5. After labelling, the cells were washed twice with phosphate-buffered saline (PBS) containing fatty acid-free bovine serum albumin (BSA) (2 mg/mL) in order to remove unincorporated radioactivity. After stimulation (EGF 100 ng/mL, 60 min), the supernatants were cleared of detached cells by centrifugation (13,000 rpm, 10 min). The release of ^{3}H-AA from the cells was assessed by liquid scintillation counting in a LS 6500 Multi-Purpose Scintillation Counter (Beckman Coulter, Brea, CA, USA). Adherent cells were dissolved in 1M NaOH in order to determine incorporated ^{3}H-AA in the cells by liquid scintillation counting. The results are given as released ^{3}H-AA in the supernatants relative to total ^{3}H-AA incorporated into the cells.

2.6. *Resazurin Assay*

HaCaT were seeded in 96-well plates in DMEM-5 at a density of 3000 cells per well. Following 72 h of cultivation, when cells reached a density of approximately 50%, we replaced the medium with DMEM-0.5. The following day, the cells were treated with AVX001, in a series of eight wells per treatment, for 24h. Resazurin (RnD systems, Abingdon, United Kingdom) was added according to the manufacturer's instructions and left to incubate for 2h at 37 °C with 5% CO_2 in a humidified atmosphere. Fluorescence was read at 544 nm excitation and 590 nm emission wavelengths using the Cytation 5 cell imaging multimode reader (Biotek Instruments, Winooski, VT, USA).

2.7. High Throughput Microscopy Assay for Population Analysis of Cell Cycle and Apoptosis

Cells were seeded in Greiner Bio-one CELLSTAR 96-well flat clear flat-bottomed plates (BioNordika, Oslo, Norway) in DMEM-5 at a density of 3000 cells per well. After 72 h, when cells reached a density of approximately 50%, we replaced the medium with DMEM-0.5. The following day, cells were treated with vehicle, AVX001, or etoposide (10 μM) in DMEM-0.5 for 24 h. We then followed the manufacturer's guidelines for the Click-iT 5-ethyl-2′-deoxyuridine (EdU) Alexa Fluor 594 imaging kit (ThermoFisher Scientific, Waltham, MA, USA) using a final concentration of 10 μM EdU per well incubated for a further 2 h at 37 °C, 5% CO_2. Following incubation with EdU, we removed the media and replaced it with the CellEvent Caspase 3/7 Green detection reagent (ThermoFisher Scientific) prepared at 2 μM in Dulbecco's (D)-PBS +5% FBS. The cells were incubated for a further 45 min, then the reagent was removed, and the cells were immediately fixed using 4% formaldehyde in D-PBS for 20 min on ice. Permeabilization was carried out using 0.1% Triton-X 100 in D-PBS and the Click-iT reaction was performed according to the manufacturer's guidelines using the Alexa-594 picoyl azide to label incorporated EdU. Finally, the cells were counterstained by incubation with 1 μg/mL 4′,6-diamidino-2-phenylindole (DAPI) (ThermoFisher Scientific) in D-PBS for 5 min. DAPI solution was removed and replaced with D-PBS for imaging. Plates were stored in the dark at 4 °C. All steps were performed at room temperature unless otherwise stated. Automated imaging was carried out on the Cytation5 cell imaging multimode reader (Biotek Instruments) at 4× magnification using DAPI, TexasRed, and GFP filter sets to image the DAPI, Alexa-594, and CellEvent Green signals, respectively. Four images were taken per well and 3 wells per treatment were used for the analysis.

Image analysis was performed in the freeware CellProfiler version 3.1.9 [48]. Firstly, nuclei were segmented from DAPI images using an Ostu 2-class thresholding approach, and were then counted. In further analyses, filters were employed to remove images with fewer than 50 cells. For cell cycle analysis, the total DAPI intensity and total EdU staining were then measured per nuclei from 12 images per treatment group and the freeware Flowing version 2.5.1 (Perttu Terho, Turku Centre for Biotechnology) was used to identify cells in G1, G2, and S-phases of the cell cycle, with gating based on $\log_{10}$ total EdU intensity vs. total DNA intensity. Apoptotic cells were identified on the basis of robust-background thresholding of the CellEvent Green signal and reported as a percentage of the total number of cells per image. Four images were taken per well and data were based on 3 wells per treatment group. To calculate the proliferative index, we segmented EdU-positive cells using an Ostu 2-class thresholding approach and reported them as a proportion of the total number of cells per image. Four images were taken per well and data were based on 3 wells per treatment group.

2.8. RNA Extraction and Real-Time Quantitative PCR

Cells were seeded in 6-well plates in DMEM-5. Following 72 h of cultivation, when cells reached a density of approximately 50%, we replaced the medium with DMEM-0.5. The following day, the cells were preincubated with AVX001 for 2 h prior to stimulation with EGF (30 ng/mL, 4 h). Total RNA was extracted with Total RNA kit I from Omega BIO-TEK (Norcross, GA, USA) according to the manufacturer's protocol. The amount and purity of the RNA samples were quantified using a Nanodrop One/One^{C} Microvolume UV–VIS Spectrophotometer (ND-ONE-W) from ThermoFisher Scientific. RNA samples with absorbance (A) A260/A230 between 1.8 and 2.1 and A260/280 between 2.0 and 2.2 were accepted. Reverse transcription was carried out using the QuantiTect Reverse Transcription Kit (Qiagen, Hilden, Germany) with 1 μg of RNA per sample, according to the manufacturer's protocol.

Real-time PCR analysis was performed using the LightCycler 480 SYBR Green I Master MIX and LightCycler 96 instrument from Roche (Basel, Switzerland), according to the manufacturer's protocol.

2.9. 3D Culture of HaCaT Keratinocytes

3D stratified HaCaT cultures were grown in Nunc cell culture inserts (0.4 μm pore size) using the 24-well carrier plates system (Thermo Fisher Scientific #141002) The culture inserts were coated using the Coating Matrix kit (Thermofisher Scientific #R-011-K) according to the manufacturer's protocol. HaCaT were plated at a density of 0.3×10^5 cells per insert in 0.5 mL DMEM-5 and incubated for 24 h before being lifted to the air–liquid interface. The media in the lower chamber was replaced with DMEM-5 (without antibiotics) + 1 ng/mL EGF, and 5 μg/mL L-ascorbic acid in the absence or presence of AVX001 (5 μM). The media in the lower chambers and treatments were changed every 3rd day for 12 days.

The cultures were fixed in 4% paraformaldehyde (PFA) overnight, before processing for paraffin embedding. Briefly, the membranes were removed from the inserts and prepared in Tissue Clear (Sakura, Osaka, Japan) for paraffin wax embedding using the Excelsior AS Tissue processor (ThermoFisher Scientific). Paraffin embedded sections (4 μm) were cut onto SuperFrost Plus slides (ThermoFisher scientific), dried over night at 37 °C, and then baked for 60 min at 60 °C. The sections were dewaxed in Tissue Clear and rehydrated through graded alcohols to water in an automatic slide stainer (Tissue-Tek Prisma, Sakura). Next, the sections were pretreated in Target Retrieval Solution, High pH (Dako, Glostrup, Denmark, K8004) in PT Link (Dako) for 20 min at 97 °C to facilitate antigen retrieval. The staining was performed according to the manufacturer's procedure with EnVision G|2 Doublestain System Rabbit/Mouse (DAB+/Permanent Red) kit (Dako/Agilent K5361) on the Dako Autostainer. Following their soaking in wash buffer, we quenched endogenous peroxidase and alkaline phosphatase activity with Dual Endogenous Enzyme Block (Dako). Sections were then rinsed in wash buffer and incubated with primary antibody against Ki67 (MIB1 (Dako M7240) diluted 1:300) for 40 min. The slides were rinsed before incubating in horseradish peroxidase (HRP) - polymer and 3,3′-Diaminobenzidine (DAB) to develop the stain. After a double stain block, the sections were incubated in antibody against cytokeratin 10 (Invitrogen #MA5-13705 diluted 1:100) for 60 min. After incubation in the mouse/rabbit linker, the sections were incubated in AP- polymer and the corresponding red substrate buffer with washing between each step. Tris-buffered saline (TBS; Dako K8007) was used throughout for the washing steps. The slides were lightly counterstained with hematoxylin, completely dried, and coverslipped. Appropriate negative controls were performed; both mouse monoclonal isotype control (Biolegend, San Diego, CA, USA) and omitting the primary antibody (negative method control).

2.10. Statistical Analysis

Statistical analysis was carried out in GraphPad Prism Software, version 7, using one-way ANOVA with Dunnet's post-analysis. For normalized data, we used the Kruskal–Wallis test with Dunn's post-analysis.

3. Results

3.1. Inhibition of cPLA2α Using AVX001 Resulted in a Balanced Reduction of Eicosanoids

Several eicosanoids including PGE_2, prostaglandin F_2 (PGF_2), LTB_4, and 12S-HETE are known to be elevated in psoriatic lesions [20–23,49], and their direct roles in disease progression are supported by animal models of inflammatory skin disease [24,31,50,51]. Hence, it has been proposed that generating a balanced reduction in overall eicosanoid production is a promising therapeutic strategy [12,31]. To test whether the cPLA2α inhibitor AVX001 can normalize a broader range of eicosanoids than known COX- and LOX inhibitors, we treated PBMC with AVX001, the non-specific lipo-oxygenase (LOX) inhibitor nordihydroguaiaretic acid (NDGA), the cyclooxygenase (COX)-2 selective inhibitor celecoxib, or the dual COX1/COX2 inhibitor naproxen, and stimulated the mixture with the Ca^{++} ionophore A23178. The levels of PGE_2, TxB_2, LTB_4, and 12S-HETE were assayed by ELISA and inhibition was calculated as a percentage of the A23178 stimulation. The maximum inhibition and concentration at which this was reached is presented in Table 1, and the absolute eicosanoid levels are available in Appendix A (Figure A1). We observed that AVX001 treatment dose-dependently inhibited the release of PGE_2, TxB_2, and LTB_4. AVX001 also reduced 12S-HETE release at the maximum dose tested (10 μM), although it did not reach significance ($p = 0.12$, $n = 3$). As expected, NDGA dose-dependently inhibited the release of LOX metabolites LTB_4 and 12S-HETE, but not COX metabolites PGE_2 and TxB_2. Celecoxib treatment gave a dose-dependent reduction of PGE_2, TxB_2, and LTB_4 but did not inhibit 12S-HETE. The finding that the COX-2 selective inhibitor celecoxib reduced LTB_4 release was surprising, however, it was consistent with the following study that demonstrated inhibition of 5-LOX but not 12- or 15-LOX by celecoxib in human blood [52]. Naproxen treatment, on the other hand, gave dose-dependent inhibition of the COX metabolites PGE_2 and TxB_2 without affecting LOX metabolites LTB_4 or 12S-HETE. AVX001 treatment was therefore shown to result in a broader inhibition of eicosanoid release than the COX or LOX inhibitors tested, having a similar efficacy to reduce both COX and 5-LOX metabolites, and possibly also the 12-LOX metabolite 12S-HETE, albeit at a higher dose.

Table 1. Inhibition of eicosanoid release in A23178-stimulated peripheral blood mononuclear cells (PBMC). Eicosanoid levels were measured in the supernatants of PBMC pre-incubated with AVX001, naproxen, celecoxib, or nordihydroguaiaretic acid (NDGA) in the range of 0.5-10 μM and stimulated with A23178 (30 μM, 15 min). The percentage inhibition of the A23178-stimulated release was calculated for each eicosanoid and is shown as the mean ± standard error of the mean (SEM) for $n = 4$ (prostaglandin E_2 (PGE_2)), $n = 4$ (leukotriene B_4 (LTB_4)), and $n = 3$ (12S-hydroxyeicosatetranoic acid (HETE)) individuals. The inhibitor concentration at which maximal inhibition was reached is given below. Statistical significance was calculated using Kruskal–Wallis tests with Dunn's post-analysis; * $p < 0.05$, ** $p < 0.01$. A graphical presentation of absolute eicosanoid levels is available in Figure A1.

Inhibitor	PGE_2	6-keto $PGF_1\alpha$	LTB_4	TxB_2	12S-HETE
AVX001	93 ± 9 5 μM *	114 ± 16 5 μM	95 ± 6 5 μM *	106 ± 3 5 μM *	90 ± 6 10 μM
Naproxen	96 ± 6 5 μM **	102 ± 18 5 μM	48 ± 7 5 μM	106 ± 2 5 μM **	−18 ± 24 10 μM
Celecoxib	97 ± 5 5 μM *	108 ± 20 5 μM	91 ± 3 5 μM *	107 ± 2 5 μM **	42 ± 35 10 μM
NDGA	49 ± 7 10 μM	34.3 ± 7 10 μM	99 ± 1 10 μM *	42 ± 12 10 μM	105 ± 1 10 μM *

3.2. *AVX001 Inhibited PGE_2 Release in Response to Inflammatory Stimuli*

AVX001 inhibited eicosanoid release from A23187-stimulated PBMC, however, the use of the Ca^{++} ionophore does not represent a physiological stimulus, and therefore it was important to confirm the effects using biologically relevant stimuli. For these experiments, we measured the PGE_2 release. This was selected because PGE_2 is one of the main eicosanoids produced in the skin, being released by epidermal keratinocytes, dermal fibroblasts, and immune cells. It has proinflammatory and immuno-modulatory properties and promotes keratinocyte proliferation (reviewed in [53]). We measured PGE_2 levels in response to inflammatory stimuli and epidermal growth factor (EGF) and investigated the role of $cPLA_2\alpha$ in these responses.

To investigate the role of $cPLA_2\alpha$ in the response to pro-inflammatory stimuli, we preincubated human PBMC or HaCaT with AVX001 and stimulated them with lipopolysaccharide (LPS) or tumor necrosis factor (TNF)-α, respectively. LPS stimulated PGE_2 release in PBMC by an average of 91-fold. Pretreatment with AVX001 dose-dependently inhibited LPS-stimulated PGE_2 release with an IC_{50} of

5 μM (Figure 1A). HaCaT released a low but detectable level of PGE_2 that was stimulated by TNF-α treatment by an average of 67-fold. AVX001 also dose-dependently inhibited TNF-α-stimulated PGE_2 release (Figure 1B). These data support a role for cPLA$_2$α in mediating pro-inflammatory eicosanoid release from immune cells and keratinocytes, suggesting that AVX001 would have anti-inflammatory properties in the skin.

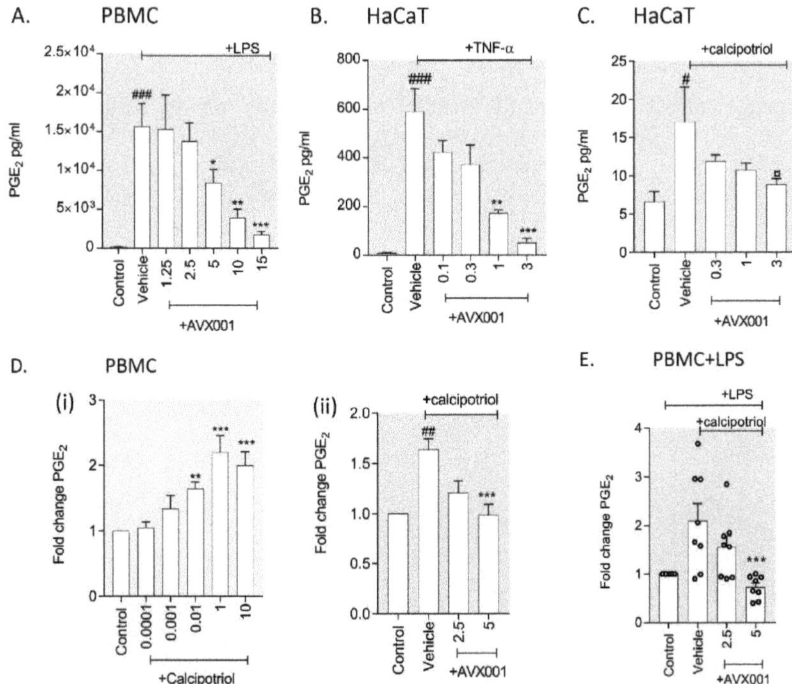

Figure 1. Effect of AVX001 on pro-inflammatory eicosanoid production. (**A**) PBMC were pre-incubated with AVX001 (μM) and stimulated with lipopolysaccharide (LPS) (10 ng/mL, 72 h). (**B**) Immortalized keratinocytes (HaCaT) were pre-incubated with AVX001 (μM) and stimulated with tumor necrosis factor (TNF)-α (10 ng/mL, 72 h). (**C**) HaCaT were pre-incubated with AVX001 (μM) and stimulated with calcipotriol (10 nM, 72 h). (**D**) (**i**) PBMC were treated with calcipotriol (μM) for 72 h, (**ii**) PBMC were pre-treated with AVX001 (μM) and stimulated with calcipotriol (10 nM, 72 h). (**E**) PBMC were pre-incubated with AVX001 (μM) and stimulated with calcipotriol (10 nM, 72 h) or vehicle in the presence of LPS (10 ng/mL). PGE_2 levels in supernatants were determined by ELISA and reported as either pg/mL (**A**–**C**), or as the fold change in PGE_2 levels over unstimulated (**D**) or LPS-stimulated (**E**) controls. Data are the mean ± SEM of ≥ 4 individuals (PBMC) or 3 replicates (HaCaT). Statistical significance was calculated by one-way ANOVA with Dunnett's post-analysis, or, for normalized data, using the Kruskal–Wallis test with Dunn's post-analysis; # $p < 0.05$, ## $p < 0.01$, or ### $p < 0.005$ versus unstimulated control and * $p < 0.05$, ** $p < 0.01$ versus vehicle-treated control.

Skin toxicity is a common side-effect of a variety of drug types including the vitamin D receptor (VDR) agonist calcitriol and its analogue calcipotriol, both of which are used to treat psoriasis [54]. The pro-inflammatory effects of VDR agonists are suggested to involve PGE_2 [55,56]. To investigate the role of cPLA$_2$α in VDR agonist-mediated PGE_2 release, we treated PBMC and HaCaT with calcipotriol in the absence or presence of AVX001. We observed a small but significant increase in PGE_2 in HaCaT exposed to calcipotriol; pre-incubation with AVX001 reduced the stimulatory effect, reaching marginal significance ($p < 0.1$) at the highest dose used (3 μM) (Figure 1C). Exposure to calcipotriol also stimulated the release of PGE_2 from PBMC by approximately twofold; pre-incubation with AVX001 (5 μM) blocked the calcipotriol-induced increase in PGE_2 (Figure 1D). Calcipotriol also augmented the LPS-stimulated release of PGE_2 in PBMC from a subset of individuals, although the high variability between the responses meant that this was not significant overall. In cases where PGE_2 levels were increased, AVX001 (5 μM) inhibited the response (Figure 1E). Our data support the hypothesis that adverse skin reactions resulting from calcipotriol treatment could result from increased PGE_2 and suggest that AVX001 may be useful to abrogate these reactions by limiting PGE_2.

3.3. *AVX001 Inhibited EGF-Stimulated Release of AA and PGE_2.*

Epidermal growth factor receptor (EGFR) activation is a well-known mitogenic signal for epidermal keratinocytes [57] and can activate $cPLA_2\alpha$ via mitogen-activated protein kinase (MAPK)-dependent phosphorylation [3,58]. To investigate the effect of AVX001 on growth factor-mediated $cPLA_2\alpha$ activation and eicosanoid release in keratinocytes, we stimulated HaCaT with EGF in the absence or presence of AVX001 and measured AA release and PGE_2 levels. Consistent with activation of $cPLA_2\alpha$, EGF stimulated the release of both AA and PGE_2 in HaCaT (Figure 2.) Treatment with AVX001 (≥ 0.3 μM) significantly and dose-dependently inhibited EGF-stimulated AA (Figure 2A) and PGE_2 (Figure 2B) release.

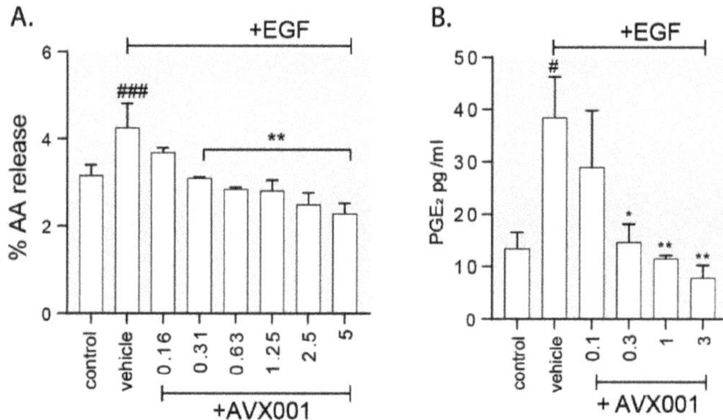

Figure 2. Effect of $cPLA_2\alpha$ inhibition on epidermal growth factor (EGF)-dependent arachidonic acid (AA) and PGE_2 release. (**A**) AA release was measured by the [3H]-arachidonic acid release assay in HaCaT treated with AVX001 and stimulated with EGF (100 ng/mL, 60 min). Data are the mean ± SD for three replicates from a representative experiment repeated twice. (**B**) PGE_2 levels measured in the supernatant of cells preincubated with AVX001 at the indicated concentrations and stimulated with EGF (30 ng/mL, 24 h). Data are mean ± SEM for three replicates. Statistical significance was calculated by one-way ANOVA with Dunnett's post-analysis; # $p < 0.05$, ### $p < 0.005$ versus unstimulated control and * $p < 0.05$, ** $p < 0.01$ versus vehicle-treated control.

3.4. *AVX001 Inhibited Keratinocyte Proliferation*

We showed that inhibition of $cPLA_2\alpha$ activity using AVX001 attenuated both AA and PGE_2 release stimulated by the addition of exogenous growth factor in HaCaT. In fibroblasts, this signaling cascade was required for cell cycle progression [38]. We thus hypothesized that the treatment of keratinocytes with AVX001 could inhibit proliferation. To test this, we first treated actively proliferating HaCaT cells with AVX001 for 24 h and measured cell viability using resazurin as an indicator of metabolically active cells. We then quantified cell number, cell cycle distribution, and apoptosis using a high content image-based assay based on the fluorescent labeling and single-cell quantification of DNA, incorporated EdU, and caspase-3/7 activity, as detailed in Section 2.7. AVX001 inhibited HaCaT viability with an IC_{50} of 8.5 μM (Figure 3A). The reduced viability observed using 10 μM AVX001 was associated with a reduction in the total cell count (Figure 3B) (i) and a reduction in the proportion of the cells in S-phase (Figure 3B) (ii); there was no significant difference in the number of apoptotic cells (Figure 3B) (iii). The data also suggested an accumulation of cells in G1, although this finding was not statistically significant ($p = 0.06$). As a positive control, cells were treated with etoposide, a chemotherapy drug known to block cell cycle progression in HaCaT cells [59–61]. As expected, treatment with the etoposide (10 µM) caused a decrease in cell numbers that, in contrast to AVX001, was associated with an accumulation of cells in G2/M, and an increased proportion of apoptotic cells, consistent with its known effect as a blocker of the G2/M transition [62,63].

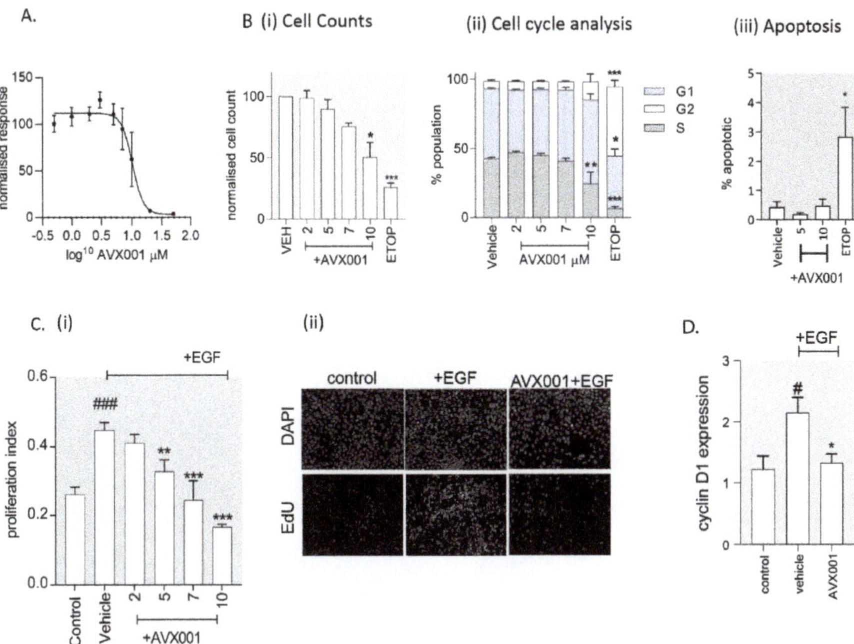

Figure 3. AVX001 inhibited proliferation in HaCaT monolayers. (**A**) Cell viability was measured using the resazurin assay in proliferating HaCaT treated with AVX001 for 24 h at the concentrations indicated. Measurements were normalized to the vehicle-treated control, and the mean ± SEM for three replicates is shown. (**B**) Automated microscopy and analysis of fluorescently labeled DNA, incorporated EdU, and caspase-3/7 activity was used to quantify (**i**) cell number, (**ii**) cell cycle distribution, and (**iii**) apoptosis in proliferating HaCaT treated with AVX001 for 24 h at the concentrations indicated. Data are mean ± SEM for ≥4 replicates. (**C**) (**i**) The proportion of cells in S-phase of the cell cycle (proliferation index) was determined by counting the total and EdU-positive nuclei in proliferating HaCaT pre-incubated with vehicle or AVX001 and stimulated with EGF (30 ng/mL, 24 h). Data are the mean ± SEM for three replicates from a representative experiment, repeated twice. (**ii**) Representative images showing 4′,6-diamidino-2-phenylindole (DAPI)-stained nuclei (DAPI) and fluorescently labelled EdU (EdU) for unstimulated HaCaT (control), and HaCaT stimulated with EGF (30 ng/mL, 24 h) in the absence (+EGF) or presence of 7 µM AVX001 (AVX001+EGF). (**D**) The relative expression of cyclin D1 measured by quantitative PCR in proliferating HaCaT pre-treated with AVX001 and stimulated with EGF (30 ng/mL, 4h). Data were normalized to the unstimulated control (control) and are the mean ± SEM from six replicates. Statistical significance was calculated by one-way ANOVA with Dunnett's post-analysis, or, for normalized data, the Kruskal–Wallis test with Dunn's post-analysis; # $p < 0.05$, ## $p < 0.01$, or ### $p < 0.005$ versus unstimulated control and * $p < 0.05$, ** $p < 0.01$, *** $p < 0.005$ versus vehicle-treated control.

Treatment with EGF for 24 h did not significantly impact cell viability (not shown), but rather led to an increase in the proportion of cells in S-phase of the cell cycle (proliferation index), which was inhibited by AVX001 (≥5 μM) (Figure 3C). Progression of the cell cycle from G1 to S-phase in response to growth factors is typically associated with increased levels of cyclin D1 [64]. Following 4 h of EGF treatment, cyclin D1 transcript levels were increased approximately twofold in HaCaT, and this was inhibited by AVX001 (5 μM) (Figure 3D). These findings support cPLA$_2$α as a target for inhibiting growth factor-dependent proliferation by halting cells in G1.

AVX001 inhibited cell proliferation in HaCaT grown in monolayers; however, experiments performed in 3D culture systems are often considered to have more physiological relevance. HaCaT retain the ability to stratify and differentiate in culture [65,66], and this is dependent on the presence of exogenous growth factors [67]. To test whether AVX001 is an effective inhibitor of proliferation in stratified keratinocytes, we cultured HaCaT at the air–liquid interface for 12 days in the absence or presence of AVX001. We measured PGE$_2$ levels and the thickness of the stratified epithelia. The proportions of proliferating and differentiating cells were determined based on Ki-67 and cytokeratin (CK) 10 positivity, respectively. AVX001-treated cultures had reduced levels of PGE$_2$, indicating the compound maintains the ability to suppress PGE$_2$ levels in stratified cultures (Figure 4A.) There was no significant difference in the thickness of the cultures (Figure 4B,E), however, immunohistochemical analysis of Ki67 showed strikingly fewer proliferating cells in the AVX001-treated cultures compared to the vehicle-treated controls (Figure 4C,E). This was not, however, accompanied by an increase in the proportion of CK10-positive cells, as might be expected, but rather by an accompanying reduction in the proportion of CK10-positive cells (Figure 4D,E). These findings give support for cPLA$_2$α being primarily a regulator of keratinocyte proliferation.

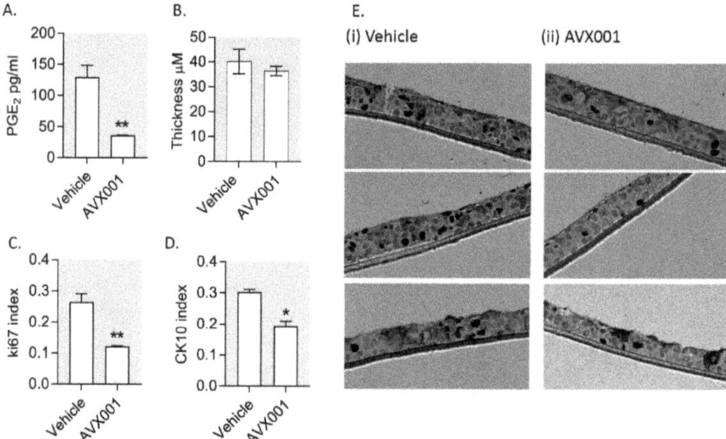

Figure 4. AVX001 inhibited proliferation in a stratified epithelium. HaCaT grown on porous collagen-coated membranes were maintained at the air–liquid interface for 12 days with exposure to exogenous EGF (1 ng/mL) in the absence or presence of AVX001 (5 μM). (**A**) PGE$_2$ levels measured in the supernatants at day 12 by ELISA. (**B–D**) Immunohistochemistry using anti-Ki-67 MIB1 (DAB+) and anti-cytokeratin (CK)10 (Permanent Red) antibodies with hematoxylin counterstain. Brightfield images were taken at 40X magnification and used to quantify (**B**) the thickness (stratification) of the cultures, (**C**) the proportion of proliferating cells (proliferation index = Ki-67-positive cells/total cells), and (**D**) the proportion of CK10-positive cells (CK10 index = CK10 positive cells/total cells.) Typically, >20 images were collected at 40X magnification per replicate, with data being the mean ± SEM for three replicates. Statistical significance was calculated by one-way ANOVA with Dunnett's post-analysis; * $p < 0.05$, ** $p < 0.01$, *** $p < 0.005$ versus vehicle-treated control. (**E**) Three representative images are shown for (i) vehicle and (ii) AVX001-treated cultures.

4. Discussion

In this study, we investigated the effects of the cPLA$_2\alpha$ inhibitor AVX001 on inflammatory eicosanoid release and epidermal proliferation to understand its mode of action for treating psoriatic skin disease.

We demonstrate for the first time that AVX001 can significantly and dose-dependently suppress the production of both COX and LOX AA metabolites in stimulated human PBMC. The findings are consistent with the use of the cPLA$_2\alpha$ inhibitors pyrrophenone and WAY-196025, which similarly inhibited both PGE$_2$ and LTB$_4$ release from PBMC stimulated with A23178 [68,69]. Our data thus support the fact that targeting the cPLA$_2\alpha$ enzyme results in a balanced suppression of inflammatory eicosanoid release.

We further demonstrate the inhibitory effect of AVX001 on eicosanoid released in response to pro-inflammatory stimuli. The Toll-like receptor (TLR) 4 agonist LPS induced PGE$_2$ release from PBMC, which was inhibited by AVX001. Our data using human PBMC supports the previously described involvement of cPLA$_2\alpha$ in LPS-stimulated PGE$_2$ production in THP-1 monocytes [35,70] reported to result from the induction of both the levels and activity of cPLA$_2\alpha$ [5]. TNF-α is a pro-inflammatory cytokine and a key contributor to the pathogenesis of psoriasis [71]. TNF-α induced a robust production of PGE$_2$ in HaCaT, which was inhibited by AVX001. These findings support cPLA2α as a mediator of the pro-inflammatory effects of TNF-α, as proposed by Sjursen et al. [36], however, while we demonstrated TNF-α-induced PGE$_2$ production, Sjursen et al. reported that 6 h treatment with TNF-α preferentially induced HETE and not PGE$_2$ production in HaCaT. It is therefore likely that stimulation of PGE$_2$ release involves additional transcriptional upregulation of COX pathway enzymes in addition to cPLA$_2\alpha$ activation in these cells, as demonstrated by Seo et al. [72].

Calcipotriol is a topical therapeutic for psoriasis and is known to cause skin irritation [54]. We demonstrate that calcipotriol stimulates the release of PGE$_2$ both in PBMC and keratinocytes, which is in agreement with the following studies [55,56,73] and further supports the involvement of VDR/PGE$_2$ signaling in drug-induced skin toxicity, as proposed by Shah et al. [56]. The mechanism by which calcipotriol stimulates PGE$_2$ production is unclear. In AVX001-treated cells, calcipotriol was unable to stimulate PGE$_2$ release, implicating that cPLA2α activation is required. This is in contrast to studies in keratinocytes by Ravid et al. [55], who suggest that the upregulation of COX-2 as opposed to increased AA production is responsible for the stimulation of PGE$_2$ production. Doroudi et al. [74] present a VDR-independent mechanism for activation of cPLA$_2\alpha$ by calcitriol via Ca^{2+}/calmodulin-dependent protein kinase II (CAMKII)-dependent phosphorylation. It will be interesting to determine how PGE$_2$ is regulated by calcipotriol in PBMC and keratinocytes and whether direct activation of cPLA$_2\alpha$ by CAMKII is involved. For the treatment of psoriasis, calcipotriol is commonly combined with the potent corticosteroid betamethasone dipropionate (Daivobet), resulting in improved efficacy and tolerance [75]. This poses the possibility that the use of AVX001 could be an interesting non-steroidal alternative combination partner for reducing inflammation and improving tolerance to calcipotriol.

Our finding that EGF-stimulated PGE$_2$ release in keratinocytes is reduced by inhibition of cPLA$_2\alpha$ is in line with several reports linking EGF stimulation with AA release [76–79]. Furthermore, Naini et al. describe a requirement for intact cPLA$_2\alpha$/PGE2 signaling in growth factor-dependent cell cycle progression in both mouse embryonic fibroblasts (MEFs) and mesangial cells [38]. Collectively, this puts regulation of the cPLA$_2\alpha$ enzyme, by means of its level and activity, in a central position to modulate growth factor-dependent responses. Thus, cPLA$_2\alpha$ may control both inflammatory and mitogenic processes, which are hallmarks of the pathogenesis of psoriasis.

We further show that treatment with AVX001 inhibits EGF-stimulated S-phase entry and reduces the proliferation of HaCaT keratinocytes grown both in monolayers and stratified cultures. Our findings are in agreement with the established role of cPLA$_2\alpha$ and eicosanoid signaling molecules as drivers of proliferation in several cancerous and non-cancerous cell types (reviewed in [80] and [81]). The described role of PGE$_2$ as an autacoid growth factor [82,83] and effector of EGF responses in keratinocytes [84] make it a good candidate for mediating the effects of cPLA$_2$α inhibition on keratinocyte proliferation. Knockout of the PGE$_2$ receptor, EP2, also supports a role for PGE$_2$ in regulating keratinocyte proliferation [85,86]. However, PGE$_2$ is certainly not the only candidate, and a weakness of this study was our focus on the effects of AVX001 on AA metabolites. It is likely that cPLA$_2$α inhibition with AVX001 would also suppress the production of LPC and its metabolites, e.g. platelet-activating factor (PAF). Like the eicosanoids, PAF has pro-inflammatory and proliferative effects in the epidermis [46,87,88], and PAF inhibition was found to suppress psoriasis-like skin disease progression in mice [89]. In future studies, it will therefore be important to determine whether AVX001 can also suppress the formation of LPC metabolites, as well as to determine which lipid mediators are the most critical effectors of keratinocyte proliferation under conditions of chronic inflammation.

5. Conclusions

In summary, we show that inhibition of cPLA$_2$α with AVX001 inhibits eicosanoid release from primary human PBMC, limits the release of PGE$_2$ in response to inflammatory mediators and EGF, and inhibits the proliferation of keratinocytes by preventing S-phase entry in response to growth factor stimulation. These findings suggest that the therapeutic mode of action of AVX001 in psoriasis could depend both on reducing inflammatory eicosanoid production and on inhibition of the hyperproliferative state of keratinocytes. We additionally propose that AVX001 could improve tolerance to calcipotriol, which could be relevant for developing a combination therapy to treat psoriasis.

Author Contributions: Conceptualization, B.J., A.J.F., and F.J.A.; methodology, A.J.M., F.J.A., N.M., and H.M.F.; software, F.J.A.; validation, F.J.A., N.M., and A.J.F.; formal analysis, F.J.A., and N.M.; investigation, F.J.A., N.M., and H.M.F.; resources, B.J. and A.J.F.; data curation, F.J.A.; writing—original draft preparation, F.J.A.; writing—review and editing, F.J.A., A.J.F., and B.J.; visualization, F.J.A. and A.J.F.; supervision, B.J., A.J.F., and F.J.A.; project administration, B.J. and A.J.F.; funding acquisition, B.J. and A.J.F. All authors have read and agreed to the published version of the manuscript.

Funding: This research was funded by the Research Council of Norway grant no. 269792 and Coegin Pharma.

Acknowledgments: The immunohistochemistry was carried out at the Cellular and Molecular Imaging Core Facility (CMIC), Norwegian University of Science and Technology (NTNU), with the help of Ingunn Nervik. CMIC is funded by the Faculty of Medicine at NTNU and Central Norway Regional Health Authority.

Conflicts of Interest: B.J. is a shareholder of Coegin Pharma. A.J.F. and N.M. are employees of Coegin Pharma. The other authors declare that they have no competing interests. The funders had no role in the design of the study; in the collection, analyses, or interpretation of data; in the writing of the manuscript; or in the decision to publish the results.

Appendix A

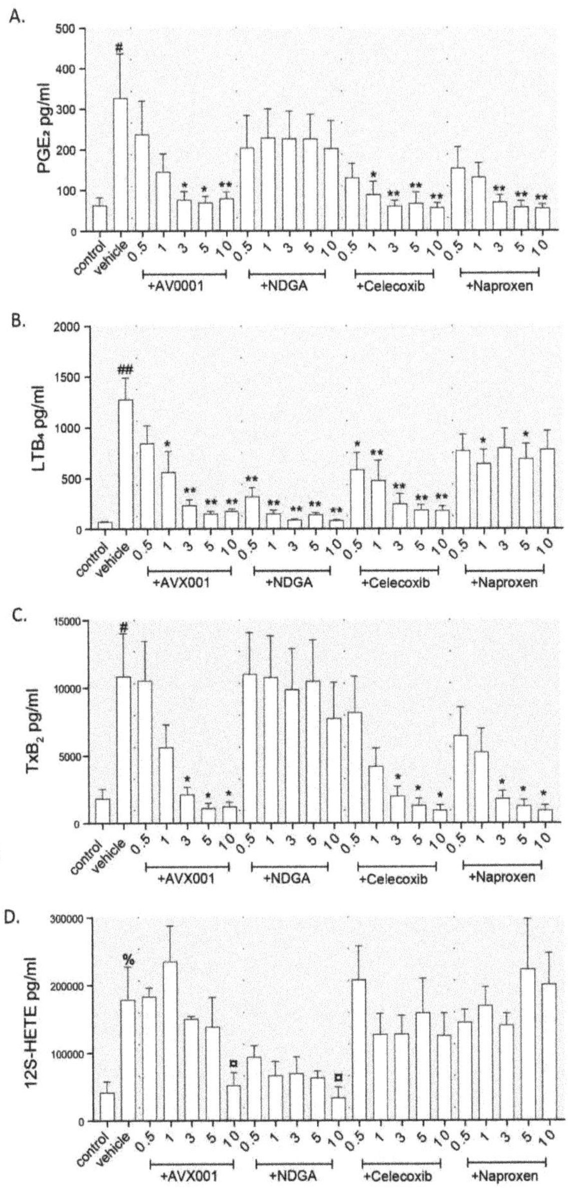

Figure A1. Inhibition of eicosanoid release in A23178-stimulated PBMC. Eicosanoid levels were measured in the supernatants of PBMC pre-incubated with vehicle, AVX001, naproxen, celecoxib, or NDGA in the range of 0.5–10 μM, and stimulated with A23178 (30 μM, 15 min). Data are the mean ± SEM for (**A**) PGE_2, (**B**) LTB_4, (**C**) TxB2, and (**D**) 12S-HETE from four individuals (PGE_2, LTB_4, TxB2) or three individuals (12S-HETE). Statistical significance was calculated using ANOVA with Dunnett's post-analysis, whereby % $p < 0.1$, # $p < 0.05$, ## $p < 0.01$ versus unstimulated control and ¤ $p < 0.1$, * $p < 0.05$, ** $p < 0.01$ versus vehicle-treated control.

References

1. Leslie, C.C. Cytosolic phospholipase A2: Physiological function and role in disease. *J. Lipid Res.* **2015**, *56*, 1386–1402. [CrossRef] [PubMed]
2. Clark, J.D.; Lin, L.-L.; Kriz, R.W.; Ramesha, C.S.; Sultzman, L.A.; Lin, A.Y.; Milona, N.; Knopf, J.L. A novel arachidonic acid-selective cytosolic PLA2 contains a Ca2+-dependent translocation domain with homology to PKC and GAP. *Cell* **1991**, *65*, 1043–1051. [CrossRef]
3. Schalkwijk, C.G.; Spaargaren, M.; Defize, L.H.; Verkleij, A.J.; van den Bosch, H.; Boonstra, J. Epidermal growth factor (EGF) induces serine phosphorylation-dependent activation and calcium-dependent translocation of the cytosolic phospholipase A2. *Eur. J. Biochem.* **1995**, *231*, 593–601. [CrossRef] [PubMed]
4. Schievella, A.R.; Regier, M.K.; Smith, W.L.; Lin, L.L. Calcium-mediated translocation of cytosolic phospholipase A2 to the nuclear envelope and endoplasmic reticulum. *J. Biol. Chem.* **1995**, *270*, 30749–30754. [CrossRef]
5. Roshak, A.; Sathe, G.A. Marshall, L. Suppression of monocyte 85-kDa phospholipase A2 by antisense and effects on endotoxin-induced prostaglandin biosynthesis. *J. Biol. Chem.* **1994**, *269*, 25999–26005.
6. Diaz, B.L.; Fujishima, H.; Mejia, R.S.; Sapirstein, A.; Bonventre, J.V.; Austen, K.F.; Arm, J.P. Cytosolic phospholipase A2 is essential for both the immediate and the delayed phases of eicosanoid generation in mouse bone marrow-derived mast cells. *Prostaglandins Other Lipid Mediat.* **1999**, *59*, 39. [CrossRef]
7. Raichel, L.; Berger, S.; Hadad, N.; Kachko, L.; Karter, M.; Szaingurten-Solodkin, I.; Williams, R.O.; Feldmann, M.; Levy, R. Reduction of cPLA2alpha overexpression: An efficient anti-inflammatory therapy for collagen-induced arthritis. *Eur. J. Immunol.* **2008**, *38*, 2905–2915. [CrossRef]
8. Rosengarten, M.; Hadad, N.; Solomonov, Y.; Lamprecht, S.; Levy, R. Cytosolic phospholipase A2α has a crucial role in the pathogenesis of DSS-induced colitis in mice. *Eur. J. Immunol.* **2015**, *46*, 400–408. [CrossRef]
9. Uozumi, N.; Kume, K.; Nagase, T.; Nakatani, N.; Ishii, S.; Tashiro, F.; Komagata, Y.; Maki, K.; Ikuta, K.; Ouchi, Y.; et al. Role of cytosolic phospholipase A2 in allergic response and parturition. *Nature* **1997**, *390*, 618–622. [CrossRef]
10. Malaviya, R.; Ansell, J.; Hall, L.; Fahmy, M.; Argentieri, R.L.; Olini, G.C.; Pereira, D.W.; Sur, R.; Cavender, D. Targeting cytosolic phospholipase A2 by arachidonyl trifluoromethyl ketone prevents chronic inflammation in mice. *Eur. J. Pharmacol.* **2006**, *539*, 195–204. [CrossRef]
11. Bonventre, J.V.; Huang, Z.; Taheri, M.R.; O'Leary, E.; Li, E.; Moskowitz, M.A.; Sapirstein, A. Reduced fertility and postischaemic brain injury in mice deficient in cytosolic phospholipase A2. *Nat.* **1997**, *390*, 622–625. [CrossRef] [PubMed]
12. Nikolaou, A.; Kokotou, M.G.; Vasilakaki, S.; Kokotos, G. Small-molecule inhibitors as potential therapeutics and as tools to understand the role of phospholipases A2. *Biochim. Biophys. Acta Mol. Cell. Biol. Lipids* **2019**, *1864*, 941–956. [CrossRef] [PubMed]
13. Hewson, C.A.; Patel, S.; Calzetta, L.; Campwala, H.; Havard, S.; Luscombe, E.; Clarke, P.A.; Peachell, P.T.; Matera, M.G.; Cazzola, M.; et al. Preclinical Evaluation of an Inhibitor of Cytosolic Phospholipase A2α for the Treatment of Asthma. *J. Pharmacol. Exp. Ther.* **2011**, *340*, 656–665. [CrossRef]
14. Bhowmick, R.; Clark, S.; Bonventre, J.V.; Leong, J.M.; McCormick, B.A. Cytosolic Phospholipase A2α Promotes Pulmonary Inflammation and Systemic Disease during Streptococcus pneumoniae Infection. *Infect. Immun.* **2017**, *85*. [CrossRef] [PubMed]
15. Feuerherm, A.J.; Dennis, E.A.; Johansen, B. Cytosolic group IVA phospholipase A2 inhibitors, AVX001 and AVX002, ameliorate collagen-induced arthritis. *Arthritis Res. Ther.* **2019**, *21*, 29. [CrossRef]
16. Albanesi, C.; Madonna, S.; Gisondi, P.; Girolomoni, G. The Interplay Between Keratinocytes and Immune Cells in the Pathogenesis of Psoriasis. *Front. Immunol.* **2018**, *9*, 1549. [CrossRef]
17. Chiba, H.; Michibata, H.; Wakimoto, K.; Seishima, M.; Kawasaki, S.; Okubo, K.; Mitsui, H.; Torii, H.; Imai, Y. Cloning of a gene for a novel epithelium-specific cytosolic phospholipase A2, cPLA2delta, induced in psoriatic skin. *J. Biol. Chem.* **2004**, *279*, 12890–12897. [CrossRef]
18. Andersen, S.; Sjursen, W.; Volden, G.; Johansen, B. Elevated expression of human nonpancreatic phospholipase A2 in psoriatic tissue. *Inflammation* **1994**, *18*, 1–12. [CrossRef]
19. Quaranta, M.; Knapp, B.; Garzorz, N.; Mattii, M.; Pullabhatla, V.; Pennino, D.; Andres, C.; Traidl-Hoffmann, P.-D.D.C.; Cavani, A.; Theis, F.J.; et al. Intraindividual genome expression analysis reveals a specific molecular signature of psoriasis and eczema. *Sci. Transl. Med.* **2014**, *6*, 244ra90. [CrossRef]

20. Hammarstrom, S.; Hamberg, M.; Samuelsson, B.; Duell, E.A.; Stawiski, M.; Voorhees, J.J. Increased concentrations of nonesterified arachidonic acid, 12L-hydroxy-5,8,10,14-eicosatetraenoic acid, prostaglandin E2, and prostaglandin F2alpha in epidermis of psoriasis. *Proc. Natl. Acad. Sci. USA* **1975**, *72*, 5130–5134. [CrossRef]
21. Ryborg, A.; Grøn, B.; Kragballe, K. Increased lysophosphatidylcholine content in lesional psoriatic skin. *Br. J. Dermatol.* **1995**, *133*, 398–402. [CrossRef] [PubMed]
22. Barr, R.; Wong, E.; Mallet, A.; Olins, L.; Greaves, M. The analysis of arachidonic acid metabolites in normal, uninvolved and lesional psoriatic skin. *Prostaglandins* **1984**, *28*, 57–65. [CrossRef]
23. Ruzicka, T.; Simmet, T.; Peskar, B.A.; Ring, J. Skin levels of arachidonic acid-derived inflammatory mediators and histamine in atopic dermatitis and psoriasis. *J. Investig. Dermatol.* **1986**, *86*, 105–108. [CrossRef] [PubMed]
24. Sumida, H.; Yanagida, K.; Kita, Y.; Abe, J.; Matsushima, K.; Nakamura, M.; Ishii, S.; Sato, S.; Shimizu, T. Interplay between CXCR2 and BLT1 Facilitates Neutrophil Infiltration and Resultant Keratinocyte Activation in a Murine Model of Imiquimod-Induced Psoriasis. *J. Immunol.* **2014**, *192*, 4361–4369. [CrossRef]
25. Ueharaguchi, Y.; Honda, T.; Kusuba, N.; Hanakawa, S.; Adachi, A.; Sawada, Y.; Otsuka, A.; Kitoh, A.; Dainichi, T.; Egawa, G. Thromboxane A2 facilitates IL-17A production from Vgamma4(+) gammadelta T cells and promotes psoriatic dermatitis in mice. *J. Allergy Clin. Immunol.* **2018**, *142*, 680–683. [CrossRef]
26. Sawada, Y.; Honda, T.; Nakamizo, S.; Otsuka, A.; Ogawa, N.; Kobayashi, Y.; Nakamura, M.; Kabashima, K. Resolvin E1 attenuates murine psoriatic dermatitis. *Sci. Rep.* **2018**, *8*, 11873. [CrossRef]
27. Lee, J.; Aoki, T.; Thumkeo, D.; Siriwach, R.; Yao, C.; Narumiya, S. T cell-intrinsic prostaglandin E2-EP2/EP4 signaling is critical in pathogenic TH17 cell-driven inflammation. *J. Allergy Clin. Immunol.* **2018**, *43*, 631–643. [CrossRef]
28. Conaghan, P. A turbulent decade for NSAIDs: Update on current concepts of classification, epidemiology, comparative efficacy, and toxicity. *Rheumatol. Int.* **2011**, *32*, 1491–1502. [CrossRef]
29. Fry, L.; Baker, B.S. Triggering psoriasis: The role of infections and medications. *Clin. Dermatol.* **2007**, *25*, 606–615. [CrossRef]
30. Kim, G.K.; Del Rosso, J.Q. Drug-Provoked Psoriasis: Is It Drug Induced or Drug Aggravated? *J. Clin. Aesthetic Dermatol.* **2010**, *3*, 32–38.
31. Liaras, K.; Fesatidou, M.; Geronikaki, A. Thiazoles and Thiazolidinones as COX/LOX Inhibitors. *Molecules* **2018**, *23*, 685. [CrossRef]
32. Huwiler, A.; Feuerherm, A.J.; Sakem, B.; Pastukhov, O.; Filipenko, I.; Nguyen, T.; Johansen, B. The omega3-polyunsaturated fatty acid derivatives AVX001 and AVX002 directly inhibit cytosolic phospholipase A(2) and suppress PGE(2) formation in mesangial cells. *Br. J. Pharmacol.* **2012**, *167*, 1691–1701. [CrossRef]
33. Omland, S.; Habicht, A.; Damsbo, P.; Wilms, J.; Johansen, B.; Gniadecki, R. A randomized, double-blind, placebo-controlled, dose-escalation first-in-man study (phase 0) to assess the safety and efficacy of topical cytosolic phospholipase A2 inhibitor, AVX001, in patients with mild to moderate plaque psoriasis. *J. Eur. Acad. Dermatol. Venereol.* **2017**, *31*, 1161–1167. [CrossRef] [PubMed]
34. Anthonsen, M.W.; Solhaug, A.; Johansen, B. Functional Coupling between Secretory and Cytosolic Phospholipase A2Modulates Tumor Necrosis Factor-α- and Interleukin-1β-induced NF-κB Activation. *J. Biol. Chem.* **2001**, *276*, 30527–30536. [CrossRef]
35. Oestvang, J.; Anthonsen, M.W.; Johansen, B. Role of secretory and cytosolic phospholipase A(2) enzymes in lysophosphatidylcholine-stimulated monocyte arachidonic acid release. *FEBS Lett.* **2003**, *555*, 257–262. [CrossRef]
36. Sjursen, W.; Brekke, O.-L.; Johansen, B. Secretory and cytosolic phospholipase A2regulate the long-term cytokine-induced eicosanoid production in human keratinocytes. *Cytokine* **2000**, *12*, 1189–1194. [CrossRef] [PubMed]
37. Thommesen, L.; Sjursen, W.; Gåsvik, K.; Hanssen, W.; Brekke, O.L.; Skattebøl, L.; Holmeide, A.K.; Espevik, T.; Johansen, B.; Laegreid, A. Selective inhibitors of cytosolic or secretory phospholipase A2 block TNF-induced activation of transcription factor nuclear factor-kappa B and expression of ICAM-1. *J. Immunol.* **1998**, *161*, 3420–3430.
38. Naini, S.M.; Choukroun, G.J.; Ryan, J.R.; Hentschel, D.M.; Shah, J.V.; Bonventre, J.V. Cytosolic phospholipase A 2 α regulates G 1 progression through modulating FOXO1 activity. *FASEB J.* **2015**, *30*, 1155–1170. [CrossRef]

39. Boukamp, P.; Petrussevska, R.T.; Breitkreutz, D.; Hornung, J.; Markham, A.E.; Fusenig, N. Normal keratinization in a spontaneously immortalized aneuploid human keratinocyte cell line. *J. Cell Biol.* **1988**, *106*, 761–771. [CrossRef]
40. Farkas, A.; Kemeny, L.; Szell, M.; Dobozy, A.; Bata-Csorgo, Z. Ethanol and acetone stimulate the proliferation of HaCaT keratinocytes: The possible role of alcohol in exacerbating psoriasis. *Arch. Dermatol. Res.* **2003**, *295*, 56–62. [CrossRef]
41. Lehmann, B. HaCaT Cell Line as a Model System for Vitamin D3 Metabolism in Human Skin. *J. Investig. Dermatol.* **1997**, *108*, 78–82. [CrossRef] [PubMed]
42. Ziv, E.; Rotem, C.; Miodovnik, M.; Ravid, A.; Koren, R. Two modes of ERK activation by TNF in keratinocytes: Different cellular outcomes and bi-directional modulation by vitamin D. *J. Cell. Biochem.* **2008**, *104*, 606–619. [CrossRef] [PubMed]
43. George, S.E.; Anderson, R.J.; Cunningham, A.C.; Donaldson, M.; Groundwater, P.W. Evaluation of a Range of Anti-Proliferative Assays for the Preclinical Screening of Anti-Psoriatic Drugs: A Comparison of Colorimetric and Fluorimetric Assays with the Thymidine Incorporation Assay. *ASSAY Drug Dev. Technol.* **2010**, *8*, 384–395. [CrossRef] [PubMed]
44. Yang, X.; Yan, H.; Zhai, Z.; Hao, F.; Ye, Q.; Zhong, B. Pharmacology and therapeutics: Neutrophil elastase promotes proliferation of HaCaT cell line and transwell psoriasis organ culture model. *Int. J. Dermatol.* **2010**, *49*, 1068–1074. [CrossRef]
45. Feuerherm, A.J.; Jørgensen, K.M.; Sommerfelt, R.M.; Eidem, L.E.; Lægreid, A.; Johansen, B. Platelet-activating factor induces proliferation in differentiated keratinocytes. *Mol. Cell. Biochem.* **2013**, *384*, 83–94. [CrossRef]
46. Pozzi, G.; Guidi, M.; Laudicina, F.; Marazzi, M.; Falcone, L.; Betti, R.; Crosti, C.; Muller, E.E.; DiMattia, G.E.; Locatelli, V.; et al. IGF-I stimulates proliferation of spontaneously immortalized human keratinocytes (HACAT) by autocrine/paracrine mechanisms. *J. Endocrinol. Investig.* **2004**, *27*, 142–149. [CrossRef]
47. Carpenter, A.E.; Jones, T.R.; Lamprecht, M.R.; Clarke, C.; Kang, I.H.; Friman, O.; A. Guertin, D.; Chang, J.H.; Lindquist, R.A.; Moffat, J.; et al. CellProfiler: Image analysis software for identifying and quantifying cell phenotypes. *Genome Biol.* **2006**, *7*, R100. [CrossRef]
48. Sorokin, A.V.; Domenichiello, A.F.; Dey, A.K.; Yuan, Z.-X.; Goyal, A.; Rose, S.M.; Playford, M.P.; Ramsden, C.E.; Mehta, N.N. Bioactive Lipid Mediator Profiles in Human Psoriasis Skin and Blood. *J. Investig. Dermatol.* **2018**, *138*, 1518–1528. [CrossRef]
49. Kawahara, K.; Hohjoh, H.; Inazumi, T.; Tsuchiya, S.; Sugimoto, Y. Prostaglandin E2-induced inflammation: Relevance of prostaglandin E receptors. *Biochim. Biophys. Acta (BBA) Mol. Cell Biol. Lipids* **2015**, *1851*, 414–421. [CrossRef]
50. Rundhaug, J.E.; Simper, M.S.; Surh, I.; Fischer, S.M. The role of the EP receptors for prostaglandin E2 in skin and skin cancer. *Cancer Metastasis Rev.* **2011**, *30*, 465–480. [CrossRef]
51. Wu, S.; Han, J.; Qureshi, A. Use of aspirin, non-steroidal anti-inflammatory drugs, and acetaminophen (paracetamol), and risk of psoriasis and psoriatic arthritis: A cohort study. *Acta Derm. Venereol.* **2015**, *95*, 217–223. [CrossRef] [PubMed]
52. Maier, T.J.; Tausch, L.; Hoernig, M.; Coste, O.; Schmidt, R.; Angioni, C.; Metzner, J.; Groesch, S.; Pergola, C.; Steinhilber, D. Celecoxib inhibits 5-lipoxygenase. *Biochem. Pharmacol.* **2008**, *76*, 862–872. [CrossRef] [PubMed]
53. Nicolaou, A. Eicosanoids in skin inflammation. *Prostaglandins Leukot. Essent. Fat. Acids* **2013**, *88*, 131–138. [CrossRef] [PubMed]
54. Bruner, C.R.; Feldman, S.; Ventrapragada, M.; Fleischer, A.B. A systematic review of adverse effects associated with topical treatments for psoriasis. *Dermatol. Online J.* **2003**, *9*, 2.
55. Ravid, A.; Shenker, O.; Buchner-Maman, E.; Rotem, C.; Koren, R. Vitamin D Induces Cyclooxygenase 2 Dependent Prostaglandin E2Synthesis in HaCaT Keratinocytes. *J. Cell. Physiol.* **2015**, *231*, 837–843. [CrossRef]
56. Shah, F.; Stepan, A.F.; O'Mahony, A.; Velichko, S.; Folias, A.E.; Houle, C.; Shaffer, C.L.; Marcek, J.; Whritenour, J.; Stanton, R.; et al. Mechanisms of Skin Toxicity Associated with Metabotropic Glutamate Receptor 5 Negative Allosteric Modulators. *Cell Chem. Biol.* **2017**, *24*, 858–869. [CrossRef]
57. Schneider, M.R.; Werner, S.; Paus, R.; Wolf, E. Beyond wavy hairs: The epidermal growth factor receptor and its ligands in skin biology and pathology. *Am. J. Pathol.* **2008**, *173*, 14–24. [CrossRef]
58. Lin, L.L.; Wartmann, M.; Lin, A.Y.; Knopf, J.L.; Seth, A.; Davis, R.J. cPLA2 is phosphorylated and activated by MAP kinase. *Cell* **1993**, *72*, 269–278. [CrossRef]

59. Bowen, A.R.; Hanks, A.N.; Allen, S.M.; Alexander, A.; Diedrich, M.J.; Grossman, D. Apoptosis Regulators and Responses in Human Melanocytic and Keratinocytic Cells. *J. Investig. Dermatol.* **2003**, *120*, 48–55. [CrossRef]
60. Calay, D.; Vind-Kezunovic, D.; Frankart, A.; Lambert, S.; Poumay, Y.; Gniadecki, R. Inhibition of Akt Signaling by Exclusion from Lipid Rafts in Normal and Transformed Epidermal Keratinocytes. *J. Investig. Dermatol.* **2010**, *130*, 1136–1145. [CrossRef]
61. Lee, E.-R.; Kang, Y.-J.; Kim, J.-H.; Lee, H.T.; Cho, S.-G. Modulation of Apoptosis in HaCaT Keratinocytes via Differential Regulation of ERK Signaling Pathway by Flavonoids. *J. Biol. Chem.* **2005**, *280*, 31498–31507. [CrossRef] [PubMed]
62. Nam, C.; Doi, K. Etoposide induces G2/M arrest and apoptosis in neural progenitor cells via DNA damage and an ATM/p53-related pathway. *Histol. Histopathol.* **2010**, *25*, 485–493. [CrossRef] [PubMed]
63. Kim, K.Y.; Cho, Y.J.; Jeon, G.A.; Ryu, P.D.; Myeong, J.N. Membrane-bound alkaline phosphatase gene induces antitumor effect by G2/M arrest in etoposide phosphate-treated cancer cells. *Mol. Cell. Biochem.* **2003**, *252*, 213–221. [CrossRef]
64. Duronio, R.J.; Xiong, Y. Signaling pathways that control cell proliferation. *Cold Spring Harb. Perspect. Biol.* **2013**, *5*, a008904. [CrossRef] [PubMed]
65. Hinitt, C.; Benn, T.; Threadgold, S.; Wood, J.; Williams, C.; Hague, A. BAG-1L promotes keratinocyte differentiation in organotypic culture models and changes in relative BAG-1 isoform abundance may lead to defective stratification. *Exp. Cell Res.* **2011**, *317*, 2159–2170. [CrossRef]
66. Schoop, V.M.; Fusenig, N.E.; Mirancea, N. Epidermal Organization and Differentiation of HaCaT Keratinocytes in Organotypic Coculture with Human Dermal Fibroblasts. *J. Investig. Dermatol.* **1999**, *112*, 343–353. [CrossRef]
67. Maas-Szabowski, N.; Stärker, A.; Fusenig, N.E. Epidermal tissue regeneration and stromal interaction in HaCaT cells is initiated by TGF-alpha. *J. Cell Sci.* **2003**, *116*, 116. [CrossRef]
68. Flamand, N.; Picard, S.; Lemieux, L.; Pouliot, M.; Bourgoin, S.G.; Borgeat, P. Effects of pyrrophenone, an inhibitor of group IVA phospholipase A2, on eicosanoid and PAF biosynthesis in human neutrophils. *Br. J. Pharmacol.* **2006**, *149*, 385–392. [CrossRef]
69. Whalen, K.A.; Legault, H.; Hang, C.; Hill, A.; Kasaian, M.; Donaldson, D.; Bensch, G.W.; Baker, J.; Reddy, P.S.; Wood, N.; et al. In vitro allergen challenge of peripheral blood induces differential gene expression in mononuclear cells of asthmatic patients: Inhibition of cytosolic phospholipase A2α overcomes the asthma-associated response. *Clin. Exp. Allergy* **2008**, *38*, 1590–1605. [CrossRef]
70. de Carvalho, M.G.; McCormack, A.L.; Olson, E.; Ghomashchi, F.; Gelb, M.H.; Yates, J.R., 3rd; Leslie, C.C. Identification of phosphorylation sites of human 85-kDa cytosolic phospholipase A2 expressed in insect cells and present in human monocytes. *J. Biol. Chem.* **1996**, *271*, 6987–6997. [CrossRef]
71. Conrad, C.; Gilliet, M. Psoriasis: From Pathogenesis to Targeted Therapies. *Clin. Rev. Allergy Immunol.* **2018**, *54*, 102–113. [CrossRef] [PubMed]
72. Seo, S.-H.; Jeong, G.-S. Fisetin inhibits TNF-α-induced inflammatory action and hydrogen peroxide-induced oxidative damage in human keratinocyte HaCaT cells through PI3K/AKT/Nrf-2-mediated heme oxygenase-1 expression. *Int. Immunopharmacol.* **2015**, *29*, 246–253. [CrossRef] [PubMed]
73. Zarrabeitia, M.T.; Riancho, J.A.; Amado, J.A.; Olmos, J.M.; Gonzalez-Macias, J. Effect of calcitriol on the secretion of prostaglandin E2, interleukin 1, and tumor necrosis factor alpha by human monocytes. *Bone* **1992**, *13*, 185–189. [CrossRef]
74. Doroudi, M.; Plaisance, M.C.; Boyan, B.D.; Schwartz, Z. Membrane actions of 1alpha,25(OH)2D3 are mediated by Ca(2+)/calmodulin-dependent protein kinase II in bone and cartilage cells. *J. Steroid Biochem. Mol. Biol.* **2015**, *145*, 65–74. [CrossRef] [PubMed]
75. Papp, K.A.; Guenther, L.; Boyden, B.; Larsen, F.G.; Harvima, R.J.; Guilhou, J.J.; Kaufmann, R.; Rogers, S.; Van De Kerkhof, P.; Hanssen, L.I.; et al. Early onset of action and efficacy of a combination of calcipotriene and betamethasone dipropionate in the treatment of psoriasis. *J. Am. Acad. Dermatol.* **2003**, *48*, 48–54. [CrossRef] [PubMed]
76. Goldman, R.; Moshonov, S.; Chen, X.; Berchansky, A.; Furstenberger, G.; Zor, U. Crosstalk between elevation of [Ca2+]i, reactive oxygen species generation and phospholipase A2 stimulation in a human keratinocyte cell line. *Adv. Exp. Med. Biol.* **1997**, *433*, 41–45. [CrossRef]

77. Goldman, R.; Zor, U.; Meller, R.; Moshonov, S.; Fürstenberger, G.; Seger, R. Activation of Map Kinases, cPLA2 and Reactive Oxygen Species Formation by EGF and Calcium Mobilizing Agonists in a Human Keratinocyte Cell Line. *Adv. Exp. Med. Biol.* **1997**, *407*, 289–293. [CrossRef]
78. Sakaguchi, M.; Huh, N.-H. S100A11, a dual growth regulator of epidermal keratinocytes. *Amino Acids* **2010**, *41*, 797–807. [CrossRef]
79. Sakaguchi, M.; Murata, H.; Sonegawa, H.; Sakaguchi, Y.; Futami, J.-I.; Kitazoe, M.; Yamada, H.; Huh, N.-H. Truncation of Annexin A1 Is a Regulatory Lever for Linking Epidermal Growth Factor Signaling with Cytosolic Phospholipase A2 in Normal and Malignant Squamous Epithelial Cells. *J. Biol. Chem.* **2007**, *282*, 35679–35686. [CrossRef]
80. Nakanishi, M.; Rosenberg, D.W. Roles of cPLA2α and arachidonic acid in cancer. *Biochim. Biophys. Acta (BBA) Mol. Cell Biol. Lipids* **2006**, *1761*, 1335–1343. [CrossRef]
81. Kim, W.; Son, B.; Lee, S.; Do, H.; Youn, B. Targeting the enzymes involved in arachidonic acid metabolism to improve radiotherapy. *Cancer Metastasis Rev.* **2018**, *37*, 213–225. [CrossRef]
82. Pentland, A.P.; Needleman, P. Modulation of keratinocyte proliferation in vitro by endogenous prostaglandin synthesis. *J. Clin. Investig.* **1986**, *77*, 246–251. [CrossRef] [PubMed]
83. Conconi, M.; Bruno, P.; Bonali, A.; De Angeli, S.; Parnigotto, P. Relationship between the proliferation of keratinocytes cultured in vitro and prostaglandin E2. *Ann. Anat. Anat. Anz.* **1996**, *178*, 229–236. [CrossRef]
84. Loftin, C.D.; Eling, T.E. Prostaglandin Synthase 2 Expression in Epidermal Growth Factor-Dependent Proliferation of Mouse Keratinocytes. *Arch. Biochem. Biophys.* **1996**, *330*, 419–429. [CrossRef] [PubMed]
85. Ansari, K.M.; Sung, Y.M.; He, G.; Fischer, S.M. Prostaglandin receptor EP2 is responsible for cyclooxygenase-2 induction by prostaglandin E2 in mouse skin. *Carcinogenesis* **2007**, *28*, 2063–2068. [CrossRef] [PubMed]
86. Sung, Y.M.; He, G.; Fischer, S.M. Lack of Expression of the EP2 but not EP3 Receptor for Prostaglandin E2 Results in Suppression of Skin Tumor Development. *Cancer Res.* **2005**, *65*, 9304–9311. [CrossRef]
87. Marques, S.A.; Dy, L.C.; Southall, M.D.; Yi, Q.; Smietana, E.; Kapur, R.; Marques, M.; Travers, J.B.; Spandau, D.F. The platelet-activating factor receptor activates the extracellular signal-regulated kinase mitogen-activated protein kinase and induces proliferation of epidermal cells through an epidermal growth factor-receptor-dependent pathway. *J. Pharmacol. Exp. Ther.* **2002**, *300*, 1026–1035. [CrossRef]
88. Sato, S.; Kume, K.; Ito, C.; Ishii, S.; Shimizu, T. Accelerated proliferation of epidermal keratinocytes by the transgenic expression of the platelet-activating factor receptor. *Arch. Dermatol. Res.* **1999**, *291*, 614–621. [CrossRef]
89. Singh, T.P.; Huettner, B.; Koefeler, H.; Mayer, G.; Bambach, I.; Wallbrecht, K.; Schon, M.P.; Wolf, P. Platelet-activating factor blockade inhibits the T-helper type 17 cell pathway and suppresses psoriasis-like skin disease in K5.hTGF-beta1 transgenic mice. *Am. J. Pathol.* **2011**, *178*, 699–708. [CrossRef]

© 2020 by the authors. Licensee MDPI, Basel, Switzerland. This article is an open access article distributed under the terms and conditions of the Creative Commons Attribution (CC BY) license (http://creativecommons.org/licenses/by/4.0/).

MDPI

Article

Lipid Profile of Activated Macrophages and Contribution of Group V Phospholipase A_2

Masaya Koganesawa, Munehiro Yamaguchi, Sachin K. Samuchiwal and Barbara Balestrieri *

Department of Medicine, Division of Allergy and Clinical Immunology, Brigham and Women's Hospital, Harvard Medical School, Boston, MA 02115, USA; mkoganesawa@bwh.harvard.edu (M.K.); munehiro1006@yahoo.com (M.Y.); ssamuchiwal@bwh.harvard.edu (S.K.S.)

* Correspondence: bbalestrieri@bwh.harvard.edu

Abstract: Macrophages activated by Interleukin (IL)-4 (M2) or LPS+ Interferon (IFN)γ (M1) perform specific functions respectively in type 2 inflammation and killing of pathogens. Group V phospholipase A_2 (Pla2g5) is required for the development and functions of IL-4-activated macrophages and phagocytosis of pathogens. Pla2g5-generated bioactive lipids, including lysophospholipids (LysoPLs), fatty acids (FAs), and eicosanoids, have a role in many diseases. However, little is known about their production by differentially activated macrophages. We performed an unbiased mass-spectrometry analysis of phospholipids (PLs), LysoPLs, FAs, and eicosanoids produced by Wild Type (WT) and *Pla2g5*-null IL-4-activated bone marrow-derived macrophages (IL-4)BM-Macs (M2) and (LPS+IFNγ)BM-Macs (M1). Phosphatidylcholine (PC) was preferentially metabolized in (LPS+IFNγ)BM-Macs and Phosphatidylethanolamine (PE) in (IL-4)BM-Macs, with Pla2g5 contributing mostly to metabolization of selected PE molecules. While Pla2g5 produced palmitic acid (PA) in (LPS+IFNγ)BM-Macs, the absence of Pla2g5 increased myristic acid (MA) in (IL-4)BM-Macs. Among eicosanoids, Prostaglandin E_2 (PGE_2) and prostaglandin D_2 (PGD_2) were significantly reduced in (IL-4)BM-Macs and (LPS+IFNγ)BM-Macs lacking Pla2g5. Instead, the IL-4-induced increase in 20-carboxy arachidonic acid (20CooH AA) was dependent on Pla2g5, as was the production of 12-hydroxy-heptadecatrienoic acid (12-HHTrE) in (LPS+IFNγ)BM-Macs. Thus, Pla2g5 contributes to PE metabolization, PGE_2 and PGD_2 production independently of the type of activation, while in (IL-4)BM-Macs, Pla2g5 regulates selective lipid pathways and likely novel functions.

Keywords: Group V phospholipase A_2; macrophages; lipids

itation: Koganesawa, M.; amaguchi, M.; Samuchiwal, S.K.; alestrieri, B. Lipid Profile of ctivated Macrophages and ontribution of Group V hospholipase A_2. *Biomolecules* **2021**, *1*, 25. https://doi.org/biom 010025

eceived: 30 November 2020
ccepted: 23 December 2020
ublished: 29 December 2020

ublisher's Note: MDPI stays neu- al with regard to jurisdictional claims published maps and institutional filiations.

opyright: © 2020 by the authors. Li- ensee MDPI, Basel, Switzerland. This rticle is an open access article distributed nder the terms and conditions of the reative Commons Attribution (CC BY) cense (https://creativecommons.org/ censes/by/4.0/).

1. Introduction

Macrophages are heterogeneous cells that contribute to the pathogenesis of infectious disease, type 2 immune responses, and metabolic disorders [1–3]. Macrophages exposed to pathogen-associated molecular patterns (PAMPS), endogenously formed danger-associated molecular patterns (DAMPs), and to the microenvironmental milieu rearrange their repertoire of cytokines and chemokines and contribute to the immune responses [4,5]. In vitro, the immune responses of macrophages are simplified by LPS+Interferon (IFN)γ (M1) and Interleukin (IL)-4 (M2) polarization [6,7]. Although it is unlikely that macrophages in vivo exist in mutually exclusive phenotypes [2,4], the polarization paradigm has helped to dissect critical immune features of macrophages. M2 macrophages develop during type 2 inflammation, prompted by Type 2 cytokines including IL-4 and IL-13, and they are characterized by the production of CCL22, CCL17, Arginase-1, and Transglutaminase 2 (TGM2) [7–9]. M1 macrophages develop during infection and are equipped with an array of cytokines, including IL-12, TNFα, and reactive oxygen species (ROS), to fight pathogens. Metabolically, macrophage polarization is distinguished by activation of aerobic glycolysis in M1 macrophages, particularly when M1 activation is achieved by IFNγ, in addition to LPS [10] and fatty acid oxidation in M2 [11]. Fatty acids (FAs) are involved in the remodeling of membrane phospholipid (mPLs), macrophage development and polarization [12–14], and they are essential for phagocytosis by fully differentiated macrophages [14]. Fu thermore, polyunsaturated fatty acids (PUFAs) are the precursors of eicosanoids, prc or anti-inflammatory lipid mediators, which are generated through three main path ways, cyclooxygenase (COX), lipoxygenase (LOX), and monooxygenases like CYP45 (CYP). Even before the identification of macrophage polarization and their spectrum o phenotypes, stimuli like Granulocyte macrophage colony-stimulating factor (GM-CSF and IL-4 reportedly induced 5-lipoxygenase (5-LOX) activation in macrophages [15,16 the limiting step in the generation of Cysteinyl leukotrienes (CysLTs), major mediator of inflammation [17]. COX-derived Prostaglandin E_2 (PGE_2) and PGD_2 may have prc or anti-inflammatory functions associated with LPS or IL-4 immune responses [18–20 CYP2J2-derived epoxyeicosatrienoic acids (EETs) have opposite roles in human monocyte macrophages depending on the activation state of the macrophages [21], while CYP omega oxidase 20-hydroxy-eicosatetraenoic acid (20-HETE) and its metabolite 20-carboxy arach donic acid (20CooH AA) may have anti-inflammatory properties due to the ability to activate Peroxisome Proliferator-Activated Receptor α (PPARα) and PPARγ [22] while also activating G Protein-Coupled Receptor 75 (GPR75) [23]. However, whether M1 or M2 activated macrophages differentially produce FAs and eicosanoids has not been fully eluci dated, but is potentially critical to understand how lipids produced by macrophages may contribute alone or in combination to the many diseases involving activated macrophages

The release of FAs and eicosanoids requires the activity of at least one of the phos pholipases A_2 (PLA_2s) enzymes. PLA_2s hydrolyze membrane phospholipids to generate FAs, including saturated fatty acids, monounsaturated fatty acids (MUFAs), and polyur saturated fatty acids (PUFAs) and their metabolites eicosanoids, and lysophospholipids (LysoPLs). As PLA_2s have substrate specificities and cell-expression preferences, likely, ac tivated macrophages would preferentially express one or more PLA_2s and, thus, contribute to the generation of selective bioactive lipids. It is worth noting that the constitutive and ubiquitous group IVA cytosolic PLA_2 (c$PLA_2\alpha$ or Pla2g4a), which preferentially releases Arachidonic Acid (AA), is expressed more abundantly in M1 macrophages than in M2 macrophages [24]. Of the secretory PLA_2 family members, group V PLA_2 (Pla2g5) i expressed in macrophages [25] and other cells [26]. Several studies, including ours, have reported the presence of Pla2g5 mRNA and protein in bone marrow-derived macrophages (BM-Macs), peritoneal macrophages, human monocyte-derived macrophages, macrophage cell lines [27]. Pla2g5 can potentiate Pla2g4a activation and AA release in macrophages activated with Toll-Like Receptor (TLR) agonists [28] and is required for phagocytosis and killing of pathogens [27,29]. In vivo, its absence results in increased LPS inflammation and mortality following Candida infection [30,31]. On the other hand, Pla2g5 is induced by IL-4 in mouse and human macrophages [20,32–34] and is expressed together with TGM in macrophages present in human nasal polyps of subjects with chronic rhinosinusitis [20 Furthermore, Pla2g5 contributes to M2 macrophage development and functions [20,33,34 Indeed, in allergic type 2 inflammation, Pla2g5 potentiates inflammation and activation of target cells, including T-cells and innate lymphoid cells type 2 (ILC2s) through the generation of the linoleic acid (LA) and oleic acid (OA) [32,34]. However, the contribution of Pla2g5 to lipid generated during M1 and M2 polarization is still not known, but is potentially critical to account for the functions of M1 and M2 macrophages.

To investigate the lipid generated during macrophage polarization, we activated BM-Macs with IL-4 or LPS+IFNγ and analyzed FAs, PLs, LysoPLs, and eicosanoids by gas chromatography coupled to mass spectrometry (GC-MS) and liquid chromatogra phy coupled to mass spectrometry (LC-MS). We found that macrophages activated by LPS+IFNγ or IL-4 rearrange their membrane phospholipid composition by decreasing phospholipid content, particularly Phosphatidylethanolamines (PE) during IL-4 activa tion and Phosphatidylcholine (PC) during LPS+IFNγ activation. Furthermore, in M2 macrophages, the absence of Pla2g5 increased PE 34:2, PE 36:2, PE 38:4, and PE 38:5, while in M1 macrophages, there was a significant increase in PC 34:1 and PC 36:2. LysoPLs were not significantly modified during macrophage activation in BM-Macs lacking Pla2g5. We also found that Pla2g5 regulated the release of selective saturated fatty acids in activated BM-Macs. Furthermore, PGD_2 and PGE_2 were significantly induced by Pla2g5 in both M1 and M2 activation, while 20CooH AA production was dependent on Pla2g5 only in M2. Therefore, it is likely that the combined action of the lipids produced by M1 and M2 macrophages may contribute to the functions of macrophages in different settings. Identifying potential lipid signatures in M1 and M2 macrophages could uncover new pathways critical for the development, persistence, or reduction of inflammation. Some of these functions may rely on the presence or absence of Pla2g5 in macrophages.

2. Materials and Methods

2.1. Macrophage Cultures

BM-Macs were generated as previously described [34]. Briefly, BM cells were collected from femurs and tibiae of mice of Wild Type (WT) and *Pla2g5*-null mice [35]. The disaggregated cells were counted and suspended in complete medium (Dulbecco's Modified Eagle Medium (DMEM) F12, 5% Fetal Bovine Serum (FBS), 100 U mL^{-1} penicillin, 100 mg mL^{-1} streptomycin, 0.1 mM nonessential amino acids, 2 mM L-glutamine and 0.05 mM 2-mercaptoethanol (2-ME)) at a concentration of 4.0×10^6 cells per mL in 10 mL in 100 mm Petri dish. WT and *Pla2g5*-null BM-macrophages were cultured for 7 days in 50 ng mL^{-1} murine recombinant-Macrophage Stimulating Factor (rM-CSF) (PeproTech). Macrophages were activated with IL-4 or LPS+IFNγ (PeproTech) for 24 h, as previously described [32]. Supernatants and adherent cells were collected [20,34], frozen, and shipped for analysis by mass spectrometry.

Human monocyte-derived macrophages were generated as previously described [20]. Briefly, leukocyte-enriched buffy coat from healthy donors was overlaid on Ficoll-Paque Plus (GE Healthcare, Buckinghamshire, UK) and centrifuged at 600× *g* for 20 min. The mononuclear layer at the interface was collected, washed, and counted. Monocytes were isolated by negative selection (Miltenyi Biotec, Auburn, CA, USA) and plated at 2.2×10^5 cells/cm^2 in 100 mm Petri dishes. Monocytes were cultured for 13 d in complete medium (RPMI 1640, 10% FBS, 2 mM L-glutamine, 100 U/mL penicillin, 100 mg/mL streptavidin, 10% nonessential amino acids, 1% 4-(2-hydroxyethyl)-1-piperazineethanesulfonic acid (HEPES), 1% sodium pyruvate, and 50 mM 2-ME supplemented with 50 ng mL^{-1} human rGM-CSF (R&D Systems, Minneapolis, MN, USA) [36]. To knock down PLA2G5, monocytes were cultured for 13 d in rGM-CSF, then they were transfected with human PLA2G5 ON-TARGET Plus SMART Pool siRNA or nontargeting vector ONTARGET Plus Control Pool (1000 nM; GE Dharmacon, Lafayette, CO, USA) as previously described [20]. To activate macrophages, cells were polarized for 24 h in complete medium, supplemented with 40 ng mL^{-1} human IL-4 (R&D Systems). Cell-free supernatants were collected, frozen and shipped for analysis by mass spectrometry.

2.2. Mass Spectrometry of Lipids

Lipid analysis was performed at the University of California San Diego Lipidomics Core [37]. For analysis of eicosanoids samples, the amount of sample used was 200 μL, and 100 μL internal standard was added. Lipids were extracted with Solid Phase Extraction (SPE): Strata-x polymeric reverse phase columns (8B-S100-UBJ; Phenomenex, Torrance, CA, USA). The following was added to each column: 100% MeOH, 100% H_2O, sample, 10% MeOH, and 100% MeOH for elution. Samples were dried with a Speed-Vac (Thermo Scientific) and taken up in 100 mL buffer A (63% H_2O, 37% acetonitrile, 0.02% acetic acid). Five microliters were injected into the ultra-high performance liquid chromatography system. The analysis was performed with a mass spectrometer (6500 Qtrap; Sciex, Framingham, MA, USA) [38]. For analysis of phospholipids, samples were extracted via Bleigh Dyer. Samples were dried down and taken up in 50 μL buffer A (59/40/1 isopropanol (IPA)/hexane (HEX)/H_20 with 10 mM NH_4OAC). Then, 10 μL were injected into UPLC-MS/MS. The first number of the phospholipids designates the total number of carbons and the second number after the colon indicates the total number of double bonds. For example, PC 36:4 indicates a phospholipid molecule with a total of 36 carbons and 4 doub bonds. Lysophosphatidylcholine (LysoPC)-O and PC-O indicate ether-linked fatty acids i the sn-1 position. We measure isobaric species. For example, PC 36:4 can consist of eithe PC (18:2; 18:2) or PC (16:0; 20:4). Similarly, we did not discriminate between 34:0 PC an 36:7 PC-O—both have identical molecular masses and are not separated by LC-MS.

FA analysis was performed according to a previously published method [39]. Briefl the cell pellet was homogenized in 500 mL of PBS/10% methanol. An aliquot of 200 m corresponding to about 0.5×10^6 cells was withdrawn and a cocktail of internal standard consisting of 15 deuterated fatty acids was added. The extraction was initiated wit 500 mL of methanol and 25 mL of 1 N HCl and a bi-phasic solution is formed by additio of 1.5 mL of isooctane. The phases are separated by centrifugation and the isooctan phase containing the FFAs fraction was removed. The extraction is repeated once an the combined extracts were evaporated to dryness. The FFAs were derivatized wit pentafluorobenzyl (PFB) bromide and the resulting fatty acid PFB esters were analyze by gas chromatography/mass spectrometry using a negative chemical ionization mod (Agilent 6890N gas chromatograph equipped with an Agilent 5973 mass selective detecto Agilent, Santa Clara, CA, USA). Standard curves for each of the fatty acids were acquired i parallel using identical conditions. The quantitative assessment of fatty acids in a sampl was achieved by comparison of the mass spectrometric ion signal of the target molecul normalized to the internal standard with the matching standard curve according to th isotope dilution method and by protein content.

2.3. *qPCR*

Total RNA was isolated from lysate with the RNeasy Micro Kit (Qiagen, Louisvill KY, USA), reverse transcribed into cDNA (High-Capacity cDNA Reverse Transcription Ki Thermo ARTICLES science-Applied Biosystems, Foster City, CA, USA) and measured b real-time PCR with the use of SYBR® Green/ROX master mix (SABiosciences, Frederic MD, USA) on an Mx3005P thermal cycler (Stratagene, Santa Clara, CA, USA). The ratio c each mRNA relative to the Glyceraldehyde-3-Phosphate Dehydrogenase (GAPDH) mRN was calculated with the $\Delta\Delta$Ct threshold cycle method. The mouse primers used wer GAPDH F: 5′-TCAACAGCAACTCCCACTCTTCCA-3′; R: 5′-ACCCTGTTGCTGTAGCCG ATTCA-30′ Pla2g5 F: 5′-TGGTTCCTGGCTTGCAGTGTG-3′; R: 5′-TTCGCAGATGACTAG CCATT-3′ [34]. Real-time PCR products were run on a 1.5% agarose gel and visualize using chemiImager 4400 fluorescence system (Alpha Innotech, Missouri, TX, USA).

2.4. *Statistical Analysis*

Comparisons between two groups were made by using an unpaired Student's *t* tes and other comparisons were made with two-way ANOVA with Tukey, Sidak, or Dunnet correction for multiple comparisons. Comparisons were performed with Prism softwar (GraphPad, San Diego, CA, USA). Data are expressed as mean ± standard error of th mean (SEM); significance was set at $p < 0.05$.

3. Results

3.1. *Phospholipid Metabolism in Bone Marrow-Derived Macrophages Activated by IL-4 or LPS+IFNγ*

To investigate the production of bioactive lipids in activated BM-Macs, we first aske whether, during macrophage activation, there were changes in the composition of men brane phospholipids (PLs), substrates of PLA_2. We cultured mouse BM-Macs for 7 day in rM-CSF and 24 h with IL-4 or LPS+IFNγ, hereafter referred to as (IL-4)BM-Macs c (LPS+IFNγ)BM-Macs, respectively, for M2 or M1 (Figure 1a, inset), to distinguish then from other in vitro derived M2 or M1 macrophages [40]. The supernatants were co lected and the phospholipid species analyzed by mass spectrometry. Five PL specie were detected in resting WT BM-Macs. Phosphatidylcholine (PC) was the most abur dant, followed by phosphatidylethanolamines (PE), while phosphatidylinositol (PI), pho phatidylserine (PS), and phosphatidic acid (PA) contributed minimally to membrane composition (Figure 1a, black columns). Compared to WT unstimulated (U)BM-Macs, PE was significantly decreased in WT (IL-4)BM-Macs, while PC was significantly reduced in (LPS+IFNγ)BM-Macs (Figure 1a). PC, PS, PI, and PA did not change following activation. PG species were undetectable. These data suggest that PE could be the prevalent substrate for a PLA_2 in (IL-4)BM-Macs and PC in (LPS+IFNγ)BM-Macs, although other phospholipases could contribute to remodeling of membrane phospholipids.

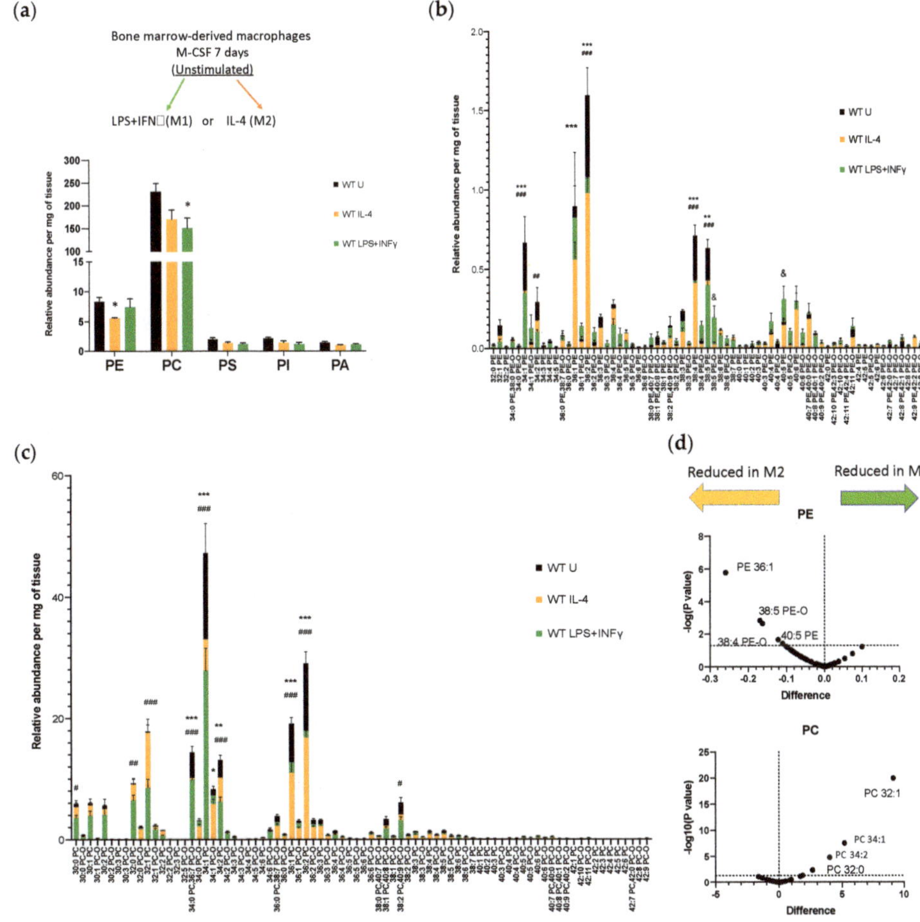

Figure 1. Phospholipid metabolism in bone marrow-derived macrophages (BM-Macs). Wild Type (WT) BM-Macs were Unstimulated (U) or stimulated with Interleukin (IL)-4 or LPS+ Interferon (IFN)γ for 24 h, supernatants were removed and collected for analysis by liquid chromatography coupled to mass spectrometry (LC-MS). (**a**) * $p < 0.05$ of (U) vs. IL-4 or LPS+IFNγ BM-Macs. (**b**,**d**) Phosphatidylethanolamine (PE) molecules or (**c**,**d**) phosphatidylcholine (PC) molecules are identified by the binding of fatty acids which are described by the numbers of carbons and double bonds. PE-O and PC-O indicate ether linked fatty acids in the sn-1 position. The data are shown volcano plot of mean (**d**) or graph (**a**–**c**) of mean and standard error of three independent determinations. *** $p < 0.0001$, ** $p < 0.005$ (IL-4)BM-Macs reduction vs (U)BM-Macs; ### $p < 0.0001$, ## $p < 0.005$, # $p < 0.05$ (LPS+IFNγ)BM-Macs reduction vs. (U)BM-Macs; & $p < 0.05$ increase in (LPS+IFNγ)BM-Macs vs. (U)BM-Macs. *p*-values were obtained by two-way ANOVA with Dunnett's correction for multiple comparisons.

Next, we wanted to ascertain whether, among PE or PC species, there were pre ferred phospholipid molecules decreased during each of the macrophage activation mod els and, therefore, preferred bioactive lipid produced. We analyzed 69 PE molecule In WT (U)BM-Macs, PE 36:2 was the most abundant. Compared to WT (U)BM-Mac PE 34:1, PE 36:2, PE 38:4, and PE 38:5 were significantly decreased in both (IL-4) and (LPS+IFNγ)BM-Macs, and PE 34:2 was significantly reduced in (LPS+IFNγ)BM-Mac while PE 36:1 was significantly reduced in (IL-4)BM-Macs (Figure 1b). These data in dicate that long-chain PUFAs, including AA, are liberated from PE in both IL-4 and LPS+IFNγ activation. Instead, PE-O 38:5 and PE 40:5 were significantly increased in W (LPS+IFNγ)BM-Macs compared to WT (U)BM-Macs, likely indicating re-acylation. Of the 80 PC species analyzed (Figure 1c), PC 34:1 was the most abundant in WT (U)BM-Mac and was reduced in both (IL-4) and (LPS+IFNγ)BM-Macs as were PC 34:2, PC 36:1, PC 36: PC-O 34:0/PC-O 36:7. LPS+IFNγ activation resulted in the selective reduction of PC 30: PC 32:0, PC 32:1, and PC 38:2/PC-O 40:9 compared to (U)BM-Macs. A direct comparison of PE or PC molecules in (IL-4)BM-Macs versus (LPS+IFNγ)BM-Macs confirmed tha PE species were preferentially reduced and likely metabolized in (IL-4)BM-Macs and PC species in (LPS+IFNγ)BM-Macs (Figure 1d). These results suggest that one or more PLA may hydrolyze preferred substrates in IL-4 or LPS+IFNγ activated macrophages.

Since Pla2g5 is induced by IL-4 [32,34] and prefers PE as a substrate at least in human monocyte-macrophages activated by IL-4 [33], we wanted to understand whether the changes in membrane phospholipid were due at least partially to Pla2g5, particularly in IL-4 BM-Macs. We confirmed that Pla2g5 is induced in mouse BM-Macs by IL-4, bu not LPS+IFNγ (Figure 2a) as previously shown [32]. In *Pla2g5*-null BM-Macs, PE wa significantly increased in unstimulated, IL-4 and LPS+IFNγ stimulated BM-Macs compare to equally treated WT BM-Macs (Figure 2b) while PC was similar in both WT and *Pla2g5* null BM-Macs independently of the activation state. Furthermore, compared to equally treated WT BM-Macs, the percentage increase of PE was 71.7 ± 21.1% in (IL-4)BM-Macs 41.1 ± 2.4% in (U)BM-Macs and 59.7 ± 26.1% in (LPS+IFNγ)BM-Macs (data not shown These data suggests that PE is the preferred substrate of Pla2g5 in (IL-4)BM-Macs and tha the low expression level of Pla2g5 in (U)BM-Macs and (LPS+IFNγ)BM-Macs is sufficient to mobilize PE.

To understand whether Pla2g5 was active on selective PE or PC molecules, we an alyzed PE and PC species in *Pla2g5*-null BM-Macs compared to WT BM-Macs. PE 36: (18:0, 18:1), PE 36:2 (18:0, 18:2), and PE 38:4 (18:0, 20:4) were increased in BM-Macs lacking Pla2g5 independently of the activation state. PE 34:1(16:0; 18:1), PE 34:2 (16:0,18:2), and PE 38:5 (16:0, 22:5; 18:1, and 20:4) were significantly increased in *Pla2g5*-null (IL-4)BM-Mac while PE 34:1 (16:0; 18:1) was increased in *Pla2g5*-null (LPS+IFNγ)BM-Macs, compared to their respective controls (Figure 2c). In the PC group, PC 30:2 and PC 36:2 were in creased in *Pla2g5*-null (U)BM-Macs compared to WT (U)BM-Macs (Figure 2d), while PC 34:1 and PC 36:2 were increased in *Pla2g5*-null (LPS+IFNγ)BM-Macs compared to rela tive controls. These data suggest that Pla2g5 preferentially metabolized PE molecules in both (IL-4)BM-Macs and (LPS+IFNγ)BM-Macs and that its activity on PC is restricted to (LPS+IFNγ)BM-Macs.

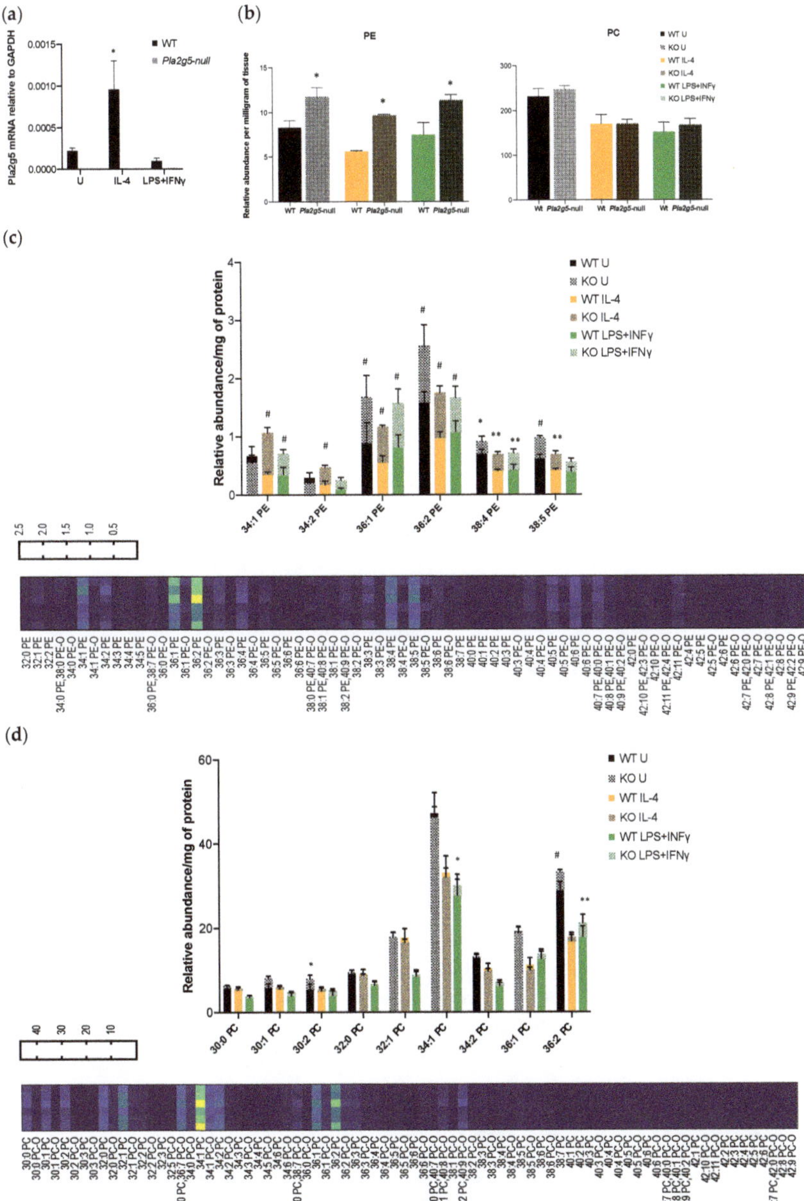

Figure 2. PE and PC molecules from WT and *Group V phospholipase* A_2 *(Pla2g5)*-null BM-Macs determined by LC-MS. Expression of Pla2g5 mRNA relative to Glyceraldehyde-3-Phosphate Dehydrogenase (GAPDH) measured by qPCR in WT (black bars) or *Pla2g5*-null BM-Macs (grey bars), Unstimulated (U) or stimulated with IL-4 or LPS+IFNγ (**a**). BM-Macs isolated from WT (solid bars) or *Pla2g5*-null (dotted bars) mice were Unstimulated (U) or treated with IL-4 or LPS+IFNγ for 24 h and (**b**) total PC and PE, (**c**) PE, or (**d**) PC molecules are reported. The data are shown as a heatmap of mean or graph of mean and standard error of three independent determinations. # $p < 0.0001$, ** $p < 0.005$, * $p < 0.05$ by two-way ANOVA with Sidak's or Tukey's correction for multiple comparisons.

3.2. *Lysophospholipid and Fatty Acid Generation in Activated BM-Macs*

To further investigate the changes in bioactive lipids during macrophage activatio we analyzed LysoPLs by LC-MS. A heat map of LysoPLs generated in BM-Macs showe that compared to unstimulated WT BM-Macs, LysoPC and Lysophosphatidylethanolamin (LysoPE) are substantially reduced following activation with either IL-4 or LPS+IFN (Supplemental Figure S1). This is likely due to LysoPL being rapidly metabolized c re-acylated into membrane phospholipids [41]. Furthermore, there was no significan difference between WT and *Pla2g5*-null BM-Macs in either LysoPE or LysoPC molecules i any stimulation (Supplemental Figure S1).

FAs are bioactive lipids and the second product of PLA_2 activity on membrane pho pholipids. We analyzed 33 FAs by GC-MS [42]. WT BM-Macs showed that FAs generate in (LPS+IFNγ)BM-Macs and (IL-4)BM-Macs were reduced compared to (U)BM-Macs (dat not shown), likely because they are being metabolized to their final products, eicosanoid However, to understand whether Pla2g5 differentially contributes to FAs generated b activated macrophages, we compared FAs produced by WT BM-Macs and *Pla2g5*-nul BM-Macs activated by IL-4 or LPS+IFNγ (Figure 3a,b). There was a trend toward reductio in LA (18:2) in *Pla2g5*-null (IL-4)BM-Macs compared to WT (IL-4)BM-Macs (Figure 3a However, myristic acid (MA; 14:0) was significantly higher in *Pla2g5*-null (IL-4)BM-Mac compared to WT (IL-4)BM-Macs. Instead, compared to WT (LPS+IFNγ)BM-Macs, *Pla2g* null (LPS+IFNγ)BM-Macs had a significant reduction palmitic acid (PA; 16:0), and trend toward a reduction in AA (20:4) (Figure 3b).

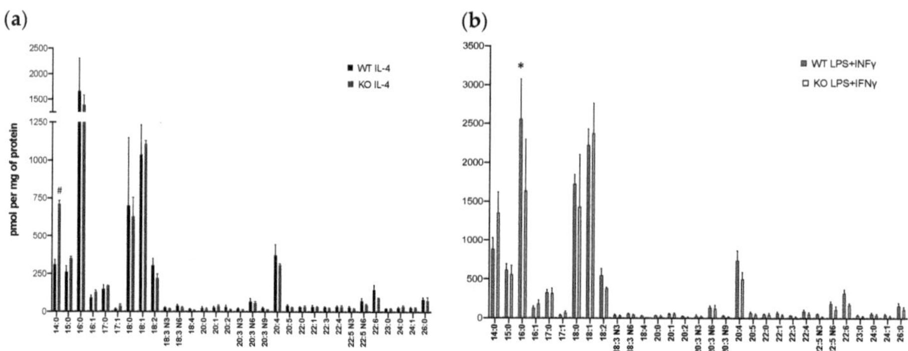

Figure 3. FAs release by activated WT and *Pla2g5*-null BM-Macs. BM-Macs isolated from WT or *Pla2g5*-null mice were Unstimulated (U) or treated with IL-4 (**a**) or LPS+IFNγ (**b**) for 24 h. The production of FAs measured by gas chromatography coupled to mass spectrometry (GC-MS) in WT and *Pla2g5*-null BM-Macrophages. Data are from three independent experiments. Two Way ANOVA followed by Fisher's LSD post-test. # $p < 0.005$, * $p < 0.01$.

3.3. *Differential Eicosanoid Generation in Bone Marrow-Derived Macrophages Activated by IL-4 or LPS+IFNγ*

Because LysoPLs and FAs were reduced in activated BM-Macs compared to unstimu lated BM-Macs, we reasoned that eicosanoid could be increased. To verify the effects c IL-4 and LPS+IFNγ on the eicosanoids generated in BM-Macs, we performed an unbiase lipid profile by LC-MS/MS. We detected 32 of the 154 analyzed lipids. The eicosanoid produced by WT BM-Macs unstimulated and following IL-4 and LPS+IFNγ activatio originated from AA (C20:4), Eicosapentaenoic acid (EPA, 20:5), Dihomo-γ linolenic aci (DGLA, 20:3), linoleic acid (LA, 18:2), Docosahexaenoic acid (DHA, 22:6), and Adreni acid (AdA, 22:4) (Figure 4). WT unstimulated BM-Macs produced AA-derived metab lites generated through the three major enzymatic pathways COX, LOX, CYP, and nor enzymatically (n.e.). PGD_2, Thromboxane B_2 (TxB_2), 12-Hydroxy-eicosatetraenoic aci (12-HETE), 20CooH AA, 12-hydroxy-heptadecatrienoic acid (12-HHTrE), 13,14-dihydr

15-keto prostaglandin D_2 (dhk PGD_2), and Prostaglandin F metabolite (PGFM) were the most abundant (Figure 4). Compared to WT (U)BM-Macs, both (IL-4) and (LPS+IFNγ)BM-Macs had a significant increase in the production of the COX metabolites PGE_2 and PGD_2. Instead, compared to WT (U)BM-Macs, in (LPS+IFNγ)BM-Macs, there were increased amounts of TxB_2, PGA_2, 12-HHTrE (COX metabolites), 11-HETE (non-enzymatic product), and 12-HETE (LOX product). Furthermore, in (LPS+IFNγ)BM-Macs, there was a trend in increasing 13-hydroxy-docosahexaenoic acid (13-HDoHE), 14-HDoHE, and dihomo-$PGF_2\alpha$. Also, 20CooH AA, a metabolic product of CYP-generated 20-HETE, was significantly increased in (IL-4)BM-Macs compared to (U)BM-Macs. These data suggest that while LPS+IFNγ robustly activates COX and LOX pathways, IL-4 activation of BM-Macs produces selective COX metabolites while also sustaining CYP metabolism as indicated by the increase in 20CooH AA.

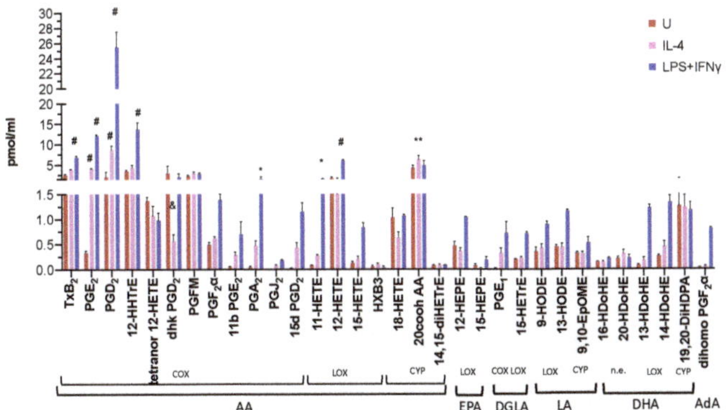

Figure 4. Eicosanoids produced by activated WT BM-Macs. Analysis by LC-MS/MS of cell-free supernatants obtained from WT BM-Macs unstimulated (U) and stimulated for 24 h with IL-4 or LPS+IFNγ. Shown is mean and standard error of eicosanoids produced by WT BM-Macs unstimulated (U, Purple bars) or activated with IL-4 (Red bars) or LPS+IFNγ (Blue bars). Eicosanoids were identified based on matching chromatographic retention times (RTs), fragmentation patterns, and six characteristic and diagnostic ions. Cyclooxygenase (Cox), Lipoxygenase (LOX) Cytochrome P450 (CYP), non-enzymatically (n.e.), Thromboxane B_2 (TxB_2), Prostaglandin E_2 (PGE_2), prostaglandin D_2 (PGD_2), 13,14-dihydro-15-keto prostaglandin D_2 (dhk PGD_2), 20-Carboxy-arachidonic acid (20CooH AA), 12-Hydroxyheptadecatrenoic acid (12-HHTrE), Prostaglandin F metabolite (PGFM), Hepoxilin B3 (HXB3), HODE (Hydroxy-octadecadienoic acid), EpHOME (epoxy-octadecenoic acid), HETrE (hydroxy-eicosatrienoic acid), HDoHE (hydroxy-docosahexaenoic acid), DiHDPA (dihydroxydocosapentaenoic acid), DiHETrE (dihydroxy-eicosatrienoic acid), HEPE (Hydroxy-eicosapentaenoic acid), HETE (Hydroxy-eicosatetraenoic acid), HPETE (Hydroperoxy-eicosatetraenoic acid), arachidonic acid (AA), LA (linoleic acid), DGLA (Dihomo-gamma linolenic acid) EPA (eicosapentaenoic acid), DHA (docosahexaenoic acid), AdA (Adrenic acid). # $p < 0.0001$, ** $p < 0.001$ * $p < 0.05$ by two-way ANOVA with Dunnett's correction for multiple comparisons of three independent determinations.

3.4. *Pla2g5 Contributes to Selective Eicosanoid Generation in Unstimulated And Activated Macrophages*

To better understand the contribution of Pla2g5 to eicosanoid generation during macrophage polarization, we compared eicosanoid generation in WT BM-Macs and *Pla2g5*-null (U)BM-Macs and (IL-4) or (LPS+IFNγ)BM-Macs (Figure 5). Compared to WT (U)BM-Macs, in *Pla2g5*-null (U)BM-Macs there was a reduction in PGD_2, and dhk PGD_2 (Figure 5b,c upper panels). Furthermore, compared to WT (IL-4)BM-Macs, in *Pla2g5*-null (IL-4)BM-Macs, there was a significant reduction of PGE_2, PGD_2, and 20CooH AA (Figure 5b,c middle panels), the three metabolites that were significantly induced by IL-4 in WT BM-Macs. Stimulation of macrophages with LPS+IFNγ revealed that *Pla2g5*-null

(LPS+IFNγ)BM-Macs have a significant reduction in PGE_2, PGD_2, and 12-HHTrE, con pared to WT (LPS+IFNγ)BM-Macs (Figure 5b,c lower panels). Unexpectedly, the absenc of Pla2g5 in BM-Macs resulted in an increase of selected lipids: in unstimulated and i IL-4 activated BM-Macs lacking Pla2g5, PGFM was increased compared to equally treate WT BM-Macs (Figure 5b), while in (LPS+IFNγ)BM-Macs lacking Pla2g5, 20CooH AA wa increased compared to equally stimulated WT BM-Macs, likely as a result of reduce metabolism through the COX and LOX pathways. These data suggest that Pla2g5 is in volved in the generation of selected metabolites of the COX pathway (IL-4)BM-Macs an metabolites of the COX and LOX pathways in (LPS+IFNγ)BM-Macs, while CYP-induce production of 20CooH AA seemed to be associated with Pla2g5 function in IL-4 activate macrophages. Cysteinyl-leukotrienes (Cys-LTs) were not detected in any conditions i either WT or *Pla2g5*-null BM-Macs.

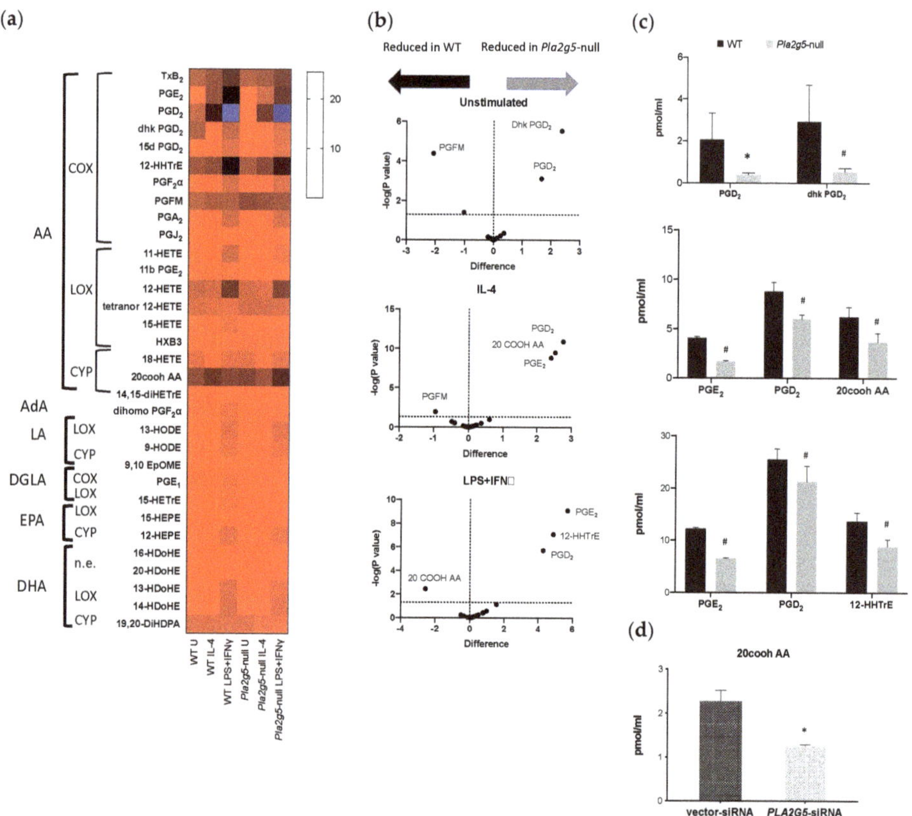

Figure 5. Contribution of Pla2g5 to eicosanoids produced by activated Macs. Analysis by LC-MS/MS of cell-free supernatants obtained from (**a–c**) WT and *Pla2g5*-null BM-Macs unstimulated (U) and stimulated for 24 h with IL-4 or LPS+IFNγ or (**d**) 20CooH AA measured in the supernatants collected from vector- and *PLA2G5*-siRNA transfected human monocyte-derived macrophages (h-Macs) stimulated with IL-4. Shown are (**a**) heatmap (**b**) volcano plots of mean values (**c**,**d**) graphs of mean and standard error of eicosanoids produced. # $p < 0.0001$, * $p < 0.05$ by two-way ANOVA with Sidak's (**c**) correction for multiple comparisons or *t*-Test (**d**) of three independent determinations.

To ascertain the contribution of human group V PLA_2 (*PLA2G5*) to eicosanoid genera tion in human monocyte-derived macrophages (h-Macs) activated by IL-4, we analyzed th lipidomic data set previously generated that demonstrated reduced PGE_2 generation in th absence of *PLA2G5* [20]. When compared with vector-treated (IL-4)h-Macs, *PLA2G5*-siRNA-treated (IL-4)h-Macs showed significant reduction of 20CooH AA ($p = 0.014$) (Figure 5d). Thus, Pla2g5 supports 20CooH AA production in mouse BM-Macs and h-Macs activated by IL-4.

4. Discussion

The study of macrophage activation has received increasing attention because of its potential implications in the development and treatment of multiple diseases. Macrophage activation is exemplified by the polarization of macrophages with IL-4 or LPS+IFNγ, each leading to differential gene expression, metabolism, cytokine, and chemokine production [13]. Although PLA_2-generated bioactive lipids, which include FAs, PUFA-derived eicosanoids, and LysoPLs are central to many critical macrophage functions, they have not been extensively studied concerning macrophage activation and functions. Because Pla2g5 is expressed in macrophages and induced by IL-4, it is an attractive target to understand the contribution of bioactive lipids to macrophage polarization and functions. Here we performed a comprehensive analysis of PL, LysoPLs, FAs, and eicosanoids produced by BM-Macs under polarizing conditions, namely LPS+IFNγ and IL-4 (Figure 6).

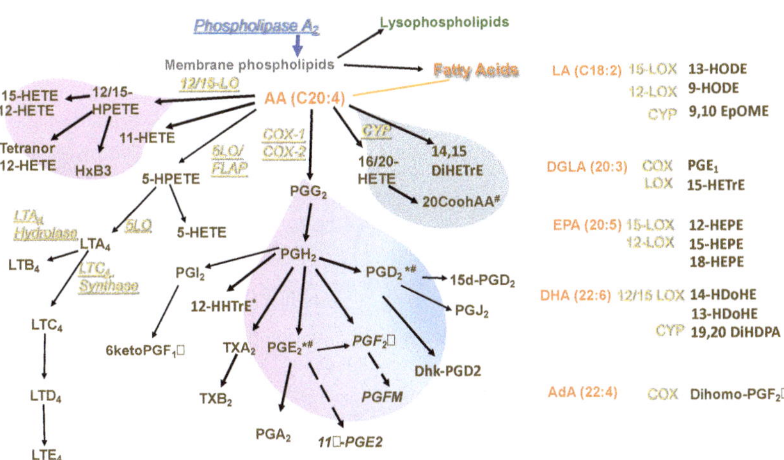

Figure 6. Flow-chart of bioactive lipids produced in activated BM-Macs and contribution of Pla2g5. Phospholipase A_2 hydrolysis of membrane glycerophospholipids generates lysophospholipids, and free fatty acids including polyunsaturated fatty acids (PUFAs) which are then metabolized to eicosanoids through 3 enzymatic pathways: COX, LOX, and monooxygenases like CYP. Depicted in yellow are the enzymes; in brown the metabolites; and in orange are the fatty acids. The purple bubble depicts eicosanoids generate mainly through activation by LPS+IFNγ in blue those generated mainly through IL-4 activation, and in purple–blue are those in common. Metabolites reduced in *Pla2g5*-null (LPS+IFNγ)BM-Macs (*) or (IL-4)BM-Macs (#). FLAP (5-lipoxygenase-activating protein).

Analysis of phospholipids in activated BM-Macs showed that PC and PE in BM-Macs are significantly reduced respectively in (LPS+IFNγ)BM-Macs and (IL-4)BM-Macs. Major species including 34:1 PC and 36:2 PE were metabolized in both M1 (LPS+IFNγ) and M2 (IL-4) macrophages, which underscores the importance of lipid metabolism in macrophage activation. LPS+IFNγ was more effective than IL-4 in reducing saturated PC species, including PC 30:0 and PC 32:0. These data are in line with reports in RAW264.7 mouse cell line and human monocyte-macrophages [39,43], although the overall composition of membrane phospholipids in RAW264.7 showed a robust content of PA and PI at baseline, while in human monocyte-derived macrophages PS, PI, and PG contributed to PL composition and changes after activation [44]. Furthermore, IL-4 was more effective on PE metabolites than LPS+IFNγ, while both LPS+IFNγ and IL-4 metabolized PE 38:4 PE and PE 38:5, tw molecules that generate AA (20:4) by the action of PLA_2s.

Macrophage functions are regulated by several PLA_2s, including Pla2g5, Pla2g4 Pla2g10, and Pla2g2d which may contribute to PL remodeling during activation [26,43,45 Notably, Pla2g5 mRNA is induced in macrophages by IL-4 but not LPS+IFNγ [20,32,33 Indeed, *Pla2g5*-null BM-Macs had a significant increase in several PE molecules in (IL-4)BM Macs. During LPS+IFNγ activation, *Pla2g5*-null BM-Macs had an increase only in PC 34: PC 36:2, and few PE molecules increased during IL-4 activation. Pla2g5 protein is expresse in resting macrophages and translocates following stimulation with pathogens [28]. Th effects of Pla2g5 activity in (LPS+IFNγ)BM-Macs are likely due to direct action of preexi tent Pla2g5 on membrane PLs or through activation of Pla2g4a, constitutively expressed b macrophages [28] or other PLAs expressed in BM-Macs.

The analysis of LysoPLs did not show increased production of any LysoPLs in a tivated BM-Macs or a reduction in BM-Macs lacking Pla2g5. In our experiments, IL- did not increase the amount of LysoPL molecules in WT BM-Macs. LysoPLs could likel be re-acetylated during activation [46]. Alternatively, Pla2g5 could exert its functions o membrane lipids earlier than 24 h, which is the time point used for macrophage activa tion [6]. However, it has been reported that macrophages derived from peripheral bloo mononuclear cells and activated by IL-4 have reduced LysoPE but not LysoPC when de pleted of PLA2G5 by siRNA and activated by IL-4 [33]. Our results also show a preferenc for Pla2g5 to target PE rather than PC during IL-4 activation of BM-Macs, but we did no detect a reduction of LysoPE molecules in *Pla2g5*-null BM-Macs. Differences in the type c macrophages and depletion of Pla2g5 may account for the discrepancies.

Given our previous data showing that LA and OA generated from macrophages in duce pulmonary inflammation [34], we would have expected a decrease at least in selecte PUFAs and MUFAs in BM-Macs lacking Pla2g5, particularly in IL-4 activation. Instead, i WT BM-Macs, despite the reduction in PC and PE species containing PUFAs and MUFA we could not detect an increase in FAs, making it difficult to detect a decrease in FA in *Pla2g5*-null BM-Macs. However, FAs can serve as second messengers by binding t cognate receptors, they can be metabolized to supply energy to the cell, re-acylated into th membrane, or serve as substrate to generate eicosanoids [11,12,14,46,47]. Therefore, in cell activated in vitro, FAs are likely heavily used. However, our data showed that PA, the mos abundant FA produced in either IL-4 or LPS+IFNγ activation, is reduced in *Pla2g5*-nul (LPS+IFNγ)BM-Macs compared to equally treated WT BM-Macs. These results agree wit the release of PA by Pla2g5 from other cell types during type 1 inflammation [48,49]. Su prisingly, (IL-4)BM-Macs lacking Pla2g5 had a significant increase in MA (14:0) compare to WT (IL-4)BM-Macs. As MA can be incorporated into a protein with consequences o signal transduction and AA metabolism, future studies await to ascertain the relevance c *Pla2g5*-generated MA in macrophage functions.

The reduction of PE 38:4 and 38:5 in *Pla2g5*-null BM-Macs suggests that, althoug the analysis of FAs does not support a contribution of Pla2g5 to the generation of A or other PUFAs, it is likely that eicosanoid generation requires Pla2g5. In macrophage prostanoids are abundantly produced following TLR-4 stimulation [50]. Furthermor LPS stimulation increases eicosanoid generation by increasing $cPLA_2\alpha$ activation an COX-2 expression [51–53]. Our data show that PGD_2 was the most abundant eicosanoi produced by (U)BM-Macs and (LPS+IFNγ)BM-Macs, as reported for RAW264.7 cells [53 Furthermore, the absence of Pla2g5 significantly reduced PGE_2 and PGD_2 in IL-4 an (LPS+IFNγ)BM-Macs, although the effect of LPS+IFNγ was more robust. However, i (LPS+IFNγ)BM-Macs, there was a reduction in 12-HHTrE, an eicosanoid reportedly pro duced by LPS activated macrophages [54]. These data suggest that the overarching effect o Pla2g5-generated eicosanoids during LPS+IFNγ or IL-4 activation depends, in the forme on the combined action of at least PGD_2, PGE_2, and 12-HHTrE, in the latter on PGD PGE_2, and 20CooH AA. Indeed, 20CooH AA was significantly induced (IL-4)BM-Mac and reduced in *Pla2g5*-null (IL-4)BM-Macs. 20CooH AA is produced from 20-HETE, product of CYP ω-oxidation. It is a vasoactive metabolite and activator of PPARγ and PPARα and, therefore, could play a role in Pla2g5-mediated IL-4-induced activation of macrophages [55]. These data indicate that compared with LPS+IFNγ, IL-4 stimulation of BM-Macs results in a weaker COX induction while also activating lipid ω-oxidation, at least partially. The fact that 20CooH AA is increased in *Pla2g5*-null (LPS+IFNγ)BM-Macs underscores the importance of Pla2g5 in the balance between lipid metabolites generated during LPS+IFNγ activation of BM-Macs (COX and 12-15LOX derived) and IL-4 activation (COX and CYP derived). The importance of 20CooH AA in macrophages expressing Pla2g5 is also supported by our data showing that (IL-4)h-Macs lacking PLA2G5 have reduced production of 20CooH AA. In another report, human monocyte-derived macrophages activated with IL-4 and deprived of PLA2G5, did not show any eicosanoid reduction. Our protocol involves monocytes' isolation by negative selection and culture of the cells in GM-CSF for 10 days to achieve high TGM2 expression [7] and PLA2G5 expression [20]. Therefore, the discrepancies may depend on protocols adopted. However, the relevance of 20CooH AA in any macrophage functions related to Pla2g5 needs to be validated likely by in vivo experiments using mouse models.

Lipoxygenase derived lipids, including Cys-LTs, are potent proinflammatory mediators which are generated by dendritic cells, mast cells, macrophages and Th2 cells. In BM-Macs, we could not detect Cys-LTs (data not shown), likely because the delayed stimulation of 24 h necessary for macrophage polarization in vitro prevents the detection of Cys-LTs. Our data are in line with reports in RAW264.7 cells in which stimulation with Kdo2-Lipid A, a lipopolysaccharide, induces the COX pathway and downregulates the 5-LOX pathway [56]. Furthermore, it cannot be excluded that exogenous Pla2g5 may induce CysLTs generation [57].

The profile of lipids of BM-Macs in one-time point and in *Pla2g5*-null BM-Macs provides a snapshot of the function of a lipid-packed, heterogeneous cell, like macrophages. It does not consider other PLA_2s expressed in macrophages, including Pla2g4a, Pla2g2d, and Pla2g10, or other sources of exogenous Pla2g5. However, an imbalance between pro and anti-inflammatory lipids could account for the different role of Pla2g5 in the pathogenesis of several pathologies. Additionally, the expression of Pla2g5 in hematopoietic vs. non-hematopoietic cells could also determine the function of Pla2g5 in different diseases [25,26]. A recent study showed that in endothelial cells, the expression of Pla2g5 mRNA is higher than other PLA_2s (Pla2g1b, 2a, 2d, 2e, 2f, and 10) then its expression is reduced by Angiotensin II stimulation while Pla2g5 protein is still present on the cell surface, likely linked to proteoglycans [49]. On the other hand, in resting macrophages, Pla2g5 mRNA is almost undetectable and is induced by IL-4, while the protein, located intracellularly in resting macrophages, is secreted upon activation [20,29]. As macrophages are hematopoietic derived cells while endothelial cells are derived from mesodermal cells, it is likely that subcellular location and cell ontogeny could also predict Pla2g5 pro or anti-inflammatory functions.

5. Conclusions

The combined proinflammatory or anti-inflammatory properties of the lipids produced by M1 (LPS+IFNγ) or M2 (IL-4) activated macrophages may contribute to the functions of macrophages in different diseases. The identification of a potential lipid signature in M1 and M2 macrophages could identify new pathways critical for the development, persistence, or reduction of inflammation. Some of these functions may rely on the presence or absence of Pla2g5 in macrophages.

Supplementary Materials: The following are available online at https://www.mdpi.com/2218-273X/11/1/25/s1, Figure S1: LysoPL produced by activates BM-Macs.

Author Contributions: Conceptualization, B.B.; methodology, M.K., M.Y., and S.K.S.; validation, M.Y., M.K., and B.B.; formal analysis, M.K., M.Y., and S.K.S.; investigation, M.K., S.K.S., and M.Y.; resources, B.B.; data curation, M.K., S.K.S., and M.Y.; writing—original draft preparation, B.B.;

writing—review and editing, M.K., M.Y., S.K.S., and B.B.; supervision, B.B.; funding acquisition, B.
All authors have read and agreed to the published version of the manuscript.

Funding: This research was funded by National Heart, Lung, and Blood Institute, grant number 11307

Conflicts of Interest: The authors declare no conflict of interest.

References

1. Gordon, S.; Pluddemann, A. The mononuclear phagocytic system. Generation of diversity. *Front Immunol.* **2019**, *10*, 189; [CrossRef]
2. Locati, M.; Curtale, G.; Mantovani, A. Diversity, mechanisms, and significance of macrophage plasticity. *Annu. Rev. Patho* **2020**, *15*, 123–147. [CrossRef] [PubMed]
3. Ying, W.; Fu, W.; Lee, Y.S.; Olefsky, J.M. The role of macrophages in obesity-associated islet inflammation and beta-cell abnormal ties. *Nat. Rev. Endocrinol.* **2020**, *16*, 81–90. [CrossRef] [PubMed]
4. Gosselin, D.; Link, V.M.; Romanoski, C.E.; Fonseca, G.J.; Eichenfield, D.Z.; Spann, N.J.; Stender, J.D.; Chun, H.B.; Garner, H Geissmann, F.; et al. Environment drives selection and function of enhancers controlling tissue-specific macrophage identitie *Cell* **2014**, *159*, 1327–1340. [CrossRef]
5. Ginhoux, F.; Schultze, J.L.; Murray, P.J.; Ochando, J.; Biswas, S.K. New insights into the multidimensional concept of macrophag ontogeny, activation and function. *Nat. Immunol.* **2016**, *17*, 34–40. [CrossRef] [PubMed]
6. Martinez, F.O.; Gordon, S. The M1 and M2 paradigm of macrophage activation: Time for reassessment. *F1000Prime Rep.* **2014**, [CrossRef]
7. Martinez, F.O.; Helming, L.; Milde, R.; Varin, A.; Melgert, B.N.; Draijer, C.; Thomas, B.; Fabbri, M.; Crawshaw, A.; Ho, L.P.; et a Genetic programs expressed in resting and IL-4 alternatively activated mouse and human macrophages: Similarities and diffe ences. *Blood* **2013**, *121*, e57–e69. [CrossRef]
8. Gieseck, R.L., III; Wilson, M.S.; Wynn, T.A. Type 2 immunity in tissue repair and fibrosis. *Nat. Rev. Immunol.* **2018**, *18*, 62–7 [CrossRef]
9. Lackey, D.E.; Olefsky, J.M. Regulation of metabolism by the innate immune system. *Nat. Rev. Endocrinol.* **2016**, *12*, 15–2 [CrossRef]
10. Wang, F.; Zhang, S.; Jeon, R.; Vuckovic, I.; Jiang, X.; Lerman, A.; Folmes, C.D.; Dzeja, P.D.; Herrmann, J. Interferon gamma induce reversible metabolic reprogramming of M1 macrophages to sustain cell viability and pro-inflammatory activity. *EBioMedicin* **2018**, *30*, 303–316. [CrossRef]
11. Vats, D.; Mukundan, L.; Odegaard, J.I.; Zhang, L.; Smith, K.L.; Morel, C.R.; Wagner, R.A.; Greaves, D.R.; Murray, P.J.; Chawla, A Oxidative metabolism and PGC-1beta attenuate macrophage-mediated inflammation. *Cell Metab.* **2006**, *4*, 13–24. [CrossRef [PubMed]
12. Ecker, J.; Liebisch, G.; Grandl, M.; Schmitz, G. Lower SCD expression in dendritic cells compared to macrophages leads t membrane lipids with less mono-unsaturated fatty acids. *Immunobiology* **2010**, *215*, 748–755. [CrossRef] [PubMed]
13. Jha, A.K.; Huang, S.C.; Sergushichev, A.; Lampropoulou, V.; Ivanova, Y.; Loginicheva, E.; Chmielewski, K.; Stewart, K.M Ashall, J.; Everts, B.; et al. Network integration of parallel metabolic and transcriptional data reveals metabolic modules tha regulate macrophage polarization. *Immunity* **2015**, *42*, 419–430. [CrossRef] [PubMed]
14. Ecker, J.; Liebisch, G.; Englmaier, M.; Grandl, M.; Robenek, H.; Schmitz, G. Induction of fatty acid synthesis is a key requireme for phagocytic differentiation of human monocytes. *Proc. Natl. Acad. Sci. USA* **2010**, *107*, 7817–7822. [CrossRef]
15. Lukic, A.; Larssen, P.; Fauland, A.; Samuelsson, B.; Wheelock, C.E.; Gabrielsson, S.; Radmark, O. GM-CSF- and M-CSF-prime macrophages present similar resolving but distinct inflammatory lipid mediator signatures. *FASEB J.* **2017**, *31*, 4370–438 [CrossRef]
16. Stein, M.; Keshav, S.; Harris, N.; Gordon, S. Interleukin 4 potently enhances murine macrophage mannose receptor activity A marker of alternative immunologic macrophage activation. *J. Exp. Med.* **1992**, *176*, 287–292. [CrossRef]
17. Kanaoka, Y.; Austen, K.F. Roles of cysteinyl leukotrienes and their receptors in immune cell-related functions. *Adv. Immuno* **2019**, *142*, 65–84. [CrossRef]
18. Samuchiwal, S.K.; Balestrieri, B.; Raff, H.; Boyce, J.A. Endogenous prostaglandin E2 amplifies IL-33 production by macrophage through an E prostanoid (EP)2/EP4-cAMP-EPAC-dependent pathway. *J. Biol. Chem.* **2017**, *292*, 8195–8206. [CrossRef]
19. Samuchiwal, S.K.; Boyce, J.A. Role of lipid mediators and control of lymphocyte responses in type 2 immunopathology. *J. Allerg Clin. Immunol.* **2018**, *141*, 1182–1190. [CrossRef]
20. Yamaguchi, M.; Zacharia, J.; Laidlaw, T.M.; Balestrieri, B. PLA2G5 regulates transglutaminase activity of human IL-4-activate M2 macrophages through PGE2 generation. *J. Leukoc. Biol.* **2016**, *100*, 131–141. [CrossRef]
21. Bystrom, J.; Wray, J.A.; Sugden, M.C.; Holness, M.J.; Swales, K.E.; Warner, T.D.; Edin, M.L.; Zeldin, D.C.; Gilroy, D.W Bishop-Bailey, D. Endogenous epoxygenases are modulators of monocyte/macrophage activity. *PLoS ONE* **2011**, *6*, e2659 [CrossRef] [PubMed]
22. Fang, X.; Dillon, J.S.; Hu, S.; Harmon, S.D.; Yao, J.; Anjaiah, S.; Falck, J.R.; Spector, A.A. 20-carboxy-arachidonic acid is dual activator of peroxisome proliferator-activated receptors alpha and gamma. *Prostaglandins Lipid Mediat.* **2007**, *82*, 175–18 [CrossRef] [PubMed]

3. Garcia, V.; Gilani, A.; Shkolnik, B.; Pandey, V.; Zhang, F.F.; Dakarapu, R.; Gandham, S.K.; Reddy, N.R.; Graves, J.P.; Gruzdev, A.; et al. 20-HETE signals through g-protein-coupled receptor GPR75 (Gq) to affect vascular function and trigger hypertension. *Circ. Res.* **2017**, *120*, 1776–1788. [CrossRef] [PubMed]
4. Huang, S.C.; Everts, B.; Ivanova, Y.; O'Sullivan, D.; Nascimento, M.; Smith, A.M.; Beatty, W.; Love-Gregory, L.; Lam, W.Y.; O'Neill, C.M.; et al. Cell-intrinsic lysosomal lipolysis is essential for alternative activation of macrophages. *Nat. Immunol.* **2014**, *15*, 846–855. [CrossRef] [PubMed]
5. Samuchiwal, S.K.; Balestrieri, B. Harmful and protective roles of group V phospholipase A2: Current perspectives and future directions. *Biochim. Biophys. Acta Mol. Cell Biol. Lipids* **2019**, *1864*, 819–826. [CrossRef]
6. Murakami, M.; Sato, H.; Taketomi, Y. Updating phospholipase A2 biology. *Biomolecules* **2020**, *10*, 1457. [CrossRef]
7. Balestrieri, B.; Hsu, V.W.; Gilbert, H.; Leslie, C.C.; Han, W.K.; Bonventre, J.V.; Arm, J.P. Group V secretory phospholipase A2 translocates to the phagosome after zymosan stimulation of mouse peritoneal macrophages and regulates phagocytosis. *J. Biol. Chem.* **2006**, *281*, 6691–6698. [CrossRef]
8. Ruiperez, V.; Astudillo, A.M.; Balboa, M.A.; Balsinde, J. Coordinate regulation of TLR-mediated arachidonic acid mobilization in macrophages by group IVA and group V phospholipase A2s. *J. Immunol.* **2009**, *182*, 3877–3883. [CrossRef]
9. Balestrieri, B.; Maekawa, A.; Xing, W.; Gelb, M.H.; Katz, H.R.; Arm, J.P. Group V secretory phospholipase A2 modulates phagosome maturation and regulates the innate immune response against Candida albicans. *J. Immunol.* **2009**, *182*, 4891–4898. [CrossRef]
0. Lapointe, S.; Brkovic, A.; Cloutier, I.; Tanguay, J.F.; Arm, J.P.; Sirois, M.G. Group V secreted phospholipase A2 contributes to LPS-induced leukocyte recruitment. *J. Cell. Physiol.* **2010**, *224*, 127–134. [CrossRef]
1. Munoz, N.M.; Meliton, A.Y.; Meliton, L.N.; Dudek, S.M.; Leff, A.R. Secretory group V phospholipase A2 regulates acute lung injury and neutrophilic inflammation caused by LPS in mice. *Am. J. Physiol. Lung Cell Mol. Physiol.* **2009**, *296*, L879–L887. [CrossRef] [PubMed]
2. Ohta, S.; Imamura, M.; Xing, W.; Boyce, J.A.; Balestrieri, B. Group V secretory phospholipase A2 is involved in macrophage activation and is sufficient for macrophage effector functions in allergic pulmonary inflammation. *J. Immunol.* **2013**, *190*, 5927–5938. [CrossRef] [PubMed]
3. Rubio, J.M.; Rodriguez, J.P.; Gil-de-Gomez, L.; Guijas, C.; Balboa, M.A.; Balsinde, J. Group V secreted phospholipase A2 is upregulated by IL-4 in human macrophages and mediates phagocytosis via hydrolysis of ethanolamine phospholipids. *J. Immunol.* **2015**, *194*, 3327–3339. [CrossRef] [PubMed]
4. Yamaguchi, M.; Samuchiwal, S.K.; Quehenberger, O.; Boyce, J.A.; Balestrieri, B. Macrophages regulate lung ILC2 activation via Pla2g5-dependent mechanisms. *Mucosal. Immunol.* **2017**. [CrossRef]
5. Satake, Y.; Diaz, B.L.; Balestrieri, B.; Lam, B.K.; Kanaoka, Y.; Grusby, M.J.; Arm, J.P. Role of group V phospholipase A2 in zymosan-induced eicosanoid generation and vascular permeability revealed by targeted gene disruption. *J. Biol. Chem.* **2004**, *279*, 16488–16494. [CrossRef]
6. Xue, J.; Schmidt, S.V.; Sander, J.; Draffehn, A.; Krebs, W.; Quester, I.; de Nardo, D.; Gohel, T.D.; Emde, M.; Schmidleithner, L.; et al. Transcriptome-based network analysis reveals a spectrum model of human macrophage activation. *Immunity* **2014**, *40*, 274–288. [CrossRef]
7. Quehenberger, O.; Armando, A.M.; Brown, A.H.; Milne, S.B.; Myers, D.S.; Merrill, A.H.; Bandyopadhyay, S.; Jones, K.N.; Kelly, S.; Shaner, R.L.; et al. Lipidomics reveals a remarkable diversity of lipids in human plasma. *J. Lipid Res.* **2010**, *51*, 3299–3305. [CrossRef]
8. Wang, Y.; Armando, A.M.; Quehenberger, O.; Yan, C.; Dennis, E.A. Comprehensive ultra-performance liquid chromatographic separation and mass spectrometric analysis of eicosanoid metabolites in human samples. *J. Chromatogr. A* **2014**, *1359*, 60–69. [CrossRef]
9. Quehenberger, O.; Armando, A.; Dumlao, D.; Stephens, D.L.; Dennis, E.A. Lipidomics analysis of essential fatty acids in macrophages. *Prostaglandins Leukot. Essent. Fat. Acids* **2008**, *79*, 123–129. [CrossRef]
0. Murray, P.J.; Allen, J.E.; Biswas, S.K.; Fisher, E.A.; Gilroy, D.W.; Goerdt, S.; Gordon, S.; Hamilton, J.A.; Ivashkiv, L.B.; Lawrence, T.; et al. Macrophage activation and polarization: Nomenclature and experimental guidelines. *Immunity* **2014**, *41*, 14–20. [CrossRef]
1. Shindou, H.; Hishikawa, D.; Harayama, T.; Eto, M.; Shimizu, T. Generation of membrane diversity by lysophospholipid acyltransferases. *J. Biochem.* **2013**, *154*, 21–28. [CrossRef] [PubMed]
2. Quehenberger, O.; Armando, A.M.; Dennis, E.A. High sensitivity quantitative lipidomics analysis of fatty acids in biological samples by gas chromatography-mass spectrometry. *Biochim. Biophys. Acta* **2011**, *1811*, 648–656. [CrossRef] [PubMed]
3. Dennis, E.A.; Deems, R.A.; Harkewicz, R.; Quehenberger, O.; Brown, H.A.; Milne, S.B.; Myers, D.S.; Glass, C.K.; Hardiman, G.; Reichart, D.; et al. A mouse macrophage lipidome. *J. Biol. Chem.* **2010**, *285*, 39976–39985. [CrossRef] [PubMed]
4. Montenegro-Burke, J.R.; Sutton, J.A.; Rogers, L.M.; Milne, G.L.; McLean, J.A.; Aronoff, D.M. Lipid profiling of polarized human monocyte-derived macrophages. *Prostaglandins Other Lipid Mediat.* **2016**, *127*, 1–8. [CrossRef] [PubMed]
5. Nolin, J.D.; Lai, Y.; Ogden, H.L.; Manicone, A.M.; Murphy, R.C.; An, D.; Frevert, C.W.; Ghomashchi, F.; Naika, G.S.; Gelb, M.H.; et al. Secreted PLA2 group X orchestrates innate and adaptive immune responses to inhaled allergen. *JCI Insight* **2017**, 2. [CrossRef] [PubMed]
6. Shindou, H.; Hishikawa, D.; Harayama, T.; Yuki, K.; Shimizu, T. Recent progress on acyl CoA: Lysophospholipid acyltransferase research. *J. Lipid Res.* **2009**, *50*, S46–S51. [CrossRef]

47. Alvarez-Curto, E.; Milligan, G. Metabolism meets immunity: The role of free fatty acid receptors in the immune systen *Biochem. Pharm.* **2016**, *114*, 3–13. [CrossRef]
48. Sato, H.; Taketomi, Y.; Ushida, A.; Isogai, Y.; Kojima, T.; Hirabayashi, T.; Miki, Y.; Yamamoto, K.; Nishito, Y.; Kobayashi, T.; et a The adipocyte-inducible secreted phospholipases PLA2G5 and PLA2G2E play distinct roles in obesity. *Cell Metab.* **2014**, *20*, 119–13 [CrossRef]
49. Watanabe, K.; Taketomi, Y.; Miki, Y.; Kugiyama, K.; Murakami, M. Group V secreted phospholipase A2 plays a protective rol against aortic dissection. *J. Biol. Chem.* **2020**, *295*, 10092–10111. [CrossRef]
50. Norris, P.C.; Reichart, D.; Dumlao, D.S.; Glass, C.K.; Dennis, E.A. Specificity of eicosanoid production depends on the TLR-stimulated macrophage phenotype. *J. Leukoc. Biol.* **2011**, *90*, 563–574. [CrossRef]
51. Qi, H.Y.; Shelhamer, J.H. Toll-like receptor 4 signaling regulates cytosolic phospholipase A2 activation and lipid generation i lipopolysaccharide-stimulated macrophages. *J. Biol. Chem.* **2005**, *280*, 38969–38975. [CrossRef] [PubMed]
52. Ueno, N.; Takegoshi, Y.; Kamei, D.; Kudo, I.; Murakami, M. Coupling between cyclooxygenases and terminal prostanoid synthase *Biochem. Biophys. Res. Commun.* **2005**, *338*, 70–76. [CrossRef] [PubMed]
53. Buczynski, M.W.; Stephens, D.L.; Bowers-Gentry, R.C.; Grkovich, A.; Deems, R.A.; Dennis, E.A. TLR-4 and sustained calcium agonis synergistically produce eicosanoids independent of protein synthesis in RAW264. *7 cells. J. Biol. Chem.* **2007**, *282*, 22834–2284 [CrossRef] [PubMed]
54. Fromel, T.; Kohlstedt, K.; Popp, R.; Yin, X.; Awwad, K.; Barbosa-Sicard, E.; Thomas, A.C.; Lieberz, R.; Mayr, M.; Fleming, Cytochrome P4502S1: A novel monocyte/macrophage fatty acid epoxygenase in human atherosclerotic plaques. *Basic Res. Cardic* **2013**, *108*, 319. [CrossRef]
55. Odegaard, J.I.; Ricardo-Gonzalez, R.R.; Goforth, M.H.; Morel, C.R.; Subramanian, V.; Mukundan, L.; Eagle, A.R.; Vats, D.; Bron bacher, F.; Ferrante, A.W.; et al. Macrophage-specific PPARgamma controls alternative activation and improves insulin resistanc *Nature* **2007**, *447*, 1116–1120. [CrossRef]
56. Buczynski, M.W.; Dumlao, D.S.; Dennis, E.A. Thematic review series: Proteomics. An integrated omics analysis c eicosanoid biology. *J. Lipid Res.* **2009**, *50*, 1015–1038. [CrossRef]
57. Boilard, E.; Lai, Y.; Larabee, K.; Balestrieri, B.; Ghomashchi, F.; Fujioka, D.; Gobezie, R.; Coblyn, J.S.; Weinblatt, M.E Massarotti, E.M.; et al. A novel anti-inflammatory role for secretory phospholipase A2 in immune complex-mediated arthriti *EMBO Mol. Med.* **2010**, *2*, 172–187. [CrossRef]

Article

A Lipidomic Perspective of the Action of Group IIA Secreted Phospholipase A_2 on Human Monocytes: Lipid Droplet Biogenesis and Activation of Cytosolic Phospholipase $A_2\alpha$

Juan P. Rodríguez [1,2,†], Elbio Leiguez [1,3,†], Carlos Guijas [1,4], Bruno Lomonte [5], José M. Gutiérrez [5], Catarina Teixeira [3], María A. Balboa [1,4] and Jesús Balsinde [1,4,*]

1 Instituto de Biología y Genética Molecular, Consejo Superior de Investigaciones Científicas (CSIC), Universidad de Valladolid, 47003 Valladolid, Spain; rodriguezcasco@me.com (J.P.R.); elbio.leiguez@butantan.gov.br (E.L.); cguijas@ibgm.uva.es (C.G.); mbalboa@ibgm.uva.es (M.A.B.)

2 Laboratorio de Investigaciones Bioquímicas de la Facultad de Medicina (LIBIM), Instituto de Química Básica y Aplicada del Nordeste Argentino (IQUIBA-NEA), Universidad Nacional del Nordeste, Consejo Nacional de Investigaciones Científicas y Técnicas (UNNE-CONICET), Corrientes 3400, Argentina

3 Laboratorio de Farmacologia, Instituto Butantan, Sao Paulo 01000, Brazil; catarina.teixeira@butantan.gov.br

4 Centro de Investigación Biomédica en Red de Diabetes y Enfermedades Metabólicas Asociadas (CIBERDEM), 28029 Madrid, Spain

5 Instituto Clodomiro Picado, Facultad de Microbiología, Universidad de Costa Rica, San José 11501–2060, Costa Rica; bruno.lomonte@ucr.ac.cr (B.L.); jose.gutierrez@ucr.ac.cr (J.M.G.)

* Correspondence: jbalsinde@ibgm.uva.es; Tel.: +34-983-423-062

† These authors contributed equally to this work.

Received: 14 May 2020; Accepted: 10 June 2020; Published: 10 June 2020

Abstract: Phospholipase A_2s constitute a wide group of lipid-modifying enzymes which display a variety of functions in innate immune responses. In this work, we utilized mass spectrometry-based lipidomic approaches to investigate the action of Asp-49 Ca^{2+}-dependent secreted phospholipase A_2 (sPLA_2) (MT-III) and Lys-49 sPLA2 (MT-II), two group IIA phospholipase A_2s isolated from the venom of the snake *Bothrops asper*, on human peripheral blood monocytes. MT-III is catalytically active, whereas MT-II lacks enzyme activity. A large decrease in the fatty acid content of membrane phospholipids was detected in MT III-treated monocytes. The significant diminution of the cellular content of phospholipid-bound arachidonic acid seemed to be mediated, in part, by the activation of the endogenous group IVA cytosolic phospholipase $A_2\alpha$. MT-III triggered the formation of triacylglycerol and cholesterol enriched in palmitic, stearic, and oleic acids, but not arachidonic acid, along with an increase in lipid droplet synthesis. Additionally, it was shown that the increased availability of arachidonic acid arising from phospholipid hydrolysis promoted abundant eicosanoid synthesis. The inactive form, MT-II, failed to produce any of the effects described above. These studies provide a complete lipidomic characterization of the monocyte response to snake venom group IIA phospholipase A_2, and reveal significant connections among lipid droplet biogenesis, cell signaling and biochemical pathways that contribute to initiating the inflammatory response.

Keywords: phospholipase A_2; lipidomics; mass spectrometry; lipid signaling; inflammation; monocytes/macrophages

1. Introduction

The phospholipase A_2 (PLA_2) superfamily consists of a broad range of enzymes defined by their ability to catalyze the hydrolysis of the ester bond at the sn-2 position of glycerophospholipids.

The hydrolysis products of this reaction, free fatty acid and lysophospholipid, serve as precursors for a variety of bioactive lipid mediators with important biological roles [1]. The PLA_2s are systematically classified according to sequence homology criteria, and include 16 groups (I-XVI), most of them with several subgroups, comprising more than 30 proteins [1]. An alternative classification also exists that groups these enzymes into six major classes on the basis of biochemical similarities and/or cell regulation properties. These are the Ca^{2+}-dependent cytosolic PLA_2s, the Ca^{2+}-dependent secreted PLA_2s ($sPLA_2$), the Ca^{2+}-independent cytosolic PLA_2s, the platelet-activating factor acetyl hydrolases, the lysosomal PLA_2, and the adipose-specific PLA_2 [2].

The $sPLA_2$ family represents the largest class of PLA_2 enzymes and possesses, as a common motif, a conserved His-Asp catalytic dyad [3]. $sPLA_2$s are widely distributed in pancreatic secretions, inflammatory exudates, and also in arthropod and snake venoms. A variety of biological activities have been described for $sPLA_2$s, including digestive actions, toxic activities (neurotoxic, myotoxic, hypotensive, etc.) and immune roles. In this regard, group IIA $sPLA_2$ was defined as a pro-inflammatory PLA_2, since its gene induction and synthesis were observed after cell stimulation by endotoxin and cytokines [3–5]. In contrast, another member of the family, the group V enzyme, is described as anti-inflammatory in some models [6–8].

$sPLA_2$s have been often observed to cooperate with other PLA_2s in eliciting certain biological responses. A prominent example of this is the mobilization of arachidonic acid (AA) and attendant eicosanoid production by innate immune cells responding to inflammatory stimuli [9–12]. The Ca^{2+}-dependent cytosolic group IVA PLA_2 ($cPLA_2\alpha$) is the essential enzyme in this process [12–14]. Depending on cell type and stimulation conditions, regulatory crosstalk mechanisms exist between $cPLA_2\alpha$ and other $sPLA_2$ enzymes present in the cells—in particular, those belonging to groups IIA, V and X—which results in the amplification of the AA mobilization response [15–20].

Previous work from our laboratory has utilized advanced mass spectrometry approaches to characterize multiple aspects of PLA_2-mediated phospholipid fatty acid remodeling in cells of the innate immune system such as monocytes and macrophages [21–28]. In these studies, the activation mechanisms of multiple PLA_2 enzymes expressed by the cells were characterized [21–28]. However, no approaches were undertaken to characterize the cellular responses to exogenously added $sPLA_2$ enzymes. It has been suggested that the response of cells exposed to exogenous $sPLA_2$ is dependent upon the nature of the lipid mediator generated on the membrane where the $sPLA_2$ acts [29]. However, it is also known that some $sPLA_2$s lack catalytic activity but still exert potent biological actions [30]. This has led to the proposal that some $sPLA_2$ effects depend on protein–protein or protein–glycan interactions [3,31]. Once bound to its target(s) on the membrane, $sPLA_2$s may exert their actions via activity-independent mechanisms that affect cellular functions or trigger a cellular response [3,31,32]. Additionally, the existence of a $sPLA_2$ receptor for these enzymes has long been proposed [33]. Whether activity-based effects prevail over activity-independent effects, or both occur simultaneously, remains unclear.

Exposure of immune cells to exogenous $sPLA_2$ occurs in numerous pathological situations such as inflammatory syndromes, sepsis, autoimmune diseases, and even bite or sting envenomations [3–5]. In this work, we have used advanced mass spectrometry-based lipidomics to analyze the effect of two different group IIA $sPLA_2$, with and without catalytic activity, on human monocytes. The $sPLA_2$ enzymes utilized in this study, termed Asp-49 $sPLA_2$ (MT-III) and Lys-49 $sPLA_2$ (MT-II), were purified from the venom of the Central American snake *Bothrops asper*. MT-III is a catalytically active enzyme similar to human synovial group IIA PLA_2. MT-II is identical to MT-III, except for the replacement of Asp49 with Lys49 within the active site, which renders it catalytically inactive [34–38]. The combined use of these two PLA_2s thus constitutes an excellent tool to distinguish between the activity-dependent and -independent actions of $sPLA_2$ enzymes. Consistent with the previously described proinflammatory properties of MT-III, our data show that it promotes remarkable changes in the lipid composition of cell membranes, triggers lipid droplet biogenesis, and induces eicosanoid synthesis. None of these actions are induced by the inactive form MT-II. These data agree with previous work demonstrating that MT-III, but not MT-II induces phospholipid hydrolysis in murine muscle cells [38].

2. Materials and Methods

2.1. Enzymes

Asp-49 $sPLA_2$ (MT-III) and Lys-49 $sPLA_2$ (MT-II) from Bothrops asper venom were purified by ion-exchange chromatography on CM-Sephadex C-50, using a KCl gradient from 0 to 0.75 M [36]. The complete amino acid sequence and toxicological profile of these enzymes have been previously described in detail [37]. The absence of endotoxin contamination in the batches used was demonstrated by performing the quantitative Limulus amebocyte lysate assay [39], which revealed no detectable levels of endotoxin (< 0.125 EU/mL).

2.2. Cell Culture

Human monocytes were isolated from buffy coats of healthy volunteer donors obtained from the Centro de Hemoterapia y Hemodonación de Castilla y León (Valladolid, Spain). Written informed consent was obtained from each donor. Briefly, blood cells were diluted 1:1 with phosphate-buffered saline, layered over a cushion of Ficoll-Paque, and centrifuged at 750 g for 30 min. The mononuclear cellular layer was recovered and washed three times, resuspended in RPMI 1640 medium supplemented with 40 µg/mL gentamicin, and allowed to adhere in sterile dishes for 2 h at 37 °C in a humidified atmosphere of CO_2/air (1:19). Nonadherent cells were removed by washing extensively with phosphate-buffered saline, and the remaining attached monocytes were used the following day [40,41]. For experiments, subconfluent cell monolayers were incubated with serum-free medium for 1 h before the addition of $sPLA_2$. After stimulation, the monocyte monolayers were washed twice with phosphate-buffered saline, scraped with a cell scraper, sonicated with a tip homogenizer twice for 15 s, and prepared for their further analysis by mass spectrometry, as described below. For eicosanoid determinations, supernatants were collected and prepared for mass spectrometry analysis as described below.

2.3. Cellular Staining and Fluorescence Microscopy

For these experiments, the cells were plated on coverslips on the bottom of 6-well dishes in a volume of 2 mL. The cells were fixed with 1 mL of 4% paraformaldehyde in phosphate-buffered saline containing 3% sucrose for 20 min. Afterward, paraformaldehyde was removed by washing the cells three times with phosphate-buffered saline, and Nile Red and 4′,6′-diamidino-2-phenylindole (DAPI) stainings were carried out by treating cells with these dyes at concentrations of 5 µg/mL and 1 µg/mL, respectively, in phosphate-buffered saline for 10 min. Coverslips were mounted on microscopy slides with 25 µL of a polyvinyl alcohol solution until analysis by fluorescence microscopy. Fluorescence was monitored by microscopy using a NIKON Eclipse 90i device equipped with a CCD camera (model DS-Ri1; Nikon, Tokyo, Japan). A mercury HBO excitation lamp (Osram, Munich, Germany) was used, and the fluorescence was recovered using the combination of a UV-2A (Ex 330–380; DM 400; BA 420) and a B-2A (Ex 450–490; DM 505; BA 520) filter, respectively. Images were analyzed with the software NIS-Elements (Nikon). Red and blue channels were merged with the Image-J software (version 1.52a).

2.4. Gas Chromatography/Mass Spectrometry (GC/MS) Analysis of Fatty Acid Methyl Esters

Total lipids from approximately 10^7 cells were extracted according to Bligh and Dyer [42]. For separation of total phospholipids from neutral lipids, the following internal standards were added: 10 nmol of 1,2-diheptadecanoyl-sn-glycero-3-phosphocholine, 10 nmol of 1,2,3-trihepta-decanoylglycerol, 20 nmol of nonadecanoic acid, and 30 nmol of cholesteryl tridecanoate. Phospholipids were separated from neutral lipids by thin-layer chromatography, using *n*-hexane/diethyl ether/acetic acid (70:30:1, *v*/*v*/*v*) as the mobile phase [43]. For separation of phospholipid classes, the following internal standards were added: 20 pmol each of 1,2-diheptadecanoyl-sn-glycero-3-phosphoethanolamine, 1,2-diheptadecanoyl-sn-glycero-3-phospho-choline, and 1,2-dinonadecanoyl-sn-glycero-phosphoinositol. Phospholipids were separated twice with chloroform/methanol/28% (*w/w*) ammonium hydroxide (60:37.5:4, *v/v/v*) as the mobile phase, using plates impregnated with boric acid [44]. The bands corresponding to the different lipid classes were scraped from the plate, and fatty acid methyl esters were obtained from the various lipid fractions by transmethylation with 0.5 M KOH in methanol for 60 min at 37 °C [45–49]. Analyses were carried out using an Agilent 7890A gas chromatograph coupled to an Agilent 5975C mass-selective detector operated in an electron impact mode (EI, 70 eV). (Agilent Technologies, Santa Clara, CA, USA). Data acquisition was carried out both in scan and selected ion monitoring mode. Scan mode was used for compound identification, comparing with authentic fatty acid methyl ester standards, and the National Institute of Standards and Technology MS library spectra. Selected ion monitoring mode was used for quantitation, using 74 and 87 fragments for saturated, 83 for monounsaturated, 67 and 81 for diunsaturated, and 79 and 91 for polyunsaturated fatty acid methyl esters. A 37-component mixture (Supelco, Sigma-Aldrich, Madrid, Spain) was used for calibration curves.

2.5. Mass Spectrometry Analysis of Free Fatty Acids

The thin layer chromatography spots corresponding to the non-esterified free fatty acid fraction were scraped, redissolved in *n*-hexane, and analyzed separately in an Agilent 1260 Infinity high-performance liquid chromatograph coupled to an API2000 triple quadrupole mass spectrometer (Applied Biosystems, Carlsbad, CA, USA). The column was a Supelcosil LC-8 (150 × 3 mm, 3 µm particle size), protected with a Supelguard LC-8 (20 × 3 mm) guard cartridge (Sigma-Aldrich). The mobile phase was used on a gradient of solvent A (methanol with 0.01% ammonium hydroxide) and solvent B (water with 0.01% ammonium hydroxide). The gradient was started at 60% solvent A and 40% solvent B. The former was linearly increased to 95% at 10 min, and held at 95% solvent A until 18 min. The initial solvent mixture (60%, 40%B) was recovered at 20 min and the column was re-equilibrated for an additional 5 min before the injection of next sample. The flow rate was fixed at 400 µL/min. Non-esterified fatty acid fraction, extracted from silica plates and filtered, was re-dissolved in 100 µL of methanol/water 60:40 *v/v* and 90 µL were injected into the high-performance liquid chromatograph. The parameters for electrospray ionization source of mass spectrometer were set as follows: Ion spray voltage, −4500 V; CUR, 20 psi; GS1, 40 psi; GS2, 80 psi; TEM, 525 °C. The analyzer mode was set to Q1MS (DP, −70 V; EP, −10 V; FP, −300 V) performing a m/z scan between 100 and 400 with a step size of 0.1 amu. Non-esterified fatty acids were detected as $[M-H]^-$ ions using the Analyst 1.5.2 software version (Applied Biosystems, Carlsbad, CA, USA), and chromatographic peaks were quantified by comparison with peaks of authentic analytical standards.

2.6. Liquid Chromatography/Mass Spectrometry (LC/MS) Analyses of Phospholipids

This was carried out exactly as described elsewhere [21–26,50,51], using a high-performance liquid chromatograph equipped with a binary pump Hitachi LaChrom Elite L-2130 and a Hitachi Autosampler L-2200 (Merck), coupled on-line to a Bruker esquire6000 ion-trap mass spectrometer (Bruker Daltonics, Bremen, Germany). Ethanolamine-containing phospholipids (PE) and phosphatidylinositol (PI) species were detected in negative ion mode as $[M-H]^-$ ions in MS experiments. Choline-containing phospholipids (PC) species were detected in positive ion mode, as $[M+H]^+$ ions by MS. Acyl chains in PI and PE species were identified by multiple reaction monitoring MS^2 experiments on chromatographic effluent by comparison to previously published data [21–26,50,51]. For the identification of acyl chains in PC species, ionization was carried out in negative mode with the post-column addition of acetic acid at a flow rate of 100 mL/h as $[M+CH_3CO_2]^-$ adducts, and acyl chains were identified by MS^3 experiments. Quantification was carried out by integrating the chromatographic peaks of the previously identified phospholipid species and comparing with an external calibration curve made with authentic standards.

2.7. Liquid Chromatography/Mass Spectrometry (LC/MS) Analyses of Eicosanoids

Analysis of eicosanoids by LC/MS was carried out exactly as described elsewhere [6,25,52], using an Agilent 1260 Infinity high-performance liquid chromatograph coupled to an API2000 triple quadrupole mass spectrometer (Applied Biosystems, Carlsbad, CA, USA). Quantification was carried out by integrating the chromatographic peaks of each species and by comparing with an external calibration curve made with analytical standards [6,25,52].

2.8. Immunoblot

Cells were lysed with 20 mM Tris-HCl (pH 7.4), containing 150 mM NaCl, 0.5% Triton X-100, 1 mM Na_3VO_4, 150 mM NaF, 1 mM phenylmethylsulfonyl fluoride, and a protease inhibitor mixture (Sigma-Aldrich, Madrid, Spain) at 4 °C. Homogenates were then clarified by centrifugation at 13,000× *g* for 10 min. Protein from the supernatants was quantified according to Bradford [53], and 100 µg of protein was analyzed by immunoblot using an antibody specific for the phosphorylated form of cPLA$_2\alpha$ at Ser505 (Cell Signaling, Danvers, MA, USA) [54,55]. The detection of immunoreactive bands was conducted by chemiluminescence (ECLTM, Amersham Biosciences, Little Chalfont, UK).

3. Results

The two group IIA sPLA$_2$s utilized in this study are from *Bothrops asper* venom and differ in a natural mutation at position 49. Asp49-sPLA$_2$, with catalytic activity, was named MT-III; Lys-49-sPLA$_2$, devoid of catalytic activity, was named MT-II [34–38]. When added to human monocytes, catalytically active MT-III, at concentrations not compromising cell viability (0.4 µg/mL), promoted an extensive loss of phospholipid-bound fatty acids, as measured by GC/MS (Figure 1).

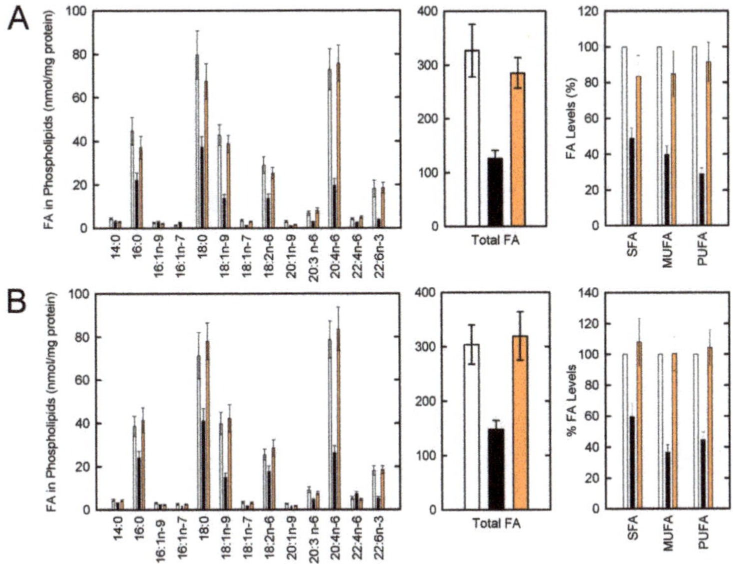

Figure 1. Phospholipid fatty acid content of human monocytes. The cells were either untreated (open bars) or treated with Asp-49 Ca^{2+}-dependent secreted phospholipase A$_2$ (sPLA$_2$) (MT-III) (black bars) or Lys-49 sPLA2 (MT-II) (orange bars) for 1 h (**A**) or 6 h (**B**).

Afterward, phospholipids were isolated and their fatty acid content was measured by GC/MS. The profile of fatty acids, the total phospholipid fatty acid amount, and distribution according to the number of double bonds is given. Fatty acids are designated by their number of carbon atoms and, after a colon, their number of double bonds. To differentiate isomers, the n−x (n minus x) nomenclature is used, where n is the number of carbons of a given fatty acid, and x is an integer which, subtracted from n, gives the position of the last double bond of the molecule. The data are expressed as mean values ± standard error of three independent determinations. Fatty acid (FA); saturated fatty acid (SFA); monounsaturated fatty acid (MUFA); polyunsaturated fatty acid (PUFA).

Despite the relatively high phospholipid hydrolysis rates detected in these experiments, cell viability always remained above 90%, as assessed by the MTT assay [56–58]. The action of MT-III was prominent on all kinds of fatty acids, including saturated, monounsaturated, and polyunsaturated. Importantly, no significant differences were observed between unstimulated control cells and MT-II-treated cells at any time tested. Since there was little difference in phospholipid hydrolysis between 1 h and 6 h, a 1 h time point was chosen to be employed in all subsequent experiments. The marked decrease in cellular AA (20:4n−6) levels after treating the monocytes with MT-III, as shown in Figure 1, was striking. Given the key role of AA in inflammatory reactions as a precursor of eicosanoids, we set out to characterize further the effect of MT-III on this particular fatty acid. Figure 2 shows the profile of AA-containing glycerophospholipid species of human monocytes, as measured by LC/MS.

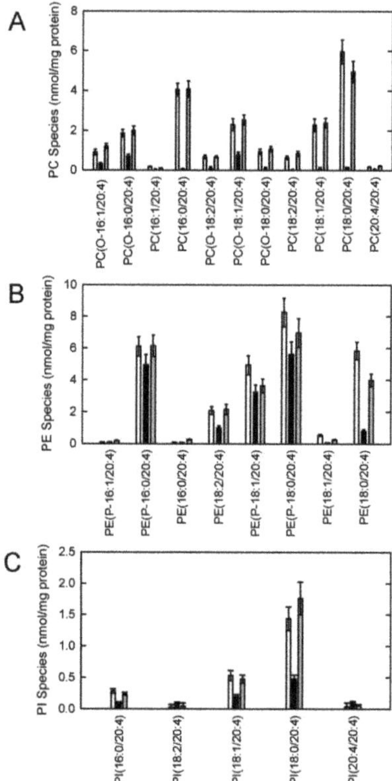

Figure 2. Arachidonic acid (AA)-containing phospholipid species of human monocytes. The cells were either untreated (open bars) or treated with MT-III (black bars) or MT-II (orange bars) for 1 h. Afterward, the distribution profile of AA between choline-containing phospholipids (PC) (**A**), ethanolamine-containing phospholipids (PE) (**B**), and phosphatidylinositol (PI) (**C**) was determined by LC/MS.

Fatty chains within the different phospholipid species are designated by their numbers of carbons and double bonds. A designation of O- before the first fatty chain indicates that the sn-1 position is ether linked, whereas a *p*- designation indicates a plasmalogen form (sn-1 vinyl ether linkage). The data are expressed as mean values ± standard error of three independent determinations.

Treatment of the monocytes with MT-III, but not MT-II, resulted in a marked decrease in the total cellular content of AA-containing PC and AA-containing PI (Figure 2). Note that some of the most abundant species such as the diacyl species PC(18:0/20:4) or PC(18:1/20:4) almost disappeared after treating the cells with MT-III. Regarding PE species, it was noted that the diacyl species also experienced dramatic decreases, similar to their PC and PI counterparts; however, the plasmalogen forms were much less affected (Figure 2). Although this could indicate that plasmalogen species may not be within the reach of MT-III, it seems likely that, immediately after hydrolysis, these species were rapidly replenished with AA via CoA-dependent transacylation reactions at the expense of diacyl PC species [21,25,26,52,59–61].

Figure 3 shows the LC/MS analysis of major phospholipid species not containing AA. In many cases, fragmentation of the *m*/*z* peaks detected in MS analyses yielded fragments corresponding to several species, which made it not possible to unequivocally assign structures to these *m*/*z* peaks. Thus, the data are given in abbreviated form, indicating phospholipid class and number of carbon atoms and double bonds of the two lateral chains together. In Table S1, the fatty acid combinations detected for each *m*/*z* are shown. For example, PI (34:1) represents a mix of PI (18:0/16:1) plus PI (16:0/18:1), and PI (36:2) represents a mix of PI (18:0/18:2) and PI (18:1/18:1).

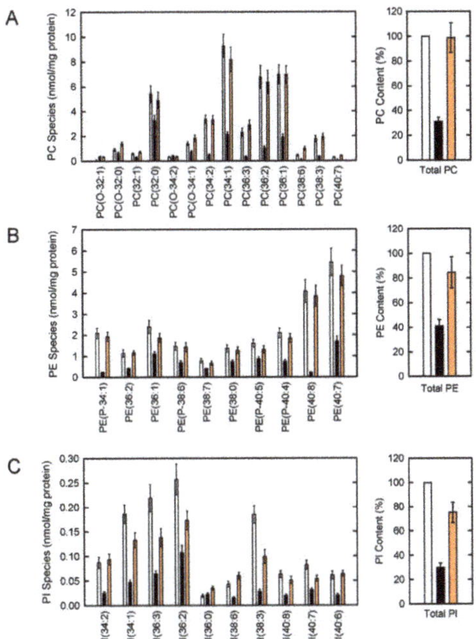

Figure 3. Phospholipid species not containing AA of human monocytes. The cells were either untreated (open bars) or treated with MT-III (black bars) or MT-II (orange bars) for 1 h. Afterward, the cellular content of PC (**A**), PE (**B**), and PI (**C**) molecular species was determined by LC/MS. Species are given in abbreviated form, indicating phospholipid class and number of carbon atoms and double bonds of the two lateral chains together. The data are expressed as mean values ± standard error of three independent determinations.

Marked reductions in all phospholipid classes were observed in the MT-III-treated monocytes. PE plasmalogen species not containing AA were hydrolyzed to much more extent that their AA-containing counterparts (*cf.* Figures 2B and 3B). Notably, some PI species such as PI (34:1), PI (36:3), PI (36:2) and PI (38:3) showed clear decreases also on stimulation with MT-II (Figure 4C). This finding represents the only positive effect observed in this study for MT-II with regard to lipid turnover. Its significance is unclear at this time but we speculate that it could constitute a ligand-like effect of MT-II related to activation of PI-dependent signaling (e.g., PI 3-kinase or intracellular Ca^{2+}-mediated pathways) via ligand binding to elements of the cell surface [31].

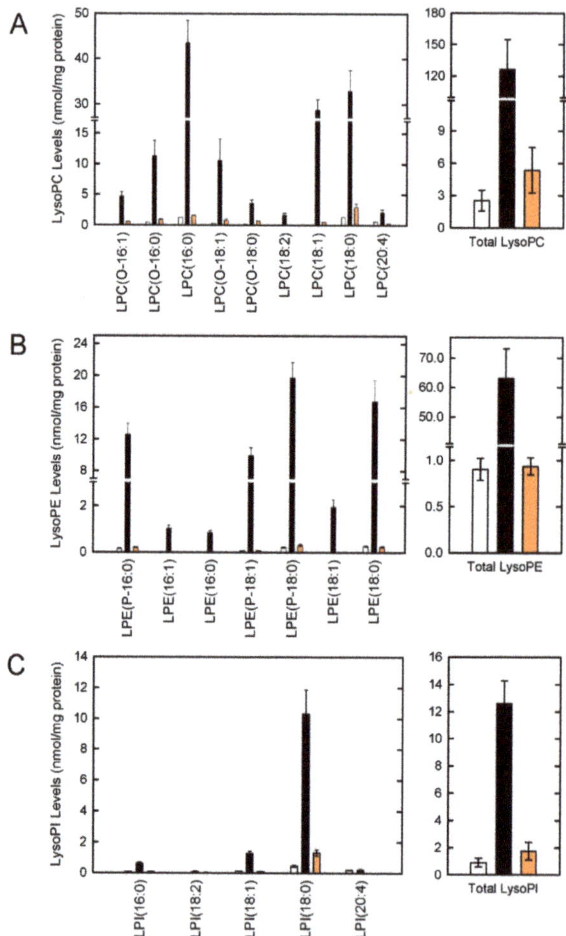

Figure 4. Lysophospholipid molecular species of human monocytes. The cells were either untreated (open bars) or treated with MT-III (black bars) or MT-II (orange bars) for 1 h. The cellular content of lysoPC (**A**), lysoPE (**B**), and lysoPI (**C**) molecular species was determined by LC/MS. Fatty chains within the different lysophospholipid species are designated by their numbers of carbons and double bonds. A designation of O- before the fatty chain indicates an sn-1 ether linkage, whereas a P- designation indicates an sn-1 vinyl ether linkage. The data are expressed as mean values ± standard error of three independent determinations.

To complete the overall lipidomic picture of changes occurring via phospholipid deacylation in the human monocytes, the measurement of lysophospholipid species was also carried out, and the results are shown in Figure 4. We detected a large lysophospholipid production after cellular treatment with MT-III, but not with MT-II, as measured by LC/MS. Consistent with the phospholipid composition of human monocytes [62], lysoPC species were detected in greater abundance, followed by lysoPE and lysoPI.

In the next series of experiments, we assessed the metabolic fate of the free fatty acids produced upon MT-III treatment. The data shown in Figure 5 indicate that a very significant incorporation of fatty acids did occur into neutral lipid classes, i.e., cholesterol esters and triacylglycerol. Consistent with the results of Figure 1, four of the five the major fatty acids lost from phospholipids—palmitic acid (16:0), stearic acid (18:0), oleic acid (18:1n−9) and linoleic acid (18:2n−6)—were also those that were incorporated in the highest proportion in neutral lipids. Very noticeably, however, the incorporation of AA into neutral lipids was negligible, suggesting other specific metabolic fates for this particular fatty acid.

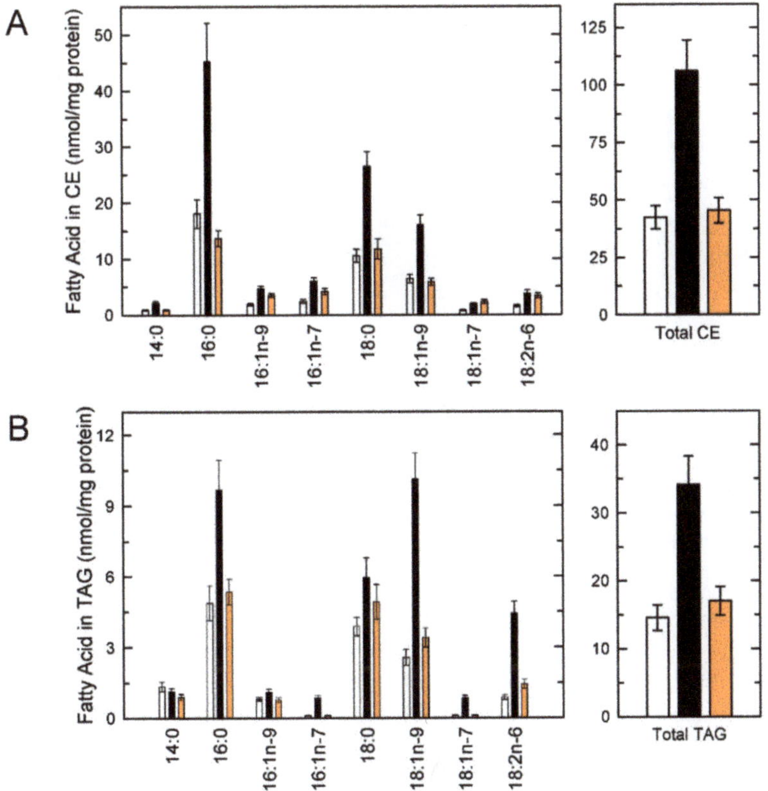

Figure 5. Fatty acid content of neutral lipids in human monocytes. The cells were either untreated (open bars) or treated with MT-III (black bars) or MT-II (orange bars) for 1 h. Afterward, cholesterol esters (CE) (**A**) and triacylglycerol (TAG) (**B**) fractions were isolated and their fatty acid content was measured by GC/MS. The profile of fatty acids, and the total CE or TAG amount are given. The data are expressed as mean values ± standard error of three independent determinations.

To investigate whether the increased synthesis of neutral lipids in the MT-III-treated cells resulted in the formation of lipid droplets, experiments were carried out to visualize these cytoplasmic organelles. Unlike human macrophages, resting human monocytes contain few lipid droplets [46]. Treatment of the cells with MT-III induced a very significant increase in the number of lipid droplets in comparison with control cells, incubated with culture medium alone (Figure 6). Mammalian cells contain five long-chain acyl-CoA synthetases, termed ACSL-1, -3, -4, -5, and -6, and human monocytes express all five of them [46]. The presence of triacsin C, a general inhibitor of long-chain acyl-CoA synthetases [63,64], quantitatively inhibited lipid droplet formation and fatty acid incorporation into neutral lipids (Figure 6). Collectively, these results suggest that lipid droplet production in MT-III-treated monocytes occurs as a consequence of increased availability of intracellular free fatty acids, which are converted into acyl-CoAs, acylated into neutral lipids, and stored in lipid droplets.

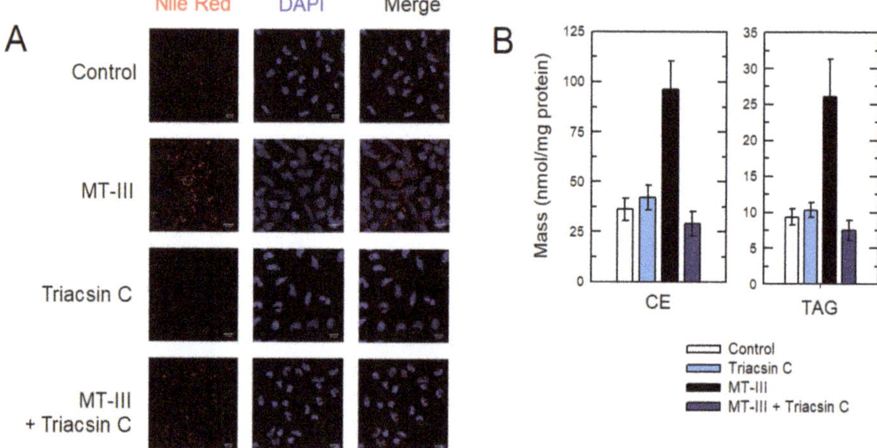

Figure 6. Lipid droplet formation in human monocytes. The cells, pretreated without or with triacsin, were exposed to MT-III as indicated. (**A**) After fixation, cells were stained with Nile Red to visualize lipid droplets (red; left panels) and DAPI to mark the nuclei (blue; central panels). Right panels show the merge. (**B**) Total fatty acid content in cholesterol esters (CE) and triacylglycerol (TAG) was analyzed in cells pretreated without (open bars and black bars) or with (light blue bars and dark blue bars) 3 μM triacsin C, and exposed to MT-III (black bars and dark blue bars) or left otherwise untreated (open bars and light blue bars). The data are expressed as mean values ± standard error of three independent determinations.

The absence of incorporation of AA into neutral lipids was unexpected, and prompted us to determine free fatty acid levels in the MT-III-treated cells. The data demonstrated the abundant presence of free AA as well as palmitic, stearic and oleic acids. Lower levels of the polyunsaturated fatty acids linoleic acid and docosahexaenoic acid were also detected (Figure 7). Importantly, only free AA levels were significantly blunted when the analyses were conducted with cells that had been pretreated with pyrrophenone prior to MT-III exposure. Pyrrophenone is a well-established inhibitor of intracellular cytosolic group IVA phospholipase $A_2\alpha$ (c$PLA_2\alpha$), and exhibits more than 1000-fold selectivity for the inhibition of c$PLA_2\alpha$ versus other types of PLA_2s, including the group IIA enzymes such as MT-III [65–68]. There was a tendency for other fatty acids—e.g., oleic acid and linoleic acid—to also decrease after pyrrophenone treatment; however, the differences failed to reach statistical significance. Collectively, these data suggest that MT-III activates c$PLA_2\alpha$; therefore, AA mobilization under these conditions would be a composite of the actions of both MT-III and c$PLA_2\alpha$. Cross-talk between c$PLA_2\alpha$ and sPLA_2 in AA release has often been described during the activation of innate immune cells [14–20,69–77]. Supporting our view that MT-III activates c$PLA_2\alpha$, MT-III-treated cells showed increased phosphorylation of cPLA$_2\alpha$ at Ser505, a hallmark of cPLA$_2\alpha$ activation [12,13] (Figure 7, inset).

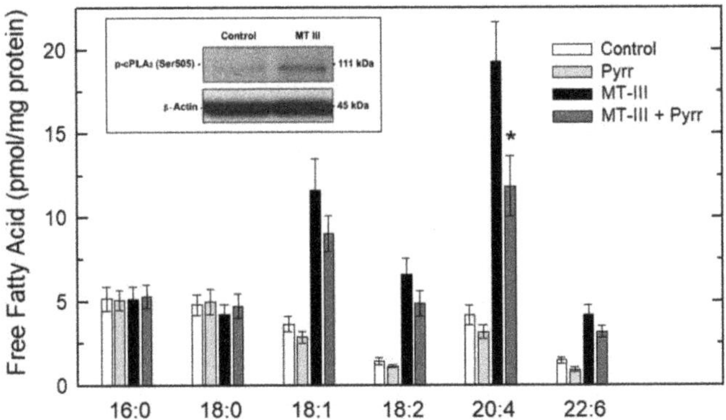

Figure 7. Free fatty acid release by human monocytes. The cells, treated without (open bars and black bars) or with (light gray bars and dark gray bars) 1 μM pyrrophenone (pyrr), were exposed to MT-III (black bars and dark gray bars) or left otherwise untreated (open bars and light gray bars). Free fatty acids were isolated and analyzed by LC/MS. The data are expressed as mean values ± standard error of three individual replicates. * $p < 0.05$, significantly different from cells not treated with pyrrophenone (Student's t-test). Inset: cell protein was separated by SDS-PAGE and the phosphorylation of cPLA$_2\alpha$ at Ser^{505} was analyzed by immunoblot using a specific antibody. β-actin was used as a load control.

Figure 8 shows that a substantial part of the AA lost from phospholipids by the action of MT-III was metabolized to a variety of eicosanoids, mostly from the cyclooxygenase pathway. Prostaglandin E$_2$ and thromboxane B$_2$ were the major metabolites produced by the monocytes, in agreement with previous estimates [78–80]. Lower amounts of products of the lipoxygenase and cytochrome P450 pathways were also detected upon MT-III stimulation (Figure 8).

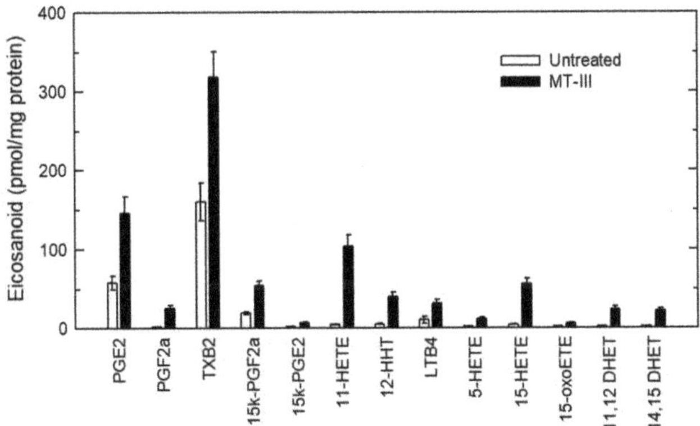

Figure 8. Eicosanoid production by stimulated monocytes. The cells were either untreated (open bars) or treated with MT-III (black bars) for 1h. Afterward, the eicosanoid content in the supernatants was analyzed by LC/MS. The data are expressed as mean values ± standard error of three individual replicates. Prostaglandin E$_2$ (PGE2); prostaglandin F$_{2\alpha}$ (PGF2a); thromboxane B$_2$ (TXB2); 15-ketoprostaglandin F$_{2\alpha}$ (15k-PGF2a); 15-ketoprostaglandin E$_2$ (15k-PGE2); 11-hydroxyeicosatetraenoic acid (11-HETE); 12- hydroxyheptadecatrienoic acid (12-HHT); leukotriene B$_4$ (LTB4); 5-hydroxyeicosatetraenoic acid (5-HETE); 15-hydroxyeicosatetraenoic acid (15-HETE); 15-oxoeicosatetraenoic acid (15-oxoETE); 11,12-dihydroxyeicosatrienoic acid (11,12-DHET); 14,15-dihydroxyeicosatrienoic acid (14,15-DHET).

4. Discussion

This work provides a mass spectrometry-based lipidomic analysis of the actions of group IIA secreted phospholipase A_2 on human peripheral blood monocytes. Although previous work has dealt with the interactions of this class of enzymes with cell surface structures and subsequent signaling [3,31], no reports, to the best of our knowledge, have characterized the global changes in the cellular lipidome as done in this study.

Group IIA $sPLA_2$ is synthesized and secreted by a variety of cells in response to inflammatory cytokines, and is found at large amounts in fluids from inflammatory exudates. Furthermore, group IIA $sPLA_2$ enzymes are also present in the venom of scorpions, wasps and, more abundantly, snake venoms, where they behave as relevant inducers of acute inflammation reactions [81]. However, it remains to be clarified how this secreted protein acts on the outer surface of the plasma membrane of mammalian cells to activate immune cells and trigger inflammation. To shed light on this issue, we first analyzed the complete lipidomic profile of metabolites produced by the action of the enzyme. Importantly, AA, a major player in inflammation reactions, is the fatty acid showing the largest decrease in monocyte membranes after MT III exposure, followed by palmitic acid, stearic acid and oleic acid. Other polyunsaturated fatty acids such as linoleic acid, dihomo-γ-linoleic acid and docosahexaenoic acid are also released in smaller quantities and do not contribute significantly to the pool of bioactive oxygenated metabolites produced under these conditions.

We observed that several PC species disappear almost entirely from the membrane. Intriguingly, it has been suggested that membranes enriched in PC may behave as poor substrates for $sPLA_2$-IIA [32]. Several studies also indicated that the phospholipid preference may be partially explained by the number of positively charged amino acids and the lack of tryptophan in the interfacial binding site. These residues may constitute key structural determinants that permit binding and hydrolysis to whole membranes [3]. Other $sPLA_2$ family members such as the -IB and -V proteins possess tryptophan residues on their putative interfacial binding surfaces; therefore, they show an enhanced capacity to bind to PC-rich vesicles [82]. Our results, using a pathophysiologically relevant setting, raise the concept that in addition to sequence differences, the molecular composition of the membrane to which the $sPLA_2$ binds—including protein components—may influence the subsequent hydrolytic steps. Our results may also provide an appropriate experimental frame to relate the catalytic activity of various $sPLA_2$ on different classes of phospholipids with their sequences for future studies.

Lipid droplet biogenesis has been demonstrated to be associated with signaling events triggered by inflammation and metabolic stress [83–87]. We show here that the catalytic activity of group IIA $sPLA_2$ is required for lipid droplet formation to occur, it probably being the only factor involved, since inactive MT-II does not reproduce the effect. Our results suggest that the extensive hydrolysis of membrane phospholipids promoted by MT-III generates free fatty acids that are channeled to neutral lipids and the formation of cytoplasmic lipid droplets. In support of this view, neutral lipid formation is strongly blunted by the acyl-CoA synthetase inhibitor triacsin C, indicating that the activation of the carboxyl group of a free fatty acid is a required event. In turn, this implicates the participation of CoA-dependent acyltransferase reactions utilizing free fatty acids, not the direct transfer of fatty acids between lipids via CoA-independent transacylation reactions.

A striking feature of the present work is that, of all major fatty acids released by MT-III, AA was excluded from incorporating into neutral lipids. We have recently shown that human monocytes exposed to micromolar amounts of AA do incorporate the fatty acid into neutral lipids, implying that this pathway is fully functional in these cells [88]. This is an interesting concept because recent work has suggested a link between lipid droplets and AA metabolism in mast cells and neutrophils [89,90]. These studies showed that AA recently incorporated into neutral lipids of lipid droplets may be mobilized under activation conditions, thus providing an alternative source of free fatty acid. Therefore, our finding that AA does not incorporate into neutral lipids during exposure of the monocytes to $sPLA_2$ clearly suggests that the pathway described in mast cells and neutrophils is not operative, and the fatty acid is used to fulfill other important cellular functions. The most immediate is the direct channeling of AA to the production of proinflammatory eicosanoids that help establish a strong inflammatory reaction. Consistent with this view, we have detected abundant production of proinflammatory mediators, especially those arising from the cyclooxygenase pathway. We did not detect significant formation of putatively anti-inflammatory eicosanoids such as lipoxins or $n-3$ fatty acid derivatives.

In previous work we showed that lipid droplet formation by cells exposed to various stimulants, including MT-III, is blunted by the cPLA$_2\alpha$ selective inhibitor pyrrophenone [46,58,91]. In this work, we show that pyrrophenone significantly inhibits the accumulation of free AA in the supernatants of MT-III-treated cells. Moreover, MT-III-treated cells demonstrate increased phosphorylation of cPLA$_2\alpha$ at Ser505. Collectively, the data are suggestive of the possibility that crosstalk exists between cPLA$_2\alpha$ and MT-IIII. As a matter of fact, evidence has accumulated to suggest that the high AA specificity of cPLA$_2\alpha$ and the lack of fatty acid selectivity in sPLA$_2$s can be combined to achieve specific cellular responses [1,3,14]. Since MT-III causes extensive phospholipid hydrolysis, we speculate that the ensuing membrane disruption may favor Ca^{2+} fluxes that activate intracellular enzymes such as cPLA$_2\alpha$. A scenario such as this has even been proposed for catalytically inactive sPLA$_2$s, acting via receptor-like mechanisms [3,31]. However, in our studies, inactive MT-II does not induce phospholipid hydrolysis; thus, cPLA$_2\alpha$ has no active participation in the signaling mediated by this sPLA$_2$.

Collectively, our results highlight important actions of catalytically active group IIA sPLA$_2$ on the surface of innate immune cells. These actions may trigger different cellular responses, depending on the lipid mediator released (Scheme 1). Clearly, further research will be needed to define the role of MT-III in supplying AA for eicosanoid biosynthesis, the mechanism of crosstalk between MT-III and intracellular cPLA$_2\alpha$, and a possible role for the latter in regulating lipid droplet biogenesis, as suggested elsewhere [87,91,92].

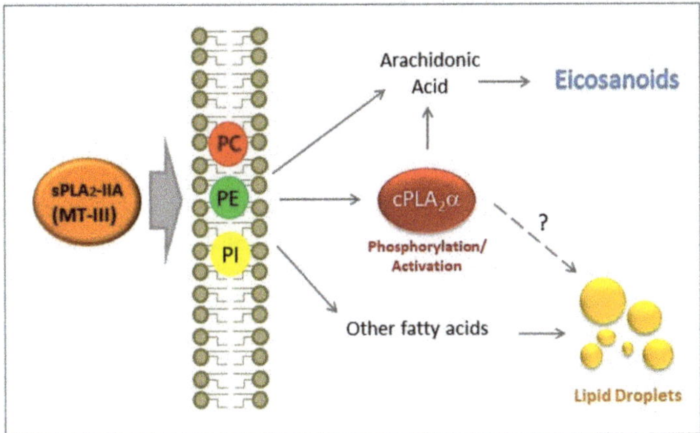

Scheme 1. Lipid mediators and cellular responses triggered by catalytically active group IIA sPLA$_2$ (MT-III) on human monocytes.

5. Conclusions

A complex network of chemical mediators including cytokines or eicosanoids characterizes the inflammation process triggered in many diseases or envenomations, involving the hydrolytic action of sPLA$_2$s. Regardless of their catalytic activity, it has been demonstrated that PLA$_2$s isolated from snake venoms (myotoxins) induce a marked local inflammatory reaction. This is characterized by an early increase in plasma extravasation, edema and a conspicuous infiltration of leukocytes followed by hyperalgesia. All these processes are the result of local and/or systemic concerted action of cytokines such as interleukin-1β, interleukin-6, tumor necrosis factor-α or interferon-γ. Despite this, lipid profiling of the changes induced by these sPLA$_2$s on circulating blood cells had not been documented.

This study provides an in depth lipidomic profiling of the monocyte response to the direct action of a group IIA $sPLA_2$, i.e., MT-III. The data reveal significant connections among lipid droplets biogenesis, cellular signaling, and biochemical pathways that contribute to initiating the inflammatory response.

Supplementary Materials: The following are available online at http://www.mdpi.com/2218-273X/10/6/891/s1, Table S1: Fatty acid compositions of phospholipid species not containing AA in human monocytes.

Author Contributions: Conceptualization, J.P.R.; formal analysis, J.P.R., E.L., C.G. and J.B.; funding acquisition, M.A.B. and J.B.; investigation, J.P.R., E.L. and C.G.; methodology, J.P.R., E.L., C.G., B.L., J.M.G. and C.T.; project administration, M.A.B. and J.B.; resources, B.L., J.M.G., C.T., M.A.B. and J.B.; supervision, M.A.B. and J.B.; writing—original draft, J.P.R., E.L., C.G. and J.B.; writing—review & editing, J.P.R., C.T., M.A.B. and J.B. All authors have read and agreed to the published version of the manuscript.

Funding: This research was funded by the Spanish Ministry of Economy, Industry, and Competitiveness, grant number SAF2016-80883-R. CIBERDEM is an initiative of Instituto de Salud Carlos III. E.L. was the recipient of a BEPE/PhD fellowship from the Fundação de Amparo a Pesquisa do Estado de São Paulo (FAPESP), Brazil (grant 12/06730-8).

Acknowledgments: We thank Montse Duque for excellent technical assistance.

Conflicts of Interest: The authors declare no conflict of interest. The funders had no role in the design of the study; in the collection, analyses, or interpretation of data; in the writing of the manuscript, or in the decision to publish the results.

Abbreviations

AA	arachidonic acid
CE	cholesterol esters
$cPLA_2\alpha$	group IVA cytosolic phospholipase $A_2\alpha$
GC/MS	gas chromatography coupled to mass spectrometry
LC/MS	liquid chromatography coupled to mass spectrometry
PC	choline-containing phospholipids
PE	ethanolamine-containing phospholipids
PI	phosphatidylinositol
$sPLA_2$	secreted phospholipase A_2
TAG	triacylglycerol

References

1. Dennis, E.A.; Cao, J.; Hsu, Y.H.; Magrioti, V.; Kokotos, G. Phospholipase A_2 enzymes: Physical structure, biological function, disease implication, chemical inhibition, and therapeutic intervention. *Chem. Rev.* **2011**, *111*, 6130–6185. [CrossRef] [PubMed]
2. Murakami, M. Novel functions of phospholipase A_2s: Overview. *Biochim. Biophys. Acta* **2019**, *1864*, 763–765. [CrossRef] [PubMed]
3. Lambeau, G.; Gelb, M.H. Biochemistry and physiology of mammalian secreted phospholipases A_2. *Annu. Rev. Biochem.* **2008**, *77*, 495–520. [CrossRef] [PubMed]
4. Triggiani, M.; Granata, F.; Giannattasio, G.; Marone, G. Secretory phospholipases A_2 in inflammatory and allergic diseases: Not just enzymes. *J. Allergy Clin. Immunol.* **2005**, *116*, 1000–1006. [CrossRef] [PubMed]
5. Rosenson, R.S.; Gelb, M.H. Secretory Phospholipase A_2: A multifaceted family of proatherogenic enzymes. *Curr. Cardiol. Rep.* **2009**, *11*, 445–451. [CrossRef] [PubMed]
6. Rubio, J.M.; Rodríguez, J.P.; Gil-de-Gómez, L.; Guijas, C.; Balboa, M.A.; Balsinde, J. Group V secreted phospholipase A_2 is upregulated by IL-4 in human macrophages and mediates phagocytosis via hydrolysis of ethanolamine phospholipids. *J. Immunol.* **2015**, *194*, 3327–3339. [CrossRef] [PubMed]
7. Balestrieri, B.; Maekawa, A.; Xing, W.; Gelb, M.H.; Katz, H.R.; Arm, J.P. Group V secretory phospholipase A_2 modulates phagosome maturation and regulates the innate immune response against *Candida albicans*. *J. Immunol.* **2009**, *182*, 4891–4898. [CrossRef] [PubMed]
8. Boilard, E.; Lai, Y.; Larabee, K.; Balestrieri, B.; Ghomashchi, F.; Fujioka, D.; Gobezie, R.; Coblyn, J.S.; Weinblatt, M.E.; Massarotti, E.M.; et al. A novel anti-inflammatory role for secretory phospholipase A_2 in immune complex-mediated arthritis. *EMBO Mol. Med.* **2010**, *2*, 172–187. [CrossRef] [PubMed]

9. Dennis, E.A.; Norris, P.C. Eicosanoid storm in infection and inflammation. *Nat. Rev. Immunol.* **2015**, *15*, 511–523. [CrossRef] [PubMed]
10. Astudillo, A.M.; Balgoma, D.; Balboa, M.A.; Balsinde, J. Dynamics of arachidonic acid mobilization by inflammatory cells. *Biochim. Biophys. Acta* **2012**, *1821*, 249–256. [CrossRef] [PubMed]
11. Astudillo, A.M.; Balboa, M.A.; Balsinde, J. Selectivity of phospholipid hydrolysis by phospholipase A_2 enzymes in activated cells leading to polyunsaturated fatty acid mobilization. *Biochim. Biophys. Acta* **2019**, *1864*, 772–783. [CrossRef] [PubMed]
12. Pérez-Chacón, G.; Astudillo, A.M.; Balgoma, D.; Balboa, M.A.; Balsinde, J. Control of free arachidonic acid levels by phospholipases A_2 and lysophospholipid acyltransferases. *Biochim. Biophys. Acta* **2009**, *1791*, 1103–1113. [CrossRef] [PubMed]
13. Leslie, C.C. Cytosolic phospholipase A_2: Physiological function and role in disease. *J. Lipid Res.* **2015**, *56*, 1386–1402. [CrossRef] [PubMed]
14. Balsinde, J.; Winstead, M.V.; Dennis, E.A. Phospholipase A_2 regulation of arachidonic acid mobilization. *FEBS Lett.* **2002**, *531*, 2–6. [CrossRef]
15. Han, W.K.; Sapirstein, A.; Huang, C.C.; Alessandrini, A.; Bonventre, J.V. Cross-talk between cytosolic phospholipase $A_2\alpha$ ($cPLA_2\alpha$) and secretory phospholipase A_2 ($sPLA_2$) in hydrogen peroxide-induced arachidonic acid release in murine mesangial cells: $sPLA_2$ regulates $cPLA_2\alpha$ activity that is responsible for the arachidonic acid release. *J. Biol. Chem.* **2003**, *278*, 24153–24163. [PubMed]
16. Marshall, J.; Krump, E.; Lindsay, T.; Downey, G.; Ford, D.A.; Zhu, P.; Walker, P.; Rubin, B. Involvement of cytosolic phospholipase A_2 and secretory phospholipase A_2 in arachidonic acid release from human neutrophils. *J. Immunol.* **2000**, *164*, 2084–2091. [CrossRef] [PubMed]
17. Degousee, N.; Ghomashchi, F.; Stefanski, E.; Singer, A.G.; Smart, B.P.; Borregaard, N.; Reithmeier, R.; Lindsay, T.F.; Lichtenberger, C.; Reinisch, W.; et al. Groups IV, V, and X phospholipases A_2s in human neutrophils: Role in eicosanoid production and gram-negative bacterial phospholipid hydrolysis. *J. Biol. Chem.* **2002**, *277*, 5061–5073. [CrossRef] [PubMed]
18. Ruipérez, V.; Casas, J.; Balboa, M.A.; Balsinde, J. Group V phospholipase A_2-derived lysophosphatidyl-choline mediates cyclooxygenase-2 induction in lipopolysaccharide-stimulated macrophages. *J. Immunol.* **2007**, *179*, 631–638. [CrossRef] [PubMed]
19. Ruipérez, V.; Astudillo, M.A.; Balboa, M.A.; Balsinde, J. Coordinate regulation of TLR-mediated arachidonic acid mobilization in macrophages by group IVA and group V phospholipase A_2s. *J. Immunol.* **2009**, *182*, 3877–3883. [CrossRef] [PubMed]
20. Balboa, M.A.; Pérez, R.; Balsinde, J. Amplification mechanisms of inflammation: Paracrine stimulation of arachidonic acid mobilization by secreted phospholipase A2 is regulated by cytosolic phospholipase A_2-derived hydroperoxyeicosatetraenoic acid. *J. Immunol.* **2003**, *171*, 989–994. [CrossRef] [PubMed]
21. Balgoma, D.; Astudillo, A.M.; Pérez-Chacón, G.; Montero, O.; Balboa, M.A.; Balsinde, J. Markers of monocyte activation revealed by lipidomic profiling of arachidonic acid-containing phospholipids. *J. Immunol.* **2010**, *184*, 3857–3865. [CrossRef] [PubMed]
22. Astudillo, A.M.; Pérez-Chacón, G.; Meana, C.; Balgoma, D.; Pol, A.; del Pozo, M.A.; Balboa, M.A.; Balsinde, J. Altered arachidonate distribution in macrophages from caveolin-1 null mice leading to reduced eicosanoid synthesis. *J. Biol. Chem.* **2011**, *286*, 35299–35307. [CrossRef] [PubMed]
23. Valdearcos, M.; Esquinas, E.; Meana, C.; Gil-de-Gómez, L.; Guijas, C.; Balsinde, J.; Balboa, M.A. Subcellular localization and role of lipin-1 in human macrophages. *J. Immunol.* **2011**, *186*, 6004–6013. [CrossRef] [PubMed]
24. Gil-de-Gómez, L.; Astudillo, A.M.; Meana, C.; Rubio, J.M.; Guijas, C.; Balboa, M.A.; Balsinde, J. A phosphatidylinositol species acutely generated by activated macrophages regulates innate immune responses. *J. Immunol.* **2013**, *190*, 5169–5177. [CrossRef] [PubMed]
25. Gil-de-Gómez, L.; Astudillo, A.M.; Guijas, C.; Magrioti, V.; Kokotos, G.; Balboa, M.A.; Balsinde, J. Cytosolic group IVA and calcium-independent group VIA phospholipase A_2s act on distinct phospholipid pools in zymosan-stimulated mouse peritoneal macrophages. *J. Immunol.* **2014**, *192*, 752–762. [CrossRef]
26. Gil-de-Gómez, L.; Astudillo, A.M.; Lebrero, P.; Balboa, M.A.; Balsinde, J. Essential role for ethanolamine plasmalogen hydrolysis in bacterial lipopolysaccharide priming of macrophages for enhanced arachidonic acid release. *Front. Immunol.* **2017**, *8*, 1251. [CrossRef] [PubMed]

27. Rubio, J.M.; Astudillo, A.M.; Casas, J.; Balboa, M.A.; Balsinde, J. Regulation of phagocytosis in macrophages by membrane ethanolamine plasmalogens. *Front. Immunol.* **2018**, *9*, 1723. [CrossRef] [PubMed]
28. Monge, P.; Garrido, A.; Rubio, J.M.; Magrioti, V.; Kokotos, G.; Balboa, M.A.; Balsinde, J. The contribution of cytosolic group IVA and calcium-independent group VIA phospholipase A_2s to adrenic acid mobilization in murine macrophages. *Biomolecules* **2020**, *10*, 542. [CrossRef]
29. Dore, E.; Boilard, E. Roles of secreted phospholipase A_2 group IIA in inflammation and host defense. *Biochim. Biophys. Acta* **2019**, *1864*, 789–802. [CrossRef] [PubMed]
30. Mora-Obando, D.; Fernández, J.; Montecucco, C.; Gutiérrez, J.M.; Lomonte, B. Synergism between basic Asp49 and Lys49 phospholipase A_2 myotoxins of viperid snake venom in vitro and in vivo. *PLoS ONE* **2014**, *9*, e109846. [CrossRef] [PubMed]
31. Birts, C.N.; Barton, C.H.; Wilton, D.C. Catalytic and non-catalytic functions of human IIA phospholipase A_2. *Trends Biochem. Sci.* **2010**, *35*, 28–35. [CrossRef]
32. Koduri, R.S.; Baker, S.F.; Snitko, Y.; Han, S.K.; Cho, W.; Wilton, D.C.; Gelb, M.H. Action of human group IIA secreted phospholipase A_2 on cell membranes. Vesicle but not heparinoid binding determines rate of fatty acid release by exogenously added enzyme. *J. Biol. Chem.* **1999**, *273*, 32142–32153. [CrossRef] [PubMed]
33. Valentin, E.; Lambeau, G. Increasing molecular diversity of secreted phospholipases A_2 and their receptors and binding proteins. *Biochim. Biophys. Acta* **2000**, *1488*, 59–70. [CrossRef]
34. Lomonte, B.; Angulo, Y.; Calderón, L. An overview of lysine-49 phospholipase A_2 myotoxins from crotalid snake venoms and their structural determinants of myotoxic action. *Toxicon* **2003**, *42*, 885–901. [CrossRef] [PubMed]
35. Gutiérrez, J.M.; Lomonte, B. Phospholipases A_2: Unveiling the secrets of a functionally versatile group of snake venom toxins. *Toxicon* **2013**, *62*, 27–39. [CrossRef] [PubMed]
36. Mora-Obando, D.; Díaz, C.; Angulo, Y.; Gutiérrez, J.M.; Lomonte, B. Role of enzymatic activity in muscle damage and cytotoxicity induced by Bothrops asper Asp49 phospholipase A_2 myotoxins: Are there additional effector mechanisms involved? *Peer J.* **2014**, *2*, e569. [CrossRef] [PubMed]
37. Kaiser, I.I.; Gutiérrez, J.M.; Plummer, D.; Aird, S.D.; Odell, G.V. The amino acid sequence of a myotoxic phospholipase from the venom of Bothrops asper. *Arch. Biochem. Biophys.* **1990**, *278*, 319–325. [CrossRef]
38. Fernández, J.; Caccin, P.; Koster, G.; Lomonte, B.; Gutiérrez, J.M.; Montecucco, C.; Postle, A.D. Muscle phospholipid hydrolysis by Bothrops asper Asp49 and Lys49 phospholipase A_2 myotoxins – distinct mechanisms of action. *FEBS J.* **2013**, *280*, 3878–3886. [CrossRef] [PubMed]
39. Takayama, K.; Mitchell, D.H.; Din, Z.Z.; Mukerjee, P.; Li, C.; Coleman, D.L. Monomeric Re lipopolysaccharide from Escherichia coli is more active than the aggregated form in the Limulus amebocyte lysate assay and in inducing Egr-1 mRNA in murine peritoneal macrophages. *J. Biol. Chem.* **1994**, *269*, 2241–2244. [PubMed]
40. Pérez-Chacón, G.; Astudillo, A.M.; Ruipérez, V.; Balboa, M.A.; Balsinde, J. Signaling role for lysophospholipid acyltransferase 3 in receptor-regulated arachidonic acid reacylation reactions in human monocytes. *J. Immunol.* **2010**, *184*, 1071–1078. [CrossRef] [PubMed]
41. Casas, J.; Meana, C.; Esquinas, E.; Valdearcos, M.; Pindado, J.; Balsinde, J.; Balboa, M.A. Requirement of JNK-mediated phosphorylation for translocation of group IVA phospholipase A_2 to phagosomes in human macrophages. *J. Immunol.* **2009**, *183*, 2767–2774. [CrossRef] [PubMed]
42. Bligh, E.G.; Dyer, W.J. A rapid method of total lipid extraction and purification. *Can. J. Biochem. Physiol.* **1959**, *37*, 911–917. [CrossRef] [PubMed]
43. Diez, E.; Balsinde, J.; Aracil, M.; Schüller, A. Ethanol induces release of arachidonic acid but not synthesis of eicosanoids in mouse peritoneal macrophages. *Biochim. Biophys. Acta* **1987**, *921*, 82–89. [CrossRef]
44. Fine, J.B.; Sprecher, H. Unidimensional thin-layer chromatography of phospholipids on boric acid-impregnated plates. *J. Lipid Res.* **1982**, *23*, 660–663. [PubMed]
45. Astudillo, A.M.; Pérez-Chacón, G.; Balgoma, D.; Gil-de-Gómez, L.; Ruipérez, V.; Guijas, C.; Balboa, M.A.; Balsinde, J. Influence of cellular arachidonic acid levels on phospholipid remodeling and CoA-independent transacylase activity in human monocytes and U937 cells. *Biochim. Biophys. Acta* **2011**, *1811*, 97–103. [CrossRef] [PubMed]
46. Guijas, C.; Pérez-Chacón, G.; Astudillo, A.M.; Rubio, J.M.; Gil-de-Gómez, L.; Balboa, M.A.; Balsinde, J. Simultaneous activation of p38 and JNK by arachidonic acid stimulates the cytosolic phospholipase A_2-dependent synthesis of lipid droplets in human monocytes. *J. Lipid Res.* **2012**, *53*, 2343–2354. [CrossRef] [PubMed]

47. Guijas, C.; Meana, C.; Astudillo, A.M.; Balboa, M.A.; Balsinde, J. Foamy monocytes are enriched in cis-7-hexadecenoic fatty acid (16:1n-9), a possible biomarker for early detection of cardiovascular disease. *Cell Chem. Biol.* **2016**, *23*, 689–699. [CrossRef] [PubMed]
48. Rodríguez, J.P.; Guijas, C.; Astudillo, A.M.; Rubio, J.M.; Balboa, M.A.; Balsinde, J. Sequestration of 9-hydroxystearic acid in FAHFA (fatty acid esters of hydroxy fatty acids) as a protective mechanism for colon carcinoma cells to avoid apoptotic cell death. *Cancers* **2019**, *11*, 524. [CrossRef] [PubMed]
49. Astudillo, A.M.; Meana, C.; Guijas, C.; Pereira, L.; Lebrero, R.; Balboa, M.A.; Balsinde, J. Occurrence and biological activity of palmitoleic acid isomers in phagocytic cells. *J. Lipid Res.* **2018**, *59*, 237–249. [CrossRef] [PubMed]
50. Balgoma, D.; Montero, O.; Balboa, M.A.; Balsinde, J. Calcium-independent phospholipase A_2-mediated formation of 1,2-diarachidonoyl-glycerophosphoinositol in monocytes. *FEBS J.* **2008**, *275*, 6180–6191. [CrossRef] [PubMed]
51. Guijas, C.; Astudillo, A.M.; Gil-de-Gómez, L.; Rubio, J.M.; Balboa, M.A.; Balsinde, J. Phospholipid sources for adrenic acid mobilization in RAW 264.7 macrophages: Comparison with arachidonic acid. *Biochim. Biophys. Acta* **2012**, *1821*, 1386–1393. [CrossRef] [PubMed]
52. Lebrero, P.; Astudillo, A.M.; Rubio, J.M.; Fernández-Caballero, J.; Kokotos, G.; Balboa, M.A.; Balsinde, J. Cellular plasmalogen content does not influence arachidonic acid levels or distribution in macrophages: A role for cytosolic phospholipase $A_2\gamma$ in phospholipid remodeling. *Cells* **2019**, *8*, 799. [CrossRef] [PubMed]
53. Bradford, M.M. A rapid and sensitive method for the quantitation of microgram quantities of protein utilizing the principle of protein-dye binding. *Anal. Biochem.* **1976**, *72*, 248–254. [CrossRef]
54. Pindado, J.; Balsinde, J.; Balboa, M.A. TLR3-dependent induction of nitric oxide synthase in RAW 264.7 macrophage-like cells via a cytosolic phospholipase A_2/cyclooxygenase-2 pathway. *J. Immunol.* **2007**, *179*, 4821–4828. [CrossRef] [PubMed]
55. Valdearcos, M.; Esquinas, E.; Meana, C.; Peña, L.; Gil-de-Gómez, L.; Balsinde, J.; Balboa, M.A. Lipin-2 reduces proinflammatory signaling induced by saturated fatty acids in macrophages. *J. Biol. Chem.* **2012**, *287*, 10894–10904. [CrossRef] [PubMed]
56. Leiguez, E.; Zuliani, J.P.; Cianciarullo, A.M.; Fernandes, C.M.; Gutiérrez, J.M.; Teixeira, C. A group IIA-secreted phospholipase A_2 from snake venom induces lipid body formation in macrophages: The roles of intracellular phospholipases A_2 and distinct signaling pathways. *J. Leukoc. Biol.* **2011**, *90*, 155–166. [CrossRef] [PubMed]
57. Zuliani, J.P.; Gutiérrez, J.M.; Casais e Silva, L.L.; Sampaio, S.C.; Lomonte, B.; Teixeira, C.F.P. Activation of cellular functions in macrophages by venom secretory Asp-49 and Lys-49 phospholipases A_2. *Toxicon* **2005**, *46*, 523–532. [CrossRef] [PubMed]
58. Leiguez, E.; Giannotti, K.C.; Moreira, V.; Matsubara, M.H.; Gutiérrez, J.M.; Lomonte, B.; Rodríguez, J.P.; Balsinde, J.; Teixeira, C. Critical role of TLR2 and MyD88 for functional response of macrophages to a group IIA-secreted phospholipase A_2 from snake venom. *PLoS ONE* **2014**, *9*, e93741. [CrossRef] [PubMed]
59. Rouzer, C.A.; Ivanova, P.T.; Byrne, M.O.; Milne, S.B.; Brown, H.A.; Marnett, L.J. Lipid profiling reveals glycerophospholipid remodeling in zymosan-stimulated macrophages. *Biochemistry* **2007**, *46*, 6026–6042. [CrossRef] [PubMed]
60. Chilton, F.H.; Fonteh, A.N.; Surette, M.E.; Triggiani, M.; Winkler, J.D. Control of arachidonate levels within inflammatory cells. *Biochim. Biophys. Acta* **1996**, *1299*, 1–15. [CrossRef]
61. Yamashita, A.; Hayashi, Y.; Matsumoto, N.; Nemoto-Sasaki, Y.; Koizumi, T.; Inagaki, Y.; Oka, S.; Tanikawa, T.; Sugiura, T. Coenzyme-A-independent transacylation system; possible involvement of phospholipase A_2 in transacylation. *Biology* **2017**, *6*, 23. [CrossRef] [PubMed]
62. Kennett, F.F.; Schenkein, H.A.; Ellis, T.M.; Rutherford, R.B. Phospholipid composition of human monocytes and alterations occurring due to culture and stimulation by C3b. *Biochim. Biophys. Acta* **1984**, *804*, 301–307. [CrossRef]
63. Hartman, E.J.; Omura, S.; Laposata, M. Triacsin C: A differential inhibitor of arachidonoyl-CoA synthetase and nonspecific long chain acyl-CoA synthetase. *Prostaglandins* **1989**, *37*, 655–671. [CrossRef]
64. Coleman, R.A.; Lewin, T.M.; Van Horn, C.G.; González-Baró, M.R. Do long-chain acyl-CoA synthetases regulate fatty acid entry into synthetic versus degradative pathways? *J. Nutr.* **2002**, *132*, 2123–2126. [CrossRef] [PubMed]

65. Ono, T.; Yamada, K.; Chikazawa, Y.; Ueno, M.; Nakamoto, S.; Okuno, T.; Seno, K. Characterization of a novel inhibitor of cytosolic phospholipase $A_2\alpha$, pyrrophenone. *Biochem. J.* **2002**, *363*, 727–735. [CrossRef] [PubMed]
66. Balboa, M.A.; Balsinde, J. Involvement of calcium-independent phospholipase A_2 in hydrogen peroxide-induced accumulation of free fatty acids in human U937 cells. *J. Biol. Chem.* **2002**, *277*, 40384–40389. [CrossRef] [PubMed]
67. Ghomashchi, F.; Stewart, A.; Hefner, Y.; Ramanadham, S.; Turk, J.; Leslie, C.C.; Gelb, M.H. A pyrrolidine-based specific inhibitor of cytosolic phospholipase $A_2\alpha$ blocks arachidonic acid release in a variety of mammalian cells. *Biochim. Biophys. Acta* **2001**, *1513*, 160–166. [CrossRef]
68. Balboa, M.A.; Sáez, Y.; Balsinde, J. Calcium-independent phospholipase A_2 is required for lysozyme secretion in U937 promonocytes. *J. Immunol.* **2003**, *170*, 5276–5280. [CrossRef] [PubMed]
69. Balestrieri, B.; Hsu, V.W.; Gilbert, H.; Leslie, C.C.; Han, W.K.; Bonventre, J.V.; Arm, J.P. Group V secretory phospholipase A_2 translocates to the phagosome after zymosan stimulation of mouse peritoneal macrophages and regulates phagocytosis. *J. Biol. Chem.* **2006**, *281*, 6691–6698. [CrossRef] [PubMed]
70. Casas, J.; Gijón, M.A.; Vigo, A.G.; Crespo, M.S.; Balsinde, J.; Balboa, M.A. Phosphatidylinositol 4,5-bisphosphate anchors cytosolic group IVA phospholipase A_2 to perinuclear membranes and decreases its calcium requirement for translocation in live cells. *Mol. Biol. Cell* **2006**, *17*, 155–162. [CrossRef] [PubMed]
71. Satake, Y.; Diaz, B.L.; Balestrieri, B.; Lam, B.K.; Kanaoka, Y.; Grusby, M.J.; Arm, J.P. Role of group V phospholipase A_2 in zymosan-induced eicosanoid generation and vascular permeability revealed by targeted gene disruption. *J. Biol. Chem.* **2004**, *279*, 16488–16494. [CrossRef] [PubMed]
72. Balboa, M.A.; Shirai, Y.; Gaietta, G.; Ellisman, M.H.; Balsinde, J.; Dennis, E.A. Localization of group V phospholipase A_2 in caveolin-enriched granules in activated P388D1 macrophage-like cells. *J. Biol. Chem.* **2003**, *278*, 48059–48065. [CrossRef] [PubMed]
73. Balsinde, J.; Shinohara, H.; Lefkowitz, L.J.; Johnson, C.A.; Balboa, M.A.; Dennis, E.A. Group V phospholipase A_2-dependent induction of cyclooxygenase-2 in macrophages. *J. Biol. Chem.* **1999**, *274*, 25967–25970. [CrossRef] [PubMed]
74. Balsinde, J.; Balboa, M.A.; Yedgar, S.; Dennis, E.A. Group V phospholipase A_2-mediated oleic acid mobilization in lipopolysaccharide-stimulated $P388D_1$ macrophages. *J. Biol. Chem.* **2000**, *275*, 4783–4786. [CrossRef] [PubMed]
75. Balboa, M.A.; Balsinde, J.; Dillon, D.A.; Carman, G.M.; Dennis, E.A. Proinflammatory macrophage-activating properties of the novel phospholipid diacylglycerol pyrophosphate. *J. Biol. Chem.* **1999**, *274*, 522–526. [CrossRef] [PubMed]
76. Balsinde, J.; Balboa, M.A.; Dennis, E.A. Inflammatory activation of arachidonic acid signaling in murine $P388D_1$ macrophages via sphingomyelin synthesis. *J. Biol. Chem.* **1997**, *272*, 20373–20377. [CrossRef] [PubMed]
77. Balsinde, J.; Balboa, M.A.; Insel, P.A.; Dennis, E.A. Differential regulation of phospholipase D and phospholipase A_2 by protein kinase C in $P388D_1$ macrophages. *Biochem. J.* **1997**, *321*, 805–809. [CrossRef] [PubMed]
78. Hoffman, T.; Lizzio, E.F.; Suissa, J.; Rotrosen, D.; Sullivan, J.A.; Mandell, G.L.; Bonvini, E. Dual stimulation of phospholipase activity in human monocytes. Role of calcium-dependent and calcium-independent pathways in arachidonic acid release and eicosanoid formation. *J. Immunol.* **1988**, *140*, 3912–3918. [PubMed]
79. Hoffman, T.; Brando, C.; Lizzio, E.F.; Lee, Y.L.; Hansen, M.; Tripathi, A.K.; Taplits, M.; Puri, J.; Bonvini, E.; Abrahamsen, T.G.; et al. Calcium-dependent eicosanoid metabolism by concanavalin A-stimulated human monocytes in vitro. Synergism with phorbol ester indicates separate regulation of leukotriene B_4 synthesis and release. *J. Immunol.* **1991**, *146*, 692–700. [PubMed]
80. Penglis, P.S.; Cleland, L.G.; Demasi, M.; Caughey, G.E.; James, M.J. Differential regulation of prostaglandin E_2 and thromboxane A_2 production in human monocytes: Implications for the use of cyclooxygenase inhibitors. *J. Immunol.* **2000**, *165*, 1605–1611. [CrossRef] [PubMed]
81. Teixeira, C.; Cury, Y.; Moreira, V.; Picolo, G.; Chaves, F. Inflammation induced by Bothrops asper venom. *Toxicon* **2009**, *54*, 67–76. [CrossRef] [PubMed]
82. Han, S.K.; Kim, K.P.; Koduri, R.; Bittova, L.; Muñoz, N.M.; Leff, A.R.; Wilton, D.C.; Gelb, M.H.; Cho, W. Roles of Trp31 in high membrane binding and proinflammatory activity of human group V phospholipase A_2. *J. Biol. Chem.* **1999**, *274*, 11881–11888. [CrossRef] [PubMed]

83. Melo, R.C.N.; Weller, P.F. Lipid droplets in leukocytes: Organelles linked to inflammatory responses. *Exp. Cell Res.* **2016**, *340*, 193–197. [CrossRef] [PubMed]
84. Cruz, A.L.S.; Barreto, E.A.; Fazolini, N.P.B.; Viola, J.P.B.; Bozza, P.T. Lipid droplets: Platforms with multiple functions in cancer hallmarks. *Cell Death Dis.* **2020**, *11*, 105. [CrossRef] [PubMed]
85. Jarc, E.; Petan, T. A twist of FATe: Lipid droplets and inflammatory lipid mediators. *Biochimie* **2020**, *169*, 69–87. [CrossRef] [PubMed]
86. Guijas, C.; Rodríguez, J.P.; Rubio, J.M.; Balboa, M.A.; Balsinde, J. Phospholipase A_2 regulation of lipid droplet formation. *Biochim. Biophys. Acta* **2014**, *1841*, 1661–1671. [CrossRef] [PubMed]
87. Chapman, K.D.; Aziz, M.; Dyer, J.M.; Mullen, R.T. Mechanisms of lipid droplet biogenesis. *Biochem. J.* **2019**, *476*, 1929–1942. [CrossRef] [PubMed]
88. Guijas, C.; Bermúdez, M.A.; Meana, C.; Astudillo, A.M.; Pereira, L.; Fernández-Caballero, L.; Balboa, M.A.; Balsinde, J. Neutral lipids are not a source of arachidonic acid for lipid mediator signaling in human foamy monocytes. *Cells* **2019**, *8*, 941. [CrossRef] [PubMed]
89. Dichlberger, A.; Schlager, S.; Maaninka, K.; Schneider, W.J.; Kovanen, P.T. Adipose triglyceride lipase regulates eicosanoid production in activated human mast cells. *J. Lipid Res.* **2014**, *55*, 2471–2478. [CrossRef] [PubMed]
90. Schlager, S.; Goeritzer, M.; Jandl, K.; Frei, R.; Vujic, N.; Kolb, D.; Strohmaier, H.; Juliane, J.; Dorow, J.; Eichmann, T.O.; et al. Adipose triglyceride lipase acts on neutrophil lipid droplets to regulate substrate availability for lipid mediator synthesis. *J. Leukoc. Biol.* **2015**, *98*, 837–850. [CrossRef] [PubMed]
91. Gubern, A.; Casas, J.; Barceló-Torns, M.; Barneda, D.; de la Rosa, X.; Masgrau, R.; Picatoste, F.; Balsinde, J.; Balboa, M.A.; Claro, E. Group IVA phospholipase A_2 is necessary for the biogenesis of lipid droplets. *J. Biol. Chem.* **2008**, *283*, 27369–27382. [CrossRef] [PubMed]
92. Gubern, A.; Barceló-Torns, M.; Barneda, D.; López, J.M.; Masgrau, R.; Picatoste, F.; Chalfant, C.E.; Balsinde, J.; Balboa, M.A.; Claro, E. JNK and ceramide kinase govern the biogenesis of lipid droplets through activation of group IVA phospholipase A_2. *J. Biol. Chem.* **2009**, *284*, 32359–32369. [CrossRef] [PubMed]

© 2020 by the authors. Licensee MDPI, Basel, Switzerland. This article is an open access article distributed under the terms and conditions of the Creative Commons Attribution (CC BY) license (http://creativecommons.org/licenses/by/4.0/).

Article

A Representative GIIA Phospholipase A_2 Activates Preadipocytes to Produce Inflammatory Mediators Implicated in Obesity Development

Elbio Leiguez [1,*], Priscila Motta [1], Rodrigo Maia Marques [1], Bruno Lomonte [2], Suely Vilela Sampaio [3] and Catarina Teixeira [1,*]

[1] Laboratório de Farmacologia, Instituto Butantan, São Paulo 05503-900, Brazil; priscila.motta@butantan.gov.br (P.M.); rodrigo.marques@butantan.gov.br (R.M.-M.)

[2] Instituto Clodomiro Picado, Facultad de Microbiología, Universidad de Costa Rica, San José 11501-2060, Costa Rica; bruno.lomonte@ucr.ac.cr

[3] Faculdade de Ciências Farmacêuticas de Riberão Preto, Universidade de São Paulo, Ribeirão Preto 14040-903, SP, Brazil; suvilela@fcfrp.usp.br

* Correspondence: elbio.junior@butantan.gov.br (E.L.); catarina.teixeira@butantan.gov.br (C.T.)

Received: 11 September 2020; Accepted: 18 November 2020; Published: 24 November 2020

Abstract: Adipose tissue secretes proinflammatory mediators which promote systemic and adipose tissue inflammation seen in obesity. Group IIA (GIIA)-secreted phospholipase A_2 ($sPLA_2$) enzymes are found to be elevated in plasma and adipose tissue from obese patients and are active during inflammation, generating proinflammatory mediators, including prostaglandin E_2 (PGE_2). PGE_2 exerts anti-lipolytic actions and increases triacylglycerol levels in adipose tissue. However, the inflammatory actions of GIIA $sPLA_2$s in adipose tissue cells and mechanisms leading to increased PGE_2 levels in these cells are unclear. This study investigates the ability of a representative GIIA $sPLA_2$, MT-III, to activate proinflammatory responses in preadipocytes, focusing on the biosynthesis of prostaglandins, adipocytokines and mechanisms involved in these effects. Our results showed that MT-III induced biosynthesis of PGE_2, PGI_2, MCP-1, IL-6 and gene expression of leptin and adiponectin in preadipocytes. The MT-III-induced PGE_2 biosynthesis was dependent on cytosolic PLA_2 ($cPLA_2$)-α, cyclooxygenases (COX)-1 and COX-2 pathways and regulated by a positive loop via the EP_4 receptor. Moreover, MT-III upregulated COX-2 and microsomal prostaglandin synthase (mPGES)-1 protein expression. MCP-1 biosynthesis induced by MT-III was dependent on the EP4 receptor, while IL-6 biosynthesis was dependent on EP3 receptor engagement by PGE_2. These data highlight preadipocytes as targets for GIIA $sPLA_2$s and provide insight into the roles played by this group of $sPLA_2$s in obesity.

Keywords: phospholipase A_2; preadipocytes; prostaglandins; adipokines; cytokines; EP receptors

---

1. Introduction

Obesity is a chronic low-grade inflammatory condition in which adipose tissue serves as the source of inflammatory mediators. In obesity and associated diseases, such as diabetes and cardiovascular disease, high plasma and tissue activities of secreted phospholipase A_2 ($sPLA_2$) enzymes, especially group IIA (GIIA) $sPLA_2$s, have been demonstrated [1,2].

Phospholipases A_2s (PLA_2s) are lipolytic enzymes with important physiological functions, including cell membrane remodelling and lipid metabolism. These enzymes are classified according to their cellular localization as either intracellular PLA_2 ($iPLA_2$) enzymes with high molecular weight or $sPLA_2$s, enzymes with low molecular weight. $sPLA_2$s hydrolyze glycerophospholipids at the sn-2 position of the glycerol backbone, releasing fatty acids and lysophospholipids in a calcium-dependent manner. $sPLA_2s$ are classified into 11 groups and possess, as a common motif, a conserved His-Asp catalytic dyad. Group IIA $sPLA_2$ comprises mammalian $sPLA_2s$ found in the inflammatory fluid of mammals and $sPLA_2s$ from Viperidae snake venoms. Besides their role in cell membrane physiology, mammalian group IIA $sPLA_2$ are known as important autocrine and paracrine players in inflammatory processes by releasing fatty acids from cell membranes leading to production of pro-inflammatory mediators such as leukotrienes and prostaglandins [3–5]. Their role in metabolic diseases, such as obesity, has also been shown [1,4,6]. It is known that inhibition of $sPLA_2s$, using pharmacological intervention, reduced lipid mediator's synthesis and inflammatory parameters linked to obesity. In this sense, prostaglandin E_2 (PGE_2) is the most abundant lipid mediator produced by the body. This mediator is constitutively produced in all tissues by the cyclooxygenases (COX) enzymatic system and terminal PGE-synthases [7,8]. PGE_2 is a powerful molecule carrying multiple biological effects, which are mediated by four subtypes of G protein-coupled receptors, named EP1, EP2, EP3 and EP4, depending on the tissue or cell type [7,8]. PGE_2 is recognized as an important mediator of inflammation, pain and fever. In addition, PGE_2 plays important roles in the regulation of proliferation and cell differentiation, and exerts anti-lipolytic actions and increases triacylglycerol levels in adipose tissue cells, contributing to lipid accumulation in these cells [9,10]. However, the molecular mechanisms triggered by GIIA $sPLA_2s$ that lead to the biosynthesis of PGE_2 by adipose tissue cells are poorly known.

Preadipocytes correspond to a greater cellular fraction present in white adipose tissue and contribute significantly to the production and secretion of inflammatory mediators, such as PGE_2 and adipokines, involved in the pathogenesis of obesity [11–14]. It has been shown that preadipocytes are target cells for a variety of inflammatory factors secreted by macrophages, which are the main cells involved in establishing an inflammatory environment in adipose tissue. In addition, when compared to mature adipocytes, preadipocytes are more responsive to inflammatory stimuli, as they offer a greater activation of the transcription nuclear factor kappa B (NF-kB) and related protein kinases [15]. Therefore, these cells may be used as a cell model for the understanding of the inductors and mechanisms involved in the development of inflammatory processes linked to obesity.

Myotoxin-III (MT-III) is a representative GIIA $sPLA_2$ isolated from *Bothrops asper* snake venom that shares functional and structural similarities with mammalian pro-inflammatory $sPLA_2s$ of the same group [16–18]. MT-III is known to trigger inflammatory events in both in vivo and in vitro experimental models. Our group has previously shown that MT-III activates macrophages' functions and induces the accumulation of lipids into these cells [19]. In addition, this enzyme is able to upregulate the differentiation of macrophages into foam cells [20], which are closely associated with diseases linked to lipid imbalance, including obesity [21,22]. On these bases, in this study, the ability of MT-III to activate proinflammatory responses in preadipocytes focusing on the biosynthesis of lipid mediators, cytokines and adipokines and the mechanisms involved in this process were investigated. In this study, we show for the first time that preadipocytes are target cells for the action of MT-III, a representative GIIA $sPLA_2$, which triggers inflammatory pathways implicated in the development of obesity. The effect of MT-III involves the biosynthesis of PGE_2, MCP-1 and IL-6 and gene expression of leptin and adiponectin. PGE_2 biosynthesis is dependent upon the activation of cytosolic PLA_2 ($cPLA_2$)-α, COX-1, COX-2 and mPGES-1 pathways. EP3 and EP4 receptors play key roles in the release of PGE_2 and cytokines.

2. Materials and Methods

2.1. Chemicals and Reagents

(3-(4,5-dimethylthiazol-2-yl)-2,5-diphenyltetrazolium bromide) MTT and L-glutamine were obtained from USB (Cleveland, OH, USA). Mouse mAb anti-β-actin was purchased from Sigma-Aldrich (St. Louis, MO, USA). The PGE_2 enzyme immunoassay kit, Valeryl Salicylate, compounds NS-398, AH6809, AH23848, SC-19220, L-798106 and polyclonal antibodies against COX-1, COX-2, mPGES-1 and the EP4 receptor were purchased from Cayman Chemical Company (Ann Arbor, MI, USA). Pyrrolidine-2

(Pyr-2) was purchased from Calbiochem-Novabiochem Corp. (La Jolla, CA, USA). Secondary antibodies, anti-mouse and anti-rabbit, conjugated to HRP and nitrocellulose membrane, were obtained from GE Healthcare (Buckinghamshire, UK). The Cytometric Bead Assay (CBA) kit was purchased from BD Bioscience (San Jose, CA, USA). Gentamicin was purchased from Schering-Plough (Whitehouse Station, NJ, USA), DMSO from Amresco (Solon, OH, USA) and Dulbecco's Modified Eagle Medium (DMEM), fetal bovine serum and real-time polymerase chain reaction (PCR) assay kit from Life Technologies (São Paulo, SP, Brazil).

2.2. *Phospholipase A_2 (PLA_2)*

Aspartate-49 $sPLA_2$, named MT-III (Uniprot accession no.: P20474), from *B. asper* venom was purified by ion-exchange chromatography on CM Sephadex C-25 using a KCl gradient from 0 to 0.75 M at pH 7.0 as described [23], followed by RP-HPLC on a semipreparative C8 column (Vydac; 106,250 mm, 5 mm particle size), eluted at a flow rate of 2.5 mL/min with a gradient of acetonitrile (0–70%, containing 0.1% trifluoroacetic acid) over 30 min. Homogeneity was verified by SDS-PAGE, run under reducing conditions, in which a single band of 14 kDa was observed. The complete amino acid sequence of this enzyme has been described previously [23,24]. The absence of endotoxin contamination in the MT-III batches used was demonstrated by a quantitative LAL test [25], which revealed undetectable levels of endotoxin (0.125 EU/mL).

2.3. *Cytotoxicity Assay*

The cytotoxicity of MT-III and toward the 3T3-L1 preadipocyte was evaluated using the MTT assay previously described [19]. In brief, 4×10^3 preadipocytes per well in DMEM, supplemented with 40 μg/mL gentamicin sulfate and 2 mM L-glutamine, were plated in 96-well plates and incubated with MT-III (0.4 μM), COX inhibitors or PGE_2 antagonist receptors, diluted in medium or with the same volume of medium alone (control) for 1, 3, 6, 12, 24 and 48 h at 37 °C in a humidified atmosphere (5% CO_2). MTT (5 mg/mL) was dissolved in PBS and filtered for sterilization and removal of insoluble residues. Stock MTT solution (10% in culture medium) was added to all wells in each assay, and plates were incubated for 3 h at 37 °C. Dimethyl sulfoxide (DMSO) (100 μL) was added to all wells and mixed thoroughly for 30 min, at room temperature. Absorbances were then recorded in a microtiter plate reader, at 540 nm. Results were expressed as percentages of viable cells, considering control cells incubated with medium alone as 100% viable.

2.4. *3T3-L1 Cell Culture and Stimulation*

3T3-L1 preadipocytes obtained from the American Type Culture Collection were cultured as described [26]. Cells were processed according to the experimental protocol, in which 5×10^3 preadipocytes per well were seeded in 12-wells culture plates and maintained in culture medium for 48 h before stimulation. Preadipocytes were serum-starved in DMEM with 1% (*v*/*v*) gentamicin sulfate supplemented with 1% (*v*/*v*) *L*-glutamine for 18 h prior to all treatments. Cellular homogenates were used for the Western blotting analysis of COX-1, COX-2, EP1–EP4 receptors and mPGES-1 protein expression, and supernatants of each treatment were used to measure lipid mediators PGE_2, PGI_2, LTB_4 and TXA_2 by Enzyme Immunoassay (EIA) and cytokines MCP-1, IL-6, IL-10, IL-12, TNF-α and INF-γ by CBA. Cells were stimulated with MT-III (0.4 mM) diluted in DMEM (serum free) or DMEM alone (control) for selected periods of time and maintained at 37 °C in a humidified atmosphere (5% CO_2). To investigate the mechanism involved in the PGE_2 and cytokine biosynthesis, selective inhibitors or antagonists were used at concentrations previously tested: 10 μM valeryl salicylate (COX-1 inhibitor) and NS-398 (COX-2 inhibitor); 10 μM SC-19220 (EP1 receptor antagonist); AH6809 (EP2 receptor antagonist) and AH23848 (EP4 receptor antagonist); and 1 μM L-798106 (EP3 receptor antagonist) [27–31]. All of the stock solutions were prepared in DMSO and stored at −20 °C. Aliquots were diluted in DMEM immediately before use. DMSO concentration was always lower than 1%. The viability of cells treated with inhibitors or antagonists was evaluated with MTT assay. No significant changes in cell viability were registered with any of the above agents or the vehicle at the concentrations used (data not shown).

2.5. Western Blotting

COX-1, COX-2, EP1–EP4 receptors and mPGES-1 protein expression from homogenate cells were detected by Western blotting. Briefly, MT-III-stimulated and non-stimulated cells were lysed with 100 mL of a sample buffer (0.5 M Tris-HCl, pH 6.8, 20% SDS, 1% glycerol, 1 M β-mercaptoethanol, 0.1% bromophenol blue) and boiled for 10 min. Samples were resolved by SDS-PAGE on 10% bis-acrylamide gels overlaid with a 5% stacking gel. Proteins were then transferred to nitrocellulose membranes using a Mini Trans-Blot (Bio-Rad Laboratories, Richmond, CA, USA). Membranes were blocked for 1 h with 5% albumin in Tris-buffered saline (20 mM Tris, 100 mM NaCl and 0.5% Tween 20, pH 7.2) and incubated overnight with primary antibodies against COX-1 and COX-2; EP1, EP2, EP3, and EP4 receptors; and mPGES-1 (1:500 dilution) or β-actin (1:3000 dilution) for 1 h at room temperature. Membranes were then washed and incubated with the appropriate secondary antibody conjugated to horseradish peroxidase. Immunoreactive bands were detected by the entry-level peroxidase substrate for enhanced chemiluminescence, according to the instructions of the manufacturer (GE Healthcare). Band densities were quantified with an ImageQuant LAS 4000 mini densitometer (GE Healthcare) using the image analysis software ImageQuant TL (GE Healthcare).

2.6. Eicosanoid and Cytokines Quantification

PGE_2, PGI_2, LTB_4 and TXA_2 were measured using an EIA kit, while cytokines (MCP-1, IL-6, IL-10, IL-12, TNF-α, INF-γ) were quantified using a CBA kit from supernatants of preadipocytes incubated with each treatment. Kits were used according to the instructions of the manufacturer.

2.7. Adipocytokines Expression by Quantitative Real-Time PCR

Quantitative polymerase chain reaction was performed as described [32]. Briefly, the total RNA from preadipocytes, incubated MT-III or DMEM alone (control) for 1, 3 and 6 h was isolated using TRIzol reagent (Invitrogen, Carlsbad, CA, USA) and reverse transcription with oligo (dT) priming was performed from 2 μg of total RNA using Superscript III (Invitrogen, Carlsbad, CA, USA). The relative expression of each transcript was determined by quantitative real-time PCR in an ABI 7000 Sequence Detection System (Applied Biosystems, Forrest City, CA, USA). Each well of the 96-well reaction plate contained a total volume of 25 μL of Power SYBR Green PCR Master Mix (Applied Biosystems). The threshold cycle (Ct) was used to determine the relative expression level of each gene by normalizing to the Ct of Glyceraldehyde-3-Phosphate Dehydrogenase (GAPDH). The method of delta–delta cycle threshold (ddCt) was used to calculate the relative fold change of each gene. Data are represented as mean + SEM.

2.8. Statistical Analysis

Data are expressed as mean ± SEM ($n = 4$). Multiple comparisons among groups were performed using the one-way ANOVA and, as a post-test, the Bonferroni test. Differences between experimental groups were considered significant for p-values < 0.05. All statistical tests were performed using Prism version 5 software (GraphPad, San Diego, CA, USA).

3. Results

3.1. MT-III Induces the Release of Lipid Mediators by Preadipocytes

Lipid mediators are involved in lipid abnormalities and contribute to the triggering of inflammatory processes in adipose tissue [33]. Therefore, we investigated the ability of MT-III to induce the release of lipid mediators linked to inflammatory processes, such as PGE_2, PGI_2, TXA_2 and LTB_4, by cultured preadipocytes. From preliminary studies (data not shown), the submaximal concentration of 0.4 μM of

MT-III was chosen for these studies as it would allow potential inhibition or exacerbation of its effects by drug treatment to be detected. As shown in Figure 1, the incubation of preadipocytes with MT-III induced a significant release of PGE_2 (A) from 1 to 24 h and of PGI_2 (B) from 12 to 48 h when compared with controls. However, the incubation of cells with MT-III did not alter TXA_2 (C) or LTB_4 (D) levels in any of the time periods evaluated. These results indicate the ability of MT-III to activate preadipocytes for the production of PGE_2 and PGI_2.

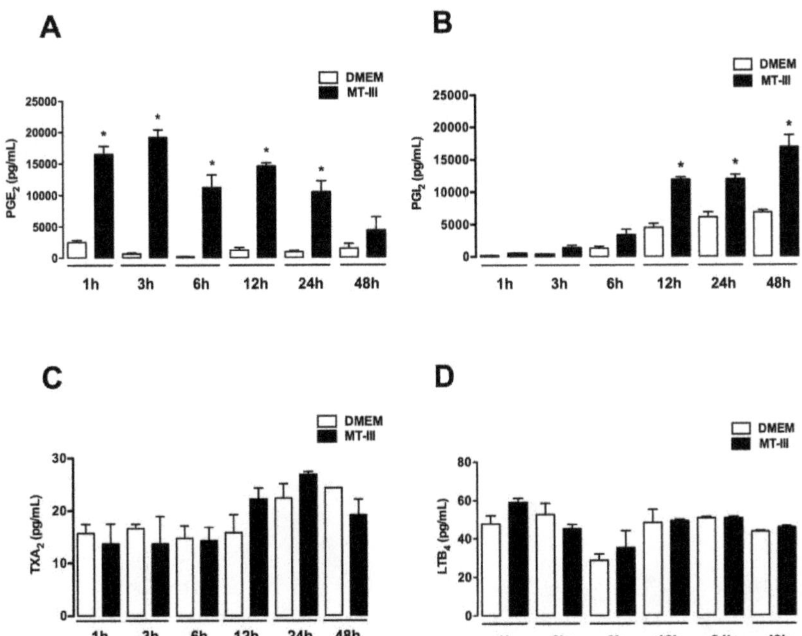

Figure 1. MT-III induces production of PGE_2, PGI_2, TXA_2 and LTB_4 by 3T3-L1 preadipocytes. Cells were incubated with MT-III (0.4 μM) or DMEM (control) for 1 to 48 h. Bar graphs show the MT-III-induced release of PGE_2 (**A**), PGI_2 (**B**), TXA_2 (**C**) and LTB_4 (**D**) by preadipocytes. Concentrations were quantified in culture supernatants by EIA commercial kit. Results are expressed as mean ± SEM from 3 independent experiments. * $p < 0.05$ as compared with control group (two-way ANOVA and Bonferroni posttest).

3.2. MT-III-Induced Release of PGE_2 Is Dependent on COX-1 and COX-2 in Preadipocytes

COX-1 and COX-2 are enzymes responsible for the metabolization of arachidonic acid–generating prostanoids, such as PGE_2 [8,34]. In order to verify the mechanism involved in the MT-III-induced biosynthesis of PGE_2, we investigated the participation of COX-1 and COX-2 in this effect. As seen in Figure 2A, preadipocytes incubated with MT-III, in the presence of vehicle (DMSO), showed a significant release of PGE_2 after 6 h when compared with controls. Preadipocytes treated either with COX-1 inhibitor (valerylsalicylate) or COX-2 inhibitor (NS-398) before the MT-III stimulus showed a reduction in PGE_2 release which was statistically significant when compared to the positive control. Treatment of cells with both valerylsalicylate and NS-398 compounds abolished the MT-III-induced release of PGE_2 when compared to the positive control. These results indicate that COX-1 and COX-2 are key enzymes involved in the release of PGE_2, induced by MT-III, in preadipocytes. Having shown that both COX isoforms participate in the signalling pathway triggered by MT-III that leads to PGE_2 production, we next investigated whether MT-III is able to upregulate the protein expression of COX-1 and COX-2 in preadipocytes. Our results show that preadipocytes constitutively expressed both isoforms of COX. Figure 2B,C show that COX-1 protein expression did not differ significantly between control cells and cells treated with MT-III. However, the protein expression of COX-2 was higher in cells incubated with the phospholipase A_2 after 6 and 12 h (Figure 2D,E). Therefore, although the COX-2 isoform is constitutively expressed by preadipocytes [35], our results show that MT-III upregulates the protein expression of COX-2 but not COX-1 in preadipocytes.

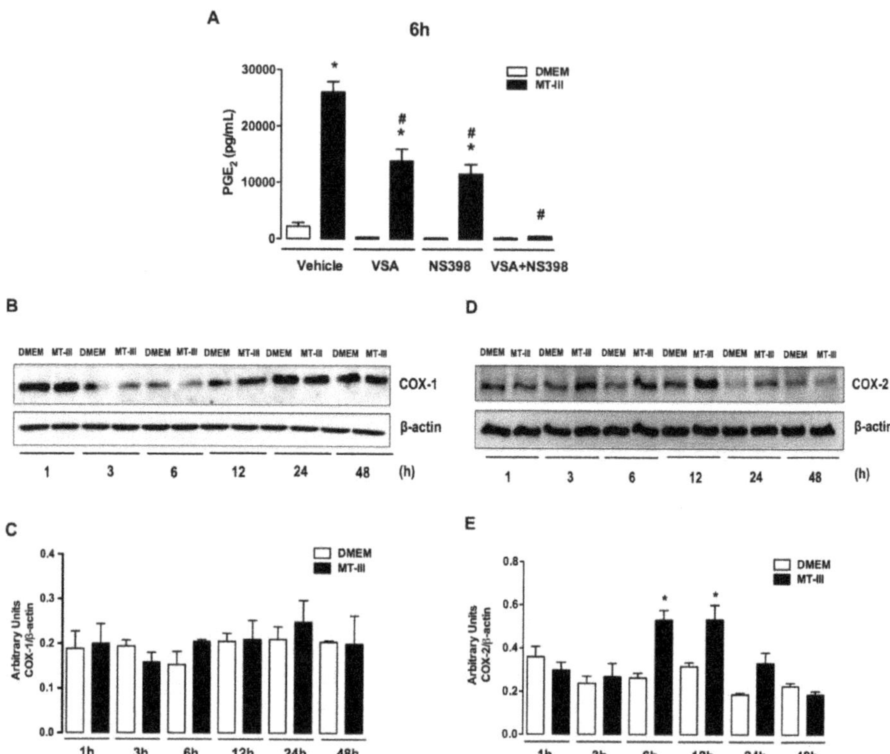

Figure 2. MT-III activates COX-1 and COX-2 pathways for release of PGE_2 by 3T3-L1 preadipocytes. (**A**) Cells were incubated with either valerylsalicylate (VSA) (10 μM), or NS-398 (10 μM), or both for 1 h, followed by incubation with MT-III (0.4 μM) for 6 h. PGE_2 concentrations were quantified in culture supernatants by EIA commercial kit. (**B–E**) 3T3-L1 preadipocytes were incubated with MT-III (0.4 μM) or DMEM (control) for 1 up to 48 h. (**B**) Western blotting of COX-1 and β-actin (loading control) showing immunoreactive bands. (**D**) Western blotting of COX-2 and β-actin (loading control) showing immunoreactive bands. Densitometric analysis of immunoreactive (**C**) COX-1 and (**E**) COX-2 bands. Density data (in arbitrary units) were normalized with those of β-actin. Results are expressed as mean ± SEM from 3 independent experiments. * $p < 0.05$ as compared with control group and # $p < 0.05$ as compared with MT-III group (two-way ANOVA and Bonferroni posttest).

3.3. MT-III Upregulates Protein Expression of mPGES-1 by Preadipocytes

An inducible synthase responsible for the terminal synthesis of PGE_2, mPGES-1 is upregulated in inflammatory conditions [36,37]. Based on this, we evaluated the ability of MT-III to upregulate the protein expression of this enzyme in preadipocytes. Our results show that MT-III upregulated the protein expression of mPGES-1 after 1 h of stimulation when compared with the control (Figure 3). The phospholipase A_2 did not alter the protein expression of mPGES-1 at other time intervals evaluated. These results demonstrate that MT-III induces protein expression of mPGES-1 in preadipocytes.

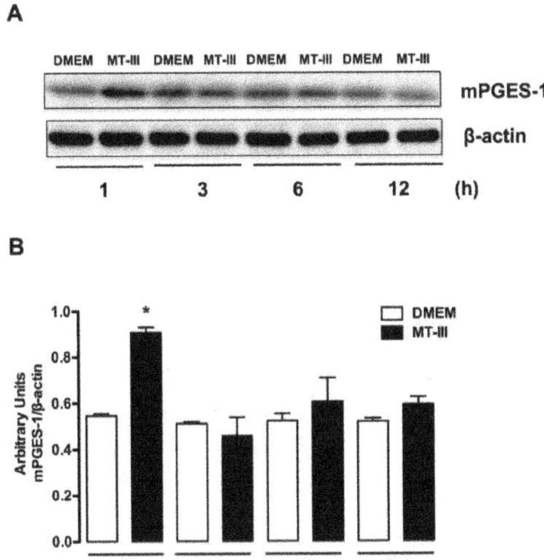

Figure 3. MT-III upregulates protein expression of mPGES-1 in 3T3-L1 preadipocyte. Cells were incubated with MT-III (0.4 μM) or DMEM (control) for 1 up to 12 h. (**A**) Western blotting of mPGES-1 and β-actin (loading control) showing immunoreactive bands. (**B**) Densitometric analysis of immunoreactive mPGES-1 bands. Density data (in arbitrary units) were normalized with those of β-actin. Results are expressed as mean ± SEM from 3 experiments. * $p < 0.05$ as compared with the control group (two-way ANOVA and Bonferroni posttest).

3.4. MT-III-Induced Release of PGE_2 Is Dependent on Cytosolic PLA_2-α in Preadipocytes

It is known that $sPLA_2$s cross-talk with $cPLA_2$ for the synthesis of inflammatory mediators [38–40]. Based on this, we investigated the participation of $cPLA_2$-α in the release of PGE_2 induced by MT-III. Our results show that preadipocytes incubated with MT-III, in the presence of vehicle (positive cotrol), showed a significant release of PGE_2 after 3 h when compared with respective control. However, pre-treatment of cells with $cPLA_2$-α inhibitor (Pyr-2) abrogated the MT-III-induced release of PGE_2 when compared to the positive control (Figure 4). These results indicate that the MT-III-induced production of PGE_2 is dependent on $cPLA_2$-α in preadipocytes.

3.5. MT-III-Induced Release of PGE_2 Is Dependent on the EP4 Receptor in Preadipocytes

PGE_2 exerts its effects through activation of four subtypes of G protein-coupled receptors, named EP1, EP2, EP3 and EP4, and these receptors are able to regulate PGE_2 biosynthesis [34,37,41]. It is known that the activation of the EP4 receptor by PGE_2 may lead to the increased expression of key enzymes of biosynthesis cascade of this prostaglandin, such as COX-2 and mPGES-1 [8,41,42]. Therefore, we investigated whether PGE_2 biosynthesis, induced by MT-III, was dependent on the activation of these receptors. As shown in Figure 5A, the incubation of preadipocytes with DMEM plus vehicle or antagonists did not cause a significant release of PGE_2 after 6 h of incubation. Pre-treatment of cells with DMEM plus vehicle followed by incubation with MT-III (0.4 μM), for the same time period, induced a significant increase in PGE_2 release, relative to baseline control. However, pre-treatment of cells with antagonists of EP1 (SC-19220), EP2 (AH6809) or EP3 (L-798106) receptors did not alter MT-III-induced PGE_2 release when compared with controls. In contrast, pre-treatment of cells with the EP4 receptor antagonist (AH23848) abolished the MT-III-induced release of PGE_2 when compared to the positive control. To better understand the involvement of the EP receptors in the effects induced by MT-III, we next analysed the protein expression of these receptors in preadipocytes stimulated with MT-III and in control cells incubated with culture medium alone. Our results show that there was no alteration in the protein expression of EP receptors in preadipocytes incubated with MT-III (0.4 μM) in any of the time periods evaluated when compared with controls (Figure 5B–I). These results indicate that PGE_2 biosynthesis, induced by MT-III, in preadipocytes is dependent on the engagement of the EP4 receptor by PGE_2, but not on increased protein expression of this receptor.

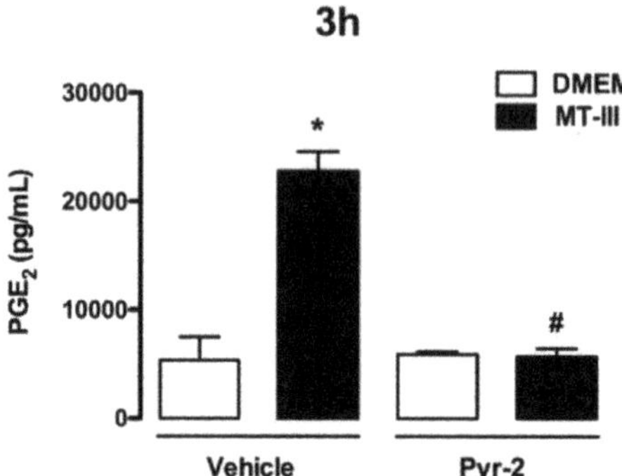

Figure 4. MT-III-induced PGE_2 release is dependent on cPLA2-α in 3T3-L1 preadipocytes. Cells were incubated with Pyr-2 (1 μM) for 1 h followed by incubation with MT-III (0.4 μM) for 3 h. PGE_2 concentrations were quantified in culture supernatants by EIA commercial kit. * $p < 0.05$ as compared with control group and # $p < 0.05$ as compared with MT-III group (two-way ANOVA and Bonferroni posttest).

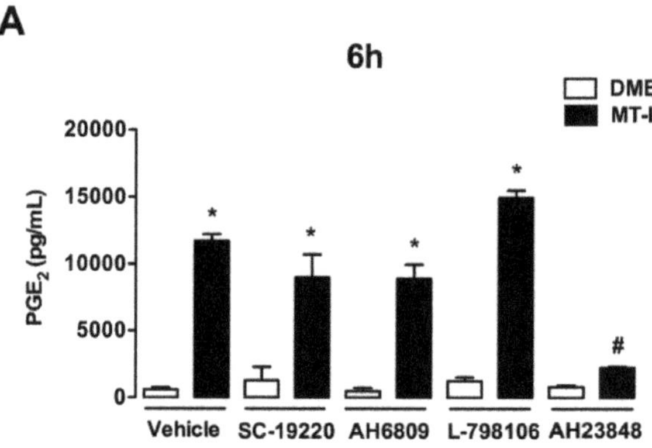

Figure 5. *Cont.*

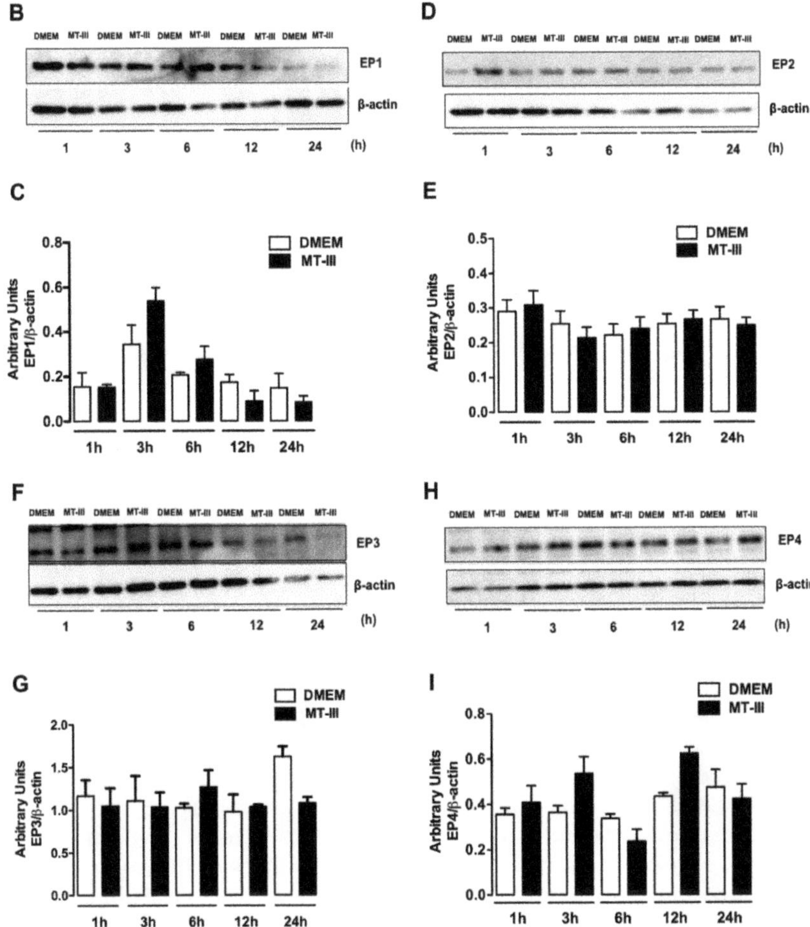

Figure 5. EP4 receptor participates in MT-III-induced PGE_2 biosynthesis in 3T3-L1 preadipocytes. (**A**) Preadipocytes were incubated with SC-19220 (10 µM), AH6809 (10 µM), L-798106 (1 µM) or AH23848 (10 µM) for 1 h followed by incubation with MT-III (0.4 µM) for 6 h. PGE_2 concentrations were quantified in culture supernatants by EIA commercial kit. (**B–I**) 3T3-L1 cells were incubated with MT-III (0.4 µM) or DMEM (control) for 1 up to 48 h. (**B**,**D**,**F**,**H**) Western blotting of EP1, EP2, EP3 and EP4 receptors, respectively, and β-actin (loading control), showing immunoreactive bands. (**C**,**E**,**G**,**I**) Densitometric analysis of immunoreactive bands for EP1, EP2, EP3 and EP4 receptors, respectively. Results are expressed as mean ± SEM from 3 independent experiments. * $p < 0.05$ as compared with control group and # $p < 0.05$ as compared with MT-III group (one-way ANOVA and Bonferroni posttest in (**A**) and two-way ANOVA and Bonferroni posttest in (**C**,**E**,**G**,**I**)).

3.6. MT-III Induces Release of Inflammatory Cytokines by Preadipocytes

Inflammatory cytokines are found in high levels in obesity inflammatory processes and contribute to the development and maintenance of this inflammatory state [43–45]. On these bases, we investigated the capacity of MT-III to induce the release of the inflammatory cytokines MCP-1, IL-6, IL-10, IL-12, TNF-α and IFN-γ by preadipocytes. Figure 6A shows that MT-III induced significant release of MCP-1 from 30 min up to 24 h of incubation when compared with controls. In addition, MT-III induced significant release of IL-6 after 12 h of incubation when compared to the respective controls (Figure 6B).

However, MT-III did not alter the release of IL-10, IL-12, TNF-α or IFN-γ (data not shown). In this sense, these results evidence the capacity of MT-III to induce the release of MCP-1 and IL-6 by 3T3-L1 in preadipocytes.

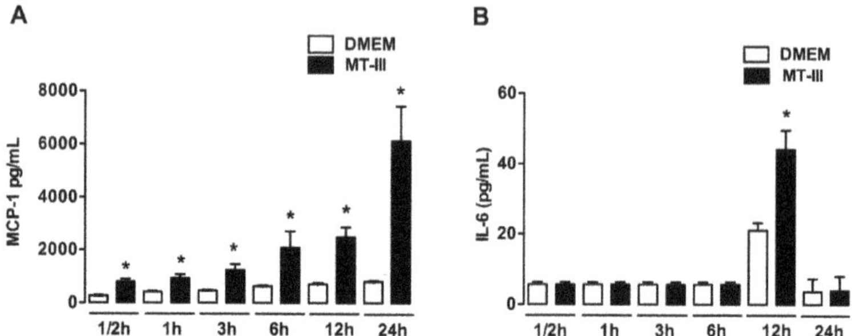

Figure 6. MT-III induces MCP-1 and IL-6 production by 3T3-L1 preadipocytes. Cells were incubated with MT-III (0.4 μM), or DMEM (control) for $\frac{1}{2}$ up to 48 h. Bar graphs show concentrations of (**A**) MCP-1 and (**B**) IL-6 released by cells incubated with MT-III. Cytokines concentrations were quantified in culture supernatants by Cytometric Bead Array (CBA). Results are expressed as mean ± SEM from 5 experiments. * $p < 0.05$ as compared with control group (two-way ANOVA and Bonferroni posttest).

3.7. EP3 and EP4 Receptors Participate in the MT-III-Induced Release of IL-6 and MCP-1 by Preadipocytes

Previous studies have shown that EP3 and EP4 PGE_2 receptors regulate the release of proinflammatory cytokines [46,47]. Therefore, we investigated the participation of EP3 and EP4 receptors in the MT-III-induced release of IL-6 and MCP-1, respectively, by preadipocytes. Figure 7A shows that the stimulation of preadipocytes with MT-III, in the presence of vehicle, significantly increased MCP-1 release after 24 h when compared with the control. Pre-treatment of preadipocytes with the EP4 antagonist (AH238481) significantly reduced the release of MCP-1 in cells stimulated with MT-III in comparison with the positive control. Similarly, pre-treatment of cells with the EP3 antagonist (L-798106) reduced the MT-III-induced release of IL-6 after 12 h, which was significant in comparison with the positive control (Figure 7B). These results indicate that EP3 and EP4 receptors participate in the release of IL-6 and MCP-1, respectively, in preadipocytes stimulated with MT-III.

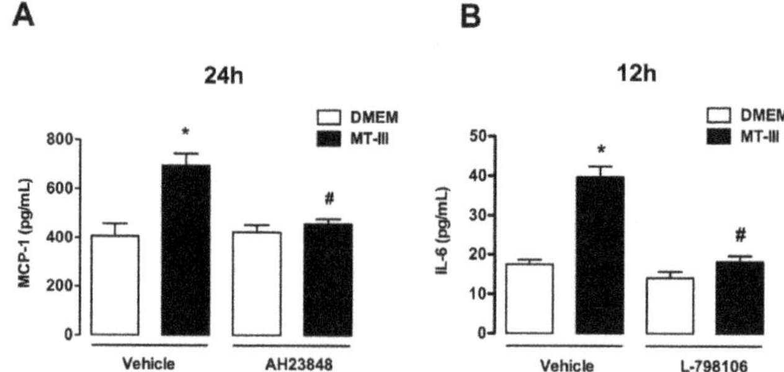

Figure 7. EP3 and EP4 receptors participate in the MT-III-induced release of IL-6 and MCP-1, respectively, by 3T3-L1 preadipocytes. Cells were incubated with AH23848 (10 μM) or L-798106 (1 μM) or vehicle for 1 h followed by incubation with MT-III (0,4 μM) for 12 or 24 h. Graphs show participation of EP4 receptor in the MT-III-induced release of MCP-1 (**A**) and participation of the EP3 receptor in the MT-III-induced release of IL-6 (**B**). Concentration of cytokines were quantified from culture supernatants by Cytometric Bead Array (CBA). Results are expressed as mean ± SEM from 5 experiments. * $p < 0.05$ as compared with control group and # $p < 0.05$ as compared with MT-III group (one-way ANOVA and Bonferroni posttest).

3.8. MT-III Upregulates Gene Expression of Adipokines in Preadipocytes

Adipokines are produced by white adipose tissue and are involved in a wide variety of physiological and pathological processes. Proinflammatory adipokines contribute to the development and maintenance of the inflammatory state in obese individuals [35,48,49]. In light of this, we investigated the ability of MT-III to induce gene expression of the adipokines leptin, resistin and adiponectin by preadipocytes. As demonstrated in Figure 8A, preadipocytes incubated with MT-III showed a significant increase in the gene expression of leptin from 1 to 3 h when compared with controls. In addition, preadipocytes incubated with MT-III showed a significant increase in the gene expression of adiponectin after 3 h (Figure 8B). However, the phospholipase A_2 did not affect the gene expression of resistin in any of the time periods evaluated (Figure 8C). These results indicate that preadipocytes can respond to MT-III with the production of leptin and adiponectin but not resistin.

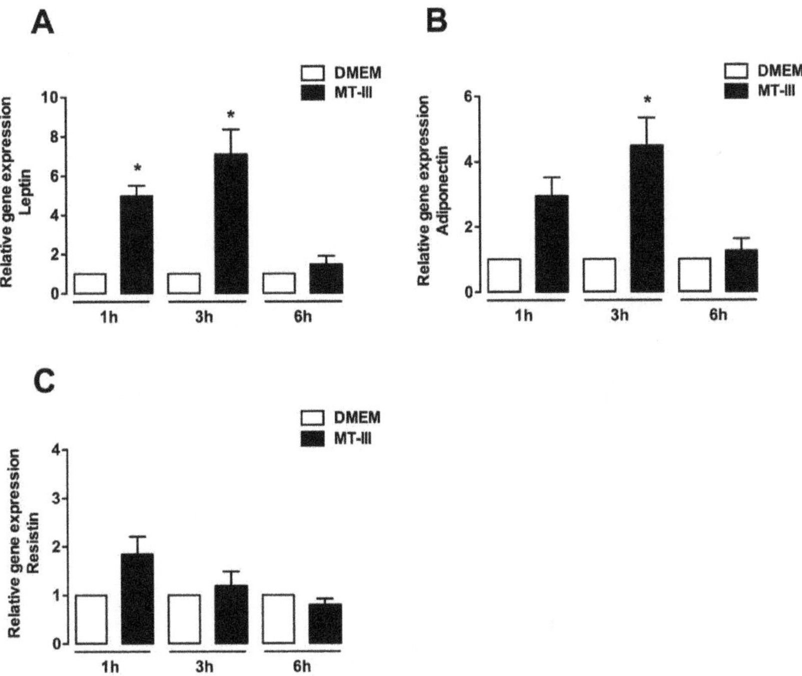

Figure 8. MT-III upregulates gene expression of leptin and adiponectin by 3T3-L1 preadipocytes. Cells were incubated with MT-III (0.4 μM) or DMEM (control) for 1, 3 or 6 h. Graphs show gene expression of leptin (**A**), adiponectin (**B**) and resistin (**C**) in the presence of MT-III. Concentrations of adipokines were quantified in cell lysates by qPCR. Results are expressed as mean ± SEM from 5 experiments. * $p < 0.05$ as compared with the control group (two-way ANOVA and Bonferroni posttest).

4. Discussion

Levels of $sPLA_2$ are elevated in the serum of obese patients as well as in inflamed fat tissue [1,2,50,51]. Previous studies have implicated $sPLA_2$s in metabolic diseases, including obesity [1,4,6]. However, the direct effects and mechanisms triggered by this class of enzymes on adipose tissue cells are not completely known. We herein report the ability of MT-III, a representative GIIA $sPLA_2$, to activate proinflammatory pathways in preadipocytes.

Prostanoids are produced from the metabolism of arachidonic acid by the cyclooxygenases system (COX-1 and COX-2) and are implicated in events related to the development of obesity, including inflammation and the differentiation of preadipocytes into mature adipocytes [33,52]. Our results demonstrate that MT-III induced an early and sustained release of PGE_2, followed by a late release of PGI_2. Taking into account the marked biosynthesis of PGE_2 in preadipocytes stimulated by MT-III and the contribution of this mediator to the inflammation process in the adipose tissue, the mechanisms involved in PGE_2 biosynthesis, induced by MT-III, were investigated. Our findings with a pharmacological approach indicated that PGE_2 production induced by MT-III is dependent upon the activation of COX-1 and COX-2 in preadipocytes. As an additional mechanism, MT-III upregulated COX-2 protein expression, but not COX-1 protein expression in preadipocytes. To our knowledge, this is the first demonstration that a GIIA $sPLA_2$ directly activates PGE_2 biosynthesis in preadipocytes, the precursor cells of mature adipocytes. Furthermore, our data evidence preadipocytes as target cells for GIIA $sPLA_2$ action.

PGE_2 synthases, including mPGES-1, mPGES-2 and cPGES, participate in the terminal step of PGE_2 biosynthesis by converting PGH_2 into PGE_2 [53]. In contrast to mPGES-2 and cPGES, the mPGES-1 isoform is upregulated in inflammatory conditions [54,55]. During obesity, upregulation of mPGES-1 expression has been shown in the adipose tissue during adipogenic processes [56]. Accordingly, we found that MT-III increased the expression of mPGES-1 in preadipocytes. The early release of PGE_2 correlated with mPGES-1 expression, indicating the participation of this terminal synthase in PGE_2 biosynthetic cascade triggered by MT-III, probably via functional coupling of COX-1 and mPGES-1 in the early stage of PLA_2 stimulation. These data reinforce the ability of GIIA $sPLA_2$s to activate mechanisms in preadipocytes that contribute to the development of obesity.

It is now well recognized that mammalian GIIA $sPLA_2$s do not exert their biological actions through their catalytic mechanism alone [57]. Several reports evidence that human GIIA $sPLA_2$s lead to eicosanoid production by means other than by directly providing arachidonic acid (AA) through catalysis. Among the non-catalytic mechanisms described is the crosstalk between mammalian GIIA sPLA2 and cPLA2 [58–62]. Several lines of evidence point out that the high AA specificity of $cPLA_2$-α and the lack of fatty acid selectivity in $sPLA_2$s can be combined to achieve specific cellular responses [38–40]. In this context, our finding that inhibition of the cytosolic ($cPLA_2$)-α by compound Pyr-2 abrogated the release of PGE_2 induced by MT-III indicates that the $cPLA_2$-α is a crucial partner for the effect triggered by MT-III in preadipocytes. This finding is in line with our previous data showing that MT-III increased phosphorylation of $cPLA_2$-α at Ser505, a hallmark of $cPLA_2$-alpha activation, in human monocytes [63]. Furthermore, although MT-III has the ability to release arachidonic acid from membrane phosphatidylcholine [63], our results evidence that the catalytic activity of MT-III does not play a role in production of PGE_2 in preadipocytes. A similar mechanism is widely accepted for mammalian GIIA sPLA2s [57].

We further extended our knowledge of the mechanisms involved in the generation of PGE_2 induced by MT-III by focusing on the participation of the EP4 receptor, which was shown to regulate the expression of key enzymes involved in PGE_2 biosynthesis, including COX-2 and PGESm-1 [2,41,52,64–66]. Our results, showing that EP4 antagonism by compound AH23848 abolished PGE_2 release induced by MT-III, indicate a critical role of this receptor in the effect of MT-III. These results strongly suggest that engagement of the EP4 receptor by PGE_2 triggers a positive feedback loop regulating the biosynthetic cascade of this mediator in preadipocytes stimulated by MT-III. Activation of this positive loop likely contributes to increased levels of PGE_2 observed throughout the period of stimulation with MT-III. In accordance with our data, studies using siRNA for knockdown EP4 gene in macrophages have shown reduced COX-2 expression upon stimuli by lipopolysaccharide [64]. In addition, our data evidenced a late release of PGI_2, which is considered a biomarker of adipocyte differentiation [67,68], in cells stimulated by MT-III. This suggests the involvement of GIIA $sPLA_2$s in the differentiation of preadipocytes. Although not investigated in this study, this hypothesis is currently being investigated in our laboratory.

Development of inflammation in the adipose tissue involves an early migration of leukocytes, mainly monocytes, into this tissue, followed by the secretion of several pro-inflammatory mediators by these cells, including cytokines, thus establishing an inflammatory environment [69–72]. Our findings showing a long-lasting release of MCP-1 in preadipocytes stimulated by MT-III strongly suggest that GIIA $sPLA_2s$ are implicated in the infiltration of monocytes and macrophages into adipose tissue and contribute to an inflammatory response in this tissue since MCP-1 is the key chemoattractant for monocytes during inflammatory conditions [70,72,73].

In addition, the release of IL-6 seen in preadipocytes stimulated by MT-III may contribute to the establishment of an inflammatory environment in the adipose tissue. These findings are in accordance with previous reports that levels of IL-6 are elevated in inflamed adipose tissue of obese patients that was associated with the induction of insulin resistance [74–76]. Furthermore, using pharmacological interference, we found that the MT-III-induced release of MCP-1 and IL-6 was dependent on EP4 or EP3 activation, respectively, in preadipocytes stimulated by MT-III. In view of previous evidence that the engagement of distinct EP receptors by PGE_2 triggers signalling pathways linked to the biosynthesis of proinflammatory cytokines [77,78], our findings indicate that PGE_2 biosynthesis, induced by MT-III, is an essential step for the activation of proinflammatory pathways linked to cytokine production in preadipocytes stimulated by the phospholipase A_2.

Adipose tissue produces specific cytokines known as adipokines, which are pivotal mediators that maintain a low-grade inflammation, which characterizes obesity [79,80]. These mediators have been described as exerting autocrine and paracrine effects and regulating appetite and satiety, glucose and lipid metabolism, blood pressure regulation, inflammation and immune functions [81–83]. Accordingly, we found that MT-III upregulated the expression of the adipokines leptin and adiponectin in preadipocytes. Previous reports have demonstrated that leptin is able to stimulate the production of proinflammatory cytokines by macrophages and expression of adhesion molecules by endothelial cells, thus contributing to the development of the inflammatory process in the adipose tissue [84]. Therefore, this mediator may be critical for the inflammatory effects triggered by MT-III in adipose tissue by promoting key inflammatory events. Moreover, in light of the modulatory effects of adiponectin in biological systems and inflammation [85], our findings suggest that this mediator may control the inflammatory response induced by MT-III leading to a low-grade inflammation environment, which characterizes obesity. In contrast, MT-III did not affect resistin expression in preadipocytes. This may be due to the predominance of mature adipocytes over preadipocytes for the production of this mediator [86]. Although the mechanisms related to the release of adipokines by MT-III have not been presently investigated, participation of PGE_2 and MCP-1 in the expression of leptin can be suggested since PGE_2 and MCP-1 have been described as activators of signalling pathways leading to leptin biosynthesis [87,88]. To the best of our knowledge, this is the first demonstration that a GIIA PLA_2 has the ability to induce the expression of adipokines in preadipocytes.

5. Conclusions

In this study, we demonstrate for the first time the ability of a representative GIIA phospholipase A_2, MT-III, to directly activate preadipocytes to release PGE_2 and the critical role of this mediator, acting via receptors EP3 and EP4, in inflammatory responses induced by this $sPLA_2$. MT-III also induced release of PGI_2, MCP-1 and IL-6 but not TNF-α, INF-γ, IL-12 or IL-10, and upregulated the expression of leptin and adiponectin. The MT-III-induced PGE_2 biosynthesis was dependent on the activation of $cPLA_2$-α, COX-1 and COX-2 pathways and positively regulated by the EP4 receptor. As an additional mechanism, MT-III upregulated COX-2 and mPGES-1 protein expression. MCP-1 biosynthesis induced by this $sPLA_2$ was dependent on the activation of the EP4 receptor, while IL-6 biosynthesis was dependent on the EP3 receptor in preadipocytes. Taken together, these findings provide evidence of a new target cell of the action of GIIA $sPLA_2s$, extending the knowledge of the effect of this class of enzymes in the adipose tissue (Scheme 1) given new insights into the roles of GIIA $sPLA_2s$ in obesity and associated disorders.

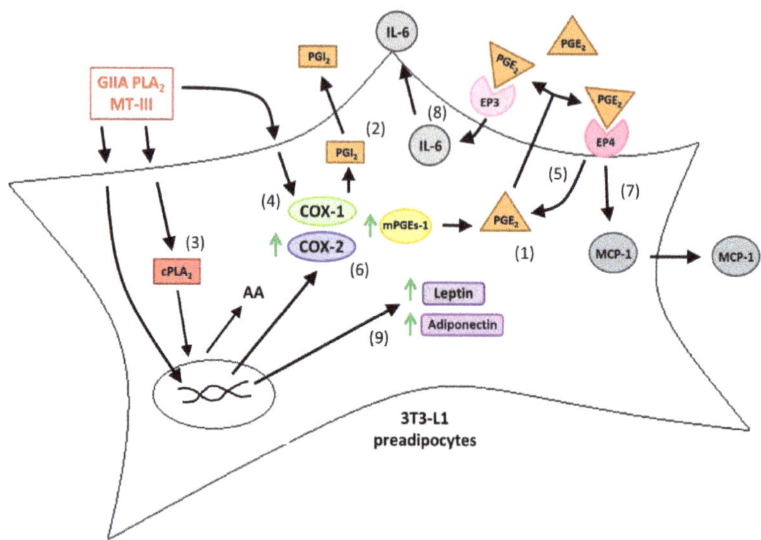

Scheme 1. Proinflammatory pathways activated by MT-III, a GIIA snake venom $sPLA_2$, in 3T3-L1 preadipocytes. (1) MT-III stimulates preadipocytes to release PGE_2 and (2) PGI_2; (3) PGE_2 release induced by MT-III is dependent on $cPLA_2$-α, (4) COX-1, COX-2, mPGES-1 and (5) EP4 receptor, which triggers a positive loop for PGE_2 production; (6) MT-III up-regulates COX-2 and mPGES-1, key enzymes involved in PGE biosynthesis. Moreover, MT-III induces the release of (7) MCP-1, dependent on the EP4 receptor, and (8) IL-6, dependent on the EP3 receptor. Furthermore, (9) MT-III up-regulates leptin and adiponectin gene expression

Author Contributions: Conceptualization: E.L. and C.T.; methodology, E.L., P.M., R.M.M.; formal analysis, E.L., P.M., R.M.M.; investigation, E.L., P.M., R.M.M.; B.L. isolated and purified MT-III; writing—original draft preparation, E.L., P.M., R.M.M. and C.T.; writing—review and editing, E.L., P.M., R.M.M. and C.T.; supervision, C.T.; project administration, S.V.S. and C.T. All authors have read and agreed to the published version of the manuscript.

Funding: This research was funded by Fundação de Amparo à Pesquisa do Estado de São Paulo (FAPESP), grant numbers 2011/23236-4, 2015/24701-3, 2017/197339.

Acknowledgments: The authors thank Renata Hage do Amaral Hernandez for providing technical assistance.

Conflicts of Interest: The authors declare no conflict of interest. The funders had no role in the design of the study; in the collection, analyses, or interpretation of data; in the writing of the manuscript, or in the decision to publish the results.

References

1. Garces, F.; López, F.; Niño, C.; Fernandez, A.; Chacin, L.; Hurt-Camejo, E.; Camejo, G.; Apitz-Castro, R. High Plasma Phospholipase A2 Activity, Inflammation Markers, and LDL Alterations in Obesity with or Without Type 2 Diabetes. *Obesity* **2010**, *18*, 2023–2029. [CrossRef] [PubMed]
2. Dutour, A.; Achard, V.; Sell, H.; Naour, N.; Collart, F.; Gaborit, B.; Silaghi, A.; Eckel, J.; Alessi, M.-C.; Henegar, C.; et al. Secretory Type II Phospholipase A2 Is Produced and Secreted by Epicardial Adipose Tissue and Overexpressed in Patients with Coronary Artery Disease. *J. Clin. Endocrinol. Metab.* **2010**, *95*, 963–967. [CrossRef] [PubMed]
3. Murakami, M.; Taketomi, Y.; Miki, Y.; Sato, H.; Yamamoto, K.; Lambeau, G. Emerging roles of secreted phospholipase A2 enzymes: The 3rd edition. *Biochimie* **2014**, *107*, 105–113. [CrossRef] [PubMed]
4. Murakami, M.; Taketomi, Y.; Sato, H.; Yamamoto, K. Secreted phospholipase A2 revisited. *J. Biochem.* **2011**, *150*, 233–255. [CrossRef]
5. Murakami, M.; Taketomi, Y. Secreted phospholipase A 2 and mast cells. *Allergol. Int.* **2015**, *64*, 4–10. [CrossRef]

6. Iyer, A.; Lim, J.; Poudyal, H.; Reid, R.C.; Suen, J.Y.; Webster, J.; Prins, J.B.; Whitehead, J.P.; Fairlie, D.P.; Brown, L. An Inhibitor of Phospholipase A2 Group IIA Modulates Adipocyte Signaling and Protects Against Diet-Induced Metabolic Syndrome in Rats. *Diabetes* **2012**, *61*, 2320–2329. [CrossRef]
7. Jacob, P.J.; Manju, S.L.; Ethiraj, K.R.; Elias, G. Safer anti-inflammatory therapy through dual COX-2/5-LOX inhibitors: A structure-based approach. *Eur. J. Pharm. Sci.* **2018**, *121*, 356–381. [CrossRef]
8. Park, J.Y.; Pillinger, M.H.; Abramson, S.B. Prostaglandin E2 synthesis and secretion: The role of PGE2 synthases. *Clin. Immunol.* **2006**, *119*, 229–240. [CrossRef]
9. Pérez, S.; Aspichueta, P.; Ochoa, B.; Chico, Y.; Ochoa, B. The 2-series prostaglandins suppress VLDL secretion in an inflammatory condition-dependent manner in primary rat hepatocytes. *Biochim. Et Biophys. Acta (BBA) Mol. Cell Biol. Lipids* **2006**, *1761*, 160–171. [CrossRef]
10. Enomoto, N.; Ikejima, K.; Yamashina, S.; Enomoto, A.; Nishiura, T.; Nishimura, T.; Brenner, D.A.; Schemmer, P.; Bradford, B.U.; Rivera, C.A.; et al. Kupffer cell-derived prostaglandin E2is involved in alcohol-induced fat accumulation in rat liver. *Am. J. Physiol. Liver Physiol.* **2000**, *279*, G100–G106. [CrossRef]
11. O'Hara, A.; Lim, F.-L.; Mazzatti, D.J.; Trayhurn, P. Stimulation of inflammatory gene expression in human preadipocytes by macrophage-conditioned medium: Upregulation of IL-6 production by macrophage-derived IL-1β. *Mol. Cell. Endocrinol.* **2012**, *349*, 239–247. [CrossRef] [PubMed]
12. Wood, I.; Trayhurn, P. Signalling role of adipose tissue: Adipokines and inflammation in obesity. *Biochem. Soc. Trans.* **2005**, *33*, 1078. [CrossRef]
13. Mafra, D.; Mafra, D. Adipokines in obesity. *Clin. Chim. Acta* **2013**, *419*, 87–94. [CrossRef]
14. Karastergiou, K.; Mohamed-Ali, V. The autocrine and paracrine roles of adipokines. *Mol. Cell. Endocrinol.* **2010**, *318*, 69–78. [CrossRef] [PubMed]
15. Chung, S.; LaPoint, K.; Martinez, K.; Kennedy, A.; Sandberg, M.B.; McIntosh, M.K. Preadipocytes Mediate Lipopolysaccharide-Induced Inflammation and Insulin Resistance in Primary Cultures of Newly Differentiated Human Adipocytes. *Endocrinology* **2006**, *147*, 5340–5351. [CrossRef] [PubMed]
16. Davidson, F.F.; Dennis, E.A. Evolutionary relationships and implications for the regulation of phospholipase A2 from snake venom to human secreted forms. *J. Mol. Evol.* **1990**, *31*, 228–238. [CrossRef] [PubMed]
17. Gutiérrez, J.M.; Lomonte, B. Phospholipase A2 myotoxins from Bothrops snake venoms. *Toxicon* **1995**, *33*, 1405–1424. [CrossRef]
18. Schaloske, R.H.; Dennis, E.A. The phospholipase A2 superfamily and its group numbering system. *Biochim. Et Biophys. Acta (BBA) Mol. Cell Biol. Lipids* **2006**, *1761*, 1246–1259. [CrossRef]
19. Leiguez, E.; Zuliani, J.P.; Cianciarullo, A.M.; Fernandes, C.M.; Gutiérrez, J.M.; Teixeira, C.D.F.P. A group IIA-secreted phospholipase A2 from snake venom induces lipid body formation in macrophages: The roles of intracellular phospholipases A2 and distinct signaling pathways. *J. Leukoc. Biol.* **2011**, *90*, 155–166. [CrossRef]
20. Leiguez, E.; Giannotti, K.C.; Viana, M.D.N.; Matsubara, M.H.; Fernandes, C.M.; Gutiérrez, J.M.; Lomonte, B.; Teixeira, C. A Snake Venom-Secreted Phospholipase A2Induces Foam Cell Formation Depending on the Activation of Factors Involved in Lipid Homeostasis. *Mediat. Inflamm.* **2018**, *2018*, 1–13. [CrossRef]
21. Xie, Z.; Wang, X.; Liu, X.; Du, H.; Sun, C.; Shao, X.; Tian, J.; Gu, X.; Wang, H.; Tian, J.; et al. Adipose-Derived Exosomes Exert Proatherogenic Effects by Regulating Macrophage Foam Cell Formation and Polarization. *J. Am. Heart Assoc.* **2018**, *7*, e007442. [CrossRef] [PubMed]
22. Shapiro, H.; Pecht, T.; Shaco-Levy, R.; Harman-Boehm, I.; Kirshtein, B.; Kuperman, Y.; Chen, A.; Blüher, M.; Shai, I.; Rudich, A. Adipose Tissue Foam Cells Are Present in Human Obesity. *J. Clin. Endocrinol. Metab.* **2013**, *98*, 1173–1181. [CrossRef] [PubMed]
23. Kaiser, I.I.; Gutiérrez, J.M.; Plummer, D.; Aird, S.D.; Odell, G.V. The amino acid sequence of a myotoxic phospholipase from the venom of Bothrops asper. *Arch. Biochem. Biophys.* **1990**, *278*, 319–325. [CrossRef]
24. Díaz-Oreiro, C.; Gutiérrez, J.M. Chemical modification of histidine and lysine residues of myotoxic phospholipases A2 isolated from Bothrops asper and Bothrops godmani snake venoms: Effects on enzymatic and pharmacological properties. *Toxicon* **1997**, *35*, 241–252. [CrossRef]
25. Takayama, K.; Mitchell, D.H.; Din, Z.Z.; Mukerjee, P.; Li, C.; Coleman, D.L. Monomeric Re lipopolysaccharide from Escherichia coli is more active than the aggregated form in the Limulus amebocyte lysate assay and in inducing Egr-1 mRNA in murine peritoneal macrophages. *J. Biol. Chem.* **1994**, *269*, 2241–2244.
26. Dordevic, A.L.; Konstantopoulos, N.; Cameron-Smith, D. 3T3-L1 Preadipocytes Exhibit Heightened Monocyte-Chemoattractant Protein-1 Response to Acute Fatty Acid Exposure. *PLoS ONE* **2014**, *9*, e99382. [CrossRef]

27. Chang, Y.-H.; Lee, S.T.; Lin, W.-W. Effects of cannabinoids on LPS-stimulated inflammatory mediator release from macrophages: Involvement of eicosanoids. *J. Cell. Biochem.* **2001**, *81*, 715–723. [CrossRef]
28. Choi, H.C.; Kim, H.S.; Lee, K.Y.; Chang, K.C.; Kang, Y.J. NS-398, a selective COX-2 inhibitor, inhibits proliferation of IL-1β-stimulated vascular smooth muscle cells by induction of HO-1. *Biochem. Biophys. Res. Commun.* **2008**, *376*, 753–757. [CrossRef]
29. Lin, Y.-S.; Hsieh, M.; Lee, Y.-J.; Liu, K.-L.; Lin, T.-H. AH23848 accelerates inducible nitric oxide synthase degradation through attenuation of cAMP signaling in glomerular mesangial cells. *Nitric Oxide* **2008**, *18*, 93–104. [CrossRef]
30. Lin, C.-C.; Lin, W.-N.; Wang, W.-J.; Sun, C.-C.; Tung, W.-H.; Wang, H.-H.; Yang, C.-M. Functional coupling expression of COX-2 and cPLA2 induced by ATP in rat vascular smooth muscle cells: Role of ERK1/2, p38 MAPK, and NF-κB. *Cardiovasc. Res.* **2009**, *82*, 522–531. [CrossRef]
31. Chen, L.; Miao, Y.; Zhang, Y.; Dou, D.; Liu, L.; Tian, X.; Yang, G.; Pu, D.; Zhang, X.; Kang, J.; et al. Inactivation of the E-Prostanoid 3 Receptor Attenuates the Angiotensin II Pressor Response via Decreasing Arterial Contractility. *Arter. Thromb. Vasc. Biol.* **2012**, *32*, 3024–3032. [CrossRef] [PubMed]
32. Ratchford, A.M.; Esguerra, C.R.; Moley, K.H. Decreased Oocyte-Granulosa Cell Gap Junction Communication and Connexin Expression in a Type 1 Diabetic Mouse Model. *Mol. Endocrinol.* **2008**, *22*, 2643–2654. [CrossRef] [PubMed]
33. Iyer, A.; Fairlie, D.P.; Prins, J.B.; Hammock, B.D.; Brown, L. Inflammatory lipid mediators in adipocyte function and obesity. *Nat. Rev. Endocrinol.* **2010**, *6*, 71–82. [CrossRef] [PubMed]
34. Kawahara, K.; Hohjoh, H.; Inazumi, T.; Tsuchiya, S.; Sugimoto, Y. Prostaglandin E2-induced inflammation: Relevance of prostaglandin E receptors. *Biochim. Et Biophys. Acta (BBA) Mol. Cell Biol. Lipids* **2015**, *1851*, 414–421. [CrossRef] [PubMed]
35. Yan, H.; Kermouni, A.; Abdel-Hafez, M.; Lau, D.C. Role of cyclooxygenases COX-1 and COX-2 in modulating adipogenesis in 3T3-L1 cells. *J. Lipid Res.* **2003**, *44*, 424–429. [CrossRef] [PubMed]
36. Yang, G.; Chen, L. An Update of Microsomal Prostaglandin E Synthase-1 and PGE2Receptors in Cardiovascular Health and Diseases. *Oxid. Med. Cell. Longev.* **2016**, *2016*, 1–9. [CrossRef] [PubMed]
37. McCoy, J.M.; Wicks, J.R.; Audoly, L.P. The role of prostaglandin E2 receptors in the pathogenesis of rheumatoid arthritis. *J. Clin. Investig.* **2002**, *110*, 651–658. [CrossRef]
38. Dennis, E.A.; Cao, J.; Hsu, Y.-H.; Magrioti, V.; Kokotos, G. Phospholipase A2Enzymes: Physical Structure, Biological Function, Disease Implication, Chemical Inhibition, and Therapeutic Intervention. *Chem. Rev.* **2011**, *111*, 6130–6185. [CrossRef]
39. Balsinde, J.; Winstead, M.V.; A Dennis, E. Phospholipase A2 regulation of arachidonic acid mobilization. *FEBS Lett.* **2002**, *531*, 2–6. [CrossRef]
40. Lambeau, G.; Gelb, M.H. Biochemistry and Physiology of Mammalian Secreted Phospholipases A2. *Annu. Rev. Biochem.* **2008**, *77*, 495–520. [CrossRef]
41. Viana, M.N.; Leiguez, E.; Gutiérrez, J.M.; Rucavado, A.; Markus, R.P.; Marçola, M.; Teixeira, C.; Fernandes, C.M. A representative metalloprotease induces PGE2 synthesis in fibroblast-like synoviocytes via the NF-κB/COX-2 pathway with amplification by IL-1β and the EP4 receptor. *Sci. Rep.* **2020**, *10*, 3269. [CrossRef] [PubMed]
42. Fantuzzi, G. Adipose tissue, adipokines, and inflammation. *J. Allergy Clin. Immunol.* **2005**, *115*, 911–919. [CrossRef] [PubMed]
43. Wang, T.; He, C. Pro-inflammatory cytokines: The link between obesity and osteoarthritis. *Cytokine Growth Factor Rev.* **2018**, *44*, 38–50. [CrossRef] [PubMed]
44. Stolarczyk, E. Adipose tissue inflammation in obesity: A metabolic or immune response? *Curr. Opin. Pharmacol.* **2017**, *37*, 35–40. [CrossRef]
45. Yamane, H.; Sugimoto, Y.; Tanaka, S.; Ichikawa, A. Prostaglandin E2 Receptors, EP2 and EP4, Differentially Modulate TNF-α and IL-6 Production Induced by Lipopolysaccharide in Mouse Peritoneal Neutrophils. *Biochem. Biophys. Res. Commun.* **2000**, *278*, 224–228. [CrossRef]
46. Li, X.; Ellman, M.; Muddasani, P.; Wang, J.H.-C.; Cs-Szabo, G.; Van Wijnen, A.J.; Im, H.-J. Prostaglandin E2and its cognate EP receptors control human adult articular cartilage homeostasis and are linked to the pathophysiology of osteoarthritis. *Arthritis Rheum.* **2009**, *60*, 513–523. [CrossRef]
47. Tilg, H.; Moschen, A.R. Adipocytokines: Mediators linking adipose tissue, inflammation and immunity. *Nat. Rev. Immunol.* **2006**, *6*, 772–783. [CrossRef]

48. Fasshauer, M.; Blüher, M. Adipokines in health and disease. *Trends Pharmacol. Sci.* **2015**, *36*, 461–470. [CrossRef]
49. Golia, E.; Limongelli, G.; Natale, F.; Fimiani, F.; Maddaloni, V.; Russo, P.E.; Riegler, L.; Bianchi, R.; Crisci, M.; Di Palma, G.; et al. Adipose tissue and vascular inflammation in coronary artery disease. *World J. Cardiol.* **2014**, *6*, 539. [CrossRef]
50. Paradis, M.-E.; Hogue, M.-O.; Mauger, J.-F.; Couillard, C.; Couture, P.; Bergeron, N.; Lamarche, B. Visceral adipose tissue accumulation, secretory phospholipase A2-IIA and atherogenecity of LDL. *Int. J. Obes.* **2006**, *30*, 1615–1622. [CrossRef]
51. Hui, D.Y. Phospholipase A2 enzymes in metabolic and cardiovascular diseases. *Curr. Opin. Lipidol.* **2012**, *23*, 235–240. [CrossRef] [PubMed]
52. Rahman, M.S. Prostacyclin: A major prostaglandin in the regulation of adipose tissue development. *J. Cell. Physiol.* **2018**, *234*, 3254–3262. [CrossRef] [PubMed]
53. Maione, F.; Casillo, G.M.; Raucci, F.; Iqbal, A.J.; Mascolo, N. The functional link between microsomal prostaglandin E synthase-1 (mPGES-1) and peroxisome proliferator-activated receptor γ (PPARγ) in the onset of inflammation. *Pharmacol. Res.* **2020**, *157*, 104807. [CrossRef] [PubMed]
54. Ikeda-Matsuo, Y. The Role of mPGES-1 in Inflammatory Brain Diseases. *Biol. Pharm. Bull.* **2017**, *40*, 557–563. [CrossRef]
55. Samuelsson, B.; Morgenstern, R.; Jakobsson, P. Membrane Prostaglandin E Synthase-1: A Novel. *Pharmacol. Rev.* **2007**, *59*, 207–224. [CrossRef]
56. Michaud, A.; Lacroix-Pepin, N.; Pelletier, M.; Daris, M.; Biertho, L.; Fortier, M.A.; Tchernof, A. Expression of Genes Related to Prostaglandin Synthesis or Signaling in Human Subcutaneous and Omental Adipose Tissue: Depot Differences and Modulation by Adipogenesis. *Mediat. Inflamm.* **2014**, *2014*, 1–13. [CrossRef]
57. Kim, R.R.; Chen, Z.; Mann, T.J.; Bastard, K.; Scott, K.F.; Church, W.B. Structural and Functional Aspects of Targeting the Secreted Human Group IIA Phospholipase A_2. *Molecules* **2020**, *25*, 4459. [CrossRef]
58. Suga, H.; Murakami, M.; Kudo, I.; Inoue, K. Participation in cellular prostaglandin synthesis of type-II phospholipase A2 secreted and anchored on cell-surface heparan sulfate proteoglycan. *JBIC J. Biol. Inorg. Chem.* **1993**, *218*, 807–813. [CrossRef]
59. Lee, L.K.; Bryant, K.J.; Bouveret, R.; Lei, P.-W.; Duff, A.P.; Harrop, S.J.; Huang, E.P.; Harvey, R.P.; Gelb, M.H.; Gray, P.P.; et al. Selective Inhibition of Human Group IIA-secreted Phospholipase A2 (hGIIA) Signaling Reveals Arachidonic Acid Metabolism Is Associated with Colocalization of hGIIA to Vimentin in Rheumatoid Synoviocytes. *J. Biol. Chem.* **2013**, *288*, 15269–15279. [CrossRef]
60. Perlson, E.; Michaelevski, I.; Kowalsman, N.; Ben-Yaakov, K.; Shaked, M.; Seger, R.; Eisenstein, M.; Fainzilber, M. Vimentin Binding to Phosphorylated Erk Sterically Hinders Enzymatic Dephosphorylation of the Kinase. *J. Mol. Biol.* **2006**, *364*, 938–944. [CrossRef]
61. Sales, T.A.; Marcussi, S.; Da Cunha, E.F.; Kuca, K.; Kuca, K. Can Inhibitors of Snake Venom Phospholipases A2 Lead to New Insights into Anti-Inflammatory Therapy in Humans? A Theoretical Study. *Toxins* **2017**, *9*, 341. [CrossRef] [PubMed]
62. Murakami, M.; Kambe, T.; Shimbara, S.; Yamamoto, S.; Kuwata, H.; Kudo, I. Functional Association of Type IIA Secretory Phospholipase A2with the Glycosylphosphatidylinositol-anchored Heparan Sulfate Proteoglycan in the Cyclooxygenase-2-mediated Delayed Prostanoid-biosynthetic Pathway. *J. Biol. Chem.* **1999**, *274*, 29927–29936. [CrossRef] [PubMed]
63. Rodríguez, J.P.; Leiguez, E.; Guijas, C.; Lomonte, B.; Gutiérrez, J.M.; Teixeira, C.; Balboa, M.; Balsinde, J. A Lipidomic Perspective of the Action of Group IIA Secreted Phospholipase A2 on Human Monocytes: Lipid Droplet Biogenesis and Activation of Cytosolic Phospholipase A2α. *Biomolecules* **2020**, *10*, 891. [CrossRef] [PubMed]
64. Khan, K.M.F.; Kothari, P.; Du, B.; Dannenberg, A.J.; Falcone, D.J. Matrix Metalloproteinase-Dependent Microsomal Prostaglandin E Synthase-1 Expression in Macrophages: Role of TNF-α and the EP4 Prostanoid Receptor. *J. Immunol.* **2012**, *188*, 1970–1980. [CrossRef] [PubMed]
65. Kojima, F.; Naraba, H.; Sasaki, Y.; Beppu, M.; Aoki, H.; Kawai, S. Prostaglandin E2is an enhancer of interleukin-1β-induced expression of membrane-associated prostaglandin E synthase in rheumatoid synovial fibroblasts. *Arthritis Rheum.* **2003**, *48*, 2819–2828. [CrossRef]

66. Khan, H.; Rengasamy, K.R.; Pervaiz, A.; Nabavi, S.M.; Atanasov, A.G.; Kamal, M.A.; Perviaz, A. Plant-derived mPGES-1 inhibitors or suppressors: A new emerging trend in the search for small molecules to combat inflammation. *Eur. J. Med. Chem.* **2018**, *153*, 2–28. [CrossRef]
67. Rahman, M.S.; Khan, F.; Syeda, P.K.; Nishimura, K.; Jisaka, M.; Nagaya, T.; Shono, F.; Yokota, K. Endogenous synthesis of prostacyclin was positively regulated during the maturation phase of cultured adipocytes. *Cytotechnology* **2013**, *66*, 635–646. [CrossRef] [PubMed]
68. Darimont, C.; Vassaux, G.; Ailhaud, G.; Negrel, R. Differentiation of preadipose cells: Paracrine role of prostacyclin upon stimulation of adipose cells by angiotensin-II. *Endocrinology* **1994**, *135*, 2030–2036. [CrossRef]
69. Weinstock, A.; Silva, H.M.; Moore, K.J.; Schmidt, A.M.; Fisher, E.A. Leukocyte Heterogeneity in Adipose Tissue, Including in Obesity. *Circ. Res.* **2020**, *126*, 1590–1612. [CrossRef]
70. Mraz, M.; Haluzik, M. The role of adipose tissue immune cells in obesity and low-grade inflammation. *J. Endocrinol.* **2014**, *222*, R113–R127. [CrossRef]
71. Russo, L.; Lumeng, C.N. Properties and functions of adipose tissue macrophages in obesity. *Immunology* **2018**, *155*, 407–417. [CrossRef] [PubMed]
72. Crop, M.; Baan, C.; Korevaar, S.; Ijzermans, J.; Pescatori, M.; Stubbs, A.; Van Ijcken, W.; Dahlke, M.; Eggenhofer, E.; Weimar, W.; et al. Inflammatory conditions affect gene expression and function of human adipose tissue-derived mesenchymal stem cells. *Clin. Exp. Immunol.* **2010**, *162*, 474–486. [CrossRef] [PubMed]
73. Christiansen, T.; Richelsen, B.; Bruun, J.M. Monocyte chemoattractant protein-1 is produced in isolated adipocytes, associated with adiposity and reduced after weight loss in morbid obese subjects. *Int. J. Obes.* **2005**, *29*, 146–150. [CrossRef]
74. Hagman, E.; Besor, O.; Hershkop, K.; Santoro, N.; Pierpont, B.; Mata, M.; Caprio, S.; Weiss, R. Relation of the degree of obesity in childhood to adipose tissue insulin resistance. *Acta Diabetol.* **2019**, *56*, 219–226. [CrossRef] [PubMed]
75. Bastard, J.-P.; Maachi, M.; Lagathu, C.; Kim, M.J.; Caron, M.; Vidal, H.; Capeau, J.; Fève, B. Recent advances in the relationship between obesity, inflammation, and insulin resistance. *Eur. Cytokine Netw.* **2006**, *17*, 4–12. [PubMed]
76. Eder, K.; Baffy, N.; Falus, A.; Fulop, A.K. The major inflammatory mediator interleukin-6 and obesity. *Inflamm. Res.* **2009**, *58*, 727–736. [CrossRef] [PubMed]
77. Józefowski, S.; Bobek, M.; Marcinkiewicz, J. Exogenous but not endogenous prostanoids regulate cytokine secretion from murine bone marrow dendritic cells: EP2, DP, and IP but not EP1, EP3, and FP prostanoid receptors are involved. *Int. Immunopharmacol.* **2003**, *3*, 865–878. [CrossRef]
78. Jiang, J.; Dingledine, R. Role of Prostaglandin Receptor EP2 in the Regulations of Cancer Cell Proliferation, Invasion, and Inflammation. *J. Pharmacol. Exp. Ther.* **2013**, *344*, 360–367. [CrossRef]
79. Sam, S.; Mazzone, T. Adipose tissue changes in obesity and the impact on metabolic function. *Transl. Res.* **2014**, *164*, 284–292. [CrossRef]
80. Kershaw, E.E.; Flier, J.S. Adipose Tissue as an Endocrine Organ. *J. Clin. Endocrinol. Metab.* **2004**, *89*, 2548–2556. [CrossRef]
81. Lago, F.; Diéguez, C.; Gómez-Reino, J.; Gualillo, O. Adipokines as emerging mediators of immune response and inflammation. *Nat. Clin. Pract. Rheumatol.* **2007**, *3*, 716–724. [CrossRef] [PubMed]
82. Lago, F.; Gómez, R.; Gómez-Reino, J.J.; Dieguez, C.; Gualillo, O. Adipokines as novel modulators of lipid metabolism. *Trends Biochem. Sci.* **2009**, *34*, 500–510. [CrossRef] [PubMed]
83. Wozniak, S.E.; Gee, L.L.; Wachtel, M.S.; Frezza, E.E. Adipose Tissue: The New Endocrine Organ? A Review Article. *Dig. Dis. Sci.* **2009**, *54*, 1847–1856. [CrossRef] [PubMed]
84. Fernández-Riejos, P.; Najib, S.; Santos-Alvarez, J.; Martín-Romero, C.; Pérez-Pérez, A.; González-Yanes, C.; Sánchez-Margalet, V. Role of Leptin in the Activation of Immune Cells. *Mediat. Inflamm.* **2010**, *2010*, 1–8. [CrossRef] [PubMed]
85. Balistreri, C.R.; Caruso, C.; Candore, G. The Role of Adipose Tissue and Adipokines in Obesity-Related Inflammatory Diseases. *Mediat. Inflamm.* **2010**, *2010*, 802078. [CrossRef]
86. Schäffler, A.; Schölmerich, J.; Salzberger, B. Adipose tissue as an immunological organ: Toll-like receptors, C1q/TNFs and CTRPs. *Trends Immunol.* **2007**, *28*, 393–399. [CrossRef]

87. Fain, J.N.; Bahouth, S.W. Regulation of Leptin Release by Mammalian Adipose Tissue. *Biochem. Biophys. Res. Commun.* **2000**, *274*, 571–575. [CrossRef]
88. Powell, K. The two faces of fat. *Nat. Cell Biol.* **2007**, *447*, 525–527. [CrossRef]

Publisher's Note: MDPI stays neutral with regard to jurisdictional claims in published maps and institutional affiliations.

© 2020 by the authors. Licensee MDPI, Basel, Switzerland. This article is an open access article distributed under the terms and conditions of the Creative Commons Attribution (CC BY) license (http://creativecommons.org/licenses/by/4.0/).

biomolecules

Article

Crotoxin-Induced Mice Lung Impairment: Role of Nicotinic Acetylcholine Receptors and COX-Derived Prostanoids

Marco Aurelio Sartim [1], Camila O. S. Souza [1], Cassiano Ricardo A. F. Diniz [2], Vanessa M. B. da Fonseca [3], Lucas O. Sousa [1], Ana Paula F. Peti [1], Tassia Rafaella Costa [1], Alan G. Lourenço [4], Marcos C. Borges [3], Carlos A. Sorgi [1], Lucia Helena Faccioli [1] and Suely Vilela Sampaio [1,*]

1 Department of Clinical Analysis, Toxicology and Food Sciences, School of Pharmaceutical Sciences of Ribeirão Preto, University of São Paulo, Ribeirão Preto 14040-903, SP, Brazil; marcosartim@hotmail.com (M.A.S.); camila.oliveirasilva@usp.br (C.O.S.S.); losousa@usp.br (L.O.S.); anaferranti@usp.br (A.P.F.P.); tassiarcosta@yahoo.com.br (T.R.C.); sorgi@fcfrp.usp.br (C.A.S.); faccioli@fcfrp.usp.br (L.H.F.)

2 Department of Pharmacology, Ribeirão Preto Medical School, University of São Paulo, Ribeirão Preto 14049-900, SP, Brazil; crafd87@gmail.com

3 Department of Internal Medicine, Ribeirão Preto Medical School, University of São Paulo, Ribeirão Preto 14049-900, SP, Brazil; van.macielfonseca@gmail.com (V.M.B.d.F.); marcosborges@fmrp.usp.br (M.C.B.)

4 Department of Basic and Oral Biology, School of Dentistry of Ribeirão Preto, University of São Paulo, Ribeirão Preto 14040-904, SP, Brazil; lourenco@forp.usp.br

* Correspondence: suvilela@usp.br; Tel.: +55-16-3315-4287

Received: 10 February 2020; Accepted: 14 March 2020; Published: 20 May 2020

Abstract: Respiratory compromise in *Crotalus durissus terrificus* (C.d.t.) snakebite is an important pathological condition. Considering that crotoxin (CTX), a phospholipase A_2 from C.d.t. venom, is the main component of the venom, the present work investigated the toxin effects on respiratory failure. Lung mechanics, morphology and soluble markers were evaluated from Swiss male mice, and mechanism determined using drugs/inhibitors of eicosanoids biosynthesis pathway and autonomic nervous system. Acute respiratory failure was observed, with an early phase (within 2 h) characterized by enhanced presence of eicosanoids, including prostaglandin E2, that accounted for the increased vascular permeability in the lung. The alterations of early phase were inhibited by indomethacin. The late phase (peaked 12 h) was marked by neutrophil infiltration, presence of pro-inflammatory cytokines/chemokines, and morphological alterations characterized by alveolar septal thickening and bronchoconstriction. In addition, lung mechanical function was impaired, with decreased lung compliance and inspiratory capacity. Hexamethonium, a nicotinic acetylcholine receptor antagonist, hampered late phase damages indicating that CTX-induced lung impairment could be associated with cholinergic transmission. The findings reported herein highlight the impact of CTX on respiratory compromise, and introduce the use of nicotinic blockers and prostanoids biosynthesis inhibitors as possible symptomatic therapy to *Crotalus durissus terrificus* snakebite.

Keywords: crotoxin; snake venom; lung impairment; inflammatory response; lipid mediators; neuromuscular blocker

---

1. Introduction

Rattlesnakes are native to the Americas and are responsible for several cases of envenomation in the continent [1]. They caused 56.3% of the ophidic accidents in North America, while snakes from genus *Crotalus*—the major rattlesnake genus in Central and South Americas—caused less than 10% of the 50,000 cases of snakebite per year in the region [1–4]. In Brazil, the 2016 report from the Ministry of Health [5] recorded that snakes from genus *Crotalus* participated in approximately 10% of the notified accidents caused by venomous snakes, and accounted for the highest mortality rate.

Typical clinical manifestations during envenomation are related to severe systemic disturbances, such as neurotoxicity, coagulation alterations, and respiratory and renal failure associated with myotoxicity, leading to failure of end-organs and death [6,7]. Although rarely reported, respiratory impairment induced by rattlesnake bite is a potential lethal complication associated with severe cases of envenomation [8–12], and it is characterized by airway obstruction, bronchospasm, soft tissue edema, or subjective symptoms including throat tightening and nasal congestion [11]. *Crotalus durissus* snakebite causes other respiratory abnormalities within the first 48 h, such as dyspnea, tachypnea, use of accessory muscles of respiration and flaring of the nostrils, followed by decreased blood pH and pO_2, and increased pCO_2 levels [9]. *Crotalus durissus terrificus* (C.d.t.) [13] and *Crotalus durissus cascavella* [14] crude venom induces similar respiratory disturbances in a mice model of envenomation, in addition to (i) mechanical alterations in lung tissues characterized by increased lung static- and dynamic-elastance, and resistive- and viscoelastic-pressure; and (ii) morphological alterations including increased leukocyte infiltration, hemorrhage, and edema [13,14].

Crotoxin (CTX) is the main toxic component of the venom from the South American rattlesnake C.d.t. This toxin is isolated as a heterodimeric complex composed of a basic enzymatically active phospholipase A_2 (CB) non-covalently bound to an acidic non-enzymatic domain (crotapotin) [15–17]. CTX has been associated with several pathological conditions such as neurotoxicity, myotoxicity, and immune alterations [18–22], but its participation in respiratory disturbances is poorly reported and remains controversial. The CTX complex (CB/crotapotin), but not its components alone (CB or crotapotin), causes complete respiratory arrest associated with decreased blood pH and pO_2, and increased pCO_2 in rabbits [23]. In contrast, CTX do not modulate respiration frequency and amplitude in dogs [24].

The reported clinical and experimental data on C.d.t. effects on respiratory function stress the importance of investigating how CTX, the most abundant venom toxin, participates in the impairment of lung physiology. Literature reports are limited, do not show lung pathological alterations in depth, and do not elucidate the mechanism by which CTX acts. In this sense, the present work investigated the pathophysiology of CTX-induced lung disturbances in mice, in particular the morphological and functional alterations, as well as the participation of peripheral nervous system and production of lipid mediators during respiratory failure.

2. Materials and Methods

2.1. Animals

Male 8–9 week-old Swiss mice (35–40 g) were provided by the Central Animal Facility of the University of São Paulo, Campus of Ribeirão Preto (Ribeirão Preto, SP, Brazil). The animals were housed at Animal Facility at Pharmaceutical Sciences School of Ribeirão Preto (FCFRP–USP) under controlled conditions of temperature (23 °C) and brightness (12 h light/dark cycles), and with free access to food and water. The experiments were performed at FCFRP-USP following animal care procedures, which experimental protocols are in accordance with the COBEA (Brazilian College of Animal Experimentation) guidelines and were approved by the Ethics Committee on Animal Use (CEUA) from the University of São Paulo, Campus of Ribeirão Preto (protocol number: 15.1.807.60.1).

2.2. Crotoxin

Crotoxin (CTX) was isolated from C.d.t crude venom as described by Muller and colleagues [1]. To eliminate endotoxin contaminants, CTX sample was purified using Affi-Prep Polymyxin Resin according to the manufacturer's instructions (Bio-Rad—Hercules, CA, USA). The endotoxin levels were lower than 0.01 EU/µg of CTX (1 EU = 0.1 ng of endotoxin), as determined using the limulus amoebocyte lysate kit (Lonza Biosciences—Walkersville, MD, USA). Protein concentration in CTX samples was quantified using the BCA kit, according to the manufacturer's instructions (Thermo Scientific—Rockford, IL, USA).

2.3. CTX and Drug Treatments

2.3.1. CTX Dose- and Time-Response Experiments

To select a suitable CTX dose for the in vivo assays, a dose-response experiment was carried out using subcutaneous injection (s.c.) of CTX at 10–300 µg/Kg or saline (control). After 6 h, mice were anesthetized with intraperitoneal (i.p.) injection of ketamine/xylazine solution (80/10 mg/kg), their lung and heart were removed for analysis, and their blood was collected by cardiac puncture for analysis of whole blood and serum. Based on the survival profile of the animals, the CTX dose of 300 µg/Kg s.c. and time treatments of 2, 6, and 12 h were selected for further experiments. Animals treated with saline under the same conditions were used as the control group.

2.3.2. Drug Treatments

Drugs that act as antagonists or inhibitors of eicosanoids production and peripheral neuronal pathways were used to investigate the toxicological mechanisms of CTX action in mice. Indomethacin (Sigma-Aldrich—St. Louis, MO, USA) was administered (3 mg/Kg i.p.) 4.5 h before CTX s.c. injection [2,3]. MK-591 (AdooQ Bioscience—Irvine, CA, USA) was administered (40 mg/Kg i.p.) 30 min before CTX s.c. injection [4]. Hexamethonium bromide (Sigma-Aldrich) was administered intravenously (i.v. tail vein) at a dose of 10 mg/Kg, 15 min before CTX s.c. injection [5]. Methyl-atropine (Sigma-Aldrich) was administered (30 mg/Kg i.p.) 30 min before CTX s.c. injection [6]. Neostigmine (Sigma-Aldrich) was administered (0.1 mg/Kg i.p.) 10 min before CTX s.c. injection [7]. Propranolol (Tocris Bioscience—Bristol, UK) was administered (5 mg/Kg i.p.) 30 min before CTX s.c. injection [8]. Indomethacin was prepared in Tris-HCl 100 mM pH 8.2, while the other drugs were prepared in saline (0.9% NaCl).

2.4. In Vivo Experiments

In order to better illustrate the experimental protocol rationale, a scheme was performed (Scheme 1).

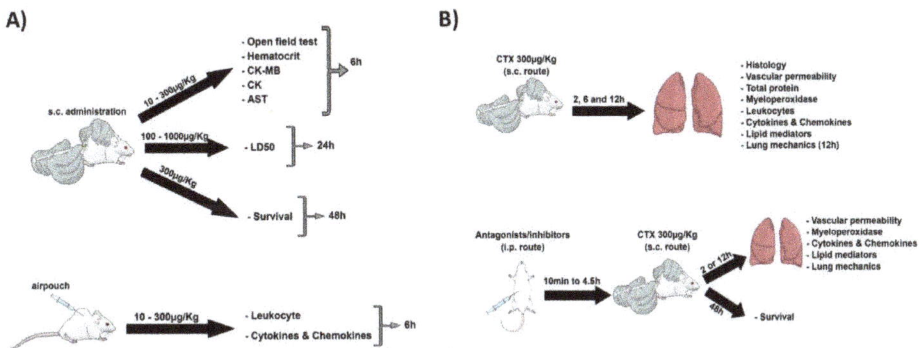

Scheme 1. Experimental protocol ratilonale. (**A**) Determination of crotoxin (CTX) working dose. (**B**) CTX-induced lung alterations.

2.4.1. Lethality

To determine the median lethal dose (LD_{50}), CTX was administered to mice at doses ranging from 100 to 1000 µg/Kg (s.c.), and the animal survival rate was analyzed after 24 h. To analyze the survival rate profile, the toxin was administered at a dose of 300 µg/Kg (s.c.) and the survival rate was monitored for 144 h (6 days). To examine the effect of pharmacological antagonists on the survival rate of CTX-treated animals, the drugs were administered before CTX (300 μg/Kg s.c.) as described in the previous section, and the survival rate was monitored every 12 h during 48 h (2 days).

2.4.2. Open Field Test

The open field test was performed in an independent experimental group to evaluate the CTX-induced locomotor effects, as previously described [9]. Mice were placed individually at the center of a circular open-field arena (40 cm diameter) divided into quadrants where the exploratory activity was videotaped during 6 min. A trained experimenter manually counted the number of quadrants crossed in the last 4 min.

2.4.3. Vascular Permeability

The Evans blue permeability assay was performed as described elsewhere [10]. Thirty minutes before the end of the experimental period, 200 μL of 0.5% Evans blue solution in sterile saline were injected into the mice tail vein. Next, the animals were euthanized and their left lung lobe and heart were collected, weighted, and immersed in 500 μL of formamide for 24 h, at 55 °C, to extract Evans blue. Finally, absorbance of the supernatant was recorded at 610 nm and the amount of extravasated dye per mg of organ was calculated from the Evans blue standard curve.

2.4.4. Air Pouch Model to Examine Local Inflammation

To examine the CTX-induced local inflammation, we used the dorsal air pouch model. First, mice were anesthetized with ketamine/xylazine solution (80/10 mg/kg, i.p.), and the midline of their dorsal region was shaved. Approximately 3 mL of sterile air were injected subcutaneously with a 25 G needle, through a sterile 0.22 μm filter (Millex, Merck Millipore—Burlington, VT, USA). At the third day, a second boost of 2 mL of sterile air was injected into the pre-existing air pouch. At the sixth day, 0.5 mL of CTX (10–300 μg/Kg) or saline (control) were injected into the air pouch, followed by injection of 2 mL of incomplete RPMI medium. The pouch fluid was collected and stained with Trypan Blue for total leukocyte counting, using the Countess II automated cell counter (Life Technologies—Carlsbad, CA, USA). Differential leukocyte counts were performed on cytospin preparations of pouch fluid stained with Panoptic kit (Laborclin—Pinhais, Brazil). Supernatant was stored at −80 °C for further quantification of cytokines and total proteins.

2.4.5. Lung Mechanics

A tracheal cannula connected to a small animal FlexiVent® ventilator (Scireq—Montreal, QC, Canada) was inserted into mice anesthetized with ketamine/xylazine (100/10 mg/kg, i.p.), and further ventilated with respiratory frequency of 150 breaths/minute and positive end-expiratory pressure of 3 cmH_2O. Pancuronium bromide (1.2 mg/kg i.p.) was administered for total paralysis before analysis of lung mechanical functions using the forced oscillation technique, in particular the single compartment and the constant phase model. In addition, a respiratory pressure–volume curve was built and the quasi-static and dynamic respiratory compliance were calculated by fitting the Salazar–Knowles equation to pressure-volume curves. Results were expressed as respiratory system elastance, tissue elastance, quasi-static and dynamic respiratory compliance, respiratory system compliance, inspiratory capacity, and tissue resistance.

2.5. Biological Parameters and Markers

2.5.1. Biochemical Markers

Mice blood samples collected without anticoagulant were kept at room temperature for 30 min to allow clotting, and centrifuged at 1300× *g* for 15 min. The resulting serum supernatant was collected and stored at −80 °C. The damage-associated serum biomarkers aspartate aminotransferase (AST), creatine kinase (CK), and creatine kinase MB (CK-MB) were quantified according to the manufacturer's instructions (Wiener Lab—Rosário, Argentina).

2.5.2. Pro-Inflammatory Cytokines

The pro-inflammatory mediators chemokine (CXC motif) ligand 1 (CXCL-1), interleukin-1β (IL-1 β), interleukin-6 (IL-6), and tumor necrosis factor-α (TNF-α) were quantified in dorsal air pouch fluid, lung homogenates and bronchoalveolar fluid (BALF) using enzyme-linked immunosorbent assay (ELISA) kits, as recommended by the manufacturer (R&D Systems—Minneapolis, MN, USA).

2.5.3. Hematocrit

Hematocrit was determined in EDTA-anticoagulated whole blood samples, using the automated hematology analyzer Cell Dyn 3700 (Abbott—Chicago, IL, USA).

2.5.4. Gene Expression

Expression of genes of inflammatory cytokines and enzymes involved in eicosanoid metabolism was analyzed in lung after 2 h of treatment with CTX or saline. The right lower lung lobule was harvested, weighed, and homogenized. Total RNA was extracted using PureLink RNA Mini Kit according to the manufacturer's specifications (Invitrogen—Carlsbad, CA, USA), quantified using NanoDrop 2000 (Thermo Scientific)—considering the absorbance ratios A260/280 and A260/230 between 1.8–2.2—and treated with DNase I amplification grade (Invitrogen). Next, cDNA was synthesized from 2 µg of the total RNA extracted, using the High Capacity cDNA Reserve Transcription Kit (Applied Biosystems—Foster City, CA, USA).

Aliquots (40 ng) of the total cDNA were amplified by quantitative reverse transcriptase-polymerase chain reaction (qRT–PCR). Custom plates for RT^2 PCR analysis were acquired from Applied Biosystems and contained genes of arachidonate 5-lipoxygenase-activating protein (*Alox5ap*–Mm00802100_m1), cyclooxygenase-2 (*Ptgs2*–Mm00478374_m1), cytosolic phospholipase A_2 (*Pla2g4a*–Mm00447040_m1), interleukin-1β (*IL1b*–Mm01336189_m1), interleukin-6 (*IL6*–Mm00446191_m1), leukotriene A_4 hydrolase (*Lta4h*–Mm01246216_m1), 5-lipoxygenase (*Alox5*–Mm01182743_m1), 12-lipoxygenase (*Alox12*–Mm00545833 _m1), 15-lipoxygenase (*Alox15*–Mm01250458_m1), and tumor necrosis factor (*Tnf*–Mm00443258_m1). Glucuronidase beta (*Gusb*–Mm00446953_m1), glyceraldehyde-3-phosphate dehydrogenase (*Gapdh*–Mm99999915_g1), and hypoxanthine phosphoribosyltransferase 1 (*Hprt1*–Mm00446968_m1) were used as reference genes. PCR reactions were performed using the TaqMan Fast Universal PCR Mastermix 2X (Applied Biosystems—Austin, TX, USA) in a StepOnePlus Real-Time PCR System (Applied Biosystems), according to the manufacturer's instructions.

Results were analyzed using the DataAssist™ v3.01 software (Applied Biosystems), and the threshold cycle (Ct) cut-off value was set up as 40. Normalization was done by subtracting the Ct mean value of the gene of interest from the Ct mean value of the three reference genes (*GAPDH, GUSB* and *HPRT1*). The values obtained for the negative control were used as reference for comparison. The relative expression of each gene was calculated by the $2^{\Delta\Delta Ct}$ method [11].

2.5.5. Eicosanoids

The eicosanoids 11-hydroxyeicosatetraenoic acid (11-HETE), 12-hydroxyeicosatetraenoic acid (12-HETE), 15-hydroxyeicosatetraenoic acid (15-HETE), and 5-oxo-eicosatetraenoic acid (5-oxo-ETE), leukotriene B_4 (LTB_4), 6-*trans*-leukotriene B_4 (6-*trans*-LTB_4), prostaglandin B_2 (PGB_2), prostaglandin D_2 (PGD_2), prostaglandin E_2 (PGE_2), 6-keto-prostaglandin E_2 (6-keto-PGE_2), prostaglandin $F_2\alpha$ ($PGF_2\alpha$), 6-keto-prostaglandin $F_1\alpha$ (6-keto-$PGF_1\alpha$, the stable prostacyclin (PGI_2) metabolite), and tromboxane B_2 (TXB_2) were quantified in lungs from mice treated with CTX or saline.

Left lung lobules were collected, weighed, and homogenized in incomplete RPMI 1640 medium (1 mL/100 mg lung). The lung homogenate supernatant was mixed with methanol (1:1 *v/v* final) and submitted to solid phase extraction in a C18 column for lipids extraction. A 10 µL aliquot of each sample extracted was analyzed using the liquid chromatography-tandem mass spectrometry (LC-MS/MS) system TripleTOF® 5600+ (AB Sciex—Foster, CA, USA), as previously described [12]. Data were acquired using the Analyst software (SCIEX—Framingham, MA, USA), reviewed using the PeakView™ software (SCIEX), and quantified using the MultiQuant™ software (SCIEX).

2.5.6. Myeloperoxidase Activity

Myeloperoxidase (MPO) activity was determined as described elsewhere [13]. Animals heart and mid right lung lobule were removed, weighed, and homogenized in a tissue homogenizer. MPO activity in the supernatant was determined using 3,3′,5,5′-tetramethylbenzidine (TMB) as substrate (BD Bioscience—San Jose, USA), and recording absorbance at 450 nm. The results were reported as Units (1 Unit = Δ0.1 Abs 450 nm) per mg of tissue.

2.6. Lung Histology

The left lung lobule was collected and fixed in a 10% formaldehyde solution in PBS pH 7.4 for 24 h. Afterwards, the samples were immersed in alcohol and xylol solutions, and included in paraffin. Sections of 4 µm were prepared using the RM-2125 microtome (Leica—Wetzlar, Germany) and stained with hematoxylin and eosin (HE). Morphological analysis was performed using the DM LB2light microscope (Leica) coupled to the DC 300F camera (Leica). The captured images were analyzed using the Leica QWin software (Leica). Quantitative histopathologic analysis of lung injury was performed using a score system based on the following criteria: leukocyte infiltration, vascular congestion, alveolar hemorrhage, and edema. Each criterion was graded on a scale from 0 to 3 (0, absent; 1, mild; 2, moderate; and 3, severe). Lung injury score was calculated for each specimen and treatment period. The pathologist who performed histological analysis was blinded to the intervention. The alveolar sac area represented by the empty space (white area) in HE-stained lung histological images was calculated using the IM-50 software (Leica).

2.7. Analysis of Lung Leukocyte Population

Infiltrating leukocytes were isolated from mice lung using the protocol reported by Souza and colleagues [14]. Briefly, the upper right lung lobule was collected, minced with sterile scissors in RPMI 1640 medium, and treated with digestion buffer containing 0.05 mg/mL liberase (Roche—Basel, Switzerland) and 0.5 mg/mL DNase (Sigma-Aldrich) for 45 min, at 37 °C, under shaking at 2000 rpm. Tissue debris were removed using a 100 µm cell strainer. Next, red blood cells were lysed and the remaining cells were washed with PBS, centrifuged, and suspended in RPMI 1640 containing 10% FBS. Cells were fixed using cytospin slides and stained with Panoptic kit (Laborclin—Pinhais, Brazil) to perform differential leukocyte counting. Next, neutrophil phenotypes were analyzed by flow cytometry using the Ly6G PE-Cy7 conjugated (Cat#560601, RRID:AB_1727562) and CD62L BB515 conjugated (Clone MEL-14) antibodies (BD Bioscience—San Jose, CA, USA). Data from 20,000 events were acquired using a FACSCanto II flow cytometer equipped with the FACSDiva software (BD Biosciences), and further plotted and analyzed using the FlowJo software v.10.0.7 (Tree Star, Inc.—Ashland, OR, USA).

2.8. Statistical Analysis

The GraphPad Prism software version 5.01 (GraphPad Software Inc.—San Diego, CA, USA) was used to plot graphics and perform statistical data analysis. The unpaired Student's t-test was used to analyze differences between two groups, while one-way analysis of variance (ANOVA) followed by the Bonferroni's post-test was used for comparison of multiple groups. The survival rate was expressed as percentage of live animals, and the Mantel–Cox log-rank test was used to compare the survival curves. Two-way ANOVA followed by the Bonferroni's post-test was used to analyze the time-course plots. Differences with $p < 0.05$ were considered statistically significant.

3. Results

3.1. CTX Working Dose

Crotoxin induces several harmful effects, which vary according to the target organ studied, the experimental design, animal species, and toxin dose and route of administration [16–20]. To determine a CTX dose capable of promoting tissue damage, in the present study we performed dose-response experiments where mice were treated with CTX at 10–300 μg/Kg (s.c.) for 6 h.

Compared with control mice, animals treated with 300 μg/Kg CTX exhibited reduced exploratory activity, as evidenced by the open field test analysis of locomotor alterations (Figure 1A), and a slightly increased hematocrit level, as evidenced by hematological analysis (Figure 1B). The serum levels of CK and CK-MB (biomarkers of muscle and heart tissue damage, respectively) and AST (a marker of liver damage) were increased in mice treated with 100 and 300 μg/Kg CTX (Figure 1C–E). Analysis of local inflammation in the dorsal air pouch cavity fluid evidenced increased infiltration of polymorphonuclear cells and production of the inflammatory mediators IL-6 and CXCL-1 in mice treated with the highest CTX dose (300 μg/Kg) (Figure 1F–H). In general, the CTX dose of 300 μg/Kg induced alterations in all the biological parameters evaluated; this dose is 1.3-fold greater than the determined LD_{50} of 229.6 μg/Kg (Figure 1i). The survival profile revealed that the toxin was lethal to approximately 20%, 60%, and 70% of the animals at 12, 24, 48 h, respectively (Figure 1I). Based on the set of results obtained so far, the CTX dose of 300 μg/Kg s.c. and the time treatment of 12 h—when the survival rate was ~80% (Figure 1J)—were selected for further investigations on how the toxin affects the respiratory system.

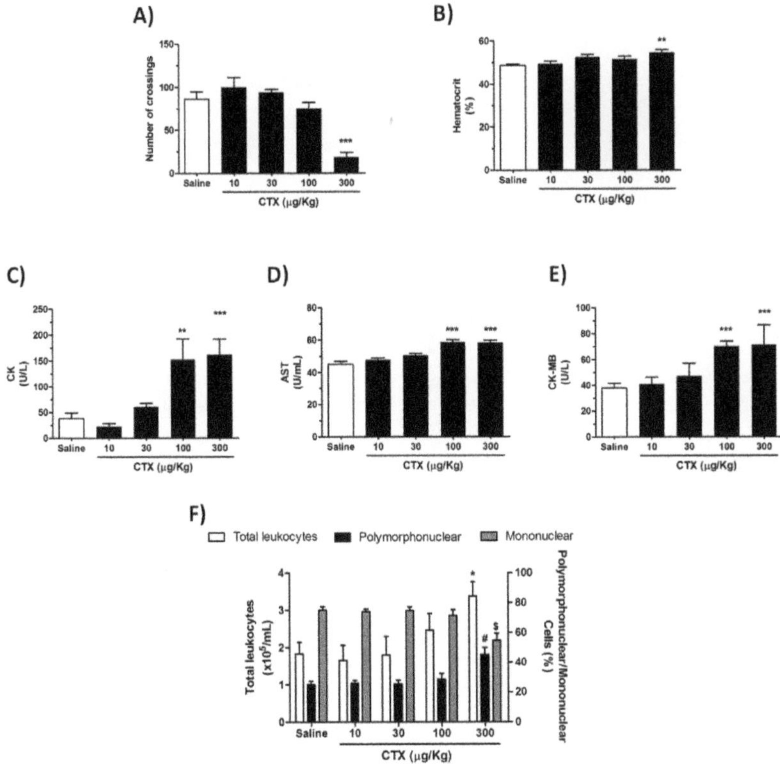

Figure 1. *Cont.*

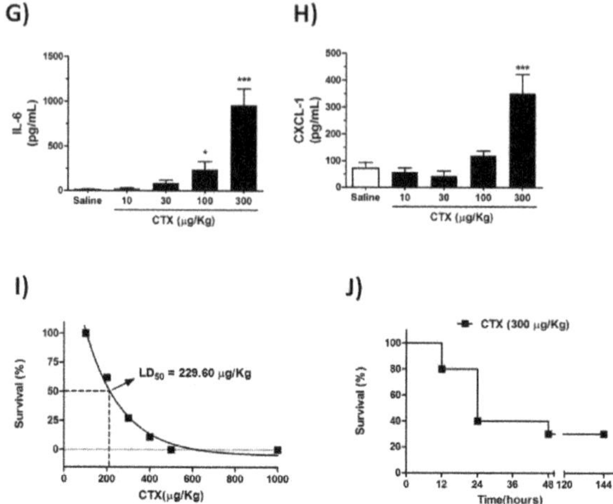

Figure 1. Dose-response toxicological effects of CTX. Mice were treated with different CTX doses (10–300 μg/Kg s.c.) and the biological parameters were evaluated at 6 h. (**A**) Open field test. (**B**) Hematocrit. (**C**) Serum creatine kinase (CK). (**D**) Serum creatine kinase-MB (CK-MB). (**E**) Serum aspartate aminotransferase (AST). (**F**) Total and differential leukocyte counting in dorsal air pouch fluid. (**G**) IL-6 levels in dorsal air pouch fluid. (**H**) CXCL-1 levels in dorsal air pouch fluid. (**I**) Determination of LD_{50}. (**J**) Time-course survival profile. The results are representative from two-independent experiments ($n = 6$–7). $^{\#}\, p < 0.05$, $^{\$}\, p < 0.05$, $^{*}\, p < 0.05$, $^{**}\, p < 0.01$, $^{***}\, p < 0.001$ compared with animals treated with saline (control)—One-way ANOVA followed by Tukey's multiple comparison test multiple comparison test.

3.2. CTX Induces Lung Alterations

Morphological analysis of lungs from animals treated with CTX for 2, 6, and 12 h revealed reduction of alveolar sac area and increased septum wall thickness, as compared with saline-treated animals (control) (Figure 2A–D). The lung histological score increased with time and was significantly different from the control at 12 h after CTX injection, indicating that lung damage was time-dependent (Figure 2E). At this treatment time, the presence of edema, vascular congestion, and –alveolar hemorrhage (Figure 2F–H), as well as leukocyte infiltration (especially of polymorphonuclear leukocytes) (Figure 2H), foamy macrophages, and hyperemia (data not shown) was also more evident.

The findings from morphological analysis guided determination of vascular permeability, total protein concentration, myeloperoxidase activity, and leukocyte infiltration in lung homogenates. Lung vascular permeability increased only at 2 h after CTX administration (Figure 3A). At 6 and 12 h of treatment with CTX, total protein concentration (Figure 3B) increased as a function of time and were significantly different from those detected in the control group. The time course myeloperoxidase activity increasing (Figure 3C) was followed by augmentation of leukocyte infiltration of polymorphonuclear cells (Figure 3D) and reflected by increased percentage of single $Ly6G^+$ neutrophil population (Figure 3E), corroborating data from the increased number of granulocytes. Treatment with CTX did not alter the percentage of double $Ly6G^+$ $CD62L^+$ cells (data not shown).

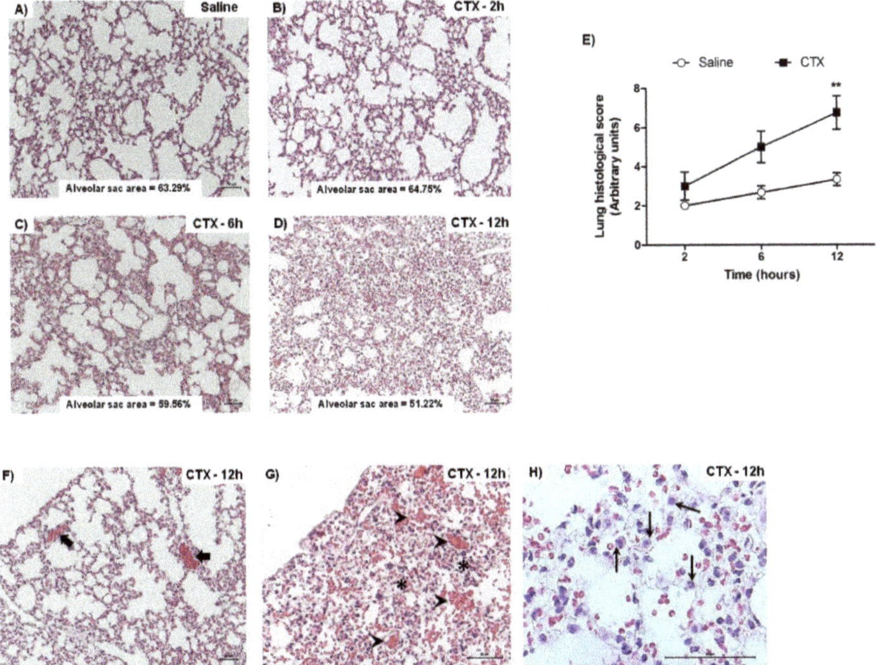

Figure 2. Histological alterations in mice lung induced by CTX. Representative sections of lungs from mice treated with saline or CTX (300 µg/Kg s.c.) for 2, 6, or 12 h. (**A–D**) Images from HE-stained sections were captured at 20X magnification (scale is represented in the image) and alveolar sac area was calculated using the IM-50 software. (**E**) Lung injury score was graded from 0–3 for each animal, and the result is representative of the mean of injury score. (**F–H**) Lung alterations after 12 h of CTX treatment: vascular congestion (thick arrow), edema (asterisk), alveolar hemorrhage (arrow head), and leukocyte infiltrate (thin arrow). ** $p < 0.01$ vs. saline-treated mice (control)—two-way ANOVA followed by the Bonferroni's post-test ($n = 3$ animals/group).

Next, we analyzed the kinetics of leukocyte infiltration profile in bronchoalveolar fluid (BALF) and dorsal skin pouch fluid (PF). No neutrophil infiltration was detected in BALF from CTX-treated mice, at all experimental periods (data not shown), indicating that leukocyte infiltration was restricted to lung parenchyma. Leukocyte infiltration into PF increased at 6 h after CTX administration, but returned to basal levels at 12 h of treatment; this infiltrate was mainly composed of Ly6G$^+$ polymorphonuclear cells (Supplementary Figure S1A–C).

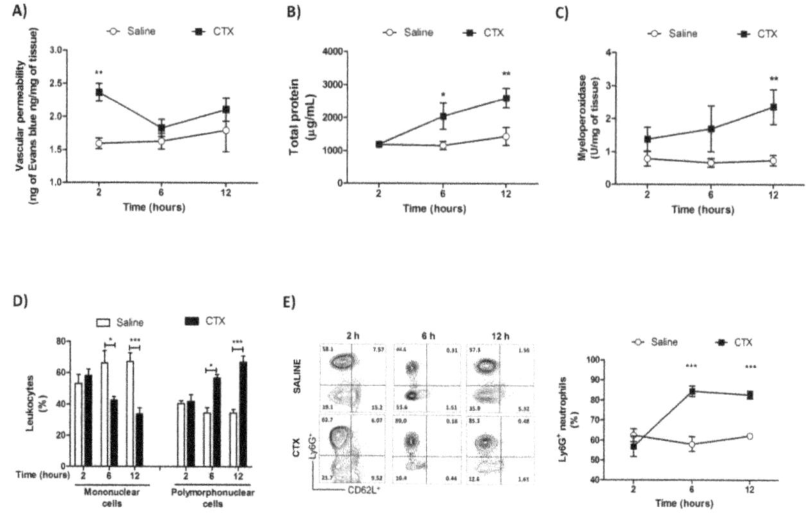

Figure 3. Lung alterations detected in lung homogenates from CTX-treated mice. The animals were treated with CTX (300 µg/Kg s.c.) for 2, 6, and 12 h, euthanized, and their lungs were removed and homogenized for the assays. (**A**) Vascular permeability was evaluated by injecting Evans blue solution 30 min before euthanasia. (**B**) Total protein concentration. (**C**) Myeloperoxidase activity. (**D**) Differential leukocyte counting from cytospin smears. (**E**) Flow cytometry data are summarized in the representative contour plots and line plots showing kinetics of increase of Ly6G$^+$ cells. The results from (**A–C**) are representative from two independent experiments (n = 6–7 animals/group), while the results from (**D–E**) are representative from one independent experiment (n = 5 animals/group). * $p < 0.05$, ** $p < 0.01$, and *** $p < 0.001$ vs. saline-treated animals (control) from the respective time group—two-way ANOVA followed by the Bonferroni's post-test.

3.3. CTX Elevates the Levels of Inflammatory Mediators and Impairs Pulmonary Function

We analyzed the kinetics of cytokine, chemokine, and lipid mediator release in mice lung homogenates. The levels of the pro-inflammatory cytokines IL-1β, IL-6, and TNF-α increased at 12 h of treatment with CTX (Figure 4A–C). The levels of CXCL-1, which is a crucial mediator of neutrophil recruitment, increased at 6 and 12 h after CTX injection (Figure 4D).

Quantification of cytokines/chemokines in mice BALF and PF revealed that CTX injection did not alter the levels of IL-1β, IL-6, TNF-α, and CXCL-1 in BALF during the studied period of 12 h (data not shown). Considering that lung homogenates are composed of both lung tissue and BALF, and that the inflammatory markers remained unaltered in BALF, our findings indicate that the aforementioned lung alterations are exclusive to lung parenchyma, i.e., they do not occur in bronchoalveolar cavity. IL-6 levels in PF raised in a time-dependent manner up to 6 h after CTX administration, but returned to basal levels at 12 h of treatment. In contrast, the highest CXCL-1 levels in PF were detected at 2 h of CTX injection, and gradually declined up to basal levels at 12 h of treatment. The PF levels of TNF-α and IL-1β were not altered by CTX injection (Supplementary Figure S1D–G). Together, these findings illustrate a contrasting profile of cytokine/chemokine production in lungs and local dorsal skin in CTX-treated animals.

As the enzymatic phospholipase A_2 activity of CTX mediates the production of signaling molecules, including eicosanoids that are associated with several biological effects induced by the toxin [21–23], in this study we examined the kinetics of lipid mediator production in mice lungs using a LC-MS/MS approach. Compared with the control group, concentration of the eicosanoids LTB_4, 6-*trans*-LTB_4, PGE_2, and 12-HETE increased at 2 h of CTX injection (Figure 4E–H), while concentration of 15-HETE increased at 12 h of CTX injection (Figure 4I) but concentration of the mediators 6-keto-PGE_2, PGD_2,

PGB_2, PGF_2a, 6-keto-PGF_1a, TXB_2, 11-HETE, and 5-oxo-ETE remained unaltered after CTX injection (data not shown). Gene transcripts for eicosanoid metabolism enzymes were analyzed in random lung samples from mice treated with CTX for 2 h. The heatmap (Figure 4J) illustrates a tendency of increase in expression of genes *Ptgs-2* (COX-2), *Lta4h*, *Alox15*, *Alox5*, and protein FLAP (*Alox5ap* gene), which play crucial roles during production of the abovementioned eicosanoids.

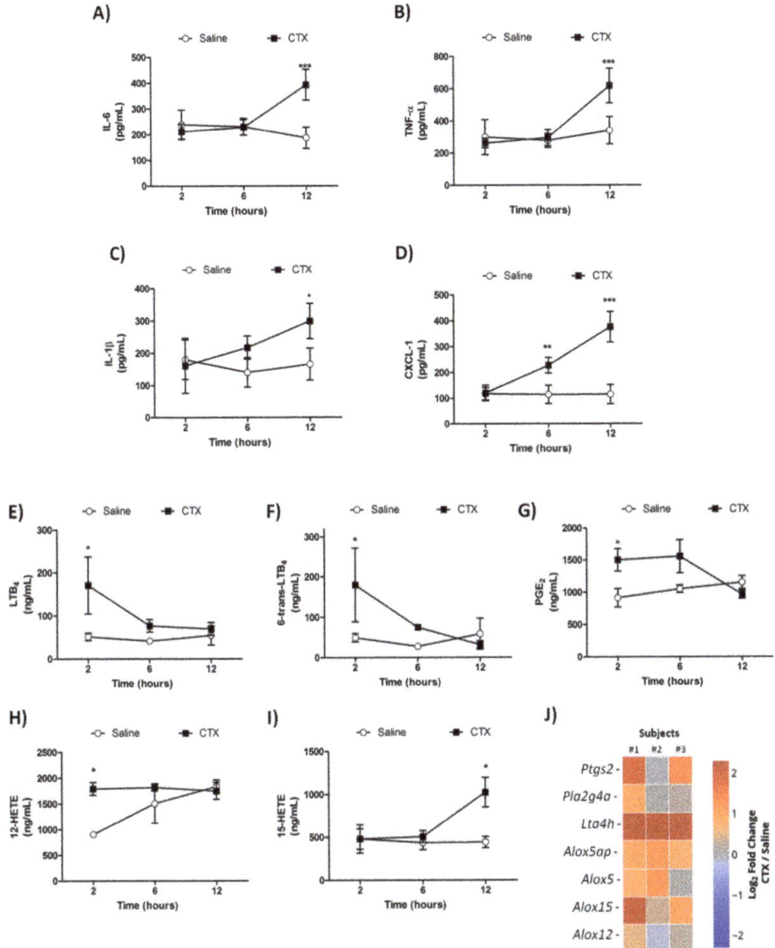

Figure 4. CTX-induced production of inflammatory mediators in mice lung. The animals were treated with CTX (300 μg/Kg s.c.) for 2, 6, and 12 h, euthanized, and their lungs were removed and homogenized for the quantification of pro-inflammatory cytokines/chemokines (**A–D**) and eicosanoids (**E–I**) using ELISA and LC-MS/MS, respectively. (**A**) Interleukin 6 (IL-6). (**B**) Tumor necrosis factor α (TNF-α). (**C**) Interleukin 1β (IL-1β). (**D**) Keratinocyte-derived chemokine (CXCL-1). (**E**) Leukotriene B_4 (LTB_4). (**F**) 6-*trans*-Leukotriene B_4 (6-*trans*-LTB_4). (**G**) Prostaglandin E_2 (PGE_2). (**H**) 15-Hydroxyeicosatetraenoic acid (15-HETE). (**I**) 12-Hydroxyeicosatetraenoic acid (12-HETE). (**J**) Gene expression profile in mice lung after 2 h of treatment with CTX, assessed by qRT–PCR. Data were expressed as log2 fold-change compared with the control (saline) ($n = 3$). The results from (**A–D**) are representative from two independent experiments (n = 6–7), while the results from (**E–I**) are representative from one independent experiment (n = 4 animals/group). * $p < 0.05$, *** $p < 0.001$ vs. saline treated animals (control) from the respective time group—two-way ANOVA followed by the Bonferroni's post-test.

Considering that CTX induced pathological alterations in lung, next we examined how the toxin affected lung mechanics in mice 12 h after its injection. Compared with the control group (saline), the respiratory system of CTX-treated mice exhibited reduced dynamic and quasi-static compliances (Figure 5A–B), indicating lung stiffness, as well as decreased inspiratory capacity (Figure 5C). In addition, the respiratory system and tissue elastance, and tissue resistance tended to increase in CTX-treated mice, but these parameters did not significantly differ from the control (p = 0.061, 0.110, and 0.108 respectively) (Figure 5D–F).

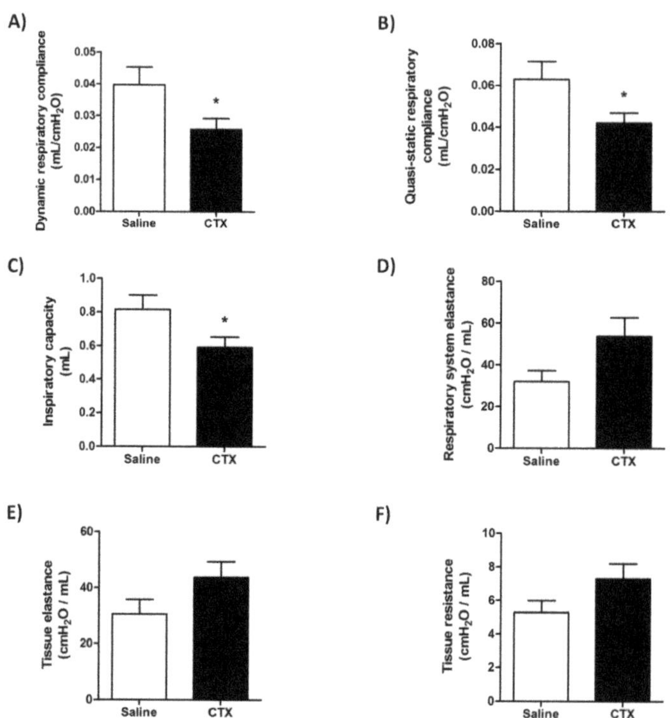

Figure 5. Lung mechanics in mice treated with CTX for 12 h. The respiratory mechanics parameters (**A**) Dynamic respiratory compliance, (**B**) Quasi-static respiratory compliance, (**C**) Inspiratory capacity, (**D**) Respiratory system elastance, (**E**) Tissue elastance, and (**F**) Tissue resistance were measured in vivo using a small animal ventilator, 12 h after CTX (300 μg/Kg s.c.) or saline (control) administration. The results are representative from two independent experiments (n = 8–10 animals/group). * $p < 0.05$ vs. control group—unpaired Student's t-test.

3.4. *Indomethacin and Hexamethonium Mitigate CTX-Induced Lethality*

Here we used drugs that act as antagonists or inhibitors of lipid mediator metabolism or peripheral nervous system pathways as the pharmacological approach to investigate the mechanisms by which CTX induced lethality. The COX-1 and -2 inhibitor indomethacin, but not the 5-lipoxygenase pathway inhibitor MK-591, increased the survival rate of CTX-treated animals (Figure 6A,B). Analysis of the peripheral neural effects involving nicotine pharmacology evidenced that the blocker of ganglionic nicotinic receptors hexamethonium, but not the acetylcholinesterase inhibitor neostigmine, increased the survival rate of CTX-treated animals (Figure 6C,D). In addition, neither the peripheral muscarinic antagonist methyl-atropine nor the adrenergic antagonist propranolol altered the survival rate of CTX-treated mice (Figure 6E,F). Therefore, these results clearly indicate that both COX-derived prostanoids and peripheral nicotinic receptors are associated with CTX-induced lethality.

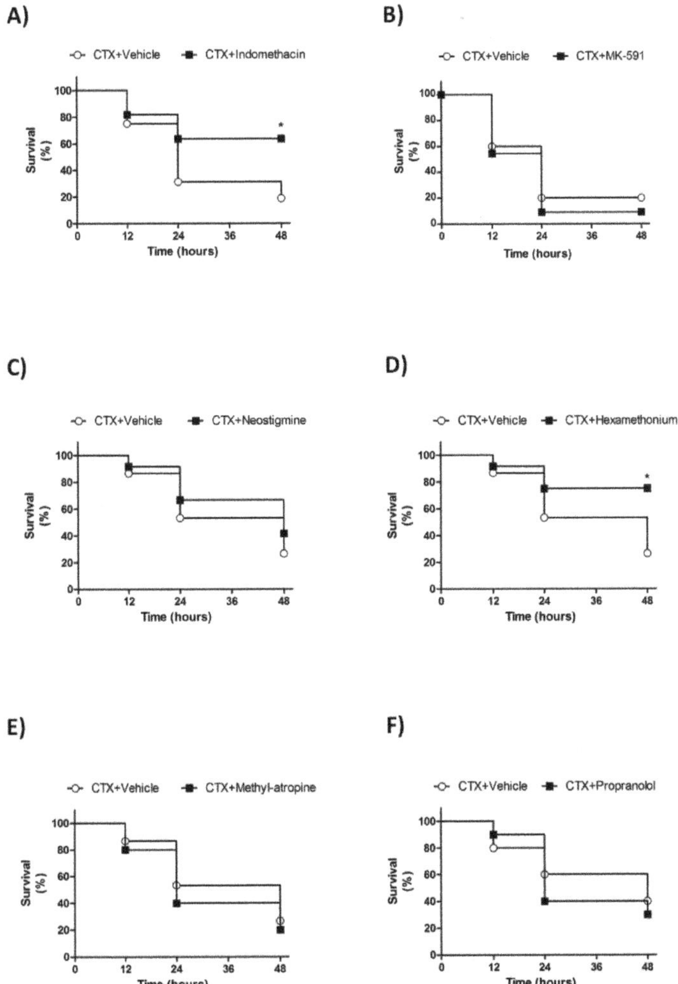

Figure 6. Role of eicosanoids metabolism and peripheral nervous system on survival of CTX-treated mice. Animals were treated with (**A**) Indomethacin (3 mg/Kg i.p.), (**B**) MK-591 (40 mg/Kg i.p.), (**C**) Neostigmine (0.1 mg/Kg i.p.), (**D**) Hexamethonium (10 mg/Kg i.v.), (**E**) Methyl-atropine (30 mg/Kg i.p.), and (**F**) Propranolol (5 mg/Kg i.p.) or their respective vehicles before CTX administration (300 µg/Kg). Survival rate was determined every 12 h for 48 h. Results are expressed as percentage of survival and are representative from two independent experiments (n = 13–16 animals/group). * $p < 0.05$—Mantel–Cox log-rank test.

3.5. Role of Hexamethonium And Indomethacin on CTX-Induced Lung Alterations

Our previous results revealed that CTX induced several morphological and functional alterations in lung parenchyma associated with inflammatory and vascular responses. Analysis of kinetics of CTX action evidenced that some alterations were clearly detected at the early (at 2 h) or late (at 12 h) phases of response to toxin administration. Considering that COX-derived prostanoids and nicotinic acetylcholine receptors were involved in CTX pathogenesis, here we examined how their antagonists indomethacin and hexamethonium, respectively, interfered with early and late phase lung alterations.

3.5.1. Early Phase

The early phase lung alterations, which were analyzed 2 h after CTX treatment, were marked by increased vascular permeability and release of eicosanoids (previous Figures 3A and 4E–I, respectively). Indomethacin, but not hexamethonium, mitigated the CTX-induced enhancement of vascular permeability (Figure 7A,B)—as assessed by the Evans blue extravasation assay—and strongly lowered basal (saline/vehicle) and CTX-induced increase of lung PGE_2 levels (Figure 7C,D). Altogether, these findings indicated that PGE_2 participated in the early phase lung alterations induced by CTX.

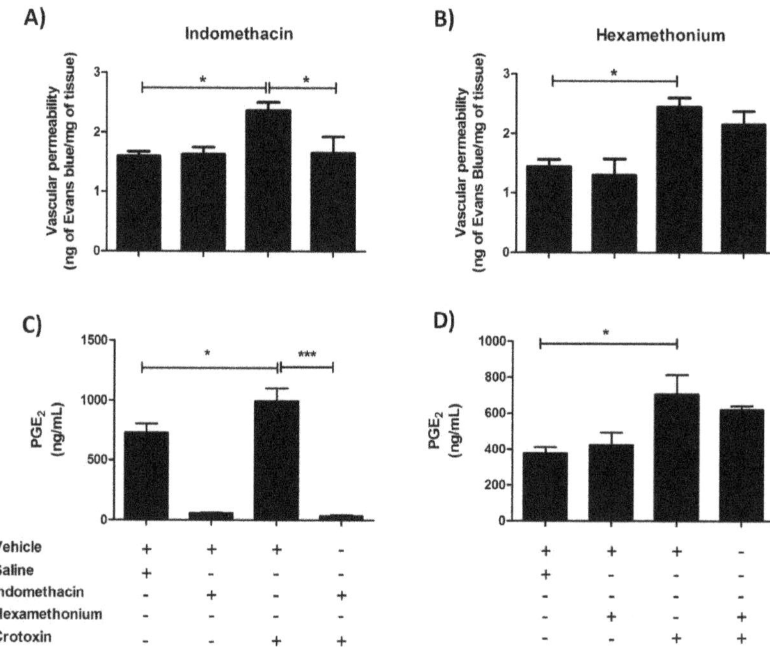

Figure 7. Effect of indomethacin and hexamethonium on CTX-induced early phase lung alterations. Mice were treated with indomethacin (3 mg/Kg i.p.; panels (**A**,**C**)), hexamethonium (10 mg/Kg i.v.; panels (**B**,**D**)), or their respective vehicles, before CTX (300 μg/Kg s.c.) administration. After 2 h, lungs were collected for (**A**,**B**) analysis of vascular permeability using the Evans blue extravasation assay and (**C**,**D**) quantification of prostaglandin E_2 (PGE_2) by ELISA. The results are representative from one independent experiment (n = 4 animals/group). * $p < 0.05$ and *** $p < 0.001$—one-way ANOVA followed by the Tukey's multiple comparison test.

3.5.2. Late Phase

The late phase lung alterations, analyzed 12 h after CTX treatment, were characterized by increased myeloperoxidase activity and increased levels of pro-inflammatory cytokines/chemokines (previous Figure 3C and Figure 4A–D, respectively). The effects of indomethacin and hexamethonium were the opposite of those detected on the early phase lung alterations: the latter but not the former drug mitigated the CTX-induced enhancement of myeloperoxidase activity (Figure 8A,B) and levels of the inflammatory markers IL-1β (Figure 8C,D), TNF-α (Figure 8E,F), IL-6 (Figure 8G,H), and CXCL-1 (Figure 8I,J). These results pointed out the participation of peripheric nicotinic receptors in the late phase lung alterations.

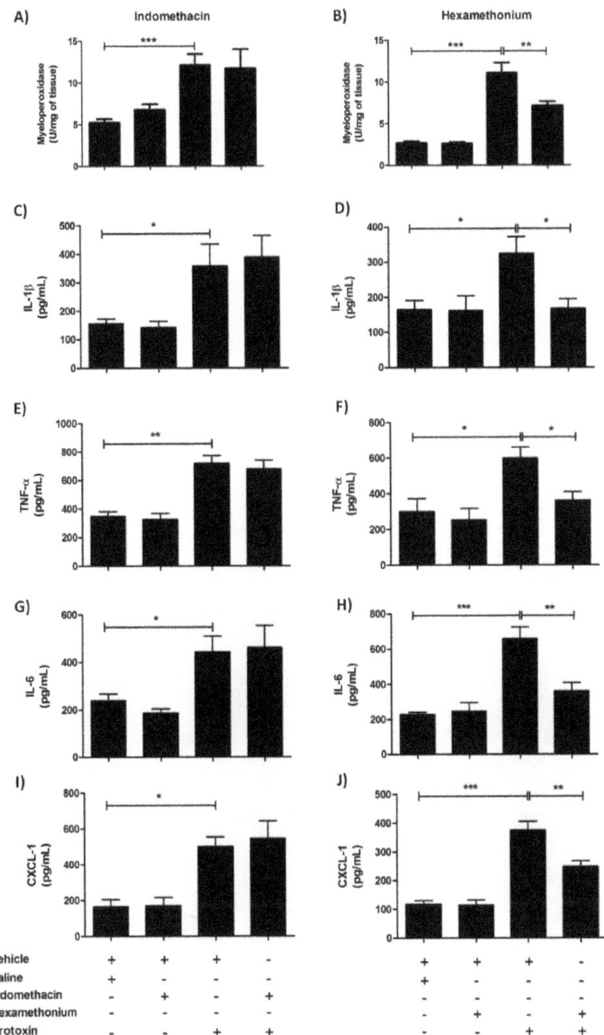

Figure 8. Effect of indomethacin and hexamethonium on ctx-induced late phase lung alterations. Animals were treated with indomethacin (3 mg/Kg i.p.) or hexamethonium (10 mg/Kg i.v.), or their respective vehicles, before CTX (300 μg/Kg s.c.) administration. After 12 h, lungs were collected for quantification of (**A**,**B**) Myeloperoxidase activity, (**C**,**D**) Interleukin 1β (IL-1β), (**E**,**F**) Tumor necrosis factor α (TNF-α), (**G**,**H**) Interleukin 6 (IL-6), and (**I**,**J**) Keratinocyte-derived chemokine (CXCL-1). The results are representative from two independent experiments (n = 6–7 animals/group). * $p < 0.05$, ** $p < 0.01$, and *** $p < 0.001$—one-way ANOVA followed by the Tukey's multiple comparison test.

We also investigated whether indomethacin and hexamethonium altered the levels of leukocyte infiltration and inflammatory mediators in PF. Compared with CTX-treated mice, animals treated with indomethacin prior to CTX injection exhibited increased levels of CXCL-1 and IL-6, and an almost significantly increased leukocyte infiltration level (Supplementary Figure S2A,C,E). Administration of hexamethonium before CTX injection did not alter the levels of the inflammatory parameters analyzed (Supplementary Figure S2B,D,F).

To continue analyzing the late phase inflammatory parameters, we determined the serum CK-MB levels at 12 h after CTX administration in mice pretreated with hexamethonium and indomethacin. Both drugs mitigated the CTX-induced increase in CK-MB levels (Supplementary Figure S3), suggesting that both COX metabolism and peripheral nicotinic receptors are associated with cardiovascular alterations induced by CTX.

3.6. Hexamethonium Mitigates CTX-Induced Lung Mechanics Impairment

Based on the previous results that clearly demonstrated that hexamethonium effectively dampened CTX-induced late phase lung alterations, here we examined how the nicotinic blocker modulated the toxin-induced impairment of pulmonary function after 12 h of treatment. Pre-treatment with hexamethonium prevented the CTX-induced alterations in pulmonary mechanics, i.e., it improved quasi-static and respiratory system compliances (Figure 9A,B) and inspiratory capacity (Figure 9C) when compared with mice treated with CTX alone.

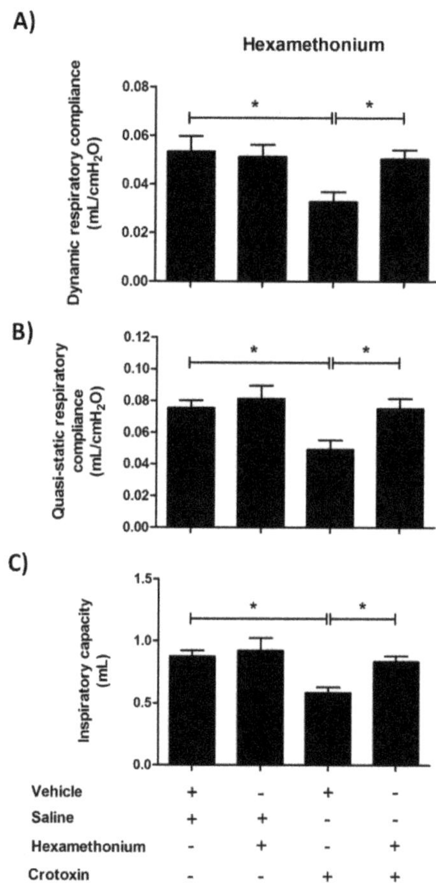

Figure 9. Effect of hexamethonium on lung mechanics of CTX-treated mice. Mice were treated with hexamethonium (10 mg/Kg i.v.) 30 min before CTX (300 μg/Kg s.c.) administration. After 12 h, we used a small animal ventilator to determine the respiratory mechanics parameters in vivo. (**A**) Respiratory system compliance. (**B**) Quasi-static compliance. (**C**) Inspiratory capacity. The results are representative from two independent experiments ($n = 7$–8 animals/group). * $p < 0.05$—one-way ANOVA followed Tukey's multiple comparison test.

4. Discussion

The systemic effects of *C. durissus* venom are stronger than its local effects, and are mainly characterized by neurotoxicity, systemic myotoxicity, and respiratory and acute renal failure [24–26]. Respiratory failure only occurs in severe cases of envenomation by rattlesnake bite [27–31]. As the major venom component, CTX plays a significant role on *Crotalus* accidents, including neuromuscular blockade and systemic myotoxicity as the main toxicological effects, associated with other alterations.

To investigate how CTX affects respiratory physiology, we selected a dose that causes significant intoxication by evaluating some biological alterations, based on previous findings. CTX-induced locomotor alterations are associated with neuromuscular disturbances caused by neuromuscular blockade [16,17,32,33] and myotoxicity [20,34]. We found that subcutaneous injection of CTX at a dose of 300 μg/Kg significantly decreased mice exploratory activity at the open field test and increased serum CK, which indicate locomotor impairment and systemic myotoxicity. This toxin dose also effectively induces other previously reported biological alterations such as increased blood hematocrit [35], dorsal skin local inflammation [19], and increased levels of AST and CK-MB associated with liver and heart damage, respectively [18,36–38]. The LD_{50} found in the present work (229.6 μg/Kg s.c.) was relatively close to that reported by Brazil and colleagues [39] (177.5 μg/Kg, s.c.) in mice. The selected dose for the following experiments (300 μg/Kg) represented 1.3 LD_{50}, and enabled a reliable characterization of the toxicological effects of CTX on the respiratory system.

The respiratory system is a complex arrangement of organs that promote the respiration, coordinated by neural control of respiratory muscles of pump (diaphragm, intercostals and abdominal) and bronchomotor tone (airway smooth muscles), and which the respiratory tract (composed of specialized tissues and cells) responsible for the maintenance of the architecture and gas exchange [40]. Lung disorders are characterized by alterations that impair airway, vessel or lung function, and are accompanied by changes in tissue morphology [41]. In the present study, histological analysis of lung from CTX-treated mice evidenced progressive tissue damage characterized by the presence of edema, vascular congestion, alveolar hemorrhage, and leukocyte infiltration, with the highest histological scores at 12 h after CTX injection. These results agree with previous findings on lung morphological changes after C.d.t. and *C. durissus cascavella* whole venom administration, especially the presence of perivascular edema, diffuse hemorrhage, and leukocyte infiltration up to 24 h after venom administration [42,43].

We used lung homogenates to investigate deep tissue alterations, including vascular permeability, total protein concentration, myeloperoxidase activity, leukocyte infiltration profile, and levels of pro-inflammatory cytokines and lipid mediators. We also analyzed the leukocyte infiltration profile and concentration of pro-inflammatory cytokines in BALF, but we did not detect alterations in these inflammatory markers (data not shown), indicating that CTX selectively acted on lung parenchyma rather than bronchoalveolar cavity.

An overall analysis of the time-course of lung injuries evidenced two patterns: early phase alterations within the first two hours, and late phase alterations that begun (or not) at 6 h and peaked after 12 h. The early phase was marked by a transient increase in lung vascular permeability, identified by tissue accumulation of Evans blue. Acute lung injuries are characterized by an early phase increase in vascular permeability, whose evolution can impair respiratory function [44,45]. CTX also upregulated the cyclo- and lipoxygenase pathways of eicosanoids production. The levels of the enzymatic product PGE_2 were augmented. Expression of the 15-lipoxygenase (*Alox15*), but not the 12-lipoxygenase (*Alox12*) gene was upregulated and associated with increased 12- and 15-HETE levels. Elevation of LTB_4 levels was accompanied by a rise in gene expression of the 5-lipoxygenase complex enzyme (*ALOX5*), its associated protein 5-lipoxygenase activated protein (FLAP—*ALOX5ap*)—which is required for 5-lipoxygenase activity [46], and LTA_4 hydrolase (*Lta4h*); both enzymes, 5-lipoxygenase and LTA_4 hydrolase, catalyze LTB_4 biosynthesis [47].

We detected increased levels of 6-*trans*-LTB_4—a non-enzymatic product from LTA_4—along with gene expression of cytosolic PLA_2 (*Pla2g4a*)—an important enzyme responsible for arachidonic acid release [48]. The aforementioned eicosanoids are common mediators of inflammatory response that can participate in lung impairment [49–52], but this hypothesis will be discussed further. Another plausible assumption for the participation of eicosanoids in CTX-induced lung alteration relies on the PGE_2-mediated increase of vascular permeability [53]. Additionally, PGE_2 is a known bronchoconstrictor agent with a direct effect on airways smooth muscle [54], and possibly being responsible for minor effect on airway impairment.

The involvement of eicosanoids in the biological effects of CTX has been widely investigated. For instance, the literature reports that PLA_2 enzymatic activity of CTX towards cell membrane phospholipids releases free arachidonic acid, and mediates the toxin-induced biosynthesis of lipoxin A_4 via the lipoxygenase pathway, and of the prostanoids PGE_2, PGD_2, and 15-d-PGJ_2 via participation of COX-1 and Ca^{2+}-independent PLA_2. These mediators are associated with several biological effects of the toxin, such as modulation of leukocyte function and anti-inflammatory and immunosuppressive responses [21,55–57].

The late phase lung injuries were characterized by inflammatory alterations that peaked 12 h after CTX administration, including neutrophil infiltrate composed of $Ly6G^+$ cells, associated with increased levels of myeloperoxidase activity, an indirect indicator of tissue neutrophil content [58], total proteins, and the pro-inflammatory cytokines/chemokines IL-6, TNF-α, IL-1β, and CXCL-1. High CTX doses promote local and systemic pro-inflammatory effects, such as paw edema, local and systemic muscle necrosis with neutrophil infiltration, and blood neutrophilia associated with increased serum levels of IL-6 and IL-10 [31,59–61]. Although CTX induced an inflammatory pattern with increased leukocyte infiltration and production of pro-inflammatory cytokines and lipid mediators, its intensity was not as strong as that found in infectious disease [62] and was restricted to parenchyma; these findings unveil a new pathophysiologic scenario in the experimental model studied herein.

Lung mechanics is an important feature in respiratory physiology that is associated with elastic and resistive properties of lung tissue [63]. Hence, alterations in parenchymal tissue can induce biomechanical loss of function and result in respiratory impairment [64]. Considering the CTX-induced injury and morphological alterations in mice lungs, we used the forced oscillation technique [65] to assess invasive lung function. We found that CTX-treated mice exhibited diminished dynamic and quasi-static respiratory compliances—two parameters that are associated with the lung ability to expand during inspiration and active expiration, and whose decrease is associated with lung stiffness [66]. Furthermore, tissue changes that lead to lung stiffness and make the respiratory system to work harder culminate in reduced inspiratory capacity [67,68]. We hypothesize that the lung tissue morphological injuries caused by CTX accounted for the lung stiffness and reduced inspiratory capacity in our study.

Next, we investigated the mechanisms by which CTX caused pulmonary impairment. CTX is a β-neurotoxin that induces neuromuscular blockade by inhibiting presynaptic acetylcholine release and postsynaptic desensitization of nicotinic receptors in neuromuscular junction, resulting in a flaccid paralysis [16,69–71]. Additionally, the PLA_2 enzymatic activity of CTX is associated with (i) several toxicological effects, such as myotoxicity characterized by degradation of cell membrane phospholipids and muscle tissue necrosis; and (ii) production of lipid mediators like PGE_2, which are associated with myotoxicity, neurotoxicity, and activation of immune responses [21,55,72–75]. In this sense, we examined the participation of peripheral nervous system and lipid mediators in CTX-induced pulmonary alterations and lethality. Pretreatment of mice with indomethacin and hexamethonium prior to CTX injection reduced the toxin lethality rate. Indomethacin is a COX-1 and -2 inhibitor that suppresses the biosynthesis of prostaglandins [76], while hexamethonium is a nicotinic acetylcholine receptor antagonist acting promiscuously on both ganglionic and less effectively on neuromuscular junction from peripheral nervous system, [77,78]. We also observed that methyl-atropine (muscarinic acetylcholine receptor antagonist) and propranolol (β-adrenergic receptor antagonist) did not change the CTX-induced lethality rate, as well as neostigmine (acetylcholinesterase inhibitor), probably due to the toxin capacity to deplete acetylcholine vesicles in nerve cholinergic endings [79]. These results suggest that the toxicological effects could be associated with the modulation of cholinergic transmission involving nicotinic receptors, but not muscarinic. Another harmful effect of CTX related to participation of both prostanoids and peripheral nervous system is associated with the toxin capacity to induce cardiovascular alterations. The toxin causes systemic hypotension in dogs and rabbits when the vagus nerve response is stimulated, since it is a parasympathetic-related controller of heart and lung function [35,80]. An in vitro study using the Langendorff model has demonstrated that CTX weakens heart contractile force and increases CK release [72]. Indomethacin reverts both effects, indicating the participation of COX-derived mediators, such as PGE_2, in these events [72]. In the present work, CTX increased serum CK-MB levels, and pretreatment with hexamethonium and indomethacin mitigated such rise. As CK-MB is a biomarker of heart damage, our findings strongly indicate that CTX-induced cardiovascular alterations are associated with production of COX-derived prostanoids and modulation of nicotinic receptors, and can account for the toxin-induced respiratory compromise.

Analysis of the interference of hexamethonium and indomethacin on both early and late phase lung alterations revealed that they have divergent modulatory actions. Indomethacin, but not hexamethonium, lowered the CTX-induced lung vascular permeability and PGE_2 production. Vascular permeability occurs in the early phase of lung diseases and is associated with lung tissue alterations such as edema [44,45,81]. PGE_2 also elicits vasodilation and thereby increases vascular permeability by acting on EP2 and EP4 receptors [53]. Considering previous studies on CTX biodistribution that report that the toxin reaches lung tissues after 10 min of i.v. administration [82,83], we can assume that, in the early phase, (i) CTX directly acted on lung tissues in order to induce the production of lipid mediators; and (i) PGE_2 mediated the increased lung vascular permeability. Regarding the late phase lung alterations, characterized by an inflammatory response, hexamethonium but not indomethacin mitigated the CTX-induced increase in lung myeloperoxidase activity and in the levels of the pro-inflammatory cytokines/chemokines IL-1β, IL-6, TNF-α, and CXCL-1. Altogether, these findings indicate that CTX-induced PGE_2 production was not involved in lung inflammation, and that hexamethonium acted as peripheral nervous system nicotinic blocker and did not have a direct anti-inflammatory action; hence, the CTX-induced lung inflammatory response seemed to be associated with a secondary effect of the toxin.

To examine whether CTX induces lung inflammation directly or indirectly, we compared data from local inflammatory profile using the dorsal air pouch model with data from lung tissue homogenates. Both local and tissue inflammatory responses were characterized by the presence of neutrophil infiltration and pro-inflammatory cytokines/chemokines, which peaked at 6 and 12 h in pouch fluid and lung tissue, respectively. Such kinetic difference can be explained by CTX biodistribution, but it does not exclude the possibility of a direct toxin action on the lungs. The increased levels of TNF-α and IL-1β in lung tissue but not in pouch fluid indicated that these compartments had different inflammation patterns. In addition, hexamethonium had no effect on total leukocyte counting and IL-6 and CXCL-1 levels in pouch fluid, but mitigated the CTX-induced increase in these parameters in mice lungs. Interestingly, pre-treatment with indomethacin worsened the CTX-induced local inflammatory response, suggesting that PGE_2 favors resolution of the inflammatory process. Although it is known that PGE_2 plays an anti-inflammatory role [50] and CTX also presents anti-inflammatory and immunosuppressive properties [84], the resolution effect of PGE_2 associated with CTX has never been described. Therefore, lung inflammation in CTX-treated mice was not associated with a direct effect of the toxin on the lungs.

Treatment with hexamethonium also significantly mitigated the CTX-induced impairment of lung function, which was another late phase alteration; in particular, it restored the dynamic and quasi-static respiratory compliances and inspiratory capacity. The neurophysiology of respiration involves the participation of cholinergic transmission from i) parasympathetic nerves that activates mAChRs, present on airway smooth muscle and blood vessels, causing bronchoconstriction and vasodilatation, and ii) and motor nerve fibers responsible for promoting diaphragm and intercostals muscles activation mediated by nAChRs [85,86]. Consequently, a possible mechanism of CTX lung toxicity would involve the modulation of nicotinic receptors, at ganglionar and/or neuromuscular junction levels, of cholinergic fibers of respiratory muscles. Our data clearly support the hypothesis that CTX induced respiratory impairment by triggering peripheral paralysis of airway muscles, including the diaphragm, which can be associated with development of lung hypoxia. CTX-induced respiratory paralysis in rabbits promotes severe acidosis and hypoxia, as demonstrated by decreased blood pH and pO_2, and increased pCO_2 [80]. Some alterations that occur in several pathologies associated with neuromuscular paralysis and hypoxia are the development of an inflammatory response comprising neutrophil infiltration, release of pro-inflammatory mediators, lung tissue fibrosis, decreased lung compliance, and increased lung resistance [87–89]. Therefore, the CTX-induced late phase alterations were not associated with a direct toxin effect on the lungs, but with a neuromuscular blockade that triggered airway muscle paralysis and promoted a hypoxemic condition.

5. Conclusions

Here we report for the first time the CTX-induced lung alterations and their implications on the respiratory function, as well as the mechanisms involved. The toxin causes acute respiratory failure characterized by early and late phase lung alterations. The early phase is marked by a direct CTX action on lung tissue that increases the production of lipid mediators, among which PGE_2 is supposed to mediate the increased vascular permeability. Additionally, the capacity to modulate lung cholinergic transmission via nicotinic receptors, possible at ganglionic and neuromuscular levels, induced a set of late phase alterations characterized by a moderate, but consistent, inflammatory response associated with morphological alterations characterized by tissue septum wall thickening, edema, hemorrhage, and decreased alveolar sac areas. Together, these effects impair lung mechanical function—more specifically, lung compliance and inspiratory capacity are decreased—and lead to animal death. Considering the CTX-induced mice lung impairment, we conclude that respiratory failure is the determinant factor for murine death. The findings reported herein highlight the impact of CTX on respiratory compromise, and introduce the use of nicotinic blockers and prostaglandin biosynthesis pathway inhibitors as possible symptomatic therapy to envenomed patients.

The present work deeply investigated the CTX-induced respiratory impairment, which is an important pathological condition of rattlesnake-envenomed patients. Additional studies on the toxicological effects of CTX are required due to its dual role: (1) as the major component, the toxin is a key player of pathological events induced by C.d.t. venom, whose understanding may help to develop therapeutic interventions to envenomed patients thru nicotinic acetylcholine antagonists drugs; (2) the toxin has promising medicinal applications due to its anti-inflammatory, immunosuppressive, antitumor, and analgesic properties [84,90–92]. The potential medicinal application of CTX, as demonstrated in phase I clinical trials against cancer [18,93], stresses the importance of analyzing the side effects of the toxin that may occur in possible cases of intoxication during its therapeutic use.

Supplementary Materials: The following are available online at http://www.mdpi.com/2218-273X/10/5/794/s1, Figure S1: Inflammatory parameters in air pouch fluid from CTX-treated mice; Figure S2: Effects of indomethacin and hexamethonium on air pouch inflammation in CTX-treated mice; Figure S3: Modulation of serum CK-MB levels by hexamethonium and indomethacin in CTX-treated mice.

Author Contributions: M.A.S. idealized and designed the study, performed the experiments and analyzed the data and discussed the overall research; S.V.S. idealized and designed the study, analyzed the data and discussed the overall research; C.R.A.F.D. designed and performed the dose-response experiments; C.O.S.S. designed, performed and analyzed the lung leukocyte phenotyping and flow cytometry analysis; V.M.B.d.F. and M.C.B. designed, performed, analyzed and discussed the lung mechanical experiments; L.O.S. performed and analyzed the microscopy experiments; T.R.C. designed, performed and analyzed the PCR experiment; A.G.L. performed, analyzed and discussed lung histological data; A.P.F.P., C.A.S. and L.H.F. designed, performed, analyzed and discussed mass spectrometry eicosanoids quantifications. All authors have read and agreed to the published version of the manuscript.

Funding: This study was supported by the São Paulo Research Foundation [FAPESP, grants n. 2011/23236-4 and 2015/06290-6].

Conflicts of Interest: The authors declare no conflict of interest.

References

1. Muller, V.D.M.; Russo, R.R.; Oliveira Cintra, A.C.; Sartim, M.A.; De Melo Alves-Paiva, R.; Figueiredo, L.T.M.; Sampaio, S.V.; Aquino, V.H. Crotoxin and phospholipases A 2 from Crotalus durissus terrificus showed antiviral activity against dengue and yellow fever viruses. *Toxicon* **2012**, *59*, 507–515. [CrossRef] [PubMed]
2. Zoccal, K.F.; Sorgi, C.A.; Hori, J.I.; Paula-Silva, F.W.G.; Arantes, E.C.; Serezani, C.H.; Zamboni, D.S.; Faccioli, L.H. Opposing roles of LTB4 and PGE2 in regulating the inflammasome-dependent scorpion venom-induced mortality. *Nat. Commun.* **2016**, *7*, 10760. [CrossRef] [PubMed]
3. da Silva, N.G.; Sampaio, S.C.; Gonçalves, L.R.C. Inhibitory effect of Crotalus durissus terrificus venom on chronic edema induced by injection of bacillus Calmette-Guérin into the footpad of mice. *Toxicon* **2013**, *63*, 98–103. [CrossRef] [PubMed]
4. Giannopoulos, P.F.; Chu, J.; Joshi, Y.B.; Sperow, M.; Li, J.G.; Kirby, L.G.; Praticò, D. 5-lipoxygenase activating protein reduction ameliorates cognitive deficit, synaptic dysfunction, and neuropathology in a mouse model of Alzheimer's disease. *Biol. Psychiatry* **2013**, *74*, 348–356. [CrossRef] [PubMed]
5. Chaisakul, J.; Ahmad Rusmili, M.R.; Hodgson, W.C.; Hatthachote, P.; Suwan, K.; Inchan, A.; Chanhome, L.; Othman, I.; Chootip, K. A pharmacological examination of the cardiovascular effects of Malayan krait (Bungarus candidus) venoms. *Toxins* **2017**, *9*, E122. [CrossRef] [PubMed]
6. de Sousa Nogueira-Neto, F.; Amorim, R.L.; Brigatte, P.; Picolo, G.; Ferreira, W.A.; Gutierrez, V.P.; Conceição, I.M.; Della-Casa, M.S.; Takahira, R.K.; Nicoletti, J.L.M.; et al. The analgesic effect of crotoxin on neuropathic pain is mediated by central muscarinic receptors and 5-lipoxygenase-derived mediators. *Pharmacol. Biochem. Behav.* **2008**, *91*, 252–260. [CrossRef]
7. Lee, C.Y.; Chen, Y.M.; Joubert, F.J. Protection by atropine against synergistic lethal effects of the Angusticeps-type toxin F7 from eastern green mamba venom and toxin I from black mamba venom. *Toxicon* **1982**, *20*, 665–667. [CrossRef]
8. Durand, M.T.; Becari, C.; Tezini, G.C.S.V.; Fazan, R.; Oliveira, M.; Guatimosim, S.; Prado, V.F.; Prado, M.A.M.; Salgado, H.C. Autonomic cardiocirculatory control in mice with reduced expression of the vesicular acetylcholine transporter. *Am. J. Physiol. Heart Circ. Physiol.* **2015**, *309*, H655–H662. [CrossRef]
9. Zanelati, T.V.; Biojone, C.; Moreira, F.A.; Guimarães, F.S.; Joca, S.R.L. Antidepressant-like effects of cannabidiol in mice: Possible involvement of 5-HT 1A receptors. *Br. J. Pharmacol.* **2010**, *159*, 122–128. [CrossRef]
10. Radu, M.; Chernoff, J. An in vivo assay to test blood vessel permeability. *J. Vis. Exp.* **2013**, *73*, e50062. [CrossRef]
11. Pfaffl, M.W.; Tichopad, A.; Prgomet, C.; Neuvians, T.P. Determination of stable housekeeping genes, differentially regulated target genes and sample integrity: BestKeeper—Excel-based tool using pair-wise correlations. *Biotechnol. Lett.* **2004**, *26*, 509–515. [CrossRef] [PubMed]
12. Sorgi, C.A.; Peti, A.P.F.; Petta, T.; Meirelles, A.F.G.; Fontanari, C.; Moraes, L.A.B.; Faccioli, L.H. Comprehensive high-resolution multiple-reaction monitoring mass spectrometry for targeted eicosanoid assays. *Sci. Data* **2018**, *21*, 180167. [CrossRef] [PubMed]
13. Cardoso, C.R.; Provinciatto, P.R.; Godoi, D.F.; Ferreira, B.R.; Teixeira, G.; Rossi, M.A.; Cunha, F.Q.; Silva, J.S. IL-4 regulates susceptibility to intestinal inflammation in murine food allergy. *Am. J. Physiol. Gastrointest. Liver Physiol.* **2009**, *296*, G593–G600. [CrossRef] [PubMed]
14. Souza, C.O.S.; Espíndola, M.S.; Fontanari, C.; Prado, M.K.B.; Frantz, F.G.; Rodrigues, V.; Gardinassi, L.G.; Faccioli, L.H. CD18 regulates monocyte hematopoiesis and promotes resistance to experimental schistosomiasis. *Front. Immunol.* **2018**, *9*, 1970. [CrossRef]
15. Diz Filho, E.B.S.; Marangoni, S.; Toyama, D.O.; Fagundes, F.H.R.; Oliveira, S.C.B.; Fonseca, F.V.; Calgarotto, A.K.; Joazeiro, P.P.; Toyama, M.H. Enzymatic and structural characterization of new PLA2 isoform isolated from white venom of Crotalus durissus ruruima. *Toxicon* **2009**, *53*, 104–114. [CrossRef]
16. Cavalcante, W.L.G.; Noronha-Matos, J.B.; Timóteo, M.A.; Fontes, M.R.M.; Gallacci, M.; Correia-de-Sá, P. Neuromuscular paralysis by the basic phospholipase A2 subunit of crotoxin from Crotalus durissus terrificus snake venom needs its acid chaperone to concurrently inhibit acetylcholine release and produce muscle blockage. *Toxicol. Appl. Pharmacol.* **2017**, *334*, 8–17. [CrossRef]
17. Araújo, D.A.; Beirão, P.S. Effects of crotoxin on the action potential kinetics of frog skeletal muscle. *Braz. J. Med. Biol. Res.* **1993**, *26*, 1111–1121.

18. Cura, J.E.; Blanzaco, D.P.; Brisson, C.; Cura, M.A.; Cabrol, R.; Larrateguy, L.; Mendez, C.; Sechi, J.C.; Silveira, J.S.; Theiller, E.; et al. Phase I and pharmacokinetics study of crotoxin (cytotoxic PLA2, NSC-624244) in patients with advanced cancer. *Clin. Cancer Res.* **2002**, *8*, 1033–1041.
19. Câmara, P.R.S.; Esquisatto, L.C.M.; Camargo, E.A.; Ribela, M.T.C.P.; Toyama, M.H.; Marangoni, S.; De Nucci, G.; Antunes, E. Inflammatory oedema induced by phospholipases A2 isolated from Crotalus durissus sp. in the rat dorsal skin: A role for mast cells and sensory C-fibers. *Toxicon* **2003**, *41*, 823–829. [CrossRef]
20. Gutiérrez, J.M.; Alberto Ponce-Soto, L.; Marangoni, S.; Lomonte, B. Systemic and local myotoxicity induced by snake venom group II phospholipases A2: Comparison between crotoxin, crotoxin B and a Lys49 PLA2 homologue. *Toxicon* **2008**, *51*, 80–92. [CrossRef]
21. Moreira, V.; Gutiérrez, J.M.; Soares, A.M.; Zamunér, S.R.; Purgatto, E.; de Fátima Pereira Teixeira, C. Secretory phospholipases A2 isolated from Bothrops asper and from Crotalus durissus terrificus snake venoms induce distinct mechanisms for biosynthesis of prostaglandins E2 and D2 and expression of cyclooxygenases. *Toxicon* **2008**, *52*, 428–459. [CrossRef] [PubMed]
22. Wei, S.; Ong, W.Y.; Thwin, M.M.; Fong, C.W.; Farooqui, A.A.; Gopalakrishnakone, P.; Hong, W. Group IIA secretory phospholipase A2 stimulates exocytosis and neurotransmitter release in pheochromocytoma-12 cells and cultured rat hippocampal neurons. *Neuroscience* **2003**, *121*, 891–898. [CrossRef]
23. Xavier, C.V.; da S. Setúbal, S.; Lacouth-Silva, F.; Pontes, A.S.; Nery, N.M.; de Castro, O.B.; Fernandes, C.F.C.; Soares, A.M.; Fortes-Dias, C.L.; Zuliani, J.P. Phospholipase A2 Inhibitor from Crotalus durissus terrificus rattlesnake: Effects on human peripheral blood mononuclear cells and human neutrophils cells. *Int. J. Biol. Macromol.* **2017**, *105*, 1117–1125. [PubMed]
24. Azevedo-Marques, M.M.; Cupo, P.; Coimbra, T.M.; Hering, S.E.; Rossi, M.A.; Laure, C.J. Myonecrosis, myoglobinuria and acute renal failure induced by south american rattlesnake (Crotalus durissus terrificus) envenomation in brazil. *Toxicon* **1985**, *23*, 631–636. [CrossRef]
25. Azevedo-Marques, M.M.; Cupo, P.; Amaral, C.F.S.; Hering, S.E. Rattlesnake bites. Clinical features and complementary tests. *Mem. Inst. Butantan* **1990**, *52*, 27–30.
26. Bucaretchi, F.; De Capitani, E.M.; Branco, M.M.; Fernandes, L.C.R.; Hyslop, S. Coagulopathy as the main systemic manifestation after envenoming by a juvenile South American rattlesnake (Crotalus durissus terrificus): Case report. *Clin. Toxicol.* **2013**, *51*, 505–508. [CrossRef] [PubMed]
27. Hardy, D.L. Envenomation by the Mojave rattlesnake (Crotalus scutulatus scutulatus) in southern Arizona, USA. *Toxicon* **1983**, *21*, 11–118. [CrossRef]
28. Amaral, C.F.; Magalhães, R.A.; de Rezende, N.A. Respiratory involvement secondary to crotalid ophidian bite (Crotalus durissus). *Rev. Inst. Med. Trop. Sao Paulo* **1991**, *33*, 251–255. [CrossRef]
29. Kerns, W.; Tomaszewski, C. Airway obstruction following canebrake rattlesnake envenomation. *J. Emerg. Med.* **2001**, *20*, 377–380. [CrossRef]
30. Brooks, D.E.; Graeme, K.A.; Ruha, A.M.; Tanen, D.A. Respiratory compromise in patients with rattlesnake envenomation. *J. Emerg. Med.* **2002**, *23*, 329–332. [CrossRef]
31. Baum, R.A.; Bronner, J.; Akpunonu, P.D.S.; Plott, J.; Bailey, A.M.; Keyler, D.E. Crotalus durissus terrificus (viperidae; crotalinae) envenomation: Respiratory failure and treatment with antivipmyn TRI ® antivenom. *Toxicon* **2019**, *163*, 32–35. [CrossRef] [PubMed]
32. Moreira, E.G.; Nascimento, N.; Rogero, J.R.; Vassilieff, V.S. Gabaergic-benzodiazepine system is involved in the crotoxin-induced anxiogenic effect. *Pharmacol. Biochem. Behav.* **2000**, *65*, 7–13. [CrossRef]
33. De Barros Ribeiro, G.; De Almeida, H.C.; Velarde, D.T.; De Melo Sá, M.L.V. Study of crotoxin on the induction of paralysis in extraocular muscle in animal model. *Arq. Bras. Oftalmol.* **2012**, *75*, 307–312. [CrossRef] [PubMed]
34. Rangel-Santos, A.; Dos-Santos, E.C.; Lopes-Ferreira, M.; Lima, C.; Cardoso, D.F.; Mota, I. A comparative study of biological activities of crotoxin and CB fraction of venoms from Crotalus durissus terrificus, Crotalus durissus cascavella and Crotalus durissus collilineatus. *Toxicon* **2004**, *43*, 801–810. [CrossRef] [PubMed]
35. Brazil, O.V.; Fariña, R.; Yoshida, L.; De Oliveira, V.A. Pharmacology of crystalline crotoxin. 3. Cardiovascular and respiratory effects of crotoxin and Crotalus durissus terrificus venom. *Mem. Inst. Butantan* **1966**, *33*, 1000.
36. Cupo, P.; Azevedo-Marques, M.M.; Hering, S.E. Acute myocardial infarction-like enzyme profile in human victims of Crotalus durissus terrificus envenoming. *Trans. R. Soc. Trop. Med. Hyg.* **1990**, *84*, 447–451. [CrossRef]

37. Sano-Martins, I.S. Coagulopathy following lethal and non-lethal envenoming of humans by the South American rattlesnake (Crotalus durissus) in Brazil. *QJM* **2001**, *94*, 551–559. [CrossRef]
38. de Sousa-e-Silva, M.C.C.; Tomy, S.C.; Tavares, F.L.; Navajas, L.; Larsson, M.H.M.A.; Lucas, S.R.R.; Kogika, M.M.; Sano-Martins, I.S. Hematological, hemostatic and clinical chemistry disturbances induced by Crotalus durissus terrificus snake venom in dogs. *Hum. Exp. Toxicol.* **2003**, *22*, 491–500. [CrossRef]
39. Brazil, O.V. Pharmacology of crystalline crotoxin. II. Neuromuscular blocking action. *Mem. Inst. Butantan* **1966**, *33*, 981–992.
40. Haschek, W.M.; Rousseaux, C.G.; Wallig, M.A.; Bolon, B.; Ochoa, R. *Haschek and Rousseaux's Handbook of Toxicologic Pathology*, 3rd ed.; Elsevier: Waltham, MA, USA, 2013; ISBN 9780124157590.
41. Sanderson, M.J. Exploring lung physiology in health and disease with lung slices. *Pulm. Pharmacol. Ther.* **2011**, *24*, 452–465. [CrossRef]
42. Nonaka, P.N.; Amorim, C.F.; Paneque Peres, A.C.; e Silva, C.A.M.; Zamuner, S.R.; Ribeiro, W.; Cogo, J.C.; Vieira, R.P.; Dolhnikoff, M.; de Oliveira, L.V.F. Pulmonary mechanic and lung histology injury induced by Crotalus durissus terrificus snake venom. *Toxicon* **2008**, *51*, 1158–1166. [CrossRef] [PubMed]
43. de Oliveira Neto, J.; de Moraes Silveira, J.A.; Serra, D.S.; de Araújo Viana, D.; Borges-Nojosa, D.M.; Sampaio, C.M.S.; Monteiro, H.S.A.; Cavalcante, F.S.Á.; Evangelista, J.S.A.M. Pulmonary mechanic and lung histology induced by Crotalus durissus cascavella snake venom. *Toxicon* **2017**, *137*, 144–149. [CrossRef] [PubMed]
44. Parker, J.C. Acute lung injury and pulmonary vascular permeability: Use of transgenic models. *Compr. Physiol.* **2011**, *1*, 835–882. [PubMed]
45. Mammoto, A.; Mammoto, T.; Kanapathipillai, M.; Yung, C.W.; Jiang, E.; Jiang, A.; Lofgren, K.; Gee, E.P.S.; Ingber, D.E. Control of lung vascular permeability and endotoxin-induced pulmonary oedema by changes in extracellular matrix mechanics. *Nat. Commun.* **2013**, *4*, 1759. [CrossRef]
46. Peters-Golden, M.; Brock, T.G. 5-Lipoxygenase and FLAP. *Prostaglandins Leukot. Essent. Fat. Acids* **2003**, *69*, 99–109. [CrossRef]
47. Wan, M.; Tang, X.; Stsiapanava, A.; Haeggström, J.Z. Biosynthesis of leukotriene B4. *Semin. Immunol.* **2017**, *33*, 3–15. [CrossRef]
48. Nakamura, H.; Wakita, S.; Suganami, A.; Tamura, Y.; Hanada, K.; Murayama, T. Modulation of the activity of cytosolic phospholipase A2α (cPLA2α) by cellular sphingolipids and inhibition of cPLA2α by sphingomyelin. *J. Lipid Res.* **2010**, *51*, 720–728. [CrossRef]
49. Fretland, D.J.; Widomski, D.L.; Anglin, C.P.; Gaginella, T.S. The antiinflammatory agent SC-41930 inhibits granulocyte infiltration of the rodent dermis induced by 6-trans-leukotriene B4. *Prostaglandins Leukot. Essent. Fat. Acids* **1991**, *44*, 61–65. [CrossRef]
50. Kalinski, P. Regulation of immune responses by prostaglandin E 2. *J. Immunol.* **2012**, *188*, 21–28. [CrossRef]
51. Powell, W.S.; Rokach, J. Biosynthesis, biological effects, and receptors of hydroxyeicosatetraenoic acids (HETEs) and oxoeicosatetraenoic acids (oxo-ETEs) derived from arachidonic acid. *Biochim. Biophys. Acta Mol. Cell Biol. Lipids* **2015**, *1851*, 340–355. [CrossRef]
52. Miyabe, Y.; Miyabe, C.; Luster, A.D. LTB4 and BLT1 in inflammatory arthritis. *Semin. Immunol.* **2017**, *33*, 52–57. [CrossRef] [PubMed]
53. Omori, K.; Kida, T.; Hori, M.; Ozaki, H.; Murata, T. Multiple roles of the PGE2-EP receptor signal in vascular permeability. *Br. J. Pharmacol.* **2014**, *171*, 4879–4889. [CrossRef] [PubMed]
54. Walters, E.H.; Davies, B.H. Dual effect of prostaglandm E2 on normal airways smooth muscle in vivo. *Thorax* **1982**, *37*, 918–922. [CrossRef] [PubMed]
55. Giannotti, K.C.; Leiguez, E.; Carvalho, A.E.Z.D.; Nascimento, N.G.; Matsubara, M.H.; Fortes-Dias, C.L.; Moreira, V.; Teixeira, C. A snake venom group IIA PLA2 with immunomodulatory activity induces formation of lipid droplets containing 15-d-PGJ2 in macrophages. *Sci. Rep.* **2017**, *7*, 4098. [CrossRef]
56. Sampaio, S.C.; Alba-Loureiro, T.C.; Brigatte, P.; Landgraf, R.G.; Santos, E.C.D.; Curi, R.; Cury, Y. Lipoxygenase-derived eicosanoids are involved in the inhibitory effect of Crotalus durissus terrificus venom or crotoxin on rat macrophage phagocytosis. *Toxicon* **2006**, *47*, 313–321. [CrossRef]
57. De Almeida, C.S.; Andrade-Oliveira, V.; Câmara, N.O.S.; Jacysyn, J.F.; Faquim-Mauro, E.L. Crotoxin from Crotalus durissus terrificus is able to down-modulate the acute intestinal inflammation in mice. *PLoS ONE* **2015**, *10*, e0121427. [CrossRef]

58. Pulli, B.; Ali, M.; Forghani, R.; Schob, S.; Hsieh, K.L.C.; Wojtkiewicz, G.; Linnoila, J.J.; Chen, J.W. Measuring myeloperoxidase activity in biological samples. *PLoS ONE* **2013**, *8*, e67976. [CrossRef]
59. Kouyoumdjian, J.A.; Harris, J.B.; Johnson, M.A. Muscle necrosis caused by the sub-units of crotoxin. *Toxicon* **1986**, *24*, 575–583. [CrossRef]
60. Landucci, E.C.T.; Antunes, E.; Donato, J.L.; Faro, R.; Hyslop, S.; Marangoni, S.; Oliveira, B.; Cirino, G.; de Nucci, G. Inhibition of carrageenin-induced rat paw oedema by crotapotin, a polypeptide complexed with phospholipase A2. *Br. J. Pharmacol.* **1995**, *114*, 578–583. [CrossRef]
61. Cardoso, D.F.; Lopes-Ferreira, M.; Faquim-Mauro, E.L.; Macedo, M.S.; Farsky, S.H.P. Role of crotoxin, a phospholipase A2 isolated from Crotalus durissus terrificus snake venom, on inflammatory and immune reactions. *Mediat. Inflamm.* **2001**, *10*, 125–133. [CrossRef]
62. Massis, L.M.; Assis-Marques, M.A.; Castanheira, F.V.S.; Capobianco, Y.J.; Balestra, A.C.; Escoll, P.; Wood, R.E.; Manin, G.Z.; Correa, V.M.A.; Alves-Filho, J.C.; et al. Legionella longbeachae is immunologically silent and highly virulent in vivo. *J. Infect. Dis.* **2017**, *215*, 440–451. [CrossRef] [PubMed]
63. Mitzner, W. Mechanics of the lung in the 20th century. *Compr. Physiol.* **2011**, *1*, 2009–2027. [PubMed]
64. Faffe, D.S.; Zin, W.A. Lung parenchymal mechanics in health and disease. *Physiol. Rev.* **2009**, *89*, 759–775. [CrossRef] [PubMed]
65. McGovern, T.K.; Robichaud, A.; Fereydoonzad, L.; Schuessler, T.F.; Martin, J.G. Evaluation of respiratory system mechanics in mice using the forced oscillation technique. *J. Vis. Exp.* **2013**, e50172. [CrossRef]
66. Grinnan, D.C.; Truwit, J.D. Clinical review: Respiratory mechanics in spontaneous and assisted ventilation. *Crit. Care* **2005**, *9*, 472–484. [CrossRef] [PubMed]
67. Lutfi, M.F. The physiological basis and clinical significance of lung volume measurements. *Multidiscip. Respir. Med.* **2017**, *12*, 3. [CrossRef]
68. Fernandez, I.E.; Amarie, O.V.; Mutze, K.; Königshoff, M.; Yildirim, A.Ö.; Eickelberg, O. Systematic phenotyping and correlation of biomarkers with lung function and histology in lung fibrosis. *Am. J. Physiol. Lung Cell. Mol. Physiol.* **2016**, *310*, L919–L927. [CrossRef]
69. Vulfius, C.A.; Kasheverov, I.E.; Kryukova, E.V.; Spirova, E.N.; Shelukhina, I.V.; Starkov, V.G.; Andreeva, T.V.; Faure, G.; Zouridakis, M.; Tsetlin, V.I.; et al. Pancreatic and snake venom presynaptically active phospholipases A2 inhibit nicotinic acetylcholine receptors. *PLoS ONE* **2017**, *12*, e0186206. [CrossRef]
70. Hawgood, B.J.; de Sa, S.S. Changes in spontaneous and evoked release of transmitter induced by the crotoxin complex and its component phospholipase A2 at the frog neuromuscular junction. *Neuroscience* **1979**, *4*, 293–303. [CrossRef]
71. Brazil, O.V.; Fontana, M.D.; Heluany, N.F. Nature of the postsynaptic action of crotoxin at guinea-pig diaphragm end-plates. *J. Nat. Toxins* **2000**, *9*, 33–42.
72. Santos, P.E.B.; Souza, S.D.; Freire-Maia, L.; Almeida, A.P. Effects of crotoxin on the isolated guinea pig heart. *Toxicon* **1990**, *28*, 215–224. [CrossRef]
73. Muniz, Z.M.; Diniz, C.R. The effect of crotoxin on the longitudinal muscle-myenteric plexus preparation of the guinea pig ileum. *Neuropharmacology* **1989**, *28*, 741–747. [CrossRef]
74. Radvanyi, F.; Keil, A.; Saliou, B.; Lembezat, M.P.; Bon, C. Binding of divalent and trivalent cations with crotoxin and with its phospholipase and its non-catalytic subunits: Effects on enzymatic activity and on the interaction of phospholipase component with phospholipids. *Biochim. Biophys. Acta (BBA) Lipids Lipid Metab.* **1989**, *1006*, 183–192. [CrossRef]
75. Yen, C.H.; Tzeng, M.C. Identification of a new binding protein for crotoxin and other neurotoxic phospholipase A 2 s on brain synaptic membranes. *Biochemistry* **1991**, *30*, 11473–11477. [CrossRef] [PubMed]
76. Lucas, S. The Pharmacology of Indomethacin. *Headache* **2016**, *56*, 436–446. [CrossRef] [PubMed]
77. Gurney, A.M.; Rang, H.P. The channel-blocking action of methonium compounds on rat submandibular ganglion cells. *Br. J. Pharmacol.* **1984**, *82*, 623–642. [CrossRef]
78. Rang, H.P.; Rylett, R.J. The interaction between hexamethonium and tubocurarine on the rat neuromuscular junction. *Br. J. Pharmacol.* **1984**, *81*, 519–531. [CrossRef]
79. Hernández, M.; Scannone, H.; Finol, H.J.; Pineda, M.E.; Fernández, I.; Vargas, A.M.; Girón, M.E.; Aguilar, I.; Rodríguez-Acosta, A. Alterations in the ultrastructure of cardiac autonomic nervous system triggered by crotoxin from rattlesnake (Crotalus durissus cumanensis) venom. *Exp. Toxicol. Pathol.* **2007**, *59*, 129–137. [CrossRef]

80. Breithaupt, H. Neurotoxic and myotoxic effects of crotalus phospholipase A and its complex with crotapotin. *Naunyn Schmiedebergs Arch. Pharmacol.* **1976**, *292*, 271–278. [CrossRef]
81. Huaringa, A.J.; Leyva, F.J.; Glassman, A.B.; Haro, M.H.; Arellano-Kruse, A.; Kim, E.E. The lung permeability index: A feasible measurement of pulmonary capillary permeability. *Respir. Med.* **2011**, *105*, 230–235. [CrossRef]
82. Habermann, E.; Walsch, P.; Breithaupt, H. Biochemistry and pharmacology of the crotoxin complex-II. Possible interrelationships between toxicity and organ distribution of phospholipase A, crotapotin and their combination. *Naunyn Schmiedebergs Arch. Pharmacol.* **1972**, *273*, 313–330. [CrossRef] [PubMed]
83. Soares, M.A.; Silveira, M.B.; Simal, C.; Forte-Dias, C.L.; dos Santos, R.G. Biodistibution and spect imaging of crotoxin on mice bearing Ehrlich tumor. In Proceedings of the International Nuclear Atlantic Conference, Aben, Associação Brasileira De Energia Nuclear, Rio de Janeiro, Brazil, 27 September–2 October 2009.
84. Sartim, M.A.; Menaldo, D.L.; Sampaio, S.V. Immunotherapeutic potential of Crotoxin: Anti-inflammatory and immunosuppressive properties. *J. Venom. Anim. Toxins Incl. Trop. Dis.* **2018**, *24*, 39. [CrossRef] [PubMed]
85. Shao, X.M.; Feldman, J.L. Central cholinergic regulation of respiration: Nicotinic receptors. *Acta Pharmacol. Sin.* **2009**, *30*, 761–770. [CrossRef]
86. Scott, G.D.; Fryer, A.D. Role of parasympathetic nerves and muscarinic receptors in allergy and asthma. *Chem. Immunol. Allergy* **2012**, *98*, 48–69. [PubMed]
87. Hughes, R.A.; Bihari, D. Acute neuromuscular respiratory paralysis. *J. Neurol. Neurosurg. Psychiatry* **1993**, *56*, 334–343. [CrossRef] [PubMed]
88. Nicolls, M.R.; Voelkel, N.F. Hypoxia and the lung: Beyond hypoxic vasoconstriction. *Antioxid. Redox Signal.* **2007**, *9*, 741–743. [CrossRef]
89. Araneda, O.F.; Tuesta, M. Lung oxidative damage by hypoxia. *Oxid. Med. Cell. Longev.* **2012**, *2012*, 856918. [CrossRef]
90. Wolz-Richter, S.; Esser, K.H.; Hess, A. Antinociceptive activity of crotoxin in the central nervous system: A functional Magnetic Resonance Imaging study. *Toxicon* **2013**, *74*, 44–55. [CrossRef]
91. Wang, J.; Qin, X.; Zhang, Z.; Chen, M.; Wang, Y.; Gao, B. Crotoxin suppresses the tumorigenic properties and enhances the antitumor activity of Iressa® (gefinitib) in human lung adenocarcinoma SPCA-1 cells. *Mol. Med. Rep.* **2014**, *10*, 3009–3014. [CrossRef]
92. Muller, S.P.; Silva, V.A.O.; Silvestrini, A.V.P.; de Macedo, L.H.; Caetano, G.F.; Reis, R.M.; Mazzi, M.V. Crotoxin from Crotalus durissus terrificus venom: In vitro cytotoxic activity of a heterodimeric phospholipase A2 on human cancer-derived cell lines. *Toxicon* **2018**, *156*, 13–22. [CrossRef]
93. Medioni, J.; Brizard, M.; Elaidi, R.; Reid, P.F.; Benlhassan, K.; Bray, D. Innovative design for a phase 1 trial with intra-patient dose escalation: The Crotoxin study. *Contemp. Clin. Trials Commun.* **2017**, *7*, 186–188. [CrossRef] [PubMed]

© 2020 by the authors. Licensee MDPI, Basel, Switzerland. This article is an open access article distributed under the terms and conditions of the Creative Commons Attribution (CC BY) license (http://creativecommons.org/licenses/by/4.0/).

Review

Rescue of Hepatic Phospholipid Remodeling Defect in iPLA$_2$β-Null Mice Attenuates Obese but Not Non-Obese Fatty Liver

Walee Chamulitrat [1,*], Chutima Jansakun [1], Huili Li [1] and Gerhard Liebisch [2]

1 Department of Internal Medicine IV, University of Heidelberg Hospital, Im Neuenheimer Feld 410, 69120 Heidelberg, Germany; Chutima.Jansakun@med.uni-heidelberg.de (C.J.); Huili.Li@med.uni-heidelberg.de (H.L.)

2 Institute of Clinical Chemistry and Laboratory Medicine, University of Regensburg, Franz-Josef-Strauss-Allee 11, 93053 Regensburg, Germany; gerhard.liebisch@klinik.uni-regensburg.de

* Correspondence: Walee.Chamulitrat@med.uni-heidelberg.de; Tel.: +49-6221-56-38731

Received: 24 August 2020; Accepted: 15 September 2020; Published: 17 September 2020

Abstract: Polymorphisms of group VIA calcium-independent phospholipase A2 (iPLA$_2$β or PLA2G6) are positively associated with adiposity, blood lipids, and Type-2 diabetes. The ubiquitously expressed iPLA$_2$β catalyzes the hydrolysis of phospholipids (PLs) to generate a fatty acid and a lysoPL. We studied the role of iPLA$_2$β on PL metabolism in non-alcoholic fatty liver disease (NAFLD). By using global deletion iPLA$_2$β-null mice, we investigated three NAFLD mouse models; genetic Ob/Ob and long-term high-fat-diet (HFD) feeding (representing obese NAFLD) as well as feeding with methionine- and choline-deficient (MCD) diet (representing non-obese NAFLD). A decrease of hepatic PLs containing monounsaturated- and polyunsaturated fatty acids and a decrease of the ratio between PLs and cholesterol esters were observed in all three NAFLD models. iPLA$_2$β deficiency rescued these decreases in obese, but not in non-obese, NAFLD models. iPLA$_2$β deficiency elicited protection against fatty liver and obesity in the order of Ob/Ob › HFD » MCD. Liver inflammation was not protected in HFD NAFLD, and that liver fibrosis was even exaggerated in non-obese MCD model. Thus, the rescue of hepatic PL remodeling defect observed in iPLA$_2$β-null mice was critical for the protection against NAFLD and obesity. However, iPLA$_2$β deletion in specific cell types such as macrophages may render liver inflammation and fibrosis, independent of steatosis protection.

Keywords: PLA2G6; fatty liver; phospholipid remodeling; diet-induced obesity; morbidly obesity; choline and methionine deficiency

1. Obesity and NAFLD

Obesity is an epidemic with a prevalence rate of 13% of the world's population [1] and has become a major public health problem resulting in decreased quality of life, reduced working ability, and early death. Obesity-associated co-morbidity and diseases include atherosclerosis, diabetes, non-alcoholic fatty liver disease (NAFLD), and non-alcoholic steatohepatitis (NASH) [2]. A significant proportion of the risk of obesity is due to genetic variance [3–5]. Demographics (ethnicity, age, and gender) and behavior (eating behavior, physical activity, and smoking) are environmental factors contributing to obesity as well [6–8]. Increased consumption of high-fat-diet (HFD) contributes in a major way to obesity in a genetic variance-dependent manner [7,8]. One example is that C57BL/6J mice are more vulnerable to diet-induced obesity compared to other genetic backgrounds [9–11]. The gene-by-diet interactions may be highly heritable and they could significantly have a large impact on obesity in human offspring [12].

A study using more than 100 inbred strains of mice revealed that a high-fat/high-sucrose diet promotes strain-specific changes in obesity that is not accounted for by food intake [13]. This provides evidence for a genetically determined set-point for obesity at least for the case of high-fat/high-sucrose feeding [13]. The genome-wide association studies (GWAS) of obesity data have been used to elucidate the function of genetic variants. The GWAS analyses of knockout mouse phenotypes have provided PLA2G6 association with body weights, lipids, energy, and nervous system [14]. In human studies, the failure to explain a larger fraction of the genetic basis of obesity alone highlights the gene-by-diet interactions, genetic determinants of habitual dietary intake, as well as the interplay between diet, genes, and obesity [13,14].

Hepatic manifestation of obesity is NAFLD [15]. NAFLD is one of the most common causes of chronic liver disease worldwide [16]. NAFLD pathogenesis has a spectrum covering from steatosis through NASH to cirrhosis, which may progress to primary liver cancer [17]. NAFLD prevalence is 27–34% of the general population in the USA, and 40–90% of global obese populations have this disease [16,17]. Similar to obesity, genetic variances [18–20], hormones [20], sex [21,22], ethnicity [23] combined with age [24], as well as dietary and physical activity habits [25] are important factors and traits for NAFLD development.

While NAFLD is commonly seen in obese subjects, it is however not rare among non-obese and lean individuals [26,27] particularly those with specific ethnic backgrounds, such as, Asia-Pacific [28,29]. Genetic predispositions, fructose- and cholesterol-rich diet, visceral adiposity, and dyslipidaemia play an important role in the pathogenesis of lean NAFLD [26,27]. Lean-NAFLD patients show less severe histological features as compared to overweight and obese NAFLD patients. For the latter, a significant ~25% increment of mean fibrosis score is found suggesting that obesity could predict a worse long-term prognosis [30]. Lean subjects with evidence of NAFLD have clinically relevant impaired glucose tolerance, low adiponectin concentrations, and a distinct metabolite profile with an increased rate of patatin-like phospholipase containing lipase 3 (PNPLA3) risk allele carriage [31]. Cardiovascular events are the main cause of mortality and morbidity in non-obese NAFLD [26,27]; which is similar to obese NAFLD [32]. As atherogenic dyslipidaemia arises from hepatic steatosis [32], the metabolism of intrahepatic fat in NAFLD is also recognized to contribute to complications of obesity [33]. While lifestyle changes that include physical activity and weight loss are the mainstay of NAFLD treatment, the understanding of hepatic lipid metabolism may provide some clues for specific interactions between nutrients and dietary needs [34]. Thus, the understanding of the balanced biomolecules and nutrients in the diets would become important in providing insights for an alternative strategy to treat and alleviate NAFLD and obesity [35].

2. Animal Models of Obese and Non-Obese NAFLD/NASH

In order to identify important biomolecules and nutrients involved in NAFLD, we have used mouse NAFLD models because mice have shorter lifespan and provide research results in a relatively short period of time. We performed our studies using three different mouse models of NAFLD/NASH. They included leptin-deficient Ob/Ob and long-term HFD-fed mice (representing obese NAFLD), and mice fed with a methionine-choline deficient (MCD) diet (representing non-obese NAFLD). All of these mice had C57BL/6 background which is prone for obesity [11]. Ob/Ob and HFD feeding represent over-nutrition NAFLD model with metabolic perturbations, glucose intolerance, and insulin resistance that are common in humans with mild NASH [36]. MCD diet feeding of mice causes no increase in weight and obesity and no insulin resistance thus representing a non-obese NAFLD model with pathological mechanisms that lead to NASH [37].

The induction of NAFLD/NASH by MCD diet is based on an impaired synthesis of phosphatidylcholine (PC) and the subsequent reduced production of very low-density lipoproteins (VLDLs), and this leads to accumulation of hepatic triglycerides (TGs) and hepatic steatosis development [36,37]. Although histological features and inflammatory response of MCD model reflect human NAFLD/NASH, this model does not however resemble human metabolic physiology;

in that the levels of serum TGs, cholesterol, insulin, glucose, and leptin are not increased. As MCD-fed mice lose their bodyweight and do not exhibit insulin resistance, these mice may represent a model of non-obese NAFLD/NASH. Interestingly, non-obese NAFLD has also been described in mice deficient with phosphatidylethanolamine *N*-methyltransferase (PEMT), which is the enzyme that converts phosphatidylethanolamine (PE) to PC [38]. This bolsters the notion that altered phospholipid (PL) metabolism and changes in the composition of PC and PE are linked to NAFLD pathogenesis.

In this review, we investigated the extent of hepatic fatty acid (FA) and PL metabolism in livers of male Ob/Ob mice at six months old [39], male C57BL/6 mice at six months old fed with HFD (60 kcal % fat, Research Diet, USA) for six months [40], and female C57BL/6 mice at 12 months old fed with MCD diet (ssniff GmbH, Germany) for four weeks [41].

3. Phospholipids in NAFLD/NASH

In NAFLD, hepatic TG contents are the bulk vesicular fat stored in lipid droplets, thus the alteration in TG metabolism has been a focus for NAFLD prevention [42]. However, hepatic PLs could also play a role in NAFLD by three major mechanisms because PLs and their metabolism are important for (1) the formation of lipid droplets [43,44], (2) the regulation of de novo lipogenesis via sterol regulatory element-binding proteins (SREBPs), a family of membrane-bound transcription factors that regulate synthesis of cholesterol and unsaturated FAs [45,46], and (3) the metabolism and secretion of VLDLs [36,37].

In (1), cytosolic lipid droplets are the sites for storage of neutral lipids including TGs, which are surrounded by a monolayer of PLs [43]. It is known that relative abundance of PC and PE on the surface of lipid droplets is important for their dynamics [44]. An inhibition of PC biosynthesis during conditions that promote TG storage increases the size of the lipid droplets [44].

For (2), the disturbance of PC [47] or PE [48] synthesis by respective genetic deletion of *C. elegans* and *Drosophila* leads to an activation of SREBPs. Thus, decreased PL mass due to suppressed synthesis could lead to a compensatory upregulation of SREBPs eventually resulting in an increase in de novo lipogenic lipid synthesis, steatohepatitis, and metabolic syndrome. This notion could be supported by the data using transgenic mice with deletion of PC [47] and PE [48] in the liver; whereby these knockout mice exhibit propensity to develop NAFLD.

For the last (3) case, it has been long known that PLs are required for the formation and stability of lipoproteins [49]. Depletion of PC can affect the endoplasmic reticulum (ER) and protein trafficking in the Golgi [50]. Moreover, a block of the ER-to-Golgi trafficking associated with a decrease in PC synthesis is shown to induce TG accumulation and subsequent lipoprotein secretion [51]. Consistently, ω-3 FA-induced PL remodeling can alter the utilization of TGs in the form of TG-rich lipoproteins [52]. While the ratio of PC/PE that influence hepatocyte membrane integrity can regulate NAFLD [53], hepatocellular PC is shown to exhibit protective effects on hepatic steatosis, however this PC does not protect liver inflammation in NASH [54]. This may indicate a differential role of PC and perhaps other PLs in hepatocytes *versus* in immune cells. Thus, the alteration of hepatic PL metabolism during NAFLD in mice is linked to the syntheses and trafficking of TGs and FAs, as well as the synthesis and secretion of TG-rich lipoproteins.

Accordingly, patients with NAFLD contain a decrease in liver total PC and PE levels, and the contents of arachidonate (20:4)-containing PC and docosahexaenate (22:6)-containing TG are also decreased in livers of NASH patients [55]. The decrease of polyunsaturated fatty acid (PUFA)-containing lipids in NASH livers indicates that there is an impairment of PL remodeling, which could be due to the down-regulation of PL synthesis genes by pro-inflammatory cytokines, such as tumor necrosis factor-α [56]. It is therefore essential to determine PL profiles as a function of unsaturation of not only PC and PE, but also other lipids including sphingomyelin (SM), ceramides (Cer), and cholesteryl esters (CEs). These results will help identify the relative significance of these types of biomolecules and their different roles in obese and non-obese NAFLD.

4. Phospholipid-Metabolizing Genes and Phospholipases A2 (PLA_2) in Obesity and NAFLD

Since hepatocellular phospholipids play important role in NAFLD/NASH, phospholipid-metabolizing genes have thus been inherently subjected to research investigations. These genes may include phospholipase A2 (PLA_2) such as group IVA PLA_2 (or cytosolic $PLA_2\alpha$), group IIA PLA_2 (or PLA2G2A or secretory PLA_2), as well as lipid hydrolases with specificities for diverse substrates such as TGs, PLs, and retinol esters. These lipid hydrolases include PNPLA family consisting of six enzymes, namely PNPLA2 (ATGL or $iPLA_2\xi$), PNPLA3 (adiponutrin or $iPLA_2\varepsilon$), PNPLA4 ($iPLA_2\eta$), PNPLA6 ($iPLA_2\delta$), PNPLA8 (group VIB $iPLA_2$ or $iPLA_2\gamma$), and PNPLA9 (group VIA $iPLA_2$ or PLA2G6 or $iPLA_2\beta$) [57]. An ablation of group IVA PLA_2 [58] or or $iPLA_2\gamma$ [59,60] in mice leads to strong and partial protection against diet-induced obesity, respectively. The attenuation of obesity by group IVA PLA_2 deficiency could likely be due to the reduction of adipocyte differentiation [61], as well as attenuation of neutrophil infiltration and hepatic insulin resistance [62]. BL/6 mice expressing the human PLA2G2A gene when fed with a fat diet showed more insulin sensitivity and glucose tolerance with a mechanism of mitochondrial uncoupling activation in brown adipose tissues [63]. Inhibitor of secretory PLA_2 reduces obesity-induced inflammation in Beagle dogs [64], and protects diet-induced metabolic syndrome in rats [65]. Inactivation of the group 1B PLA_2 (PLA2G1B), a gut digestive enzyme, suppresses diet-induced obesity, hyperglycemia, insulin resistance, and hyperlipidemia in C57BL/6 mice [66,67] and attenuates atherosclerosis and metabolic diseases in LDL receptor-deficient mice [68]. Conversely, transgenic mice with pancreatic acinar cell-specific overexpression of the human PLA2G1B gene gain more weight and display elevated insulin resistance when challenged with a high-fat/carbohydrate diet [69]. Moreover, two secreted PLA_2s, PLA2G5 and PLA2G2E capable of hydrolysis of lipoproteins, are robustly induced in adipocytes of obese mice, and PLA2G5 prevents palmitate-induced M1 macrophage polarization and PLA2G2E moderately facilitates lipid accumulation in adipose tissue and liver [70]. On the contrary to previously mentioned PLA_2s, mice deficient with ATGL, $iPLA_2\xi$, or PNPLA2 when fed with MCD diet show exacerbated hepatic steatosis and inflammation [71]. Interestingly, mice deficient with adiponutrin or PNPLA3 show no protection against HFD or Ob/Ob background [72], but on the other hand show protection under ER stress [73]. Taken together, these publications show a growing list of lipolytic enzymes that act as metabolic coordinators of obesity and NALFD in mice.

Consistent with mouse data, obese human subjects with or without Type-2 diabetes show high activities of total PLA_2 and of Ca^{2+}-dependent and Ca^{2+}-independent enzymes; and that Ca^{2+}-dependent secretory $sPLA_2$ are the main enzyme responsible of obesity-associated high activity [74]. Moreover, sPLA2 activity is increased with high correlation with sensitive C-reactive proteins in morbidly obese patients [75]. Lastly, plasma PLA2 activity is increased in asthma patients and associated with high plasma cholesterol and body mass index [76].

Genome-wide (GWAS) and candidate gene association studies have identified several variants that predispose individuals to developing NAFLD. A study in mouse GWAS has identified 11 genome-wide significant loci to be associated with obesity traits, and a PL-metabolizing enzyme lysophospholipase-like 1 (LYPLAL1) was among these loci identified in the epididymal adipose tissues of diet-induced obese mice [13]. These results are consistent with association of this gene with human NAFLD [77]. When NASH/fibrosis was assessed histologically and non-invasive computed tomography (CT) was used for hepatic steatosis, it is reported that three variants near PNPLA3 are associated with CT hepatic steatosis, and variants in or near LYPLAL1 and adiponutrin or PNPLA3 are associated with histologic lobular inflammation/fibrosis [77]. NAFLD progression has a strong genetic component, and the most robust contributor is PNPLA3 rs738409 encoding the 148M protein sequence variant [78]. Moreover, antisense oligonucleotides-mediated silencing of Pnpla3 reduces liver steatosis in homozygous Pnpla3 148M/M knock-in mutant mice, but not in wild-type littermates fed a steatogenic high-sucrose diet [79]. As the variation in PNPLA3 contributes to ancestry-related differences in hepatic fat content and susceptibility to NAFLD, consistently the weight loss is effective in decreasing liver fat in subjects who are homozygous for the rs738409 PNPLA3 G or C allele [80].

Hence, current data bolster the notion that PL-metabolizing enzymes, particularly PNPLA3, may be involved in NAFLD development and thus may be used as targets for development of drugs for NAFLD/NASH treatment and prevention.

5. iPLA$_2$β in Obesity and NAFLD and Use of iPLA$_2$β-Null Mice

GWAS in >100,000 individuals of primarily European ancestry have identified group VIA calcium-independent PLA$_2$ (iPLA$_2$β, PLA2G6, or PNPLA9) as one of the 12 loci to be associated with human body fat percentage (BFP) [81]. This is consistent with PLA2G6 association with bodyweight in mice [14]. Further extended GWAS also identified a strong association of PLA2G6 to BFP in metabolically healthy obesity [82]. In this study, the BFP-increasing allele in the locus near PLA2G6 is associated with lower plasma TG levels in men and women, with lower insulin levels and risk of Type-2 diabetes particularly in men, and higher visceral adipose tissue in men [82]. Another GWAS for plasma lipids in >100,000 individuals also identified SNP rs5756931 of PLA2G6 as one of the 95 loci to be associated with plasma TGs [83]. These results were also recently reviewed [84]. While PLA2G6 association with plasma TG is shown to have no effects on cardiovascular disease (CAD) risk [85], PLA2G6 together with PLA2G2 and PLA2G5 levels are however increased in subgroups of patients with CAD [86]. Furthermore, genetic variants at or near PLA2G6 are associated with Type-2 diabetes in European-Americans [87], European-American women [88], and a Chinese population [89]. This is in line with the reported suppressed insulin secretion by islets in response to glucose and forskolin upon global iPLA$_2$β deletion in mice [90]. Hence, iPLA$_2$β or PLA2G6 may represent a key PL-metabolizing enzyme being critical in the development of obesity and Type-2 diabetes.

iPLA$_2$s are lipolytic enzymes not requiring calcium for catalysis in hydrolyzing ester bond of PL at sn-2 position to release a 2-lysoPL and a free FA [91–93]. iPLA$_2$β is a prototypic iPLA$_2$ that is ubiquitously expressed and plays a house-keeping role in PL metabolism and PL remodeling [91–93]. iPLA$_2$β mediates PL remodeling by regulating the composition of PUFA in PL pools, for example, an increase of PUFA-containing PLs was observed upon treatment of cultured cells with an iPLA$_2$β inhibitor [94–96].

In 2011, we obtained global-deficient iPLA$_2$β-null (KO) mice with exon 9 deletion [90,97] from Dr. John Turk, Washington University School of Medicine, St. Louis, MO, USA. Our first publication in 2016 revealed the functional role of iPLA$_2$β inactivation in morbidly obese NAFLD [39]. Male Ob/Ob mice were cross-bred with KO mice. Compared to Ob/Ob mice, the double Ob/Ob-iPLA$_2$β KO mice showed protection with significant reduction of body and liver weights, improved glucose tolerance, and reduction in islet hyperplasia [39]. The improvement in hepatic steatosis was also seen by attenuation of liver TG, FA, and CE contents in double Ob/Ob-iPLA$_2$β KO mice.

Work from Dr. Turk's laboratory showed that HFD feeding of iPLA$_2$β-null mice for six months did not improve, but rather further impaired glucose intolerance likely due to an impairment of insulin secretion by pancreatic islets [90]. Moreover, the global deletion of exon 2 in the iPLA$_2$β gene in mice fed with HFD for eight weeks also did not show any improvement in serum and liver TGs [98]. The lack of effects could be due to relative short HFD feeding such that hepatic PLs were not yet affected. We therefore attempted to define the conditions among the three NAFLD models that an inactivation of iPLA$_2$β is effective in alleviating obesity and NAFLD. We were particularly interested in comparing hepatic PL profiles in iPLA$_2$β-null mice in obese and non-obese NAFLD. We considered that hepatic PLs and TGs may be affected by the metabolism in adipose tissues of iPLA$_2$β-null mice, since iPLA$_2$β is shown to regulate adipocyte differentiation [99]. Interestingly unlike PLA2G6, group IVA PLA$_2$ or group VIB iPLA$_2$ (iPLA$_2$γ) are not included in human GWAS data on obesity/adiposity and blood lipids as discussed above [14,81–89]. Thus, PLA2G6 or iPLA$_2$β may exhibit a unique activity with a preference toward obesity and hence obese NAFLD [39].

We performed HFD feeding of WT and iPLA$_2$β KO mice for six months as another model of obese NAFLD [40]. We showed that protection was observed in iPLA$_2$β KO mice with an attenuation of HFD-induced body and liver-weight gains, liver enzymes, serum-free FAs, as well as hepatic TGs and steatosis scores. However, this deficiency did not attenuate hepatic ER stress, fibrosis, and inflammation markers. No protection was observed after short-term 3–5 week HFD feeding when hepatic PL contents were not yet depleted.

Since PL syntheses are disturbed by MCD feeding of mice [36,37], we tested whether iPLA$_2$β KO mice could still be protected from fatty liver in this non-obese NAFLD model. MCD feeding of female wild-type (WT) for four weeks induced hepatic steatosis with a severe reduction of body and visceral fat weights, which were not altered in MCD-fed iPLA$_2$β-KO mice [41]. However, iPLA$_2$β deficiency attenuated MCD-induced elevation of serum transaminase activities and hepatic expression of FA translocase Cd36, fatty-acid binding protein-4, peroxisome-proliferator activated receptorγ, and HDL-uptake gene scavenger receptor B type 1 (SR-B1). The reduction of lipid uptake genes was consistent with a decrease of hepatic esterified and un-esterified FAs and CEs [41]. On the contrary, iPLA$_2$β deficiency under MCD did not have any effects on inflammasomes and pro-inflammatory markers but rather exacerbated hepatic expression of myofibroblast α-smooth muscle actin and vimentin [41].

Taken together, iPLA$_2$β deficiency elicited protection against hepatic steatosis in an order of Ob/Ob › HFD » MCD; or that protection was better in obese NAFLD compared to non-obese NAFLD model.

6. Metabolic Lipid Changes in Ob/Ob Mice and Modulation by iPLA$_2$β Deficiency

Because the liver does not serve as a storage depot for fat, the steady-state concentration of hepatic TGs is low under physiological conditions. There is nevertheless a considerable trafficking of both TGs and FAs into and out of the liver during NAFLD development induced by genetic alterations, increased fat intake, and/or alteration in hepatic metabolism [34,42]. Hepatic steatosis in NAFLD develops when the rate of FA input is greater than that of FA output. Thus, the mechanisms for gene-to-diet interactions on the extent of steatosis are very complex since many genes are involved in the regulation of TG, FA, and lipoprotein syntheses; and that some of these lipids have been identified as obese [19] and non-obese [100] NAFLD modifier genes.

We first compared hepatic steatosis among WT and Ob/Ob mice. Liver TG and total FA contents were increased in Ob/Ob mice (Figure 1A). Metabolomic profiling has been used to study hepatic lipid metabolism [39–41]. Gas chromatography mass spectrometry (GC/MS) method was used to measure un-esterified and esterified FA species present in all lipids [101]. Regarding FA composition (% Mol), Ob/Ob mice showed an increase of hepatic FAs containing monounsaturated FAs (MUFA) but a decrease in those containing di- and >2 unsaturated FAs (Figure 1B).

PC and PE metabolism is important in pathogenesis of fatty liver owing to disrupted membrane integrity and suppressed PC syntheses as well as altered PC and PE composition [102]. Hepatic steatosis is associated with the reduction of hepatic PC or PLs as demonstrated in experiments in transgenic mice with a deletion of a PC synthesis gene [47,48,53]. Here, an electrospray ionization tandem mass spectrometry (ESI-MS/MS) method was utilized to profile PL species including PC, SM, lysoPC (LPC), LPE, PE, phosphatidylserine (PS), phosphatidylinositol (PI), plasmalogens (Pla), Cer, CE, and free cholesterol (FC) [103–106]. The composition among these PL subclasses (% Mol) showed a decrease of PC, PE, and PI, but an increase of CE in genetic Ob/Ob mice (Figure 1C). This indicates a shift from polar PLs to neutral lipids, namely, TGs and CEs (Figure 1A,C). The increase of CEs in Ob/Ob livers may reflect diabetes and hyperinsulinemia in these mice. Genetically obese Ob/Ob livers showed a significant increase in MUFA-containing CEs concomitant with a decrease of PUFA-containing PC, PE, PS, and PI as well as SM contents (Figure 1D–F).

PL contents and composition in livers of WT, Ob/Ob, and Ob/Ob-iPLA$_2$β KO mice were determined by ESI-MS/MS. Liver histology showed marked steatosis attenuation in Ob/Ob-iPLA$_2$β KO mice (Figure 2A). Here, PL composition (% Mol) (Figure 2B), MUFA-PL (Figure 2C), and PUFA-PL (Figure 2D) contents were analyzed showing the suppression of PUFA-PC, PUFA-PE, and PUFA-PS contents in Ob/Ob livers. This suppression was reversed in Ob/Ob-iPLA$_2$β KO mice. Moreover, the elevation of

MUFA-CEs and PUFA-CEs in Ob/Ob livers was also attenuated in Ob/Ob-iPLA$_2$β KO mice (Figure 2C,D). These changes were associated with an attenuation of bodyweight gains, hepatic steatosis, as well as the reduction of hepatic and plasma TGs [39]. These results showed that iPLA$_2$β has a pathophysiological function by depleting PUFA concentrations in Ob/Ob liver PLs. iPLA$_2$β inactivation re-establishes PL remodeling to return to normal homeostasis.

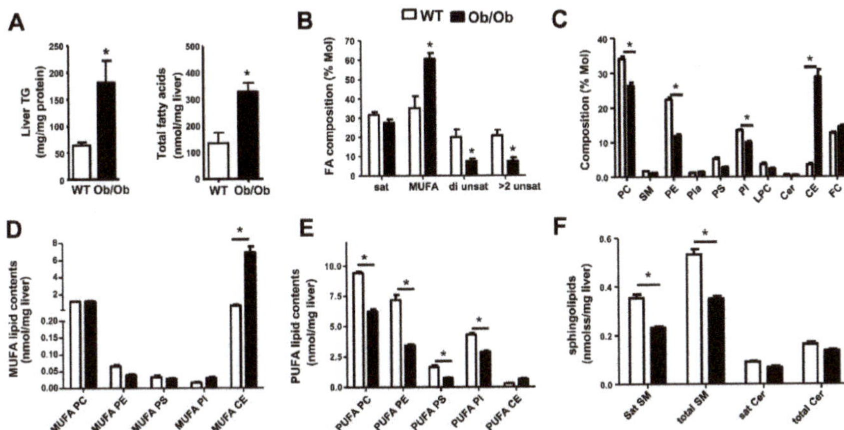

Figure 1. Lipid contents and composition show a defect of hepatic PL remodeling in Ob/Ob mice. Male wild-type (WT) and Ob/Ob mice at six months old were used. (**A**) The contents of hepatic triglycerides (TG), total fatty acids (FAs). (**B**) FA composition of saturated, monounsaturated fatty acids (MUFA), di unsaturated, and > 2 unsaturated FA. (**C**) The composition of phospholipids (PLs) and cholesterol esters (CEs). (**D**) The contents of PLs and CEs containing monounsaturated fatty acids (MUFA). (**E**) The contents of PLs and CEs containing polyunsaturated fatty acids (PUFA). (**F**) The contents of sphingolipids. Data are mean ± SEM, N = 5–7; *, $p < 0.05$ *versus* WT.

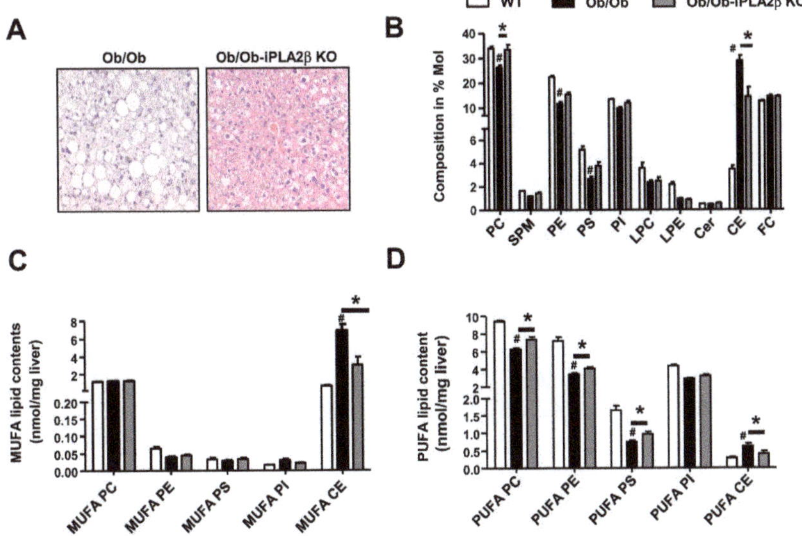

Figure 2. iPLA$_2$β inactivation in Ob/Ob mice rescues the defect of hepatic PL remodeling. Male WT,

Ob/Ob, and double knockout Ob/Ob-iPLA$_2\beta$ KO at six months old were used. (**A**) Representative photographs of hematoxylin- and eosin-stained livers of Ob/Ob mice and Ob/Ob-iPLA$_2\beta$ KO. (**B**) The composition of PLs and CEs. (**C**) The contents of PLs and CEs containing MUFA. (**D**) The contents of PLs and CEs containing PUFA. Data are mean ± SEM, N = 5–7; #, $p < 0.05$, *versus* WT; *, $p < 0.05$, Ob/Ob *versus* Ob/Ob-iPLA$_2\beta$ KO.

7. Metabolic Lipid Changes in HFD-Fed Mice and Modulation by iPLA$_2\beta$ Deficiency

Livers of WT mice fed with HFD showed a significant increase of TGs and FAs (Figure 3A) [40]. On the contrary to Ob/Ob mice (Figure 1B), hepatic FA composition showed an increase of di-unsaturated FAs by HFD feeding (Figure 3B). For HFD-fed mice, PL composition plot (% Mol) showed a weaker shift from PC to CEs (Figure 3C) when compared with Ob/Ob mice (Figure 1C). There was a decrease in liver MUFA-PC (Figure 3D) concomitant with an increase in SM (Figure 3F). Hence, hepatic PC particularly PUFA-PC was the key PL that was suppressed in HFD obese NAFLD model. This suppression indicated a defect in hepatic fatty-acyl PC remodeling. It appears that PUFA-PE contents were modulated differently between Ob/Ob (Figure 1E) and HFD (Figure 3E) obese NAFLD.

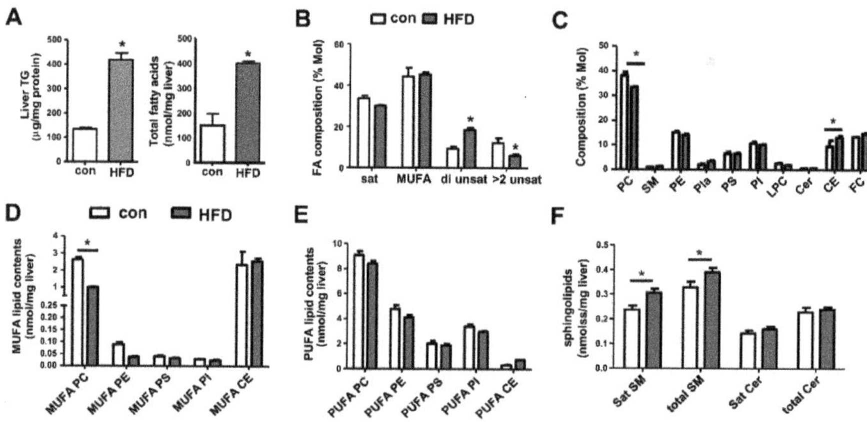

Figure 3. Lipid contents and composition show a defect of hepatic PL remodeling of WT mice fed with high-fat diet (HFD) for six months. Male WT mice at six months old were used. (**A**) The contents of hepatic triglycerides (TG), total fatty acids (FA). (**B**) FA composition of saturated, MUFA, di unsaturated, and > 2 unsaturated FA. (**C**) The composition of PLs and CEs. (**D**) The contents of PLs and CEs containing MUFA. (**E**) The contents (nmol/mg liver) of PLs and CEs containing PUFA. (**F**) The contents of sphingolipids SMs and Cers. Data are mean ± SEM, N = 5–12; *, $p < 0.05$, *versus* WT.

PL contents and composition in livers of WT, HFD-fed WT, and HFD-fed iPLA$_2\beta$ KO mice were determined by ESI-MS/MS. Liver histology showed marked steatosis attenuation in HFD-fed iPLA$_2\beta$ KO mice (Figure 4A). Analyses of hepatic PL composition showed that the elevation of liver CEs in HFD-fed WT mice was attenuated by iPLA$_2\beta$ deficiency (Figure 4B). In HFD model, iPLA$_2\beta$ deficiency was rescued by loss of PUFA-PC and PUFA-PE (Figure 4D) and only a trend rescue of MUFA-PE and MUFA-PS (Figure 4C). Hence, similar to Ob/Ob mice (Figure 2D), the loss of PUFA-PC and PUFA-PE was rescued by iPLA$_2\beta$ deficiency in HFD obese model (Figure 4D).

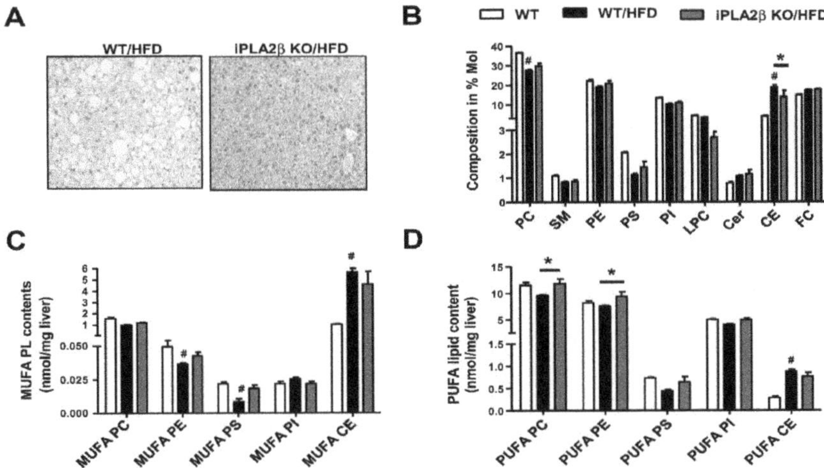

Figure 4. Inactivation in HFD-fed mice rescues the defect of hepatic PL remodeling. Male WT and iPLA$_2\beta$ KO mice at six months old were fed with HFD for six months. (**A**) Representative photographs of hematoxylin- and eosin-stained livers of HFD-fed WT and iPLA$_2\beta$ KO mice. (**B**) The composition of PLs and CEs. (**C**) The contents of PLs and CEs containing MUFA. (**D**) The contents of PLs and CEs containing PUFA. Data are mean ± SEM, N = 5–12; #, $p < 0.05$, *versus* WT; *, $p < 0.05$, WT/HFD *versus* iPLA$_2\beta$ KO/HFD.

8. Metabolic Lipid Changes in MCD-Fed Mice and Modulation by iPLA$_2\beta$ Deficiency

Unlike Ob/Ob and HFD obese models, hepatic steatosis was not protected by iPLA$_2\beta$ deficiency in non-obese mice fed with MCD diet [41] (Figure 5A). There was a decrease in composition of liver PC but an increase of CEs by MCD feeding (Figure 5B). iPLA$_2\beta$ deficiency attenuated the elevation of CEs in composition plot (Figure 5B) and CE contents (Figure 5C) associated with attenuation of SR-B1 by iPLA$_2\beta$ deficiency [41]. MCD feeding of WT mice decreased the contents of MUFA-PC, MUFA-PE (Figure 5C), and PUFA-PC (Figure 5D) concomitant with a significant increase in total Cer (Figure 5E), and all these changes were not altered by iPLA$_2\beta$ deficiency.

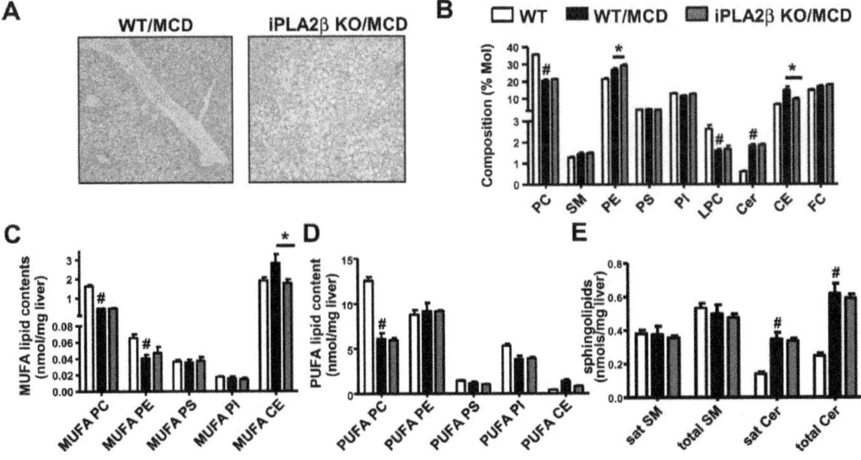

Figure 5. Inactivation does not rescue the defect of hepatic phospholipid remodeling in methionine-

and choline-deficient (MCD) diet-fed mice. Female WT and iPLA$_2$β KO mice at 12 months old were fed with MCD diet for four weeks. (**A**) Representative photographs of hematoxylin- and eosin-stained livers of MCD-fed WT and iPLA$_2$β KO mice. (**B**) The composition of hepatic PLs, CEs, and free cholesterol (FC). (**C**) The contents of PLs and CEs containing MUFA. (**D**) The contents of PLs and CEs containing PUFA. (**E**) The contents of sphingolipids SMs and Cers. Data are mean ± SEM, N = 5–6; #, $p < 0.05$, *versus* WT; *, $p < 0.05$, WT/MCD *versus* iPLA$_2$β KO/MCD.

9. PL in Liver Endoplasmic Reticulum of HFD- or MCD-Fed Mice and Modulation by iPLA$_2$β Deficiency

Because iPLA$_2$β is localized in the ER [107] where PL syntheses take place [102], we surmise that PLs in the ER membrane during obese and non-obese NAFLD could be modulated by iPLA$_2$β inactivation. In support of this notion, NAFLD induced by HFD feeding of PEMT-knockout mice [108] and genetic obese Ob/Ob mice [109] are associated with changes of PC and PE in liver ER fractions. We determined whether PL contents in the ER could be affected by HFD [40] or MCD diet [110] feeding and in combination with iPLA$_2$β deficiency. Our ER preparations from livers led to an enrichment of a resident ER protein calnexin in ER fractions (but not in liver homogenates) [110]; thus confirming the purity of ER membranes for lipidomic measurements.

PL profiles of liver ER fractions of HFD-fed mice were analyzed as PL subclasses (Figure 6A). HFD feeding of WT mice depleted ER PC contents and iPLA$_2$β deficiency showed a rescue trend. A similar pattern of a rescue-trend effect of iPLA$_2$β deficiency could be observed for ER PE and ER PS. Due to substrate depletion of PC synthesis [37], MCD feeding of WT mice caused a strong reduction of ER PC and ER PE (Figure 6B) [110]. iPLA$_2$β deficiency under MCD further suppressed ER PE contents, particularly, those containing PUFA. This deficiency did not, however, have any effects on ER PC contents suggesting specificity iPLA$_2$β towards PE in the ER. Hence, MCD-induced defect of ER PL remodeling became more severe by iPLA$_2$β deficiency [110].

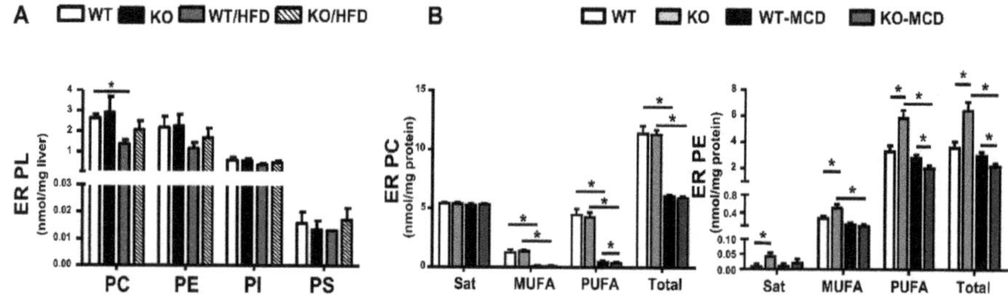

Figure 6. iPLA$_2$β inactivation on PL profiles of liver ER fractions of HFD- and MCD-fed mice. Feeding with HFD or MCD diet was described in Figures 4 and 5, respectively. ER fractions were isolated from mouse livers and ER proteins subjected to PL profiling by LC-MS/MS. (**A**) The contents of PC, PE, PI, and PS in the ER of WT, iPLA$_2$β-KO, WT/HFD, and iPLA$_2$β-KO/HFD livers. (**B**) Saturated, MUFA, PUFA, and total contents of PC and PE in liver ER fractions of WT, iPLA$_2$β-KO, WT/MCD, and iPLA$_2$β-KO/MCD livers. Data are mean ± SEM, N = 5–12 for (A) and 5–6 for (B); *, $p < 0.05$, between indicated pairs.

10. Hepatic PL Ratio among Obese and Non-Obese NAFLD and Modulation by iPLA$_2$β Deficiency

The importance of maintaining an appropriate hepatic PC/PE ratio has been extensively studied by D. Vance's research group using PEMT-knockout mice [53,54,102,108]. The clinical relevance of this ratio has been shown that the proportion of patients with NASH have a lower hepatic PC/PE ratio compared to healthy subjects [53]. Interestingly, both low and high hepatic PC/PE ratios in different NAFLD models are associated with an increase in NAFLD scores [102]. The lower PC/PE ratio seen with a deficiency of PEMT, betaine:homocysteine methyltransferase, or CPT:phosphocholine cytidyltransferaseα, correlates with increased NAFLD severity [102]. In contrast, mice with the deficiency of glycine *N*-methyltransferase [102], Ob/Ob mice [109], and mice fed with high fat/high cholesterol/cholate diet [111] show a higher hepatic PC/PE ratio. Hence, hepatic changes in PL composition and PC/PE ratio may be dependent on the experimental models distinguishing between genetic *versus* diet or obese *versus* non-obese NAFLD.

To this end, we investigated whether the ratios among PL subclasses and CEs would indicate hepatic steatosis among our three NAFLD mouse models. The ratios were calculated and separated into groups with the indicated PL subclasses used in the calculation including total PLs, saturated (sat) PLs, MUFA-PLs, and PUFA-PLs (Figure 7). Consistent with previous report in Ob/Ob mice [109], PC/PE ratio among total PLs in our Ob/Ob livers was increased from 1.5 to 2.3; and this increase was seen among sat and PUFA-PLs (Figure 7A). With a significant increase in CEs in Ob/Ob livers (Figure 1D), PC/CE and PE/CE ratios were therefore decreased with genetic obesity (Figure 7A). For HFD-fed WT mice, PC/PE was decreased among total and sat PLs, but a marked decrease was observed in PC/CE and PE/CE among MUFA-PLs and PUFA-PLs (Figure 7B). For MCD-fed WT mice, PC/PE was decreased from 1.5 to 0.5 among total and PUFA-PLs (Figure 7C). MCD feeding caused a marked decrease in PC/PS and PC/PI seen in MUFA-PLs. With a significant decrease in CEs in MCD-fed livers (Figure 5C), PC/CE and PE/CE ratios were therefore decreased in sat- and PUFA PLs (Figure 7C).

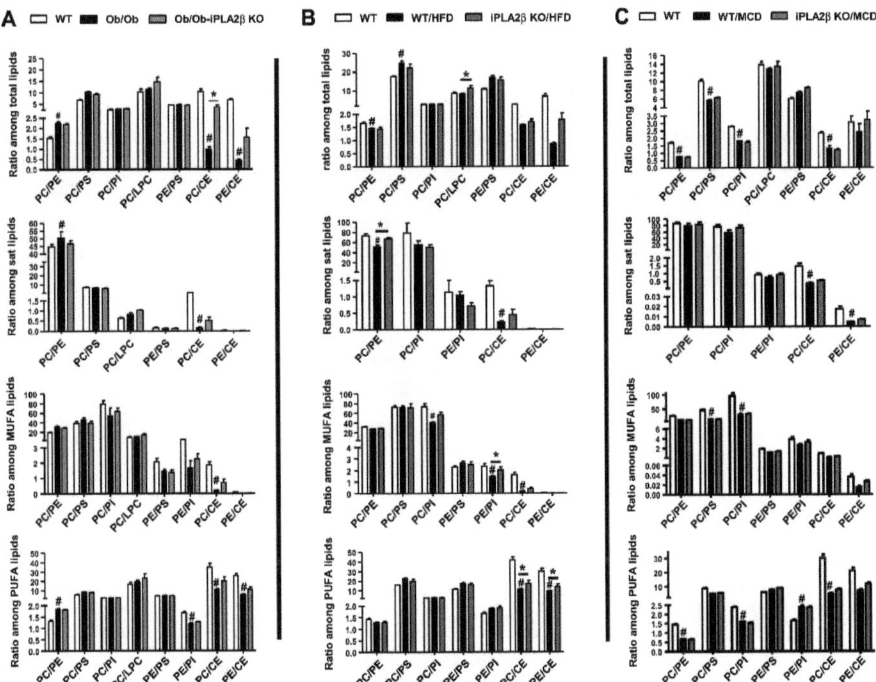

Figure 7. Alters the ratios among phospholipid subclasses in 3 NAFLD models: Ob/Ob mice, HFD-, and MCD-fed mice. Ob/Ob mice, HFD-, and MCD-fed mice are described in Figures 1, 3 and 5, respectively. The ratio among PLs in livers of (**A**) Ob/Ob, (**B**) HFD-fed mice, and (**C**) MCD-fed mice. Data are mean ± SEM, N = 5–7 for (**A**); N = 4–5 for (**B**), and N = 5–6 for (**C**). #, $p < 0.05$, *versus* WT; *, $p < 0.05$, Ob/Ob *versus* Ob/Ob-iPLA$_2\beta$ KO or WT/HFD *versus* iPLA$_2\beta$ KO/HFD.

Our data showed PC/PE among total PLs increased in Ob/Ob, but on the other hand, decreased in MCD livers. Thus, PC/PE ratios are changed in a U-shape curve from genetic obesity to non-obese

NAFLD/NASH [102]. In addition to PC/PE, ratios among other PLs were also altered with fatty liver. A weak decrease in PE/PI was observed in Ob/Ob livers (Figure 7A). A decrease in PE/PI was observed in HFD-fed WT mice (Figure 7B). A weak decrease of PC/PS and PC/PI ratios was observed in MCD-induced NAFLD (Figure 7C). Such changes during NAFLD can support the changes in electrostatics of PL bilayers and the existence of asymmetric lipid membranes due to charged anionic PLs, such as PS and PE relative to PC [112]. This may correlate with the least extent of hepatic inflammatory status in Ob/Ob mice as compared with HFD- and MCD-fed mice [39–41].

iPLA$_2$β deficiency did not alter PC/PE ratio in Ob/Ob (Figure 7A) and MCD-fed mice (Figure 7C). This deficiency however reversed the suppression of PC/PE and PE/PI in HFD-fed mice (Figure 7B). Since marked exacerbation of cholesterol metabolism is reported in Ob/Ob [113] and diabetic mice [114], the elevation of CEs caused a decrease of PC/CE and PE/CE in all three NAFLD models. iPLA$_2$β deficiency reversed the suppression of PC/CE and PE/CE ratios in Ob/Ob and HFD-fed mice, but not in MCD-fed mice (Figure 7A–C). This may suggest that the remodeling with a shuttling of PUFA and MUFA could occur between PC, PE, and CEs via acylation and transacylation in choline/methionine-rich Ob/Ob and HFD livers [115]. Taken together, we have demonstrated a difference between genetic and diet (HFD and MCD) NAFLD regarding the ratios among PL subclasses, a contribution of PS and PI relative to PC and PE, as well as CE metabolism.

11. iPLA$_2$β and De Novo Lipogenesis Gene Expression in Livers of Mice in 3 NAFLD Models

Associated with hepatic steatosis protection in obese but in non-obese NAFLD models [39–41], we further compared expression of iPLA$_2$β protein and de novo lipogenesis in livers of Ob/Ob, HFD- and MCD-fed WT mice. Rather than an increase, a slight decrease in iPLA$_2$β protein expression was observed in fatty livers of obese models (Figure 8A,B) and a strong decrease in MCD model (Figure 8C). iPLA$_2$β mRNA expression was not markedly altered (not shown) [39–41]. These data imply that iPLA$_2$β protein may be subjected to degradation at post translational levels during NASH. It is shown that iPLA$_2$ expression is decreased in rat cirrhotic livers [116], and may support iPLA$_2$β as a target for degradation during severe liver injury. Currently, no published data on iPLA$_2$β or PLA2G6 expression in livers of NAFLD/NASH patients are available. We could not correlate iPLA$_2$β protein expression observed in our results with human data.

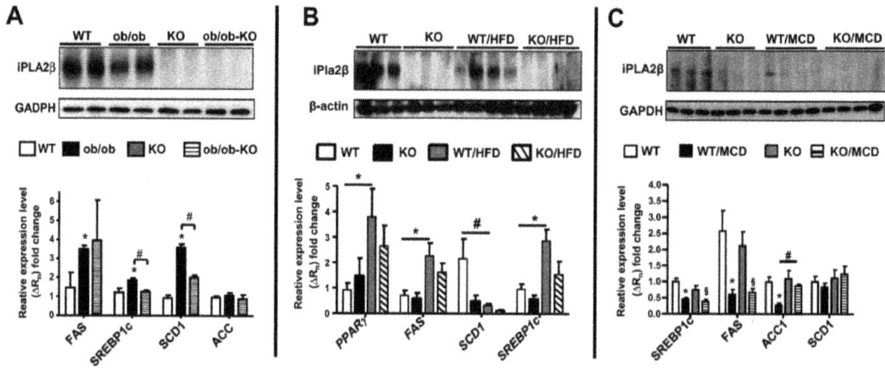

Figure 8. Expression of iPLA$_2$β protein and de novo lipogenesis mRNA in livers of WT, Ob/Ob mice, HFD-, and MCD-fed mice. Ob/Ob mice, HFD-, and MCD-fed mice are described in Figures 1, 3 and 5, respectively. Expression of (**A**) iPLA$_2$β protein, (**B**) HFD-fed mice, and (**C**) MCD-fed mice. Data are mean ± SEM, N = 5–7 for PCR data; #, $p < 0.05$; §, $p < 0.05$, KO *versus* KO/MCD; *, $p < 0.05$ between indicated groups.

With obesity and fatty liver, livers of Ob/Ob and HFD-fed mice showed marked elevation of de novo lipogenesis genes including fatty acid synthase (FAS) and transcription factor SREBP1c

(Figure 8A,B). It was shown that the transcription of iPLA$_2$β is regulated by SREBP-1 [117], consistently attenuated expression of FAS and SREBP1c was observed in livers of iPLA$_2$β-deficient obese mice. This attenuation was in indeed correlated with hepatic steatosis protection. On the other hand, MCD feeding of WT mice caused suppressed expression of these genes, which was not altered by iPLA$_2$β deficiency (Figure 8C). This was associated with no steatosis protection by iPLA$_2$β deficiency in this non-obese model.

12. Summarized PL Characteristics in Ob/Ob, HFD-, and MCD-Fed Mice and Effects of iPLA$_2$β Deficiency

Because PC and PE are the two major zwitterionic PLs in cells and their metabolism has been a focus as a key mechanistic base for healthy liver, the alterations in PC and PE are critical in liver disease development and NAFLD [53,54,102,108]. iPLA$_2$β deficiency protects obesity and NAFLD with an order of - Ob/Ob [39] › HFD [40] » MCD [41]; and the latter showed no steatosis protection (Figure 9).

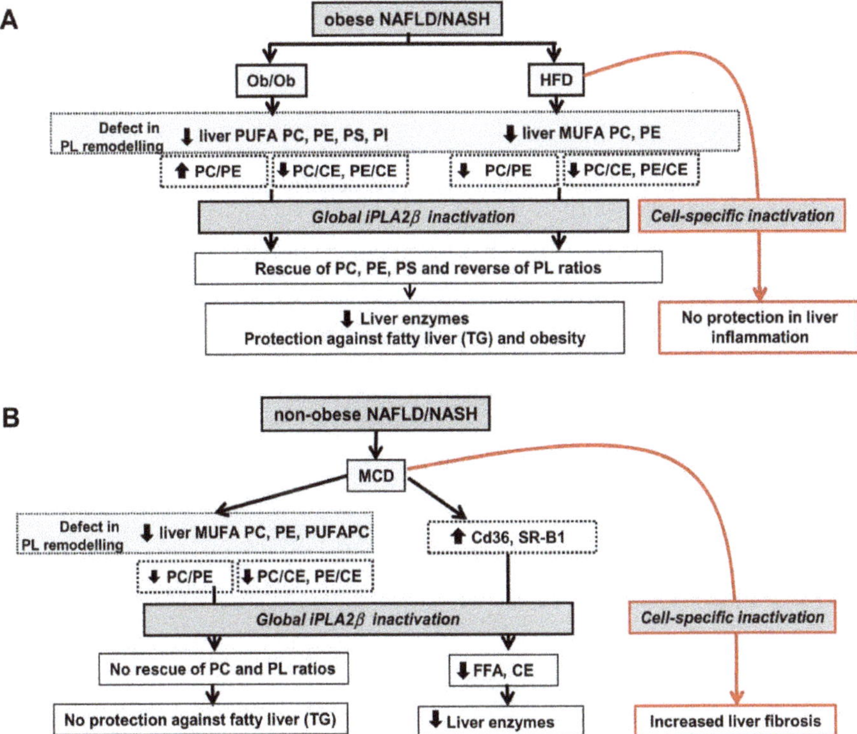

Figure 9. Role of iPLA$_2$β deficiency in obese and non-obese NAFLD/NASH mouse models. (**A**) Livers of genetic Ob/Ob and chronic HFD-fed mice exhibited a defect in PL remodeling with suppressed contents of PUFA PLs. PC/PE ratio was increased in Ob/Ob mice while that of HFD-fed mice was decreased. iPLA$_2$β inactivation replenished PLs associated with fatty liver protection. (**B**) Livers of MCD-fed mice exhibited a defect in PL remodeling with suppressed PUFA PLs as well as PC/PE ratio. iPLA$_2$β inactivation in MCD-fed mice did not rescue this defect with no protection. We propose that iPLA$_2$β deficiency in specific cell types may lead to no protection in liver inflammation and liver fibrosis in HFD and MCD NAFLD model, respectively (marked in red).

Human NAFLD and NASH are associated with numerous changes in the lipid composition of the liver. A decrease of the total PC and a decrease of arachidonic acid (20:4n-6) in FFA, TGs, and PC are reported in both NAFLD and NASH [55]. The contents of eicosapentanoic acid (20:5n-3) and docosahexanoic acid (22:6n-3) are decreased in NASH livers. In another study, PUFA-PLs are decreased in NASH livers compared to normal livers, and liver CEs are increased in NAFLD and NASH livers compared to normal livers [118]. Thus, a defect in PL remodeling and increased CEs could be observed in livers of human NAFLD and NASH. It is reported that the activity of the desaturase FADS1 is decreased in NAFLD liver biopsies [119]. This decrease in desaturation of FFA would likely lead to a depletion of MUFA and PUFA lipids, particularly PLs in NAFLD/NASH. It is shown that the hepatic PC/PE ratio is decreased in human NASH livers [53]. In another study, this ratio is lower in simple steatosis and NASH patients compared with controls, but it was not different between SS and NASH [120]. PC was lower and PE higher in the liver of simple patients compared with controls, whereas in NASH patients, only PE was higher [120]. Thus, the decrease of hepatic PC/PE ratio is a key parameter for human NAFLD and progression to NASH.

Livers of C57b/S129J mice fed a high-fat/high-cholesterol diet show an increase of hepatic CEs while hepatic PC, PE, and PS contents are decreased in NAFLD and a further decrease in PC and PE are observed in NASH [121]. Not only MUFA-PLs, hepatic PUFA-PC, PUFA-PE and PUFA-PI are reported to be decreased in MCD-fed mice compared with chow-fed or HFD-fed mice [122]. Hepatic 16:0 CEs are also increased in MCD-fed mice compared to chow-fed mice.

Results on liver lipids in our studies overall are consistent with those reported in human [55,118–120] and mouse [121,122] NAFLD/NASH. In our studies, a decrease of PUFA-PLs and MUFA-PLs was observed in livers of Ob/Ob and HFD-fed mice (Figure 9A) and MCD-fed mice (Figure 9B). The elevation of CEs led to a decrease of PC/CE and PE/CE ratios in obese and non-obese lives. iPLA$_2\beta$ deficiency rescued not only the defect of liver PL remodeling, but also reversed the suppression of PC/CE and PE/CE in Ob/Ob and HFD-fed mice (Figure 9A). iPLA$_2\beta$ deficiency in MCD-fed mice did not alter these parameters (Figure 9B). In our models, iPLA$_2\beta$ deficiency did not interfere with liver sphingolipids. Our work suggests the functions of iPLA$_2\beta$ on the hepatocyte PL remodeling in obese NAFLD models. Interestingly, hepatic PC/PE ratio is increased in Ob/Ob mice but decreased in HFD- and MCD-fed mice (Figure 9). Thus, this ratio is a marker in discriminating genetic *versus* diet-induced NASH, which is in a similar manner to human NASH [53,120]. iPLA$_2\beta$ deficiency reversed PC/PE ratio in obese but not non-obese model, rendering this ratio as a marker for phenotypic changes in NASH.

Besides PC and PE, our current data present an additional evidence for changes in other PLs including anionic PS and PI in Ob/Ob mice (Figure 9A). iPLA$_2\beta$ deficiency elicited full protection against fatty liver, obesity, and elevation of liver enzymes in Ob/Ob and HFD-fed mice [39,40]. In these obese models, we propose that iPLA$_2\beta$ specifically hydrolyzes PUFA-PLs and MUFA-PLs for generation of FAs subsequently utilized for TG and CE syntheses for hepatic steatosis. This process may be coordinated with other PLA$_2$ enzymes. By this way, PUFA-PL and MUFA-PL contents in Ob/Ob and HFD livers are suppressed, and this suppression is rescued or replenished by iPLA$_2\beta$ deficiency. In line with this, administration of n-3 essential FAs [123] and PUFAs [124] have been shown to ameliorate hepatic steatosis in obese mice likely by increasing membrane fluidity [125]. Unlike Ob/Ob mice [39], iPLA$_2\beta$ deficiency during HFD feeding did not protect mice from liver inflammation [40]. Consistently, PLs, such as PC, are shown to elicit protection of hepatic steatosis in NAFLD without attenuating liver inflammation in NASH [54]. Alternatively, iPLA$_2\beta$ deletion by birth under the background of leptin deficiency may render a complete protection [39]; possibly due to adaptation in different cell types upon gene deletions throughout the mouse lifetime. Chronic HFD feeding on the other hand would represent an external stress to mice [40]. As iPLA$_2\beta$ KO mice were global deletion, we therefore proposed that iPLA$_2\beta$ deletion may affect specific cell types, such as hepatocytes, immune cells, and adipocytes, differently in response to HFD feeding (Figure 9A).

Due to the lack of choline and methionine in the diet, MCD feeding limits PC synthesis thus resulting in a defect of PL remodeling (Figure 9B). iPLA$_2\beta$ deficiency did not protect mice from MCD-induced fatty liver, but attenuated elevation of liver enzymes (Figure 9B). This attenuation could be due to an inhibition of the uptake of FA as determined by Cd36 expression and FA contents, as well as an inhibition of HDL reverse transport as determined by SR-B1 expression and CE contents [41]. The latter may indicate an involvement of iPLA$_2$β in cholesterol esterification to CEs [114,115]. Despite attenuation of liver enzymes, iPLA$_2$β inactivation during MCD however showed an increase of α-smooth muscle actin and vimentin expression [41]. Such increased liver fibrosis could be due to iPLA$_2$β inactivation in specific cell types such as macrophages or hepatocytes associated with stress induced by MCD feeding (Figure 9B).

Taken together, our results highlight the significance of cross-talk between the metabolism of PLs and neutral lipids, i.e., CEs [115,126] and TGs in lipid droplets [43,127], which can be modulated by iPLA$_2$β deficiency. In the latter case, iPLA$_2$β may co-function with other PLA$_2$s such as group IVA PLA2 in the shuttling of FAs toward TG synthesis [127]. iPLA$_2$β inactivation was effective in attenuating obese (Ob/Ob and chronic-HFD) NAFLD indicating specific involvement of iPLA$_2$β in obesity pathogenesis. As iPLA$_2$β inactivation was ineffective to treat non-obese MCD NAFLD, this indicates that choline and methionine in hepatic PC and PL synthesis and metabolism were necessary for protection in obese models. Our data also emphasize the contributions and involvement of hepatic PL, TG, and CE metabolism in the development of NAFLD.

13. Perspectives

13.1. Consideration of Cell-Type Specificity of iPLA$_2$β

The phenotypes of iPLA$_2$β-null mice have been recently reviewed [92,93]. On one hand, iPLA$_2$β is detrimental in mediating ER stress and cell death of pancreatic β-cells. On the other hand, iPLA$_2$β plays a homeostatic role and the loss of iPLA$_2$β in mice leads to ageing-related diseases, such as male infertility, bone-density loss, and neurological disorders. Accordingly, PLA2G6 mutations lead to the pathogenesis of infantile neuroaxonal dystrophy and PARK14-linked Parkinson's disease [92,93]. These data have highlighted various and often opposing functions of iPLA$_2$β in a cell-type specific manner.

In our investigations of obese and non-obese NAFLD/NASH, global deletion iPLA$_2$β-null mice were used. Thus, we could not identify whether the observed effects of iPLA$_2$β inactivation were due to altered functions in adipocytes [99], immune cells such as macrophages [128,129], and Kupffer cells [130], as well as hepatocytes as shown by our work [39–41]. Concurrently, we have also reported that global deletion of iPLA$_2$β was able to sensitize hepatocellular damage induced by concanavalin A [131] or during ageing [132]. It is thought that such epithelial damage caused by iPLA$_2$β deficiency may secrete mediators that in turn activate inflammatory macrophages leading to sensitized injury. The observed sensitization of liver injury [131,132] could therefore be the combined effects of iPLA$_2$β deficiency in hepatocytes and macrophages.

The effects of iPLA$_2$β inactivation in macrophages have been reported [128,129]. On one hand, macrophages from iPLA$_2$β-null mice exhibited suppressed pro-inflammatory M1 response [128,130] and showed enhanced IL-4-induced M2 polarization in vitro [128]. Consistently, iPLA$_2$β-null mice treated with anti-CD95 antibody primed Kupffer cells for attenuated release of TNF-α but enhanced release of interleukin-6 in vitro [130]. iPLA$_2$β-null mice showed the inability to phagocytose infected parasites in vivo [133]; thus iPLA$_2$β deficiency could lead to a defect in innate immunity. As iPLA$_2$β-null mice showed propensity for increased liver inflammation and fibrosis during HFD and MCD NAFLD (Figure 9), this could be due to the ability of iPLA$_2$β-deficient macrophages to differentially regulate M1 and M2 cytokines during NAFLD.

Our work has demonstrated iPLA$_2$β activity in the hepatocytes by a decrease of products lysoPL and an accumulation of substrates PC and PE observed in livers of iPLA$_2$β-null mice fed with chow [39,40]. Such accumulation could therefore lead to replenishment of PLs and steatosis protection in obese Ob/Ob and HFD NAFLD (Figure 9A). On the other hand, the alterations of hepatocellular PL composition and a decrease of PC/PE in chow-fed iPLA$_2$β-null mice may render altered PL membranes leading to susceptibility for previously observed liver injury [131,132]. Further studies are warranted to determine whether iPLA$_2\beta$ inactivation specifically in hepatocytes could affect hepatic steatosis and inflammation/fibrosis in HFD (Figure 9A) and MCD (Figure 9B) models.

While knockdown of iPLA$_2\beta$ inhibits hormone-induced differentiation of adipocytes in vitro [99], iPLA$_2\beta$ inactivation in adipocytes may inhibit adipocyte expansion, thus attenuating adipogenesis and obesity observed in our obese NAFLD models [39,40]. It has recently been shown that protection of hepatic steatosis by an ablation of adipocyte PLA$_2$ is mediated by adipocyte hormone leptin [134]. By unknown mechanisms, iPLA$_2\beta$ inactivation may lead to an increase secretion of leptin, and may thus elicit protection in a similar way as adipocyte PLA$_2$. Accordingly, leptin has biological activity in depleting liver TG [135] as well as activating Kupffer cells and thereby altering hepatic lipid metabolism [136]. To figure out the contribution of adipocytes, macrophages, and hepatocytes on the NAFLD/NASH pathogenesis, the generation of tissue-specific iPLA$_2\beta$-deficient mice is therefore warranted. These results will help us understand that iPLA$_2\beta$ deficiency in which cell type is responsible for steatosis protection in HFD NAFLD and for increased liver fibrosis in MCD NAFLD. Results from tissue-specific iPLA$_2\beta$ KO mice will also help with the designs of iPLA$_2\beta$ inhibitors and formulations for specific-tissue delivery for effective treatment of obese and non-obese NAFLD.

13.2. Use of PLs or iPLA$_2\beta$ Antagonists for Steatosis Protection in Obese Versus Non-Obese NAFLD

Because there are no approved drugs for treatment of NAFLD/NASH, our research results may provide some insights for further development toward NAFLD treatment. Here, our data support the idea for repletion of PL loss by use of PLs themselves or use of iPLA$_2\beta$ inhibitors for treatment of obese NAFLD. Some iPLA$_2\beta$ inhibitors have been found to be effective for treatment of diabetes at least in experimental animals [137]. Potent and selective inhibitors of iPLA$_2\beta$ have been developed [138]. Pending the results on the phenotypes of hepatocyte-, macrophage-, and adipose-specific iPLA$_2\beta$ KO mice in obese NAFLD models, investigators may design iPLA$_2\beta$ inhibitors in combination with specific delivery to hepatocytes [139], macrophages [140], or adipocytes [141] for better treatment of the common disease obese NAFLD.

For non-obese MCD NAFLD, iPLA$_2\beta$ inhibitors may be found to be partially effective in attenuating liver enzymes but not hepatic steatosis (Figure 9B). Choline supplementation in patients under parenteral nutrition reverses hepatic steatosis [142], and the strategy for non-obese NAFLD treatment may involve the use of choline and methyl donors as co-treatment with iPLA$_2\beta$ inhibitors. Choline deficiency in humans has been shown to induce hepatic steatosis [143,144] causing liver dysfunctions [145]. Hence, supplementation of choline [142] and methyl donors [146] may be effective in attenuating hepatic steatosis under non-obese NAFLD/NASH. Further investigations in experimental animals are necessary to evaluate the long-term use of iPLA$_2\beta$ inhibitors alone for obese NAFLD and in combination with choline/methyl donor supplementation for non-obese NAFLD models. Nonetheless, the hierarchical mode of action by iPLA$_2\beta$ deficiency indicates that an iPLA$_2\beta$ inhibitor may be designed perhaps with tissue-specific delivery for therapeutic development to treat metabolic syndromes due to obese NAFLD, and may not be suitable for non-obese NASH.

14. Conclusions

We demonstrated a pivotal role of iPLA$_2\beta$ in the development of hepatic steatosis and inflammation in obese and non-obese NAFLD models. iPLA$_2\beta$ inactivation rescued the defect in PL remodeling and elicited steatosis protection in obese NAFLD models, but not in non-obese MCD model. While our study suggests the use of iPLA$_2\beta$ inhibitors for therapy of obese NAFLD due to genetics or chronic HFD intake, further investigations using tissue-specific iPLA$_2\beta$-deficient mice are still warranted. Here, the usefulness of the lipidomics methodology is shown in deciphering the alterations in hepatic PL pools and ratios among three NAFLD models.

Funding: This research was funded by Deutsche Forschungsgemeinschaft (CH288/6-1 and CH288/6-2).

Acknowledgments: We thank John Turk who provided us with iPLA$_2$β-null mice, and previous lab members who were involved in this project. We thank Wolfgang Stremmel and Uta Merle for research insights.

Conflicts of Interest: Authors declare no conflict of interest.

Abbreviations

BFP	body fat percentage
CAD	cardiovascular disease
CE	cholesterol esters
Cer	ceramides
CT	computed tomography
ER	endoplasmic reticulum
ESI-MS/MS	electrospray ionization tandem mass spectrometry
FA	fatty acid
FC	free cholesterol
FAS	fatty acid synthase
GC/MS	gas chromatography mass spectrometry
GWAS	genome-wide association studies
HFD	high fat diet
iPLA$_2$β	group VIA calcium-independent PLA$_2$
LPC	lysophosphatidylcholine
LPE	lysophosphatidylethanolamine
MCD	methionine- and choline-deficient diet
MUFA	monounsaturated fatty acids
NAFLD	non-alcoholic fatty liver disease
NASH	non-alcoholic steatohepatitis
Ob/Ob mice	leptin-deficient mice
PC	phosphatidylcholine
PE	phosphatidylethanolamine
PEMT	phosphatidylethanolamine *N*-methyltransferase
PI	phosphatidylinositol
PL	phospholipid
PLA$_2$	phospholipase A2
Pla	plasmalogens
PNPLA	patatin-like phospholipase containing lipase
PS	phosphatidylserine
PUFA	polyunsaturated fatty acids
SR-B1	scavenger receptor B type 1
SREBP	sterol regulatory element-binding protein
SM	sphingomyelin
TG	triglyceride
WT	wild-type
VLDL	very low-density lipoproteins

References

1. Lissner, L.; Visscher, T.L.; Rissanen, A.; Heitmann, B.L.; Prevention and Public Health Task Force of the European Association for the Study of Obesity. Monitoring the obesity epidemic into the 21st century—Weighing the evidence. *Obes. Facts* **2013**, *6*, 561–565. [CrossRef] [PubMed]
2. Gallagher, E.J.; LeRoith, D. Epidemiology and molecular mechanisms tying obesity, diabetes, and the metabolic syndrome with cancer. *Diabetes Care* **2013**, *36* (Suppl. 2), S233–S239. [CrossRef] [PubMed]
3. Astrup, A.; Buemann, B.; Western, P.; Toubro, S.; Raben, A.; Christensen, N.J. Obesity as an adaptation to a high-fat diet: Evidence from a cross-sectional study. *Am. J. Clin. Nutr.* **1994**, *59*, 350–355. [CrossRef] [PubMed]

4. Clement, K.; Garner, C.; Hager, J.; Philippi, A.; LeDuc, C.; Carey, A.; Harris, T.J.; Jury, C.; Cardon, L.R.; Basdevant, A.; et al. Indication for linkage of the human OB gene region with extreme obesity. *Diabetes* **1996**, *45*, 687–690. [CrossRef] [PubMed]
5. York, B.; Truett, A.A.; Monteiro, M.P.; Barry, S.J.; Warden, C.H.; Naggert, J.K.; Maddatu, T.P.; West, D.B. Gene-environment interaction: A significant diet-dependent obesity locus demonstrated in a congenic segment on mouse chromosome 7. *Mamm. Genome* **1999**, *10*, 457–462. [CrossRef]
6. Hill, J.O.; Peters, J.C. Environmental contributions to the obesity epidemic. *Science* **1998**, *280*, 1371–1374. [CrossRef] [PubMed]
7. Lissner, L.; Heitmann, B.L.; Bengtsson, C. Population studies of diet and obesity. *Br. J. Nutr.* **2000**, *83* (Suppl. 1), S21–S24. [CrossRef]
8. Hasselbalch, A.L. Genetics of dietary habits and obesity—A twin study. *Dan. Med. Bull.* **2010**, *57*, B4182.
9. West, D.B.; York, B. Dietary fat, genetic predisposition, and obesity: Lessons from animal models. *Am. J. Clin. Nutr.* **1998**, *67*, 505S–512S. [CrossRef]
10. Collins, S.; Martin, T.L.; Surwit, R.S.; Robidoux, J. Genetic vulnerability to diet-induced obesity in the C57BL/6J mouse: Physiological and molecular characteristics. *Physiol. Behav.* **2004**, *81*, 243–248. [CrossRef]
11. Alexander, J.; Chang, G.Q.; Dourmashkin, J.T.; Leibowitz, S.F. Distinct phenotypes of obesity-prone AKR/J, DBA2J and C57BL/6J mice compared to control strains. *Int. J. Obes.* **2006**, *30*, 50–59. [CrossRef] [PubMed]
12. Wu, Q.; Suzuki, M. Parental obesity and overweight affect the body-fat accumulation in the offspring: The possible effect of a high-fat diet through epigenetic inheritance. *Obes. Rev.* **2006**, *7*, 201–208. [CrossRef] [PubMed]
13. Parks, B.W.; Nam, E.; Org, E.; Kostem, E.; Norheim, F.; Hui, S.T.; Pan, C.; Civelek, M.; Rau, C.D.; Bennett, B.J.; et al. Genetic control of obesity and gut microbiota composition in response to high-fat, high-sucrose diet in mice. *Cell Metab.* **2013**, *17*, 141–152. [CrossRef]
14. Ghosh, S.; Bouchard, C. Convergence between biological, behavioural and genetic determinants of obesity. *Nat. Rev. Genet.* **2017**, *18*, 731–748. [CrossRef] [PubMed]
15. Moore, J.B. Non-alcoholic fatty liver disease: The hepatic consequence of obesity and the metabolic syndrome. *Proc. Nutr. Soc.* **2010**, *69*, 211–220. [CrossRef]
16. Younossi, Z.; Anstee, Q.M.; Marietti, M.; Hardy, T.; Henry, L.; Eslam, M.; George, J.; Bugianesi, E. Global burden of NAFLD and NASH: Trends, predictions, risk factors and prevention. *Nat. Rev. Gastroenterol. Hepatol.* **2018**, *15*, 11–20. [CrossRef]
17. Kim, G.A.; Lee, H.C.; Choe, J.; Kim, M.J.; Lee, M.J.; Chang, H.S.; Bae, I.Y.; Kim, H.K.; An, J.; Shim, J.H.; et al. Association between non-alcoholic fatty liver disease and cancer incidence rate. *J. Hepatol.* **2017**. [CrossRef]
18. Merriman, R.B.; Aouizerat, B.E.; Bass, N.M. Genetic influences in nonalcoholic fatty liver disease. *J. Clin. Gastroenterol.* **2006**, *40* (Suppl. 1), S30–S33. [CrossRef]
19. Anstee, Q.M.; Daly, A.K.; Day, C.P. Genetic modifiers of non-alcoholic fatty liver disease progression. *Biochim. Biophys. Acta* **2011**, *1812*, 1557–1566. [CrossRef]
20. Norheim, F.; Hui, S.T.; Kulahcioglu, E.; Mehrabian, M.; Cantor, R.M.; Pan, C.; Parks, B.W.; Lusis, A.J. Genetic and hormonal control of hepatic steatosis in female and male mice. *J. Lipid Res.* **2017**, *58*, 178–187. [CrossRef]
21. Pramfalk, C.; Pavlides, M.; Banerjee, R.; McNeil, C.A.; Neubauer, S.; Karpe, F.; Hodson, L. Sex-Specific Differences in Hepatic Fat Oxidation and Synthesis May Explain the Higher Propensity for NAFLD in Men. *J. Clin. Endocrinol. Metab.* **2015**, *100*, 4425–4433. [CrossRef] [PubMed]
22. Wang, L.; Guo, J.; Lu, J. Risk factor compositions of nonalcoholic fatty liver disease change with body mass index in males and females. *Oncotarget* **2016**, *7*, 35632–35642. [CrossRef] [PubMed]
23. Chan, W.K.; Bahar, N.; Razlan, H.; Vijayananthan, A.; Sithaneshwar, P.; Goh, K.L. Non-alcoholic fatty liver disease in a young multiracial Asian population: A worrying ethnic predilection in Malay and Indian males. *Hepatol. Int.* **2014**, *8*, 121–127. [CrossRef] [PubMed]
24. Summart, U.; Thinkhamrop, B.; Chamadol, N.; Khuntikeo, N.; Songthamwat, M.; Kim, C.S. Gender differences in the prevalence of nonalcoholic fatty liver disease in the Northeast of Thailand: A population-based cross-sectional study. *F1000Research* **2017**, *6*, 1630. [CrossRef] [PubMed]
25. Katsagoni, C.N.; Georgoulis, M.; Papatheodoridis, G.V.; Fragopoulou, E.; Ioannidou, P.; Papageorgiou, M.; Alexopoulou, A.; Papadopoulos, N.; Deutsch, M.; Kontogianni, M.D. Associations Between Lifestyle Characteristics and the Presence of Nonalcoholic Fatty Liver Disease: A Case-Control Study. *Metab. Syndr. Relat. Disord.* **2017**, *15*, 72–79. [CrossRef]

26. Kumar, R.; Mohan, S. Non-alcoholic Fatty Liver Disease in Lean Subjects: Characteristics and Implications. *J. Clin. Transl. Hepatol.* **2017**, *5*, 216–223. [CrossRef]
27. Yousef, M.H.; Al Juboori, A.; Albarrak, A.A.; Ibdah, J.A.; Tahan, V. Fatty liver without a large "belly": Magnified review of non-alcoholic fatty liver disease in non-obese patients. *World J. Gastrointest. Pathophysiol.* **2017**, *8*, 100–107. [CrossRef]
28. Liu, C.J. Prevalence and risk factors for non-alcoholic fatty liver disease in Asian people who are not obese. *J. Gastroenterol. Hepatol.* **2012**, *27*, 1555–1560. [CrossRef]
29. Feng, R.N.; Du, S.S.; Wang, C.; Li, Y.C.; Liu, L.Y.; Guo, F.C.; Sun, C.H. Lean-non-alcoholic fatty liver disease increases risk for metabolic disorders in a normal weight Chinese population. *World J. Gastroenterol.* **2014**, *20*, 17932–17940. [CrossRef]
30. Sookoian, S.; Pirola, C.J. Systematic review with meta-analysis: The significance of histological disease severity in lean patients with nonalcoholic fatty liver disease. *Aliment. Pharmacol. Ther.* **2018**, *47*, 16–25. [CrossRef]
31. Feldman, A.; Eder, S.K.; Felder, T.K.; Kedenko, L.; Paulweber, B.; Stadlmayr, A.; Huber-Schonauer, U.; Niederseer, D.; Stickel, F.; Auer, S.; et al. Clinical and Metabolic Characterization of Lean Caucasian Subjects With Non-alcoholic Fatty Liver. *Am. J. Gastroenterol.* **2017**, *112*, 102–110. [CrossRef] [PubMed]
32. Bril, F.; Sninsky, J.J.; Baca, A.M.; Superko, H.R.; Portillo Sanchez, P.; Biernacki, D.; Maximos, M.; Lomonaco, R.; Orsak, B.; Suman, A.; et al. Hepatic Steatosis and Insulin Resistance, But Not Steatohepatitis, Promote Atherogenic Dyslipidemia in NAFLD. *J. Clin. Endocrinol. Metab.* **2016**, *101*, 644–652. [CrossRef] [PubMed]
33. Fabbrini, E.; Sullivan, S.; Klein, S. Obesity and nonalcoholic fatty liver disease: Biochemical, metabolic, and clinical implications. *Hepatology* **2010**, *51*, 679–689. [CrossRef] [PubMed]
34. Zivkovic, A.M.; German, J.B.; Sanyal, A.J. Comparative review of diets for the metabolic syndrome: Implications for nonalcoholic fatty liver disease. *Am. J. Clin. Nutr.* **2007**, *86*, 285–300. [CrossRef]
35. Veena, J.; Muragundla, A.; Sidgiddi, S.; Subramaniam, S. Non-alcoholic fatty liver disease: Need for a balanced nutritional source. *Br. J. Nutr.* **2014**, *112*, 1858–1872. [CrossRef]
36. Larter, C.Z.; Yeh, M.M. Animal models of NASH: Getting both pathology and metabolic context right. *J. Gastroenterol. Hepatol.* **2008**, *23*, 1635–1648. [CrossRef]
37. Machado, M.V.; Michelotti, G.A.; Xie, G.; Almeida Pereira, T.; Boursier, J.; Bohnic, B.; Guy, C.D.; Diehl, A.M. Mouse models of diet-induced nonalcoholic steatohepatitis reproduce the heterogeneity of the human disease. *PLoS ONE* **2015**, *10*, e0127991. [CrossRef]
38. Nakatsuka, A.; Matsuyama, M.; Yamaguchi, S.; Katayama, A.; Eguchi, J.; Murakami, K.; Teshigawara, S.; Ogawa, D.; Wada, N.; Yasunaka, T.; et al. Insufficiency of phosphatidylethanolamine N-methyltransferase is risk for lean non-alcoholic steatohepatitis. *Sci. Rep.* **2016**, *6*, 21721. [CrossRef]
39. Deng, X.; Wang, J.; Jiao, L.; Utaipan, T.; Tuma-Kellner, S.; Schmitz, G.; Liebisch, G.; Stremmel, W.; Chamulitrat, W. iPLA2beta deficiency attenuates obesity and hepatic steatosis in ob/ob mice through hepatic fatty-acyl phospholipid remodeling. *Biochim. Biophys. Acta* **2016**, *1861*, 449–461. [CrossRef]
40. Otto, A.C.; Gan-Schreier, H.; Zhu, X.; Tuma-Kellner, S.; Staffer, S.; Ganzha, A.; Liebisch, G.; Chamulitrat, W. Group VIA phospholipase A2 deficiency in mice chronically fed with high-fat-diet attenuates hepatic steatosis by correcting a defect of phospholipid remodeling. *Biochim. Biophys. Acta Mol. Cell Biol. Lipids* **2019**, *1864*, 662–676. [CrossRef]
41. Zhu, X.; Gan-Schreier, H.; Otto, A.C.; Cheng, Y.; Staffer, S.; Tuma-Kellner, S.; Ganzha, A.; Liebisch, G.; Chamulitrat, W. iPla2beta deficiency in mice fed with MCD diet does not correct the defect of phospholipid remodeling but attenuates hepatocellular injury via an inhibition of lipid uptake genes. *Biochim. Biophys. Acta Mol. Cell Biol. Lipids* **2019**, *1864*, 677–687. [CrossRef] [PubMed]
42. Donnelly, K.L.; Smith, C.I.; Schwarzenberg, S.J.; Jessurun, J.; Boldt, M.D.; Parks, E.J. Sources of fatty acids stored in liver and secreted via lipoproteins in patients with nonalcoholic fatty liver disease. *J. Clin. Investig.* **2005**, *115*, 1343–1351. [CrossRef] [PubMed]
43. Tauchi-Sato, K.; Ozeki, S.; Houjou, T.; Taguchi, R.; Fujimoto, T. The surface of lipid droplets is a phospholipid monolayer with a unique Fatty Acid composition. *J. Biol. Chem.* **2002**, *277*, 44507–44512. [CrossRef] [PubMed]
44. Krahmer, N.; Guo, Y.; Wilfling, F.; Hilger, M.; Lingrell, S.; Heger, K.; Newman, H.W.; Schmidt-Supprian, M.; Vance, D.E.; Mann, M.; et al. Phosphatidylcholine synthesis for lipid droplet expansion is mediated by localized activation of CTP:phosphocholine cytidylyltransferase. *Cell Metab.* **2011**, *14*, 504–515. [CrossRef] [PubMed]

45. Walker, A.K.; Jacobs, R.L.; Watts, J.L.; Rottiers, V.; Jiang, K.; Finnegan, D.M.; Shioda, T.; Hansen, M.; Yang, F.; Niebergall, L.J.; et al. A conserved SREBP-1/phosphatidylcholine feedback circuit regulates lipogenesis in metazoans. *Cell* **2011**, *147*, 840–852. [CrossRef]
46. Lim, H.Y.; Wang, W.; Wessells, R.J.; Ocorr, K.; Bodmer, R. Phospholipid homeostasis regulates lipid metabolism and cardiac function through SREBP signaling in Drosophila. *Genes Dev.* **2011**, *25*, 189–200. [CrossRef]
47. Fullerton, M.D.; Hakimuddin, F.; Bonen, A.; Bakovic, M. The development of a metabolic disease phenotype in CTP:phosphoethanolamine cytidylyltransferase-deficient mice. *J. Biol. Chem.* **2009**, *284*, 25704–25713. [CrossRef]
48. Leonardi, R.; Frank, M.W.; Jackson, P.D.; Rock, C.O.; Jackowski, S. Elimination of the CDP-ethanolamine pathway disrupts hepatic lipid homeostasis. *J. Biol. Chem.* **2009**, *284*, 27077–27089. [CrossRef]
49. Higgins, J.A. Evidence that during very low density lipoprotein assembly in rat hepatocytes most of the triacylglycerol and phospholipid are packaged with apolipoprotein B in the Golgi complex. *FEBS Lett.* **1988**, *232*, 405–408. [CrossRef]
50. Testerink, N.; van der Sanden, M.H.; Houweling, M.; Helms, J.B.; Vaandrager, A.B. Depletion of phosphatidylcholine affects endoplasmic reticulum morphology and protein traffic at the Golgi complex. *J. Lipid Res.* **2009**, *50*, 2182–2192. [CrossRef]
51. Gaspar, M.L.; Jesch, S.A.; Viswanatha, R.; Antosh, A.L.; Brown, W.J.; Kohlwein, S.D.; Henry, S.A. A block in endoplasmic reticulum-to-Golgi trafficking inhibits phospholipid synthesis and induces neutral lipid accumulation. *J. Biol. Chem.* **2008**, *283*, 25735–25751. [CrossRef]
52. Tran, K.; Sun, F.; Cui, Z.; Thorne-Tjomsland, G.; St Germain, C.; Lapierre, L.R.; McLeod, R.S.; Jamieson, J.C.; Yao, Z. Attenuated secretion of very low density lipoproteins from McA-RH7777 cells treated with eicosapentaenoic acid is associated with impaired utilization of triacylglycerol synthesized via phospholipid remodeling. *Biochim. Biophys. Acta* **2006**, *1761*, 463–473. [CrossRef]
53. Li, Z.; Agellon, L.B.; Allen, T.M.; Umeda, M.; Jewell, L.; Mason, A.; Vance, D.E. The ratio of phosphatidylcholine to phosphatidylethanolamine influences membrane integrity and steatohepatitis. *Cell Metab.* **2006**, *3*, 321–331. [CrossRef]
54. Niebergall, L.J.; Jacobs, R.L.; Chaba, T.; Vance, D.E. Phosphatidylcholine protects against steatosis in mice but not non-alcoholic steatohepatitis. *Biochim. Biophys. Acta* **2011**, *1811*, 1177–1185. [CrossRef] [PubMed]
55. Puri, P.; Baillie, R.A.; Wiest, M.M.; Mirshahi, F.; Choudhury, J.; Cheung, O.; Sargeant, C.; Contos, M.J.; Sanyal, A.J. A lipidomic analysis of nonalcoholic fatty liver disease. *Hepatology* **2007**, *46*, 1081–1090. [CrossRef] [PubMed]
56. Mallampalli, R.K.; Ryan, A.J.; Salome, R.G.; Jackowski, S. Tumor necrosis factor-alpha inhibits expression of CTP:phosphocholine cytidylyltransferase. *J. Biol. Chem.* **2000**, *275*, 9699–9708. [CrossRef] [PubMed]
57. Kienesberger, P.C.; Oberer, M.; Lass, A.; Zechner, R. Mammalian patatin domain containing proteins: A family with diverse lipolytic activities involved in multiple biological functions. *J. Lipid Res.* **2009**, *50*, S63–S68. [CrossRef]
58. Ii, H.; Yokoyama, N.; Yoshida, S.; Tsutsumi, K.; Hatakeyama, S.; Sato, T.; Ishihara, K.; Akiba, S. Alleviation of high-fat diet-induced fatty liver damage in group IVA phospholipase A2-knockout mice. *PLoS ONE* **2009**, *4*, e8089. [CrossRef]
59. Song, H.; Wohltmann, M.; Bao, S.; Ladenson, J.H.; Semenkovich, C.F.; Turk, J. Mice deficient in group VIB phospholipase A2 (iPLA2gamma) exhibit relative resistance to obesity and metabolic abnormalities induced by a Western diet. *Am. J. Physiol. Endocrinol. Metab.* **2010**, *298*, E1097–E1114. [CrossRef]
60. Mancuso, D.J.; Sims, H.F.; Yang, K.; Kiebish, M.A.; Su, X.; Jenkins, C.M.; Guan, S.; Moon, S.H.; Pietka, T.; Nassir, F.; et al. Genetic ablation of calcium-independent phospholipase A2gamma prevents obesity and insulin resistance during high fat feeding by mitochondrial uncoupling and increased adipocyte fatty acid oxidation. *J. Biol. Chem.* **2010**, *285*, 36495–36510. [CrossRef]
61. Pena, L.; Meana, C.; Astudillo, A.M.; Lorden, G.; Valdearcos, M.; Sato, H.; Murakami, M.; Balsinde, J.; Balboa, M.A. Critical role for cytosolic group IVA phospholipase A2 in early adipocyte differentiation and obesity. *Biochim. Biophys. Acta* **2016**, *1861*, 1083–1095. [CrossRef] [PubMed]
62. Hadad, N.; Burgazliev, O.; Elgazar-Carmon, V.; Solomonov, Y.; Wueest, S.; Item, F.; Konrad, D.; Rudich, A.; Levy, R. Induction of cytosolic phospholipase a2alpha is required for adipose neutrophil infiltration and hepatic insulin resistance early in the course of high-fat feeding. *Diabetes* **2013**, *62*, 3053–3063. [CrossRef] [PubMed]

63. Kuefner, M.S.; Pham, K.; Redd, J.R.; Stephenson, E.J.; Harvey, I.; Deng, X.; Bridges, D.; Boilard, E.; Elam, M.B.; Park, E.A. Secretory phospholipase A2 group IIA modulates insulin sensitivity and metabolism. *J. Lipid Res.* **2017**, *58*, 1822–1833. [CrossRef]
64. Xu, J.; Bourgeois, H.; Vandermeulen, E.; Vlaeminck, B.; Meyer, E.; Demeyere, K.; Hesta, M. Secreted phospholipase A2 inhibitor modulates fatty acid composition and reduces obesity-induced inflammation in Beagle dogs. *Vet. J.* **2015**, *204*, 214–219. [CrossRef] [PubMed]
65. Iyer, A.; Lim, J.; Poudyal, H.; Reid, R.C.; Suen, J.Y.; Webster, J.; Prins, J.B.; Whitehead, J.P.; Fairlie, D.P.; Brown, L. An inhibitor of phospholipase A2 group IIA modulates adipocyte signaling and protects against diet-induced metabolic syndrome in rats. *Diabetes* **2012**, *61*, 2320–2329. [CrossRef]
66. Hui, D.Y.; Cope, M.J.; Labonte, E.D.; Chang, H.T.; Shao, J.; Goka, E.; Abousalham, A.; Charmot, D.; Buysse, J. The phospholipase A(2) inhibitor methyl indoxam suppresses diet-induced obesity and glucose intolerance in mice. *Br. J. Pharmacol.* **2009**, *157*, 1263–1269. [CrossRef]
67. Hollie, N.I.; Hui, D.Y. Group 1B phospholipase A(2) deficiency protects against diet-induced hyperlipidemia in mice. *J. Lipid Res.* **2011**, *52*, 2005–2011. [CrossRef]
68. Hollie, N.I.; Konaniah, E.S.; Goodin, C.; Hui, D.Y. Group 1B phospholipase A(2) inactivation suppresses atherosclerosis and metabolic diseases in LDL receptor-deficient mice. *Atherosclerosis* **2014**, *234*, 377–380. [CrossRef]
69. Cash, J.G.; Kuhel, D.G.; Goodin, C.; Hui, D.Y. Pancreatic acinar cell-specific overexpression of group 1B phospholipase A2 exacerbates diet-induced obesity and insulin resistance in mice. *Int. J. Obes.* **2011**, *35*, 877–881. [CrossRef]
70. Sato, H.; Taketomi, Y.; Ushida, A.; Isogai, Y.; Kojima, T.; Hirabayashi, T.; Miki, Y.; Yamamoto, K.; Nishito, Y.; Kobayashi, T.; et al. The adipocyte-inducible secreted phospholipases PLA2G5 and PLA2G2E play distinct roles in obesity. *Cell Metab.* **2014**, *20*, 119–132. [CrossRef]
71. Jha, P.; Claudel, T.; Baghdasaryan, A.; Mueller, M.; Halilbasic, E.; Das, S.K.; Lass, A.; Zimmermann, R.; Zechner, R.; Hoefler, G.; et al. Role of adipose triglyceride lipase (PNPLA2) in protection from hepatic inflammation in mouse models of steatohepatitis and endotoxemia. *Hepatology* **2014**, *59*, 858–869. [CrossRef] [PubMed]
72. Chen, W.; Chang, B.; Li, L.; Chan, L. Patatin-like phospholipase domain-containing 3/adiponutrin deficiency in mice is not associated with fatty liver disease. *Hepatology* **2010**, *52*, 1134–1142. [CrossRef] [PubMed]
73. Ochi, T.; Munekage, K.; Ono, M.; Higuchi, T.; Tsuda, M.; Hayashi, Y.; Okamoto, N.; Toda, K.; Sakamoto, S.; Oben, J.A.; et al. Patatin-like phospholipase domain-containing protein 3 is involved in hepatic fatty acid and triglyceride metabolism through X-box binding protein 1 and modulation of endoplasmic reticulum stress in mice. *Hepatol. Res.* **2016**, *46*, 584–592. [CrossRef] [PubMed]
74. Garces, F.; Lopez, F.; Nino, C.; Fernandez, A.; Chacin, L.; Hurt-Camejo, E.; Camejo, G.; Apitz-Castro, R. High plasma phospholipase A2 activity, inflammation markers, and LDL alterations in obesity with or without type 2 diabetes. *Obesity* **2010**, *18*, 2023–2029. [CrossRef] [PubMed]
75. Konukoglu, D.; Uzun, H.; Firtina, S.; Cigdem Arica, P.; Kocael, A.; Taskin, M. Plasma adhesion and inflammation markers: Asymmetrical dimethyl-L-arginine and secretory phospholipase A2 concentrations before and after laparoscopic gastric banding in morbidly obese patients. *Obes. Surg.* **2007**, *17*, 672–678. [CrossRef]
76. Misso, N.L.; Petrovic, N.; Grove, C.; Celenza, A.; Brooks-Wildhaber, J.; Thompson, P.J. Plasma phospholipase A2 activity in patients with asthma: Association with body mass index and cholesterol concentration. *Thorax* **2008**, *63*, 21–26. [CrossRef]
77. Speliotes, E.K.; Yerges-Armstrong, L.M.; Wu, J.; Hernaez, R.; Kim, L.J.; Palmer, C.D.; Gudnason, V.; Eiriksdottir, G.; Garcia, M.E.; Launer, L.J.; et al. Genome-wide association analysis identifies variants associated with nonalcoholic fatty liver disease that have distinct effects on metabolic traits. *PLoS Genet.* **2011**, *7*, e1001324. [CrossRef]
78. Romeo, S.; Kozlitina, J.; Xing, C.; Pertsemlidis, A.; Cox, D.; Pennacchio, L.A.; Boerwinkle, E.; Cohen, J.C.; Hobbs, H.H. Genetic variation in PNPLA3 confers susceptibility to nonalcoholic fatty liver disease. *Nat. Genet.* **2008**, *40*, 1461–1465. [CrossRef]

79. Linden, D.; Ahnmark, A.; Pingitore, P.; Ciociola, E.; Ahlstedt, I.; Andreasson, A.C.; Sasidharan, K.; Madeyski-Bengtson, K.; Zurek, M.; Mancina, R.M.; et al. Pnpla3 silencing with antisense oligonucleotides ameliorates nonalcoholic steatohepatitis and fibrosis in Pnpla3 I148M knock-in mice. *Mol. Metab.* **2019**, *22*, 49–61. [CrossRef]
80. Sevastianova, K.; Kotronen, A.; Gastaldelli, A.; Perttila, J.; Hakkarainen, A.; Lundbom, J.; Suojanen, L.; Orho-Melander, M.; Lundbom, N.; Ferrannini, E.; et al. Genetic variation in PNPLA3 (adiponutrin) confers sensitivity to weight loss-induced decrease in liver fat in humans. *Am. J. Clin. Nutr.* **2011**, *94*, 104–111. [CrossRef]
81. Lu, Y.; Day, F.R.; Gustafsson, S.; Buchkovich, M.L.; Na, J.; Bataille, V.; Cousminer, D.L.; Dastani, Z.; Drong, A.W.; Esko, T.; et al. New loci for body fat percentage reveal link between adiposity and cardiometabolic disease risk. *Nat. Commun.* **2016**, *7*, 10495. [CrossRef] [PubMed]
82. Huang, L.O.; Loos, R.J.F.; Kilpelainen, T.O. Evidence of genetic predisposition for metabolically healthy obesity and metabolically obese normal weight. *Physiol. Genom.* **2018**, *50*, 169–178. [CrossRef] [PubMed]
83. Teslovich, T.M.; Musunuru, K.; Smith, A.V.; Edmondson, A.C.; Stylianou, I.M.; Koseki, M.; Pirruccello, J.P.; Ripatti, S.; Chasman, D.I.; Willer, C.J.; et al. Biological, clinical and population relevance of 95 loci for blood lipids. *Nature* **2010**, *466*, 707–713. [CrossRef] [PubMed]
84. Bauer, R.C.; Khetarpal, S.A.; Hand, N.J.; Rader, D.J. Therapeutic Targets of Triglyceride Metabolism as Informed by Human Genetics. *Trends Mol. Med.* **2016**, *22*, 328–340. [CrossRef]
85. Johansen, C.T.; Hegele, R.A. Allelic and phenotypic spectrum of plasma triglycerides. *Biochim. Biophys. Acta* **2012**, *1821*, 833–842. [CrossRef] [PubMed]
86. Peng, X.Y.; Wang, Y.; Hu, H.; Zhang, X.J.; Li, Q. Identification of the molecular subgroups in coronary artery disease by gene expression profiles. *J. Cell. Physiol.* **2019**. [CrossRef]
87. Klimentidis, Y.C.; Chougule, A.; Arora, A.; Frazier-Wood, A.C.; Hsu, C.H. Triglyceride-Increasing Alleles Associated with Protection against Type-2 Diabetes. *PLoS Genet.* **2015**, *11*, e1005204. [CrossRef] [PubMed]
88. Ahmad, S.; Mora, S.; Ridker, P.M.; Hu, F.B.; Chasman, D.I. Gene-Based Elevated Triglycerides and Type 2 Diabetes Mellitus Risk in the Women's Genome Health Study. *Arterioscler. Thromb. Vasc. Biol.* **2019**, *39*, 97–106. [CrossRef]
89. Yan, J.; Hu, C.; Jiang, F.; Zhang, R.; Wang, J.; Tang, S.; Peng, D.; Chen, M.; Bao, Y.; Jia, W. Genetic variants of PLA2G6 are associated with Type 2 diabetes mellitus and triglyceride levels in a Chinese population. *Diabet. Med. J. Br. Diabet. Assoc.* **2015**, *32*, 280–286. [CrossRef] [PubMed]
90. Bao, S.; Song, H.; Wohltmann, M.; Ramanadham, S.; Jin, W.; Bohrer, A.; Turk, J. Insulin secretory responses and phospholipid composition of pancreatic islets from mice that do not express Group VIA phospholipase A2 and effects of metabolic stress on glucose homeostasis. *J. Biol. Chem.* **2006**, *281*, 20958–20973. [CrossRef]
91. Dennis, E.A.; Cao, J.; Hsu, Y.H.; Magrioti, V.; Kokotos, G. Phospholipase A2 enzymes: Physical structure, biological function, disease implication, chemical inhibition, and therapeutic intervention. *Chem. Rev.* **2011**, *111*, 6130–6185. [CrossRef] [PubMed]
92. Ramanadham, S.; Ali, T.; Ashley, J.W.; Bone, R.N.; Hancock, W.D.; Lei, X. Calcium-independent phospholipases A2 and their roles in biological processes and diseases. *J. Lipid Res.* **2015**, *56*, 1643–1668. [CrossRef]
93. Turk, J.; White, T.D.; Nelson, A.J.; Lei, X.; Ramanadham, S. iPLA2beta and its role in male fertility, neurological disorders, metabolic disorders, and inflammation. *Biochim. Biophys. Acta Mol. Cell Biol. Lipids* **2019**, *1864*, 846–860. [CrossRef] [PubMed]
94. Balsinde, J.; Bianco, I.D.; Ackermann, E.J.; Conde-Frieboes, K.; Dennis, E.A. Inhibition of calcium-independent phospholipase A2 prevents arachidonic acid incorporation and phospholipid remodeling in P388D1 macrophages. *Proc. Natl. Acad. Sci. USA* **1995**, *92*, 8527–8531. [CrossRef] [PubMed]
95. Balsinde, J. Roles of various phospholipases A2 in providing lysophospholipid acceptors for fatty acid phospholipid incorporation and remodelling. *Biochem. J.* **2002**, *364*, 695–702. [CrossRef]
96. Zhang, X.H.; Zhao, C.; Ma, Z.A. The increase of cell-membranous phosphatidylcholines containing polyunsaturated fatty acid residues induces phosphorylation of p53 through activation of ATR. *J. Cell Sci.* **2007**, *120*, 4134–4143. [CrossRef]
97. Bao, S.; Miller, D.J.; Ma, Z.; Wohltmann, M.; Eng, G.; Ramanadham, S.; Moley, K.; Turk, J. Male mice that do not express group VIA phospholipase A2 produce spermatozoa with impaired motility and have greatly reduced fertility. *J. Biol. Chem.* **2004**, *279*, 38194–38200. [CrossRef] [PubMed]

98. Zhang, L.; Zhong, S.; Li, Y.; Ji, G.; Sundaram, M.; Yao, Z. Global Inactivation of the Pla2g6 Gene in Mice Does Not Cause Dyslipidemia under Chow or High-fat Diet Conditions. *J. Cancer Prev.* **2013**, *18*, 235–248. [CrossRef]
99. Su, X.; Mancuso, D.J.; Bickel, P.E.; Jenkins, C.M.; Gross, R.W. Small interfering RNA knockdown of calcium-independent phospholipases A2 beta or gamma inhibits the hormone-induced differentiation of 3T3-L1 preadipocytes. *J. Biol. Chem.* **2004**, *279*, 21740–21748. [CrossRef]
100. Honda, Y.; Yoneda, M.; Kessoku, T.; Ogawa, Y.; Tomeno, W.; Imajo, K.; Mawatari, H.; Fujita, K.; Hyogo, H.; Ueno, T.; et al. Characteristics of non-obese non-alcoholic fatty liver disease: Effect of genetic and environmental factors. *Hepatol. Res.* **2016**, *46*, 1011–1018. [CrossRef]
101. Ecker, J.; Scherer, M.; Schmitz, G.; Liebisch, G. A rapid GC-MS method for quantification of positional and geometric isomers of fatty acid methyl esters. *J. Chromatogr. Banal. Technol. Biomed. Life Sci.* **2012**, *897*, 98–104. [CrossRef] [PubMed]
102. Van der Veen, J.N.; Kennelly, J.P.; Wan, S.; Vance, J.E.; Vance, D.E.; Jacobs, R.L. The critical role of phosphatidylcholine and phosphatidylethanolamine metabolism in health and disease. *Biochim. Biophys. Acta Biomembr.* **2017**, *1859*, 1558–1572. [CrossRef] [PubMed]
103. Liebisch, G.; Lieser, B.; Rathenberg, J.; Drobnik, W.; Schmitz, G. High-throughput quantification of phosphatidylcholine and sphingomyelin by electrospray ionization tandem mass spectrometry coupled with isotope correction algorithm. *Biochim. Biophys. Acta* **2004**, *1686*, 108–117. [CrossRef] [PubMed]
104. Liebisch, G.; Binder, M.; Schifferer, R.; Langmann, T.; Schulz, B.; Schmitz, G. High throughput quantification of cholesterol and cholesteryl ester by electrospray ionization tandem mass spectrometry (ESI-MS/MS). *Biochim. Biophys. Acta* **2006**, *1761*, 121–128. [CrossRef]
105. Liebisch, G.; Drobnik, W.; Lieser, B.; Schmitz, G. High-throughput quantification of lysophosphatidylcholine by electrospray ionization tandem mass spectrometry. *Clin. Chem.* **2002**, *48*, 2217–2224. [CrossRef]
106. Liebisch, G.; Drobnik, W.; Reil, M.; Trumbach, B.; Arnecke, R.; Olgemoller, B.; Roscher, A.; Schmitz, G. Quantitative measurement of different ceramide species from crude cellular extracts by electrospray ionization tandem mass spectrometry (ESI-MS/MS). *J. Lipid Res.* **1999**, *40*, 1539–1546.
107. Song, H.; Bao, S.; Lei, X.; Jin, C.; Zhang, S.; Turk, J.; Ramanadham, S. Evidence for proteolytic processing and stimulated organelle redistribution of iPLA(2)beta. *Biochim. Biophys. Acta* **2010**, *1801*, 547–558. [CrossRef]
108. Gao, X.; van der Veen, J.N.; Vance, J.E.; Thiesen, A.; Vance, D.E.; Jacobs, R.L. Lack of phosphatidylethanolamine N-methyltransferase alters hepatic phospholipid composition and induces endoplasmic reticulum stress. *Biochim. Biophys. Acta* **2015**, *1852*, 2689–2699. [CrossRef] [PubMed]
109. Fu, S.; Yang, L.; Li, P.; Hofmann, O.; Dicker, L.; Hide, W.; Lin, X.; Watkins, S.M.; Ivanov, A.R.; Hotamisligil, G.S. Aberrant lipid metabolism disrupts calcium homeostasis causing liver endoplasmic reticulum stress in obesity. *Nature* **2011**, *473*, 528–531. [CrossRef]
110. Ming, Y.; Zhu, X.; Tuma-Kellner, S.; Ganzha, A.; Liebisch, G.; Gan-Schreier, H.; Chamulitrat, W. iPla2beta Deficiency Suppresses Hepatic ER UPR, Fxr, and Phospholipids in Mice Fed with MCD Diet, Resulting in Exacerbated Hepatic Bile Acids and Biliary Cell Proliferation. *Cells* **2019**, *8*, 879. [CrossRef]
111. Tu, L.N.; Showalter, M.R.; Cajka, T.; Fan, S.; Pillai, V.V.; Fiehn, O.; Selvaraj, V. Metabolomic characteristics of cholesterol-induced non-obese nonalcoholic fatty liver disease in mice. *Sci. Rep.* **2017**, *7*, 6120. [CrossRef] [PubMed]
112. Gurtovenko, A.A.; Vattulainen, I. Membrane potential and electrostatics of phospholipid bilayers with asymmetric transmembrane distribution of anionic lipids. *J. Phys. Chem. B* **2008**, *112*, 4629–4634. [CrossRef] [PubMed]
113. Sena, A.; Rebel, G.; Bieth, R.; Hubert, P.; Waksman, A. Lipid composition in liver and brain of genetically obese (ob/ob), heterozygote (ob/+)and normal (+/+) mice. *Biochim. Biophys. Acta* **1982**, *710*, 290–296. [CrossRef]
114. Van Rooyen, D.M.; Larter, C.Z.; Haigh, W.G.; Yeh, M.M.; Ioannou, G.; Kuver, R.; Lee, S.P.; Teoh, N.C.; Farrell, G.C. Hepatic free cholesterol accumulates in obese, diabetic mice and causes nonalcoholic steatohepatitis. *Gastroenterology* **2011**, *141*, 1393–1403.e5. [CrossRef] [PubMed]
115. Patelski, J.; Piorunska-Stolzmann, M. Effects of substrate fatty acids on products of lecithin hydrolysis and acyl-CoA-independent transacylation with cholesterol by aortic enzyme preparations. *Enzyme* **1985**, *34*, 217–219. [CrossRef]

116. Zhao, G.; Wakabayashi, R.; Shimoda, S.; Fukunaga, Y.; Kumagai, M.; Tanaka, M.; Nakano, K. Impaired activities of cyclic adenosine monophosphate-responsive element binding protein, protein kinase A and calcium-independent phospholipase A2 are involved in deteriorated regeneration of cirrhotic liver after partial hepatectomy in rats. *Hepatol. Res.* **2011**, *41*, 1110–1119. [CrossRef]
117. Lei, X.; Zhang, S.; Barbour, S.E.; Bohrer, A.; Ford, E.L.; Koizumi, A.; Papa, F.R.; Ramanadham, S. Spontaneous development of endoplasmic reticulum stress that can lead to diabetes mellitus is associated with higher calcium-independent phospholipase A2 expression: A role for regulation by SREBP-1. *J. Biol. Chem.* **2010**, *285*, 6693–6705. [CrossRef]
118. Gorden, D.L.; Myers, D.S.; Ivanova, P.T.; Fahy, E.; Maurya, M.R.; Gupta, S.; Min, J.; Spann, N.J.; McDonald, J.G.; Kelly, S.L.; et al. Biomarkers of NAFLD progression: A lipidomics approach to an epidemic. *J. Lipid Res.* **2015**, *56*, 722–736. [CrossRef]
119. Chiappini, F.; Coilly, A.; Kadar, H.; Gual, P.; Tran, A.; Desterke, C.; Samuel, D.; Duclos-Vallee, J.C.; Touboul, D.; Bertrand-Michel, J.; et al. Metabolism dysregulation induces a specific lipid signature of nonalcoholic steatohepatitis in patients. *Sci. Rep.* **2017**, *7*, 46658. [CrossRef]
120. Arendt, B.M.; Ma, D.W.; Simons, B.; Noureldin, S.A.; Therapondos, G.; Guindi, M.; Sherman, M.; Allard, J.P. Nonalcoholic fatty liver disease is associated with lower hepatic and erythrocyte ratios of phosphatidylcholine to phosphatidylethanolamine. *Appl. Physiol. Nutr. Metab.* **2013**, *38*, 334–340. [CrossRef]
121. Sanyal, A.J.; Pacana, T. A Lipidomic Readout of Disease Progression in A Diet-Induced Mouse Model of Nonalcoholic Fatty Liver Disease. *Trans. Am. Clin. Climatol. Assoc.* **2015**, *126*, 271–288. [PubMed]
122. Chiappini, F.; Desterke, C.; Bertrand-Michel, J.; Guettier, C.; Le Naour, F. Hepatic and serum lipid signatures specific to nonalcoholic steatohepatitis in murine models. *Sci. Rep.* **2016**, *6*, 31587. [CrossRef] [PubMed]
123. Cunnane, S.C.; McAdoo, K.R.; Horrobin, D.F. n-3 Essential fatty acids decrease weight gain in genetically obese mice. *Br. J. Nutr.* **1986**, *56*, 87–95. [CrossRef]
124. Sekiya, M.; Yahagi, N.; Matsuzaka, T.; Najima, Y.; Nakakuki, M.; Nagai, R.; Ishibashi, S.; Osuga, J.; Yamada, N.; Shimano, H. Polyunsaturated fatty acids ameliorate hepatic steatosis in obese mice by SREBP-1 suppression. *Hepatology* **2003**, *38*, 1529–1539. [CrossRef]
125. Stubbs, C.D.; Smith, A.D. The modification of mammalian membrane polyunsaturated fatty acid composition in relation to membrane fluidity and function. *Biochim. Biophys. Acta* **1984**, *779*, 89–137. [CrossRef]
126. Ii, H.; Oka, M.; Yamashita, A.; Waku, K.; Uozumi, N.; Shimizu, T.; Sato, T.; Akiba, S. Inhibition of cytosolic phospholipase A(2) suppresses production of cholesteryl ester through the reesterification of free cholesterol but not formation of foam cells in oxidized LDL-stimulated macrophages. *Biol. Pharm. Bull.* **2008**, *31*, 6–12. [CrossRef]
127. Gubern, A.; Barcelo-Torns, M.; Casas, J.; Barneda, D.; Masgrau, R.; Picatoste, F.; Balsinde, J.; Balboa, M.A.; Claro, E. Lipid droplet biogenesis induced by stress involves triacylglycerol synthesis that depends on group VIA phospholipase A2. *J. Biol. Chem.* **2009**, *284*, 5697–5708. [CrossRef] [PubMed]
128. Ashley, J.W.; Hancock, W.D.; Nelson, A.J.; Bone, R.N.; Tse, H.M.; Wohltmann, M.; Turk, J.; Ramanadham, S. Polarization of Macrophages toward M2 Phenotype Is Favored by Reduction in iPLA2beta (Group VIA Phospholipase A2). *J. Biol. Chem.* **2016**, *291*, 23268–23281. [CrossRef]
129. Monge, P.; Garrido, A.; Rubio, J.M.; Magrioti, V.; Kokotos, G.; Balboa, M.A.; Balsinde, J. The Contribution of Cytosolic Group IVA and Calcium-Independent Group VIA Phospholipase A2s to Adrenic Acid Mobilization in Murine Macrophages. *Biomolecules* **2020**, *10*, 542. [CrossRef]
130. Inhoffen, J.; Tuma-Kellner, S.; Straub, B.; Stremmel, W.; Chamulitrat, W. Deficiency of iPLA(2)beta Primes Immune Cells for Proinflammation: Potential Involvement in Age-Related Mesenteric Lymph Node Lymphoma. *Cancers* **2015**, *7*, 2427–2442. [CrossRef]
131. Jiao, L.; Gan-Schreier, H.; Tuma-Kellner, S.; Stremmel, W.; Chamulitrat, W. Sensitization to autoimmune hepatitis in group VIA calcium-independent phospholipase A2-null mice led to duodenal villous atrophy with apoptosis, goblet cell hyperplasia and leaked bile acids. *Biochim. Biophys. Acta* **2015**, *1852*, 1646–1657. [CrossRef] [PubMed]
132. Jiao, L.; Gan-Schreier, H.; Zhu, X.; Wei, W.; Tuma-Kellner, S.; Liebisch, G.; Stremmel, W.; Chamulitrat, W. Ageing sensitized by iPLA2beta deficiency induces liver fibrosis and intestinal atrophy involving suppression of homeostatic genes and alteration of intestinal lipids and bile acids. *Biochim. Biophys. Acta Mol. Cell Biol. Lipids* **2017**, *1862*, 1520–1533. [CrossRef] [PubMed]

133. Sharma, J.; Blase, J.R.; Hoft, D.F.; Marentette, J.O.; Turk, J.; McHowat, J. Mice with Genetic Deletion of Group VIA Phospholipase A2beta Exhibit Impaired Macrophage Function and Increased Parasite Load in Trypanosoma cruzi-Induced Myocarditis. *Infect. Immun.* **2016**, *84*, 1137–1142. [CrossRef] [PubMed]
134. Jaworski, K.; Ahmadian, M.; Duncan, R.E.; Sarkadi-Nagy, E.; Varady, K.A.; Hellerstein, M.K.; Lee, H.Y.; Samuel, V.T.; Shulman, G.I.; Kim, K.H.; et al. AdPLA ablation increases lipolysis and prevents obesity induced by high-fat feeding or leptin deficiency. *Nat. Med.* **2009**, *15*, 159–168. [CrossRef]
135. Shimabukuro, M.; Koyama, K.; Chen, G.; Wang, M.Y.; Trieu, F.; Lee, Y.; Newgard, C.B.; Unger, R.H. Direct antidiabetic effect of leptin through triglyceride depletion of tissues. *Proc. Natl. Acad. Sci. USA* **1997**, *94*, 4637–4641. [CrossRef] [PubMed]
136. Metlakunta, A.; Huang, W.; Stefanovic-Racic, M.; Dedousis, N.; Sipula, I.; O'Doherty, R.M. Kupffer cells facilitate the acute effects of leptin on hepatic lipid metabolism. *Am. J. Physiol. Endocrinol. Metab.* **2017**, *312*, E11–E18. [CrossRef]
137. Bone, R.N.; Gai, Y.; Magrioti, V.; Kokotou, M.G.; Ali, T.; Lei, X.; Tse, H.M.; Kokotos, G.; Ramanadham, S. Inhibition of Ca2+-independent phospholipase A2beta (iPLA2beta) ameliorates islet infiltration and incidence of diabetes in NOD mice. *Diabetes* **2015**, *64*, 541–554. [CrossRef]
138. Mouchlis, V.D.; Limnios, D.; Kokotou, M.G.; Barbayianni, E.; Kokotos, G.; McCammon, J.A.; Dennis, E.A. Development of Potent and Selective Inhibitors for Group VIA Calcium-Independent Phospholipase A2 Guided by Molecular Dynamics and Structure-Activity Relationships. *J. Med. Chem.* **2016**, *59*, 4403–4414. [CrossRef]
139. Guo, H.H.; Feng, C.L.; Zhang, W.X.; Luo, Z.G.; Zhang, H.J.; Zhang, T.T.; Ma, C.; Zhan, Y.; Li, R.; Wu, S.; et al. Liver-target nanotechnology facilitates berberine to ameliorate cardio-metabolic diseases. *Nat. Commun.* **2019**, *10*, 1981. [CrossRef]
140. Sharma, Y.; Ahmad, A.; Yavvari, P.S.; Kumar Muwal, S.; Bajaj, A.; Khan, F. Targeted SHP-1 Silencing Modulates the Macrophage Phenotype, Leading to Metabolic Improvement in Dietary Obese Mice. *Mol. Ther. Nucleic Acids* **2019**, *16*, 626–636. [CrossRef]
141. Cao, J.; Peterson, S.J.; Sodhi, K.; Vanella, L.; Barbagallo, I.; Rodella, L.F.; Schwartzman, M.L.; Abraham, N.G.; Kappas, A. Heme oxygenase gene targeting to adipocytes attenuates adiposity and vascular dysfunction in mice fed a high-fat diet. *Hypertension* **2012**, *60*, 467–475. [CrossRef] [PubMed]
142. Buchman, A.L.; Dubin, M.D.; Moukarzel, A.A.; Jenden, D.J.; Roch, M.; Rice, K.M.; Gornbein, J.; Ament, M.E. Choline deficiency: A cause of hepatic steatosis during parenteral nutrition that can be reversed with intravenous choline supplementation. *Hepatology* **1995**, *22*, 1399–1403.
143. Corbin, K.D.; Zeisel, S.H. Choline metabolism provides novel insights into nonalcoholic fatty liver disease and its progression. *Curr. Opin. Gastroenterol.* **2012**, *28*, 159–165. [CrossRef] [PubMed]
144. Spencer, M.D.; Hamp, T.J.; Reid, R.W.; Fischer, L.M.; Zeisel, S.H.; Fodor, A.A. Association between composition of the human gastrointestinal microbiome and development of fatty liver with choline deficiency. *Gastroenterology* **2011**, *140*, 976–986. [CrossRef]
145. Sha, W.; da Costa, K.A.; Fischer, L.M.; Milburn, M.V.; Lawton, K.A.; Berger, A.; Jia, W.; Zeisel, S.H. Metabolomic profiling can predict which humans will develop liver dysfunction when deprived of dietary choline. *FASEB J.* **2010**, *24*, 2962–2975. [CrossRef] [PubMed]
146. Cordero, P.; Gomez-Uriz, A.M.; Campion, J.; Milagro, F.I.; Martinez, J.A. Dietary supplementation with methyl donors reduces fatty liver and modifies the fatty acid synthase DNA methylation profile in rats fed an obesogenic diet. *Genes Nutr.* **2013**, *8*, 105–113. [CrossRef] [PubMed]

© 2020 by the authors. Licensee MDPI, Basel, Switzerland. This article is an open access article distributed under the terms and conditions of the Creative Commons Attribution (CC BY) license (http://creativecommons.org/licenses/by/4.0/).

Article

Metabolic Effects of Selective Deletion of Group VIA Phospholipase A2 from Macrophages or Pancreatic Islet Beta-Cells

John Turk [1,*,†], Haowei Song [1,2,†], Mary Wohltmann [1,†], Cheryl Frankfater [1], Xiaoyong Lei [3] and Sasanka Ramanadham [3,*]

1 Mass Spectrometry Facility, Division of Endocrinology, Metabolism, and Lipid Research, Department of Medicine, Department of Pathology and Immunology, Washington University School of Medicine, St. Louis, MO 63110, USA; songhw@gmail.com (H.S.); wohltmann@wustl.edu (M.W.); c.frankf@wustl.edu (C.F.)

2 Sigma-Aldrich Corporation, St. Louis, MO 63118, USA

3 Department of Cell, Developmental, and Integrative Biology, Comprehensive Diabetes Center, University of Alabama at Birmingham, Birmingham, AL 35294, USA; xlei@uab.edu

* Correspondence: jturk@wustl.edu (J.T.); sramvem@uab.edu (S.R.); Tel.: +1-314-852-8554 (J.T.); +1-205-996-5973 (S.R.)

† All three of these authors contributed equally to this manuscript and are co-primary authors.

Received: 30 August 2020; Accepted: 13 October 2020; Published: 17 October 2020

Abstract: To examine the role of group VIA phospholipase A_2 (iPLA$_2$β) in specific cell lineages in insulin secretion and insulin action, we prepared mice with a selective iPLA$_2$β deficiency in cells of myelomonocytic lineage, including macrophages (MØ-iPLA$_2$β-KO), or in insulin-secreting β-cells (β-Cell-iPLA$_2$β-KO), respectively. MØ-iPLA$_2$β-KO mice exhibited normal glucose tolerance when fed standard chow and better glucose tolerance than floxed-iPLA$_2$β control mice after consuming a high-fat diet (HFD). MØ-iPLA$_2$β-KO mice exhibited normal glucose-stimulated insulin secretion (GSIS) in vivo and from isolated islets ex vivo compared to controls. Male MØ-iPLA$_2$β-KO mice exhibited enhanced insulin responsivity vs. controls after a prolonged HFD. In contrast, β-cell-iPLA$_2$β-KO mice exhibited impaired glucose tolerance when fed standard chow, and glucose tolerance deteriorated further when introduced to a HFD. β-Cell-iPLA$_2$β-KO mice exhibited impaired GSIS in vivo and from isolated islets ex vivo vs. controls. β-Cell-iPLA$_2$β-KO mice also exhibited an enhanced insulin responsivity compared to controls. These findings suggest that MØ iPLA$_2$β participates in HFD-induced deterioration in glucose tolerance and that this mainly reflects an effect on insulin responsivity rather than on insulin secretion. In contrast, β-cell iPLA$_2$β plays a role in GSIS and also appears to confer some protection against deterioration in β-cell functions induced by a HFD.

Keywords: pancreatic islets; β-cells; insulin secretion; glucose tolerance; insulin resistance; group VIA phospholipase A_2

---

1. Introduction

Glycerophospholipids are the most abundant molecular components of biological membrane bilayers and are both critical determinants of membrane structure and the source of signaling molecules produced from their hydrolysis by phospholipase enzymes. Glycerophospholipids consist of a glycerol backbone with a phosphate ester in the *sn*-3 position that may form a phosphodiester linkage to a polar head-group, such as choline, ethanolamine, serine, inositol, or glycerol, inter alia. A fatty acid is esterified to the glycerol backbone in the *sn*-2 position of phospholipids, and in the *sn*-1 position there is an ester, ether, or vinyl ether linkage to a fatty acid, fatty alcohol, or fatty aldehyde residue, respectively. Phospholipase A_2 (PLA_2) enzymes hydrolyze the phospholipid *sn*-2 ester bond to yield a free fatty acid and a 2-lysophospholipid as products [1,2]. The PLA_2 superfamily consists of at least 16 groups of structurally and functionally diverse enzymes that include secreted ($sPLA_2$), cytosolic ($cPLA_2$), calcium-independent ($iPLA_2$), lipoprotein-associated ($Lp\text{-}PLA_2$), and adipose-PLA_2 (AdPLA). These enzymes play central roles in cellular lipid metabolism and signaling [1].

The enzyme that is now designated as Group VIA PLA_2 was the first recognized mammalian member of the Ca^{2+}-independent PLA_2 enzymes [3–5], and its abbreviated designation is $iPLA_2\beta$ [6,7]. Akin to the plant lipase patatin, $iPLA_2\beta$ contains a GXSXG lipase consensus sequence in the enzyme catalytic center, in which the central Ser is a component of a Ser-Asp catalytic dyad [8]. Enzymes with patatin-like phospholipase domains comprise the PNPLA family, and the human genome expresses nine members of this family, of which $iPLA_2\beta$ is PNPLA9 [9,10]. The phenotypic properties of experimental mouse models with induced mutations in the genes that encode PNPLA family members and the clinical phenotypes of patients with corresponding mutations indicate that several of these lipases play important metabolic roles in mammalian lipid and energy homeostasis [9,10].

Global $iPLA_2\beta$-null mice produced by homologous recombination exhibit several phenotypic abnormalities, including greatly impaired male fertility [11] and the development of a neurodegenerative condition [12] that is similar to the human genetic disease Infantile Neuroaxonal Dystrophy (INAD), which arises from mutations in the Group VIA PLA_2 gene [13,14]. Consistent with previous evidence that $iPLA_2\beta$ participates in signaling events leading to glucose-stimulated insulin secretion (GSIS) from pancreatic islet β-cells [15–22], islets isolated from male [23] or female [24] $iPLA_2\beta$-null mice exhibit impaired GSIS ex vivo, and male $iPLA_2\beta$-null mice exhibit impaired glucose tolerance in vivo [23]. In contrast, female $iPLA_2\beta$-null mice exhibit normal glucose tolerance in the unstressed state [24] but develop a more severe glucose intolerance than wild-type littermates after exposure to the β-cell toxin streptozotocin or after the introduction of a high-fat diet (HFD) [24]. Surprisingly, female $iPLA_2\beta$-null mice experienced less deterioration in insulin sensitivity than did wild-type littermates after being introduced to a HFD, although pancreatic islets isolated from HFD-fed female $iPLA_2\beta$-null mice exhibited a much more severe impairment of GSIS than did islets isolated from HFD-fed wild-type littermates [24].

This discordance of the effects of a HFD on insulin secretion and insulin sensitivity in global $iPLA_2\beta$-null mice suggests that $iPLA_2\beta$ plays distinct roles in the molecular mechanisms underlying insulin secretion and insulin action and in the impact of a HFD on these processes. Evidence indicates that $iPLA_2\beta$ amplifies glucose-induced Ca^{2+} entry into β-cells [23], suggesting that this is one component of the mechanism(s) through which $iPLA_2\beta$ participates in signaling events underlying GSIS, and $iPLA_2\beta$ also participates in the repair of β-cell mitochondrial membranes that are oxidized upon exposure to high concentrations of palmitic acid [25]. This might represent a mechanism whereby $iPLA_2\beta$ mitigates β-cell injury in HFD-fed mice and accounts for the fact that loss of $iPLA_2\beta$ results in a greater impairment of insulin secretion in HFD-fed $iPLA_2\beta$-null mice compared to HFD-fed wild-type littermates [24].

Macrophages and their precursor monocytes also express $iPLA_2\beta$ [26–29], and its expression level affects the macrophage phenotype [30]. Migration of monocytes into extravascular sites, including adipose tissue, and their differentiation into macrophages that elaborate cytokines, including TNFα and IL-6, which impair insulin sensitivity, are thought to represent a critical series of events in the development of diet-induced insulin resistance in diabetes and obesity and to involve tissue elaboration of the cytokine Monocyte Chemoattractant-1 (MCP-1) and its interaction with the monocyte MCP-1 receptor CCR2 [31–36]. Genetic and pharmacologic evidence indicates that the monocyte chemotactic response to MCP-1 requires the action of $iPLA_2\beta$ [37–39] to generate the lipid mediator 2-lysophosphatidic acid (LPA) [29,37,39]. These observations suggest the possibility that the relative insensitivity of $iPLA_2\beta$-null mice to high-fat diet (HFD)-induced insulin resistance might reflect the failure of $iPLA_2\beta$-null monocytes to migrate into adipose tissue and other extravascular sites in response to HFD-induced tissue elaboration of MCP-1.

A plausible hypothesis is thus that the net metabolic effects of global deletion of iPLA$_2$β might reflect opposing effects on β-cells and monocyte-macrophages. A loss of iPLA$_2$β in β-cells would result in impaired insulin secretion and increased sensitivity to lipid-induced injury, but loss of iPLA$_2$β in macrophages might provide relative protection against the HFD-induced deterioration of insulin sensitivity because of an impaired migration of monocytes into extravascular tissues, differentiation into macrophages, and elaboration of cytokines that result in insulin resistance. To test this hypothesis, we generated mice with floxed-iPLA$_2$β alleles and mated them with mice that express Cre recombinase under control of LysM or RIP2 promoters to produce mice with selective iPLA$_2$β deficiency in macrophages (MØ-iPLA$_2$β-KO) or insulin-secreting β-cells (β-cell-iPLA$_2$β-KO), respectively. The metabolic phenotypes of these mice were then characterized and are described in this report.

2. Materials and Methods

2.1. Materials

Enhanced chemiluminescence reagents were obtained from Amersham Biosciences (Piscataway, NJ, USA); SDS-PAGE supplies from Bio-Rad (Richmond, CA, USA); ATP, common reagents, and salts from Sigma (St. Louis, MO, USA); culture media, penicillin, streptomycin, Hanks' balanced salt solution, L-glutamine, agarose, and RT-PCR reagents from Invitrogen (Carlsbad, CA, USA); fetal bovine serum from Hyclone (Logan, UT); Pentex bovine serum albumin (BSA, fatty acid-free, fraction V) from ICN Biomedical (Aurora, OH, USA); forskolin from Calbiochem (La Jolla, CA, USA). Krebs–Ringer bicarbonate (KRB) buffer contained (in mM) 25 HEPES (pH 7.4), 115 NaCl, 24 $NaHCO_3$, 5 KCl, 1 $MgCl_2$, and 2.5 $CaCl_2$.

2.2. Preparation of Mice with Selective Deletion of iPLA$_2$β from Restricted Cell Lineages

The Washington University Animal Studies Committee approved all animal studies. Mice with floxed-iPLA$_2$β alleles were prepared and mated with mice expressing Cre recombinase under control of cell type-restricted promoters to generate conditionally iPLA$_2$β-deficient mice. European Conditional Mouse Mutagenesis (EUCOMM) embryonic stem (ES) cells with an iPLA$_2$β-targeting construct incorporated by homologous recombination [40] were purchased (Figure 1A). In this construct, LoxP sites L2 and L3 were recognized by Cre recombinase [41,42] flank critical iPLA$_2$β gene exons 6–8 (Figure 1A). Removing this "floxed" segment results in a truncated mRNA species eliminated by nonsense mediated decay. The 5′ fragment with the neo cassette is flanked by FRT (flippase recognition target) sites F1 and F2 (Figure 1A) that are recognized by FLP (flippase) recombinase [43,44]. Correct integration of 5′ and 3′ arms was confirmed by PCR (Figure 1B) using primer sets that recognize sequences external to the construct and within the neo cassette, respectively. Primer set Raf 5 and GR3 for the 3′ arm yielded a 9.2 kb product (Figure 1A). Two clones (G05, G07) with normal karyotypes were injected into blastocysts that were then implanted into pseudo-pregnant females. Both yielded chimeric mice that transmitted the targeted allele in the germ line to yield heterozygotes for wild-type (WT) and EUCOMM iPLA$_2$β alleles [45], as verified by PCR with primers within the neo cassette and the intron between L2 and exon 6, respectively, that yield a 795 bp product (Figure 1C).

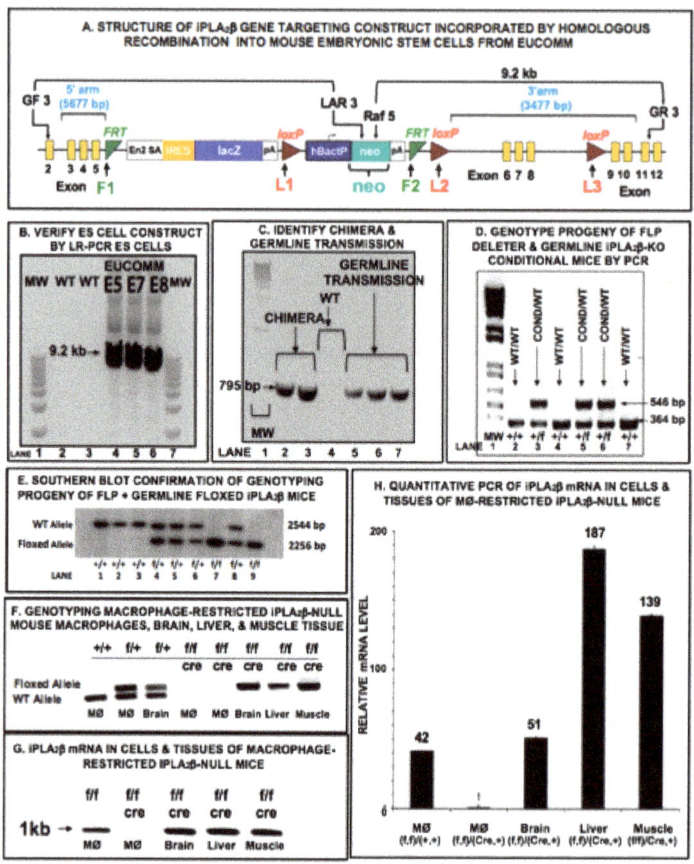

Figure 1. Cre-Lox Preparation of Mutant Mice with Tissue-Restricted Deletion of iPLA$_2\beta$. (**A**) is a schematic illustration of the structure of the gene-targeting construct that was incorporated by homologous recombination into mouse embryonic stem (ES) cells from European Conditional Mouse Mutagenesis (EUCOMM). (**B**) demonstrates correct integration of the 5′ and 3′ arms of the construct in the ES cell lines by long-range PCR using primer sets that recognize sequences external to the construct and within the neo cassette, respectively, to yield a 9.2 kb product. (**C**) illustrates that injection of the ES cells into blastocysts that were then implanted into pseudo-pregnant female mice resulted in production of chimeric mice with germline transmission of the targeted gene to yield heterozygotes for the wild-type (WT) and EUCOMM iPLA$_2\beta$ alleles. (**D**) illustrates that mating of these heterozygotes with FLP deleter mice resulted in removal of the region between F1 and F2 sites in the targeting construct that contained the neo cassette to yield an iPLA$_2\beta$ allele that contained Lox2 and Lox3 sites flanking iPLA$_2\beta$ exons 6–8 and that this allele is distinguishable from the WT allele upon PCR analyses. (**E**) is a Southern blot of gene fragments after Bam H1 endonuclease digestion and illustrates production of fragments characteristic of the WT and conditional alleles that establish the genotype of individual mice. WT and conditional iPLA$_2\beta$ heterozygotes were mated to yield conditional allele homozygotes, heterozygotes, and WT allele homozygotes in a 1:2:1 ratio. Those mice were then mated with mice that express Cre recombinase under control of LysM or RIP2 promoters to drive Cre expression selectively in macrophages or pancreatic islet β-cells, respectively. (**F**) illustrates that with conditional allele homozygotes these matings result in selective iPLA$_2\beta$ deletion in only the targeted cell line and not in other tissues. (**G**,**H**) illustrate that iPLA$_2\beta$ mRNA is not produced in the Cre-expressing cell type in mice that are iPLA$_2\beta$ conditional allele homozygotes but that iPLA$_2\beta$ mRNA is produced by non-targeted tissues.

These mice were mated with FLP deleter mice to excise the F1-F2 region that contained the neo cassette [43–45] to yield a conditional iPLA$_2\beta$ allele in which loxP sites L2 and L3 flank exons 6–8 (Figure 1A). Progeny included heterozygotes for conditional and WT iPLA$_2\beta$ alleles, as verified by PCR genotyping with primers in introns between exon 5 and F2 and between L2 and exon 6, respectively, which yielded products of 546 and 364 bp for conditional and WT alleles, respectively (Figure 1D). Genotypes were confirmed by Southern blotting after digestion with restriction endonuclease Bam H1, which cleaves at sites B1, B2 (conditional allele), and B3 (WT allele) to yield fragments of 2544 and 2256 bp for WT and conditional alleles (Figure 1E), respectively, recognized by a probe to the 3′ end of exon 4 and the following intron. Heterozygous (conditional/WT) mice were mated with mice that express Cre recombinase under control of RIP2 or LysM promoters to direct Cre expression in β-cells or macrophages [46–53], respectively. FLP-negative offspring were mated with Cre mice to produce mice homozygous for iPLA$_2\beta$ conditional alleles that express Cre recombinase (Figure 1F) and produce no iPLA$_2\beta$ mRNA in macrophages but do so in other tissues (Figure 1G,H).

Mice were housed in a specific pathogen-free barrier facility with unrestricted access to water and standard mouse chow containing 6% fat. For mice with a β-cell-specific inactivation of iPLA$_2\beta$ (β-iPLA$_2\beta$-KO), RIP2-Cre mice (The Jackson Laboratory, number 003573) were crossed with mice carrying iPLA$_2\beta$ alleles with exons 6–8 flanked by loxP recombination sites (iPLA$_2\beta^{lox/lox}$) [46]. First, generation animals hemizygous for the RIP-Cre gene and bearing one "floxed" LPL allele (iPLA$_2\beta^{lox/wt}$ Cre$^+$) were crossed with iPLA$_2\beta^{lox/lox}$ animals to generate β-cell iPLA$_2\beta$-deficient (iPLA$_2\beta^{lox/lox}$ Cre$^+$) and β-cell iPLA$_2\beta$ wild-type (iPLA$_2\beta^{lox/lox}$ Cre$^-$) littermates that were at least N5 in the C57BL/6 background with a conditional deletion of iPLA$_2\beta$ in β-cells. The following primers were used to document iPLA$_2\beta$ gene rearrangement: primer A, 5′-CCCAGCTCTGTGTCTTAGTATG-3′; primer B, 5′-TTCTTGGCCCAATGGAGTG-3′. Amplification of WT DNA yields a product of 673 bp; amplification of non-rearranged floxed DNA allele yields a product of 855 bp, whereas the amplification of appropriately rearranged DNA will not show a band since the exon 5 is deleted. For mice with a myelomonocytic cell-specific inactivation of iPLA$_2\beta$ (MØ-iPLA$_2\beta$-KO), mice with loxP-flanked iPLA$_2\beta$ alleles were mated with lysozyme M-Cre mice [51] and crossbred to yield iPLA$_2\beta$ knock-out in macrophage (MØ-iPLA$_2\beta$-KO) mice that were at least N5 in the C57BL/6 background with conditional deletion of iPLA$_2\beta$ in the myelomonocytic lineage. Mice were genotyped using iPLA$_2\beta$- and Cre-specific primer sets [54], weaned to chow providing 6% calories as fat, and subsequently fed a high-fat diet (HFD), as described below.

2.3. Analyses of iPLA$_2\beta$ mRNA in Mouse Tissues

Northern blots of iPLA$_2\beta$ mRNA were performed as described in [11]. For RT-PCR, total RNA was isolated with an RNeasy kit (Qiagen Inc.). A SuperScript First Strand Synthesis System (Invitrogen) was used to synthesize cDNA in 20 μL reactions that contained DNase I-treated total RNA (2 μg). The cDNA product was treated (20 min, 37 °C) with RNase H (2 units, Invitrogen), and was heat inactivated (70 °C for 15 min). A reaction without reverse transcriptase was performed to verify the absence of genomic DNA. The PCR performed with the pair of primers 1 and 2 was designed to amplify a fragment that spans the neomycin cassette insertion site. The PCR performed with the pair of primers 3 and 2 was designed to amplify a fragment downstream from the neomycin cassette insertion site. The sequence of primer 1 is tgtgacgtggacagcactagc; that of primer 2 is ccccagagaaacgactatgga; that of primer 3 is tatgcgtggtgtgtacttccg.

2.4. High-Fat Dietary Intervention Studies

Mice were housed in a pathogen-free barrier facility with unrestricted access to water and standard mouse chow (Purina Mills Rodent Chow 5053) with a caloric content of 13.025% fat, 62.144% carbohydrate, and 24.651% protein. For dietary intervention studies, mice were fed standard chow until 8 weeks of age and thereafter were randomized into groups that were fed either standard chow or a HFD continuously until they reached the age of three or six months, respectively, as described in [55].

The HFD (Harlan Teklad catalog TD88137) had a caloric content of 42% fat, 42.7% carbohydrate, and 15.2% protein.

2.5. Blood Glucose and Insulin Concentrations

As described previously [56], blood samples were obtained from the lateral saphenous vein in heparinized capillary tubes, and glucose concentrations were measured in whole blood with a blood-glucose monitor (Becton Dickenson) or an Ascensia ELITE XL blood-glucose meter. Plasma was prepared from heparinized blood by centrifugation, and insulin levels were determined in aliquots (5 μL) with a rat insulin ELISA kit (Crystal Chem). Fasting blood samples were obtained after an overnight fast, and fed blood samples were obtained between 9:00 and 10:00 a.m.

2.6. Glucose and Insulin Tolerance Tests

As described in [23], intraperitoneal glucose tolerance tests (IPGTTs) were performed on mice that fasted overnight from which a baseline blood sample was obtained, followed by intraperitoneal injection of D-glucose (2 mg/g body weight) and collection of blood for measurement of glucose level after 30, 60, and 120 min. Insulin tolerance tests were performed in mice with free access to water and chow that received an intraperitoneal injection (0.75 U/kg body weight) of human regular insulin (Lilly, Indianapolis, IN, USA), followed by collection of blood after 30, 60, and 120 min for glucose level determinations [24,26].

2.7. Area Under the Curve (AUC) Calculations for Glucose Tolerance Tests (GTTs)

As described previously [23,24,57], the AUC for the GTT curves was calculated by the method of Sakaguchi et al. [58], where the blood glucose concentration at t = x min is designated G(x):

$$AUC = [0.25 \times G(0)] + [0.5 \times G(30)] + [0.75 \times G(60)] + [0.5 \times G(120)] \tag{1}$$

2.8. Insulin Secretion In Vivo

Mice were fasted overnight, and baseline blood samples were obtained from the saphenous vein, followed by intraperitoneal injection of D-glucose (3 mg/kg body weight), and a blood sample was obtained 30 min thereafter for the measurement of plasma insulin levels, as described in [57].

2.9. Pancreatic Islet Isolation

Islets were isolated from pancreata removed from mice by collagenase digestion after mincing, followed by Ficoll step density gradient separation, and manual selection under stereomicroscopic visualization to exclude contaminating tissues [46,59].

2.10. Insulin Secretion Ex Vivo from Isolated Pancreatic Islets in Static Incubations

Islets were rinsed with KRB medium containing 3 mM glucose and 0.1% bovine serum albumin and placed in silanized tubes (12 × 75 mm) in the same buffer, through which 95% air/5% CO_2 was bubbled before the incubation. The tubes were capped and incubated (37 °C, 30 min) in a shaking water bath, as described in [24,46,59]. The buffer was then replaced with KRB medium containing 1, 11, or 20 mM glucose and 0.1% BSA without or with forskolin (2.5 μM), and the samples were incubated for 30 min. Insulin secreted into the medium was measured, as described in [46,57].

2.11. Other Analytical Procedures

Serum glucose was measured using reagents from Sigma, and serum insulin was measured by an enzyme-linked immunosorbent assay (Crystal Chem. Inc., Downer's Grove, IL, USA), as described in [46].

2.12. Statistical Methods

Results are presented as mean ± SEM. Data were evaluated by an unpaired, two-tailed Student's *t* test or by an analysis of variance with appropriate post-hoc tests. Significance levels are described in the figure legends.

3. Results

3.1. Mouse Genotype Characterization

As described in the experimental procedures and illustrated in Figure 1, mice homozygous for a floxed-iPLA$_2$β allele were prepared and mated with mice that express Cre recombinase in a restricted set of tissues to produce offspring with conditional iPLA$_2$β gene deletions. Such mice fail to express iPLA$_2$β in tissues that express Cre because the floxed gene is excised by the action of the recombinase, but those mice do express iPLA$_2$β in all other tissues. Two breeding lines of mice with a tissue-selective expression of Cre were used, one of which expresses Cre under control of the Rat Insulin Promoter (RIP) which is active in insulin-secreting pancreatic islet β-cells and in a limited number of other cells but not in the vast majority of cells [46–50]. When mated with mice homozygous for a floxed-iPLA$_2$β allele, some progeny, which are identified by genotyping, fail to express iPLA$_2$β in β-cells, and their genotype is designated β-cell-iPLA$_2$ β-KO. The second breeding line expresses Cre under control of the Lysozyme-M (Lys) promoter that is active in myelomonocytic lineage cells, including monocyte/macrophages [51,52]. When mated with mice homozygous for a floxed-iPLA$_2$β allele, some progeny, again identified by genotyping, fail to express iPLA$_2$β in monocyte/macrophages (MØ), and their genotype is designated MØ-iPLA$_2$β-KO. β-Cell-iPLA$_2$β-KO mice are thus selectively deficient in iPLA$_2$β in β-cells, and MØ-iPLA$_2$β-KO mice are selectively deficient in iPLA$_2$β in monocyte/macrophages. Mice homozygous for a floxed-iPLA$_2$β allele that do not express Cre are designated "Floxed-iPLA$_2$β" and serve as controls when examining the metabolic behavior of the conditional iPLA$_2$β-KO mice.

3.2. Glucose Tolerance Tests

Glucose tolerance tests (GTTs) performed with female mice 6 months of age of various genotypes after consuming food from a regular diet (RD) or high-fat diet (HFD) are illustrated in Figure 2, in which the blood glucose concentration is plotted as a function of time after an intraperitoneal administration of glucose.

Figure 2A shows that for floxed-iPLA$_2$β control mice, glucose tolerance deteriorates significantly in HFD-fed mice compared to RD-fed mice. This effect of diet was also observed in MØ-iPLA$_2$β-KO mice, but the peak glucose concentration and the area under the curve (AUC) of the GTTs were both significantly lower for MØ-iPLA$_2$ β-KO mice than for floxed-iPLA$_2$β controls, suggesting that MØ-selective iPLA$_2$β deficiency confers some protection against diet-induced glucose intolerance.

Figure 2B illustrates that GTTs performed with β-cell-iPLA$_2$ β-KO mice compared to floxed-iPLA$_2$β controls. Again, there is HFD-induced deterioration in GTTs compared to RD-fed mice for the floxed-iPLA$_2$β control mice. This dietary effect was also observed in β-cell-iPLA$_2$β-KO mice. In contrast to MØ-iPLA$_2$β-KO mice, the peak glucose concentration and the Area Under the Curve (AUC) of the GTT were both significantly higher in β-cell-iPLA$_2$β-KO mice than in floxed-iPLA$_2$β controls, suggesting that β-cell-selective iPLA$_2$β deficiency exacerbates diet-induced glucose intolerance.

Similar effects of genotype and diet were observed in male mice 6 months of age, as illustrated in Figure 3, in which the AUC of the GTT is plotted for female (F) or male (M) mice fed a regular (R) or high-fat (HF) diet. Figure 3A shows that for a given diet, males exhibit higher GTT AUC values than females and that a HFD causes deterioration in glucose tolerance, as reflected by a higher AUC, compared to RD-fed mice. For both males and females, the diet-induced rise in GTT AUC was significantly lower for MØ-iPLA$_2$β-KO mice than for floxed-iPLA$_2$β controls. In contrast, the diet-induced rise in GTT was significantly higher for β-KO (RIP) mice than for floxed-iPLA$_2$β controls for both males and females.

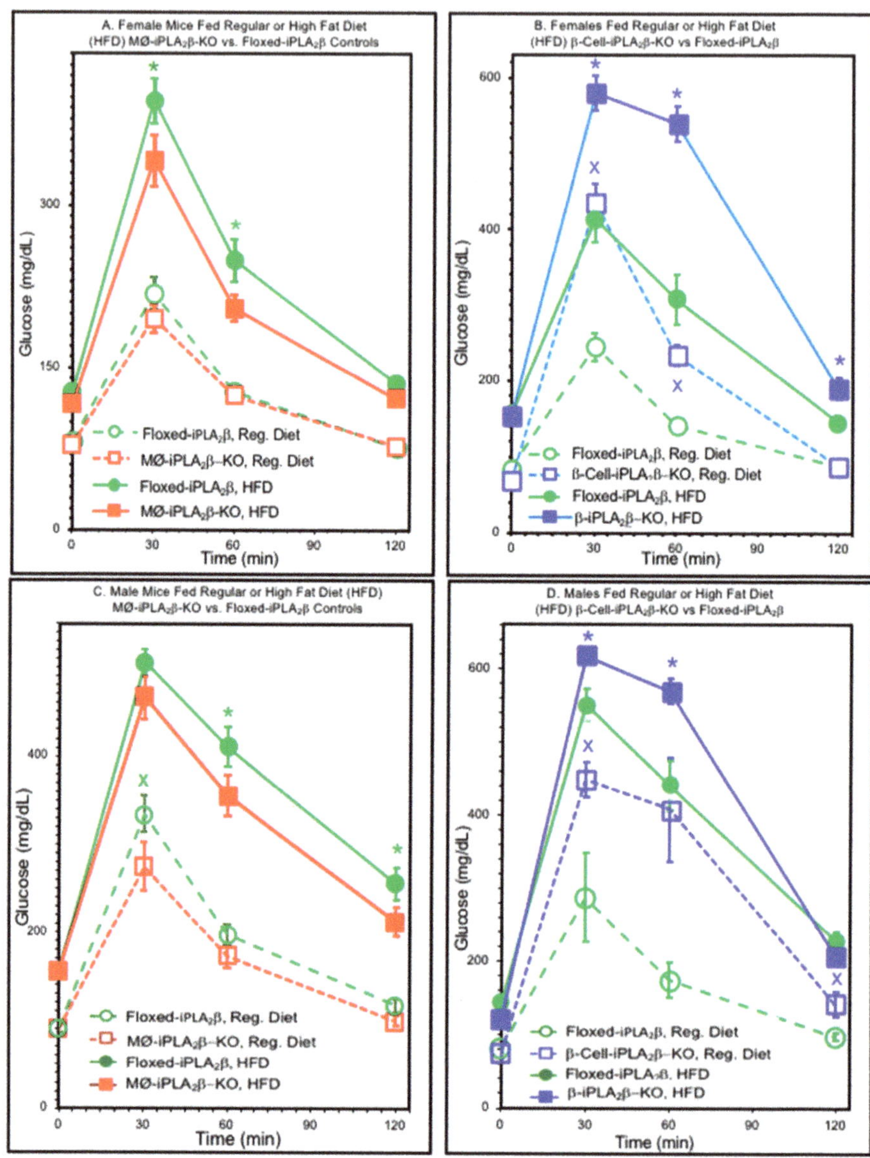

Figure 2. Glucose tolerance tests for iPLA$_2$β conditional knockout mice and floxed-iPLA$_2$β controls. D-glucose (2 mg/g body weight) was administered by intraperitoneal injection to female (**A**,**B**) or male (**C**,**D**) floxed-iPLA$_2$β control mice (circles), MØ- iPLA$_2$β-KO mice (**A**,**C**, squares), or β-cell-iPLA$_2$β-KO mice (**B**,**D**, squares) 6 months of age that had been fed a regular diet (open symbols) or high-fat diet (HFD, closed symbols) after the age of 8 weeks, and blood was collected at baseline and at 30, 60, and 120 min after glucose administration to measure blood glucose concentration. Values are displayed as means ± SEM (n = 6 to 24, as specified by condition in Table S1). An asterisk (*) denotes $p < 0.05$ for comparisons between genotypes. The symbol x denotes $p < 0.05$ for the comparison between diets.

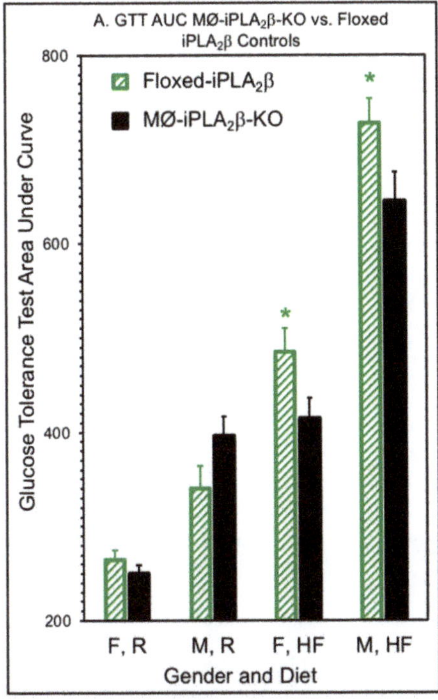

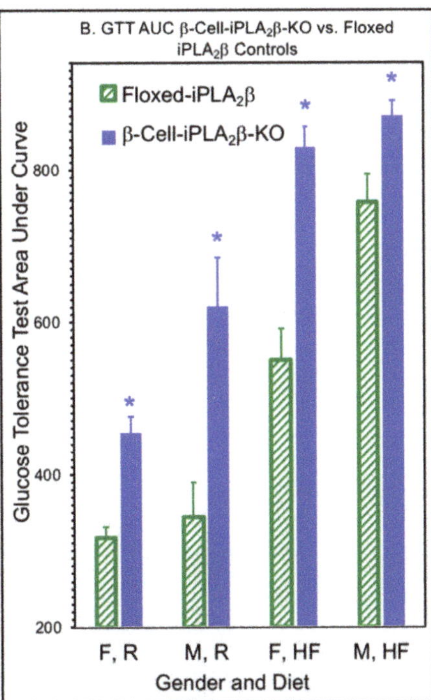

Figure 3. Areas under the curve for glucose tolerance tests for iPLA$_2$β conditional knockout mice and floxed-iPLA$_2$β controls. Glucose tolerance tests (GTTs) were performed as in Figure 2 for male (M) or female (F) MØ-iPLA$_2$β-KO mice (**A**), β-cell-iPLA$_2$β-KO mice (**B**), or floxed-iPLA$_2$β control mice 6 months of age that had been fed a regular (R) or high-fat (HF) diet, and the Areas Under the Curves (AUCs) were calculated from the measured glucose concentration values by the trapezoidal method of Sakaguchi et al. [58]. Values are displayed as means ± SEM (n = 6 to 24, as specified by condition in Table S1). An asterisk (*) denotes $p < 0.05$ for comparisons between genotypes.

Together, Figures 2 and 3 demonstrate that glucose tolerance is affected by diet, gender, and genotype, with a higher GTT AUC for males compared to females and for HFD-fed compared to RD-fed mice. Compared to flox mice, MØ-iPLA$_2$β-KO mice exhibit a significantly lower HFD-diet induced deterioration of GTT than floxed-iPLA$_2$β controls. In contrast, β-cell-iPLA$_2$β-KO mice exhibit significantly poorer glucose tolerance, as reflected by a higher GTT AUC than floxed-iPLA$_2$β controls after the introduction of an RD, and HFD-induced deterioration of glucose tolerance is greater for β-cell-iPLA$_2$β-KO mice than for floxed-iPLA$_2$β controls.

Table S1 illustrates an effect of age on glucose tolerance. Metabolic abnormalities that have developed in mice aged 6 months were found to be nascent but attenuated in mice aged 3 months. For a given condition, GTT AUC is lower for mice aged 3 months than for mice aged 6 months. A HFD also induces deterioration in glucose tolerance for both female and male mice at an age of 3 months, although the effect is smaller than with mice aged 6 months. For female mice aged 3 months, the GTT AUC was significantly lower for MØ-iPLA$_2$β-KO mice than for floxed-iPLA$_2$β control mice after RD-consumption, and the GTT AUC was significantly higher for β-cell-iPLA$_2$ β-KO mice than for floxed-iPLA$_2$β mice after RD-consumption. No other comparisons were statistically significant for female or male mice aged 3 months, although there was a strong trend for higher GTT AUC for HFD-fed β-Cell-iPLA$_2$β-KO mice than for floxed-iPLA$_2$β controls and a trend for higher GTT AUC for RD-fed male β-cell-iPLA$_2$β-KO mice than for floxed-iPLA$_2$β controls. Weaker trends were observed for the

GTT AUC to be lower in HFD-fed female MØ-iPLA$_2$β-KO mice than for floxed-iPLA$_2$β mice and for the GTT AUC to be higher in HFD-fed male β-cell-iPLA$_2$β-KO mice than for floxed-iPLA$_2$β controls.

3.3. *Ex Vivo Insulin Secretion from Isolated Pancreatic Islets*

The magnitude of GSIS from pancreatic islet β-cells has an important influence on glucose tolerance, and the secretory behavior of pancreatic islets isolated from RD-fed or HFD-fed mice of various genotypes was therefore examined ex vivo, as illustrated in Figure 4. Islets isolated from both male and female RD-fed floxed-iPLA$_2$β control mice exhibited insulin secretion that increased with medium glucose concentration over a range from 1 to 20 mM, and this response was amplified in the presence of the adenylyl cyclase activator, forskolin (Figure 4A–D), as observed in previous studies [22–24]. Insulin secretory responses from RD-fed MØ-iPLA$_2$β-KO male or female mice were not statistically different from those for floxed-iPLA$_2$β controls (Figure 4A,B). In contrast, insulin secretory responses from RD-fed β-cell-iPLA$_2$β-KO male or female mice were significantly lower from those for floxed-iPLA$_2$β controls (Figure 4C,D), and this was also observed in islets from HFD-fed β-cell-iPLA$_2$β-KO male and female mice (Figure 4D,E), which exhibited even lower insulin secretory responses relative to floxed-iPLA$_2$β controls than did islets from RD-fed mice. These observations indicate that selective iPLA$_2$β deficiency in β-cells results in impaired insulin secretion from pancreatic islets but that selective iPLA$_2$β deficiency in MØ does not. The impaired islet insulin secretion of β-cell-iPLA$_2$β-KO mice in Figure 4C–F is thus likely to contribute to the impaired glucose tolerance in these mice (Figures 2B and 3B).

3.4. *In Vivo Insulin Secretion in Mice After Intraperitoneal Glucose Administration*

To determine whether insulin secretion in vivo would reflect the effects of genotype on insulin secretion ex vivo from isolated pancreatic islets, we determined the increment in blood insulin concentration that occurred 30 min after an intraperitoneal administration of glucose to mice of various genotypes, gender, and dietary history (Figure 5). Both RD-fed male and HFD-fed female floxed-iPLA$_2$β control mice exhibited a significant increment in blood insulin concentrations after an intraperitoneal administration of glucose (Figure 5A,B and Figure S1). This was also true for MØ-iPLA$_2$β-KO mice, and the magnitude of the rise in blood insulin level was similar and not significantly different when comparing MØ-iPLA$_2$β-KO mice to floxed-iPLA$_2$β controls. This is consistent with the observation that insulin secretion from pancreatic islets isolated from MØ-iPLA$_2$β-KO mice is not impaired compared to floxed-iPLA$_2$β controls (Figure 4).

In contrast, neither RD-fed female or HFD-fed male β-cell-iPLA$_2$β-KO mice exhibited a significant increase in blood insulin levels 30 min after intraperitoneal glucose administration (Figure 5A,B and Figure S1). This is consistent with the impairment of ex vivo glucose-dependent insulin secretion that was observed with pancreatic islets isolated from β-cell-iPLA$_2$β-KO mice relative to floxed-iPLA$_2$β controls (Figure 4), suggesting that inadequate insulin secretion contributes to the substantially impaired glucose tolerance observed with β-cell-iPLA$_2$β-KO mice (Figures 2 and 3).

3.5. *Insulin Tolerance Tests*

In addition to the magnitude of GSIS from pancreatic islets, the responsivity of peripheral tissues, including skeletal muscle, to insulin is an important determinant of glucose tolerance. To evaluate insulin responsivity of mice of various genders, genotypes, and dietary history, insulin tolerance tests (ITTs) were performed by measuring the blood glucose concentrations at various times after an intraperitoneal injection of a fixed dose of insulin, as illustrated in Figure 6.

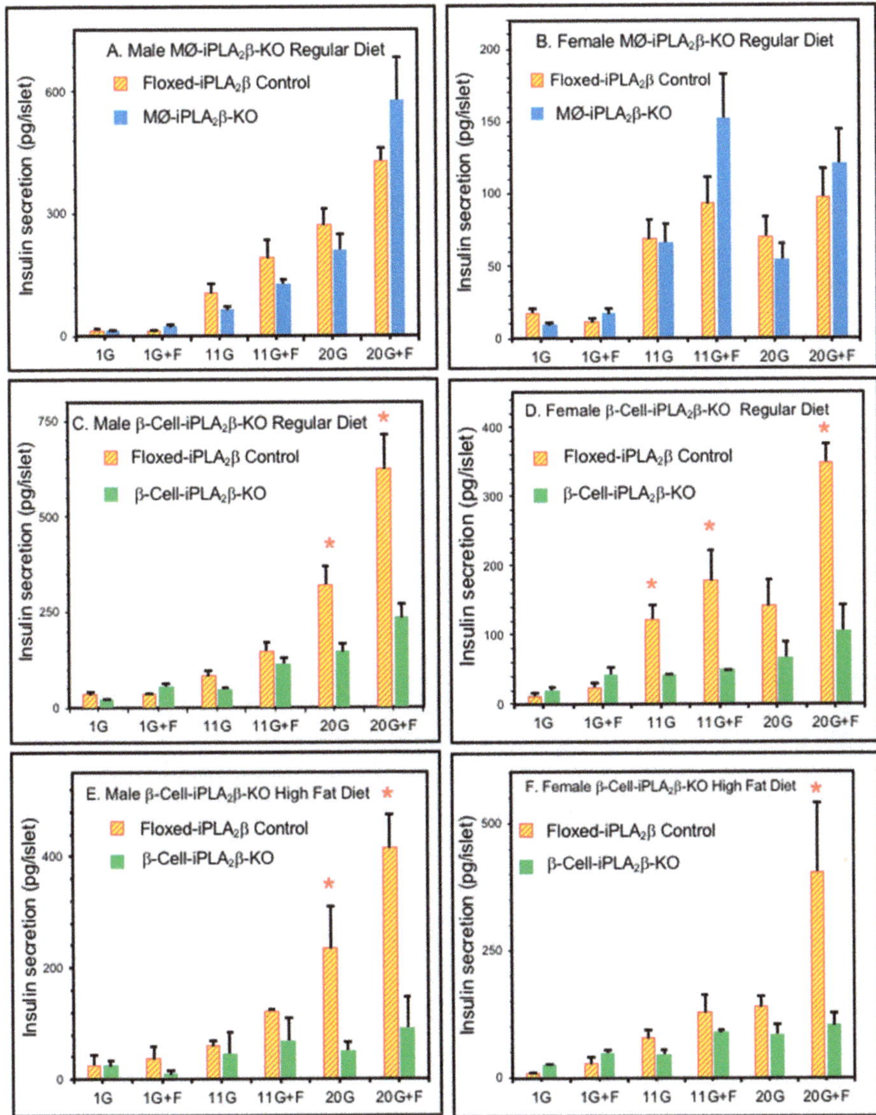

Figure 4. Insulin secretion by pancreatic islets isolated from iPLA$_2$β conditional knockout mice and floxed-iPLA$_2$β controls. Insulin secretion was stimulated by D-glucose and forskolin from pancreatic islets isolated from male (**A**,**C**,**E**) or female (**B**,**D**,**F**) MØ-iPLA$_2$β-KO mice (**A**,**B**, solid bars), β-cell-iPLA$_2$β-KO mice (**C–F**, solid bars), or floxed-iPLA$_2$β control mice (**A–F**, cross-hatched bars) 6 months of age that had been fed a RD (**A–D**) or a HFD (**E**,**F**). Incubations (30 islets per condition, 30 min, 37 °C) were performed in buffer containing 1, 11, or 20 mM D-glucose without or with 2.5 μM forskolin, and an aliquot of medium was then removed for measurement of insulin. Mean values ± SEM are displayed (n = 4, in triplicate). An asterisk (*) denotes $p < 0.05$ for the comparison between genotypes.

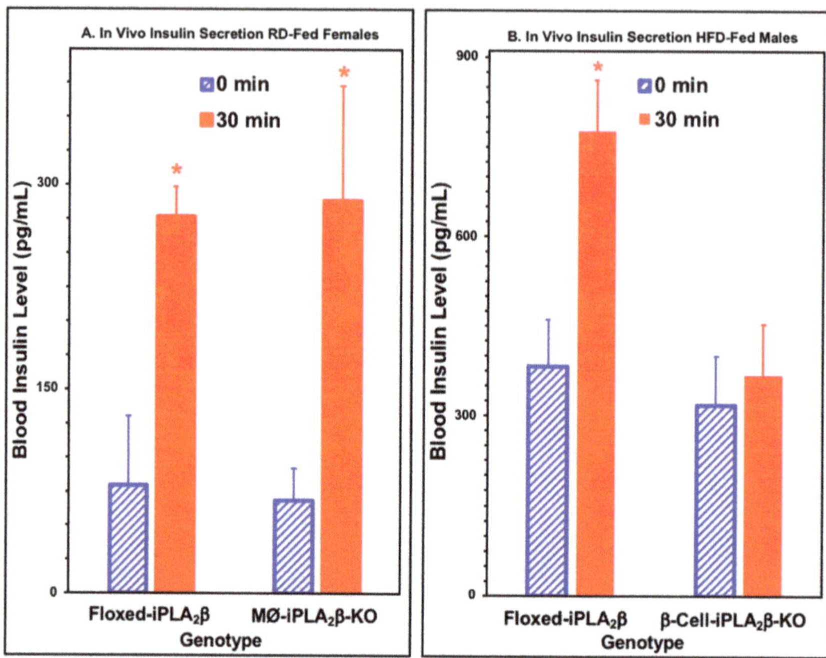

Figure 5. In vivo insulin secretion for iPLA$_2$β conditional knockout mice and floxed-iPLA$_2$β control mice fed a regular (RD) or high-fat diet (HFD). After an overnight fast, baseline blood samples were obtained from the saphenous vein of male (**A**) or female (**B**) MØ-iPLA$_2$β-KO mice, β-cell-iPLA$_2$β-KO mice, or floxed-iPLA$_2$β control mice 6 months of age that had been fed a regular diet (RD, **A**) or high-fat diet (HFD, **B**). D-glucose (3 mg/kg body weight) was administered by intraperitoneal injection, and a blood sample was obtained 30 min thereafter. The insulin contents of the baseline and 30 min samples were then measured by enzyme-linked immunosorbent assay, as described [23,24,57]. Displayed values represent mean ± SEM. An asterisk (*) denotes $p < 0.05$ for the comparison between the time 0 and 30 min values (n = 5 to 28).

For female mice aged 3 months either RD- or HFD-fed, the ITT curves for neither MØ-iPLA$_2$β-KO nor β-cell-iPLA$_2$β-KO conditional knockout mice were statistically distinguishable from those for floxed-iPLA$_2$β control mice. This was also true for 3-month-old male MØ-iPLA$_2$β-KO conditional knockout mice and for HFD-fed 3-month-old male β-cell-iPLA$_2$β-KO conditional knockout mice compared to floxed-iPLA$_2$β control mice. For 3-month-old male β-cell-iPLA$_2$β-KO conditional knockout mice fed a RD, the ITT curves showed a slightly but significantly superior insulin sensitivity compared to floxed-iPLA$_2$β control mice (not shown), and this difference was magnified further at age 6 months, as described below.

As illustrated in Figure 6, at 3 months of age the ITTs of male HFD-fed floxed-iPLA$_2$β control mice and MØ-iPLA$_2$β-KO mice (Figure 6A) did not differ significantly, and this was also true for β-cell-iPLA$_2$β-KO mice compared to floxed-iPLA$_2$β control mice aged 3 months (Figure 6C). The deterioration of insulin sensitivity with age occurs, and by 6 months of age, a statistically significant difference between the ITTs for HFD-fed male MØ-KO and floxed-iPLA$_2$β control mice had developed (Figure 6B). The 6-month-old HFD-fed MØ-iPLA$_2$β-KO mice achieved significantly lower blood glucose levels than floxed-iPLA$_2$β control mice at 30 and 60 min after insulin administration (Figure 6B). This is consistent with the hypothesis that motivated the preparation of the MØ-iPLA$_2$β-KO mice—that HFD feeding would induce a lower deterioration in glucose tolerance and insulin sensitivity in MØ-iPLA$_2$β-KO mice compared to floxed-iPLA$_2$β control mice, possibly because of impaired migration of monocytes into peripheral tissues where their differentiation into cytokine-producing macrophages ordinarily contributes to insulin resistance. Similar to the phenomenon described above for 3-month-old male β-cell-iPLA$_2$β-KO conditional knockout RD-fed mice, at 6 months of age HFD-fed β-cell-iPLA$_2$β-KO mice were also significantly more responsive to insulin than floxed-iPLA$_2$β control mice (Figure 6D), raising the unanticipated possibility that β-cell products might also contribute to the development of HFD-induced insulin resistance.

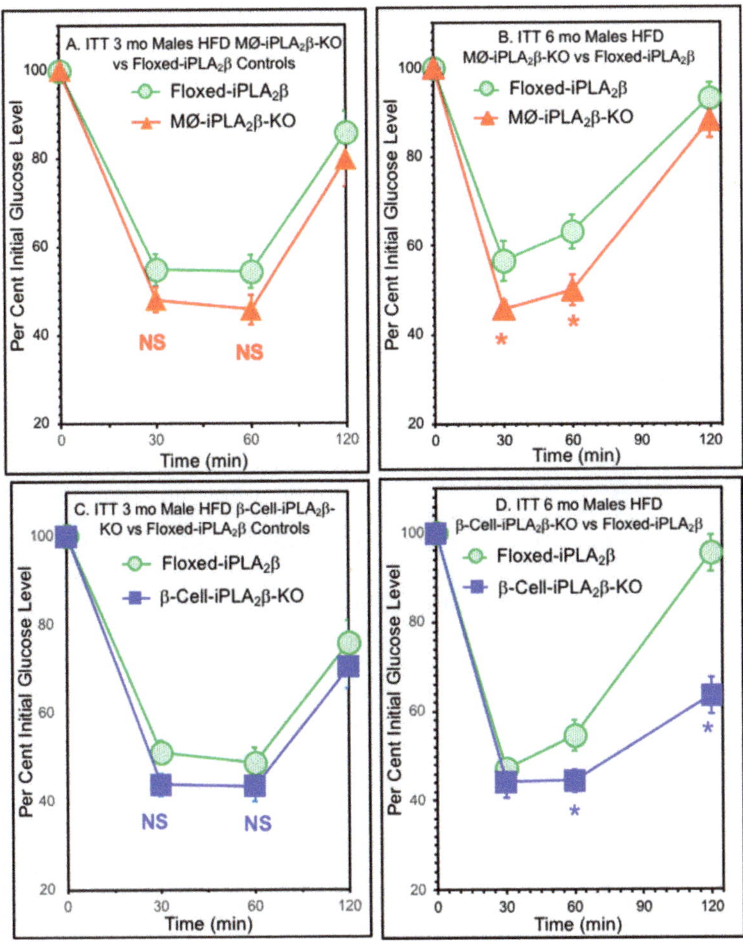

Figure 6. Insulin tolerance tests of 3- or 6-month-old male conditional iPLA$_2$β-knockout mice and floxed-iPLA$_2$β controls fed a regular or high-fat diet. Male floxed-iPLA$_2$β control mice (circles), MØ-iPLA$_2$β-KO mice (triangles), or β-cell-iPLA$_2$β-KO mice (squares) mice were fed a regular diet after weaning until they were 6 weeks of age and were then fed a HFD until aged 3 months (**A,C**) or 6 months (**B,D**). Insulin tolerance tests were then performed in mice with free access to water and chow until human regular insulin (0.75 U/kg body weight; Lilly, Indianapolis, IN) was administered by intraperitoneal injection. Blood specimens were collected at 0, 30, 60, and 120 min thereafter for glucose concentration measurements, which were expressed as a percentage of the time zero blood glucose concentration, as described in [23,24,57]. Displayed values represent mean ± SEM (n = 13 to 16 (**A**), n = 15 to 18 (**B**), n = 22 to 26 (**C**), and n = 21 to 22 (**D**). An asterisk (*) denotes $p < 0.05$ for the comparison between genotypes.

A similar phenomenon was observed in 6-month-old HFD-fed female β-cell-iPLA$_2$β-KO mice, which exhibited superior insulin sensitivity compared to floxed-β-cell-iPLA$_2$β-KO control mice, and this magnified a smaller but similar trend observed with 6-month-old female β-cell-iPLA$_2$β-KO mice fed regular chow (Figure S2A,B). For 6-month-old female MØ-iPLA$_2$β-KO mice, the ITT curves did not differ significantly from those of floxed-iPLA$_2$β control mice fed either a RD or a high-fat diet HFD (Figure S2C,D). Curiously, 6-month-old male MØ-iPLA$_2$β-KO RD-fed mice exhibited lower insulin sensitivity compared to floxed-iPLA$_2$β control mice (Figure S3), and this phenomenon has been observed previously with global iPLA$_2$β-KO mice fed RD compared to controls [23]. In contrast, 6-month-old HFD-fed MØ-iPLA$_2$β-KO male mice exhibited superior insulin sensitivity compared to floxed-iPLA$_2$β control mice (Figure S3), which may reflect less HFD-induced deterioration in insulin sensitivity for 6-month-old HFD-fed MØ-iPLA$_2$β-KO male mice than that which occurred for floxed-iPLA$_2$β control male mice. Discrepant ITT findings between male and female global iPLA$_2$β-KO mice and their responses to dietary stress have also been observed previously [23,24] and are commonplace in animal models of perturbed glucose homeostasis.

4. Discussion

Our previous studies with global iPLA$_2$β-knockout mice indicated that disturbances in glucose homeostasis, which included glucose intolerance and impaired insulin secretion by pancreatic islet β-cells, occurred as a consequence of iPLA$_2$β gene disruption [23,24]. Although iPLA$_2$β and its products might participate in multiple events in GSIS from β-cells, one of them is to amplify depolarization-induced [Ca^{2+}] entry into the β-cell. In addition, iPLA$_2$β appears to confer protection against lipid injury to β-cells that may reflect iPLA$_2$β participation in the repair of oxidative damage to β-cell mitochondrial membranes that occurs in the context of lipid toxicity [24,25]. Loss of these actions of iPLA$_2$β in β-cells provides a plausible explanation for the impaired glucose tolerance in global iPLA$_2$β-null mice, the reduced insulin secretory response to glucose of islets isolated from global iPLA$_2$β-null mice compared to their wild-type littermates, and the exaggerated deterioration in glucose tolerance for global iPLA$_2$β-null mice fed a HFD compared to wild-type controls [23,24].

More puzzling is the superior sensitivity to insulin of global iPLA$_2$β-null mice compared to wild-type controls in insulin tolerance tests [24], implying that iPLA$_2$β gene deletion has opposing effects on insulin secretion from β-cells and the insulin responsiveness of peripheral tissues, including skeletal muscle. One potential explanation for these findings is that iPLA$_2$β deficiency has distinct effects in different cells. A loss of iPLA$_2$β in β-cells would reasonably be expected to impair insulin secretion and to amplify lipid-induced β-cell injury, but a loss of iPLA$_2$β activity is some other cell type might be responsible for the amelioration of lipid-induced deterioration in insulin sensitivity of global iPLA$_2$β-null mice compared to wild-type controls. Candidates include cells of the monocyte/macrophage lineage. It is postulated that in the setting of lipid stress, the migration of blood monocytes into extravascular sites, including adipose tissue, and their differentiation into tissue macrophages results in the elaboration of cytokines, including IL-1β, IL-6 and TNFα, which impair insulin sensitivity. Because iPLA$_2$β-derived 2-lysophosphatidic acid (LPA) appears to be required for monocyte migration in response to the cytokine Monocyte Chemoattractant-1 (MCP-1) [29,37–39], we postulated that iPLA$_2$β-deficiency in cells of the monocyte/macrophage lineage might confer protection against the HFD-induced deterioration of insulin sensitivity because of a failure of iPLA$_2$β-null monocytes to migrate into peripheral tissues and differentiate into macrophages.

We therefore prepared mice that are selectively deficient in iPLA$_2$β in β-cells or cells of the monocyte/macrophage lineage. Mice with "floxed" iPLA$_2$β genes were prepared using embryonic stem cells from the EUCOMM consortium. These cells had incorporated a targeting construct by homologous recombination that replaced the wild-type iPLA$_2$β gene. In this construct, LoxP sites recognized by Cre recombinase [41,42] flanked critical iPLA$_2$β gene exons. Mice with the "floxed" iPLA$_2$β genes were then mated with mice that express Cre recombinase under the control of promoters that are expressed in only a restricted set of cells. In cells that express Cre, the action of the recombinase removes the "floxed" segment of the iPLA$_2\beta$ gene-targeting construct, and transcription of this modified gene yields a truncated mRNA that is eliminated by a nonsense mediated decay and cannot lead to production of an active iPLA$_2\beta$ protein. In this way, mice were produced that are selectively iPLA$_2\beta$-deficient in specifically targeted populations of cells that express Cre recombinase.

To direct Cre expression in β-cells, RIP-Cre mice were used that express Cre under control of the Rat Insulin 2 Promoter, which has been demonstrated to be appropriate for β-cell-specific gene deletion when used in mice on a pure C57BL/6J background [49], as is the case in this study, although anomalies may be encountered on other genetic backgrounds [53]. The use of β-cell-targeting promoters has been reviewed [60], and the use of RIP-Cre remains a popular means of directing β-cell-restricted gene deletion [46,50,61–69]. The LysM promoter is expressed in all myeloid cells, and LysM-Cre is widely used to study conditional macrophage-myeloid cell gene deletions [52,70–76].

Our expectations for the behavior of these conditional iPLA$_2\beta$-KO mouse lines was that the β-cell-iPLA$_2\beta$-KO mice would exhibit glucose intolerance, impaired insulin secretion, and increased deterioration in glucose tolerance compared to control mice in response to a HFD, and our observations largely conformed to these expectations. Expectations for the MØ-iPLA$_2\beta$-KO mice were that insulin secretion and glucose tolerance would be preserved, that deterioration in glucose tolerance induced by a HFD would be blunted compared to floxed-iPLA$_2\beta$ control mice, and that insulin sensitivity as measured in insulin tolerance tests (ITTs) would be impaired less in HFD-fed MØ-iPLA$_2\beta$-KO compared to floxed-iPLA$_2\beta$ control mice, and our observations also largely conformed to these expectations.

An unexpected finding is that HFD-fed β-cell-iPLA$_2\beta$-KO mice also had superior insulin sensitivity compared to floxed-iPLA$_2\beta$ control mice as assessed with ITTs. The explanation for this finding has not been established, but of possible relevance is the phenomenon of selective insulin resistance first proposed by McGarry [75] and subsequently discussed by Brown and Goldstein [76] and others [77,78]. McGarry proposed that hyperinsulinemia itself elicits insulin resistance in peripheral tissues [75]. During the evolution of T2D (type 2 diabetes), resistance develops to the effect of insulin to decrease hepatic gluconeogenesis but such resistance does not develop the insulin-stimulated lipogenesis of fatty acids and triacylglycerols (TAGs), which is now known to involve the activation of the transcription factor SREBP-1c [76–78]. In fact, hepatic lipogenesis and Very Low Density Lipoprotein (VLDL) secretion are amplified by the hyperinsulinemia that results from the resistance of peripheral tissues to the action of insulin, and fatty acids derived from VLDL TAG exacerbate the insulin-resistant state in muscle and adipose tissue [75–79]. These events result in the classic T2DM triad of hyperglycemia, hyperinsulinemia, and hypertriglyceridemia [75,76]. McGarry suggested that VLDL TAG play a toxic role in the evolution of T2D by their deposition in muscle, where they enhance insulin resistance, and in liver, where they can produce non-alcoholic steatohepatitis (NASH). The term lipotoxicity is used to describe the detrimental effects of accumulation of TAG and other lipids in various tissues [80], and ceramides are thought to be among the lipid mediators that contribute to insulin resistance [79].

If hyperinsulinemia is involved in inducing peripheral tissue insulin resistance, then diabetes in a mouse model caused solely by a β-cell secretory defect might fail to generate sufficiently high levels of hyperinsulinemia to drive development of insulin resistance. Compared to a control population of mice that do develop sufficient hyperinsulinemia, mice with the pure secretory defect might thus exhibit superior insulin sensitivity in insulin tolerance tests. It is of interest in this regard that knockout mice with defective K_{ATP} channel activity also have impaired insulin secretion from β-cells but enhanced insulin sensitivity of peripheral tissues in insulin tolerance tests [81], which the authors postulated might be mediated by altered extracellular hormonal or neuronal signals perturbed by disruption of K_{ATP} channels [82]. In the context of the discussion in this and the preceding paragraph, one such altered extracellular hormonal signal might be the blood insulin concentration itself.

The possibility that the insulin secretory defect of global iPLA$_2\beta$-knockout mice and of β-cell-iPLA$_2\beta$-KO mice prevents the hyperinsulinemia-driven deposition of toxic lipids in tissues that otherwise occurs in mice subjected to a HFD is of interest in the context of observations that hepatic steatosis fails to develop in global iPLA$_2\beta$-knockout mice subjected to a HFD, although it does occur in control mice [83]. Gene deletion of iPLA$_2\beta$ also greatly attenuates hepatic steatosis that otherwise occurs in the *ob/ob* mouse genetic model [84]. In addition, ceramides such as those that accumulate in tissues that develop insulin resistance [79,80], also accumulate in β-cells subjected to ER stress in an iPLA$_2\beta$-dependent process that involves activation of SREBP-1 [85–88], which is of interest in the context of involvement of SREBP-1 activation in the pathogenesis of hyperinsulinemia-driven amplification of hepatic lipogenesis in the evolution of T2D [76–78].

In contrast to the unexpected resistance to HFD-induced deterioration in insulin-responsivity of β-cell-iPLA$_2\beta$-KO mice discussed above, it was expected that HFD-fed MØ-iPLA$_2\beta$-KO mice would exhibit superior insulin sensitivity compared to HFD-fed floxed-iPLA$_2\beta$ control mice as assessed with ITTs. This expectation was based on the postulates that HFD-induced deterioration in insulin sensitivity involves migration of blood monocytes into extravascular sites, such as adipose tissue, where they differentiate into macrophages and produce cytokines, including IL-β, IL-6, and TNFα, that impair insulin sensitivity in a process that involves the tissue elaboration of the cytokine Monocyte Chemoattractant-1 (MCP1) and its interaction with the CCR2 receptor on monocytes to induce their migration into tissues [31–36]. Genetic and pharmacologic evidence indicates that the monocyte chemotactic response to MCP-1 requires the action of iPLA$_2\beta$ to produce the lipid mediator 2-lysophosphatidic acid (LPA) [37–39]. Macrophages [29,30] and other cells [89] from our global iPLA$_2\beta$ knockout mice have been demonstrated to have impaired production of LPA in response to stimuli that induce robust LPA production in wild-type cells.

Moreover, macrophages from our global iPLA$_2\beta$ knockout mice have impaired migratory responses to MCP-1 in a mouse model of diet-induced glucose intolerance and atherogenesis, and these migratory responses are restored by provision of exogenous LPA [29]. The accumulation of macrophages in atherosclerotic lesions in this mouse model of diabetic stress is also associated with increased lesion content of iPLA$_2\beta$ immunoreactive protein and enzymatic activity [29]. Migratory responses of vascular smooth muscle cells (VSMCs) in a model of vascular injury are also greatly reduced in our global iPLA$_2\beta$-null mice compared to wild-type mice [90]. The activation state of iPLA$_2\beta$-null macrophages from our global iPLA$_2\beta$ knockout mice is also shifted toward an M2 anti-inflammatory phenotype compared to the inflammatory M1 phenotype exhibited by wild-type macrophages, and this is associated with reduced TNFα production by the iPLA$_2\beta$-null macrophages [30,91–93], which suggests that in addition to participating in monocyte migration into extravascular tissues, iPLA$_2\beta$ may also be involved in their differentiation into pro-inflammatory macrophages that elaborate cytokines that impair insulin sensitivity. The pharmacologic inhibition of iPLA$_2\beta$ also ameliorates leukocyte infiltration into pancreatic islets and the onset of diabetes in NOD (non-obese diabetic) mice [92,93], which suggests that iPLA$_2\beta$ may participate in the evolution of T1D in addition to that of T2DM.

iPLA$_2\beta$ is in fact involved in regulating several fundamental aspects of monocyte/macrophage biology in addition to those discussed above, including the remodeling the fatty acid composition of macrophage phospholipids [94,95], selection of macrophage phospholipid pools from which to mobilize fatty acids upon cellular stimulation [27,96,97], governing the proliferation state of macrophage precursor cells [97], regulating macrophage apoptosis in the context of lipid stress [98], regulating the rate of transcription of the macrophage inducible nitric oxide synthase (iNOS) gene [26], and promoting macrophage adhesion and spreading in response to the engagement of the class A scavenger receptor [99]. Development of the monocyte/macrophage-selective conditional iPLA$_2\beta$-knockout mice described here may prove to be useful in further characterizing these aspects of monocyte/macrophage biology in models of diabetes, atherosclerosis, and other pathophysiologic states.

In addition, preparation of the mouse line with floxed-iPLA$_2\beta$ genes described here may prove to be useful in the development of conditional knockout mouse lines with selective deletion of iPLA$_2\beta$ in other cell lines that might participate in the development of T2DM, including skeletal muscle cells, adipocytes, and hepatocytes. All of these cell types are involved in glucose homeostasis in T2D, and all express iPLA$_2\beta$ [83,84,100,101], which is involved in regulating fatty acid oxidation in skeletal muscle [100], differentiation of adipoctyes [101], and hepatic lipid synthesis [83,84]. Conditional knockouts in skeletal muscle can be prepared from our iPLA$_2$β-floxed mice with mice that express Cre under the control of the promoter for muscle creatine kinase (MCK) [102], conditional knockouts in adipocytes can be prepared with our floxed mice and mice that express Cre under control of the promoter for adipocyte-specific fatty acid binding protein (aP2) [103]. Conditional knockouts in hepatocytes can be prepared with our floxed mice and mice that express Cre under control of the promoter for rat albumin [54]. Such lines would permit further characterization of the mechanism of resistance of the global iPLA$_2$β knockout to HFD-induced deterioration in insulin sensitivity [24].

5. Conclusions

Cre-lox technology was used to produce mice with selective iPLA$_2$β deficiency in cells of myelomonocytic lineage, including macrophages (MØ-iPLA$_2$β-KO), or in insulin-secreting β-cells (β-Cell-iPLA$_2$β-KO), respectively. MØ-iPLA$_2$β-KO mice exhibited normal glucose tolerance when fed standard chow and better glucose tolerance than floxed-iPLA$_2$β control mice after consuming a high-fat diet (HFD). MØ-iPLA$_2$β-KO mice exhibited normal GSIS in vivo and from isolated islets ex vivo compared to floxed-iPLA$_2$β controls. Male MØ-iPLA$_2$β-KO mice exhibited enhanced insulin responsivity vs. controls after a prolonged HFD. β-Cell-iPLA$_2$β-KO mice exhibited impaired glucose tolerance when fed standard chow, and glucose tolerance deteriorated further after introduced to a HFD. β-Cell-iPLA$_2$β-KO mice exhibited impaired GSIS in vivo and from isolated islets ex vivo vs. controls, and male β-cell-iPLA$_2$β-KO mice also exhibited enhanced insulin responsivity compared to controls after of introduction of the HFD. These findings suggest that MØ iPLA$_2$β participates in HFD-induced deterioration in glucose tolerance, and that this mainly reflects an effect on insulin responsivity rather than on insulin secretion. In contrast, β-cell iPLA$_2$β plays a role in GSIS and also appears to confer some protection against deterioration in β-cell function induced by a HFD.

Supplementary Materials: The following are available online at http://www.mdpi.com/2218-273X/10/10/1455/s1, Figure S1: In Vivo Insulin Secretion For iPLA$_2$β Conditional Knockout and Floxed-iPLA$_2$β Control Mice Fed a Regular or High-Fat Diet, Figure S2: Insulin Tolerance Tests of 6-month-old Female Conditional iPLA$_2$β-Knockout Mice and Floxed-iPLA$_2$β Controls Fed a Regular or High-Fat Diet (HFD), Figure S3: Insulin Tolerance Tests of 6-month-old Male Conditional iPLA$_2$β-Knockout Mice and Floxed-iPLA$_2$β Controls Fed a Regular or High-Fat Diet (HFD). Table S1: Glucose Tolerance Tests for Female and Male Mice Aged 3 or 6 months with Genotypes Floxed-iPLA$_2$β, β-Cell-iPLA$_2$β-KO, or MØ-iPLA$_2$β-KO Fed a Regular or High-Fat Diet.

Author Contributions: Conceptualization, J.T., H.S. and S.R.; methodology, H.S., M.W. and X.L.; software, H.S.; validation, H.S., M.W. and X.L.; formal analysis, J.T., H.S. and S.R.; investigation, H.S. and M.W.; resources, J.T. and S.R.; data curation, M.W., H.S., C.F., and J.T.; writing—original draft preparation, J.T.; writing—review and editing, S.R.; visualization, J.T., C.F., and S.R.; supervision, J.T. and S.R.; project administration, J.T. and S.R.; funding acquisition, J.T. and S.R. All authors have read and agreed to the published version of the manuscript.

Funding: These studies were supported in part by United States Public Health Service Grants R37-DK34388, P60-DK20579, P30 DK020579, P30 DK056341, P41 RR000954, R24GM136766, P41GM103422, DK110292, R21 AI 146743, and a grant from the Dean's fund from the Washington University School of Medicine.

Acknowledgments: The intellectual input from Zhongmin Ma and Shunzhong Bao in discussions of these data is gratefully acknowledged, as is the technical assistance of Alan Bohrer and Min Tan.

Conflicts of Interest: The authors declare no conflict of interest.

References

1. Mouchlis, V.D.; Dennis, E.A. Phospholipase A_2 catalysis and lipid mediator lipidomics. *Biochim. Biophys. Acta Mol. Cell Biol. Lipids* **2019**, *1864*, 766–771. [CrossRef] [PubMed]
2. Turk, J.; White, T.D.; Nelson, A.J.; Lei, X.; Ramanadham, S. iPLA$_2$beta and its role in male fertility, neurological disorders, metabolic disorders, and inflammation. *Biochim. Biophys. Acta Mol. Cell Biol. Lipids* **2019**, *1864*, 846–860. [CrossRef] [PubMed]
3. Balboa, M.A.; Balsinde, J.; Jones, S.S.; Dennis, E.A. Identity between the Ca^{2+}-independent phospholipase A_2 enzymes from P388D1 macrophages and Chinese hamster ovary cells. *J. Biol. Chem.* **1997**, *272*, 8576–8580. [CrossRef] [PubMed]

4. Ma, Z.; Ramanadham, S.; Kempe, K.; Chi, X.S.; Ladenson, J.; Turk, J. Pancreatic islets express a Ca^{2+}-independent phospholipase A_2 enzyme that contains a repeated structural motif homologous to the integral membrane protein binding domain of ankyrin. *J. Biol. Chem.* **1997**, *272*, 11118–11127. [CrossRef]
5. Tang, J.; Kriz, R.W.; Wolfman, N.; Shaffer, M.; Seehra, J.; Jones, S.S. A novel cytosolic calcium-independent phospholipase A_2 contains eight ankyrin motifs. *J. Biol. Chem.* **1997**, *272*, 8567–8575. [CrossRef] [PubMed]
6. Jenkins, C.M.; Mancuso, D.J.; Yan, W.; Sims, H.F.; Gibson, B.; Gross, R.W. Identification, cloning, expression, and purification of three novel human calcium-independent phospholipase A_2 family members possessing triacylglycerol lipase and acylglycerol transacylase activities. *J. Biol. Chem.* **2004**, *279*, 48968–48975. [CrossRef] [PubMed]
7. Mancuso, D.J.; Jenkins, C.M.; Gross, R.W. The genomic organization, complete mRNA sequence, cloning, and expression of a novel human intracellular membrane-associated calcium-independent phospholipase A_2. *J. Biol. Chem.* **2000**, *275*, 9937–9945. [CrossRef] [PubMed]
8. Rydel, T.J.; Williams, J.M.; Krieger, E.; Moshiri, F.; Stallings, W.C.; Brown, S.M.; Pershing, J.C.; Purcell, J.P.; Alibhai, M.F. The crystal structure, mutagenesis, and activity studies reveal that patatin is a lipid acyl hydrolase with a Ser-Asp catalytic dyad. *Biochemistry* **2003**, *42*, 6696–6708. [CrossRef]
9. Wilson, P.A.; Gardner, S.D.; Lambie, N.M.; Commans, S.A.; Crowther, D.J. Characterization of the human patatin-like phospholipase family. *J. Lipid Res.* **2006**, *47*, 1940–1949. [CrossRef]
10. Kienesberger, P.C.; Oberer, M.; Lass, A.; Zechner, R. Mammalian patatin domain containing proteins: A family with diverse lipolytic activities involved in multiple biological functions. *J. Lipid Res.* **2009**, *50*, S63–68. [CrossRef]
11. Bao, S.; Miller, D.J.; Ma, Z.; Wohltmann, M.; Eng, G.; Ramanadham, S.; Moley, K.; Turk, J. Male mice that do not express group VIA phospholipase A_2 produce spermatozoa with impaired motility and have greatly reduced fertility. *J. Biol. Chem.* **2004**, *279*, 38194–38200. [CrossRef]
12. Malik, I.; Turk, J.; Mancuso, D.J.; Montier, L.; Wohltmann, M.; Wozniak, D.F.; Schmidt, R.E.; Gross, R.W.; Kotzbauer, P.T. Disrupted membrane homeostasis and accumulation of ubiquitinated proteins in a mouse model of infantile neuroaxonal dystrophy caused by PLA2G6 mutations. *Am. J. Pathol.* **2008**, *172*, 406–416. [CrossRef] [PubMed]
13. Khateeb, S.; Flusser, H.; Ofir, R.; Shelef, I.; Narkis, G.; Vardi, G.; Shorer, Z.; Levy, R.; Galil, A.; Elbedour, K.; et al. PLA2G6 mutation underlies infantile neuroaxonal dystrophy. *Am. J. Hum. Genet.* **2006**, *79*, 942–948. [CrossRef]
14. Morgan, N.V.; Westaway, S.K.; Morton, J.E.; Gregory, A.; Gissen, P.; Sonek, S.; Cangul, H.; Coryell, J.; Canham, N.; Nardocci, N.; et al. PLA2G6, encoding a phospholipase A_2, is mutated in neurodegenerative disorders with high brain iron. *Nat. Genet.* **2006**, *38*, 752–754. [CrossRef] [PubMed]
15. Bao, S.; Bohrer, A.; Ramanadham, S.; Jin, W.; Zhang, S.; Turk, J. Effects of stable suppression of Group VIA phospholipase A_2 expression on phospholipid content and composition, insulin secretion, and proliferation of INS-1 insulinoma cells. *J. Biol. Chem.* **2006**, *281*, 187–198. [CrossRef]
16. Gross, R.W.; Ramanadham, S.; Kruszka, K.K.; Han, X.; Turk, J. Rat and human pancreatic islet cells contain a calcium ion independent phospholipase A_2 activity selective for hydrolysis of arachidonate which is stimulated by adenosine triphosphate and is specifically localized to islet beta-cells. *Biochemistry* **1993**, *32*, 327–336. [CrossRef]
17. Ma, Z.; Ramanadham, S.; Wohltmann, M.; Bohrer, A.; Hsu, F.F.; Turk, J. Studies of insulin secretory responses and of arachidonic acid incorporation into phospholipids of stably transfected insulinoma cells that overexpress group VIA phospholipase A_2 (iPLA$_2$b) indicate a signaling rather than a housekeeping role for iPLA$_2$b. *J. Biol. Chem.* **2001**, *276*, 13198–13208. [CrossRef]
18. Ma, Z.; Zhang, S.; Turk, J.; Ramanadham, S. Stimulation of insulin secretion and associated nuclear accumulation of iPLA$_2$beta in INS-1 insulinoma cells. *Am. J. Physiol. Endocrinol. Metab.* **2002**, *282*, E820–833. [CrossRef] [PubMed]
19. Ramanadham, S.; Gross, R.W.; Han, X.; Turk, J. Inhibition of arachidonate release by secretagogue-stimulated pancreatic islets suppresses both insulin secretion and the rise in beta-cell cytosolic calcium ion concentration. *Biochemistry* **1993**, *32*, 337–346. [CrossRef] [PubMed]

20. Ramanadham, S.; Hsu, F.F.; Bohrer, A.; Ma, Z.; Turk, J. Studies of the role of group VI phospholipase A_2 in fatty acid incorporation, phospholipid remodeling, lysophosphatidylcholine generation, and secretagogue-induced arachidonic acid release in pancreatic islets and insulinoma cells. *J. Biol. Chem.* **1999**, *274*, 13915–13927. [CrossRef]
21. Ramanadham, S.; Song, H.; Hsu, F.F.; Zhang, S.; Crankshaw, M.; Grant, G.A.; Newgard, C.B.; Bao, S.; Ma, Z.; Turk, J. Pancreatic islets and insulinoma cells express a novel isoform of group VIA phospholipase A_2 (iPLA_2b) that participates in glucose-stimulated insulin secretion and is not produced by alternate splicing of the iPLA_2b transcript. *Biochemistry* **2003**, *42*, 13929–13940. [CrossRef]
22. Ramanadham, S.; Wolf, M.J.; Jett, P.A.; Gross, R.W.; Turk, J. Characterization of an ATP-stimulatable Ca^{2+}-independent phospholipase A_2 from clonal insulin-secreting HIT cells and rat pancreatic islets: A possible molecular component of the beta-cell fuel sensor. *Biochemistry* **1994**, *33*, 7442–7452. [CrossRef] [PubMed]
23. Bao, S.; Jacobson, D.A.; Wohltmann, M.; Bohrer, A.; Jin, W.; Philipson, L.H.; Turk, J. Glucose homeostasis, insulin secretion, and islet phospholipids in mice that overexpress iPLA_2b in pancreatic b-cells and in iPLA_2b-null mice. *Am. J. Physiol. Endocrinol. Metab.* **2008**, *294*, 217–229. [CrossRef] [PubMed]
24. Bao, S.; Song, H.; Wohltmann, M.; Ramanadham, S.; Jin, W.; Bohrer, A.; Turk, J. Insulin secretory responses and phospholipid composition of pancreatic islets from mice that do not express Group VIA phospholipase A_2 and effects of metabolic stress on glucose homeostasis. *J. Biol. Chem.* **2006**, *281*, 20958–20973. [CrossRef]
25. Song, H.; Wohltmann, M.; Tan, M.; Ladenson, J.H.; Turk, J. Group VIA phospholipase A_2 mitigates palmitate-induced beta-cell mitochondrial injury and apoptosis. *J. Biol. Chem.* **2014**, *289*, 14194–14210. [CrossRef]
26. Gil-de-Gomez, L.; Astudillo, A.M.; Guijas, C.; Magrioti, V.; Kokotos, G.; Balboa, M.A.; Balsinde, J. Cytosolic group IVA and calcium-independent group VIA phospholipase A_2s act on distinct phospholipid pools in zymosan-stimulated mouse peritoneal macrophages. *J. Immunol.* **2014**, *192*, 752–762. [CrossRef]
27. Moran, J.M.; Buller, R.M.; McHowat, J.; Turk, J.; Wohltmann, M.; Gross, R.W.; Corbett, J.A. Genetic and pharmacologic evidence that calcium-independent phospholipase A_2beta regulates virus-induced inducible nitric-oxide synthase expression by macrophages. *J. Biol. Chem.* **2005**, *280*, 28162–28168. [CrossRef] [PubMed]
28. Perez, R.; Balboa, M.A.; Balsinde, J. Involvement of group VIA calcium-independent phospholipase A_2 in macrophage engulfment of hydrogen peroxide-treated U937 cells. *J. Immunol.* **2006**, *176*, 2555–2561. [CrossRef] [PubMed]
29. Tan, C.; Day, R.; Bao, S.; Turk, J.; Zhao, Q.D. Group VIA phospholipase A_2 mediates enhanced macrophage migration in diabetes mellitus by increasing expression of nicotinamide adenine dinucleotide phosphate oxidase 4. *Arter. Thromb. Vasc. Biol.* **2014**, *34*, 768–778. [CrossRef]
30. Ashley, J.W.; Hancock, W.D.; Nelson, A.J.; Bone, R.N.; Tse, H.M.; Wohltmann, M.; Turk, J.; Ramanadham, S. Polarization of Macrophages toward M2 Phenotype Is Favored by Reduction in iPLA_2b (Group VIA Phospholipase A_2). *J. Biol. Chem.* **2016**, *291*, 23268–23281. [CrossRef]
31. Kamei, N.; Tobe, K.; Suzuki, R.; Ohsugi, M.; Watanabe, T.; Kubota, N.; Ohtsuka-Kowatari, N.; Kumagai, K.; Sakamoto, K.; Kobayashi, M.; et al. Overexpression of monocyte chemoattractant protein-1 in adipose tissues causes macrophage recruitment and insulin resistance. *J. Biol. Chem.* **2006**, *281*, 26602–26614. [CrossRef]
32. Kanda, H.; Tateya, S.; Tamori, Y.; Kotani, K.; Hiasa, K.; Kitazawa, R.; Kitazawa, S.; Miyachi, H.; Maeda, S.; Egashira, K.; et al. MCP-1 contributes to macrophage infiltration into adipose tissue, insulin resistance, and hepatic steatosis in obesity. *J. Clin. Investig.* **2006**, *116*, 1494–1505. [CrossRef]
33. Sartipy, P.; Loskutoff, D.J. Monocyte chemoattractant protein 1 in obesity and insulin resistance. *Proc. Natl. Acad. Sci. USA* **2003**, *100*, 7265–7270. [CrossRef] [PubMed]
34. Weisberg, S.P.; Hunter, D.; Huber, R.; Lemieux, J.; Slaymaker, S.; Vaddi, K.; Charo, I.; Leibel, R.L.; Ferrante, A.W., Jr. CCR2 modulates inflammatory and metabolic effects of high-fat feeding. *J. Clin. Investig.* **2006**, *116*, 115–124. [CrossRef] [PubMed]
35. Weisberg, S.P.; McCann, D.; Desai, M.; Rosenbaum, M.; Leibel, R.L.; Ferrante, A.W., Jr. Obesity is associated with macrophage accumulation in adipose tissue. *J. Clin. Investig.* **2003**, *112*, 1796–1808. [CrossRef]
36. Lumeng, C.N.; Bodzin, J.L.; Saltiel, A.R. Obesity induces a phenotypic switch in adipose tissue macrophage polarization. *J. Clin. Investig.* **2007**, *117*, 175–184. [CrossRef]
37. Carnevale, K.A.; Cathcart, M.K. Calcium-independent phospholipase A_2 is required for human monocyte chemotaxis to monocyte chemoattractant protein 1. *J. Immunol.* **2001**, *167*, 3414–3421. [CrossRef] [PubMed]

38. Cathcart, M.K. Signal-activated phospholipase regulation of leukocyte chemotaxis. *J. Lipid Res.* **2009**, *50*, 231–236. [CrossRef]
39. Mishra, R.S.; Carnevale, K.A.; Cathcart, M.K. iPLA$_2$b: Front and center in human monocyte chemotaxis to MCP-1. *J. Exp. Med.* **2008**, *205*, 347–359. [CrossRef] [PubMed]
40. DeChiara, T.M. Gene targeting in ES cells. *Methods Mol. Biol.* **2001**, *158*, 19–45. [CrossRef]
41. Kuhn, R.; Torres, R.M. Cre/loxP recombination system and gene targeting. *Methods Mol. Biol.* **2002**, *180*, 175–204. [CrossRef] [PubMed]
42. Nagy, A. Cre recombinase: The universal reagent for genome tailoring. *Genesis* **2000**, *26*, 99–109. [CrossRef]
43. Takeuchi, T.; Nomura, T.; Tsujita, M.; Suzuki, M.; Fuse, T.; Mori, H.; Mishina, M. Flp recombinase transgenic mice of C57BL/6 strain for conditional gene targeting. *Biochem. Biophys Res. Commun.* **2002**, *293*, 953–957. [CrossRef]
44. Turan, S.; Galla, M.; Ernst, E.; Qiao, J.; Voelkel, C.; Schiedlmeier, B.; Zehe, C.; Bode, J. Recombinase-mediated cassette exchange (RMCE): Traditional concepts and current challenges. *J. Mol. Biol.* **2011**, *407*, 193–221. [CrossRef] [PubMed]
45. Friedel, R.H.; Wurst, W.; Wefers, B.; Kuhn, R. Generating conditional knockout mice. *Methods Mol. Biol.* **2011**, *693*, 205–231. [CrossRef]
46. Fex, M.; Wierup, N.; Nitert, M.D.; Ristow, M.; Mulder, H. Rat insulin promoter 2-Cre recombinase mice bred onto a pure C57BL/6J background exhibit unaltered glucose tolerance. *J. Endocrinol.* **2007**, *194*, 551–555. [CrossRef]
47. Gannon, M.; Shiota, C.; Postic, C.; Wright, C.V.; Magnuson, M. Analysis of the Cre-mediated recombination driven by rat insulin promoter in embryonic and adult mouse pancreas. *Genesis* **2000**, *26*, 139–142. [CrossRef]
48. Kalis, M.; Bolmeson, C.; Esguerra, J.L.; Gupta, S.; Edlund, A.; Tormo-Badia, N.; Speidel, D.; Holmberg, D.; Mayans, S.; Khoo, N.K.; et al. Beta-cell specific deletion of Dicer1 leads to defective insulin secretion and diabetes mellitus. *PLoS ONE* **2011**, *6*, e29166. [CrossRef]
49. Pappan, K.L.; Pan, Z.; Kwon, G.; Marshall, C.A.; Coleman, T.; Goldberg, I.J.; McDaniel, M.L.; Semenkovich, C.F. Pancreatic beta-cell lipoprotein lipase independently regulates islet glucose metabolism and normal insulin secretion. *J. Biol. Chem.* **2005**, *280*, 9023–9029. [CrossRef]
50. Sun, G.; Tarasov, A.I.; McGinty, J.A.; French, P.M.; McDonald, A.; Leclerc, I.; Rutter, G.A. LKB1 deletion with the RIP2.Cre transgene modifies pancreatic beta-cell morphology and enhances insulin secretion in vivo. *Am. J. Physiol. Endocrinol. Metab.* **2010**, *298*, 1261–1273. [CrossRef]
51. Clausen, B.E.; Burkhardt, C.; Reith, W.; Renkawitz, R.; Forster, I. Conditional gene targeting in macrophages and granulocytes using LysMcre mice. *Transgenic Res.* **1999**, *8*, 265–277. [CrossRef] [PubMed]
52. Schneider, J.G.; Yang, Z.; Chakravarthy, M.V.; Lodhi, I.J.; Wei, X.; Turk, J.; Semenkovich, C.F. Macrophage fatty-acid synthase deficiency decreases diet-induced atherosclerosis. *J. Biol. Chem.* **2010**, *285*, 23398–23409. [CrossRef] [PubMed]
53. Lee, J.Y.; Ristow, M.; Lin, X.; White, M.F.; Magnuson, M.A.; Hennighausen, L. RIP-Cre revisited, evidence for impairments of pancreatic beta-cell function. *J. Biol. Chem.* **2006**, *281*, 2649–2653. [CrossRef]
54. Chakravarthy, M.V.; Pan, Z.; Zhu, Y.; Tordjman, K.; Schneider, J.G.; Coleman, T.; Turk, J.; Semenkovich, C.F. "New" hepatic fat activates PPARalpha to maintain glucose, lipid, and cholesterol homeostasis. *Cell Metab.* **2005**, *1*, 309–322. [CrossRef] [PubMed]
55. Cheverud, J.M.; Ehrich, T.H.; Hrbek, T.; Kenney, J.P.; Pletscher, L.S.; Semenkovich, C.F. Quantitative trait loci for obesity- and diabetes-related traits and their dietary responses to high-fat feeding in LGXSM recombinant inbred mouse strains. *Diabetes* **2004**, *53*, 3328–3336. [CrossRef] [PubMed]
56. Bernal-Mizrachi, E.; Fatrai, S.; Johnson, J.D.; Ohsugi, M.; Otani, K.; Han, Z.; Polonsky, K.S.; Permutt, M.A. Defective insulin secretion and increased susceptibility to experimental diabetes are induced by reduced Akt activity in pancreatic islet beta cells. *J. Clin. Investig.* **2004**, *114*, 928–936. [CrossRef] [PubMed]
57. Song, H.; Wohltmann, M.; Bao, S.; Ladenson, J.H.; Semenkovich, C.F.; Turk, J. Mice deficient in group VIB phospholipase A_2 (iPLA$_2$g) exhibit relative resistance to obesity and metabolic abnormalities induced by a Western diet. *Am. J. Physiol. Endocrinol. Metab.* **2010**, *298*, 1097–1114. [CrossRef]
58. Sakaguchi, K.; Takeda, K.; Maeda, M.; Ogawa, W.; Sato, T.; Okada, S.; Ohnishi, Y.; Nakajima, H.; Kashiwagi, A. Glucose area under the curve during oral glucose tolerance test as an index of glucose intolerance. *Diabetol. Int.* **2016**, *7*, 53–58. [CrossRef] [PubMed]

59. McDaniel, M.L.; Colca, J.R.; Kotagal, N.; Lacy, P.E. A subcellular fractionation approach for studying insulin release mechanisms and calcium metabolism in islets of Langerhans. *Methods Enzymol.* **1983**, *98*, 182–200. [CrossRef]
60. Schwartz, M.W.; Guyenet, S.J.; Cirulli, V. The hypothalamus and ss-cell connection in the gene-targeting era. *Diabetes* **2010**, *59*, 2991–2993. [CrossRef]
61. Wu, X.; Zhang, Q.; Wang, X.; Zhu, J.; Xu, K.; Okada, H.; Wang, R.; Woo, M. Survivin is required for beta-cell mass expansion in the pancreatic duct-ligated mouse model. *PLoS ONE* **2012**, *7*, e41976. [CrossRef] [PubMed]
62. Quan, W.; Lim, Y.M.; Lee, M.S. Role of autophagy in diabetes and endoplasmic reticulum stress of pancreatic beta-cells. *Exp. Mol. Med.* **2012**, *44*, 81–88. [CrossRef] [PubMed]
63. Choi, D.; Cai, E.P.; Schroer, S.A.; Wang, L.; Woo, M. Vhl is required for normal pancreatic beta cell function and the maintenance of beta cell mass with age in mice. *Lab. Investig.* **2011**, *91*, 527–538. [CrossRef]
64. Cui, J.; Wang, Z.; Cheng, Q.; Lin, R.; Zhang, X.M.; Leung, P.S.; Copeland, N.G.; Jenkins, N.A.; Yao, K.M.; Huang, J.D. Targeted inactivation of kinesin-1 in pancreatic beta-cells in vivo leads to insulin secretory deficiency. *Diabetes* **2011**, *60*, 320–330. [CrossRef] [PubMed]
65. Wang, L.; Liu, Y.; Lu, S.Y.; Nguyen, K.T.; Schroer, S.A.; Suzuki, A.; Mak, T.W.; Gaisano, H.; Woo, M. Deletion of Pten in pancreatic ss-cells protects against deficient ss-cell mass and function in mouse models of type 2 diabetes. *Diabetes* **2010**, *59*, 3117–3126. [CrossRef] [PubMed]
66. Srinivasan, M.; Choi, C.S.; Ghoshal, P.; Pliss, L.; Pandya, J.D.; Hill, D.; Cline, G.; Patel, M.S. β-Cell-specific pyruvate dehydrogenase deficiency impairs glucose-stimulated insulin secretion. *Am. J. Physiol. Endocrinol. Metab.* **2010**, *299*, 910–917. [CrossRef] [PubMed]
67. Tokumoto, S.; Yabe, D.; Tatsuoka, H.; Usui, R.; Fauzi, M.; Botagarova, A.; Goto, H.; Herrera, P.L.; Ogura, M.; Inagaki, N. Generation and characterization of a novel mouse model that allows spatiotemporal quantification of pancreatic beta-cell proliferation. *Diabetes* **2020**. [CrossRef] [PubMed]
68. Wong, C.; Tang, L.H.; Davidson, C.; Vosburgh, E.; Chen, W.; Foran, D.J.; Notterman, D.A.; Levine, A.J.; Xu, E.Y. Two well-differentiated pancreatic neuroendocrine tumor mouse models. *Cell Death Differ.* **2020**, *27*, 269–283. [CrossRef]
69. Tauscher, S.; Nakagawa, H.; Volker, K.; Werner, F.; Krebes, L.; Potapenko, T.; Doose, S.; Birkenfeld, A.L.; Baba, H.A.; Kuhn, M. beta Cell-specific deletion of guanylyl cyclase A, the receptor for atrial natriuretic peptide, accelerates obesity-induced glucose intolerance in mice. *Cardiovasc. Diabetol.* **2018**, *17*, 103. [CrossRef]
70. Otten, J.J.; de Jager, S.C.; Kavelaars, A.; Seijkens, T.; Bot, I.; Wijnands, E.; Beckers, L.; Westra, M.M.; Bot, M.; Busch, M.; et al. Hematopoietic G-protein-coupled receptor kinase 2 deficiency decreases atherosclerotic lesion formation in LDL receptor-knockout mice. *FASEB J.* **2013**, *27*, 265–276. [CrossRef]
71. Klose, A.; Zigrino, P.; Mauch, C. Monocyte/macrophage MMP-14 modulates cell infiltration and T-cell attraction in contact dermatitis but not in murine wound healing. *Am. J. Pathol.* **2013**, *182*, 755–764. [CrossRef]
72. Pello, O.M.; Chevre, R.; Laoui, D.; De Juan, A.; Lolo, F.; Andres-Manzano, M.J.; Serrano, M.; Van Ginderachter, J.A.; Andres, V. In vivo inhibition of c-MYC in myeloid cells impairs tumor-associated macrophage maturation and pro-tumoral activities. *PLoS ONE* **2012**, *7*, e45399. [CrossRef]
73. Luiz, J.P.M.; Toller-Kawahisa, J.E.; Viacava, P.R.; Nascimento, D.C.; Pereira, P.T.; Saraiva, A.L.; Prado, D.S.; LeBert, M.; Giurisato, E.; Tournier, C.; et al. MEK5/ERK5 signaling mediates IL-4-induced M2 macrophage differentiation through regulation of c-Myc expression. *J. Leukoc. Biol.* **2020**. [CrossRef]
74. Otto, N.A.; de Vos, A.F.; van Heijst, J.W.J.; Roelofs, J.; van der Poll, T. Myeloid Liver Kinase B1 depletion is associated with a reduction in alveolar macrophage numbers and an impaired host defense during gram-negative pneumonia. *J. Infect. Dis.* **2020**. [CrossRef] [PubMed]
75. McGarry, J.D. What if Minkowski had been ageusic? An alternative angle on diabetes. *Science* **1992**, *258*, 766–770. [CrossRef]
76. Brown, M.S.; Goldstein, J.L. Selective versus total insulin resistance: A pathogenic paradox. *Cell Metab.* **2008**, *7*, 95–96. [CrossRef] [PubMed]
77. Otero, Y.F.; Stafford, J.M.; McGuinness, O.P. Pathway-selective insulin resistance and metabolic disease: The importance of nutrient flux. *J. Biol. Chem.* **2014**, *289*, 20462–20469. [CrossRef] [PubMed]
78. Chavez, J.A.; Summers, S.A. Lipid oversupply, selective insulin resistance, and lipotoxicity: Molecular mechanisms. *Biochim. Biophys. Acta* **2010**, *1801*, 252–265. [CrossRef] [PubMed]

79. Chavez, J.A.; Siddique, M.M.; Wang, S.T.; Ching, J.; Shayman, J.A.; Summers, S.A. Ceramides and glucosylceramides are independent antagonists of insulin signaling. *J. Biol. Chem.* **2014**, *289*, 723–734. [CrossRef]
80. Unger, R.H. Lipotoxicity in the pathogenesis of obesity-dependent NIDDM. Genetic and clinical implications. *Diabetes* **1995**, *44*, 863–870. [CrossRef]
81. Miki, T.; Nagashima, K.; Tashiro, F.; Kotake, K.; Yoshitomi, H.; Tamamoto, A.; Gonoi, T.; Iwanaga, T.; Miyazaki, J.; Seino, S. Defective insulin secretion and enhanced insulin action in KATP channel-deficient mice. *Proc. Natl. Acad. Sci. USA* **1998**, *95*, 10402–10406. [CrossRef] [PubMed]
82. Miki, T.; Minami, K.; Zhang, L.; Morita, M.; Gonoi, T.; Shiuchi, T.; Minokoshi, Y.; Renaud, J.M.; Seino, S. ATP-sensitive potassium channels participate in glucose uptake in skeletal muscle and adipose tissue. *Am. J. Physiol. Endocrinol. Metab.* **2002**, *283*, 1178–1184. [CrossRef] [PubMed]
83. Deng, X.; Wang, J.; Jiao, L.; Utaipan, T.; Tuma-Kellner, S.; Schmitz, G.; Liebisch, G.; Stremmel, W.; Chamulitrat, W. iPLA$_2$beta deficiency attenuates obesity and hepatic steatosis in ob/ob mice through hepatic fatty-acyl phospholipid remodeling. *Biochim. Biophys. Acta* **2016**, *1861*, 449–461. [CrossRef]
84. Otto, A.C.; Gan-Schreier, H.; Zhu, X.; Tuma-Kellner, S.; Staffer, S.; Ganzha, A.; Liebisch, G.; Chamulitrat, W. Group VIA phospholipase A$_2$ deficiency in mice chronically fed with high-fat-diet attenuates hepatic steatosis by correcting a defect of phospholipid remodeling. *Biochim. Biophys. Acta Mol. Cell Biol. Lipids* **2019**, *1864*, 662–676. [CrossRef] [PubMed]
85. Lei, X.; Zhang, S.; Bohrer, A.; Bao, S.; Song, H.; Ramanadham, S. The group VIA calcium-independent phospholipase A$_2$ participates in ER stress-induced INS-1 insulinoma cell apoptosis by promoting ceramide generation via hydrolysis of sphingomyelins by neutral sphingomyelinase. *Biochemistry* **2007**, *46*, 10170–10185. [CrossRef]
86. Lei, X.; Zhang, S.; Bohrer, A.; Ramanadham, S. Calcium-independent phospholipase A$_2$ (iPLA$_2$b)-mediated ceramide generation plays a key role in the cross-talk between the endoplasmic reticulum (ER) and mitochondria during ER stress-induced insulin-secreting cell apoptosis. *J. Biol. Chem.* **2008**, *283*, 34819–34832. [CrossRef]
87. Lei, X.; Zhang, S.; Barbour, S.E.; Bohrer, A.; Ford, E.L.; Koizumi, A.; Papa, F.R.; Ramanadham, S. Spontaneous development of endoplasmic reticulum stress that can lead to diabetes mellitus is associated with higher calcium-independent phospholipase A$_2$ expression: A role for regulation by SREBP-1. *J. Biol. Chem.* **2010**, *285*, 6693–6705. [CrossRef]
88. Lei, X.; Zhang, S.; Emani, B.; Barbour, S.E.; Ramanadham, S. A link between endoplasmic reticulum stress-induced beta-cell apoptosis and the group VIA Ca^{2+}-independent phospholipase A$_2$ (iPLA$_2$b). *Diabetes Obes. Metab.* **2010**, *12* (Suppl. 2), 93–98. [CrossRef]
89. Li, H.; Zhao, Z.; Wei, G.; Yan, L.; Wang, D.; Zhang, H.; Sandusky, G.E.; Turk, J.; Xu, Y. Group VIA phospholipase A$_2$ in both host and tumor cells is involved in ovarian cancer development. *FASEB J.* **2010**, *24*, 4103–4116. [CrossRef]
90. Moon, S.H.; Jenkins, C.M.; Mancuso, D.J.; Turk, J.; Gross, R.W. Smooth muscle cell arachidonic acid release, migration, and proliferation are markedly attenuated in mice null for calcium-independent phospholipase A2beta. *J. Biol. Chem.* **2008**, *283*, 33975–33987. [CrossRef]
91. Nelson, A.J.; Stephenson, D.J.; Cardona, C.L.; Lei, X.; Almutairi, A.; White, T.D.; Tusing, Y.G.; Park, M.A.; Barbour, S.E.; Chalfant, C.E.; et al. Macrophage polarization is linked to Ca^{2+}-independent phospholipase A$_2$beta-derived lipids and cross-cell signaling in mice. *J. Lipid Res.* **2020**, *61*, 143–158. [CrossRef] [PubMed]
92. Bone, R.N.; Gai, Y.; Magrioti, V.; Kokotou, M.G.; Ali, T.; Lei, X.; Tse, H.M.; Kokotos, G.; Ramanadham, S. Inhibition of Ca^{2+}-independent phospholipase A$_2$beta (iPLA$_2$b) ameliorates islet infiltration and incidence of diabetes in NOD mice. *Diabetes* **2015**, *64*, 541–554. [CrossRef]
93. Nelson, A.J.; Stephenson, D.J.; Bone, R.N.; Cardona, C.L.; Park, M.A.; Tusing, Y.G.; Lei, X.; Kokotos, G.; Graves, C.L.; Mathews, C.E.; et al. Lipid mediators and biomarkers associated with type 1 diabetes development. *JCI Insight* **2020**, *5*, 138034. [CrossRef]
94. Balsinde, J.; Balboa, M.A.; Dennis, E.A. Antisense inhibition of group VI Ca^{2+}-independent phospholipase A2 blocks phospholipid fatty acid remodeling in murine P388D1 macrophages. *J. Biol. Chem.* **1997**, *272*, 29317–29321. [CrossRef]

95. Balsinde, J.; Bianco, I.D.; Ackermann, E.J.; Conde-Frieboes, K.; Dennis, E.A. Inhibition of calcium-independent phospholipase A_2 prevents arachidonic acid incorporation and phospholipid remodeling in P388D1 macrophages. *Proc. Natl. Acad. Sci. USA* **1995**, *92*, 8527–8531. [CrossRef]
96. Monge, P.; Garrido, A.; Rubio, J.M.; Magrioti, V.; Kokotos, G.; Balboa, M.A.; Balsinde, J. The Contribution of Cytosolic Group IVA and Calcium-Independent Group VIA Phospholipase A_2s to Adrenic Acid Mobilization in Murine Macrophages. *Biomolecules* **2020**, *10*, 542. [CrossRef]
97. Balboa, M.A.; Perez, R.; Balsinde, J. Calcium-independent phospholipase A_2 mediates proliferation of human promonocytic U937 cells. *FEBS J.* **2008**, *275*, 1915–1924. [CrossRef]
98. Bao, S.; Li, Y.; Lei, X.; Wohltmann, M.; Jin, W.; Bohrer, A.; Semenkovich, C.F.; Ramanadham, S.; Tabas, I.; Turk, J. Attenuated free cholesterol loading-induced apoptosis but preserved phospholipid composition of peritoneal macrophages from mice that do not express group VIA phospholipase A_2. *J. Biol. Chem.* **2007**, *282*, 27100–27114. [CrossRef] [PubMed]
99. Nikolic, D.M.; Gong, M.C.; Turk, J.; Post, S.R. Class A scavenger receptor-mediated macrophage adhesion requires coupling of calcium-independent phospholipase A_2 and 12/15-lipoxygenase to Rac and Cdc42 activation. *J. Biol. Chem.* **2007**, *282*, 33405–33411. [CrossRef]
100. Carper, M.J.; Zhang, S.; Turk, J.; Ramanadham, S. Skeletal muscle group VIA phospholipase A_2 (iPLA_2b): Expression and role in fatty acid oxidation. *Biochemistry* **2008**, *47*, 12241–12249. [CrossRef] [PubMed]
101. Su, X.; Mancuso, D.J.; Bickel, P.E.; Jenkins, C.M.; Gross, R.W. Small interfering RNA knockdown of calcium-independent phospholipases A_2 beta or gamma inhibits the hormone-induced differentiation of 3T3-L1 preadipocytes. *J. Biol. Chem.* **2004**, *279*, 21740–21748. [CrossRef] [PubMed]
102. Zisman, A.; Peroni, O.D.; Abel, E.D.; Michael, M.D.; Mauvais-Jarvis, F.; Lowell, B.B.; Wojtaszewski, J.F.; Hirshman, M.F.; Virkamaki, A.; Goodyear, L.J.; et al. Targeted disruption of the glucose transporter 4 selectively in muscle causes insulin resistance and glucose intolerance. *Nat. Med.* **2000**, *6*, 924–928. [CrossRef] [PubMed]
103. Bluher, M.; Michael, M.D.; Peroni, O.D.; Ueki, K.; Carter, N.; Kahn, B.B.; Kahn, C.R. Adipose tissue selective insulin receptor knockout protects against obesity and obesity-related glucose intolerance. *Dev. Cell* **2002**, *3*, 25–38. [CrossRef]

Publisher's Note: MDPI stays neutral with regard to jurisdictional claims in published maps and institutional affiliations.

© 2020 by the authors. Licensee MDPI, Basel, Switzerland. This article is an open access article distributed under the terms and conditions of the Creative Commons Attribution (CC BY) license (http://creativecommons.org/licenses/by/4.0/).

Article

The Contribution of Cytosolic Group IVA and Calcium-Independent Group VIA Phospholipase A_2s to Adrenic Acid Mobilization in Murine Macrophages

Patricia Monge [1,2], Alvaro Garrido [1,2], Julio M. Rubio [1,2], Victoria Magrioti [3], George Kokotos [3], María A. Balboa [1,2] and Jesús Balsinde [1,2,*]

1 Instituto de Biología y Genética Molecular, Consejo Superior de Investigaciones Científicas (CSIC), Universidad de Valladolid, 47003 Valladolid, Spain; pmonge@ibgm.uva.es (P.M.); agarrido@ibgm.uva.es (A.G.); jrubio@ibgm.uva.es (J.M.R.); mbalboa@ibgm.uva.es (M.A.B.)

2 Centro de Investigación Biomédica en Red de Diabetes y Enfermedades Metabólicas Asociadas (CIBERDEM), 28029 Madrid, Spain

3 Department of Chemistry, National and Kapodistrian University of Athens, Panepistimiopolis, 15771 Athens, Greece; vmagriot@chem.uoa.gr (V.M.); gkokotos@chem.uoa.gr (G.K.)

* Correspondence: jbalsinde@ibgm.uva.es; Tel.: +34-983-423-062

Received: 16 March 2020; Accepted: 1 April 2020; Published: 3 April 2020

Abstract: Adrenic acid (AA), the 2-carbon elongation product of arachidonic acid, is present at significant levels in membrane phospholipids of mouse peritoneal macrophages. Despite its abundance and structural similarity to arachidonic acid, very little is known about the molecular mechanisms governing adrenic acid mobilization in cells of the innate immune system. This contrasts with the wide availability of data on arachidonic acid mobilization. In this work, we used mass-spectrometry-based lipidomic procedures to define the profiles of macrophage phospholipids that contain adrenic acid and their behavior during receptor activation. We identified the phospholipid sources from which adrenic acid is mobilized, and compared the data with arachidonic acid mobilization. Taking advantage of the use of selective inhibitors, we also showed that cytosolic group IVA phospholipase A_2 is involved in the release of both adrenic and arachidonic acids. Importantly, calcium independent group VIA phospholipase A_2 spared arachidonate-containing phospholipids and hydrolyzed only those that contain adrenic acid. These results identify separate mechanisms for regulating the utilization of adrenic and arachidonic acids, and suggest that the two fatty acids may serve non-redundant functions in cells.

Keywords: adrenic acid; arachidonic acid; mass spectrometry; lipid signaling; inflammation; phospholipase A_2; monocytes/macrophages

---

1. Introduction

Arachidonic acid (*cis*-5,8,11,14-eicosatetraenoic acid, AA), a fatty acid of the *n*-6 series, is the major polyunsaturated fatty acid present in cells of the innate immunity [1]. Cleavage of AA-containing membrane phospholipids by phospholipase A_2 (PLA_2) enzymes during activation results in substantial release of free AA. The free fatty acid is then metabolized by cyclooxygenases, lipoxygenases, or cytochrome P-450 enzymes into numerous oxygenated metabolites, collectively called the eicosanoids, which have key roles in inflammation [2–6]. The eicosanoids participate in immune regulation by influencing the activation of innate immune phagocytic cells at various levels, including differentiation and migration, phagocytic capacity, and cytokine production [7–10].

Innate immune cells also contain significant quantities of another *n*-6 fatty acid, adrenic acid (*cis*-7,10,13,16-docosatetraenoic acid, AdA), which is the 2-carbon elongation product of AA [1]

(Scheme 1). Similar to AA, AdA can also be metabolized via cyclooxygenase, lipoxygenase, and cytochrome P-450 pathways [11–14], giving rise to a number of oxygenated metabolites. Interestingly, a recent study using a mouse model of steatohepatitis reported elevated levels of free AdA in plasma and liver [15]. These data raised the possibility that alterations in the homeostatic mechanisms regulating AdA levels may be related to pathophysiological states [15]. Regarding the role of AdA in innate immune cells, it has been shown that the fatty acid is able to dampen inflammation by blocking leukotriene B_4 formation by neutrophils and enhancing phagocytosis by macrophages [16]. These effects are consistent with a pro-resolving mediator function for AdA in osteoarthritis [16].

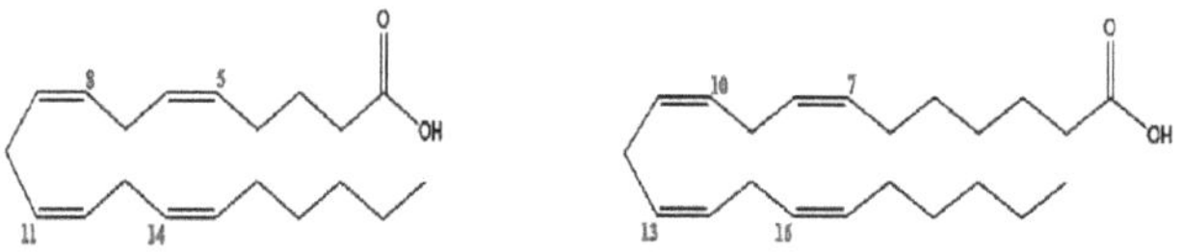

Arachidonic Acid (AA)
20:4n-6

Adrenic Acid (AdA)
22:4n-6

Scheme 1. Structure of arachidonic acid (AA) and adrenic acid (AdA). The positions of the carbons at which double bonds occur are numbered. Note that all double bonds are in the *cis* configuration.

Recent work from our laboratory has taken advantage of mass-spectrometry-based lipidomic approaches to define at a molecular level the mechanisms regulating AA mobilization from phagocytic cells responding to stimuli of the innate immune response [17–27]. Our work has highlighted not only the importance of multiple PLA_2 enzymes in the process, but also of the mechanisms regulating the reacylation of the liberated fatty acid back into phospholipids and the remodeling that places the AA in the appropriate cellular lipid pools. These findings have raised the intriguing possibility that not all AA pools may be reachable by the PLA_2 form involved in its mobilization. For instance, release of lipoxygenase products by ionophore-activated human neutrophils [28] and zymosan-stimulated mouse peritoneal macrophages [22] appears to be associated with AA mobilization from choline-containing phospholipids (PC), not ethanolamine-containing phospholipids (PE) or phosphatidylinositol (PI). Thus, depending on stimulation conditions, the cellular distribution of AA between various phospholipid locations may also limit eicosanoid synthesis [3,26].

In contrast to AA studies, very little is still known about the mechanisms governing cellular AdA availability. In this study we have analyzed the regulatory features of AdA mobilization in activated macrophages. Our work provides an in-depth examination of AdA homeostasis under pathophysiologically relevant conditions, and suggests that both cytosolic group IVA PLA_2 ($cPLA_2\alpha$) and calcium-independent group VIA PLA_2 ($iPLA_2$-VIA) participate in the process. AdA mobilization shares regulatory features with AA mobilization with regard to $cPLA_2\alpha$ involvement, but seems to be a more complex process, as it involves participation of a second PLA_2 that is apparently not involved in AA release (i.e., $iPLA_2$-VIA). In addition, our study also suggests that supplementation with AdA does not reduce AA utilization by the cells. Thus, the two fatty acids may play non-redundant biological roles that could be exploited to design selective strategies to control the production of AdA-derived products at the level of their precursor fatty acid.

2. Materials and Methods

2.1. Reagents

Cell culture medium was from Molecular Probes-Invitrogen (Carlsbad, CA, USA). Organic solvents (Optima® LC/MS grade) were from Fisher Scientific (Madrid, Spain). Lipid standards were from Avanti (Alabaster, AL, USA) or Cayman (Ann Arbor, MI, USA). Silicagel G thin-layer chromatography plates were from Macherey-Nagel (Düren, Germany). The $cPLA_2\alpha$ inhibitor pyrrophenone [29] was synthesized and provided by Dr. Alfonso Pérez (Department of Organic Chemistry, University of

Valladolid, Valladolid, Spain). The iPLA$_2$-VIA inhibitors FKGK18 and GK436 were synthesized in the Kokotos laboratory [30,31]. All other reagents were from Sigma-Aldrich (Madrid, Spain).

2.2. Cell Culture

Mouse peritoneal macrophages from Swiss mice (University of Valladolid Animal House, 10–12 weeks old) were obtained by peritoneal lavage using 5 mL cold phosphate-buffered saline and cultured in RPMI 1640 medium with 10% (*v*/*v*) fetal bovine serum, 100 U/mL penicillin, and 100 µg/mL streptomycin, as described elsewhere [32,33]. All procedures involving animals were undertaken under the supervision of the Institutional Committee of Animal Care and Usage of the University of Valladolid (No. 7406000), in accordance with the guidelines established by the Spanish Ministry of Agriculture, Food, and Environment and the European Union.

Cells were placed in serum-free medium for 1 h before addition of stimuli or inhibitors. Afterward, they were challenged by the stimuli for the time indicated. Zymosan was prepared exactly as described [32,33]. Only zymosan batches that demonstrated no measurable endogenous PLA$_2$ activity, as measured by in vitro assay under different conditions [34–37], were used in this study. Cell protein content was quantified according to Bradford [38] using a commercial kit (BioRad Protein Assay, Bio-Rad, Hercules, CA, USA).

2.3. Gas Chromatography/Mass Spectrometry (GC/MS) Analyses

Total lipids from approximately 10^7 cells were extracted according to Bligh and Dyer [39], and the following internal standards were added: 10 nmol of 1,2-diheptadecanoyl-*sn*-glycero-3-phosphocholine, 10 nmol of 1,2,3-triheptadecanoylglycerol, and 30 nmol of cholesteryl tridecanoate. Phospholipids were separated from neutral lipids by thin-layer chromatography, using *n*-hexane/diethyl ether/acetic acid (70:30:1, *v*/*v*/*v*) as the mobile phase [40]. Phospholipid classes were separated twice with chloroform/methanol/28% (*w*/*w*) ammonium hydroxide (60:37.5:4, *v*/*v*/*v*) as the mobile phase, using plates impregnated with boric acid [41]. The bands corresponding to the different lipid classes were scraped off from the plate, and fatty acid methyl esters were obtained from the various lipid fractions by transmethylation with 0.5 M KOH in methanol for 60 min at 37 °C [42–45]. Analyses were carried out using an Agilent 7890A gas chromatograph coupled to an Agilent 5975C mass-selective detector operated in an electron impact mode (EI, 70 eV). The apparatus was also equipped with an Agilent 7693 autosampler and an Agilent DB23 column (60 m length × 0.25 mm internal diameter × 0.15 µM film thickness) (Agilent Technologies, Santa Clara, CA, USA). Data analysis was carried out with the Agilent G1701EA MSD Productivity Chemstation software, revision E.02.00 [42–45].

2.4. Liquid Chromatography/Mass Spectrometry (LC/MS) Analyses of Phospholipids

Lipids were extracted according to Bligh and Dyer [39], and the following internal standards were added: 20 pmol each of 1,2-dimyristoyl-*sn*-glycero-3-phosphoglycerol, 1,2-dilauroyl-*sn*-glycero-3-phosphoethanolamine, 1,2-dimyristoyl-*sn*-glycero-3-phosphoethanolamine, 1,2-di- heptadecanoyl-*sn*-glycero-3-phosphoethanolamine, 1,2-dimyristoyl-*sn*-glycero-3-phosphoserine, 1,2-dimyristoyl-*sn*-glycero-3-phosphate, 1,2-dipentadecanoyl-*sn*-glycero-3-phosphocholine, 1,2-di-heptadecanoyl-*sn*-glycero-3-phosphocholine, and 1,2-dinonadecanoyl-*sn*-glycero-phosphocholine.

The samples were re-dissolved in hexanes/2-propanol/water (42:56:2, *v*/*v*/*v*) and injected into a Thermo Scientific Dionex Ultimate 3000 high-performance liquid chromatograph equipped with an Ultimate HPG-3400SD standard binary pump and an Ultimate ACC-3000 autosampler column compartment (Waltham, MA, USA). Separation was carried using a FORTIS HILIC column (150 × 3 mm, 3 µm particle size) (Fortis Technologies, Neston, UK). The mobile phase consisted of a gradient of solvent A (hexanes/isopropanol 30:40, *v*/*v*) and solvent B (hexanes/isopropanol/20 mM ammonium acetate in water, 30:40:7, *v*/*v*/*v*). The gradient started at 75% A from 0 to 5 min, then decreased from 75% A to 40% A at 15 min, from 40% A to 5% A at 20 min, was held at 5% until 40 min, and then increased to 75% at 41 min. Then, the column was re-equilibrated holding 75% A for an additional 14-min period before the next sample injection [46]. The flow rate through the column was fixed at 0.4 mL/min. The liquid chromatography system was coupled online to an Sciex QTRAP 4500 Mass Spectrometer equipped with a Turbo V ion source and a TurboIonSpray probe for electrospray ionization (AB Sciex, Framingham, MA, USA). Source parameters were set as follows: ion spray voltage, −4500 V; curtain gas, 30 psi; nebulizer gas, 50 psi; desolvation gas, 60 psi; temperature, 425 °C. Phospholipid species were analyzed in scheduled multiple reaction monitoring mode with negative ionization, detecting in Q3 the *m/z* of either 303.2 or 331.2, corresponding to AA and AdA, respectively, as $[M-H]^-$. Compound parameters were fixed as follows: declustering potential; −45 V (choline glycerophospholipids), −60 V (ethanolamine glycerophospholipids) −30 V (phosphatidylinositol), −50 V (phosphatidylserine), −60 V (phosphatidic acid), −50 V (phosphatidylglycerol); collision energy: −50 V (choline glycerophospholipids), −40 V (ethanolamine glycerophospholipids), −60 V (phosphatidylinositol), −50 V (phosphatidylserine), −45 V (phosphatidic acid), −45 V (phosphatidylglycerol); entrance potential, −10 V; and collision cell exit potential, −8 V. All glycerophospholipids were detected as $[M-H]^-$, ions except choline glycerophospholipids, which were detected as $[M + CH_3COO]^-$ ions. Quantification was carried out by integrating the chromatographic peaks of each species and comparing these with the peak area of the internal standard that corresponded to each class.

3. Results

3.1. *Adrenic Acid and Arachidonic Acid Contents of Murine Peritoneal Macrophages*

Lipid extracts from mouse peritoneal macrophages were analyzed for fatty acid content by GC/MS. Total AA content was 69.9 ± 4.2 nmol/mg cell protein (mean values ± standard error of the mean, $n = 5$), while AdA content was 15.1 ± 1.2 nmol/mg cell protein (mean values ± standard error of the mean, $n = 5$). Both AA and AdA were found almost exclusively in phospholipids. The distribution of AA and AdA between phospholipid classes is shown in Figure 1. Despite the difference in mass between AA and AdA, their distribution between phospholipid classes was remarkably similar, with the majority of both fatty acids being found in ethanolamine glycerophospholipids (PE), followed by choline glycerophospholipids (PC). Minor amounts of both fatty acids were found in phosphatidylinositol (PI) and phosphatidylserine (PS).

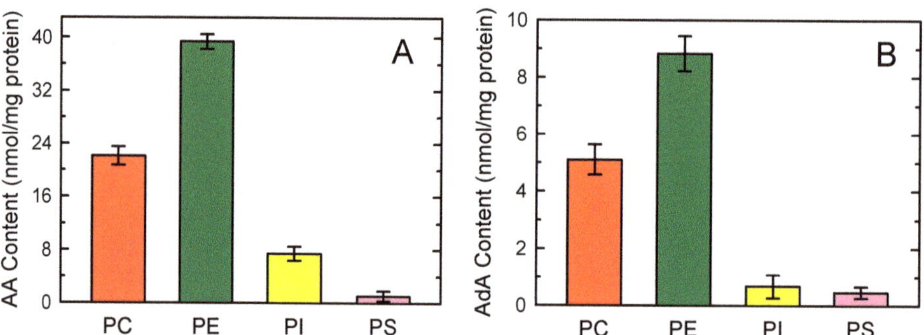

Figure 1. Distribution of AA and AdA between phospholipid classes. The various phospholipid classes were separated by thin-layer chromatography. The distribution of AA (**A**) and AdA (**B**) between choline glycerophospholipids (PC), ethanolamine glycerophospholipids (PE), phosphatidylinositol (PI), and phosphatidylserine (PS) was determined by gas chromatography/mass spectrometry (GC/MS) after converting the phospholipid-bound fatty acids into methyl esters. Results are shown as means ± standard error of the mean ($n = 3$).

Figure 2 shows the distribution of AA- and AdA-containing phospholipid molecular species, as analyzed by liquid chromatography coupled to tandem mass spectrometry (LC/MS). In agreement with previous estimates [21,22], multiple AA-containing species were detected, with the alkenylacyl and diacyl ethanolamine phospholipid species PE(P-16:0/20:4), PE(P-18:0/20:4), and PE(18:0/20:4) predominating, followed by the diacyl choline phospholipid species PC(16:0/20:4) and PC(18:0/20:4), and the unique inositol phospholipid species PI(18:0/20:4)

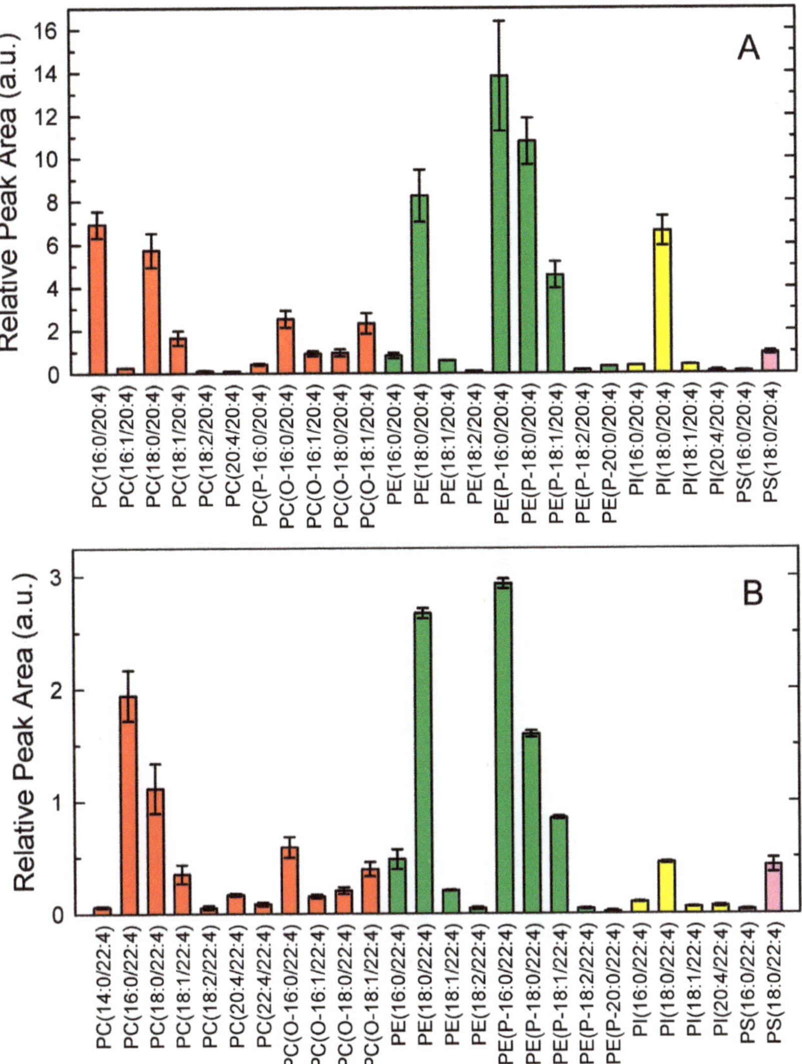

Figure 2. AA and AdA-containing phospholipid molecular species in peritoneal macrophages The profiles of AA- (**A**) or AdA- (**B**) containing PC (red), PE (green), PI (yellow), and PS (pink) species in peritoneal macrophages were determined by liquid chromatography/mass spectrometry (LC/MS). Fatty chains within the different phospholipid species are designated by their numbers of carbons and double bonds. A designation of O- before the first fatty chain indicates that the *sn*-1 position is ether linked, whereas a P- designation indicates a plasmalogen form (*sn*-1 vinyl ether linkage) [47]. Phospholipids containing two ester bonds have no designation. Results are shown as mean values ± standard error of the mean ($n = 3$).

Regarding AdA-containing species, the alkenyl acyl and diacyl ethanolamine phospholipid species PE(P-16:0/22:4), PE(P-18:0/22:4), and PE(18:0/22:4)], and the diacyl choline phospholipid species PC(16:0/22:4) and PC(18:0/22:4) also constituted the major cellular AdA reservoirs. Strikingly, the inositol phospholipid species PI(18:0/22:4) was not as prevalent as its AA equivalent, PI(18:0/20:4), was among AA-containing phospholipids. This may suggest that the acyl-CoA acyltransferase using lysoPI as the acceptor [48] shows selectivity for AA over AdA as a substrate.

Macrophage stimulation with yeast-derived zymosan markedly decreased the cellular AA content in PC and PI. Despite PE being the major AA-containing class, AA losses from PE did not reach statistical significance (Figure 3A). It should be noted in this regard that during receptor stimulation, AA is known to be transferred from AA-containing PC (1-acyl species) to PE (plasmalogen species) by CoA-independent transacylase; hence, the decline in the amount of AA-containing PE during cellular stimulation may be greatly reduced [22,24,49]. Regarding AdA, decreases in its cellular content were also observed after zymosan stimulation. However the pattern clearly differed in that PC was the only phospholipid class that contributed to AdA mobilization; AdA reductions from PE and PI did not reach statistical significance (Figure 3B).

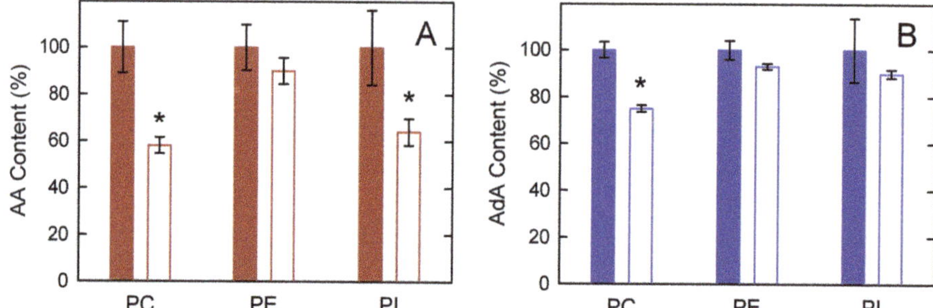

Figure 3. AA and AdA mobilization in zymosan-stimulated macrophages. The cells were unstimulated (colored bars) or stimulated (open bars) with 1 mg/mL zymosan for 1 h. Afterward, total content of AA (**A**) or AdA (**B**) in various phospholipid classes (see x-axis) was measured by GC/MS. To allow for direct comparison between AA and AdA, the fatty acid contents in phospholipid classes in unstimulated cells are given as 100%. Actual mass values can be obtained from Figure 1. Results are shown as mean values ± standard error of the mean. ($n = 3$). Note: * $p < 0.05$, significantly different from the corresponding phospholipid class.

3.2. *Defining the Role of Various PLA_2 Forms in Stimulus-Induced AA and AdA Mobilization*

The differential involvement of phospholipid classes in AA and AdA mobilization suggests the existence of separate mechanisms for regulating the availability of each fatty acid. To characterize such mechanisms, we took advantage of the use of selective inhibitors of the two major intracellular PLA_2 enzymes potentially effecting the fatty acid release in activated peritoneal macrophages, namely group IVA Ca^{2+}-dependent cytosolic PLA_2 (c$PLA_2\alpha$) and group VIA Ca^{2+}-independent PLA_2 (iPLA_2-VIA) [22, 50]. The use of selective chemical inhibitors to address the role of intracellular PLA_2s during cell activation has some advantages over other widely used methods, such as small interfering RNA or cells from knockout mice. With chemical inhibitors, inhibition develops rapidly, which reduces the impact of unspecific effects that could occur over time. Also, no compensatory mechanisms take place that might obscure the interpretation of results [22]. In addition, chemical inhibitors directly target PLA_2 effects that depend on enzymatic activity, without affecting noncatalytic functions of the enzyme [51]. The inhibitors used in this work are the most potent and selective inhibitors currently available to block c$PLA_2\alpha$ and iPLA_2-VIA in cells. Pyrrophenone potently and selectively inhibits c$PLA_2\alpha$ activity using a number of in vitro assays without detectable effects on other PLA_2 activities, blocks AA release in mammalian cells in the 0.01–1 μM range [29,52,53], and has been shown to be effective in experimental models of disease involving cPLA$_2\alpha$ [54,55]. Fluoroketone FKGK18, a selective iPLA$_2$-VIA inhibitor, is at least 200-fold more potent for inhibiting iPLA$_2$-VIA than cPLA$_2\alpha$ [30], and is useful for characterizing iPLA$_2$-VIA-mediated functions in vivo [31,56,57]. β-Lactone GK436, another iPLA$_2$-VIA inhibitor, has been found to be at least 1000-fold more potent for iPLA$_2$-VIA than for cPLA$_2\alpha$ [31]. For comparative purposes with data from the bibliography [22,58–64], we also used bromoenol lactone (BEL) to inhibit iPLA$_2$-VIA. BEL inhibits calcium-independent PLA$_2$s and exerts little or no effect on Ca^{2+}-dependent enzymes [65,66], albeit it may exhibit off-target effects depending on cell type [67–69].

In agreement with previous observations [22,24], pyrrophenone was found to almost completely inhibit zymosan-stimulated AA mobilization from phospholipids (Figure 4A). Importantly, pyrrophenone also partly inhibited AdA mobilization (Figure 4B). Conversely, none of the iPLA$_2$-VIA inhibitors tested exerted appreciable effects on AA mobilization while significantly affecting AdA mobilization (Figure 4). The simultaneous addition of inhibitors of both cPLA$_2\alpha$ and iPLA$_2$-VIA resulted in complete inhibition of AdA mobilization (Figure 4B). These results suggest that unlike AA, AdA mobilization involves the action of both cPLA$_2\alpha$ and iPLA$_2$-VIA.

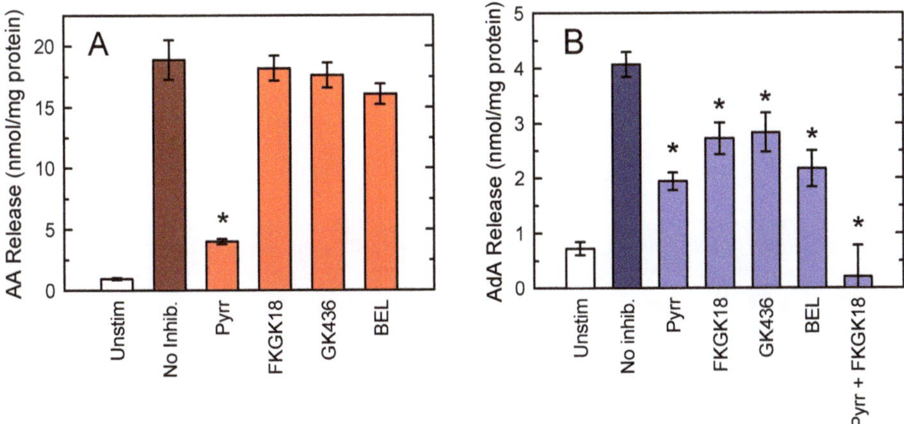

Figure 4. Release of AA and AdA by stimulated macrophages. The cells were either unstimulated (open bars) or stimulated by 1 mg/mL zymosan for 1 h (colored bars) in the absence (no inhibition) or presence of the following inhibitors: 2 μM pyrrophenone (Pyrr), 10 μM FKGK18, 5 μM GK436, 10 μM BEL, or 2 μM pyrrophenone plus 10 μM FKGK18 (see *x*-axis). Afterward, total content of AA (**A**) or AdA (**B**) was measured by GC/MS. The fatty acid release was calculated by subtracting the amount of phospholipid-bound AA or AdA in stimulated cells from that in unstimulated cells. Results are shown as mean values ± standard error of the mean. (n = 3). Note: * $p < 0.05$, significantly different from zymosan-stimulated cells in the absence of inhibitors.

3.3. Studies with AdA-Enriched Cells

Given the structural similarities between AA and AdA, it could be envisioned that in some instances, AdA competes with AA for incorporation into phospholipids, which might result in reduced amounts of "mobilizable" AA within phospholipids, and hence reduced formation of pro-inflammatory eicosanoids [16,70,71]. To address this possibility, we incubated the macrophages with exogenous AdA (10 μM, 20 h), which resulted in the cells avidly incorporating the fatty acid into cellular phospholipids. These conditions led to a 2–3-fold increase in the amount of AdA esterified into phospholipids compared with untreated cells (33.2 ± 4.6 nmol per mg protein; mean ± standard error of the mean, $n = 4$). Analysis by GC/MS of the fatty acid content of the AdA-enriched cells revealed that the fatty acids typically displaced within phospholipids by AdA upon AdA supplementation were oleic acid (18:1n-9) and the n-6 series members linoleic (18:2n-6) and dihomo-γ-linolenic (20:3n-6) acids

(Figure 5). Remarkably, little AA displacement occurred after AdA supplementation. Analysis of AA and AdA mobilization in zymosan-stimulated cells indicated that the extent of the AA response did not substantially differ from that observed from cells not treated with AdA, while AdA mobilization expectedly increased (Figure 6).

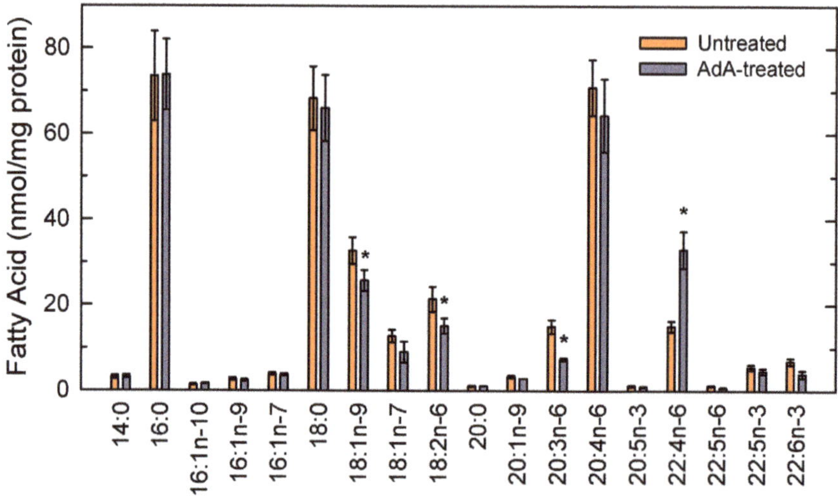

Figure 5. Fatty acid composition of murine peritoneal macrophages. The cells were either untreated (orange bars) or treated with exogenous AdA for 20 h (gray bars). Afterward, the total fatty acid profiles in cells were determined by GC/MS after converting the fatty acid glyceryl esters into fatty acid methyl esters. The fatty acids are designated by their number of carbon atoms, and after a colon, their number of double bonds. To differentiate isomers, the n − x (n minus x) nomenclature is used, where n is the number of carbons of a given fatty acid and x is an integer that gives the position of the last double bond of the molecule when subtracted from n. The data are expressed as mean values ± standard error of the mean of three individual replicates. Note: * $p < 0.05$, significantly different from incubations in the absence of AdA.

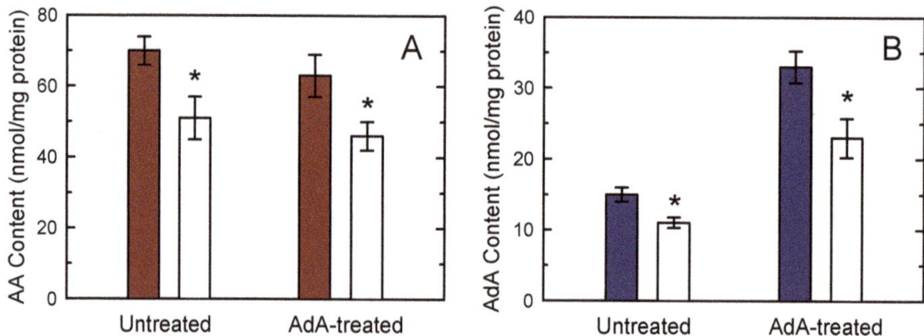

Figure 6. AA and AdA mobilization in zymosan-stimulated macrophages. The cells were either untreated or treated with exogenous AdA for 20 h (see x-axis). Afterward, the cells were unstimulated (colored bars) or stimulated (open bars) with 1 mg/mL zymosan for 1 h, and cellular AA content (**A**) or AdA content (**B**) was determined by GC/MS. Results are shown as mean values ± standard error of the mean ($n = 3$). Note: * $p < 0.05$, significantly different from unstimulated cells.

Analysis of AA-derived metabolites produced by zymosan-stimulated cells did not appreciably change whether the cells had previously been treated with AdA or not, while the levels of the only AdA metabolite detected at significant levels in the stimulated macrophages, dihomoprostaglandin E_2, increased in the AdA-treated cells (Figure 7). Collectively, these results suggest that in zymosan-activated macrophages, AdA does not influence AA metabolism leading to eicosanoid production.

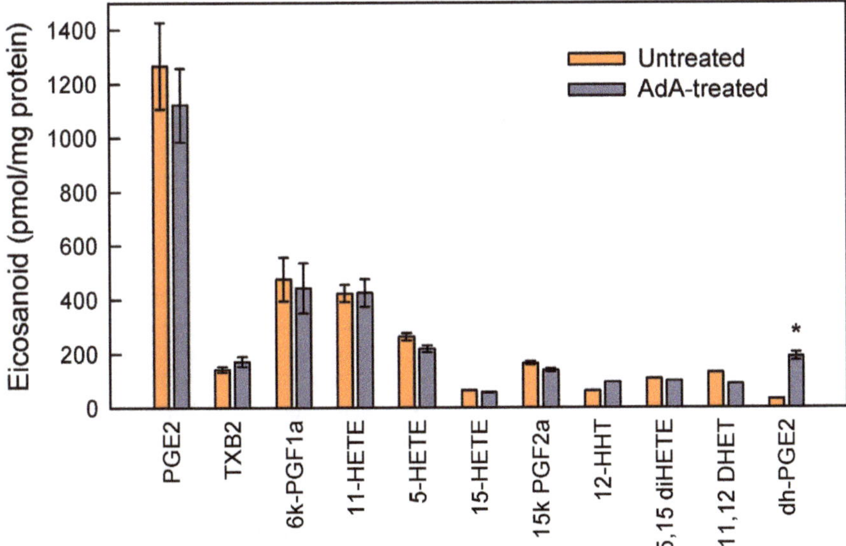

Figure 7. Eicosanoid production by stimulated macrophages. The cells were either untreated (orange bars) or treated with exogenous AdA for 20 h (gray bars). Afterward, the cells were stimulated by 1 mg/mL zymosan for 1 h, and eicosanoid content in the supernatants was analyzed by LC/MS. The data are expressed as mean values ± standard error of the mean of three individual replicates. Note: * $p < 0.05$, significantly different from incubations in the absence of AdA treatment. PGE2, prostaglandin E_2; TXB2, thromboxane B_2; 6k-PGF1a, 6-keto prostaglandin $F_{1\alpha}$ (the stable product of prostaglandin I_2); 11-HETE, 11-hydroxyeicosatetraenoic acid; 5-HETE, 5-hydroxyeicosatetraenoic acid; 15-HETE, 15-hydroxyeicosatetraenoic acid; 15k-PGF2a, 15-ketoprostaglandin $F_{2\alpha}$; 12-HHT, 12-hydroxyheptadecatrienoic acid; 5,15-diHETE, 5,15-dihydroxyeicosatetraenoic acid; 11,12-DHET, 11,12-dihydroxyeicosatrienoic acid; dh-PGE2, dihomoprostaglandin E_2.

4. Discussion

AdA is the 2-carbon elongation product of AA. Similar to AA, AdA can be mobilized from membrane phospholipids to serve as a substrate for the production of oxidized metabolites with a diverse array of biological functions [11–14,72]. We show in this work that AdA is present at significant quantities in the membrane phospholipids of murine peritoneal macrophages. The AdA-to-AA ratio in macrophages is 20–25%, which is similar to that found in other cells [13,14]. Stimulation of the macrophages with yeast-derived zymosan results in the cells releasing significant amounts of AdA, part of which is metabolized to dihomoprostaglandin E_2. While macrophages appear to utilize AA for lipid mediator synthesis more efficiently than AdA, both qualitatively and quantitatively, our results support the concept that neither fatty acid competes with the other. Further, the data suggest that the processes leading to the mobilization and metabolism of both fatty acids are independently regulated. In support of these observations, enriching the cells with exogenous AdA has little effect on endogenous AA levels. AA release and metabolism in these AdA-enriched cells proceed in essentially the same way as in cells displaying normal levels of AdA. Consistent with these observations, studies by others have highlighted that under certain settings, some AdA-derived metabolites may display higher potency than their AA-derived counterparts, suggesting that they may serve non-redundant functions on their own [13,14,72–75]. Collectively, the results suggest that separate and selective mechanisms may exist for regulating AA and AdA utilization in macrophages.

Current evidence clearly indicates that cPLA$_2\alpha$ is the major effector governing AA mobilization from membrane phospholipids in innate immune cells [55,76,77]. In contrast with the plethora of data regarding AA, no studies have been available to date identifying the PLA$_2$ enzyme(s) responsible for effecting AdA release in activated cells. This work demonstrates the involvement of two PLA$_2$ enzymes, namely cPLA$_2\alpha$ and iPLA$_2$-VIA, in agonist-induced AdA mobilization from activated macrophages, which contrasts with the sole involvement of cPLA$_2\alpha$ in AA mobilization. This is a striking finding because a large number of studies have suggested that iPLA$_2$-VIA does not participate, or plays a very limited role, in AA mobilization and attendant eicosanoid production in innate immunity and inflammation reactions [22,78–81]. The latter responses are almost completely suppressed in macrophages derived from cPLA$_2\alpha$-deficient mice compared with cells from wild-type mice [50].

In our studies, we have utilized yeast-derived zymosan particles. These are cell wall preparations from *S. cerevisiae* which have been extensively used as models for the induction of inflammatory responses to fungal infection [82]. Although zymosan activates many macrophage responses via engagement of toll-like receptor 2 (TLR2) [82], studies utilizing genetically-deficient mice have demonstrated that the receptor that couples fungal responses to cPLA$_2\alpha$ activation and enhanced AA release are dectin-1 and dectin-2, not TLR2. [83,84]. Interestingly, macrophages exposed to the Gram-positive bacterium *L. monocytogenes* do release AA via engagement of TLR2 [85]. Stimulation of macrophages via TLR4 by lipopolysaccharide from Gram-negative bacteria promotes abundant AA mobilization in macrophage cell lines [86,87], but does so very poorly in primary macrophages [88]. Thus, multiple receptors on the surface of macrophages exist that can potentially induce cPLA$_2\alpha$ activation and attendant AA release, and the involvement of one or another depends on the nature of the triggering stimulus. Future work should be aimed at defining whether, depending on stimulus, increased AdA mobilization also occurs through multiple macrophage receptors.

Although structurally related to AA, AdA does not have a double bond at C5, a potentially important feature for substrate recognition by cPLA$_2\alpha$ [89]. In our study, cPLA$_2\alpha$ inhibition by pyrrophenone blocks AA release to a larger extent than AdA release, which may reflect the higher affinity of cPLA$_2\alpha$ for AA-containing phospholipids. It has been shown that cPLA$_2\alpha$ has a deep and rigid channel-like active site that is able to accommodate a phospholipid substrate molecule in its entirety [90]. This confers the enzyme preference for phospholipids containing AA at the *sn*-2 position [90]. In contrast, iPLA$_2$-VIA contains a more flexible and versatile active site; thus, this enzyme exhibits a more permissive specificity for the fatty acid at the *sn*-2 position [90]. In spite of these observations, our data using a live cell system show that cPLA$_2\alpha$ is able to cleave both AA and AdA-containing substrates, while iPLA$_2$-VIA cleaves only AdA-containing phospholipids. This raises the intriguing concept that in cells not only the inherent substrate specificity of each PLA$_2$ determines phospholipid hydrolysis, and other factors should be taken into account. Among these, PLA$_2$ accessibility to its substrate within the cell may profoundly limit free fatty acid release and attendant generation of biologically active mediators. Analysis of phospholipid molecular species containing either AA or AdA in the macrophages reflect no major differences (i.e., the major species containing AA also tend to be the major species containing AdA). However, there is a clear difference between the phospholipid pools used for the release of AA or AdA. While free AdA appears to proceed only from the hydrolysis of PC molecules, free AA appears to come from the hydrolysis of PC and PI, and perhaps PE as well. As discussed elsewhere [3,22,24,49], the lack of AA mobilization from PE probably reflects the involvement of CoA-independent transacylases, which rapidly restore the levels of AA in PE at the expense of AA-containing PC. In a similar vein, it could be argued that the lack of hydrolysis of AdA-containing PE could also be due to the replenishing action of CoA-independent transacylation reactions restoring AdA in PE. However, to the best of our knowledge, these reactions have not been shown to involve inositol phospholipids [91,92]. Thus, the lack of AdA hydrolysis from PI may likely reflect the inability of both cPLA$_2\alpha$ and iPLA$_2$-VIA to reach this particular substrate. Hence, while the AA-containing PI pool is accessible to phospholipase attack, the AdA-containing PI pool is not. Collectively, our results highlight fundamental differences between the mobilization of AA versus AdA in activated macrophages in terms of the PLA$_2$ enzymes involved and the phospholipid class used as a substrate (Scheme 2). The data provide further support to the concept that intracellular substrate compartmentalization may limit the synthesis of fatty-acid-derived mediators.

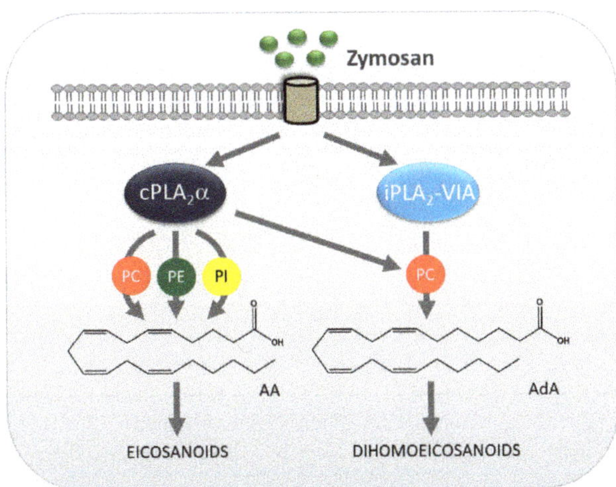

Scheme 2. The scheme shows that cPLA$_2\alpha$ and iPLA$_2$-VIA are differentially involved in AA and AdA mobilization by zymosan-stimulated mouse peritoneal macrophages.

5. Conclusions

Results from this study constitute an initial report addressing the mechanisms of AdA mobilization from membrane phospholipids in cells of innate immunity and inflammation. We demonstrated that AdA mobilization is mediated by two different PLA$_2$s, namely cPLA$_2\alpha$ and iPLA$_2$-VIA (Scheme 2). Further studies will be needed to precisely determine the contribution of each PLA$_2$ to overall AdA release under different stimulation conditions, and to assess whether cross-talk exists between the two enzymes. Elucidation of the innate immune functions of AdA and its metabolites may not prove to be an easy task, because inhibitors that block cPLA$_2\alpha$ or AdA metabolism via cyclooxygenase, lipoxygenase, or CYP450 pathways will also affect AA and its metabolism. Given the sole involvement of iPLA$_2$-VIA in AdA release but not in AA release, selective inhibition of this PLA$_2$ could be considered to evaluate the effect of altering AdA metabolism on the outcome of immunoinflammatory diseases.

Author Contributions: Conceptualization, M.A.B. and J.B.; funding acquisition, M.A.B. and J.B.; investigation, P.M., A.G., J.M.R., and V.M.; methodology, P.M., A.G., J.M.R., V.M., and G.K.; project administration, M.A.B. and J.B.; supervision, M.A.B. and J.B.; writing—original draft, J.B.; writing—review and editing, G.K., M.A.B., and J.B. All authors have read and agreed to the published version of the manuscript.

Funding: This research was funded by the Spanish Ministry of Economy, Industry, and Competitiveness, grant number SAF2016-80883-R. P.M. and A.G. were supported by predoctoral fellowships from the Regional Government of Castile and Leon. CIBERDEM is an initiative of Instituto de Salud Carlos III.

Acknowledgments: We thank Montse Duque for excellent technical assistance.

Conflicts of Interest: The authors declare no conflict of interest. The funders had no role in the design of the study; in the collection, analyses, or interpretation of data; in the writing of the manuscript; or in the decision to publish the results.

Abbreviations

AA: arachidonic acid; AdA: adrenic acid; cPLA$_2\alpha$: group IVA Ca^{2+}-dependent cytosolic phospholipase A$_2\alpha$; GC/MS: gas chromatography coupled to mass spectrometry; iPLA$_2$-VIA: group IVA Ca^{2+}-independent phospholipase A$_2$; LC/MS: liquid chromatography coupled to mass spectrometry; PC: choline-containing phospholipids; PE: ethanolamine-containing phospholipids; PI: phosphatidylinositol; PLA$_2$: phospholipase A$_2$; PS: phosphatidylserine.

References

1. Astudillo, A.M.; Meana, C.; Guijas, C.; Pereira, L.; Lebrero, R.; Balboa, M.A.; Balsinde, J. Occurrence and biological activity of palmitoleic acid isomers in phagocytic cells. *J. Lipid Res.* **2018**, *59*, 237–249. [CrossRef]
2. Dennis, E.A.; Norris, P.C. Eicosanoid storm in infection and inflammation. *Nat. Rev. Immunol.* **2015**, *15*, 511–523. [CrossRef]
3. Astudillo, A.M.; Balboa, M.A.; Balsinde, J. Selectivity of phospholipid hydrolysis by phospholipase A$_2$ enzymes in activated cells leading to polyunsaturated fatty acid mobilization. *Biochim. Biophys. Acta* **2019**, *1864*, 772–783. [CrossRef]
4. Guijas, C.; Rodríguez, J.P.; Rubio, J.M.; Balboa, M.A.; Balsinde, J. Phospholipase A$_2$ regulation of lipid droplet formation. *Biochim. Biophys. Acta* **2014**, *1841*, 1661–1671. [CrossRef]
5. Astudillo, A.M.; Balgoma, D.; Balboa, M.A.; Balsinde, J. Dynamics of arachidonic acid mobilization by inflammatory cells. *Biochim. Biophys. Acta* **2012**, *1821*, 249–256. [CrossRef]
6. Pérez-Chacón, G.; Astudillo, A.M.; Balgoma, D.; Balboa, M.A.; Balsinde, J. Control of free arachidonic acid levels by phospholipases A$_2$ and lysophospholipid acyltransferases. *Biochim. Biophys. Acta* **2009**, *1791*, 1103–1113.
7. Peters-Golden, M.; Canetti, C.; Mancuso, P.; Coffey, M.J. Leukotrienes: Underappreciated mediators of innate immune responses. *J. Immunol.* **2005**, *174*, 589–594. [CrossRef] [PubMed]
8. Serhan, C.N. Resolution phase of inflammation: Novel endogenous antiinflammatory and proresolving lipid mediators and pathways. *Ann. Rev. Immunol.* **2007**, *25*, 101–137. [CrossRef] [PubMed]
9. Ricciotti, E.; Fitzgerald, G.A. Prostaglandins and inflammation. *Arterioscler. Thromb. Vasc. Biol.* **2011**, *31*, 986–1000. [CrossRef]
10. Kalinski, P. Regulation of immune responses by prostaglandin E$_2$. *J. Immunol.* **2012**, *188*, 21–28. [CrossRef]
11. Mann, C.J.; Kaduce, T.L.; Figard, P.H.; Spector, A.A. Docosatetraenoic acid in endothelial cells: Formation, retroconversion to arachidonic acid and effect on prostacyclin production. *Arch. Biochem. Biophys.* **1986**, *244*, 813–823. [CrossRef]
12. Campbell, W.B.; Falck, J.R.; Okita, J.R.; Johnson, A.R.; Callahan, K.S. Synthesis of dihomoprostaglandins from adrenic acid (7,10,13,16-docosatetraenoic acid) by human endothelial cells in culture. *Biochim. Biophys. Acta* **1985**, *837*, 67–76. [CrossRef]
13. Yi, X.Y.; Gauthier, K.M.; Cui, L.; Nithipatikom, K.; Falck, J.R.; Campbell, W.B. Metabolism of adrenic acid to vasodilatory 1α,1β-dihomo-epoxyeicosatrienoic acids by bovine coronary arteries. *Am. J. Physiol.* **2007**, *292*, H2265–H2274. [CrossRef] [PubMed]
14. Kopf, P.G.; Zhang, D.X.; Gauthier, K.M.; Nithipatikom, K.; Yi, X.Y.; Falck, J.R.; Campbell, W.B. Adrenic acid metabolites as endogenous endothelium-derived and zona glomerulosa-derived hyperpolarizing actors. *Hypertension* **2010**, *55*, 547–554. [CrossRef] [PubMed]
15. Nababan, S.H.H.; Nishiumi, S.; Kawano, Y.; Kobayashi, T.; Yoshida, M.; Azuma, T. Adrenic acid as an inflammation enhancer in non-alcoholic fatty liver disease. *Arch Biochem. Biophys.* **2017**, *623*, 64–75. [CrossRef] [PubMed]
16. Brouwers, H.; Jonasdottir, H.; Kwekkeboom, J.; López-Vicario, C.; Claria, J.; Freysdottir, J.; Hardardottir, I.; Huizinga, T.; Toes, R.; Giera, M.; et al. Adrenic acid as a novel anti-inflammatory player in osteoarthritis. *Osteoarthr. Cartil.* **2018**, *26*, S126. [CrossRef]
17. Guijas, C.; Astudillo, A.M.; Gil-de-Gómez, L.; Rubio, J.M.; Balboa, M.A.; Balsinde, J. Phospholipid sources for adrenic acid mobilization in RAW 264.7 macrophages: Comparison with arachidonic acid. *Biochim. Biophys. Acta* **2012**, *1821*, 1386–1393. [CrossRef]

18. Balgoma, D.; Astudillo, A.M.; Pérez-Chacón, G.; Montero, O.; Balboa, M.A.; Balsinde, J. Markers of monocyte activation revealed by lipidomic profiling of arachidonic acid-containing phospholipids. *J. Immunol.* **2010**, *184*, 3857–3865. [CrossRef]
19. Astudillo, A.M.; Pérez-Chacón, G.; Meana, C.; Balgoma, D.; Pol, A.; del Pozo, M.A.; Balboa, M.A.; Balsinde, J. Altered arachidonate distribution in macrophages from caveolin-1 null mice leading to reduced eicosanoid synthesis. *J. Biol. Chem.* **2011**, *286*, 35299–35307. [CrossRef]
20. Valdearcos, M.; Esquinas, E.; Meana, C.; Gil-de-Gómez, L.; Guijas, C.; Balsinde, J.; Balboa, M.A. Subcellular localization and role of lipin-1 in human macrophages. *J. Immunol.* **2011**, *186*, 6004–6013. [CrossRef]
21. Gil-de-Gómez, L.; Astudillo, A.M.; Meana, C.; Rubio, J.M.; Guijas, C.; Balboa, M.A.; Balsinde, J. A phosphatidylinositol species acutely generated by activated macrophages regulates innate immune responses. *J. Immunol.* **2013**, *190*, 5169–5177. [CrossRef] [PubMed]
22. Gil-de-Gómez, L.; Astudillo, A.M.; Guijas, C.; Magrioti, V.; Kokotos, G.; Balboa, M.A.; Balsinde, J. Cytosolic group IVA and calcium-independent group VIA phospholipase A_2s act on distinct phospholipid pools in zymosan-stimulated mouse peritoneal macrophages. *J. Immunol.* **2014**, *192*, 752–762. [CrossRef] [PubMed]
23. Rubio, J.M.; Rodríguez, J.P.; Gil-de-Gómez, L.; Guijas, C.; Balboa, M.A.; Balsinde, J. Group V secreted phospholipase A_2 is up-regulated by interleukin-4 in human macrophages and mediates phagocytosis via hydrolysis of ethanolamine phospholipids. *J. Immunol.* **2015**, *194*, 3327–3339. [CrossRef] [PubMed]
24. Gil-de-Gómez, L.; Astudillo, A.M.; Lebrero, P.; Balboa, M.A.; Balsinde, J. Essential role for ethanolamine plasmalogen hydrolysis in bacterial lipopolysaccharide priming of macrophages for enhanced arachidonic acid release. *Front. Immunol.* **2017**, *8*, 1251. [CrossRef] [PubMed]
25. Rubio, J.M.; Astudillo, A.M.; Casas, J.; Balboa, M.A.; Balsinde, J. Regulation of phagocytosis in macrophages by membrane ethanolamine plasmalogens. *Front. Immunol.* **2018**, *9*, 1723.
26. Lebrero, P.; Astudillo, A.M.; Rubio, J.M.; Fernández-Caballero, J.; Kokotos, G.; Balboa, M.A.; Balsinde, J. Cellular plasmalogen content does not influence arachidonic acid levels or distribution in macrophages: A role for cytosolic phospholipase $A_2\gamma$ in phospholipid remodeling. *Cells* **2019**, *8*, 799.
27. Guijas, C.; Bermúdez, M.A.; Meana, C.; Astudillo, A.M.; Pereira, L.; Fernández-Caballero, L.; Balboa, M.A.; Balsinde, J. Neutral lipids are not a source of arachidonic acid for lipid mediator signaling in human foamy monocytes. *Cells* **2019**, *8*, 941.
28. Chilton, F.H. Potential phospholipid source(s) of arachidonate used for the synthesis of leukotrienes by the human neutrophil. *Biochem. J.* **1989**, *258*, 327–333. [CrossRef]
29. Ono, T.; Yamada, K.; Chikazawa, Y.; Ueno, M.; Nakamoto, S.; Okuno, T.; Seno, K. Characterization of a novel inhibitor of cytosolic phospholipase $A_2\alpha$, pyrrophenone. *Biochem. J.* **2002**, *363*, 727–735. [CrossRef]
30. Kokotos, G.; Hsu, Y.H.; Burke, J.E.; Baskakis, C.; Kokotos, C.G.; Magrioti, V.; Dennis, E.A. Potent and selective fluoroketone inhibitors of group VIA calcium-independent phospholipase A_2. *J. Med. Chem.* **2010**, *53*, 3602–3610. [CrossRef]
31. Dedaki, C.; Kokotou, M.G.; Mouchlis, V.D.; Limnios, D.; Lei, X.; Mu, C.T.; Ramanadham, S.; Magrioti, V.; Dennis, E.A.; Kokotos, G. β-Lactones: A novel class of Ca^{2+}-independent phospholipase A_2 (group VIA $iPLA_2$) inhibitors with the ability to inhibit β-cell apoptosis. *J. Med. Chem.* **2019**, *62*, 2916–2927. [CrossRef] [PubMed]
32. Balsinde, J.; Fernández, B.; Diez, E. Regulation of arachidonic acid release in mouse peritoneal macrophages. The role of extracellular calcium and protein kinase C. *J. Immunol.* **1990**, *144*, 4298–4304. [PubMed]
33. Balsinde, J.; Fernández, B.; Solís-Herruzo, J.A.; Diez, E. Pathways for arachidonic acid mobilization in zymosan-stimulated mouse peritoneal macrophages. *Biochim. Biophys. Acta* **1992**, *1136*, 75–82. [CrossRef]
34. Balboa, M.A.; Sáez, Y.; Balsinde, J. Calcium-independent phospholipase A_2 is required for lysozyme secretion in U937 promonocytes. *J. Immunol.* **2003**, *170*, 5276–5280. [CrossRef] [PubMed]
35. Balboa, M.A.; Pérez, R.; Balsinde, J. Amplification mechanisms of inflammation: Paracrine stimulation of arachidonic acid mobilization by secreted phospholipase A_2 is regulated by cytosolic phospholipase A_2-derived hydroperoxyeicosatetraenoic acid. *J. Immunol.* **2003**, *171*, 989–994. [CrossRef]
36. Balsinde, J.; Balboa, M.A.; Insel, P.A.; Dennis, E.A. Differential regulation of phospholipase D and phospholipase A_2 by protein kinase C in $P388D_1$ macrophages. *Biochem. J.* **1997**, *321*, 805–809. [CrossRef]
37. Pérez, R.; Matabosch, X.; Llebaria, A.; Balboa, M.A.; Balsinde, J. Blockade of arachidonic acid incorporation into phospholipids induces apoptosis in U937 promonocytic cells. *J. Lipid Res.* **2006**, *47*, 484–491. [CrossRef]

38. Bradford, M.M. A rapid and sensitive method for the quantitation of microgram quantities of protein utilizing the principle of protein-dye binding. *Anal. Biochem.* **1976**, *72*, 248–254. [CrossRef]
39. Bligh, E.G.; Dyer, W.J. A rapid method of total lipid extraction and purification. *Can. J. Biochem. Physiol.* **1959**, *37*, 911–917. [CrossRef]
40. Diez, E.; Balsinde, J.; Aracil, M.; Schüller, A. Ethanol induces release of arachidonic acid but not synthesis of eicosanoids in mouse peritoneal macrophages. *Biochim. Biophys. Acta* **1987**, *921*, 82–89. [CrossRef]
41. Fine, J.B.; Sprecher, H. Unidimensional thin-layer chromatography of phospholipids on boric acid-impregnated plates. *J. Lipid Res.* **1982**, *23*, 660–663. [PubMed]
42. Astudillo, A.M.; Pérez-Chacón, G.; Balgoma, D.; Gil-de-Gómez, L.; Ruipérez, V.; Guijas, C.; Balboa, M.A.; Balsinde, J. Influence of cellular arachidonic acid levels on phospholipid remodeling and CoA-independent transacylase activity in human monocytes and U937 cells. *Biochim. Biophys. Acta* **2011**, *1811*, 97–103. [CrossRef] [PubMed]
43. Guijas, C.; Pérez-Chacón, G.; Astudillo, A.M.; Rubio, J.M.; Gil-de-Gómez, L.; Balboa, M.A.; Balsinde, J. Simultaneous activation of p38 and JNK by arachidonic acid stimulates the cytosolic phospholipase A_2-dependent synthesis of lipid droplets in human monocytes. *J. Lipid Res.* **2012**, *53*, 2343–2354. [CrossRef] [PubMed]
44. Guijas, C.; Meana, C.; Astudillo, A.M.; Balboa, M.A.; Balsinde, J. Foamy monocytes are enriched in *cis*-7-hexadecenoic fatty acid (16:1*n*-9), a possible biomarker for early detection of cardiovascular disease. *Cell Chem. Biol.* **2016**, *23*, 689–699. [CrossRef] [PubMed]
45. Rodríguez, J.P.; Guijas, C.; Astudillo, A.M.; Rubio, J.M.; Balboa, M.A.; Balsinde, J. Sequestration of 9-hydroxystearic acid in FAHFA (fatty acid esters of hydroxy fatty acids) as a protective mechanism for colon carcinoma cells to avoid apoptotic cell death. *Cancers* **2019**, *11*, 524. [CrossRef] [PubMed]
46. Axelsen, P.H.; Murphy, R.C. Quantitative analysis of phospholipids containing arachidonate and docosahexaenoate chains in microdissected regions of mouse brain. *J. Lipid Res.* **2010**, *51*, 660–671. [CrossRef]
47. Fahy, E.; Subramaniam, S.; Brown, H.A.; Glass, C.K.; Merrill, A.H., Jr.; Murphy, R.C.; Raetz, C.R.; Russell, D.W.; Seyama, Y.; Shaw, W.; et al. A comprehensive classification system for lipids. *J. Lipid Res.* **2005**, *46*, 839–861. [CrossRef]
48. Murphy, R.C.; Folco, G. Lysophospholipid acyltransferases and leukotriene biosynthesis: Intersection of the Lands cycle and the arachidonate PI cycle. *J. Lipid Res.* **2019**, *60*, 219–226. [CrossRef]
49. Rouzer, C.A.; Ivanova, P.T.; Byrne, M.O.; Milne, S.B.; Brown, H.A.; Marnett, L.J. Lipid profiling reveals glycerophospholipid remodeling in zymosan-stimulated macrophages. *Biochemistry* **2007**, *46*, 6026–6042. [CrossRef]
50. Gijón, M.A.; Spencer, D.M.; Siddiqi, A.R.; Bonventre, J.V.; Leslie, C.C. Cytosolic phospholipase A_2 is required for macrophage arachidonic acid release by agonists that do and do not mobilize calcium. *J. Biol. Chem.* **2000**, *275*, 20146–20156. [CrossRef]
51. Nikolaou, A.; Kokotou, M.G.; Vasilakaki, S.; Kokotos, G. Small-molecule inhibitors as potential therapeutics and as tools to understand the role of phospholipases A_2. *Biochim. Biophys. Acta* **2019**, *1864*, 941–956. [CrossRef] [PubMed]
52. Balboa, M.A.; Balsinde, J. Involvement of calcium-independent phospholipase A_2 in hydrogen peroxide-induced accumulation of free fatty acids in human U937 cells. *J. Biol. Chem.* **2002**, *277*, 40384–40389. [CrossRef] [PubMed]
53. Ghomashchi, F.; Stewart, A.; Hefner, Y.; Ramanadham, S.; Turk, J.; Leslie, C.C.; Gelb, M.H. A pyrrolidine-based specific inhibitor of cytosolic phospholipase $A_2\alpha$ blocks arachidonic acid release in a variety of mammalian cells. *Biochim. Biophys. Acta* **2001**, *1513*, 160–166. [CrossRef]
54. Dennis, E.A.; Cao, J.; Hsu, Y.H.; Magrioti, V.; Kokotos, G. Phospholipase A_2 enzymes: Physical structure, biological function, disease implication, chemical inhibition, and therapeutic intervention. *Chem. Rev.* **2011**, *111*, 6130–6185. [CrossRef]
55. Bonventre, J.V. Cytosolic phospholipase $A_2\alpha$ reigns supreme in arthritis and bone resorption. *Trends Immunol.* **2004**, *25*, 116–119. [CrossRef]
56. Bone, R.N.; Gai, Y.; Magrioti, V.; Kokotou, M.G.; Ali, T.; Lei, X.; Tse, H.M.; Kokotos, G.; Ramanadham, S. Inhibition of Ca^{2+}-independent phospholipase $A_2\beta$ (iPL$A_2\beta$) ameliorates islet infiltration and incidence of diabetes in NOD mice. *Diabetes* **2015**, *64*, 541–554. [CrossRef]

57. Ali, T.; Kokotos, G.; Magrioti, V.; Bone, R.N.; Mobley, J.A.; Hancock, W.; Ramanadham., S. Characterization of FKGK18 as inhibitor of group VIA Ca^{2+}-independent phospholipase A_2 (iPLA$_2\beta$): Candidate drug for preventing beta-cell apoptosis and diabetes. *PLoS ONE* **2013**, *8*, e71748. [CrossRef]
58. Akiba, S.; Mizunaga, S.; Kume, K.; Hayama, M.; Sato, T. Involvement of Group VI Ca^{2+}-independent phospholipase A_2 in protein kinase C-dependent arachidonic acid liberation in zymosan-stimulated macrophage-like P388D$_1$ cells. *J. Biol. Chem.* **1999**, *274*, 19906–19912. [CrossRef]
59. Balsinde, J.; Balboa, M.A.; Dennis, E.A. Identification of a third pathway for arachidonic acid mobilization and prostaglandin production in activated P388D$_1$ macrophage-like cells. *J. Biol. Chem.* **2000**, *275*, 22544–22549. [CrossRef]
60. Balsinde, J.; Dennis, E.A. Distinct roles in signal transduction for each of the phospholipase A_2 enzymes present in P388D$_1$ macrophages. *J. Biol. Chem.* **1996**, *271*, 6758–6765. [CrossRef]
61. Balsinde, J.; Balboa, M.A.; Yedgar, S.; Dennis, E.A. Group V phospholipase A_2-mediated oleic acid mobilization in lipopolysaccharide-stimulated P388D$_1$ macrophages. *J. Biol. Chem.* **2000**, *275*, 4783–4786. [CrossRef] [PubMed]
62. Balboa, M.A.; Balsinde, J.; Dillon, D.A.; Carman, G.M.; Dennis, E.A. Proinflammatory macrophage-activating properties of the novel phospholipid diacylglycerol pyrophosphate. *J. Biol. Chem.* **1999**, *274*, 522–526. [CrossRef] [PubMed]
63. Balboa, M.A.; Shirai, Y.; Gaietta, G.; Ellisman, M.H.; Balsinde, J.; Dennis, E.A. Localization of group V phospholipase A_2 in caveolin-enriched granules in activated P388D$_1$ macrophage-like cells. *J. Biol. Chem.* **2003**, *278*, 48059–48065. [CrossRef] [PubMed]
64. Casas, J.; Gijón, M.A.; Vigo, A.G.; Crespo, M.S.; Balsinde, J.; Balboa, M.A. Phosphatidylinositol 4,5-bisphosphate anchors cytosolic group IVA phospholipase A_2 to perinuclear membranes and decreases its calcium requirement for translocation in live cells. *Mol. Biol. Cell* **2006**, *17*, 155–162. [CrossRef] [PubMed]
65. Ramanadham, S.; Ali, T.; Ashley, J.W.; Bone, R.N.; Hancock, W.D.; Lei, X. Calcium-independent phospholipases A_2 and their roles in biological processes and diseases. *J. Lipid Res.* **2015**, *56*, 1643–1668. [CrossRef]
66. Jenkins, C.M.; Mancuso, D.J.; Yan, W.; Sims, H.F.; Gibson, B.; Gross, R.W. Identification, cloning, expression and purification of three novel human calcium-independent phospholipase A_2 family members possessing triacylglycerol lipase and acylglycerol transacylase activities. *J. Biol. Chem.* **2004**, *279*, 48968–48975. [CrossRef]
67. Pindado, J.; Balsinde, J.; Balboa, M.A. TLR3-dependent induction of nitric oxide synthase in RAW 264.7 macrophage-like cells via a cytosolic phospholipase A_2/cyclooxygenase-2 pathway. *J. Immunol.* **2007**, *179*, 4821–4828. [CrossRef]
68. Balboa, M.A.; Balsinde, J.; Dennis, E.A. Involvement of phosphatidate phosphohydrolase in arachidonic acid mobilization in human amnionic WISH cells. *J. Biol. Chem.* **1998**, *273*, 7684–7690. [CrossRef]
69. Fuentes, L.; Pérez, R.; Nieto, M.L.; Balsinde, J.; Balboa, M.A. Bromoenol lactone promotes cell death by a mechanism involving phosphatidate phosphohydrolase-1 rather than calcium-independent phospholipase A_2. *J. Biol. Chem.* **2003**, *278*, 44683–44690. [CrossRef]
70. Cagen, L.M.; Baer, P.G. Adrenic acid inhibits prostaglandin synthesis. *Life Sci.* **1980**, *26*, 765–770. [CrossRef]
71. Rosenthal, M.D.; Hill, J.R. Elongation of arachidonic and eicosapentaenoic acids limits their availability for thrombin-stimulated release from the glycerolipids of vascular endothelial cells. *Biochim. Biophys. Acta* **1986**, *875*, 382–391. [CrossRef]
72. Harkewicz, R.; Fahy, E.; Andreyev, A.; Dennis, E.A. Arachidonate-derived dihomoprostaglandin production observed in endotoxin-stimulated macrophage-like cells. *J. Biol. Chem.* **2007**, *282*, 2899–2910. [CrossRef] [PubMed]
73. Leonard, A.E.; Pereira, S.L.; Sprecher, H.; Huang, Y.S. Elongation of long-chain fatty acids. *Prog. Lipid Res.* **2004**, *43*, 36–54. [CrossRef]
74. Sprecher, H.; Luthria, D.L.; Mohammed, B.S.; Baykousheva, S.P. Reevaluation of the pathways for the biosynthesis of polyunsaturated fatty acids. *J. Lipid Res.* **1995**, *36*, 2471–2477.
75. Sprecher, H.; Van Rollins, M.; Sun, F.; Wyche, A.; Needleman, P. Dihomo prostaglandin and dihomo-thromboxane. A prostaglandin family from adrenic acid that may be preferentially synthesized in the kidney. *J. Biol. Chem.* **1982**, *257*, 3912–3918.
76. Leslie, C.C. Cytosolic phospholipase A_2: Physiological function and role in disease. *J. Lipid Res.* **2015**, *56*, 1386–1402. [CrossRef]

77. Balsinde, J.; Winstead, M.V.; Dennis, E.A. Phospholipase A_2 regulation of arachidonic acid mobilization. *FEBS Lett.* **2002**, *531*, 2–6. [CrossRef]
78. Winstead, M.V.; Balsinde, J.; Dennis, E.A. Calcium-independent phospholipase A_2: Structure and function. *Biochim. Biophys. Acta* **2000**, *1488*, 28–39. [CrossRef]
79. Balsinde, J.; Balboa, M.A. Cellular regulation and proposed biological functions of group VIA calcium-independent phospholipase A_2 in activated cells. *Cell. Signal.* **2005**, *17*, 1052–1062. [CrossRef] [PubMed]
80. Shirai, Y.; Balsinde, J.; Dennis, E.A. Localization and functional interrelationships among cytosolic group IV, secreted group V, and Ca^{2+}-independent group VI phospholipase A_2s in P388D_1 macrophages using GFP/RFP constructs. *Biochim. Biophys. Acta* **2005**, *1735*, 119–129. [CrossRef]
81. Pérez-Chacón, G.; Astudillo, A.M.; Ruipérez, V.; Balboa, M.A.; Balsinde, J. Signaling role for lysophosphatidylcholine acyltransferase 3 in receptor-regulated arachidonic acid reacylation reactions in human monocytes. *J. Immunol.* **2010**, *184*, 1071–1078. [CrossRef] [PubMed]
82. Underhill, D.M. Macrophage recognition of zymosan particles. *J. Endotoxin Res.* **2003**, *9*, 176–180. [CrossRef] [PubMed]
83. Suram, S.; Brown, G.D.; Ghosh, M.; Gordon, S.; Loper, R.; Taylor, P.R.; Akira, S.; Uematsu, S.; Williams, D.L.; Leslie, C.C. Regulation of cytosolic phospholipase A_2 activation and cyclooxygenase-2 expression in macrophages by the β-glucan receptor. *J. Biol. Chem.* **2006**, *281*, 5506–5514. [CrossRef] [PubMed]
84. Suram, S.; Gangelhoff, T.A.; Taylor, P.R.; Rosas, M.; Brown, G.D.; Bonventre, J.V.; Akira, S.; Uematsu, S.; Williams, D.L.; Murphy, R.C.; et al. Pathways regulating cytosolic phospholipase A_2 activation and eicosanoid production in macrophages by Candida albicans. *J. Biol. Chem.* **2010**, *285*, 30676–30685. [CrossRef] [PubMed]
85. Noor, S.; Goldfine, H.; Tucker, D.E.; Suram, S.; Lenz, L.L.; Akira, S.; Suematsu, S.; Girotti, M.; Bonventre, J.V.; Breuel, K.; et al. Activation of cytosolic phospholipase $A_2\alpha$ in resident peritoneal macrophages by Listeria monocytogenes involves listeriolysin O and TLR2. *J. Biol. Chem.* **2008**, *283*, 4744–4755. [CrossRef] [PubMed]
86. Ruipérez, V.; Astudillo, M.A.; Balboa, M.A.; Balsinde, J. Coordinate regulation of TLR-mediated arachidonic acid mobilization in macrophages by group IVA and group V phospholipase A_2s. *J. Immunol.* **2009**, *182*, 3877–3883.
87. Shinohara, H.; Balboa, M.A.; Johnson, C.A.; Balsinde, J.; Dennis, E.A. Regulation od delayed prostaglandin production in activated P388D_1 macrophages by group IV cytosolic and group V secretory phospholipase A_2s. *J. Biol. Chem.* **1999**, *274*, 12263–12268. [CrossRef]
88. Aderem, A.A.; Cohen, D.S.; Wright, S.D.; Cohn, Z.A. Bacterial lipopolysaccharides prime macrophages for enhanced release of arachidonic acid metabolites. *J. Exp. Med.* **1986**, *164*, 165–179. [CrossRef]
89. Kramer, R.M.; Sharp, J.D. Structure, function and regulation of Ca^{2+}-sensitive cytosolic phospholipase A_2 (cPLA_2). *FEBS Lett.* **1997**, *410*, 49–53. [CrossRef]
90. Mouchlis, V.D.; Chen, Y.; McCammon, J.A.; Dennis, E.A. Membrane allostery and unique hydrophobic sites promote enzyme substrate specificity. *J. Am. Chem. Soc.* **2018**, *140*, 3285–3291. [CrossRef]
91. Chilton, F.H.; Fonteh, A.N.; Surette, M.E.; Triggiani, M.; Winkler, J.D. Control of arachidonate levels within inflammatory cells. *Biochim. Biophys. Acta* **1996**, *1299*, 1–15. [CrossRef]
92. Yamashita, A.; Hayashi, Y.; Matsumoto, N.; Nemoto-Sasaki, Y.; Koizumi, T.; Inagaki, Y.; Oka, S.; Tanikawa, T.; Sugiura, T. Coenzyme-A-independent transacylation system; possible involvement of phospholipase A_2 in transacylation. *Biology* **2017**, *6*, 23. [CrossRef] [PubMed]

© 2020 by the authors. Licensee MDPI, Basel, Switzerland. This article is an open access article distributed under the terms and conditions of the Creative Commons Attribution (CC BY) license (http://creativecommons.org/licenses/by/4.0/).

Review

Lipid Phosphate Phosphatases and Cancer

Xiaoyun Tang [1,2] and David N. Brindley [1,2,*]

1 Department of Biochemistry, University of Alberta, Edmonton, AB T6G 2S2, Canada; xtang2@ualberta.ca
2 Cancer Research Institute of Northern Alberta, University of Alberta, Edmonton, AB T6G 2E1, Canada
* Correspondence: david.brindley@ualberta.ca

Received: 12 August 2020; Accepted: 30 August 2020; Published: 2 September 2020

Abstract: Lipid phosphate phosphatases (LPPs) are a group of three enzymes (LPP1–3) that belong to a phospholipid phosphatase (PLPP) family. The LPPs dephosphorylate a wide spectrum of bioactive lipid phosphates, among which lysophosphatidate (LPA) and sphingosine 1-phosphate (S1P) are two important extracellular signaling molecules. The LPPs are integral membrane proteins, which are localized on plasma membranes and intracellular membranes, including the endoplasmic reticulum and Golgi network. LPPs regulate signaling transduction in cancer cells and demonstrate different effects in cancer progression through the breakdown of extracellular LPA and S1P and other intracellular substrates. This review is intended to summarize an up-to-date understanding about the functions of LPPs in cancers.

Keywords: PAP-2; autotaxin; lysophosphatidate; G protein-coupled receptor

---

1. Introduction

Lipid phosphate phosphatases (LPPs) consist of three enzymes (LPP1–3), which have been classified as phospholipid phosphatases (PLPP). So far, the PLPP family has seven members, PLPP1–7, in which PLPP1, PLPP2, and PLPP3 correspond to the former LPP1, LPP2, and LPP3, respectively. Mammalian LPP1–3 are encoded by three separate genes, *PLPP1*, *PLPP2*, and *PLPP3*, and they hydrolyze a wide spectrum of lipid phosphates including phosphatidate (PA), lysophosphatidate (LPA), sphingosine 1-phosphate (S1P), ceramide 1-phosphate (C1P), and diacylglycerol pyrophosphate (DGPP) in a Mg^{2+}-independent and *N*-ethylmaleimide (NEM)-insensitive manner [1,2]. PLPP4 and PLPP5 are the former diacylglycerol pyrophosphate phosphatase-like 2 (DPPL2) and 1 (DPPL1), respectively. PLPP4–5 prefer DGPP as a substrate and also hydrolyze PA and LPA [3]. The activities of PLPP4–5 are also Mg^{2+}-independent, but they can be inhibited by NEM [3]. PLPP6 is formerly known as polyisoprenyl diphosphate phosphatase 1 (PDP1) or candidate sphingomyelin synthase type 2β (CSS2β), which hydrolyzes presqualene diphosphate (PSDP), farnesyl diphosphate (FDP), S1P, LPA, and PA, but it has a preference for PSDP [4,5]. LPPs (PLPP1–3) and PLPP4–6 share highly conserved catalytic domains but show different substrate preferences. LPPs are responsible for the breakdown of extracellular LPA and S1P, which are two important signal molecules and therefore participate in many physiological and pathological processes such as vascular development [6], cell cycle regulation [7], cardiovascular disease [8], and cancer [9]. So far, there are very few reports about PLPP4–6, and their physiological functions are not clear. PLPP7, formerly known as NET39 or CSS2α, is catalytically inactive as a phosphatase due to the loss of critical amino acids in the catalytic domains [10,11].

The process of identifying LPPs dates back to the 1950s when a phosphatidate phosphatase (PAP) activity that dephosphorylates PA to form diacylglycerol (DAG) was discovered in mammalian tissue [12,13]. The PAP activity was intensively investigated as a critical regulator of lipid metabolism because the transformation from PA to DAG represents an intermediate reaction in the Kennedy pathway [14]. Early studies found that the cytosolic and membrane-bound PAPs exhibit quite different enzymological characteristics. For instance, the activity of the cytosolic PAP (PAP-1) that translocates onto membranes of the endoplasmic reticulum (ER) depends on the presence of Mg^{2+} and is sensitive to NEM [15–18]. Its activity is required for the synthesis of triacylglycerol, phosphatidylcholine, and phosphatidylethanolamine [19,20]. It was not until 2006 that PAP-1 was identified in yeast and was found to be the orthologue of a family of three mammalian proteins called lipins [21]. Then, all three of the mammalian lipins were shown to have PAP activity, which is involved in glycerolipid synthesis [22].

A Mg^{2+}-independent phosphatidate phosphatase activity (PAP-2) was also described, and this activity was found mainly in the plasma membrane fraction [23]. This activity in mammals was not inhibited by NEM, which further distinguished it from PAP-1 activity. This new class of PAP activities was characterized in liver [23–26]. Unlike PAP-1, which is specific for PA, PAP-2 degrades a wide spectrum of phospholipids including PA, LPA, S1P, C1P, and lipid pyrophosphates in vitro [1]. This observation led to the more accurate naming of the PAP-2 activity as a lipid phosphate phosphatases [27]. The identification of PAP-2 at a molecular level was achieved by the revelation of cDNA sequences of three PAP-2 isoforms (PAP-2a, PAP-2b, and PAP-2c) in human beings and other animals [28–31]. These isoforms share amino acid sequence homology, and LPP orthologs were also identified in fruit flies and yeast [32–34].

mRNA of LPP1–3 are universally expressed in different tissues of human beings including adrenal, appendix, bone marrow, brain, colon, duodenum, endometrium, esophagus, fat, gall bladder, heart, kidney, liver, lung, lymph node, ovary, pancreas, placenta, prostate, salivary gland, skin, small intestine, spleen, stomach, testis, thyroid, and urinary bladder [35]. Protein expression data from The Human Protein Atlas (http://www.proteinatlas.org) indicate that LPP1–3 are expressed in most tissues, among which LPP1 is highly expressed in the prostate and kidney. LPP2 is expressed at a higher level in the gastrointestinal tract, salivary gland, gallbladder, pancreas, kidney, urinary bladder, and brain than in other tissue, while LPP3 is high in lung, salivary gland, oral mucosa, duodenum, smooth muscle, and skin [36].

Bioactive phospholipids such as LPA and S1P in the extracellular environment signal through their families of G protein-coupled receptors to induce a plethora of effects including cell survival, migration, vascular formation, and inflammation, which play critical roles in cancer development. Functioning as integral membrane phospholipid phosphatases, LPPs hydrolyze extracellular LPA/S1P and attenuate their downstream signaling. LPPs are also present in the intracellular membranes such as the ER and Golgi network [37]. This allows LPPs to hydrolyze intracellular lipid phosphates that have access to the active sites of the LPPs, and thus, the LPPs affect intracellular signaling pathways. Considerable evidence has been accumulated about the functions of LPPs (PLPP1–3) in many physio-pathological processes, including cancer. This review is intended to summarize an up-to-date understanding of the roles of LPPs in cancer development and offer insights for the future directions of cancer treatment.

2. Structure and Membrane Topology of LPP

The mammalian LPPs are localized on the plasma membrane and intracellular network of ER and Golgi [7,37]. It has been reported that LPP1 and LPP3 are present in lipid rafts or caveolae [38,39]. There is also evidence that LPP1 can be directed to the apical surface membrane by a FDKTRL motif on the N-terminus, whereas LPP3 is accumulated at the basolateral membrane [40]. The crystal structure of the LPPs has not yet been solved. A putative topology for the LPPs was determined based on the data obtained from hydrophobicity plots and transmembrane disposition analysis of the rat Dri42 protein [41], which later proved to be rat LPP3 [28]. It has six membrane-spanning regions connected by five extramembrane loops (I–V). Both C- and N-terminal extensions and loop II and IV are located in the cytosol. Loops I, III, and V are on the extracellular side of the membrane. (Figure 1). Three conserved domains (C1, C2, and C3) that form the catalytic site are located on loops III and V outside the cells. Residues that are indispensable for the phosphatase activity in C1–C3 (Figure 2) were identified by amino acid substitution analysis [42]. LPPs inside the cells are localized in the ER [37,41] and Golgi [28]. There is an N-linked glycosylation site between C1 and C2 (Figure 1) [42], indicating that the catalytic site are on the luminal side of ER and Golgi where LPPs are glycosylated [42]. This topology enables LPPs to hydrolyze substrates outside of the cells and in the lumen of ER and Golgi [9,43].

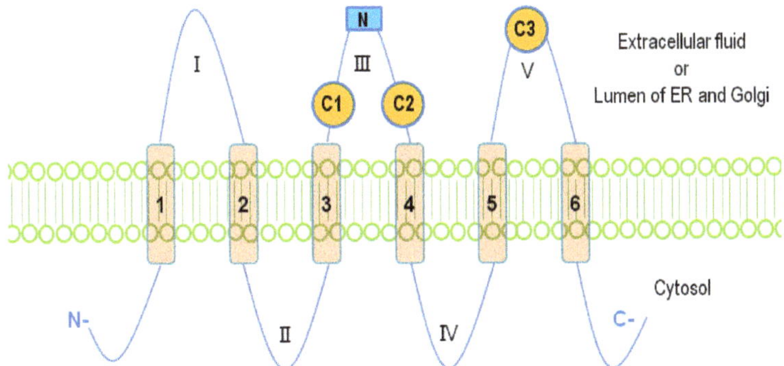

Figure 1. The membrane topology of lipid phosphate phosphatases (LPPs). Six membrane-spanning regions (1–6) are connected with five extramembrane loops (I–V). Three conserved catalytic domains, C1, C2, and C3, are located on loops III and V. The N-linked glycosylation site on the loop III is shown as a blue square.

	C 1		C 2		C 3
phospholipid phosphatase 1 *PLPP1* (LPP1, PAP-2a)	KYSIGRLRP	-----	SFYSGH	-----	GLSRVSDYKHHWSD
phospholipid phosphatase 2 *PLPP2* (LPP2, PAP-2c)	KYMIGRLRP	-----	SFYSGH	-----	GYTRVSDYKHHWSD
phospholipid phosphatase 3 *PLPP3* (LPP3, PAP-2b)	KVSIGRLRP	-----	SFFSGH	-----	GLSRVSDHKHHPSD
phospholipid phosphatase 4 *PLPP4* (DPPL2)	KLIVGRPRP	-----	SFPSIH	-----	ALSRMCDYKHHWQD
phospholipid phosphatase 5 *PLPP5* (DPPL1)	KLIVGRPRP	-----	SFPSGH	-----	ALSRTCDYKHHWQD
phospholipid phosphatase 6 *PLPP6* (PDP1, CSS2β)	KGLVRRRRP	-----	SFPSGH	-----	GLSRVMLGRHNVTD
phospholipid phosphatase 7 *PLPP7* (NET39, CSS2α)	QKLIKRRGP	-----	AFPAGH	-----	GLSRVMIGRHHVTD
sphingosine-1-phosphate phosphatase 1 *SGPP1*	KDIIRWPRP	-----	SMPSTH	-----	CLSRIYMGMHSILD
sphingosine-1-phosphate phosphatase 2 *SGPP2*	KDVLKWPRP	-----	GMPSTH	-----	CLSRLYTGMHTVLD
sphingomyelin synthase 1 *SGMS1*	---------	-----	YLYSGH	-----	IFCILLAHDHYTVD
sphingomyelin synthase 2 *SGMS2*	---------	-----	FLFSGH	-----	IICILVAHEHYTID
glucose-6-phosphatase catalytic subunit *G6PC*	KWILFGQRP	-----	GSPSGH	-----	CLSRIYLAAHFPHQ
phospholipid phosphatase related 1 *PLPPR1*	PYFLTVCKP	-----	SFPSKH	-----	GLNRVSEYRNHCSD
phospholipid phosphatase related 2 *PLPPR2*	PHFLSVCRP	-----	AFPCKD	-----	GVVRVAEYRNHWSD
phospholipid phosphatase related 3 *PLPPR3*	QLATGYHTP	-----	TFPSQH	-----	GLTQITQYRSHPVD
phospholipid phosphatase related 4 *PLPPR4*	QLSTGYQAP	-----	SFPSQH	-----	GLTRITQYKNHPVD
phospholipid phosphatase related 5 *PLPPR5*	PHFLALCKP	-----	TFPSKE	-----	GLNRVAEYRNHWSD

Figure 2. Amino acid sequences of the conserved catalytic domains, C1, C2, and C3, in human LPPs and other proteins with structure similarity. Residues critical for the catalytic activity are shown in red.

The catalytic mechanism of LPPs has been postulated and proposed through a combination of computational modeling and the crystal structure of chloroperoxidase, which is a related enzyme that also possesses the C1–3 domains [44,45]. The conserved histidine on C3 serves as the nucleophile acting on the phosphate group to form a phospho-histidine intermediate. The C2 histidine is involved in breaking the phosphate bond. The conserved lysine and arginine on C1 as well as the arginine on C3 help coordinate the substrate in the active site [43–45]. Similar domains are also found in PLPP4–7. Unlike PLPP1–3, PLPP6 only has four transmembrane helices, and C1–3 of PLPP6 are located at the cytosolic side of the membrane. This allows PLPP6 to hydrolyze polyisoprenoid diphosphates in the cytosol [46]. Sphingosine phosphate phosphatases (SPPs), sphingomyelin synthases (SMSs), phospholipid phosphatase-related proteins (PLPPRs) [43,47], glucose 6-phosphatase (G6P), and *E. coli* phosphatidylglycerol-phosphate phosphatase B (PGPB), an orthologue of human G6P [33], also have the conserved catalytic domains. It is notable that the putative structure of PGPB was established through its crystal structure, which was determined later [47]. The structure of LPPs is thought to be modeled accurately from that proposed for PGPB.

3. Ecto-Activity of LPPs

A major part of circulating LPA is generated from lysophosphatidylcholine (LPC) through the lysophospholipase D activity of autotaxin (ATX) [48,49]. LPC is abundant in circulation with a concentration (>200 μM in human beings) [50], which is much higher than the K_m of ATX for LPC (approximately 100 μM) [51]. As a secretary enzyme, ATX can readily access the LPC pool to generate LPA.

S1P is a sphingolipid analogue of LPA. The precursor for S1P synthesis is sphingosine, which is formed through the hydrolysis of ceramide by ceramidases. Sphingosine is phosphorylated by sphingosine kinase-1 and -2 (SPHK1 and 2) inside cells to generate S1P. SPHK1 is cytosolic and it interacts with the plasma membrane, whereas SPHK2 is present in the mitochondria [52] and nuclei [53]. S1P can be exported out the cells by the membrane transporters including ATP-binding cassette (ABC) transporters (ABCC1, ABCG2, and ABCA1) [54–57], spinster homolog-2 (SPNS2) [58], and major facilitator superfamily transporter 2b (Mfsd2b) [59]. This facilitates the "inside-out signaling" of S1P [54].

Both LPA and S1P outside the cells induce a plethora of cellular responses such as proliferation, migration, angiogenesis, and inflammation [51,60–62] through receptors on the cell surface. To date, six LPA receptors (LPAR1–6) and five S1P receptors (S1PR1–5) have been identified and all of them are G protein-coupled receptors (GPCRs). Plasma membrane-localized LPPs dephosphorylate extracellular LPA and S1P and thereby attenuate LPA/S1P signaling.

The ecto-phosphatase activity was established in rat2 fibroblasts where the overexpression of LPP1 increased the dephosphorylation of extracellular LPA, PA, and C1P. This action attenuated LPA-induced MAPK (mitogen-activated protein kinase) activation and inhibited cell migration [37,63,64]. Similarly, LPP1 and LPP2 inhibited the activation of MAPK that was stimulated by LPA or S1P in HEK293 cells [65]. The dephosphorylation of LPA generates monoacylglycerol (MAG), which can be transported into the cells and re-phosphorylated to form intracellular LPA [66]. This intracellular LPA can activate LPA1 receptors on the nuclear membrane and stimulate the expression of cyclooxygenase-2 (COX-2) and inducible nitric oxide synthase (iNOS) [67]. Intracellular LPA has also been reported to initiate signaling through peroxisome proliferator-activated receptor γ (PPARγ) [68].

The ecto-activity of LPPs in vivo is more complex. Exogenous LPA injected into the circulation is turned over rapidly with the half-life of approximately 1 min [69]. LPP1 knockout mice showed increased levels and a decreased turnover rate of circulating LPA [70]. A similar phenotype was observed in LPP1 hypomorph mice, which have a low expression of LPP1 in most organs except the brain [71]. Interestingly, mice that transgenically overexpressed LPP1 did not show a decrease in the circulating LPA concentrations [72], suggesting that other factors may affect the ecto-activity of LPPs in vivo. For instance, the activity of LPPs is inhibited strongly by Ca^{2+}, which is present in the extracellular environment at approximately 2 mM [37]. In addition, the physiological concentration of LPA in the plasma (0.1–1 μM) is much lower than the Km of LPP1 for LPA (approximately 36 μM) [37]. This indicates that the ecto-activity of LPPs is more important when the LPA levels are increased. In cancers, extracellular LPA levels are elevated as high as 10 μM [73–75]. We do not know if LPA concentrations in the vicinity of the LPPs are modified by other factors such as the levels of expression of the LPA receptors.

S1P concentrations in the plasma range from 100 nM to 1 μM [54]. Exogenous S1P injected into the circulation is cleared from the blood in 15–30 min [76]. S1P is dephosphorylated by SPPs and LPPs, or irreversibly cleaved by S1P lyase (SPL). SPPs and SPL are localized on the ER [77,78]. Therefore, the plasma membrane-localized LPPs have an essential role in regulating the amount of extracellular S1P. The ecto-activity of LPP1 and LPP3 against S1P has been demonstrated in cells [69,79] and animals [80,81]. LPP3 and LPP1a, a splice variant of LPP1, seem to be more efficient at hydrolyzing S1P than LPP1 and LPP2. Phospho-FTY720, an analogue of S1P, can be converted to FTY720 by LPP3 and LPP1a [82], but not by LPP1 or LPP2 [83]. Similarly, expressing exogenous LPP1, LPP2, or LPP3 in HEK293 cells enhances the ecto-activity against LPA, but only LPP3 significantly increases the degradation of extracellular S1P [69]. Sphingosine formed by the dephosphorylation of S1P can be transported into the cells and re-phosphorylated into S1P [79]. Therefore, this process represents a mechanism for the entry of S1P into cells.

4. Intracellular Activities of the LPPs

Not all of the effects of LPPs can be attributed to their ecto-activities. LPP1 is able to suppress wls-31-induced cell migration and Ca^{2+} mobilization [64,84]. Wls-31 is an isosteric phosphonate analog of LPA that activates LPAR1/2, but cannot be hydrolyzed by LPPs. LPP1 and LPP2 also inhibit MAPK activation induced by thrombin, which activates protease-activated receptors (PARs) [65]. Similarly, Ca^{2+} mobilization induced by a PAR1 peptide in MDA-MB-231 cells is inhibited by increased LPP1 expression [84]. Furthermore, LPP1 decreases the platelet derived growth factor (PDGF)-induced migration of embryonic fibroblasts through inhibiting the PDGF/PKC (protein kinase C) /MAPK pathway [63]. These effects of the LPPs are independent of their ecto-activities because these agonists cannot be degraded by LPPs. However, the effect requires LPP activity and therefore probably depends on the degradation of an intracellular lipid phosphate that is formed downstream of the activation of LPA, PAR, or PDGF receptors.

LPPs are also present on the ER and Golgi network with the catalytic domains, which should face the luminal side. As such, LPPs probably have specific access to substrates depending on the subcellular compartment. One of these possible substrates inside the cells is PA, which activates Sos (son of sevenless), Raf (rapidly accelerated fibrosarcoma), MAPK, mTOR (mammalian target of rapamycin), AKT (Ak strain transforming), and SPHK1 [85–87]. The dephosphorylation of PA generates DAG, which activates the classical and novel PKCs and Ras (rat sarcoma) guanyl nucleotide-releasing protein [88]. Increasing LPP1, LPP2, or LPP3 does decrease intracellular PA/DAG ratios [38,89]. LPP3 depletion decreases the levels of de novo synthesized DAG and the Golgi-associated DAG [90]. LPP2 decreases intracellular PA, which promotes the apoptosis of HEK293 cells in serum-deprived media [91]. However, LPP3 or LPP1 did not change intracellular DAG significantly in other studies [65,72,92].

Since the catalytic domains of LPPs are on the luminal side of ER and Golgi or the outer surface of the plasma membrane, the LPPs should not be able to dephosphorylate PA, which is formed at the cytosolic side of the membranes, unless the PA can be transported across the membranes to the catalytic sites of LPPs. However, this has yet to be shown. It should be noted that increasing LPP1 activity directly inhibits phospholipase D (PLD) activation [64], which forms a large proportion of intracellular PA. This can provide an alternative explanation for the decreased accumulation of PA. It is likely that the lipins, which are cytosolic phosphatidate phosphatases that translocate to membranes, are responsible for the degradation of most of the PA on the cytosolic surface of membranes [93].

LPPs probably also degrade intracellular C1P and S1P, both of which are involved in inflammation. C1P activates phospholipase A_2 (PLA_2) to produce arachidonate, which is converted to inflammatory eicosanoids (prostaglandins and thromboxanes) by COX-1/2 [94]. S1P helps to coordinate the metabolism of arachidonate by COX-2 to ensure the maximum production of prostaglandin E2 (PGE_2) [94]. S1P also interacts with specific intracellular target proteins such as histone deacetylase 1/2, prohibitin 2, PPARγ, and tumor necrosis factor (TNF) receptor associated factor 2, to induce cell responses [95]. The overexpression of LPP3, but not LPP2, decreases intracellular S1P in HEK293 cells [91]. The degradation of intracellular S1P can be performed by other enzymes such as S1P phosphatases and S1P lyase, which are major regulators of intracellular S1P concentrations.

5. Upregulation of LPA Signaling in Cancers

Functioning as a platelet activator, a chemoattractant, and a growth factor, LPA plays a critical role in wound healing [96]. At sites of tissue damage, LPA stimulates the proliferation of fibroblasts and endothelial cells [97], and it promotes collagen deposition [98] and angiogenesis [99,100]. Circulating LPA concentrations are normally between 0.1 and 1 μM [51], and this is regulated mainly by the balance of ATX activity versus that of the LPPs.

LPA signaling is magnified and hijacked by cancers (wounds that do not heal) [101]. Elevated ATX levels have been observed in the blood and malignant tissues from patients with thyroid [102], lung [103], breast [104], liver [105], pancreatic [106,107], kidney [108], bladder [108], and prostate cancer [109]. As a consequence, LPA levels increase in those cancers [107,110–113], which has been considered an indicator of poor prognosis [110,113]. Significantly, LPA concentrations have been reported to reach as high as 10 μM in the ascites fluid of ovarian cancer patients [73–75]. Cancer cells express high levels of LPAR1–3 [61], which are GPCRs. LPAR1–3 couple to G proteins: Gi/o, Gq/11, and G12/13 [61], and activate PI3K (phosphoinositide 3-kinase) /AKT [114,115], PLC (phospholipase C) [116], and Rho pathways [116]. LPAR1–3 are elevated in brain [117,118], pancreatic [119–121], colon [122,123], and breast cancer [124], which is associated with enhanced tumor growth and metastasis.

Introducing exogenous LPAR1 converts non-transformed MCF-10A cells into an invasive phenotype [125]. LPAR1 and/or LPAR3 activate Wnt/β-catenin and PI3K/AKT/mTOR pathways to induce the epithelial-to-mesenchymal transition (EMT) [126,127], which is an essential step during cancer cell stemness [128]. Cancer stem cell (CSC)-related genes such as ALDH1A1, OCT4, and SOX2 are upregulated by activating LPAR1 [129]. Blocking ATX or LPAR2 suppresses the growth of breast cancer stem cells [62,130] in which LPP3 expression is downregulated [131]. Transgenic mice overexpressing ATX or any of LPAR1–3 by MMTV-LTR (mouse mammary tumor virus long terminal repeat) promoter in mammary epithelial cells show an increased development of spontaneous breast tumors and subsequent metastases [132]. LPAR4–6 are closely related to purinergic receptors [61]. LPAR4 (P2Y9/GPR23) and LPAR5 (GPR92) in cancer cells demonstrate inhibitory effects on proliferation and migration/invasion [133–136], which is in contrast to the effects of LPAR1–3. It is notable that LPAR5 suppresses the function of infiltrated CD8+ cytotoxic T cells as a mediator of immune suppression in the tumor microenvironment (TME) [137]. The effects of LPAR6 (P2Y5) in cancers are uncertain [138,139] and require further investigation.

LPA induces lymphocyte homing [140] and the transformation of monocytes to macrophages [141], which provokes inflammation [102,142]. LPA is closely related to the inflammatory milieu in conditions such as pulmonary fibrosis, rheumatoid arthritis, atherosclerosis, and inflammatory bowel disease [143]. The TME is also characterized by chronic inflammation, which is one of the hallmarks of cancers [144]. Increasing evidence reveals that there is crosstalk between LPA signaling and cancer-related inflammation. TNFα increases ATX production by Huh7, HepG2, and Hep3B liver cancer cells through activating nuclear factor κB (NFκB). The subsequent increase in LPA enhanced the invasiveness of the cancer cells [145]. The secretion of IL-8 is increased by LPA in human bronchial epithelial cells, which is mediated by protein kinase Cδ (PKCδ) and NFκB [146,147]. IL-8 and IL-6 expressions in ovarian cancer cells are also increased by LPAR2 or LPAR3 activation [148]. In a colon cancer model, LPAR2 knockout mice formed smaller tumors after induction with azoxymethane (AOM)/dextran sulfate sodium (DSS). This was accompanied by decreased levels of COX2 and CCL2 and reduced macrophage infiltration [149]. Zhao et al. showed that LPP1 inhibits LPA-induced NFκB translocation, which blocks IL-8 secretion in human bronchial epithelial cells [150]. This suggests an important role of LPP1 in inflammation [50,142].

We recently proposed a model of the ATX–LPA inflammatory cycle in breast cancer [151,152]. In this model, tumor-derived inflammatory cytokines such as TNFα and IL-1β increase ATX secretion by the adjacent mammary adipose. As a consequence, LPA levels increase in the TME. The increased LPA stimulates cancer cells to produce more cytokines, which can overcome the LPA-mediated feedback inhibition of mRNA expression for ATX [153] to form a feed-forward inflammatory cycle. This ATX–LPA inflammatory cycle can be exacerbated by radiotherapy (RT), since irradiation increases COX-2 and inflammatory cytokines in cultured adipose tissues as well as in the fat pads of mice [154,155]. ATX and LPAR1/2 levels are also elevated by irradiation. Dexamethasone, an anti-inflammatory glucocorticoid, attenuates the RT-induced upregulation of the expression ATX and LPA1R and LPA2R and increases LPP1 expression [156], which together decrease LPA signaling. Pulmonary fibrosis caused by RT or bleomycin is also blocked by dexamethasone [157–159].

It is well documented that LPA signaling promotes cell survival by inhibiting the intrinsic and extrinsic apoptosis pathways [160]. LPA decreases the level of the Fas receptor and reduces the expression of the Fas ligand [161,162], which makes cancer cells less responsive to the extrinsic pro-apoptotic stimuli. LPA also attenuates the intrinsic apoptosis pathway by increasing Bcl-2 and inhibiting Bad and Bax [163,164]. These effects of LPA depend on the activation of the PI3K–Akt pathway. The decrease in the sensitivity of cancer cells to chemotherapy and RT is contributed, at least partly, by the upregulation of LPA signaling. LPA decreases the effectiveness of Taxol [165], tamoxifen [166], and doxorubicin [167] in killing breast cancer cells.

The critical role of LPAR2 in protecting cells from radiation-induced damage has been illustrated by LPAR2 knockout mice, which exhibit increased irradiation-induced apoptosis in intestinal tissue [168]. By contrast, the knockout of LPAR1 or LPAR3 does not have this effect [168]. On the other hand, LPAR2 agonists show a therapeutic potential against irradiation-induced injury [168,169]. LPA contributes to the resistance of 786-O renal cancer cells to Temsirolimus and Sunitinib by activating Arf6 GTPase through LPAR2 [170]. Similarly, blocking LPAR1/3 with Ki16425 in resistant UMRC3 renal cancer cells re-establishes the sensitivity to Sunitinib [171]. The long-term culture of PANC-1 pancreatic cancer cells in the presence of cisplatin results in an upregulation of LPAR3 [172]. LPA through the activation of LPA1R and PI3K stabilized the expression of nuclear factor erythroid 2-related factor 2 (Nrf2), a transcription factor, which through the anti-oxidant response element increases the expression of the multidrug-resistant transporters, anti-oxidant genes, and enzymes of DNA repair [166,167,173,174]. Thus, the ATX inhibitors, ONO-8430506 and GLPG1690, enhance the sensitivity of breast tumor to doxorubicin and RT [167,175]. It should be noted that the later effect of GLPG1690 involved decreased cell division in the cancer cells, and this is compatible with the major effect of RT in solid tumors being to increase cell senescence rather than apoptosis [176–178]. Effects of the upregulation of LPA signaling in cancer cells are summarized in Figure 3.

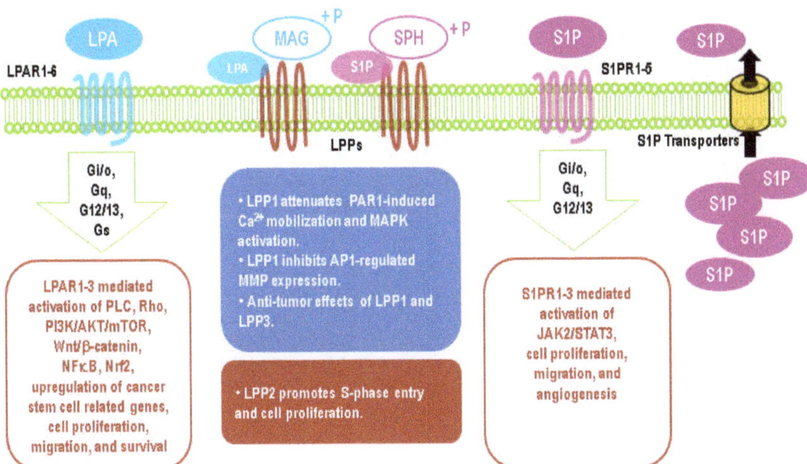

Figure 3. Major effects of upregulation of lysophosphatidate (LPA) and sphingosine 1-phosphate (S1P) signaling in cancer cells through G protein-coupled receptors and different functions of LPP1/3 and LPP2 in cancers.

6. Upregulation of S1P Signaling in Cancers

Elevated expression of SPHK, especially SPHK1, has been well documented in multiple cancers where the consequent increase in S1P promotes cell survival, growth, and invasiveness [179–181]. The overexpression of wild-type SPHK1, but not the inactive mutant, transforms NIH3T3 cells into fibrosarcoma [182]. The function of SPHK2 in cancer is unclear. Some studies indicated that SPHK2 has an opposite role to SPHK1; for instance, SPHK2 induces cell cycle arrest and promotes apoptosis [183–185]. The knockdown of SPHK2 enhances apoptosis and sensitivity to chemotherapy in lung and colon cancer cells [186,187]. However, emerging evidence has revealed the anti-tumor effect of SPHK2. Targeting SPHK2 demonstrates antitumorigenic effects in cancer cell lines and mouse models [186,188–190]. Neubauer et al. recently reported that the effect of SPHK2 on cancer depends on its expression level [191]. Moderate increases in SPHK2 promoted cell proliferation and survival, and this can be suppressed by highly overexpressed SPHK2. Interestingly, this study indicated that the highly overexpressed SPHK2 is accumulated in the nuclei, whereas at lower levels of expression, SPHK2 is on the plasma membrane. This suggests the importance of localization for the effect of SPHK2. Indeed, elevated SPHK2 has been shown in bladder, melanoma, esophageal, breast, lymphoma cancers, and leukemia [191], and this is linked to a poor prognosis in non-small cell lung cancer [192]. S1P concentrations increase in mouse and human breast tumors and in the serum of stage III breast cancer patients [193,194].

S1PRs are GPCRs. S1PR1 couples to Gi/o. It has an essential role in activating JAK2 (janus kinase 2), which causes a persistent STAT3 (signal transducer and activator of transcription 3) activation in cancers. The activated STAT3 increases the expression of S1PR1 further to form a feed-forward loop of S1PR1–JAK2–STAT3 [195]. This feed-forward loop drives tumorigenesis and metastasis [196] and contributes to the formation of the chronic inflammation milieu in colon cancer [197]. Enhanced S1PR1/STAT3 signaling has also been found in intestinal and lung cancers [197–199]. S1PR1 is required for tumor angiogenesis [200]. S1PR2 and S1PR3 couple to Gi/o, Gq/11, and G12/13. Functioning as a promoter of tumorigenesis and angiogenesis [201–203], S1PR3 is upregulated in lung cancer cells [204], and it is the most highly expressed S1PR in breast cancer cells [205]. The function of S1PR2 in cancer is uncertain, because both positive and negative impacts of S1PR2 were reported by different studies [206,207]. Compared with S1PR1–3, S1PR4 and S1PR5 have a restricted distribution and less clear functions. Effects of the upregulation of S1P signaling in cancer cells are summarized in Figure 3.

7. Alterations of LPP Expression in Cancers

The downregulation of LPPs results in an exacerbation of the excessive LPA and S1P signaling in cancers. LPP1 and LPP3 levels are significantly decreased in colon and breast tumors compared with the normal tissue [208,209]. Microarray data also demonstrated the downregulation of LPP1 or LPP3 in many other cancers [210–212]. We compared mRNA levels of LPP1–3 in all the tumors versus normal datasets of the Oncomine database, with the following threshold settings: p value, 0.05; fold change, 2; gene rank, top 10%. The results showed that LPP1 expression is significantly downregulated in melanoma, sarcoma, leukemia, bladder, breast, colorectal, kidney, lung, and ovarian cancers, and it is upregulated in lymphoma, brain and central nervous system, and prostate cancers. LPP3 is downregulated in melanoma, myeloma, sarcoma, bladder, breast, cervical, colorectal, lung, kidney, liver, and head and neck cancers, and it is upregulated in lymphoma.

In contrast, LPP2 is upregulated in 9 out of 20 categories of cancers including bladder, cervical, colorectal, esophageal, head and neck, liver, and prostate cancers, and it is downregulated in brain and central nervous system cancer, melanoma, and sarcoma (Figure 4).

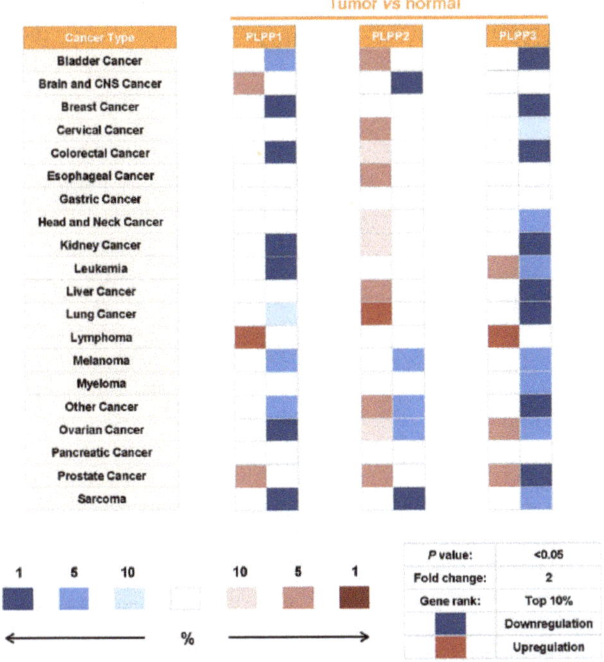

Figure 4. Alterations of mRNA levels of LPP1–3 in different tumors versus normal tissues. Values were obtained from the Oncomine database. The searching thresholds were set as follows: *p*-value, 0.05; fold change, 2; gene rank, top 10% (means 9% other genes have more significant *p*-values). The red or blue color represents the up- or downregulation of genes respectively in tumors relative to the adjacent normal tissue. The darkness of color corresponds to the gene rank; darker color means higher rank. LPP1 (PLPP1) and LPP3 (PLPP3) are downregulated, whereas LPP2 (PLPP2) is upregulated in the majority of cancers ($p < 0.05$). In some cases, such as PLPP2 in ovarian cancer and other cancers and PLPP3 in leukemia, ovarian cancer, and prostate cancer, both upregulation and downregulation are shown by different datasets. These cases are considered as neither upregulation nor downregulation.

Alterations (amplification, deletion, and mutation) of *PLPP1–3* are not common in cancers. How the expression of LPPs is regulated remains unclear so far. Several transcription factors that control the expression of LPPs have been identified. LPP1 can be induced by DAF-16, an orthologue of FOXO (class O forkhead box protein) transcription factors in *Ancylostoma caninum* [213]. The conditional knockout of SP2 in the mouse cerebral cortex leads to a decrease in LPP1 expression [214]. Oxidized low-density lipoprotein increases LPP3 expression in human macrophages through transcription factor C/EBPβ (CCAAT-enhancer-binding protein β) [215]. LPP3 expression can also be activated by NFκB through three response elements in the promoter region of *PLPP3* [216]. DNA modification is another mechanism changing the expression of LPPs. DNA methyltransferase Dnmt3a1 upregulates LPP3 transcription in mouse embryonic stem cells [217]. In addition, LPP3 expression can be elevated by androgens, EGF (epidermal growth factor), FGF (fibroblast growth factor), and VEGF (vascular endothelial growth factor) at the transcription level [28,218,219].

The ecto-activity of LPP in ovarian cancer cells can be increased by gonadotropin-releasing hormone (GnRH) [220]. So far, little is known about how the discrepant expression between LPP1/3 and LPP2 in cancers happens. Dexamethasone, an anti-inflammatory glucocorticoid, increased LPP1 expression in RT-treated breast tumors and adjacent adipose [155,157] suggesting that LPP1 could decrease in response to the inflammatory milieu created by the tumor. LPP3 expression can be decreased by hypoxia in the TME, leading to an asymmetrical redistribution of ATX and LPP1 to the leading and trailing edge of cancer cells, respectively [221]. These results suggest that the intrinsic characteristics of the TME such as inflammation and hypoxia may play an essential role in the downregulation of LPP1/3.

8. Effects of LPPs in Cancers

Increasing the low level of LPP1 or LPP3 in cancer cells leads to an inhibition in tumor growth and metastasis, which is partly caused by the ecto-activity. Ovarian cancer cells overexpressing LPP1 or LPP3 show an increased hydrolysis of extracellular LPA, resulting in impaired colony-forming ability and enhanced apoptosis [222,223]. GnRH increases the LPP ecto-activity in GnRH receptor-positive ovarian cancer cells, and this is attenuated by GnRH antagonism [220]. This effect of GnRH is associated with its antiproliferative actions on ovarian cancer cells.

We found that tetracyclines, a class of antibiotics, increase the degradation of extracellular LPA by breast cancer cells and HEK293 cells [69]. This is thought to occur through increasing the stability of LPP proteins, leading to an elevation in LPP ecto-activity. The clearance of [^{32}P]LPA from the circulation is increased from 61% to 79% at 30 s and from 75% to 85% at 60 s [69]. Doxycycline treatment delays breast tumor growth in mice and decreases LPA levels in the plasma. Furthermore, doxycycline decreases inflammation in the tumors as indicated by a decrease in 11 inflammatory cytokines/chemokines and decreased NFκB levels in the nuclei of cancer cells [224].

Overexpressing LPP1 in MDA-MB-231 breast cancer cells decreases the Ca^{2+} mobilization that is stimulated by LPA, wls-31, and a PAR1 peptide [84]. LPP1 expression decreases cell migration and also suppresses tumor growth and metastasis in both syngeneic and xenograft mouse models. The catalytically inactive mutant (R217K) of LPP1 does not have these effects. Increasing LPP1 expression in the cancer cells does not affect LPA levels in both tumors and the plasma, even though the cells have enhanced ecto-activity in vitro [84]. These results emphasize the importance of the intracellular activity of LPP1, because LPP1 does not degrade wls-31 and the PAR1 agonist. In addition, the extent of hydrolysis of optimum (10 μM) extracellular LPA concentrations is not fast enough to attenuate acute response, such as Ca^{2+} mobilization, which occur in 30 s [84].

In our recent report, increasing LPP1 in MDA-MB-231 breast cancer cells decreases the levels of c-Jun and c-Fos in nuclei and suppresses the expression of AP-1 (activator protein 1) -regulated genes including *MMPs* (matrix metalloproteinases) and *CCND1/3* (cyclin D1/D3). This effect can be partially reversed by siRNA against LPP1 [208]. In fact, human breast tumors have significantly higher protein levels of MMP-1, -7, -8, -9, -12, -13, c-Jun, and c-Fos than normal breast tissue [208], which is probably caused by the downregulation of LPP1. *PLPP1* has been recognized as one of 12 genes linked with relapse-free survival in breast cancer patients [225].

The effects of LPP1 and LPP3 in cancers are not always consistent. Nakayama et al. found a biphasic growth pattern of ovarian cancer cells in LPP1 knockout mice [70]. The high level of circulating LPA caused by the decrease in LPP1 facilitates cancer cell growth within the first 2 weeks after inoculation, showing more invasive nodules on the omentum compared with the wild-type mice. However, subsequent tumor growth after 3 weeks is slower in the LPP1 knockout mice than the wild-type mice, leading to formation of smaller tumors [70].

Chatterjee et al. reported a pro-tumorigenic action of LPP3. In their study, the knockdown of LPP3 in U87 and U118 glioblastoma cells inhibits tumor growth in mice, whereas overexpressing LPP3 in SW480 colon cancer cells promotes tumor growth [226]. WM239A melanoma cells failed to degrade extracellular LPA after the knockdown of LPP3, but not LPP1 or LPP2, leading to an impaired chemotaxis toward LPA, which was related to the loss of a self-generated LPA gradient outside cells [227].

LPP2 expression is elevated in transformed cells and a variety of cancer cell lines including MCF7, SK-LMS1, MG63, and U2OS [228]. The upregulation of LPP2 is also shown in many cancers, which is opposite to that for LPP1/3 (Figure 4). The knockdown of LPP2 impairs the anchorage-dependent growth of cancer cell lines and decreases cell proliferation [7,228]. These in vitro data validate LPP2 as a putative cancer target. Our unpublished data indicated that the knockout of LPP2 in breast cancer cells inhibits cell proliferation, but it does not affect migration. The cells with LPP2 knockdown form smaller tumors in mice than the wild-type cells. Different functions of LPPs in cancers are summarized in Figure 3.

The different effects of LPP1/3 and LPP2 on cancers may be reflected by their distinct non-redundant functions. For instance, unlike LPP1 [70] and LPP2 [229] knockout mice, which are viable, LPP3 knockout mice die between E7.5 and 9.5 and fail to form a chorioallantoic placenta and yolk sac vasculature [230]. Wunen and Wunen-2 are *Drosophila* homologues of human LPP [32]. The mutation of Wunens causes the impaired migration and death of primordial germ cells. This can be rescued by human or mouse LPP3, but not human LPP1 or mouse LPP2 [32,231]. The knockdown of LPP2 affects the cell cycle by delaying S-phase entry and cyclin A expression. Conversely, the overexpression of LPP2, but not a catalytically inactive mutant, causes premature S-phase entry, which is accompanied by premature cyclin A accumulation [7]. This effect of LPP2 is not observed with LPP1 and LPP3, where the overexpression of these two isoforms normally inhibits cell growth and migration [84,223]. Divergent subcellular distribution could be another reason. In polarized MDCK cells, LPP1 and LPP3 are differentially located on the apical and basolateral subdomains of the plasma membrane, respectively [40]. Sciorra and Morris found that LPP3, but not LPP1, increases DAG accumulation from PLD-generated PA in HEK293 cells. The authors also showed that PLD and LPP3 co-exist in caveolin-1-enriched detergent-resistant membrane microdomains where PLD and LPP3 act sequentially to generate DAG [38]. This subtle difference in localization may allow LPPs to act on specific intracellular pools of substrates and differentially regulate intracellular signaling. To understand more clearly the effects of LPPs in cancers, investigations need to be expanded to more types of cancer. It is important to understand how the LPP isoforms regulate intracellular signal transduction in addition to the degradations of extracellular LPA and S1P. The intracellular lipid phosphate targets for different LPPs need to be identified. LPP2 is a promising target because it is upregulated in many cancers. Developing LPP2 inhibitors is feasible for cancer therapy.

9. Conclusions

Plasma membrane-bound LPPs are responsible for the dephosphorylation of extracellular LPA and S1P through their ecto-activities, and this decreases the activation of the respective cell surface receptors. Intracellular LPPs, which are localized on organelles such as the ER and Golgi, also attenuate post-receptor signaling through various LPA receptors and other GPCRs such as PAR receptors. LPP1–3 have distinct and non-redundant functions in physiological processes beside the common phosphatase activities. In cancers, LPP1/3 are generally downregulated, whereas LPP2 is upregulated. LPP1/3 demonstrate antitumorigenic effects in ovarian and breast cancer cells, but they have opposite effects in melanoma and glioblastoma cells. Emerging evidence suggest that LPP2 may function as a tumor promoter, which is different from LPP1/3. The reason for these differences in LPPs are not completely understood, and they are probably caused by differences in substrate access, localization, and accessibility to intracellular substrates [232], which warrant further investigation. LPPs are promising targets for developing novel approaches to cancer therapy.

Author Contributions: X.T. wrote the first draft of the review, which was then modified by D.N.B. All authors have read and agreed to the published version of the manuscript.

Funding: D.N.B. was supported by an Innovative Grant from the Canadian Cancer Society Research Institute (703926)/Canadian Breast Cancer Foundation (300034) and the Women and Children's Health Research Institute (WCHRI) at the University of Alberta and by a project grant from the Canadian Institutes of Health Research (PJT-169140).

Conflicts of Interest: The authors declare no conflict of interest and the funders had no role in the design of the study; in the collection, analyses, or interpretation of data; in the writing of the manuscript, or in the decision to publish the results.

References

1. Brindley, D.N.; Waggoner, D.W. Phosphatidate phosphohydrolase and signal transduction. *Chem. Phys. Lipids* **1996**, *80*, 45–57. [CrossRef]
2. Dillon, D.A.; Chen, X.; Zeimetz, G.M.; Wu, W.I.; Waggoner, D.W.; Dewald, J.; Brindley, D.N.; Carman, G.M. Mammalian Mg2+-independent phosphatidate phosphatase (PAP2) displays diacylglycerol pyrophosphate phosphatase activity. *J. Biol. Chem.* **1997**, *272*, 10361–10366. [CrossRef]
3. Takeuchi, M.; Harigai, M.; Momohara, S.; Ball, E.; Abe, J.; Furuichi, K.; Kamatani, N. Cloning and characterization of DPPL1 and DPPL2, representatives of a novel type of mammalian phosphatidate phosphatase. *Gene* **2007**, *399*, 174–180. [CrossRef]
4. Fukunaga, K.; Arita, M.; Takahashi, M.; Morris, A.J.; Pfeffer, M.; Levy, B.D. Identification and functional characterization of a presqualene diphosphate phosphatase. *J. Biol. Chem.* **2006**, *281*, 9490–9497. [CrossRef]
5. Carlo, T.; Petasis, N.A.; Levy, B.D. Activation of polyisoprenyl diphosphate phosphatase 1 remodels cellular presqualene diphosphate. *Biochemistry* **2009**, *48*, 2997–3004. [CrossRef]
6. Ren, H.; Panchatcharam, M.; Mueller, P.; Escalante-Alcalde, D.; Morris, A.J.; Smyth, S.S. Lipid phosphate phosphatase (LPP3) and vascular development. *Biochim. Biophys. Acta* **2013**, *1831*, 126–132. [CrossRef]
7. Morris, K.E.; Schang, L.M.; Brindley, D.N. Lipid phosphate phosphatase-2 activity regulates S-phase entry of the cell cycle in Rat2 fibroblasts. *J. Biol. Chem.* **2006**, *281*, 9297–9306. [CrossRef]
8. Chandra, M.; Escalante-Alcalde, D.; Bhuiyan, M.S.; Orr, A.W.; Kevil, C.; Morris, A.J.; Nam, H.; Dominic, P.; McCarthy, K.J.; Miriyala, S.; et al. Cardiac-specific inactivation of LPP3 in mice leads to myocardial dysfunction and heart failure. *Redox Biol.* **2018**, *14*, 261–271. [CrossRef]
9. Tang, X.; Benesch, M.G.; Brindley, D.N. Lipid phosphate phosphatases and their roles in mammalian physiology and pathology. *J. Lipid Res.* **2015**, *56*, 2048–2060. [CrossRef]
10. Huitema, K.; van den Dikkenberg, J.; Brouwers, J.F.; Holthuis, J.C. Identification of a family of animal sphingomyelin synthases. *EMBO J.* **2004**, *23*, 33–44. [CrossRef]
11. Liu, G.H.; Guan, T.; Datta, K.; Coppinger, J.; Yates, J., 3rd; Gerace, L. Regulation of myoblast differentiation by the nuclear envelope protein NET39. *Mol. Cell Biol.* **2009**, *29*, 5800–5812. [CrossRef] [PubMed]
12. Smith, S.W.; Weiss, S.B.; Kennedy, E.P. The enzymatic dephosphorylation of phosphatidic acids. *J. Biol. Chem.* **1957**, *228*, 915–922. [PubMed]
13. Coleman, R.; Huebscher, G. Metabolism of phospholipids. V. Studies of phosphatidic acid phosphatase. *Biochim. Biophys. Acta* **1962**, *56*, 479–490. [CrossRef]
14. Kennedy, E.P.; Weiss, S.B. The function of cytidine coenzymes in the biosynthesis of phospholipides. *J. Biol. Chem.* **1956**, *222*, 193–214.
15. Lamb, R.G.; Fallon, H.J. Glycerolipid formation from sn-glycerol-3-phosphate by rat liver cell fractions. The role of phosphatidate phosphohydrolase. *Biochim. Biophys. Acta* **1974**, *348*, 166–178. [CrossRef]
16. Jamdar, S.C.; Fallon, H.J. Glycerolipid synthesis in rat adipose tissue. II. Properties and distribution of phosphatidate phosphatase. *J. Lipid Res.* **1973**, *14*, 517–524.
17. Sturton, R.G.; Brindley, D.N. Factors controlling the metabolism of phosphatidate by phosphohydrolase and phospholipase A-type activities. Effects of magnesium, calcium and amphiphilic cationic drugs. *Biochim. Biophys. Acta* **1980**, *619*, 494–505. [CrossRef]
18. Jamdar, S.C.; Osborne, L.J.; Wells, G.N. Glycerolipid biosynthesis in rat adipose tissue 12. Properties of Mg2+-dependent and -independent phosphatidate phosphohydrolase. *Arch. Biochem. Biophys.* **1984**, *233*, 370–377. [CrossRef]
19. Smith, M.E.; Hubscher, G. The biosynthesis of glycerides by mitochondria from rat liver. The requirement for a soluble protein. *Biochem. J.* **1966**, *101*, 308–316. [CrossRef]
20. Jelsema, C.L.; Morre, D.J. Distribution of phospholipid biosynthetic enzymes among cell components of rat liver. *J. Biol. Chem.* **1978**, *253*, 7960–7971.
21. Han, G.S.; Wu, W.I.; Carman, G.M. The Saccharomyces cerevisiae Lipin homolog is a Mg^{2+}-dependent phosphatidate phosphatase enzyme. *J. Biol. Chem.* **2006**, *281*, 9210–9218. [CrossRef]
22. Kok, B.P.; Venkatraman, G.; Capatos, D.; Brindley, D.N. Unlike two peas in a pod: Lipid phosphate phosphatases and phosphatidate phosphatases. *Chem. Rev.* **2012**, *112*, 5121–5146. [CrossRef] [PubMed]

23. Jamal, Z.; Martin, A.; Gomez-Munoz, A.; Brindley, D.N. Plasma membrane fractions from rat liver contain a phosphatidate phosphohydrolase distinct from that in the endoplasmic reticulum and cytosol. *J. Biol. Chem.* **1991**, *266*, 2988–2996. [PubMed]
24. Fleming, I.N.; Yeaman, S.J. Purification and characterization of N-ethylmaleimide-insensitive phosphatidic acid phosphohydrolase (PAP2) from rat liver. *Biochem. J.* **1995**, *308*, 983–989. [CrossRef] [PubMed]
25. Waggoner, D.W.; Martin, A.; Dewald, J.; Gomez-Munoz, A.; Brindley, D.N. Purification and characterization of novel plasma membrane phosphatidate phosphohydrolase from rat liver. *J. Biol. Chem.* **1995**, *270*, 19422–19429. [CrossRef] [PubMed]
26. Siess, E.A.; Hofstetter, M.M. Identification of phosphatidate phosphohydrolase purified from rat liver membranes on SDS-polyacrylamide gel electrophoresis. *FEBS Lett.* **1996**, *381*, 169–173. [CrossRef]
27. Brindley, D.N.; Waggoner, D.W. Mammalian lipid phosphate phosphohydrolases. *J. Biol. Chem.* **1998**, *273*, 24281–24284. [CrossRef]
28. Kai, M.; Wada, I.; Imai, S.; Sakane, F.; Kanoh, H. Cloning and characterization of two human isozymes of Mg^{2+}-independent phosphatidic acid phosphatase. *J. Biol. Chem.* **1997**, *272*, 24572–24578. [CrossRef]
29. Kai, M.; Wada, I.; Imai, S.; Sakane, F.; Kanoh, H. Identification and cDNA cloning of 35-kDa phosphatidic acid phosphatase (type 2) bound to plasma membranes. Polymerase chain reaction amplification of mouse H2O2-inducible hic53 clone yielded the cDNA encoding phosphatidic acid phosphatase. *J. Biol. Chem.* **1996**, *271*, 18931–18938. [CrossRef]
30. Hooks, S.B.; Ragan, S.P.; Lynch, K.R. Identification of a novel human phosphatidic acid phosphatase type 2 isoform. *FEBS Lett.* **1998**, *427*, 188–192. [CrossRef]
31. Tate, R.J.; Tolan, D.; Pyne, S. Molecular cloning of magnesium-independent type 2 phosphatidic acid phosphatases from airway smooth muscle. *Cell. Signal.* **1999**, *11*, 515–522. [CrossRef]
32. Burnett, C.; Howard, K. Fly and mammalian lipid phosphate phosphatase isoforms differ in activity both in vitro and in vivo. *EMBO Rep.* **2003**, *4*, 793–799. [CrossRef] [PubMed]
33. Stukey, J.; Carman, G.M. Identification of a novel phosphatase sequence motif. *Protein Sci.* **1997**, *6*, 469–472. [CrossRef] [PubMed]
34. Toke, D.A.; Bennett, W.L.; Oshiro, J.; Wu, W.I.; Voelker, D.R.; Carman, G.M. Isolation and characterization of the Saccharomyces cerevisiae LPP1 gene encoding a Mg^{2+}-independent phosphatidate phosphatase. *J. Biol. Chem.* **1998**, *273*, 14331–14338. [CrossRef] [PubMed]
35. Fagerberg, L.; Hallstrom, B.M.; Oksvold, P.; Kampf, C.; Djureinovic, D.; Odeberg, J.; Habuka, M.; Tahmasebpoor, S.; Danielsson, A.; Edlund, K.; et al. Analysis of the human tissue-specific expression by genome-wide integration of transcriptomics and antibody-based proteomics. *Mol. Cell Proteom.* **2014**, *13*, 397–406. [CrossRef]
36. Uhlen, M.; Fagerberg, L.; Hallstrom, B.M.; Lindskog, C.; Oksvold, P.; Mardinoglu, A.; Sivertsson, A.; Kampf, C.; Sjostedt, E.; Asplund, A.; et al. Proteomics. Tissue-based map of the human proteome. *Science* **2015**, *347*, 1260419. [CrossRef]
37. Jasinska, R.; Zhang, Q.X.; Pilquil, C.; Singh, I.; Xu, J.; Dewald, J.; Dillon, D.A.; Berthiaume, L.G.; Carman, G.M.; Waggoner, D.W.; et al. Lipid phosphate phosphohydrolase-1 degrades exogenous glycerolipid and sphingolipid phosphate esters. *Biochem. J.* **1999**, *340*, 677–686. [CrossRef]
38. Sciorra, V.A.; Morris, A.J. Sequential actions of phospholipase D and phosphatidic acid phosphohydrolase 2b generate diglyceride in mammalian cells. *Mol. Biol. Cell* **1999**, *10*, 3863–3876. [CrossRef]
39. Nanjundan, M.; Possmayer, F. Pulmonary lipid phosphate phosphohydrolase in plasma membrane signalling platforms. *Biochem. J.* **2001**, *358*, 637–646. [CrossRef]
40. Jia, Y.J.; Kai, M.; Wada, I.; Sakane, F.; Kanoh, H. Differential localization of lipid phosphate phosphatases 1 and 3 to cell surface subdomains in polarized MDCK cells. *FEBS Lett.* **2003**, *552*, 240–246. [CrossRef]
41. Barilà, D.; Plateroti, M.; Nobili, F.; Muda, A.O.; Xie, Y.; Morimoto, T.; Perozzi, G. The Dri 42 gene, whose expression is up-regulated during epithelial differentiation, encodes a novel endoplasmic reticulum resident transmembrane protein. *J. Biol. Chem.* **1996**, *271*, 29928–29936. [CrossRef] [PubMed]
42. Zhang, Q.X.; Pilquil, C.S.; Dewald, J.; Berthiaume, L.G.; Brindley, D.N. Identification of structurally important domains of lipid phosphate phosphatase-1: Implications for its sites of action. *Biochem. J.* **2000**, *345*, 181–184. [CrossRef] [PubMed]
43. Sigal, Y.J.; McDermott, M.I.; Morris, A.J. Integral membrane lipid phosphatases/phosphotransferases: Common structure and diverse functions. *Biochem. J.* **2005**, *387*, 281–293. [CrossRef] [PubMed]

44. Messerschmidt, A.; Wever, R. X-ray structure of a vanadium-containing enzyme: Chloroperoxidase from the fungus Curvularia inaequalis. *Proc. Natl. Acad. Sci. USA* **1996**, *93*, 392–396. [CrossRef]
45. Messerschmidt, A.; Prade, L.; Wever, R. Implications for the catalytic mechanism of the vanadium-containing enzyme chloroperoxidase from the fungus Curvularia inaequalis by X-ray structures of the native and peroxide form. *Biol. Chem.* **1997**, *378*, 309–315. [CrossRef]
46. Miriyala, S.; Subramanian, T.; Panchatcharam, M.; Ren, H.; McDermott, M.I.; Sunkara, M.; Drennan, T.; Smyth, S.S.; Spielmann, H.P.; Morris, A.J. Functional characterization of the atypical integral membrane lipid phosphatase PDP1/PPAPDC2 identifies a pathway for interconversion of isoprenols and isoprenoid phosphates in mammalian cells. *J. Biol. Chem.* **2010**, *285*, 13918–13929. [CrossRef]
47. Fan, J.; Jiang, D.; Zhao, Y.; Liu, J.; Zhang, X.C. Crystal structure of lipid phosphatase Escherichia coli phosphatidylglycerophosphate phosphatase B. *Proc. Natl. Acad. Sci. USA* **2014**, *111*, 7636–7640. [CrossRef]
48. Umezu-Goto, M.; Kishi, Y.; Taira, A.; Hama, K.; Dohmae, N.; Takio, K.; Yamori, T.; Mills, G.B.; Inoue, K.; Aoki, J.; et al. Autotaxin has lysophospholipase D activity leading to tumor cell growth and motility by lysophosphatidic acid production. *J. Cell Biol.* **2002**, *158*, 227–233. [CrossRef]
49. Tokumura, A.; Majima, E.; Kariya, Y.; Tominaga, K.; Kogure, K.; Yasuda, K.; Fukuzawa, K. Identification of human plasma lysophospholipase D, a lysophosphatidic acid-producing enzyme, as autotaxin, a multifunctional phosphodiesterase. *J. Biol. Chem.* **2002**, *277*, 39436–39442. [CrossRef]
50. Brindley, D.N.; Lin, F.T.; Tigyi, G.J. Role of the autotaxin-lysophosphatidate axis in cancer resistance to chemotherapy and radiotherapy. *Biochim. Biophys. Acta* **2013**, *1831*, 74–85. [CrossRef]
51. Samadi, N.; Bekele, R.; Capatos, D.; Venkatraman, G.; Sariahmetoglu, M.; Brindley, D.N. Regulation of lysophosphatidate signaling by autotaxin and lipid phosphate phosphatases with respect to tumor progression, angiogenesis, metastasis and chemo-resistance. *Biochimie* **2011**, *93*, 61–70. [CrossRef] [PubMed]
52. Strub, G.M.; Paillard, M.; Liang, J.; Gomez, L.; Allegood, J.C.; Hait, N.C.; Maceyka, M.; Price, M.M.; Chen, Q.; Simpson, D.C.; et al. Sphingosine-1-phosphate produced by sphingosine kinase 2 in mitochondria interacts with prohibitin 2 to regulate complex IV assembly and respiration. *FASEB J.* **2011**, *25*, 600–612. [CrossRef] [PubMed]
53. Hait, N.C.; Allegood, J.; Maceyka, M.; Strub, G.M.; Harikumar, K.B.; Singh, S.K.; Luo, C.; Marmorstein, R.; Kordula, T.; Milstien, S.; et al. Regulation of histone acetylation in the nucleus by sphingosine-1-phosphate. *Science* **2009**, *325*, 1254–1257. [CrossRef] [PubMed]
54. Takabe, K.; Paugh, S.W.; Milstien, S.; Spiegel, S. "Inside-out" signaling of sphingosine-1-phosphate: Therapeutic targets. *Pharmacol. Rev.* **2008**, *60*, 181–195. [CrossRef]
55. Sato, K.; Malchinkhuu, E.; Horiuchi, Y.; Mogi, C.; Tomura, H.; Tosaka, M.; Yoshimoto, Y.; Kuwabara, A.; Okajima, F. Critical role of ABCA1 transporter in sphingosine 1-phosphate release from astrocytes. *J. Neurochem.* **2007**, *103*, 2610–2619. [CrossRef]
56. Kobayashi, N.; Nishi, T.; Hirata, T.; Kihara, A.; Sano, T.; Igarashi, Y.; Yamaguchi, A. Sphingosine 1-phosphate is released from the cytosol of rat platelets in a carrier-mediated manner. *J. Lipid Res.* **2006**, *47*, 614–621. [CrossRef]
57. Ulrych, T.; Bohm, A.; Polzin, A.; Daum, G.; Nusing, R.M.; Geisslinger, G.; Hohlfeld, T.; Schror, K.; Rauch, B.H. Release of sphingosine-1-phosphate from human platelets is dependent on thromboxane formation. *J. Thromb. Haemost.* **2011**, *9*, 790–798. [CrossRef]
58. Nagahashi, M.; Kim, E.Y.; Yamada, A.; Ramachandran, S.; Allegood, J.C.; Hait, N.C.; Maceyka, M.; Milstien, S.; Takabe, K.; Spiegel, S. Spns2, a transporter of phosphorylated sphingoid bases, regulates their blood and lymph levels, and the lymphatic network. *FASEB J.* **2013**, *27*, 1001–1011. [CrossRef]
59. Vu, T.M.; Ishizu, A.N.; Foo, J.C.; Toh, X.R.; Zhang, F.; Whee, D.M.; Torta, F.; Cazenave-Gassiot, A.; Matsumura, T.; Kim, S.; et al. Mfsd2b is essential for the sphingosine-1-phosphate export in erythrocytes and platelets. *Nature* **2017**, *550*, 524–528. [CrossRef]
60. English, D.; Welch, Z.; Kovala, A.T.; Harvey, K.; Volpert, O.V.; Brindley, D.N.; Garcia, J.G. Sphingosine 1-phosphate released from platelets during clotting accounts for the potent endothelial cell chemotactic activity of blood serum and provides a novel link between hemostasis and angiogenesis. *FASEB J.* **2000**, *14*, 2255–2265. [CrossRef]
61. Yung, Y.C.; Stoddard, N.C.; Chun, J. LPA receptor signaling: Pharmacology, physiology, and pathophysiology. *J. Lipid Res.* **2014**, *55*, 1192–1214. [CrossRef] [PubMed]

62. Tigyi, G.J.; Yue, J.; Norman, D.D.; Szabo, E.; Balogh, A.; Balazs, L.; Zhao, G.; Lee, S.C. Regulation of tumor cell-Microenvironment interaction by the autotaxin-lysophosphatidic acid receptor axis. *Adv. Biol. Regul.* **2019**, *71*, 183–193. [CrossRef] [PubMed]
63. Long, J.S.; Yokoyama, K.; Tigyi, G.; Pyne, N.J.; Pyne, S. Lipid phosphate phosphatase-1 regulates lysophosphatidic acid- and platelet-derived-growth-factor-induced cell migration. *Biochem. J.* **2006**, *394*, 495–500. [CrossRef] [PubMed]
64. Pilquil, C.; Dewald, J.; Cherney, A.; Gorshkova, I.; Tigyi, G.; English, D.; Natarajan, V.; Brindley, D.N. Lipid phosphate phosphatase-1 regulates lysophosphatidate-induced fibroblast migration by controlling phospholipase D2-dependent phosphatidate generation. *J. Biol. Chem.* **2006**, *281*, 38418–38429. [CrossRef]
65. Alderton, F.; Darroch, P.; Sambi, B.; McKie, A.; Ahmed, I.S.; Pyne, N.; Pyne, S. G-protein-coupled receptor stimulation of the p42/p44 mitogen-activated protein kinase pathway is attenuated by lipid phosphate phosphatases 1, 1a, and 2 in human embryonic kidney 293 cells. *J. Biol. Chem.* **2001**, *276*, 13452–13460. [CrossRef]
66. Bektas, M.; Payne, S.G.; Liu, H.; Goparaju, S.; Milstien, S.; Spiegel, S. A novel acylglycerol kinase that produces lysophosphatidic acid modulates cross talk with EGFR in prostate cancer cells. *J. Cell Biol.* **2005**, *169*, 801–811. [CrossRef]
67. Gobeil, F., Jr.; Bernier, S.G.; Vazquez-Tello, A.; Brault, S.; Beauchamp, M.H.; Quiniou, C.; Marrache, A.M.; Checchin, D.; Sennlaub, F.; Hou, X.; et al. Modulation of pro-inflammatory gene expression by nuclear lysophosphatidic acid receptor type-1. *J. Biol. Chem.* **2003**, *278*, 38875–38883. [CrossRef]
68. Zhang, C.; Baker, D.L.; Yasuda, S.; Makarova, N.; Balazs, L.; Johnson, L.R.; Marathe, G.K.; McIntyre, T.M.; Xu, Y.; Prestwich, G.D.; et al. Lysophosphatidic acid induces neointima formation through PPARgamma activation. *J. Exp. Med.* **2004**, *199*, 763–774. [CrossRef]
69. Tang, X.; Zhao, Y.Y.; Dewald, J.; Curtis, J.M.; Brindley, D.N. Tetracyclines increase lipid phosphate phosphatase expression on plasma membranes and turnover of plasma lysophosphatidate. *J. Lipid Res.* **2016**, *57*, 597–606. [CrossRef]
70. Nakayama, J.; Raines, T.A.; Lynch, K.R.; Slack-Davis, J.K. Decreased peritoneal ovarian cancer growth in mice lacking expression of lipid phosphate phosphohydrolase 1. *PLoS ONE* **2015**, *10*, e0120071. [CrossRef]
71. Tomsig, J.L.; Snyder, A.H.; Berdyshev, E.V.; Skobeleva, A.; Mataya, C.; Natarajan, V.; Brindley, D.N.; Lynch, K.R. Lipid phosphate phosphohydrolase type 1 (LPP1) degrades extracellular lysophosphatidic acid in vivo. *Biochem. J.* **2009**, *419*, 611–618. [CrossRef] [PubMed]
72. Yue, J.; Yokoyama, K.; Balazs, L.; Baker, D.L.; Smalley, D.; Pilquil, C.; Brindley, D.N.; Tigyi, G. Mice with transgenic overexpression of lipid phosphate phosphatase-1 display multiple organotypic deficits without alteration in circulating lysophosphatidate level. *Cell. Signal.* **2004**, *16*, 385–399. [CrossRef] [PubMed]
73. Jesionowska, A.; Cecerska-Heryc, E.; Matoszka, N.; Dolegowska, B. Lysophosphatidic acid signaling in ovarian cancer. *J. Recept. Signal. Transduct. Res.* **2015**, *35*, 578–584. [CrossRef]
74. Fang, X.; Schummer, M.; Mao, M.; Yu, S.; Tabassam, F.H.; Swaby, R.; Hasegawa, Y.; Tanyi, J.L.; LaPushin, R.; Eder, A.; et al. Lysophosphatidic acid is a bioactive mediator in ovarian cancer. *Biochim. Biophys. Acta* **2002**, *1582*, 257–264. [CrossRef]
75. Baker, D.L.; Morrison, P.; Miller, B.; Riely, C.A.; Tolley, B.; Westermann, A.M.; Bonfrer, J.M.; Bais, E.; Moolenaar, W.H.; Tigyi, G. Plasma lysophosphatidic acid concentration and ovarian cancer. *JAMA* **2002**, *287*, 3081–3082. [CrossRef] [PubMed]
76. Peest, U.; Sensken, S.C.; Andreani, P.; Hanel, P.; Van Veldhoven, P.P.; Graler, M.H. S1P-lyase independent clearance of extracellular sphingosine 1-phosphate after dephosphorylation and cellular uptake. *J. Cell Biochem.* **2008**, *104*, 756–772. [CrossRef] [PubMed]
77. Mandala, S.M. Sphingosine-1-phosphate phosphatases. *Prostaglandins Other Lipid Mediat.* **2001**, *64*, 143–156. [CrossRef]
78. Le Stunff, H.; Peterson, C.; Liu, H.; Milstien, S.; Spiegel, S. Sphingosine-1-phosphate and lipid phosphohydrolases. *Biochim. Biophys. Acta* **2002**, *1582*, 8–17. [CrossRef]
79. Zhao, Y.; Kalari, S.K.; Usatyuk, P.V.; Gorshkova, I.; He, D.; Watkins, T.; Brindley, D.N.; Sun, C.; Bittman, R.; Garcia, J.G.; et al. Intracellular generation of sphingosine 1-phosphate in human lung endothelial cells: Role of lipid phosphate phosphatase-1 and sphingosine kinase 1. *J. Biol. Chem.* **2007**, *282*, 14165–14177. [CrossRef]

80. Breart, B.; Ramos-Perez, W.D.; Mendoza, A.; Salous, A.K.; Gobert, M.; Huang, Y.; Adams, R.H.; Lafaille, J.J.; Escalante-Alcalde, D.; Morris, A.J.; et al. Lipid phosphate phosphatase 3 enables efficient thymic egress. *J. Exp. Med.* **2011**, *208*, 1267–1278. [CrossRef]
81. Lopez-Juarez, A.; Morales-Lazaro, S.; Sanchez-Sanchez, R.; Sunkara, M.; Lomeli, H.; Velasco, I.; Morris, A.J.; Escalante-Alcalde, D. Expression of LPP3 in Bergmann glia is required for proper cerebellar sphingosine-1-phosphate metabolism/signaling and development. *Glia* **2011**, *59*, 577–589. [CrossRef] [PubMed]
82. Yamanaka, M.; Anada, Y.; Igarashi, Y.; Kihara, A. A splicing isoform of LPP1, LPP1a, exhibits high phosphatase activity toward FTY720 phosphate. *Biochem. Biophys. Res. Commun.* **2008**, *375*, 675–679. [CrossRef] [PubMed]
83. Mechtcheriakova, D.; Wlachos, A.; Sobanov, J.; Bornancin, F.; Zlabinger, G.; Baumruker, T.; Billich, A. FTY720-phosphate is dephosphorylated by lipid phosphate phosphatase 3. *FEBS Lett.* **2007**, *581*, 3063–3068. [CrossRef] [PubMed]
84. Tang, X.; Benesch, M.G.; Dewald, J.; Zhao, Y.Y.; Patwardhan, N.; Santos, W.L.; Curtis, J.M.; McMullen, T.P.; Brindley, D.N. Lipid phosphate phosphatase-1 expression in cancer cells attenuates tumor growth and metastasis in mice. *J. Lipid Res.* **2014**, *55*, 2389–2400. [CrossRef]
85. Nishioka, T.; Frohman, M.A.; Matsuda, M.; Kiyokawa, E. Heterogeneity of phosphatidic acid levels and distribution at the plasma membrane in living cells as visualized by a Foster resonance energy transfer (FRET) biosensor. *J. Biol. Chem.* **2010**, *285*, 35979–35987. [CrossRef]
86. Rizzo, M.A.; Shome, K.; Vasudevan, C.; Stolz, D.B.; Sung, T.C.; Frohman, M.A.; Watkins, S.C.; Romero, G. Phospholipase D and its product, phosphatidic acid, mediate agonist-dependent raf-1 translocation to the plasma membrane and the activation of the mitogen-activated protein kinase pathway. *J. Biol. Chem.* **1999**, *274*, 1131–1139. [CrossRef]
87. Andresen, B.T.; Rizzo, M.A.; Shome, K.; Romero, G. The role of phosphatidic acid in the regulation of the Ras/MEK/Erk signaling cascade. *FEBS Lett.* **2002**, *531*, 65–68. [CrossRef]
88. Brindley, D.N.; Pilquil, C. Lipid phosphate phosphatases and signaling. *J. Lipid Res.* **2009**, *50*, S225–S230. [CrossRef]
89. Alderton, F.; Rakhit, S.; Kong, K.C.; Palmer, T.; Sambi, B.; Pyne, S.; Pyne, N.J. Tethering of the platelet-derived growth factor beta receptor to G-protein-coupled receptors. A novel platform for integrative signaling by these receptor classes in mammalian cells. *J. Biol. Chem.* **2001**, *276*, 28578–28585. [CrossRef]
90. Gutierrez-Martinez, E.; Fernandez-Ulibarri, I.; Lazaro-Dieguez, F.; Johannes, L.; Pyne, S.; Sarri, E.; Egea, G. Lipid phosphate phosphatase 3 participates in transport carrier formation and protein trafficking in the early secretory pathway. *J. Cell Sci.* **2013**, *126*, 2641–2655. [CrossRef]
91. Long, J.; Darroch, P.; Wan, K.F.; Kong, K.C.; Ktistakis, N.; Pyne, N.J.; Pyne, S. Regulation of cell survival by lipid phosphate phosphatases involves the modulation of intracellular phosphatidic acid and sphingosine 1-phosphate pools. *Biochem. J.* **2005**, *391*, 25–32. [CrossRef] [PubMed]
92. Roberts, R.Z.; Morris, A.J. Role of phosphatidic acid phosphatase 2a in uptake of extracellular lipid phosphate mediators. *Biochim. Biophys. Acta* **2000**, *1487*, 33–49. [CrossRef]
93. Brindley, D.N.; Pilquil, C.; Sariahmetoglu, M.; Reue, K. Phosphatidate degradation: Phosphatidate phosphatases (lipins) and lipid phosphate phosphatases. *Biochim. Biophys. Acta* **2009**, *1791*, 956–961. [CrossRef]
94. Pettus, B.J.; Kitatani, K.; Chalfant, C.E.; Taha, T.A.; Kawamori, T.; Bielawski, J.; Obeid, L.M.; Hannun, Y.A. The coordination of prostaglandin E2 production by sphingosine-1-phosphate and ceramide-1-phosphate. *Mol. Pharmacol.* **2005**, *68*, 330–335. [CrossRef] [PubMed]
95. Pyne, S.; Adams, D.R.; Pyne, N.J. Sphingosine 1-phosphate and sphingosine kinases in health and disease: Recent advances. *Prog. Lipid Res.* **2016**, *62*, 93–106. [CrossRef]
96. Balazs, L.; Okolicany, J.; Ferrebee, M.; Tigyi, G. Topical application of LPA accelerates wound healing. *Ann. N. Y. Acad. Sci.* **2000**, *905*, 270–273. [CrossRef]
97. van Corven, E.J.; van Rijswijk, A.; Jalink, K.; van der Bend, R.L.; van Blitterswijk, W.J.; Moolenaar, W.H. Mitogenic action of lysophosphatidic acid and phosphatidic acid on fibroblasts. Dependence on acyl-chain length and inhibition by suramin. *Biochem. J.* **1992**, *281*, 163–169. [CrossRef]
98. Chabaud, S.; Marcoux, T.L.; Deschenes-Rompre, M.P.; Rousseau, A.; Morissette, A.; Bouhout, S.; Bernard, G.; Bolduc, S. Lysophosphatidic acid enhances collagen deposition and matrix thickening in engineered tissue. *J. Tissue Eng. Regen. Med.* **2015**, *9*, E65–E75. [CrossRef]

99. Van Corven, E.J.; Groenink, A.; Jalink, K.; Eichholtz, T.; Moolenaar, W.H. Lysophosphatidate-induced cell proliferation: Identification and dissection of signaling pathways mediated by G proteins. *Cell* **1989**, *59*, 45–54. [CrossRef]
100. Panetti, T.S.; Chen, H.; Misenheimer, T.M.; Getzler, S.B.; Mosher, D.F. Endothelial cell mitogenesis induced by LPA: Inhibition by thrombospondin-1 and thrombospondin-2. *J. Lab. Clin. Med.* **1997**, *129*, 208–216. [CrossRef]
101. Dvorak, H.F. Tumors: Wounds that do not heal-redux. *Cancer Immunol. Res.* **2015**, *3*, 1–11. [CrossRef] [PubMed]
102. Benesch, M.G.; Ko, Y.M.; Tang, X.; Dewald, J.; Lopez-Campistrous, A.; Zhao, Y.Y.; Lai, R.; Curtis, J.M.; Brindley, D.N.; McMullen, T.P. Autotaxin is an inflammatory mediator and therapeutic target in thyroid cancer. *Endocr. Relat. Cancer* **2015**, *22*, 593–607. [CrossRef]
103. Magkrioti, C.; Oikonomou, N.; Kaffe, E.; Mouratis, M.A.; Xylourgidis, N.; Barbayianni, I.; Megadoukas, P.; Harokopos, V.; Valavanis, C.; Chun, J.; et al. The Autotaxin-Lysophosphatidic Acid Axis Promotes Lung Carcinogenesis. *Cancer Res.* **2018**, *78*, 3634–3644. [CrossRef]
104. Shao, Y.; Yu, Y.; He, Y.; Chen, Q.; Liu, H. Serum ATX as a novel biomarker for breast cancer. *Medicine (Baltimore)* **2019**, *98*, e14973. [CrossRef] [PubMed]
105. Memet, I.; Tsalkidou, E.; Tsaroucha, A.K.; Lambropoulou, M.; Chatzaki, E.; Trypsianis, G.; Schizas, D.; Pitiakoudis, M.; Simopoulos, C. Autotaxin Expression in Hepatocellular Carcinoma. *J. Investig. Surg.* **2018**, *31*, 359–365. [CrossRef] [PubMed]
106. Auciello, F.R.; Bulusu, V.; Oon, C.; Tait-Mulder, J.; Berry, M.; Bhattacharyya, S.; Tumanov, S.; Allen-Petersen, B.L.; Link, J.; Kendsersky, N.D.; et al. A Stromal Lysolipid-Autotaxin Signaling Axis Promotes Pancreatic Tumor Progression. *Cancer Discov.* **2019**, *9*, 617–627. [CrossRef] [PubMed]
107. Nakai, Y.; Ikeda, H.; Nakamura, K.; Kume, Y.; Fujishiro, M.; Sasahira, N.; Hirano, K.; Isayama, H.; Tada, M.; Kawabe, T.; et al. Specific increase in serum autotaxin activity in patients with pancreatic cancer. *Clin. Biochem.* **2011**, *44*, 576–581. [CrossRef]
108. Xu, A.; Ahsanul Kabir Khan, M.; Chen, F.; Zhong, Z.; Chen, H.C.; Song, Y. Overexpression of autotaxin is associated with human renal cell carcinoma and bladder carcinoma and their progression. *Med. Oncol.* **2016**, *33*, 131. [CrossRef]
109. Nouh, M.A.; Wu, X.X.; Okazoe, H.; Tsunemori, H.; Haba, R.; Abou-Zeid, A.M.; Saleem, M.D.; Inui, M.; Sugimoto, M.; Aoki, J.; et al. Expression of autotaxin and acylglycerol kinase in prostate cancer: Association with cancer development and progression. *Cancer Sci.* **2009**, *100*, 1631–1638. [CrossRef]
110. Patterson, A.D.; Maurhofer, O.; Beyoglu, D.; Lanz, C.; Krausz, K.W.; Pabst, T.; Gonzalez, F.J.; Dufour, J.F.; Idle, J.R. Aberrant lipid metabolism in hepatocellular carcinoma revealed by plasma metabolomics and lipid profiling. *Cancer Res.* **2011**, *71*, 6590–6600. [CrossRef]
111. Masuda, A.; Nakamura, K.; Izutsu, K.; Igarashi, K.; Ohkawa, R.; Jona, M.; Higashi, K.; Yokota, H.; Okudaira, S.; Kishimoto, T.; et al. Serum autotaxin measurement in haematological malignancies: A promising marker for follicular lymphoma. *Br. J. Haematol.* **2008**, *143*, 60–70. [CrossRef] [PubMed]
112. Xu, Y.; Shen, Z.; Wiper, D.W.; Wu, M.; Morton, R.E.; Elson, P.; Kennedy, A.W.; Belinson, J.; Markman, M.; Casey, G. Lysophosphatidic acid as a potential biomarker for ovarian and other gynecologic cancers. *JAMA* **1998**, *280*, 719–723. [CrossRef] [PubMed]
113. Zeng, R.; Li, B.; Huang, J.; Zhong, M.; Li, L.; Duan, C.; Zeng, S.; Liu, W.; Lu, J.; Tang, Y.; et al. Lysophosphatidic Acid is a Biomarker for Peritoneal Carcinomatosis of Gastric Cancer and Correlates with Poor Prognosis. *Genet. Test. Mol. Biomark.* **2017**, *21*, 641–648. [CrossRef] [PubMed]
114. Riaz, A.; Huang, Y.; Johansson, S. G-Protein-Coupled Lysophosphatidic Acid Receptors and Their Regulation of AKT Signaling. *Int. J. Mol. Sci.* **2016**, *17*, 215. [CrossRef]
115. Zhang, G.; Cheng, Y.; Zhang, Q.; Li, X.; Zhou, J.; Wang, J.; Wei, L. ATXLPA axis facilitates estrogeninduced endometrial cancer cell proliferation via MAPK/ERK signaling pathway. *Mol. Med. Rep.* **2018**, *17*, 4245–4252.
116. Herr, D.R. Potential use of g protein-coupled receptor-blocking monoclonal antibodies as therapeutic agents for cancers. *Int. Rev. Cell Mol. Biol.* **2012**, *297*, 45–81.
117. Kishi, Y.; Okudaira, S.; Tanaka, M.; Hama, K.; Shida, D.; Kitayama, J.; Yamori, T.; Aoki, J.; Fujimaki, T.; Arai, H. Autotaxin is overexpressed in glioblastoma multiforme and contributes to cell motility of glioblastoma by converting lysophosphatidylcholine to lysophosphatidic acid. *J. Biol. Chem.* **2006**, *281*, 17492–17500. [CrossRef]

118. Manning, T.J., Jr.; Parker, J.C.; Sontheimer, H. Role of lysophosphatidic acid and rho in glioma cell motility. *Cell Motil. Cytoskelet.* **2000**, *45*, 185–199. [CrossRef]
119. Yamada, T.; Sato, K.; Komachi, M.; Malchinkhuu, E.; Tobo, M.; Kimura, T.; Kuwabara, A.; Yanagita, Y.; Ikeya, T.; Tanahashi, Y.; et al. Lysophosphatidic acid (LPA) in malignant ascites stimulates motility of human pancreatic cancer cells through LPA1. *J. Biol. Chem.* **2004**, *279*, 6595–6605. [CrossRef]
120. Gong, Y.L.; Tao, C.J.; Hu, M.; Chen, J.F.; Cao, X.F.; Lv, G.M.; Li, P. Expression of lysophosphatidic acid receptors and local invasiveness and metastasis in Chinese pancreatic cancers. *Curr. Oncol.* **2012**, *19*, eS15–eS21. [CrossRef]
121. Kato, K.; Yoshikawa, K.; Tanabe, E.; Kitayoshi, M.; Fukui, R.; Fukushima, N.; Tsujiuchi, T. Opposite roles of LPA1 and LPA3 on cell motile and invasive activities of pancreatic cancer cells. *Tumour. Biol.* **2012**, *33*, 1739–1744. [CrossRef] [PubMed]
122. Shida, D.; Watanabe, T.; Aoki, J.; Hama, K.; Kitayama, J.; Sonoda, H.; Kishi, Y.; Yamaguchi, H.; Sasaki, S.; Sako, A.; et al. Aberrant expression of lysophosphatidic acid (LPA) receptors in human colorectal cancer. *Lab. Investig.* **2004**, *84*, 1352–1362. [CrossRef] [PubMed]
123. Yun, C.C.; Sun, H.; Wang, D.; Rusovici, R.; Castleberry, A.; Hall, R.A.; Shim, H. LPA2 receptor mediates mitogenic signals in human colon cancer cells. *Am. J. Physiol. Cell Physiol.* **2005**, *289*, C2–C11. [CrossRef] [PubMed]
124. Sun, K.; Cai, H.; Duan, X.; Yang, Y.; Li, M.; Qu, J.; Zhang, X.; Wang, J. Aberrant expression and potential therapeutic target of lysophosphatidic acid receptor 3 in triple-negative breast cancers. *Clin. Exp. Med.* **2015**, *15*, 371–380. [CrossRef] [PubMed]
125. Li, T.T.; Alemayehu, M.; Aziziyeh, A.I.; Pape, C.; Pampillo, M.; Postovit, L.M.; Mills, G.B.; Babwah, A.V.; Bhattacharya, M. Beta-arrestin/Ral signaling regulates lysophosphatidic acid-mediated migration and invasion of human breast tumor cells. *Mol. Cancer Res.* **2009**, *7*, 1064–1077. [CrossRef] [PubMed]
126. Burkhalter, R.J.; Westfall, S.D.; Liu, Y.; Stack, M.S. Lysophosphatidic Acid Initiates Epithelial to Mesenchymal Transition and Induces beta-Catenin-mediated Transcription in Epithelial Ovarian Carcinoma. *J. Biol. Chem.* **2015**, *290*, 22143–22154. [CrossRef] [PubMed]
127. Xu, M.; Liu, Z.; Wang, C.; Yao, B.; Zheng, X. EDG2 enhanced the progression of hepatocellular carcinoma by LPA/PI3K/AKT/mTOR signaling. *Oncotarget* **2017**, *8*, 66154–66168. [CrossRef]
128. Shibue, T.; Weinberg, R.A. EMT, CSCs, and drug resistance: The mechanistic link and clinical implications. *Nat. Rev. Clin. Oncol.* **2017**, *14*, 611–629. [CrossRef]
129. Seo, E.J.; Kwon, Y.W.; Jang, I.H.; Kim, D.K.; Lee, S.I.; Choi, E.J.; Kim, K.H.; Suh, D.S.; Lee, J.H.; Choi, K.U.; et al. Autotaxin Regulates Maintenance of Ovarian Cancer Stem Cells through Lysophosphatidic Acid-Mediated Autocrine Mechanism. *Stem Cells* **2016**, *34*, 551–564. [CrossRef]
130. Banerjee, S.; Norman, D.D.; Lee, S.C.; Parrill, A.L.; Pham, T.C.; Baker, D.L.; Tigyi, G.J.; Miller, D.D. Highly Potent Non-Carboxylic Acid Autotaxin Inhibitors Reduce Melanoma Metastasis and Chemotherapeutic Resistance of Breast Cancer Stem Cells. *J. Med. Chem.* **2017**, *60*, 1309–1324. [CrossRef]
131. Gupta, P.B.; Onder, T.T.; Jiang, G.; Tao, K.; Kuperwasser, C.; Weinberg, R.A.; Lander, E.S. Identification of selective inhibitors of cancer stem cells by high-throughput screening. *Cell* **2009**, *138*, 645–659. [CrossRef] [PubMed]
132. Liu, S.; Umezu-Goto, M.; Murph, M.; Lu, Y.; Liu, W.; Zhang, F.; Yu, S.; Stephens, L.C.; Cui, X.; Murrow, G.; et al. Expression of autotaxin and lysophosphatidic acid receptors increases mammary tumorigenesis, invasion, and metastases. *Cancer Cell* **2009**, *15*, 539–550. [CrossRef] [PubMed]
133. Lee, Z.; Cheng, C.T.; Zhang, H.; Subler, M.A.; Wu, J.; Mukherjee, A.; Windle, J.J.; Chen, C.K.; Fang, X. Role of LPA4/p2y9/GPR23 in negative regulation of cell motility. *Mol. Biol. Cell* **2008**, *19*, 5435–5445. [CrossRef] [PubMed]
134. Ishii, S.; Hirane, M.; Fukushima, K.; Tomimatsu, A.; Fukushima, N.; Tsujiuchi, T. Diverse effects of LPA4, LPA5 and LPA6 on the activation of tumor progression in pancreatic cancer cells. *Biochem. Biophys. Res. Commun.* **2015**, *461*, 59–64. [CrossRef]
135. Jongsma, M.; Matas-Rico, E.; Rzadkowski, A.; Jalink, K.; Moolenaar, W.H. LPA is a chemorepellent for B16 melanoma cells: Action through the cAMP-elevating LPA5 receptor. *PLoS ONE* **2011**, *6*, e29260. [CrossRef]
136. Lee, S.C.; Fujiwara, Y.; Liu, J.; Yue, J.; Shimizu, Y.; Norman, D.D.; Wang, Y.; Tsukahara, R.; Szabo, E.; Patil, R.; et al. Autotaxin and LPA1 and LPA5 receptors exert disparate functions in tumor cells versus the host tissue microenvironment in melanoma invasion and metastasis. *Mol. Cancer Res.* **2015**, *13*, 174–185. [CrossRef]

137. Mathew, D.; Kremer, K.N.; Strauch, P.; Tigyi, G.; Pelanda, R.; Torres, R.M. LPA5 Is an Inhibitory Receptor That Suppresses CD8 T-Cell Cytotoxic Function via Disruption of Early TCR Signaling. *Front. Immunol.* **2019**, *10*, 1159. [CrossRef]
138. Enooku, K.; Uranbileg, B.; Ikeda, H.; Kurano, M.; Sato, M.; Kudo, H.; Maki, H.; Koike, K.; Hasegawa, K.; Kokudo, N.; et al. Higher LPA2 and LPA6 mRNA Levels in Hepatocellular Carcinoma Are Associated with Poorer Differentiation, Microvascular Invasion and Earlier Recurrence with Higher Serum Autotaxin Levels. *PLoS ONE* **2016**, *11*, e0161825. [CrossRef]
139. Takahashi, K.; Fukushima, K.; Otagaki, S.; Ishimoto, K.; Minami, K.; Fukushima, N.; Honoki, K.; Tsujiuchi, T. Effects of LPA1 and LPA6 on the regulation of colony formation activity in colon cancer cells treated with anticancer drugs. *J. Recept. Signal. Transduct. Res.* **2018**, *38*, 71–75. [CrossRef]
140. Knowlden, S.; Georas, S.N. The autotaxin-LPA axis emerges as a novel regulator of lymphocyte homing and inflammation. *J. Immunol.* **2014**, *192*, 851–857. [CrossRef]
141. Ray, R.; Rai, V. Lysophosphatidic acid converts monocytes into macrophages in both mice and humans. *Blood* **2017**, *129*, 1177–1183. [CrossRef] [PubMed]
142. Benesch, M.G.; Ko, Y.M.; McMullen, T.P.; Brindley, D.N. Autotaxin in the crosshairs: Taking aim at cancer and other inflammatory conditions. *FEBS Lett.* **2014**, *588*, 2712–2727. [CrossRef] [PubMed]
143. Benesch, M.G.K.; Yang, Z.; Tang, X.; Meng, G.; Brindley, D.N. Lysophosphatidate Signaling: The Tumor Microenvironment's New Nemesis. *Trends Cancer* **2017**, *3*, 748–752. [CrossRef] [PubMed]
144. Colotta, F.; Allavena, P.; Sica, A.; Garlanda, C.; Mantovani, A. Cancer-related inflammation, the seventh hallmark of cancer: Links to genetic instability. *Carcinogenesis* **2009**, *30*, 1073–1081. [CrossRef] [PubMed]
145. Wu, J.M.; Xu, Y.; Skill, N.J.; Sheng, H.; Zhao, Z.; Yu, M.; Saxena, R.; Maluccio, M.A. Autotaxin expression and its connection with the TNF-alpha-NF-kappaB axis in human hepatocellular carcinoma. *Mol. Cancer* **2010**, *9*, 71. [CrossRef] [PubMed]
146. Cummings, R.; Zhao, Y.; Jacoby, D.; Spannhake, E.W.; Ohba, M.; Garcia, J.G.; Watkins, T.; He, D.; Saatian, B.; Natarajan, V. Protein kinase Cdelta mediates lysophosphatidic acid-induced NF-kappaB activation and interleukin-8 secretion in human bronchial epithelial cells. *J. Biol. Chem.* **2004**, *279*, 41085–41094. [CrossRef] [PubMed]
147. Zhao, Y.; He, D.; Saatian, B.; Watkins, T.; Spannhake, E.W.; Pyne, N.J.; Natarajan, V. Regulation of lysophosphatidic acid-induced epidermal growth factor receptor transactivation and interleukin-8 secretion in human bronchial epithelial cells by protein kinase Cdelta, Lyn kinase, and matrix metalloproteinases. *J. Biol. Chem.* **2006**, *281*, 19501–19511. [CrossRef]
148. Yu, S.; Murph, M.M.; Lu, Y.; Liu, S.; Hall, H.S.; Liu, J.; Stephens, C.; Fang, X.; Mills, G.B. Lysophosphatidic acid receptors determine tumorigenicity and aggressiveness of ovarian cancer cells. *J. Natl. Cancer Inst.* **2008**, *100*, 1630–1642. [CrossRef]
149. Lin, S.; Wang, D.; Iyer, S.; Ghaleb, A.M.; Shim, H.; Yang, V.W.; Chun, J.; Yun, C.C. The absence of LPA2 attenuates tumor formation in an experimental model of colitis-associated cancer. *Gastroenterology* **2009**, *136*, 1711–1720. [CrossRef]
150. Zhao, Y.; Usatyuk, P.V.; Cummings, R.; Saatian, B.; He, D.; Watkins, T.; Morris, A.; Spannhake, E.W.; Brindley, D.N.; Natarajan, V. Lipid phosphate phosphatase-1 regulates lysophosphatidic acid-induced calcium release, NF-kappaB activation and interleukin-8 secretion in human bronchial epithelial cells. *Biochem. J.* **2005**, *385*, 493–502. [CrossRef]
151. Schmid, R.; Wolf, K.; Robering, J.W.; Strauss, S.; Strissel, P.L.; Strick, R.; Rubner, M.; Fasching, P.A.; Horch, R.E.; Kremer, A.E.; et al. ADSCs and adipocytes are the main producers in the autotaxin-lysophosphatidic acid axis of breast cancer and healthy mammary tissue in vitro. *BMC Cancer* **2018**, *18*, 1273. [CrossRef] [PubMed]
152. Benesch, M.G.; Tang, X.; Dewald, J.; Dong, W.F.; Mackey, J.R.; Hemmings, D.G.; McMullen, T.P.; Brindley, D.N. Tumor-induced inflammation in mammary adipose tissue stimulates a vicious cycle of autotaxin expression and breast cancer progression. *FASEB J.* **2015**, *29*, 3990–4000. [CrossRef] [PubMed]
153. Benesch, M.G.; Zhao, Y.Y.; Curtis, J.M.; McMullen, T.P.; Brindley, D.N. Regulation of autotaxin expression and secretion by lysophosphatidate and sphingosine 1-phosphate. *J. Lipid Res.* **2015**, *56*, 1134–1144. [CrossRef] [PubMed]
154. Meng, G.; Tang, X.; Yang, Z.; Benesch, M.G.K.; Marshall, A.; Murray, D.; Hemmings, D.G.; Wuest, F.; McMullen, T.P.W.; Brindley, D.N. Implications for breast cancer treatment from increased autotaxin production in adipose tissue after radiotherapy. *FASEB J.* **2017**, *31*, 4064–4077. [CrossRef]

155. Meng, G.; Wuest, M.; Tang, X.; Dufour, J.; Zhao, Y.; Curtis, J.M.; McMullen, T.P.W.; Murray, D.; Wuest, F.; Brindley, D.N. Repeated Fractions of X-Radiation to the Breast Fat Pads of Mice Augment Activation of the Autotaxin-Lysophosphatidate-Inflammatory Cycle. *Cancers* **2019**, *11*, 1816. [CrossRef]
156. Meng, G.; Tang, X.; Yang, Z.; Zhao, Y.; Curtis, J.M.; McMullen, T.P.W.; Brindley, D.N. Dexamethasone decreases the autotaxin-lysophosphatidate-inflammatory axis in adipose tissue: Implications for the metabolic syndrome and breast cancer. *FASEB J.* **2019**, *33*, 1899–1910. [CrossRef]
157. Meng, G.; Wuest, M.; Tang, X.; Dufour, J.; McMullen, T.P.W.; Wuest, F.; Murray, D.; Brindley, D.N. Dexamethasone Attenuates X-Ray-Induced Activation of the Autotaxin-Lysophosphatidate-Inflammatory Cycle in Breast Tissue and Subsequent Breast Fibrosis. *Cancers* **2020**, *12*, 999. [CrossRef]
158. Shi, K.; Jiang, J.; Ma, T.; Xie, J.; Duan, L.; Chen, R.; Song, P.; Yu, Z.; Liu, C.; Zhu, Q.; et al. Dexamethasone attenuates bleomycin-induced lung fibrosis in mice through TGF-beta, Smad3 and JAK-STAT pathway. *Int. J. Clin. Exp. Med.* **2014**, *7*, 2645–2650.
159. Chen, F.; Gong, L.; Zhang, L.; Wang, H.; Qi, X.; Wu, X.; Xiao, Y.; Cai, Y.; Liu, L.; Li, X.; et al. Short courses of low dose dexamethasone delay bleomycin-induced lung fibrosis in rats. *Eur. J. Pharmacol.* **2006**, *536*, 287–295. [CrossRef]
160. Deng, W.; Wang, D.A.; Gosmanova, E.; Johnson, L.R.; Tigyi, G. LPA protects intestinal epithelial cells from apoptosis by inhibiting the mitochondrial pathway. *Am. J. Physiol. Gastrointest. Liver Physiol.* **2003**, *284*, G821–G829. [CrossRef]
161. Meng, Y.; Kang, S.; Fishman, D.A. Lysophosphatidic acid inhibits anti-Fas-mediated apoptosis enhanced by actin depolymerization in epithelial ovarian cancer. *FEBS Lett.* **2005**, *579*, 1311–1319. [CrossRef] [PubMed]
162. Sui, Y.; Yang, Y.; Wang, J.; Li, Y.; Ma, H.; Cai, H.; Liu, X.; Zhang, Y.; Wang, S.; Li, Z.; et al. Lysophosphatidic Acid Inhibits Apoptosis Induced by Cisplatin in Cervical Cancer Cells. *Biomed. Res. Int.* **2015**, *2015*, 598386. [CrossRef]
163. Kang, Y.C.; Kim, K.M.; Lee, K.S.; Namkoong, S.; Lee, S.J.; Han, J.A.; Jeoung, D.; Ha, K.S.; Kwon, Y.G.; Kim, Y.M. Serum bioactive lysophospholipids prevent TRAIL-induced apoptosis via PI3K/Akt-dependent cFLIP expression and Bad phosphorylation. *Cell Death Differ.* **2004**, *11*, 1287–1298. [CrossRef] [PubMed]
164. Rusovici, R.; Ghaleb, A.; Shim, H.; Yang, V.W.; Yun, C.C. Lysophosphatidic acid prevents apoptosis of Caco-2 colon cancer cells via activation of mitogen-activated protein kinase and phosphorylation of Bad. *Biochim. Biophys. Acta* **2007**, *1770*, 1194–1203. [CrossRef] [PubMed]
165. Samadi, N.; Bekele, R.T.; Goping, I.S.; Schang, L.M.; Brindley, D.N. Lysophosphatidate induces chemo-resistance by releasing breast cancer cells from taxol-induced mitotic arrest. *PLoS ONE* **2011**, *6*, e20608. [CrossRef] [PubMed]
166. Bekele, R.T.; Venkatraman, G.; Liu, R.Z.; Tang, X.; Mi, S.; Benesch, M.G.; Mackey, J.R.; Godbout, R.; Curtis, J.M.; McMullen, T.P.; et al. Oxidative stress contributes to the tamoxifen-induced killing of breast cancer cells: Implications for tamoxifen therapy and resistance. *Sci. Rep.* **2016**, *6*, 21164. [CrossRef]
167. Venkatraman, G.; Benesch, M.G.; Tang, X.; Dewald, J.; McMullen, T.P.; Brindley, D.N. Lysophosphatidate signaling stabilizes Nrf2 and increases the expression of genes involved in drug resistance and oxidative stress responses: Implications for cancer treatment. *FASEB J.* **2015**, *29*, 772–785. [CrossRef]
168. Deng, W.; Shuyu, E.; Tsukahara, R.; Valentine, W.J.; Durgam, G.; Gududuru, V.; Balazs, L.; Manickam, V.; Arsura, M.; VanMiddlesworth, L.; et al. The lysophosphatidic acid type 2 receptor is required for protection against radiation-induced intestinal injury. *Gastroenterology* **2007**, *132*, 1834–1851. [CrossRef]
169. Tigyi, G.J.; Johnson, L.R.; Lee, S.C.; Norman, D.D.; Szabo, E.; Balogh, A.; Thompson, K.; Boler, A.; McCool, W.S. Lysophosphatidic acid type 2 receptor agonists in targeted drug development offer broad therapeutic potential. *J. Lipid Res.* **2019**, *60*, 464–474. [CrossRef]
170. Hashimoto, S.; Mikami, S.; Sugino, H.; Yoshikawa, A.; Hashimoto, A.; Onodera, Y.; Furukawa, S.; Handa, H.; Oikawa, T.; Okada, Y.; et al. Lysophosphatidic acid activates Arf6 to promote the mesenchymal malignancy of renal cancer. *Nat. Commun.* **2016**, *7*, 10656. [CrossRef]
171. Su, S.C.; Hu, X.; Kenney, P.A.; Merrill, M.M.; Babaian, K.N.; Zhang, X.Y.; Maity, T.; Yang, S.F.; Lin, X.; Wood, C.G. Autotaxin-lysophosphatidic acid signaling axis mediates tumorigenesis and development of acquired resistance to sunitinib in renal cell carcinoma. *Clin. Cancer Res.* **2013**, *19*, 6461–6472. [CrossRef] [PubMed]

172. Fukushima, K.; Takahashi, K.; Yamasaki, E.; Onishi, Y.; Fukushima, N.; Honoki, K.; Tsujiuchi, T. Lysophosphatidic acid signaling via LPA1 and LPA3 regulates cellular functions during tumor progression in pancreatic cancer cells. *Exp. Cell Res.* **2017**, *352*, 139–145. [CrossRef]
173. Nguyen, T.; Nioi, P.; Pickett, C.B. The Nrf2-antioxidant response element signaling pathway and its activation by oxidative stress. *J. Biol. Chem.* **2009**, *284*, 13291–13295. [CrossRef] [PubMed]
174. Hemmings, D.G.; Brindley, D.N. Signalling by lysophosphatidate and its health implications. *Essays Biochem.* **2020**. [CrossRef] [PubMed]
175. Tang, X.; Wuest, M.; Benesch, M.G.; Dufour, J.; Zhao, Y.; Curtis, J.M.; Monjardet, A.; Heckmann, B.; Murray, D.; Wuest, F.; et al. Inhibition of autotaxin with GLPG1690 increases the efficacy of radiotherapy and chemotherapy in a mouse model of breast cancer. *Mol. Cancer Ther.* **2019**, *19*, 63–74. [CrossRef]
176. Kim, B.C.; Yoo, H.J.; Lee, H.C.; Kang, K.A.; Jung, S.H.; Lee, H.J.; Lee, M.; Park, S.; Ji, Y.H.; Lee, Y.S.; et al. Evaluation of premature senescence and senescence biomarkers in carcinoma cells and xenograft mice exposed to single or fractionated irradiation. *Oncol. Rep.* **2014**, *31*, 2229–2235. [CrossRef]
177. Jones, K.R.; Elmore, L.W.; Jackson-Cook, C.; Demasters, G.; Povirk, L.F.; Holt, S.E.; Gewirtz, D.A. p53-Dependent accelerated senescence induced by ionizing radiation in breast tumour cells. *Int. J. Radiat. Biol.* **2005**, *81*, 445–458. [CrossRef] [PubMed]
178. Zhang, M.; Guo, X.; Gao, Y.; Lu, D.; Li, W. Tumor Cell-Accelerated Senescence Is Associated With DNA-PKcs Status and Telomere Dysfunction Induced by Radiation. *Dose Response* **2018**, *16*, 1559325818771527. [CrossRef]
179. Pyne, N.J.; El Buri, A.; Adams, D.R.; Pyne, S. Sphingosine 1-phosphate and cancer. *Adv. Biol. Regul.* **2018**, *68*, 97–106. [CrossRef] [PubMed]
180. Heffernan-Stroud, L.A.; Obeid, L.M. Sphingosine kinase 1 in cancer. *Adv. Cancer Res.* **2013**, *117*, 201–235.
181. Hatoum, D.; Haddadi, N.; Lin, Y.; Nassif, N.T.; McGowan, E.M. Mammalian sphingosine kinase (SphK) isoenzymes and isoform expression: Challenges for SphK as an oncotarget. *Oncotarget* **2017**, *8*, 36898–36929. [CrossRef]
182. Xia, P.; Gamble, J.R.; Wang, L.; Pitson, S.M.; Moretti, P.A.; Wattenberg, B.W.; D'Andrea, R.J.; Vadas, M.A. An oncogenic role of sphingosine kinase. *Curr. Biol.* **2000**, *10*, 1527–1530. [CrossRef]
183. Chipuk, J.E.; McStay, G.P.; Bharti, A.; Kuwana, T.; Clarke, C.J.; Siskind, L.J.; Obeid, L.M.; Green, D.R. Sphingolipid metabolism cooperates with BAK and BAX to promote the mitochondrial pathway of apoptosis. *Cell* **2012**, *148*, 988–1000. [CrossRef] [PubMed]
184. Hofmann, L.P.; Ren, S.; Schwalm, S.; Pfeilschifter, J.; Huwiler, A. Sphingosine kinase 1 and 2 regulate the capacity of mesangial cells to resist apoptotic stimuli in an opposing manner. *Biol. Chem.* **2008**, *389*, 1399–1407. [CrossRef] [PubMed]
185. Maceyka, M.; Sankala, H.; Hait, N.C.; Le Stunff, H.; Liu, H.; Toman, R.; Collier, C.; Zhang, M.; Satin, L.S.; Merrill, A.H., Jr.; et al. SphK1 and SphK2, sphingosine kinase isoenzymes with opposing functions in sphingolipid metabolism. *J. Biol. Chem.* **2005**, *280*, 37118–37129. [CrossRef]
186. Schnitzer, S.E.; Weigert, A.; Zhou, J.; Brune, B. Hypoxia enhances sphingosine kinase 2 activity and provokes sphingosine-1-phosphate-mediated chemoresistance in A549 lung cancer cells. *Mol. Cancer Res.* **2009**, *7*, 393–401. [CrossRef]
187. Nemoto, S.; Nakamura, M.; Osawa, Y.; Kono, S.; Itoh, Y.; Okano, Y.; Murate, T.; Hara, A.; Ueda, H.; Nozawa, Y.; et al. Sphingosine kinase isoforms regulate oxaliplatin sensitivity of human colon cancer cells through ceramide accumulation and Akt activation. *J. Biol. Chem.* **2009**, *284*, 10422–10432. [CrossRef]
188. Gao, P.; Smith, C.D. Ablation of sphingosine kinase-2 inhibits tumor cell proliferation and migration. *Mol. Cancer Res.* **2011**, *9*, 1509–1519. [CrossRef]
189. Weigert, A.; Schiffmann, S.; Sekar, D.; Ley, S.; Menrad, H.; Werno, C.; Grosch, S.; Geisslinger, G.; Brune, B. Sphingosine kinase 2 deficient tumor xenografts show impaired growth and fail to polarize macrophages towards an anti-inflammatory phenotype. *Int. J. Cancer* **2009**, *125*, 2114–2121. [CrossRef]
190. Wallington-Beddoe, C.T.; Powell, J.A.; Tong, D.; Pitson, S.M.; Bradstock, K.F.; Bendall, L.J. Sphingosine kinase 2 promotes acute lymphoblastic leukemia by enhancing MYC expression. *Cancer Res.* **2014**, *74*, 2803–2815. [CrossRef]
191. Neubauer, H.A.; Pham, D.H.; Zebol, J.R.; Moretti, P.A.; Peterson, A.L.; Leclercq, T.M.; Chan, H.; Powell, J.A.; Pitman, M.R.; Samuel, M.S.; et al. An oncogenic role for sphingosine kinase 2. *Oncotarget* **2016**, *7*, 64886–64899. [CrossRef] [PubMed]

192. Wang, Q.; Li, J.; Li, G.; Li, Y.; Xu, C.; Li, M.; Xu, G.; Fu, S. Prognostic significance of sphingosine kinase 2 expression in non-small cell lung cancer. *Tumour Biol.* **2014**, *35*, 363–368. [CrossRef] [PubMed]
193. Nagahashi, M.; Ramachandran, S.; Kim, E.Y.; Allegood, J.C.; Rashid, O.M.; Yamada, A.; Zhao, R.; Milstien, S.; Zhou, H.; Spiegel, S.; et al. Sphingosine-1-phosphate produced by sphingosine kinase 1 promotes breast cancer progression by stimulating angiogenesis and lymphangiogenesis. *Cancer Res.* **2012**, *72*, 726–735. [CrossRef] [PubMed]
194. Nagahashi, M.; Tsuchida, J.; Moro, K.; Hasegawa, M.; Tatsuda, K.; Woelfel, I.A.; Takabe, K.; Wakai, T. High levels of sphingolipids in human breast cancer. *J. Surg. Res.* **2016**, *204*, 435–444. [CrossRef]
195. Lee, H.; Deng, J.; Kujawski, M.; Yang, C.; Liu, Y.; Herrmann, A.; Kortylewski, M.; Horne, D.; Somlo, G.; Forman, S.; et al. STAT3-induced S1PR1 expression is crucial for persistent STAT3 activation in tumors. *Nat. Med.* **2010**, *16*, 1421–1428. [CrossRef]
196. Chang, Q.; Bournazou, E.; Sansone, P.; Berishaj, M.; Gao, S.P.; Daly, L.; Wels, J.; Theilen, T.; Granitto, S.; Zhang, X.; et al. The IL-6/JAK/Stat3 feed-forward loop drives tumorigenesis and metastasis. *Neoplasia* **2013**, *15*, 848–862. [CrossRef]
197. Liang, J.; Nagahashi, M.; Kim, E.Y.; Harikumar, K.B.; Yamada, A.; Huang, W.C.; Hait, N.C.; Allegood, J.C.; Price, M.M.; Avni, D.; et al. Sphingosine-1-phosphate links persistent STAT3 activation, chronic intestinal inflammation, and development of colitis-associated cancer. *Cancer Cell* **2013**, *23*, 107–120. [CrossRef]
198. Deng, J.; Liu, Y.; Lee, H.; Herrmann, A.; Zhang, W.; Zhang, C.; Shen, S.; Priceman, S.J.; Kujawski, M.; Pal, S.K.; et al. S1PR1-STAT3 signaling is crucial for myeloid cell colonization at future metastatic sites. *Cancer Cell* **2012**, *21*, 642–654. [CrossRef]
199. Theiss, A.L. Sphingosine-1-phosphate: Driver of NFkappaB and STAT3 persistent activation in chronic intestinal inflammation and colitis-associated cancer. *JAKSTAT* **2013**, *2*, e24150. [CrossRef]
200. Chae, S.S.; Paik, J.H.; Furneaux, H.; Hla, T. Requirement for sphingosine 1-phosphate receptor-1 in tumor angiogenesis demonstrated by in vivo RNA interference. *J. Clin. Investig.* **2004**, *114*, 1082–1089. [CrossRef]
201. Wang, S.; Liang, Y.; Chang, W.; Hu, B.; Zhang, Y. Triple Negative Breast Cancer Depends on Sphingosine Kinase 1 (SphK1)/Sphingosine-1-Phosphate (S1P)/Sphingosine 1-Phosphate Receptor 3 (S1PR3)/Notch Signaling for Metastasis. *Med. Sci. Monit.* **2018**, *24*, 1912–1923. [CrossRef]
202. Watson, C.; Long, J.S.; Orange, C.; Tannahill, C.L.; Mallon, E.; McGlynn, L.M.; Pyne, S.; Pyne, N.J.; Edwards, J. High expression of sphingosine 1-phosphate receptors, S1P1 and S1P3, sphingosine kinase 1, and extracellular signal-regulated kinase-1/2 is associated with development of tamoxifen resistance in estrogen receptor-positive breast cancer patients. *Am. J. Pathol.* **2010**, *177*, 2205–2215. [CrossRef] [PubMed]
203. Lee, M.J.; Thangada, S.; Claffey, K.P.; Ancellin, N.; Liu, C.H.; Kluk, M.; Volpi, M.; Sha'afi, R.I.; Hla, T. Vascular endothelial cell adherens junction assembly and morphogenesis induced by sphingosine-1-phosphate. *Cell* **1999**, *99*, 301–312. [CrossRef]
204. Zhao, J.; Liu, J.; Lee, J.F.; Zhang, W.; Kandouz, M.; VanHecke, G.C.; Chen, S.; Ahn, Y.H.; Lonardo, F.; Lee, M.J. TGF-beta/SMAD3 Pathway Stimulates Sphingosine-1 Phosphate Receptor 3 Expression: IMPLICATION OF SPHINGOSINE-1 PHOSPHATE RECEPTOR 3 IN LUNG ADENOCARCINOMA PROGRESSION. *J. Biol. Chem.* **2016**, *291*, 27343–27353. [CrossRef] [PubMed]
205. Goetzl, E.J.; Dolezalova, H.; Kong, Y.; Zeng, L. Dual mechanisms for lysophospholipid induction of proliferation of human breast carcinoma cells. *Cancer Res.* **1999**, *59*, 4732–4737.
206. Asghar, M.Y.; Kemppainen, K.; Lassila, T.; Tornquist, K. Sphingosine 1-phosphate attenuates MMP2 and MMP9 in human anaplastic thyroid cancer C643 cells: Importance of S1P2. *PLoS ONE* **2018**, *13*, e0196992. [CrossRef]
207. Liu, R.; Zhao, R.; Zhou, X.; Liang, X.; Campbell, D.J.; Zhang, X.; Zhang, L.; Shi, R.; Wang, G.; Pandak, W.M.; et al. Conjugated bile acids promote cholangiocarcinoma cell invasive growth through activation of sphingosine 1-phosphate receptor 2. *Hepatology* **2014**, *60*, 908–918. [CrossRef]
208. Tang, X.; McMullen, T.P.W.; Brindley, D.N. Increasing the low lipid phosphate phosphatase 1 activity in breast cancer cells decreases transcription by AP-1 and expressions of matrix metalloproteinases and cyclin D1/D3. *Theranostics* **2019**, *9*, 6129–6142. [CrossRef]
209. Leung, D.W.; Tompkins, C.K.; White, T. Molecular cloning of two alternatively spliced forms of human phosphatidic acid phosphatase cDNAs that are differentially expressed in normal and tumor cells. *DNA Cell Biol.* **1998**, *17*, 377–385. [CrossRef]

210. Bhattacharjee, A.; Richards, W.G.; Staunton, J.; Li, C.; Monti, S.; Vasa, P.; Ladd, C.; Beheshti, J.; Bueno, R.; Gillette, M.; et al. Classification of human lung carcinomas by mRNA expression profiling reveals distinct adenocarcinoma subclasses. *Proc. Natl. Acad. Sci. USA* **2001**, *98*, 13790–13795. [CrossRef]
211. Curtis, C.; Shah, S.P.; Chin, S.-F.; Turashvili, G.; Rueda, O.M.; Dunning, M.J.; Speed, D.; Lynch, A.G.; Samarajiwa, S.; Yuan, Y.; et al. The genomic and transcriptomic architecture of 2000 breast tumours reveals novel subgroups. *Nature* **2012**, *486*, 346–352. [CrossRef]
212. Yoshihara, K.; Tajima, A.; Komata, D.; Yamamoto, T.; Kodama, S.; Fujiwara, H.; Suzuki, M.; Onishi, Y.; Hatae, M.; Sueyoshi, K.; et al. Gene expression profiling of advanced-stage serous ovarian cancers distinguishes novel subclasses and implicates ZEB2 in tumor progression and prognosis. *Cancer Sci.* **2009**, *100*, 1421–1428. [CrossRef]
213. Gao, X.; Goggin, K.; Dowling, C.; Qian, J.; Hawdon, J.M. Two potential hookworm DAF-16 target genes, SNR-3 and LPP-1: Gene structure, expression profile, and implications of a cis-regulatory element in the regulation of gene expression. *Parasit. Vectors* **2015**, *8*, 14. [CrossRef]
214. Loziuk, P.; Meier, F.; Johnson, C.; Ghashghaei, H.T.; Muddiman, D.C. TransOmic analysis of forebrain sections in Sp2 conditional knockout embryonic mice using IR-MALDESI imaging of lipids and LC-MS/MS label-free proteomics. *Anal. Bioanal. Chem.* **2016**, *408*, 3453–3474. [CrossRef]
215. Reschen, M.E.; Gaulton, K.J.; Lin, D.; Soilleux, E.J.; Morris, A.J.; Smyth, S.S.; O'Callaghan, C.A. Lipid-induced epigenomic changes in human macrophages identify a coronary artery disease-associated variant that regulates PPAP2B Expression through Altered C/EBP-beta binding. *PLoS Genet.* **2015**, *11*, e1005061. [CrossRef]
216. Mao, G.; Smyth, S.S.; Morris, A.J. Regulation of PLPP3 gene expression by NF-kappaB family transcription factors. *J. Biol. Chem.* **2019**, *294*, 14009–14019. [CrossRef]
217. Kotini, A.G.; Mpakali, A.; Agalioti, T. Dnmt3a1 upregulates transcription of distinct genes and targets chromosomal gene clusters for epigenetic silencing in mouse embryonic stem cells. *Mol. Cell Biol.* **2011**, *31*, 1577–1592. [CrossRef]
218. Wary, K.K.; Thakker, G.D.; Humtsoe, J.O.; Yang, J. Analysis of VEGF-responsive genes involved in the activation of endothelial cells. *Mol. Cancer* **2003**, *2*, 25. [CrossRef]
219. Asirvatham, A.J.; Schmidt, M.; Gao, B.; Chaudhary, J. Androgens regulate the immune/inflammatory response and cell survival pathways in rat ventral prostate epithelial cells. *Endocrinology* **2006**, *147*, 257–271. [CrossRef]
220. Imai, A.; Furui, T.; Tamaya, T.; Mills, G.B. A gonadotropin-releasing hormone-responsive phosphatase hydrolyses lysophosphatidic acid within the plasma membrane of ovarian cancer cells. *J. Clin. Endocrinol. Metab.* **2000**, *85*, 3370–3375. [CrossRef]
221. Harper, K.; Brochu-Gaudreau, K.; Saucier, C.; Dubois, C.M. Hypoxia Downregulates LPP3 and Promotes the Spatial Segregation of ATX and LPP1 During Cancer Cell Invasion. *Cancers* **2019**, *11*, 1403. [CrossRef]
222. Tanyi, J.L.; Hasegawa, Y.; Lapushin, R.; Morris, A.J.; Wolf, J.K.; Berchuck, A.; Lu, K.; Smith, D.I.; Kalli, K.; Hartmann, L.C.; et al. Role of decreased levels of lipid phosphate phosphatase-1 in accumulation of lysophosphatidic acid in ovarian cancer. *Clin. Cancer Res.* **2003**, *9*, 3534–3545.
223. Tanyi, J.L.; Morris, A.J.; Wolf, J.K.; Fang, X.; Hasegawa, Y.; Lapushin, R.; Auersperg, N.; Sigal, Y.J.; Newman, R.A.; Felix, E.A.; et al. The human lipid phosphate phosphatase-3 decreases the growth, survival, and tumorigenesis of ovarian cancer cells: Validation of the lysophosphatidic acid signaling cascade as a target for therapy in ovarian cancer. *Cancer Res.* **2003**, *63*, 1073–1082.
224. Tang, X.; Wang, X.; Zhao, Y.Y.; Curtis, J.M.; Brindley, D.N. Doxycycline attenuates breast cancer related inflammation by decreasing plasma lysophosphatidate concentrations and inhibiting NF-kappaB activation. *Mol. Cancer* **2017**, *16*, 36. [CrossRef]
225. Mao, X.Y.; Lee, M.J.; Zhu, J.; Zhu, C.; Law, S.M.; Snijders, A.M. Genome-wide screen identifies a novel prognostic signature for breast cancer survival. *Oncotarget* **2017**, *8*, 14003–14016. [CrossRef]
226. Chatterjee, I.; Humtsoe, J.O.; Kohler, E.E.; Sorio, C.; Wary, K.K. Lipid phosphate phosphatase-3 regulates tumor growth via beta-catenin and Cyclin-D1 signaling. *Mol. Cancer* **2011**, *10*, 51. [CrossRef]
227. Susanto, O.; Koh, Y.W.H.; Morrice, N.; Tumanov, S.; Thomason, P.A.; Nielson, M.; Tweedy, L.; Muinonen-Martin, A.J.; Kamphorst, J.J.; Mackay, G.M.; et al. LPP3 mediates self-generation of chemotactic LPA gradients by melanoma cells. *J. Cell Sci.* **2017**, *130*, 3455–3466. [CrossRef]

228. Flanagan, J.M.; Funes, J.M.; Henderson, S.; Wild, L.; Carey, N.; Boshoff, C. Genomics screen in transformed stem cells reveals RNASEH2A, PPAP2C, and ADARB1 as putative anticancer drug targets. *Mol. Cancer Ther.* **2009**, *8*, 249–260. [CrossRef]
229. Zhang, N.; Sundberg, J.P.; Gridley, T. Mice mutant for Ppap2c, a homolog of the germ cell migration regulator wunen, are viable and fertile. *Genesis* **2000**, *27*, 137–140. [CrossRef]
230. Escalante-Alcalde, D.; Hernandez, L.; Le Stunff, H.; Maeda, R.; Lee, H.S.; Gang, C., Jr.; Sciorra, V.A.; Daar, I.; Spiegel, S.; Morris, A.J.; et al. The lipid phosphatase LPP3 regulates extra-embryonic vasculogenesis and axis patterning. *Development* **2003**, *130*, 4623–4637. [CrossRef]
231. Ile, K.E.; Tripathy, R.; Goldfinger, V.; Renault, A.D. Wunen, a Drosophila lipid phosphate phosphatase, is required for septate junction-mediated barrier function. *Development* **2012**, *139*, 2535–2546. [CrossRef] [PubMed]
232. Kai, M.; Sakane, F.; Jia, Y.J.; Imai, S.; Yasuda, S.; Kanoh, H. Lipid phosphate phosphatases 1 and 3 are localized in distinct lipid rafts. *J. Biochem.* **2006**, *140*, 677–686. [CrossRef] [PubMed]

© 2020 by the authors. Licensee MDPI, Basel, Switzerland. This article is an open access article distributed under the terms and conditions of the Creative Commons Attribution (CC BY) license (http://creativecommons.org/licenses/by/4.0/).

Review

Regulation of Signaling and Metabolism by Lipin-mediated Phosphatidic Acid Phosphohydrolase Activity

Andrew J. Lutkewitte and Brian N. Finck *

Center for Human Nutrition, Division of Geriatrics and Nutritional Sciences, Department of Medicine, Washington University School of Medicine, Euclid Avenue, Campus Box 8031, St. Louis, MO 63110, USA; alutkew@wustl.edu

* Correspondence: bfinck@wustl.edu; Tel: +1-3143628963

Received: 4 September 2020; Accepted: 24 September 2020; Published: 29 September 2020

Abstract: Phosphatidic acid (PA) is a glycerophospholipid intermediate in the triglyceride synthesis pathway that has incredibly important structural functions as a component of cell membranes and dynamic effects on intracellular and intercellular signaling pathways. Although there are many pathways to synthesize and degrade PA, a family of PA phosphohydrolases (lipin family proteins) that generate diacylglycerol constitute the primary pathway for PA incorporation into triglycerides. Previously, it was believed that the pool of PA used to synthesize triglyceride was distinct, compartmentalized, and did not widely intersect with signaling pathways. However, we now know that modulating the activity of lipin 1 has profound effects on signaling in a variety of cell types. Indeed, in most tissues except adipose tissue, lipin-mediated PA phosphohydrolase activity is far from limiting for normal rates of triglyceride synthesis, but rather impacts critical signaling cascades that control cellular homeostasis. In this review, we will discuss how lipin-mediated control of PA concentrations regulates metabolism and signaling in mammalian organisms.

Keywords: phosphatidic acid; diacylglycerol; lipin; signaling

---

1. Introduction

Foundational work many decades ago by the laboratory of Dr. Eugene Kennedy defined the four sequential enzymatic steps by which three fatty acyl groups were esterified onto the glycerol-3-phosphate backbone to synthesize triglyceride [1]. The penultimate step in this pathway, the dephosphorylation of phosphatidic acid (PA) to form diacylglycerol (DAG), is catalyzed by Mg^{2+}-dependent PA phosphohydrolase (PAP) enzymes; an enzymatic activity first quantified in 1957 [2]. This lipid had been measured in plants, but at that time, the existence of PA in *Animalia* was controversial. It is now known that PA is maintained at picomolar concentrations in most cells and that this glycerophospholipid constitutes a critical branching-point in the Kennedy Pathway (Figure 1). PA is the precursor of cytidine diphosphate diacylglycerol (CDP-DAG) used to make several phospholipids including phosphatidylglycerol and phosphatidylinositol, while DAG is the substrate for synthesis of other abundant phospholipids like phosphatidylcholine and phosphatidylethanolamine. Although the elegant studies of Kennedy described PAP activity in chicken liver at a biochemical level in 1957 [2], the cloning of the genes that encode proteins with PAP catalytic activity would require almost 50 years of additional study [3,4]

Convergent lines of research in multiple model organisms and serendipitous findings with freezer-archived samples would eventually lead to the identification of the mammalian lipin family of proteins as PAP enzymes [5]. In 2006, the lab of George Carman reported that the yeast *Pah* protein catalyzed the long sought Mg^{2+}-dependent PAP activity in yeast [3]. This protein was the yeast homolog of the mammalian lipin family of proteins that were identified by Dr. Karen Reue's group in 2001 [4], but at that time they had no known molecular function. Han and colleagues demonstrated that, like the yeast *Pah* protein, mammalian lipin proteins had intrinsic PAP activity, answering this enduring biological question [3]. Given a number of differences in the regulation of yeast and mammalian lipin proteins, we have elected to focus this review on the mammalian lipins.

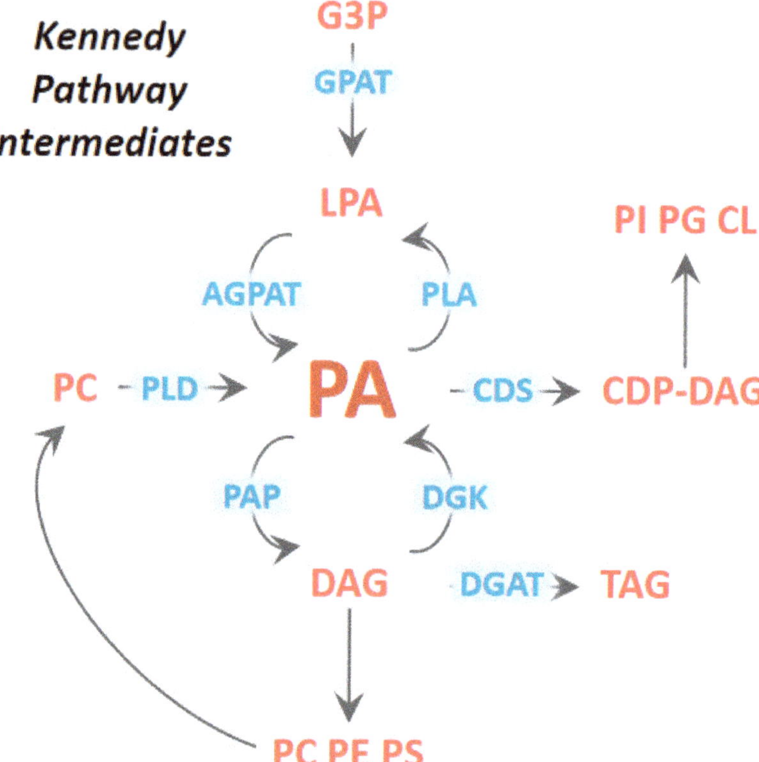

Figure 1. Phosphatidic acid as a central component in the Kennedy Pathway of lipid synthesis. Phosphatidic acid (PA) can be synthesized from and converted to numerus glycerophospholipids involved in membrane formation, cell signaling, lipid storage, and many others. Enzyme abbreviations in blue: glycerol-3-phosphate acyltransferase (GPAT), 1-acylglycerol-3-phosphate O-acyltransferase (AGPAT), phospholipase A (PLA), phospholipase D (PLD), cytidine diphosphate diacylglycerol Synthase (CDS), diacylglycerol kinase (DGK), phosphatidic acid phosphatase (PAP), diacylglycerol O-acyltransferase (DGAT). Glycerophospholipids and derivatives abbreviations in red: glycerol-3-phosphate (G3P), lysophosphatidic acid (LPA), phosphatidic acid (PA), phosphatidylcholine (PC), cytidine diphosphate diacylglycerol (CDP-DAG), phosphatidylinositol (PI), phosphatidylglycerol (PG), cardiolipin (CL), diacylglycerol (DAG), triacylglycerol (TAG), phosphatidylcholine (PC), phosphatidylethanolamine (PE), phosphatidylserine (PS).

The cloning of mammalian lipin genes resulted from another longstanding project to identify the spontaneous mutation leading to the phenotype of fatty liver dystrophic (*fld*) mice [6]. In mammals, three genes (*Lpin1, Lpin2, Lpin3*) encode lipin proteins (lipin 1, lipin 2, and lipin 3) [4,7]. Lipin family proteins exhibit distinct tissue-specific expression patterns [7]. Lipin 1 protein is enriched in adipocytes, striated muscle, and liver. Lipin 2 protein is liver-enriched and also expressed well in the intestine and central nervous system whereas lipin 3 is expressed in intestine and fat. Predictably, *fld* mice exhibit very low levels of PAP activity in most tissues where only lipin 1 is highly expressed (adipose tissue and striated muscle), but have significant PAP activity in liver, intestine, and other organs where lipin 2 is present [7–9]. While germline double deletion of lipin 1 and 3 or lipin 2 and 3 is tolerated in mice, the loss of lipin 1 and 2 is embryonic lethal [10], which is also consistent with functional redundancy of lipin 1 and 2, at least in mice. The importance of lipin 2 in human physiology is also demonstrated by the observation that mutations in lipin 2 cause Majeed's syndrome, an inflammatory syndrome of osteomyelitis [11]; the mechanistic basis for which is poorly understood.

2. Lipin Protein Structure and Regulation

Lipins are soluble proteins with conserved N- and C-terminal domains. A canonical haloacid dehalogenase catalytic site is contained in the C-terminal domain and N-terminal amphipathic helices and a polybasic domain facilitate membrane interaction [4,9] (Figure 2A). Recent crystallization studies have suggested that these conserved termini interact to form an immunoglobulin-like domain that is enabled by the variable regions in the middle of the protein [12] (Figure 2B). There is also evidence that lipin proteins form hetero- and homo-oligomers in their native state [13], although the importance of oligomer formation is still unclear. Atomic force microscopy imaging also suggested that lipin multimers may form circular structures or larger symmetrical particles [14]. Lipin proteins contain long stretches of basic amino acids (polybasic domain) that may be involved in promoting membrane localization by electrostatic interaction and also serve as a nuclear localization sequence [15,16] (Figure 2A). In the nucleus, lipin 1 interacts with DNA bound transcription factors to regulate their activity [17]. Lipin 1 has been shown to coactivate a number of nuclear receptors that regulate fatty acid metabolism [17], but can also act in a repressive manner on other transcription factors [18]. Since this activity is independent of PAP activity, the transcription regulatory function of lipin proteins will not be discussed in this review.

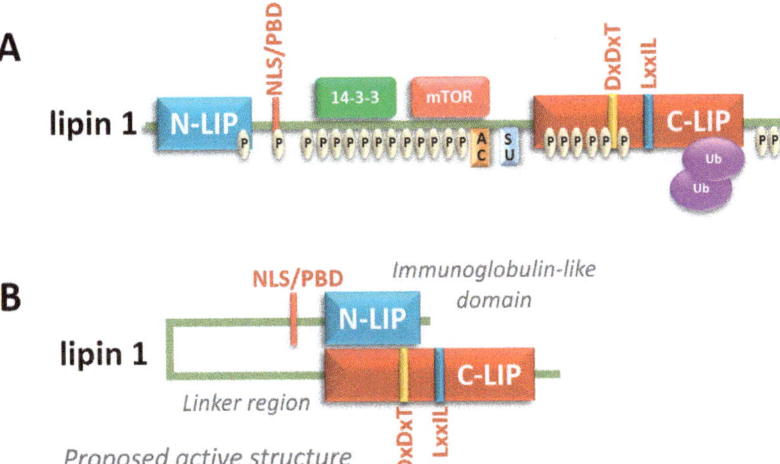

Figure 2. Lipin 1 structure and posttranslational modifications. (**A**) The lipin 1 protein contains several serine and threonine phosphorylation sites (P). Additionally, lipin 1 is acetylated (AC), sumoylated (SU), and ubiquitinated (Ub). Lipin 1 contains highly conserved N-terminal lipin (N-LIP) and C-terminal lipin (C-LIP) domains. The nuclear localization signal (NLS) is within the poly basic domain (PBD). The haloacid dehalogenase domain (DxDxT) is the catalytic motif and the LxxIL motif are contained within the C-LIP domain. (**B**) Recent crystal structure data suggests the N-LIP and C-LIP domains, which are separated by a linker region, interact to form an immunoglobulin-like domain in the native state.

Lipin activity seems to be controlled at several regulatory levels, though the control of the lipin 1 isoform is best understood compared to lipin 2 and 3. Transcription of the *Lpin1* gene is dynamically regulated in response to a variety of metabolic stimuli and disease states [17], but a great deal of lipin 1 activity is regulated post-translationally. Lipin 1 is a phospho-protein that is a target of the mechanistic target of rapamycin complex 1 (mTORC1) signaling pathway [9,19] (Figure 2A). Hyper-phosphorylation of serine/threonine residues of lipin 1 by mTORC1 drives its localization to the cytosol and away from the membrane and nuclear compartments [9,20]. Since PA is an insoluble lipid and embedded in cellular membranes, lipin 1 phosphorylation likely has the effect of reducing conversion of PA to DAG without affecting intrinsic PAP activity. mTORC1 is an important nutrient-sensing kinase and is downstream of the insulin receptor signaling cascade; thus, linking nutritional status to lipin 1 activity. In addition, the modification of lysine residues in lipin 1 by sumoylation [21], acetylation [22], and ubiquitination [23] can regulate lipin 1 localization and degradation, though it is unknown whether there is interplay among these various lysine modifications to modulate lipin 1 stability and activity (Figure 2A). Although less is known about the regulation of lipin 2, some studies have shown it is regulated both transcriptionally and translationally [24] and also post-translationally via phosphorylation [25]. Very little is known about the regulated expression and control of lipin 3 activity. The modulation of lipin expression and activity at multiple regulatory levels allows the cell to tightly control the activity of this enzyme.

3. Phosphatidic Acid and Diacylglycerol as Regulators of Signaling Pathways

For many years now, PA and DAG have been recognized as important regulators of intracellular signaling pathways and membrane biophysical properties as recently reviewed [26,27]. There are several enzymatic reactions that synthesize or catabolize these intermediates. For example, like PAP proteins, Mg^{2+}-independent lipid phosphate phosphohydrolases (LPPs) dephosphorylate PA into DAG [28]. While LPP activity is important in controlling PA- and DAG-mediated signaling, LPP activity occurs primarily at the plasma membrane. LPPs also dephosphorylate LPA, ceramide-1-phosphate, and sphingosine-1-phosphate [28]. Although many of these signaling pathways are parallel to those affected by lipin expression [29], we focus our attention to DAG- and PA-responsive pathways shown to be specifically regulated by lipin-mediated PAP activity.

Many of the effects ascribed to PA or DAG have been mechanistically demonstrated. However, it is important to note the near impossibility of modulating the abundance of one lipid without affecting levels of other related lipids. For example, PA can be rapidly converted to lysophosphatidic acid (LPA) by phospholipase A family lipases and the addition of high amounts of PA to cells in culture will likely alter abundance of LPA as well (Figure 1). Thus, caution should be taken in interpreting such results.

The mTOR signaling cascade is one of the most prominent kinases regulated by PA abundance [30,31] (Figure 3). As discussed above, mTOR is a nutrient responsive kinase that forms at least two distinct complexes of accessory proteins that regulate a multitude of downstream targets [32]. mTORC1 regulates protein synthesis, autophagy, mitochondrial metabolism and transcription of enzymes involved in lipid synthesis, whereas mTOR complex 2 (mTORC2) negatively regulates insulin signaling, controls cell stress response, apoptosis and cytoskeleton organization [32]. mTORC1 directly interacts with PA and this interaction allosterically activates mTORC1 to initiate a mitogenic response [33]. PA activation of mTOR, appears to have similar effects as insulin stimulation in myocytes [34]; yet, PA has also been shown to inhibit insulin signaling and is anti-mitogenic in adipocytes [35]. Work in lipin 1-deficient mice has demonstrated that mTORC1 activity is chronically elevated in some tissues [36]. mTORC1 has important negative regulatory effects on autophagy and mice or cells lacking lipin proteins exhibit general defects in autophagy [37,38]. Interestingly, PA accumulation seems to inhibit the activity of the mTORC2 signaling cascade [39]. In hepatocytes, lipin 1 knockdown leading to PA accumulation was associated with reduced mTORC2 activity and insulin resistance [39].

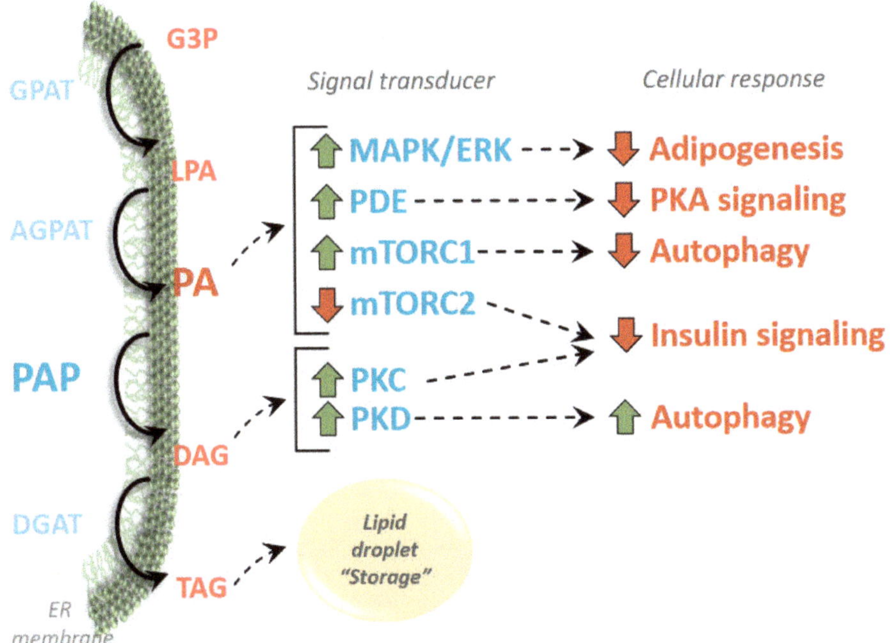

Figure 3. PAP derived phosphatidic acid activates several signaling cascades. Phosphatidic acid (PA) and diacylglycerol (DAG) synthesized from PAP activity effects several signaling modules involved in metabolism, autophagy, and differentiation. Enzyme abbreviations not listed in Figure 1 in **blue**: Mitogen Activated Protein Kinase (MAPK), Extracellular Regulated Kinase (ERK), phosphodiesterase (PDE), mechanistic Target of Rapamycin Complex 1 & 2 (mTORC1, mTORC2), Protein Kinase C (PKC), Protein Kinase D (PKD).

mTOR signaling also enhances the activity of phosphodiesterase (PDE) enzymes that degrade cAMP to control the activity of cAMP-responsive Protein Kinase A [40,41]. Additionally, PA directly activates PDE4 via allosteric interaction [42]. These dual mechanisms have been linked to impaired PKA signaling in lipin 1-deficient tissues including adipose tissue [43] and heart [36,44].

PA has also been shown to activate the extracellular signal-regulated kinase (ERK) Mitogen-Activated Protein Kinase (MAPK) signaling cascades [45–47] (Figure 3). This was first demonstrated in Schwann cells and is involved in the peripheral nerve demyelination that occurs in *fld* mice [46]. Regulation of ERK signaling may also be involved in the effects of lipin 1 on myocyte and adipocyte differentiation [45,47].

PA has also been shown to have effects on gene transcription by multiple mechanisms in cultured cells. Accumulation of some species of PA is linked to inhibition of peroxisome proliferator-activated receptor (PPAR) activity likely by effects on signaling pathways as well as cyclic phosphatidic acid possibly acting as an antagonistic ligand for this nuclear receptor [48]. It is possible that this plays a role in the regulation of adipocyte differentiation by this transcription factor and explains why nuclear-localized PAP activity is required for the induction of the adipogenic program in these cells [16]. Other work has suggested that the lipin-mediated remodeling of PA to DAG in the nuclear membrane by lipin 1 may also regulate gene expression by affecting chromatin structure and function [20].

In addition, the product of lipin 1 PAP activity, DAG, is also a significant regulator of signaling cascades including Protein Kinase C (PKC) and Protein Kinase D (PKD) (Figure 3). Activation of PKC isoforms by DAG accumulation in insulin-sensitive tissues has been linked to insulin resistance in obesity [49], and in mouse liver, lipin 1 mediated DAG production led to insulin resistance via activation of PKCε [50]. Additionally, in keratinocytes, lipin deficiency led to reduced activation of PKCα and affected the differentiation program of these cells [51]. Moreover, loss of lipin 1 and subsequent reductions in DAG levels in skeletal myocytes have been linked to inhibition of PKD activity, which led to impairments in autophagic flux and skeletal myopathy in *fld* mice [38].

Below we will detail the known connections between lipin and PA and its impact on signaling and metabolism in four tissue types. In order to focus this review, we will not discuss important findings in other types of cells and apologize for any oversights or unintentional exclusions.

4. Adipose Tissue

Mutations in lipin 1 lead to the marked lipodystrophic phenotype of *fld* mice [4], which is consistent with the effects of mutation or knockout of other enzymes involved in triglyceride synthesis also resulting in lipodystrophy [52]. This is somewhat predictable given that lipin 1 is highly expressed in adipocytes and the role of its enzymatic activity in triglyceride storage. However, in addition to an inability to store fat, lipin 1-deficient adipocytes also fail to induce the expression of canonical genes of the adipogenic program in vitro in response to a differentiation cocktail of hormones [53,54]. Accumulation of PA may explain this observation as this lipid has been shown to activate anti-adipogenic signaling, such as the ERK-MAPK pathway [53,55], and PA inhibition of differentiation is rescued by blocking ERK signaling in 3T3-L1 cells [56]. Complementation studies have also shown that both PAP activity and nuclear localization of lipin 1 are required for adipogenesis to occur in vitro, raising the possibility that this activity is required in the nucleus to induce adipogenesis [16]. This could also fit with lipin 1 transcriptional regulatory function enhancing activity of PPARγ [57], a crucial regulator of adipogenesis. In mice, knockout of lipin 3 slightly reduces PAP activity in white adipose tissue, but does not seem to affect adiposity and lipin 3 seems insufficient to compensate when lipin 1 is absent [58]. On the other hand, humans with mutations in the gene encoding lipin 1 (*LPIN1*) exhibit no defects in adipogenesis or reduction in adiposity [26,59], likely suggesting that other members of the lipin family or other PA phosphohydrolases can compensate [60].

Conditional knockout of lipin 1 after adipocyte differentiation has begun in mice has very mild effects on adiposity on a standard diet [53,61]. Fat-specific lipin 1 knockout mice have somewhat smaller fat pads on a standard diet, but a high fat diet produces a robust phenotype and fat-specific lipin 1 knockouts are highly resistant to diet-induced obesity [53]. Despite a lean phenotype, these mice are more susceptible to insulin resistance on a high fat diet likely due to accumulation of ectopic lipid in other tissues [53]. This observation is interesting in light of translational studies showing that adipose tissue lipin 1 expression in humans with obesity correlates well with insulin sensitivity. Specifically, patients with high adipose lipin 1 expression exhibit greater insulin sensitivity in skeletal muscle and liver [62,63]. Consistent with this, lipin 1 overexpression in mouse adipose tissue promotes an obese, but insulin sensitive phenotype [54]. Though PA has not been linked to this systemic effect on insulin sensitivity per se, it is possible that this lipid or other related lipids may be involved in inter-organ communication that leads to insulin resistance and that appropriate sequestration of these lipids in adipocytes protects other tissues from lipotoxicity.

Fat-specific lipin 1 knockout mice also exhibit marked reductions in Protein Kinase A (PKA) signaling that result in impaired basal and stimulated lipolysis [43]. This was due to accumulation of PA, since other methods to increase PA abundance also impaired PKA activity. Mechanistically, PA suppressed PKA activity by a two pronged mechanism involving a direct interaction with phosphodiesterase 4 (PDE4) and by activating mTOR signaling to enhance PDE activity and reduce cAMP [43]. Interestingly, it has long been known that β-adrenergic agonists increase PAP activity [64] and stimulate lipin 1 trafficking to its active site at the membrane [9]. While counterintuitive, given the role of lipin 1 in fat storage, it is possible that this effect is a mechanism to amplify PKA signaling in response to β-agonists in adipocytes. Conversely, when PA is abundant, triglyceride synthesis is favored and lipolysis is inhibited. Though first observed in mouse adipocytes, this effect on PKA activity is also observed in lipin 1-deficient mouse liver [43] and heart [36,44] and lipin 1 abundance in adipose tissue of humans with obesity is inversely correlated with basal lipolytic rates [43]. Thus, in adipose tissue, lipin 1 plays important roles in regulating fat storage and retention by regulating both triglyceride synthesis and lipolysis.

5. Skeletal Muscle

Rare mutations in the *LPIN1* gene in humans are associated with a syndrome of acute, recurrent rhabdomyolysis that usually manifests in early childhood [59,65–68]. Rhabdomyolysis is an acute syndrome of skeletal muscle injury resulting in the release of intracellular metabolites and proteins, including creatine kinase and myoglobin, into the systemic circulation that can result in death from renal, cardiac, or hematologic dysfunction. Although there are many common acquired causes of acute rhabdomyolysis in children and adults, inborn errors in intermediary metabolism are often to blame in idiopathic cases, especially in children.

To decipher the mechanisms by which loss of PAP activity leads to myocyte injury, investigators have used a variety of mouse and cell culture models. It should be noted that *fld* mice [38] or mice with muscle-specific lipin 1 deletion [37,69] exhibit a chronic and active myopathic phenotype that is not a phenocopy of the acute syndrome in humans. The myopathy is characterized by myocyte necrosis, myofibrils with central nuclei indicative of regeneration, and eventually development of fibrotic lesions. Damaged mitochondria with impaired oxidative capacity accumulate in skeletal myocytes from lipin 1-deficient mice due to impaired mitochondrial turnover through the process of mitochondrial autophagy (mitophagy) [37,38]. The phenotype of *fld* mice can be rescued by transgenic muscle specific overexpression of lipin 1 [38], which together with the skeletal muscle-specific knockout models indicate a myocyte intrinsic effect.

Loss of lipin 1 in skeletal muscle leads to very low PAP activity in muscle and both the constitutive and muscle-specific knockout of lipin 1 models all exhibit accumulation of PA and impairments in autophagy. However, the mechanistic explanations for impaired autophagy and muscle pathologic remodeling may vary upon the model used. For instance, whereas muscle-specific lipin knockouts actually exhibit increased muscle DAG content [37,69], *fld* mice exhibit depletion of DAG [38], suggesting that lipodystrophy of *fld* mice affects muscle lipid content. In *fld* mice and cells, DAG depletion leads to impaired PKD activation, which leads to defective autophagy [38] and may also affect myocyte differentiation via regulation of transcription factors that regulate developmental processes [70]. Since DAG actually accumulates in muscle of mice with conditional deletion of lipin 1 the PKD mechanism does not seem to apply to this model. Indeed, muscle-specific *LPIN1* knockout mice exhibit signs of lipotoxic and sarcoplasmic reticular stress and treatment with agents that enhance fat oxidation or chemical chaperones to alleviate stress can attenuate myopathy in these mice [69]. The mechanistic basis for myopathic remodeling in these mice and human patients with *LPIN1* mutations and optimal treatment approaches will require further study.

6. Cardiac Muscle

Much of the literature regarding patients with *LPIN1* mutations has focused on the skeletal muscle manifestation of the disease and less is known about the effects on cardiac myocytes, despite abundant expression of lipin 1 in the myocardium. Recently, it was shown that patients with *LPIN1* mutations have increased cardiac triglyceride accumulation and some patients exhibited a defects in cardiac function when challenged with exercise [71]. This may suggest that diminished mitochondrial oxidative capacity under exercise conditions impairs cardiac function with energetic challenge.

The role of lipins in regulating cardiac metabolism and function has been more extensively studied in mice. Lipin 1 and 3 seem to be expressed in the myocardium, but lipin 2 is not [8]. Despite expressing lipin 3 in heart, *fld* mice exhibit very little cardiac PAP activity and increased cardiac PA [8]. When isolated hearts from *fld* mice were perfused with ^{3}H-oleate, the radiolabeled fatty acid was more enriched in glycerophospholipids (PA, PI, PS, etc.), but cardiac triglyceride levels were not affected [36]. Hearts from *fld* mice actually exhibited cardiac triglyceride accumulation during prolonged fasting despite decreased PAP activity [72]. *Fld* hearts also exhibit mild cardiac dysfunction, but do not exhibit cardiac myocyte death or signs of fibrosis. The reasons for the different outcomes in lipin 1-deficient skeletal and cardiac myocytes is not yet clear. We have recently developed mice with cardiac-specific *LPIN1* deficiency (manuscript submitted) and like *fld* mice, cardiac-specific deletion of lipin 1 does not lead to myocyte dropout or development of myopathic remodeling [44]. However, the cardiac lipin 1 knockouts also exhibit accumulation of PA, diminished mitochondrial respiration potentially due to reduced cardiolipin content, and mild contractile dysfunction when challenged with dobutamine. Interestingly, lipin 1 expression and PAP activity are diminished and PA abundance is increased in acquired forms of heart failure in mice [8]. It is unknown whether loss of lipin 1 and the accumulation of PA may contribute to the impairments in contractile dysfunction in these models.

Loss of lipin 1 in myocardium has been shown to have several signaling effects, including activation of mTOR signaling [36]. Whereas mTOR activation is usually linked to cardiac hypertrophy, hearts from *fld* mice are actually smaller than control hearts. It is possible that the activation of mTOR is an adaptation to regulate cardiac energy metabolism. Kok and colleagues also noted reduced phosphorylation of hormone sensitive lipase [36] in *fld* mouse hearts, which is consistent with impairments in PKA activity. Indeed, cardiac-specific lipin 1 knockout also leads to impairments in PKA activity, especially in the context of β-adrenergic stimulation [44]. This likely explains the impairment in contractility observed in response to dobutamine. Further work to investigate whether these observations translate to humans and to better characterize the cardiac phenotype of patients with *LPIN1* mutations is needed.

7. Liver

Unlike striated muscle and adipose tissue, the liver highly expresses both lipin 1 and lipin 2. Although lipin 2 is more abundant than lipin 1 in normal mouse liver [7], hepatic lipin 1 expression is highly induced by fasting, diabetes, glucocorticoid administration [17], and experimental alcoholic fatty liver disease [73,74]. Lipin 2 mRNA is also induced in liver by fasting and diabetes, but lipin 1 and lipin 2 are under the control of different regulatory pathways [24]. These physiologic contexts with increased lipin expression were shown many years ago to be associated with increased hepatic PAP activity [75].

The high expression of both proteins in liver often leads to a great deal of compensation and there are limited effects of deleting only one lipin family member. For instance, neonatal *fld* mice exhibit an overt fatty liver phenotype [6,76], which is at odds with the role of lipin 1 as a PAP enzyme involved in triglyceride synthesis. However, we now know that lipin 2 protein abundance is markedly increased in *fld* liver [24] and that the fatty liver in this model is largely driven by loss of lipin 1 in adipose tissue driving a lipodystrophic phenotype [43,68]. Conversely, knockout of lipin 2 leads to increased lipin 1 protein abundance in liver and does not affect hepatic triglyceride levels [77]. Acute knockdown of lipin proteins circumvents some of these compensatory effects and has revealed pathophysiological roles for lipin 1 and 2 in mouse models of fatty liver disease.

In many models of obesity-related related fatty liver disease, lipin 1 expression is increased [17,50]. However, not all obese mouse models exhibit an induction in lipin 1 [78] and lipin 1 seems to be decreased in human subjects with obesity and hepatic steatosis [62]. Interestingly, liver-specific deletion of lipin 1 does not prevent hepatic TAG accumulation in fasted mice. Liver lipin 1 knockout mice fed a diet enriched in fat, fructose, and cholesterol were also not protected from triglyceride and DAG accumulation or insulin resistance [77], suggesting that lipin 2 may be able to compensate for loss of lipin 1. On the other hand, Ryu et al. showed that acute RNAi-mediated lipin 1 knockdown attenuated hepatic steatosis and improved insulin-stimulated AKT activation in mouse liver and primary mouse hepatocytes [50]. Mechanistically, activation of lipin 1 can increase cellular DAG; thereby activating PKCε and driving insulin resistance [50]. Similarly, activation of lipin 2 in fatty liver or by ER stress was also shown to cause insulin resistance by this mechanism [79]. However, other work in hepatocytes has suggested that PA induces insulin resistance by suppressing mTORC2 and that lipin 1 overexpression actually attenuated insulin resistance [39]. Furthermore, in a genetic model of obesity, the UCP-DTA mouse, lipin 1 expression was decreased and lipin 1 overexpression improved insulin signaling and glucose tolerance [78]. Thus, the role of lipin 1 in hepatic insulin resistance remains controversial.

It has long been known that PAP activity is induced in rodent models of alcoholic fatty liver disease (AFLD) [80] and that this coincides with a marked induction in lipin [73,74]. Surprisingly, liver-specific lipin1-KO did not attenuate, but actually exacerbated triglyceride accumulation and liver injury in a model of alcohol feeding, likely due to reduced fatty acid oxidation and triglyceride secretion [73]. This suggests that the induction of lipin 1 in hepatocytes in AFLD may actually play an adaptive or protective role by mechanisms that are still not completely clear. Interestingly, deletion of lipin 1 in myeloid cells markedly attenuated hepatic inflammation while concomitantly exacerbating hepatic steatosis in another model of AFLD [81]. This effect was attributable to altered secretion of adipokines like fibroblast growth factor 15 and adiponectin. Thus, the effects of lipin 1 PAP activity in myeloid cells may impact systemic inflammation and metabolism by altering interorgan endocrine signaling pathways.

Recently, it was demonstrated that lipin 1 and 2 expression is decreased after experimental overdose with acetaminophen (APAP) in mice coincident with a marked increase in liver and plasma PA concentrations [82]. It is possible that lipin deactivation and PA accumulation after APAP overdose is an adaptive mechanism to stimulate the hepatocyte proliferative response to regenerate liver tissue. The mechanisms by which this occurs are still emerging. However, activation of mTOR signaling after APAP treatment seems to precede the accumulation of PA, suggesting that PA is not the trigger for this response. It is possible that PA is activating other mitogenic signaling pathways.

8. Summary

In conclusion, lipin-mediated PAP activity plays important and pleiotropic roles in regulating lipid metabolism and cellular homeostasis via the metabolism of PA concentrations. Indeed, it may be that limiting the accumulation of this unabundant lipid to modulate signaling pathways and limit PA toxicity could be considered the primary function of this family of phosphohydrolases. The robust phenotypes of mice and humans with lipin deficiency underscores the important roles that lipin 1 and 2 play in regulating adipocyte differentiation, myocyte homeostasis, and whole-body metabolism. Future work will almost certainly define more important physiologic and pathophysiologic roles for lipin proteins in modulating metabolism and signaling.

Author Contributions: A.J.L. and B.N.F. wrote and edited the manuscript. All authors have read and agreed to the published version of the manuscript.

Funding: This research was funded by the Precision Nutrition and Metabolic Function, Washington University-Centene Personalized Medicine Initiative and NIH grant R01 HL119225.

Conflicts of Interest: The authors declare no conflict of interest.

References

1. Weiss, S.B.; Kennedy, E.P.; Kiyasu, J.Y. The Enzymatic Synthesis of Triglycerides. *J. Biol. Chem.* **1960**, *235*, 40–44. [CrossRef] [PubMed]
2. Smith, S.W.; Weiss, S.B.; Kennedy, E.P. The Enzymatic Dephosphorylation of Phosphatidic Acids. *J. Biol. Chem.* **1957**, *228*, 915–922. [PubMed]
3. Han, G.-S.; Wu, W.-I.; Carman, G.M. The Saccharomyces cerevisiae Lipin homolog is a Mg2+-dependent phosphatidate phosphatase enzyme. *J. Biol. Chem.* **2006**, *281*, 9210–9218. [CrossRef] [PubMed]
4. Péterfy, M.; Phan, J.; Xu, P.; Reue, K. Lipodystrophy in the *fld* mouse results from mutation of a new gene encoding a nuclear protein, lipin. *Nat. Genet.* **2001**, *27*, 121. [CrossRef] [PubMed]
5. Carman, G.M. Discoveries of the phosphatidate phosphatase genes in yeast published in the Journal of Biological Chemistry. *J. Biol. Chem.* **2019**, *294*, 1681–1689. [CrossRef]

6. Langner, C.A.; Birkenmeier, E.H.; Ben-Zeev, O.; Schotz, M.C.; Sweet, H.O.; Davisson, M.T.; Gordon, J.I. The fatty liver dystrophy (fld) mutation. A new mutant mouse with a developmental abnormality in triglyceride metabolism and associated tissue-specific defects in lipoprotein lipase and hepatic lipase activities. *J. Biol. Chem.* **1989**, *264*, 7994–8003.
7. Donkor, J.; Sariahmetoglu, M.; Dewald, J.; Brindley, D.N.; Reue, K. Three Mammalian Lipins Act as Phosphatidate Phosphatases with Distinct Tissue Expression Patterns. *J. Biol. Chem.* **2007**, *282*, 3450–3457. [CrossRef]
8. Mitra, M.S.; Schilling, J.D.; Wang, X.; Jay, P.Y.; Huss, J.M.; Su, X.; Finck, B.N. Cardiac lipin 1 expression is regulated by the peroxisome proliferator activated receptor γ coactivator 1α/estrogen related receptor axis. *J. Mol. Cell. Cardiol.* **2011**, *51*, 120–128. [CrossRef]
9. Harris, T.E.; Huffman, T.A.; Chi, A.; Shabanowitz, J.; Hunt, D.F.; Kumar, A.; Lawrence, J.C. Insulin Controls Subcellular Localization and Multisite Phosphorylation of the Phosphatidic Acid Phosphatase, Lipin 1. *J. Biol. Chem.* **2007**, *282*, 277–286. [CrossRef]
10. Dwyer, J.R.; Donkor, J.; Zhang, P.; Csaki, L.S.; Vergnes, L.; Lee, J.M.; Dewald, J.; Brindley, D.N.; Atti, E.; Tetradis, S.; et al. Mouse lipin-1 and lipin-2 cooperate to maintain glycerolipid homeostasis in liver and aging cerebellum. *PNAS* **2012**, *109*, E2486–E2495. [CrossRef]
11. Ferguson, P.J.; Chen, S.; Tayeh, M.K.; Ochoa, L.; Leal, S.M.; Pelet, A.; Munnich, A.; Lyonnet, S.; Majeed, H.A.; El-Shanti, H. Homozygous mutations in LPIN2 are responsible for the syndrome of chronic recurrent multifocal osteomyelitis and congenital dyserythropoietic anaemia (Majeed syndrome). *J. Med. Genet.* **2005**, *42*, 551–557. [CrossRef] [PubMed]
12. Khayyo, V.I.; Hoffmann, R.M.; Wang, H.; Bell, J.A.; Burke, J.E.; Reue, K.; Airola, M.V.V. Crystal structure of a lipin/Pah phosphatidic acid phosphatase. *Nat. Commun.* **2020**, *11*, 1309. [CrossRef]
13. Liu, G.-H.; Qu, J.; Carmack, A.E.; Kim, H.B.; Chen, C.; Ren, H.; Morris, A.J.; Finck, B.N.; Harris, T.E. Lipin proteins form homo- and hetero-oligomers. *Biochem J.* **2010**, *432*, 65–76. [CrossRef]
14. Creutz, C.E.; Eaton, J.M.; Harris, T.E. Assembly of High Molecular Weight Complexes of Lipin on a Supported Lipid Bilayer Observed by Atomic Force Microscopy. *Biochemistry* **2013**, *52*, 5092–5102. [CrossRef] [PubMed]
15. Boroda, S.; Takkellapati, S.; Lawrence, R.T.; Entwisle, S.W.; Pearson, J.M.; Granade, M.E.; Mullins, G.R.; Eaton, J.M.; Villén, J.; Harris, T.E. The phosphatidic acid-binding, polybasic domain is responsible for the differences in the phosphoregulation of lipins 1 and 3. *J. Biol. Chem.* **2017**, *292*, 20481–20493. [CrossRef] [PubMed]
16. Ren, H.; Federico, L.; Huang, H.; Sunkara, M.; Drennan, T.; Frohman, M.A.; Smyth, S.S.; Morris, A.J. A Phosphatidic Acid Binding/Nuclear Localization Motif Determines Lipin1 Function in Lipid Metabolism and Adipogenesis. *MBoC* **2010**, *21*, 3171–3181. [CrossRef] [PubMed]
17. Finck, B.N.; Gropler, M.C.; Chen, Z.; Leone, T.C.; Croce, M.A.; Harris, T.E.; Lawrence, J.C.; Kelly, D.P. Lipin 1 is an inducible amplifier of the hepatic PGC-1alpha/PPARalpha regulatory pathway. *Cell Metab.* **2006**, *4*, 199–210. [CrossRef] [PubMed]
18. Kim, H.B.; Kumar, A.; Wang, L.; Liu, G.-H.; Keller, S.R.; Lawrence, J.C.; Finck, B.N.; Harris, T.E. Lipin 1 Represses NFATc4 Transcriptional Activity in Adipocytes To Inhibit Secretion of Inflammatory Factors. *Mol. Cell. Biol.* **2010**, *30*, 3126–3139. [CrossRef]
19. Huffman, T.A.; Mothe-Satney, I.; Lawrence, J.C. Insulin-stimulated phosphorylation of lipin mediated by the mammalian target of rapamycin. *PNAS* **2002**, *99*, 1047–1052. [CrossRef]
20. Peterson, T.R.; Sengupta, S.S.; Harris, T.E.; Carmack, A.E.; Kang, S.A.; Balderas, E.; Guertin, D.A.; Madden, K.L.; Carpenter, A.E.; Finck, B.N.; et al. mTOR Complex 1 Regulates Lipin 1 Localization to Control the SREBP Pathway. *Cell* **2011**, *146*, 408–420. [CrossRef]
21. Liu, G.-H.; Gerace, L. Sumoylation regulates nuclear localization of lipin-1alpha in neuronal cells. *PLoS ONE* **2009**, *4*, e7031. [CrossRef]
22. Li, T.Y.; Song, L.; Sun, Y.; Li, J.; Yi, C.; Lam, S.M.; Xu, D.; Zhou, L.; Li, X.; Yang, Y.; et al. Tip60-mediated lipin 1 acetylation and ER translocation determine triacylglycerol synthesis rate. *Nat. Commun.* **2018**, *9*, 1916. [CrossRef]
23. Shimizu, K.; Fukushima, H.; Ogura, K.; Lien, E.C.; Nihira, N.T.; Zhang, J.; North, B.J.; Guo, A.; Nagashima, K.; Nakagawa, T.; et al. The SCFβ-TRCP E3 ubiquitin ligase complex targets Lipin1 for ubiquitination and degradation to promote hepatic lipogenesis. *Sci Signal.* **2017**, *10*. [CrossRef]

24. Gropler, M.C.; Harris, T.E.; Hall, A.M.; Wolins, N.E.; Gross, R.W.; Han, X.; Chen, Z.; Finck, B.N. Lipin 2 Is a Liver-enriched Phosphatidate Phosphohydrolase Enzyme That Is Dynamically Regulated by Fasting and Obesity in Mice. *J. Biol. Chem.* **2009**, *284*, 6763–6772. [CrossRef]
25. Eaton, J.M.; Mullins, G.R.; Brindley, D.N.; Harris, T.E. Phosphorylation of Lipin 1 and Charge on the Phosphatidic Acid Head Group Control Its Phosphatidic Acid Phosphatase Activity and Membrane Association. *J. Biol. Chem.* **2013**, *288*, 9933–9945. [CrossRef]
26. Zegarlinska, J.; Piaścik, M.; Sikorski, A.F.; Czogalla, A. Phosphatidic acid–a simple phospholipid with multiple faces. *Acta Biochim. Pol.* **2018**, *65*, 163–171. [CrossRef]
27. Eichmann, T.O.; Lass, A. DAG tales: The multiple faces of diacylglycerol–stereochemistry, metabolism, and signaling. *Cell. Mol. Life Sci.* **2015**, *72*, 3931–3952. [CrossRef]
28. Brindley, D.N.; Waggoner, D.W. Mammalian lipid phosphate phosphohydrolases. *J. Biol. Chem.* **1998**, *273*, 24281–24284. [CrossRef]
29. Brindley, D.N.; Pilquil, C. Lipid phosphate phosphatases and signaling. *J. Lipid Res.* **2009**, *50*, S225–S230. [CrossRef]
30. Toschi, A.; Lee, E.; Xu, L.; Garcia, A.; Gadir, N.; Foster, D.A. Regulation of mTORC1 and mTORC2 Complex Assembly by Phosphatidic Acid: Competition with Rapamycin. *Mol. Cell. Biol.* **2009**, *29*, 1411–1420. [CrossRef]
31. Yoon, M.-S.; Sun, Y.; Arauz, E.; Jiang, Y.; Chen, J. Phosphatidic Acid Activates Mammalian Target of Rapamycin Complex 1 (mTORC1) Kinase by Displacing FK506 Binding Protein 38 (FKBP38) and Exerting an Allosteric Effect. *J. Biol. Chem.* **2011**, *286*, 29568–29574. [CrossRef]
32. Laplante, M.; Sabatini, D.M. mTOR signaling at a glance. *J. Cell Sci.* **2009**, *122*, 3589–3594. [CrossRef]
33. Fang, Y.; Vilella-Bach, M.; Bachmann, R.; Flanigan, A.; Chen, J. Phosphatidic Acid-Mediated Mitogenic Activation of mTOR Signal. *Science* **2001**, *294*, 1942–1945. [CrossRef]
34. O'Neil, T.K.; Duffy, L.R.; Frey, J.W.; Hornberger, T.A. The role of phosphoinositide 3-kinase and phosphatidic acid in the regulation of mammalian target of rapamycin following eccentric contractions. *J. Physiol. (Lond.)* **2009**, *587*, 3691–3701. [CrossRef]
35. Phan, J.; Peterfy, M.; Reue, K. Biphasic expression of lipin suggests dual roles in adipocyte development. *Drug News Perspect.* **2005**, *18*, 5–11. [CrossRef]
36. Kok, B.P.C.; Kienesberger, P.C.; Dyck, J.R.B.; Brindley, D.N. Relationship of glucose and oleate metabolism to cardiac function in lipin-1 deficient (*fld*) mice. *J. Lipid Res.* **2012**, *53*, 105–118. [CrossRef]
37. Schweitzer, G.G.; Collier, S.L.; Chen, Z.; McCommis, K.S.; Pittman, S.K.; Yoshino, J.; Matkovich, S.J.; Hsu, F.-F.; Chrast, R.; Eaton, J.M.; et al. Loss of lipin 1-mediated phosphatidic acid phosphohydrolase activity in muscle leads to skeletal myopathy in mice. *Faseb J.* **2018**, *33*, 652–667. [CrossRef]
38. Zhang, P.; Verity, M.A.; Reue, K. Lipin-1 Regulates Autophagy Clearance and Intersects with Statin Drug Effects in Skeletal Muscle. *Cell Metab.* **2014**, *20*, 267–279. [CrossRef]
39. Zhang, C.; Wendel, A.A.; Keogh, M.R.; Harris, T.E.; Chen, J.; Coleman, R.A. Glycerolipid signals alter mTOR complex 2 (mTORC2) to diminish insulin signaling. *PNAS* **2012**, *109*, 1667–1672. [CrossRef]
40. Norambuena, A.; Metz, C.; Jung, J.E.; Silva, A.; Otero, C.; Cancino, J.; Retamal, C.; Valenzuela, J.C.; Soza, A.; González, A. Phosphatidic Acid Induces Ligand-independent Epidermal Growth Factor Receptor Endocytic Traffic through PDE4 Activation. *MBoC* **2010**, *21*, 2916–2929. [CrossRef]
41. Zakaroff-Girard, A.; Bawab, S.E.; Némoz, G.; Lagarde, M.; Prigent, A.-F. Relationships between phosphatidic acid and cyclic nucleotide phosphodiesterases in activated human blood mononuclear cells. *J. Leukoc. Biol.* **1999**, *65*, 381–390. [CrossRef] [PubMed]
42. Némoz, G.; Sette, C.; Conti, M. Selective Activation of Rolipram-Sensitive, cAMP-Specific Phosphodiesterase Isoforms by Phosphatidic Acid. *Mol Pharm.* **1997**, *51*, 242–249. [CrossRef] [PubMed]
43. Mitra, M.S.; Chen, Z.; Ren, H.; Harris, T.E.; Chambers, K.T.; Hall, A.M.; Nadra, K.; Klein, S.; Chrast, R.; Su, X.; et al. Mice with an adipocyte-specific lipin 1 separation-of-function allele reveal unexpected roles for phosphatidic acid in metabolic regulation. *PNAS* **2013**, *110*, 642–647. [CrossRef] [PubMed]
44. Chambers, K.T.; Cooper, M.A.; Swearingen, A.R.; Schweitzer, G.G.; Weinheirmer, C.J.; Kovacs, A.; Finck, B.N. Myocardial Lipin 1 Knockout in Mice Approximates Cardiac Effects of Human LPIN1 Mutations. 2020, Submitted.

45. Jiang, W.; Zhu, J.; Zhuang, X.; Zhang, X.; Luo, T.; Esser, K.A.; Ren, H. Lipin1 Regulates Skeletal Muscle Differentiation through Extracellular Signal-regulated Kinase (ERK) Activation and Cyclin D Complex-regulated Cell Cycle Withdrawal. *J. Biol. Chem.* **2015**, *290*, 23646–23655. [CrossRef]
46. Nadra, K.; de Preux Charles, A.-S.; Médard, J.-J.; Hendriks, W.T.; Han, G.-S.; Grès, S.; Carman, G.M.; Saulnier-Blache, J.-S.; Verheijen, M.H.G.; Chrast, R. Phosphatidic acid mediates demyelination in Lpin1 mutant mice. *Genes Dev.* **2008**, *22*, 1647–1661. [CrossRef]
47. Zhang, P.; O'Loughlin, L.; Brindley, D.N.; Reue, K. Regulation of lipin-1 gene expression by glucocorticoids during adipogenesis. *J. Lipid Res.* **2008**, *49*, 1519–1528. [CrossRef]
48. Tsukahara, T.; Hanazawa, S.; Murakami-Murofushi, K. Cyclic phosphatidic acid influences the expression and regulation of cyclic nucleotide phosphodiesterase 3B and lipolysis in 3T3-L1 cells. *Biochem. Biophys. Res. Commun.* **2011**, *404*, 109–114. [CrossRef]
49. Ter Horst, K.W.; Gilijamse, P.W.; Versteeg, R.I.; Ackermans, M.T.; Nederveen, A.J.; la Fleur, S.E.; Romijn, J.A.; Nieuwdorp, M.; Zhang, D.; Samuel, V.T.; et al. Hepatic Diacylglycerol-Associated Protein Kinase Cε Translocation Links Hepatic Steatosis to Hepatic Insulin Resistance in Humans. *Cell Rep.* **2017**, *19*, 1997–2004. [CrossRef]
50. Ryu, D.; Oh, K.-J.; Jo, H.-Y.; Hedrick, S.; Kim, Y.-N.; Hwang, Y.-J.; Park, T.-S.; Han, J.-S.; Choi, C.S.; Montminy, M.; et al. TORC2 Regulates Hepatic Insulin Signaling via a Mammalian Phosphatidic Acid Phosphatase, LIPIN1. *Cell Metab.* **2009**, *9*, 240–251. [CrossRef]
51. Chae, M.; Jung, J.-Y.; Bae, I.-H.; Kim, H.-J.; Lee, T.R.; Shin, D.W. Lipin-1 expression is critical for keratinocyte differentiation. *J. Lipid Res.* **2016**, *57*, 563–573. [CrossRef]
52. Takeuchi, K.; Reue, K. Biochemistry, physiology, and genetics of GPAT, AGPAT, and lipin enzymes in triglyceride synthesis. *Am. J. Physiol. Endocrinol. Metab.* **2009**, *296*, E1195–E1209. [CrossRef] [PubMed]
53. Nadra, K.; Médard, J.-J.; Mul, J.D.; Han, G.-S.; Grès, S.; Pende, M.; Metzger, D.; Chambon, P.; Cuppen, E.; Saulnier-Blache, J.-S.; et al. Cell Autonomous Lipin 1 Function Is Essential for Development and Maintenance of White and Brown Adipose Tissue. *Mol. Cell. Biol.* **2012**, *32*, 4794–4810. [CrossRef] [PubMed]
54. Phan, J.; Péterfy, M.; Reue, K. Lipin Expression Preceding Peroxisome Proliferator-activated Receptor-γ Is Critical for Adipogenesis in Vivo and in Vitro. *J. Biol. Chem.* **2004**, *279*, 29558–29564. [CrossRef] [PubMed]
55. Sembongi, H.; Miranda, M.; Han, G.-S.; Fakas, S.; Grimsey, N.; Vendrell, J.; Carman, G.M.; Siniossoglou, S. Distinct Roles of the Phosphatidate Phosphatases Lipin 1 and 2 during Adipogenesis and Lipid Droplet Biogenesis in 3T3-L1 Cells. *J. Biol. Chem.* **2013**, *288*, 34502–34513. [CrossRef]
56. Zhang, P.; Takeuchi, K.; Csaki, L.S.; Reue, K. Lipin-1 Phosphatidic Phosphatase Activity Modulates Phosphatidate Levels to Promote Peroxisome Proliferator-activated Receptor γ (PPARγ) Gene Expression during Adipogenesis. *J. Biol. Chem.* **2012**, *287*, 3485–3494. [CrossRef]
57. Koh, Y.-K.; Lee, M.-Y.; Kim, J.-W.; Kim, M.; Moon, J.-S.; Lee, Y.-J.; Ahn, Y.-H.; Kim, K.-S. Lipin1 Is a Key Factor for the Maturation and Maintenance of Adipocytes in the Regulatory Network with CCAAT/Enhancer-binding Protein α and Peroxisome Proliferator-activated Receptor γ2. *J. Biol. Chem.* **2008**, *283*, 34896–34906. [CrossRef]
58. Csaki, L.S.; Reue, K. Lipins: Multifunctional Lipid Metabolism Proteins. *Annu. Rev. Nutr.* **2010**, *30*, 257–272. [CrossRef]
59. Topal, S.; Köse, M.D.; Ağın, H.; Sarı, F.; Çolak, M.; Atakul, G.; Karaarslan, U.; İşgüder, R. A neglected cause of recurrent rhabdomyolysis, LPIN1 gene defect: A rare case from Turkey. *Turk. J. Pediatr.* **2020**, *62*, 647–651. [CrossRef]
60. Temprano, A.; Sembongi, H.; Han, G.-S.; Sebastián, D.; Capellades, J.; Moreno, C.; Guardiola, J.; Wabitsch, M.; Richart, C.; Yanes, O.; et al. Redundant roles of the phosphatidate phosphatase family in triacylglycerol synthesis in human adipocytes. *Diabetologia* **2016**, *59*, 1985–1994. [CrossRef]
61. Csaki, L.S.; Dwyer, J.R.; Li, X.; Nguyen, M.H.K.; Dewald, J.; Brindley, D.N.; Lusis, A.J.; Yoshinaga, Y.; de Jong, P.; Fong, L.; et al. Lipin-1 and lipin-3 together determine adiposity in vivo. *Mol. Metab.* **2014**, *3*, 145–154. [CrossRef]
62. Croce, M.A.; Eagon, J.C.; LaRiviere, L.L.; Korenblat, K.M.; Klein, S.; Finck, B.N. Hepatic Lipin 1β Expression Is Diminished in Insulin-Resistant Obese Subjects and Is Reactivated by Marked Weight Loss. *Diabetes* **2007**, *56*, 2395–2399. [CrossRef] [PubMed]

63. Yao-Borengasser, A.; Rasouli, N.; Varma, V.; Miles, L.M.; Phanavanh, B.; Starks, T.N.; Phan, J.; Spencer, H.J.; McGehee, R.E.; Reue, K.; et al. Lipin Expression Is Attenuated in Adipose Tissue of Insulin-Resistant Human Subjects and Increases With Peroxisome Proliferator–Activated Receptor γ Activation. *Diabetes* **2006**, *55*, 2811–2818. [CrossRef] [PubMed]
64. Moller, F.; Wong, K.H.; Green, P. Control of fat cell phosphohydrolase by lipolytic agents. *Can. J. Biochem.* **1981**, *59*, 9–15. [CrossRef] [PubMed]
65. Balboa, M.A.; de Pablo, N.; Meana, C.; Balsinde, J. The role of lipins in innate immunity and inflammation. *Biochim Biophys Acta Mol Cell Biol Lipids* **2019**, *1864*, 1328–1337. [CrossRef] [PubMed]
66. Michot, C.; Hubert, L.; Brivet, M.; Meirleir, L.D.; Valayannopoulos, V.; Müller-Felber, W.; Venkateswaran, R.; Ogier, H.; Desguerre, I.; Altuzarra, C.; et al. LPIN1 gene mutations: A major cause of severe rhabdomyolysis in early childhood. *Hum. Mutat.* **2010**, *31*, E1564–E1573. [CrossRef]
67. Michot, C.; Hubert, L.; Romero, N.B.; Gouda, A.; Mamoune, A.; Mathew, S.; Kirk, E.; Viollet, L.; Rahman, S.; Bekri, S.; et al. Study of LPIN1, LPIN2 and LPIN3 in rhabdomyolysis and exercise-induced myalgia. *J. Inherit. Metab. Dis.* **2012**, *35*, 1119–1128. [CrossRef]
68. Schweitzer, G.G.; Collier, S.L.; Chen, Z.; Eaton, J.M.; Connolly, A.M.; Bucelli, R.C.; Pestronk, A.; Harris, T.E.; Finck, B.N. Rhabdomyolysis-Associated Mutations in Human LPIN1 Lead to Loss of Phosphatidic Acid Phosphohydrolase Activity. *Jimd Rep.* **2015**, *23*, 113–122. [CrossRef]
69. Rashid, T.; Nemazanyy, I.; Paolini, C.; Tatsuta, T.; Crespin, P.; de Villeneuve, D.; Brodesser, S.; Benit, P.; Rustin, P.; Baraibar, M.A.; et al. Lipin1 deficiency causes sarcoplasmic reticulum stress and chaperone-responsive myopathy. *Embo J.* **2019**, *38*. [CrossRef]
70. Jama, A.; Huang, D.; Alshudukhi, A.A.; Chrast, R.; Ren, H. Lipin1 is required for skeletal muscle development by regulating MEF2c and MyoD expression. *J. Physiol.* **2019**, *597*, 889–901. [CrossRef]
71. Legendre, A.; Khraiche, D.; Ou, P.; Mauvais, F.-X.; Madrange, M.; Guemann, A.-S.; Jais, J.-P.; Bonnet, D.; Hamel, Y.; de Lonlay, P. Cardiac function and exercise adaptation in 8 children with LPIN1 mutations. *Mol. Genet. Metab.* **2018**, *123*, 375–381. [CrossRef]
72. Kok, B.P.C.; Dyck, J.R.B.; Harris, T.E.; Brindley, D.N. Differential regulation of the expressions of the PGC-1α splice variants, lipins, and PPARα in heart compared to liver. *J. Lipid Res.* **2013**, *54*, 1662–1677. [CrossRef] [PubMed]
73. Hu, M.; Yin, H.; Mitra, M.S.; Liang, X.; Ajmo, J.M.; Nadra, K.; Chrast, R.; Finck, B.N.; You, M. Hepatic-specific lipin-1 deficiency exacerbates experimental alcohol-induced steatohepatitis in mice. *Hepatology* **2013**, *58*, 1953–1963. [CrossRef] [PubMed]
74. Shen, Z.; Liang, X.; Rogers, C.Q.; Rideout, D.; You, M. Involvement of adiponectin-SIRT1-AMPK signaling in the protective action of rosiglitazone against alcoholic fatty liver in mice. *Am. J. Physiol.-Gastrointest. Liver Physiol.* **2009**, *298*, G364–G374. [CrossRef] [PubMed]
75. Brindley, D.N. Some aspects of the physiological and pharmacological control of the synthesis of triacylglycerols and phospholipids. *Int J. Obes* **1978**, *2*, 7–16. [PubMed]
76. Hall, A.M.; Brunt, E.M.; Chen, Z.; Viswakarma, N.; Reddy, J.K.; Wolins, N.E.; Finck, B.N. Dynamic and differential regulation of proteins that coat lipid droplets in fatty liver dystrophic mice. *J. Lipid Res.* **2010**, *51*, 554–563. [CrossRef]
77. Schweitzer, G.G.; Chen, Z.; Gan, C.; McCommis, K.S.; Soufi, N.; Chrast, R.; Mitra, M.S.; Yang, K.; Gross, R.W.; Finck, B.N. Liver-specific loss of lipin-1-mediated phosphatidic acid phosphatase activity does not mitigate intrahepatic TG accumulation in mice. *J. Lipid Res.* **2015**, *56*, 848–858. [CrossRef]
78. Chen, Z.; Gropler, M.C.; Norris, J.; Lawrence, J.C.; Harris, T.E.; Finck, B.N. Alterations in Hepatic Metabolism in *fld* Mice Reveal a Role for Lipin 1 in Regulating VLDL-Triacylglyceride Secretion. *Arter. Thromb. Vasc. Biol.* **2008**, *28*, 1738–1744. [CrossRef]
79. Ryu, D.; Seo, W.-Y.; Yoon, Y.-S.; Kim, Y.-N.; Kim, S.S.; Kim, H.-J.; Park, T.-S.; Choi, C.S.; Koo, S.-H. Endoplasmic reticulum stress promotes LIPIN2-dependent hepatic insulin resistance. *Diabetes* **2011**, *60*, 1072–1081. [CrossRef]
80. Brindley, D.N.; Cooling, J.; Burditt, S.L.; Pritchard, P.H.; Pawson, S.; Sturton, R.G. The involvement of glucocorticoids in regulating the activity of phosphatidate phosphohydrolase and the synthesis of triacylglycerols in the liver. Effects of feeding rats with glucose, sorbitol, fructose, glycerol and ethanol. *Biochem. J.* **1979**, *180*, 195–199. [CrossRef]

81. Wang, J.; Kim, C.; Jogasuria, A.; Han, Y.; Hu, X.; Wu, J.; Shen, H.; Chrast, R.; Finck, B.N.; You, M. Myeloid Cell-Specific Lipin-1 Deficiency Stimulates Endocrine Adiponectin-FGF15 Axis and Ameliorates Ethanol-Induced Liver Injury in Mice. *Sci. Rep.* **2016**, *6*, 34117. [CrossRef]
82. Lutkewitte, A.J.; Schweitzer, G.G.; Kennon-McGill, S.; Clemens, M.M.; James, L.P.; Jaeschke, H.; Finck, B.N.; McGill, M.R. Lipin deactivation after acetaminophen overdose causes phosphatidic acid accumulation in liver and plasma in mice and humans and enhances liver regeneration. *Food Chem. Toxicol.* **2018**, *115*, 273–283. [CrossRef] [PubMed]

© 2020 by the authors. Licensee MDPI, Basel, Switzerland. This article is an open access article distributed under the terms and conditions of the Creative Commons Attribution (CC BY) license (http://creativecommons.org/licenses/by/4.0/).

Review

Interface of Phospholipase Activity, Immune Cell Function, and Atherosclerosis

Robert M. Schilke †, Cassidy M. R. Blackburn †, Temitayo T. Bamgbose and Matthew D. Woolard *

Department of Microbiology and Immunology, Louisiana State University Health Sciences Center, Shreveport, LA 71130, USA; rschil@lsuhsc.edu (R.M.S.); cragai@lsuhsc.edu (C.M.R.B.); tbamgb@lsuhsc.edu (T.T.B.)

* Correspondence: mwoola@lsuhsc.edu; Tel.: +1-(318)-675-4160

† These authors contributed equally to this work.

Received: 12 September 2020; Accepted: 13 October 2020; Published: 15 October 2020

Abstract: Phospholipases are a family of lipid-altering enzymes that can either reduce or increase bioactive lipid levels. Bioactive lipids elicit signaling responses, activate transcription factors, promote G-coupled-protein activity, and modulate membrane fluidity, which mediates cellular function. Phospholipases and the bioactive lipids they produce are important regulators of immune cell activity, dictating both pro-inflammatory and pro-resolving activity. During atherosclerosis, pro-inflammatory and pro-resolving activities govern atherosclerosis progression and regression, respectively. This review will look at the interface of phospholipase activity, immune cell function, and atherosclerosis.

Keywords: atherosclerosis; phospholipases; macrophages; T cells; lipins

---

1. Introduction

All cellular membranes are composed mostly of phospholipids. Phospholipids are amphiphilic compounds with a hydrophilic, negatively charged phosphate group head and two hydrophobic fatty acid tail residues [1]. The glycerophospholipids, phospholipids with glycerol backbones, are the largest group of phospholipids, which are classified by the modification of the head group [1]. The negatively charged phosphate head forms an ionic bond with an amino alcohol. This bridges the glycerol backbone to the nitrogenous functional group (amino alcohol). The addition of an amino alcohol largely dictates the quaternary structure of the phospholipid [2]. A smaller but also critical family of phospholipids are the sphingolipids, which have sphingosine as a backbone [3]. The amphiphilic make-up of phospholipids allows them to create lipid bilayers, which make cellular membranes and supply structure to cells. Phospholipids also contribute to cellular responses through the binding of receptors, such as lysophosphatidic acid (LPA) binding to the family of LPA receptors, and sphingosine-1-phosphate (S1P) binding to S1P receptors [4]. Components of phospholipids, such as inositol trisphosphate (IP3), diacylglycerol (DAG), and fatty acids, are substrates for the activation of intracellular receptors (e.g., inositol trisphosphate receptors), cofactors for proteins (e.g., protein kinase C), and transcription factors (e.g., peroxisome proliferator-activated receptors (PPARs) [5]. In addition, free fatty acids are also precursors to the prostanoid family of lipid mediators, which can have a broad array of cellular and physiological effects.

Phospholipases are a group of enzymes that cleave phospholipids. Each family of phospholipases cleaves a unique site on a phospholipid or unique phospholipid family. Phospholipase A hydrolyzes the fatty acid esters from the sn-1 (PLA1) or sn-2 (PLA2) position of the glycerol backbones, generating free fatty acids [6]. Phospholipase C (PLC) hydrolyzes the glycerol linkage glycerophosphate bond of the polar head, generating DAG and IP3. Phospholipase D (PLD) hydrolyzes the head group of phospholipids, leaving phosphatide and phosphatidic acid (Figure 1). Phosphatidic acid phosphatases are a family of enzymes that can cleave phosphate heads from LPA, PA, and S1P (Figure 1). Phosphatidic acid phosphatases can be split into two families of enzymes, the LPPs, which cleave phosphate heads of lipids on the external side of the plasma membrane, and lipins, which cleave PA intracellularly. Phospholipases are critical regulators of the liberation of bioactive compounds contained within phospholipids and subsequent physiological activity of those compounds.

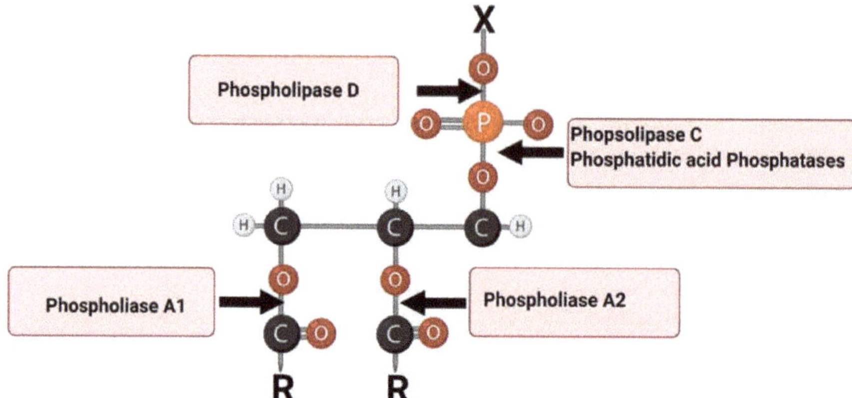

Figure 1. Schematic representation of phospholipase enzymatic sites on phospholipids. "X" represents a functional group. Red "O" represents oxygen; orange "P" represents phosphorus; grey "C" represents carbon; white "H" represents hydrogen; "R" represents fatty acid tails. Figure Created with BioRender.com.

Atherosclerosis is an immuno-metabolic disease that leads to myocardial infarction, stroke, or sudden death [7]. Excess circulating cholesterol in the form of low-density lipoproteins (LDLs) can be deposited into the arterial intima. If these LDLs are not quickly removed, they can be modified (e.g., to oxidized LDL (oxLDL)) through a variety of enzymatic and nonenzymatic modifications [8]. Oxidative stress is associated with the increased oxidation of LDL and increased atherosclerosis progression. Oxidative stress occurs when there is an imbalance in the ratio of reactive oxygen species (ROS) and antioxidants [9,10]. ROS oxidize the polyunsaturated fatty acids and apolipoprotein B-100 on the LDL [11,12]. Once formed, oxLDL leads to the recruitment and activation of inflammatory cells into the arterial intima. Monocytes/macrophages are the main cells recruited into the intimal space to clear both LDL and oxidized LDL. Macrophages engulf LDL (non-oxidized) via the low-density lipoprotein receptor (LDLR). This initiates a negative feedback inhibitory loop resulting in the downregulation and degradation of the LDLR. The negative feedback loop reduces LDL uptake to limit intracellular cholesterol. However, after oxidation, macrophage scavenger receptors, such as CD36, increase oxLDL engulfment, leading to intracellular cholesterol accumulation and lipid droplet formation.

A broad range of immune cells and immunological mediators contribute to atherosclerosis. Macrophages have long been recognized as a key component of the immune response that determine atherosclerosis severity [13]. It is now well established that neutrophils, dendritic cells, T cells, and B cells have important cellular responses within atherosclerotic plaque lesions as well [14]. Furthermore, immune mediators such as pro-inflammatory cytokines (IL-1β, TNF-α, and IL-6), anti-inflammatory cytokines (IL-10), and lipid mediators such as prostaglandins and pro-resolvins also contribute. Pro-inflammatory macrophage responses, Th1 and Th17 T cell responses, and the cytokines and prostaglandins those cells produce promote atherosclerotic progression (Figure 2) [15,16]. By contrast, pro-resolvins, macrophage efferocytosis, anti-phospholipid B cell responses, and T regulatory cell (Treg) responses promote atherosclerosis regression (Figure 2). Phospholipase activity has been documented to contribute to both pro-inflammatory and pro-resolving immune responses as well [17]. This review will concentrate on the contribution of phospholipases to atherosclerosis within immune responses.

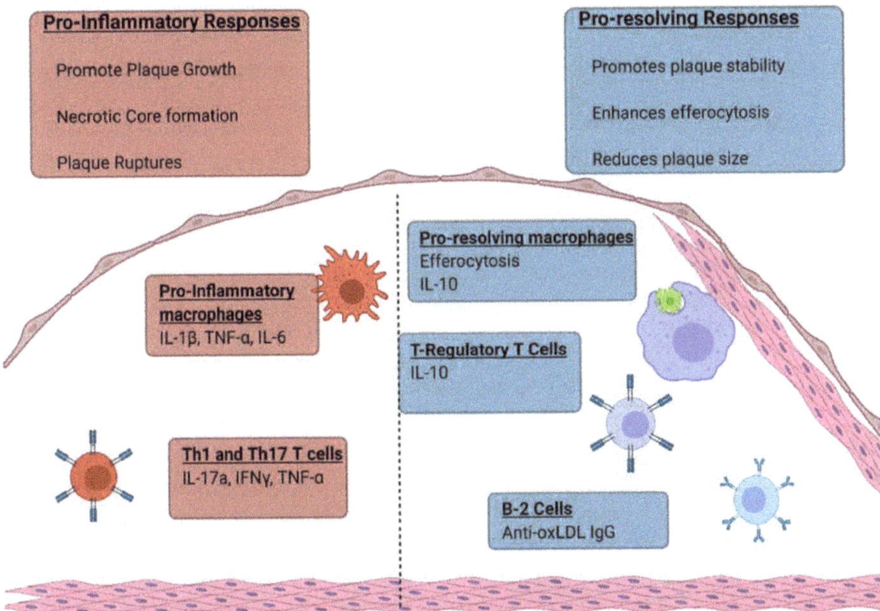

Figure 2. Immunological responses that contribute to plaque progression and plaque regression. Figure Created with BioRender.com.

2. Lipoprotein-Associated Phospholipase A2

Lipoprotein-associated phospholipase A2 (Lp-PLA2) is a 45 kDa monomeric protein and belongs to the phospholipase A2 superfamily [18]. Lp-PLA2 differs from the other phospholipase A2 members, as it does not require calcium for its enzymatic activity [13], and in its substrate specificity, as it preferentially hydrolyzes the oxidatively truncated sn-2 acyl chain of water-soluble phospholipids [19]. The enzyme is also known as platelet-activating factor acetyl hydrolase, due to its ability to hydrolyze and inactivate platelet-activating factor (PAF) [20]. Lp-PLA2 was initially suggested to play an atheroprotective role due to its enzymatic activity of hydrolyzing oxidized phospholipids in LDL and its function in degrading pro-inflammatory and atherogenesis-inducing PAF [21–24]. However, there are controversies regarding the effects of Lp-PLA2 on atherosclerosis [25].

Lp-PLA2 is encoded by the *PLA2G27* gene, which contains 12 exons. The *PLA2G27* gene is characterized by a variety of nonsynonymous polymorphisms that either attenuate Lp-PLA2 enzymatic activity or result in its complete loss [26]. Loss-of-function Lp-PLA2 is associated with an increase in cardiovascular disease, suggesting an atheroprotective role for the enzyme [22,27,28]. The loss of Lp-PLA2 activity is speculated to increase circulating PAF levels and increase the amounts of oxLDL. Lp-PLA2's proposed atheroprotective role is also attributed to the predominant association of Lp-PLA2 with high-density lipoprotein (HDL) in mice [22]. However, LDL is low in mouse species compared to humans, suggesting a potential discrepancy for the contribution of Lp-PLA2 during atherosclerosis in humans [29]. Currently, Lp-PLA2 is considered a marker of cardiovascular disease. To further support this, Singh et al. reported an increase in the number of atherosclerotic lesions in transgenic mouse models that had greater amounts of Lp-PLA2 associated with LDL [30]. However, rather than it playing an atherogenic role, it is speculated that the correlation of atherosclerosis with increased amounts of Lp-PLA2 is a result of the protective function of the enzyme [25].

Lp-PLA2 is secreted by a variety of white blood cells and other specialized cells such as hepatocytes and adipocytes [31]. Lp-PLA2 synthesis and release into the circulation have been found to predominantly occur during monocyte maturation into macrophages [32]. In humans, circulating Lp-PLA2 is bound to lipoproteins, with 70–80% of the enzyme bound to apolipoprotein B on LDL, while the remaining is carried on HDL [33]. Specific residues in the Lp-PLA2 N-terminus bind the electronegative domain of apolipoprotein B (ApoB) on the C-terminus of LDL [34]. The Lp-PLA2 association with ApoB is increased as ApoB becomes more negatively charged [34]. While Lp-PLA2 associates with LDL in the blood, its potential atherogenic activity is not observed until it is found within the arterial intima [35]. Within the arterial intima, LDLs can be oxidized, providing the oxidatively truncated sn-2 chains that Lp-PLA2 is preferentially known to hydrolyze on phospholipids [36]. Lp-PLA2 mediates the hydrolysis of oxidized LDL, yielding oxidized non-esterified fatty acids (oxNEFA) and lysophosphatidylcholine (LysoPC) [35]. These two hydrolytic products are individually and collectively pro-inflammatory and atherogenic [36]. oxNEFA and LysoPC induce the apoptosis of macrophages and increase the recruitment of leukocytes in the sub-intimal space of the artery wall [35,37]. This eventually facilitates the development of the plaque lipid core [26].

LysoPC, in particular, encompasses multiple atherogenic and pro-inflammatory activities because it acts as a monocyte chemoattractant factor, induces oxidative stress, induces endothelial dysfunction, upregulates the expression of adhesion molecules and cytokines (IL-1β, IL-6, and TNF-α), and induces apoptosis in endothelial cells, smooth muscle cells, and macrophages [35,37,38]. Increased amounts of LysoPC were found in patients with early coronary atherosclerosis when compared with control subjects [39]. Conversely, it is speculated that LysoPC does not pose much of an atherogenic threat because LysoPC is mostly found in a bound state, thus reducing its availability [25]. Consequently, the amount of LysoPC measured in the plasma is not a true representation of the amount of LysoPC that is biologically available.

Apoptotic cells are phagocytosed by neighboring macrophages in a receptor–ligand interaction called efferocytosis. Defects in efferocytosis are one of the biggest drivers of atherosclerotic plaque growth and the formation of necrotic cores that lead to destabilized plaques. The macrophage scavenger receptor CD36 recognizes exposed oxidized phosphatidylcholine and phosphatidylserine molecules on the surface of apoptotic cells. The Lp-PLA2 cleavage of oxidized phosphatidylcholine reduces the scavenger receptor recognition of apoptotic cells by macrophages [40]. The impaired clearance of apoptotic cells leads to necrosis and the subsequent expansion of the necrotic core [41]. The Lp-PLA2-induced formation of oxNEFA can also elicit monocyte and leukocyte recruitment and induce apoptosis [35,37]. The combination of enhanced leukocyte recruitment, increased apoptosis, and reduced efferocytosis are likely responsible for the expansion of the necrotic core and the thinning of the fibrous cap [35,42].

Lp-PLA2 mRNA has not only been found to be upregulated in atherosclerotic plaques but has also been shown to be strongly expressed in the macrophage populations that are found within the fibrous cap of vulnerable atherosclerotic plaques [43,44]. The presence of Lp-PLA2 substrates and products of its hydrolytic activity in lipid-laden plaques further supports the atherogenic role of Lp-PLA2 [45]. An autopsy examination study on 25 sudden coronary death patients found Lp-PLA2 to be highly upregulated in the ruptured plaques found in the human coronary arteries and their cap fibroatheromas [46]. Several large studies have continued to show that Lp-PLA2 is an independent and reliable predictor of cardiovascular diseases [47,48]. Based on these pieces of evidence and the recommendations of several major international societies, Lp-PLA2 is considered a cardiovascular disease risk factor by the Food and Drug Administration [49]. In summary, the enzymatic activity of Lp-PLA2 and the products of its hydrolytic action facilitate the continuous progression and detrimental destabilization of atherosclerotic plaques.

3. Lipid Phosphate Phosphatases

Lipid phosphate phosphatases (LPPs) are a group of enzymes that belong to the phosphatase/phosphotransferase family. LPPs dephosphorylate phosphatidic acid, lysophosphatidic acid (LPA), sphingosine-1-phosphate (S1P), ceramine-1-phosphate (C1P), and diacylglycerol pyrophosphate [50]. LPPs are typically localized on the plasma membranes, with the outer leaf containing the active site. LPPs can also be expressed on the membranes of the endoplasmic reticulum (ER) and Golgi, allowing the metabolism of internal lipid phosphates [51]. LPPs modify the concentrations of lipid phosphates and their dephosphorylated products to regulate cell signaling [52]. LPPs regulate cell signaling through the dephosphorylation of bioactive lipids. As mentioned above, LPPs dephosphorylate lipid products such as LPA, S1P, and C1P. LPA activates PPARs and nuclear LPA1 receptors, resulting in an increase in transcription and cell signaling pathways such as those involved in cell proliferation, migration, calcium mobilization, etc. [53–55]. S1P elicits calcium mobilization, ERK activity, and protection against apoptosis [56–58]. C1P promotes cell division and prevents apoptosis. The LPP-mediated degradation of LPA, S1P, and C1P will terminate the receptor-mediated activities.

LPPs have three isoforms—LPP1, LPP2, and LPP3—that each have a conserved catalytic domain to dephosphorylate lipid phosphates [51,59]. LPP3 also has noncatalytic activity that allows it to bind to integrins. This noncatalytic activity promotes endothelial cell-to-cell adhesion and depends on the arginine–glycine–aspartate recognition motif [60,61]. Each LPP contributes to different cell responses in various models of inflammation. For example, ovarian cancer cells are exposed to an elevated amount of LPA, which results in cell proliferation and survival. Ovarian cancer cells also have reduced LPP1 mRNA [62]. When LPP1 is overexpressed in ovarian cancer cells, LPA hydrolysis is increased and results in decreased cell proliferation and increased apoptosis [62]. Within platelets, LPP1 dephosphorylates LPA, which may help to recruit monocytes and macrophages after endothelial cell and vascular muscle cell stimulation [63]. Increased plasma LPA may also participate in signaling and stimulation for the growth of tumor cells and is associated with increased gynecological cancers [63]. The inducible inactivation of the LPP3 gene in endothelial and hematopoietic cells enhanced inflammation in mice after challenge with LPS or thioglycolate [64]. LPP3 overexpression in HEK293 cells increases phosphatidic acid-to-diacylglycerol conversion [51,65,66]. Altered phosphatidic acid/diacylglycerol concentrations affect different cellular processes. For example, within neutrophils, membrane-associated phosphatidic acid stimulates endothelial cell tyrosine kinases, which results in increased membrane permeability in the endothelial cells. LPP activity reduces membrane-associated phosphatidic acid and therefore stifles endothelial cell membrane permeability [64]. Overall, LPPs are involved in numerous different cell processes and are regulated by lipid phosphate availability to influence cell cycle and inflammatory responses.

Single nucleotide polymorphisms have been identified in *PLPP3* (the gene that encodes LPP3) that are associated with an increased risk of coronary artery disease [67–69]. LPP3 can be detected in human atheromas and is mainly found in foam cells [70]. Further investigation showed oxidized LDL upregulates the *PLPP3* gene and associated LPP3 protein expression within macrophages [70]. Specifically, oxidized LDL increases the enzymatic activity of LPP3. The atheroprotective role of LPP3 may be through the reduction of LPA. LPA increases plaque-associated thrombosis [71]. Multiple animal models of atherosclerosis have shown LPP3 is upregulated in endothelial cells, CD68-positive cells (monocytes/macrophages), and smooth muscle cells [68]. In mice, LPP3 is necessary during early vascular development; global deletion causes embryonic lethality [64,72]. Mice with an induced global deletion of *PLPP3* have larger atherosclerotic plaques associated with increased lesional LPA [68]. Liver-specific, conditional *PLPP3* knockout mice crossed with apolipoprotein E (ApoE) knockout mice have significantly larger plaques and necrotic cores within aortic roots compared to wild-type ApoE knockout mice. The authors show that the deletion of liver-specific LPP3 increased atherogenic lipids, such as LPA and other lysophosphatidylinositols, in the plasma [73]. The increase in atherogenic lipids correlated with increased atherosclerosis progression [73].

Oxidized LDL-treated bone marrow-derived macrophages have increased LPP3 expression, suggesting macrophage LPP3 may regulate atherosclerosis progression. However, in a model of atherosclerosis, myeloid-derived *PLPP3* does not increase LPA lesion localization or increase atherosclerosis progression. Along with macrophages, smooth muscle cells are also able to transition into foam cells during atherosclerosis. The deletion of smooth muscle cell LPP3 resulted in increased atherosclerosis plaque growth [68]. The authors demonstrated that LPP3-deficient smooth muscle cells still transition to foam cells but may have altered responses to lipids that lead to increased plaque growth and inflammation. These data suggest smooth muscle cell LPP3 is atheroprotective. The above studies demonstrate that LPP3 is involved in atherosclerosis. More work is needed to truly understand the cell-specific contributions of LPP3 and the contributions of LPP1 and LPP2 toward atherosclerosis.

4. Phospholipase C

Phospholipase C (PLC) is a calcium-dependent phosphodiesterase that regulates phosphoinositide metabolism. PLC hydrolyzes phosphatidylinositol 4,5-bis-phosphate (PI(4,5)P_2) to generate the second messengers inositol 1,4,5-trisphosphate (IP_3), and diacylglycerol (DAG) [74]. There are thirteen PLC isozymes in mammals, which are categorized into six classes based on structure. These classes include PLC β, γ, δ, ϵ, λ, and ν [74]. These structures largely dictate interactions with cell surface receptors including G-protein-coupled receptors (GPCRs), G-proteins, receptor tyrosine kinases (RTKs), and non-receptor tyrosine kinases [74]. There are numerous reviews focusing on the structure and regulation of each class of PLCs [75–77]; as such, those will not be covered here. Rather, we will review what is known about PLC and its contribution to atherosclerosis and immune responses.

Phospholipase C is known to regulate multiple immunological responses of T and B lymphocytes [78]. T cell receptor signaling results in the activation of PLC. The PLC-mediated cleavage of PI(4,5)P_2 generates IP_3 and DAG, which both have significant roles in the activation of immune cells. DAG activates protein kinase C (PKC), resulting in the initiation of NF$\kappa\beta$ signaling to promote inflammatory gene transcription [79,80]. IP3 binds to the IP3 receptor, leading to calcium release from the endoplasmic reticulum. Calcium activates calcineurin, resulting in the nuclear translocation of NFAT to promote IL-2 production and subsequent T cell proliferation [81]. In addition, PLC deficiency leads to a reduction of Treg development, which may promote chronic inflammation [82]. PLC plays a similar role in B cell activation as it does in T cells by promoting downstream NFκB- and NFAT-mediated transcription. This is accomplished through IP_3- and DAG-mediated signaling [80].

In comparison to those in lymphocytes, the functional consequences of PLC-mediated signaling in myeloid cells are diverse. PLC is required for macrophage differentiation in response to macrophage colony-stimulating factor (MCSF) [78,83]. In addition to promoting differentiation, activated macrophages and dendritic cells require PLC for appropriate cytokine production and dendritic cell migration [78,84]. Upon entry into tissue, macrophages and dendritic cells constitutively engulf surrounding antigens and present them on the cell surface. This engulfment requires the synthesis of phosphatidic acid (PA), and PLC is required for the generation of intermediates of the PA synthetic pathway, leading to subsequent RAC activation and actin polymerization [85]. PLC localizes to nascent phagosomes to promote the recruitment of PKC, leading to the uptake of IgG-opsonized antigens [85]. There are numerous studies demonstrating the critical role of PLC in immune cell activation and differentiation.

Although not extensively studied, the diverse role of PLC in immunological cells would suggest that phospholipase C likely contributes to the development of atherosclerosis. As previously mentioned, atherosclerosis is a chronic inflammatory disease, and PLC contributes to the activation and development of immune cells. Monocyte infiltration and reduced macrophage clearance exacerbate atherosclerosis [86]. PLC regulates the migration and phagocytic capacity of macrophages [78,84]. PLCβ3/ApoE-deficient mice exhibited a reduction in atherosclerotic lesion size in the aortic vessels, arches, and roots compared with littermate controls [87]. PLCβ3 deficiency also resulted in a reduction in the number of macrophages within murine atherosclerotic plaques [87]. The products of PLC enzymatic activity stimulate PKC, which is known to be atherogenic. PKC α/β positively regulates foam cell formation, and the deletion of PKCβ from ApoE KO mice reduced atherosclerotic plaque size [88,89]. Investigating the contribution of PLC within immune cells in atherosclerosis needs to be further explored.

Given that atherosclerosis is a chronic inflammatory condition, adaptive immune responses play a critical role in the progression of the disease. Immune responses from recruited T cells and B cells become the dominant factors that enhance local inflammation. Inflammatory T cell subsets (Th1) promote continued inflammation, which further exacerbates atherosclerosis. The inhibition of Th1 differentiation and cytokine production reduced the plaque area in the aortic root of atherosclerotic mice [90]. The inhibition of Th1 responses resulted in an increase in Th2 T cells, which led to a decrease in plaque area. B cell responses are largely atheroprotective, due to the production of immunoglobulins [91]. In particular, IgM and IgG directed at the epitopes of oxLDL seem to neutralize the pro-inflammatory epitopes [91]. Overall, the role of PLC in regulating T and B cell activation and function could have drastic impacts on atherosclerosis progression.

5. Phospholipase D

Phospholipase D (PLD) is a phospholipid-specific phosphodiesterase in which the enzymatic activity cleaves phosphorylcholine into phosphatidic acid and free choline [92]. PLD's enzymatic activity has pleiotropic effects on a variety of cellular pathways. Mammalian phospholipase Ds are divided into two classical isoforms, PLD1 and PLD2, which have both redundant and specific functions depending on the tissue distribution [92].

Phospholipase D is regulated transcriptionally and post-translationally. Both PLD1 and PLD2 are activated by the presence of phosphatidylinositol 4,5-bisphosphate (PtdIns(4,5)P2) [92,93]. Other lipid species also activate PLD, such as PtdIns(3,4,5)P3 and unsaturated fatty acids [92,94,95]. Not only do lipid species regulate PLD, but proteins that regulate the abundance, location, and phosphorylation state of Ptdlns(4,5)P2 are also involved in the regulation of PLD [93,94]. Various stimuli, such as PDGF, EGF, or IL-1β, result in the increased gene expression of PLD via the activation of NFκB [96]. PLD is post-translationally modified by phosphorylation and palmitoylation. Phosphorylation by GTPases, such as ARF and Rho family proteins, directly activates PLD enzymatic activity [94,97,98]. Palmitoylation has been shown to alter the localization of PLD within the cell, from perinuclear to plasma membrane regions [99,100]. This shows the highly dynamic nature of phospholipase D within the cell.

Understanding how PLD contributes to chronic inflammatory diseases, such as atherosclerosis, may have significant implications in disease progression. PLD has been shown to be present within macrophages of a human atherosclerotic plaque [101]. PLD regulates phagocytosis in macrophages through the generation of phosphatidic acid. PLD1 vesicles are recruited to both nascent and internalized phagosomes, while PLD2 is observed at nascent phagosomes [101]. The shRNA depletion of either PLD1 or PLD2 results in a reduction in the phagocytic capabilities for IgG-coated latex beads of RAW264.7 macrophages [102]. Ganesan et al. investigated the role of PLD in the phagocytosis of oxidized LDL. They show that PLD2 is critical for the uptake up of oxidized LDL through the regulation of WASP and Grb2 to polymerize actin at the phagocytic cup [103]. PLD2 is also needed for the CD36-mediated removal of aggregated oxLDL [103]. Given the importance of lipid metabolism in immunological cells, PLD activity presumably plays a greater role in the progression of atherosclerosis than the current literature suggests. Neutrophil responses are known to promote early atherogenesis. In neutrophils, FcgammaR1 binding leads to PLD activation, which is critical for the oxidative burst during degranulation [104]. PLD recruits cytochrome B to the mitochondria to increase NADPH oxidase activity and ROS generation [105]. In addition, PLD indirectly activates the p22phox subunit of cytochrome D via PA production [106]. The PLD-mediated activation of neutrophils may promote early plaque progression. Altogether, phospholipase D is critical for various immunological responses, and the contribution of PLD to atherosclerosis needs to be further investigated.

6. Cytosolic Phospholipase A2

Cytosolic phospholipase A2 (cPLA2) is one of three categories of phospholipase A2s. The other phospholipase A2s are known as secretory PLA2 and calcium-independent PLA2 [107]. Phospholipase A2s catalyze the hydrolysis of glycerophospholipids to produce arachidonic acid metabolites [107]. Of the phospholipases, cPLA2 is highly selective for arachidonic acid-containing glycerophospholipids [107]. cPLA2 is a ubiquitous enzyme that is found in most tissues and cells; however, mature T and B lymphocytes do not have any detectable levels of cPLA2 [108,109]. There are three isoforms of cPLA2: cPLA2 beta (110 kDA), cPLA2 gamma (60 kDA), and cPLA2 alpha (85 kDA). Each isoform has two catalytic domains: A and B. Catalytic domain A contains the lipase consensus sequence GXSGS [109]. Inactive cPLA2 exists in the cytosol; however, upon calcium binding to the C2 domain, cPLA2 translocates to the endoplasmic reticulum (ER), Golgi apparatus, and nuclear envelope [107]. Steady intracellular calcium greater than 100–125 nM causes cPLA2 translocation to the Golgi, whereas steady intracellular calcium greater than 210–280 nM causes cPLA2 translocation to the Golgi, ER, and nuclear envelope [109]. cPLA2 cellular localization can have effects on different lipid-mediated processes. For example, a study with renal cells demonstrated cPLA2 localization at the Golgi can change the lipid ratio and result in changes in structure and protein trafficking [110]. Along with intracellular calcium levels, the phosphorylation of cPLA2 at Ser 505, Ser 515, and Ser 727 regulates cPLA2 activity [107]. Mitogen-activated protein kinase phosphorylates the above serine residues; phosphorylation increases the enzymatic activity [107,111]. The activation of cPLA2 leads to the liberation of arachidonic acid, which can be converted into inflammatory eicosanoids including prostaglandins.

cPLA2 activity promotes pro-inflammatory immune cell activation through the production of eicosanoids, especially prostaglandin E_2 (PGE_2). PGE_2 is known to contribute to atherosclerosis and cardiovascular disease. cPLA2 hydrolyses glycerophospholipids into arachidonic acid. Cyclooxygenase (COX) enzymes then convert arachidonic acid into prostaglandins. Non-steroidal anti-inflammatory drugs inhibit COX enzymes. The inhibition of COX enzymes increases myocardial infarction risk [112]. These studies suggest cPLA2 may be involved during myocardial infarction. The contribution of cPLA2 specifically to atherosclerosis has been less studied, but there are a few studies suggesting involvement. Patients with advanced-stage cardiovascular disease had increased vascular cPLA2 expression compared to those with early-stage cardiovascular disease [113]. Treatment with the cPLA2 inhibitor AACOF3 in a cholecalciferol-overload mouse model significantly reduced vascular calcification [113]. These studies suggest cPLA2 is involved in vascular calcification during advanced atherosclerosis. There is also evidence that low-density lipoproteins increase the activity of cPLA2 by participating with secretory PLA2 to increase the release of arachidonic acid in monocytes after inflammatory stimuli [114]. Though limited, these studies do provide evidence that cPLA2 does contribute to atherosclerosis.

7. Lipin 1

Lipin-1 is a phosphatidic acid phosphatase that belongs to the evolutionarily conserved family of lipins [115]. Of the three-membered lipin family, lipin-1 exhibits the highest phosphatidate-specific phosphohydrolase activity [116]. Lipin-1 converts PA to DAG via its phosphohydrolase activity in a Mg^2+-dependent reaction [116,117]. The lipin family has two domains that are conserved from yeast to mammals [117,118]. There are sequence motifs between the N-terminal (N-LIP) and C-terminal (C-LIP) domains that mediate the functions of the lipins [117,119]. Close to the N-LIP is a nuclear localization sequence that translocates lipin-1 to the nucleus [120]. The C-LIP contains the haloacid dehalogenase (HAD)-like phosphatase motif (DXDXT) and an α-helical leucine-rich motif (LXXIL) that mediate the enzymatic and transcriptional co-regulatory activities, respectively [119,121,122]. Three isoforms (lipin1α, lipin1β, and lipin1γ) of lipin-1 are known to be present in humans as a result of the alternative mRNA splicing of the lipin-1 gene [123]. In contrast to in humans, lipin-1γ is not present in mice [122,123]. These splice variants have similar and complementary functions, even though they are differentially expressed in tissues [122,123].

Lipin-1-mediated DAG production is a key step in the biosynthesis of triacylglycerol (TAG), phosphatidylcholine (PC), and phosphatidylethanolamine (PE) [124–126]. Lipin-1 resides in the cytosol and can translocate to the endoplasmic reticulum (ER) upon dephosphorylation [127]. Lipin-1 then moves along the membrane to interact with and dephosphorylate PA to generate DAG [128]. Neither the membrane composition nor fatty acid tails of PA influence lipin-1 activity. Lipin-1's contribution to TAG, PE, and PC production is critical to lipid droplet (LD) generation, which aids in the storage of excess cholesterol, and TAG protects against lipid toxicity [129]. The shRNA depletion of lipin-1 reduced lipid droplet formation in oxLDL-fed RAW264.7 macrophages [125]. The siRNA depletion of lipin-1 in human macrophages reduces LD size and number, and TAG composition in response to fatty acid feeding [130,131]. Additionally, lipins can also protect against dietary glucose toxicity through the regulation of polyunsaturated fatty acid (PUFA) production. In *Caenorhabditis elegans*, lipin prevents dietary glucose toxicity, which leads to a shorter life span [132]. In addition to modulating lipid levels to protect against metabolite overloads, lipin-1 is important in the regulation of autophagy. Autophagy is a housekeeping mechanism of recycling nutrients and degrading dead organelles. Lipin-1-mediated DAG production regulates autophagosome formation and maturation by activating protein kinase D and subsequent VPS34 activity [133]. In support of this, CRISPR-generated lipin-1-deficient myoblasts were observed to have impaired mitochondrial function and irregular autophagic vacuoles under conditions of induced starvation [134]. Thus, lipins and especially lipin-1 are a critical regulatory node in nutrient handling within cells.

The phosphorylation of lipin-1 on multiple sites by mechanistic target of rapamycin complex-1 (mTORC-1) results in retention in the cytosol [135]. Lipin-1 acts as a transcriptional coactivator or repressor by forming a complex with transcription factors such as PPARy, PPARα, and peroxisome proliferator-activated receptor γ coactivator-1α (PGC1α) [119,136–138]. PPARs promote macrophage wound-healing activities [139]. Lipin-1 is able to coactivate these transcription factors and enhance their activity. Lipin-1's transcriptional co-regulatory activity directly facilitates the polarization of IL-4-stimulated macrophages into a wound-healing phenotype [140]. Lipin-1 also acts as a repressor for pro-inflammatory transcription factors such as sterol-response element binding protein-1 (SREBP-1) and nuclear factor of activated T cells isoform c4 (NFATc4) by preventing their binding to promoters [135,141].

Inflammatory responses contribute to the pathogenesis of various diseases. Lipin-1 facilitates the production of eicosanoids by activating cPLA2α to release arachidonic acid from phospholipids [142,143]. Several studies have shown that lipin-1 couples lipid synthesis with pro-inflammatory responses in macrophages [130,144]. Lipin-1 mediates the inflammatory response during TLR4 activation [130]. This process occurs in a diacylglycerol-dependent mechanism that regulates the activation of MAPKs and AP-1 to induce the expression of pro-inflammatory genes [130]. These findings were further supported by an in vivo experiment, which showed that mice lacking lipin-1 experienced earlier weight recovery in response to LPS treatment [130]. The faster recovery observed in lipin-1-deficient mice was due to the reduced expression of pro-inflammatory factors [130]. Lipin-1's enzymatic activity mediates macrophage pro-inflammatory responses. The uptake of oxLDLs leads to a diacylglycerol-dependent pro-inflammatory signaling cascade that is mediated by lipin-1 [144]. The activation of diacylglycerol-responsive proteins leads to the persistent activation of the pro-inflammatory PKC-MAPK-AP-1 signal transduction pathway [144]. The lipin-1-mediated production of DAG has also been shown to be implicated in colon cancer [145]. DAG increases the expression of pro-inflammatory cytokines in colon-resident macrophages to drive the transformation of dysplastic cells into cancerous cells [145].

In humans, loss-of-function mutations of lipin-1 result in fatal episodic childhood rhabdomyolysis [146,147]. Polymorphisms of LPIN1 are associated with an increased body mass index, type II diabetes, and metabolic syndrome, which are risk factors for atherosclerosis [148]. These results highlight the potential contribution of lipin-1 to cardiovascular disease in humans. In mice, the loss of lipin-1 results in lipodystrophy, although this is not seen in humans, likely due to compensatory mechanisms [149]. Additionally, in mice, lipin-1 contributes to the pathophysiology of fatty liver disease, colon cancer, and atherosclerosis through the promotion of macrophage pro-inflammatory responses [144,145,150]. In addition, lipin-1 was found to colocalize with macrophages in human atherosclerotic plaques [125]. Lipin-1's enzymatic activity has been implicated in the development of atherosclerosis, as it facilitates the formation of the lipid-laden macrophage phenotype and the production of inflammatory cytokines [144]. Mice lacking myeloid-associated lipin-1 enzymatic activity have a reduction in atherosclerosis [144]. The persistent production of DAG activates a signaling cascade that increases the production and secretion of pro-inflammatory mediators such as IL-6, IL-1, TNF-α, CCL2, and PGE_2 in response to oxLDL and LPS [125,144]. Lipin-1-deficient macrophages produce significantly less pro-inflammatory cytokines [125]. Collectively, the coupled effect of enhanced modLDL uptake and poor cholesterol efflux lead to the production of tissue-damaging inflammatory mediators that promote atherogenesis and contribute to the different stages of atherosclerosis.

The contributions of macrophage-associated lipin-1 transcriptional co-regulatory activity to atherosclerosis have not yet been published. However, there are data that suggest lipin-1 transcriptional co-regulatory activity may be involved in atherosclerosis. Lipin-1 transcriptional co-regulatory activity increases wound healing and induces macrophage wound-healing/pro-resolving polarization [140]. Macrophage wound-healing responses reduce atherosclerosis plaque growth and severity [151]. Lipin-1 transcriptional co-regulatory activity also augments PPAR promoter binding and increases PPAR-associated genes [137]. PPARs reduce early atherosclerosis progression and enhance atherosclerosis regression [136,139,152–154]. Combined, these data suggest that macrophage-associated lipin-1 transcriptional co-regulatory activity would reduce atherosclerosis severity. More work needs to be completed to understand how macrophage-associated lipin-1 transcriptional co-regulatory activity affects atherosclerosis.

8. Conclusions

Phospholipids, the components they store, and phospholipases are dynamic regulators of immune cell function. Specifically, the production and removal of bioactive lipids contributes to cellular activation, phagocytosis, ROS generation, cytokine production, and prostanoid production. Phospholipase activity is evident in almost all immune cells. The targeting of the immune system to reduce atherosclerosis is a therapeutic goal that offers a chance to reduce cardiovascular disease. We must define a mechanism of immune responses that can be targeted in atherosclerosis that does not cause global immuno-suppression. Phospholipases may represent one such target. The contribution of phospholipases to atherosclerosis must be further investigated beyond the current understanding. Future work would need to find ways to target phospholipases within the plaque. Numerous small-molecule inhibitors of phospholipase are known, and pairing with nanotechnology may be feasible [155]. The dual function of lipin-1 may also represent an interesting target for atherosclerosis therapeutics. Future work on understanding how lipin-1 is regulated in macrophages, what dictates when each lipin-1 activity will be dominant, and mechanisms to control each lipin-1 activity is needed. The further understanding of the interface of phospholipases, immune cell function, and atherosclerosis will uncover new therapeutic targets and add to our ability to better treat and prevent cardiovascular disease.

Author Contributions: All authors contributed to the collection of information and writing of this review. C.M.R.B. and R.M.S. were responsible for editing and compiling all the information. All authors have read and agreed to the published version of the manuscript.

Funding: This work was supported by the National Institutes of Health—RO1 HL131844 (M. Woolard)—and a Malcolm Feist Predoctoral Fellowship (C. Blackburn).

Conflicts of Interest: The authors declare no conflict of interest. The funders had no role in the design of the study; in the collection, analyses, or interpretation of data; in the writing of the manuscript; or in the decision to publish the results.

References

1. Yang, Y.; Lee, M.; Fairn, G.D. Phospholipid subcellular localization and dynamics. *J. Biol. Chem.* **2018**, *293*, 6230–6240. [CrossRef] [PubMed]
2. Berg, J.M.; Tymoczko, J.L.; Stryer, L. There are three common types of membrane lipids. In *Biochemistry*, 5th ed.; W H Freeman: New York, NY, USA, 2002. Available online: https://www.ncbi.nlm.nih.gov/books/NBK22361/ (accessed on 30 September 2020).
3. Albeituni, S.; Stiban, J. Roles of ceramides and other sphingolipids in immune cell function and inflammation. *Adv. Exp. Med. Biol.* **2019**, *1161*, 169–191. [PubMed]
4. Gendaszewska-Darmach, E. Lysophosphatidic acids, cyclic phosphatidic acids and autotaxin as promising targets in therapies of cancer and other diseases. *Acta Biochim. Pol.* **2008**, *55*, 227–240. [CrossRef] [PubMed]
5. Vance, D.E. Phospholipid metabolism and cell signalling in eucaryotes. *New Compr. Biochem.* **1991**, *7*, 205–240.
6. Aloulou, A.; Rahier, R.; Arhab, Y.; Noiriel, A.; Abousalham, A. Phospholipases: An Overview. *Methods Mol. Biol.* **2018**, *1835*, 69–105.
7. Gisterå, A.; Hansson, G.K. Immunol. Atherosclerosis. *Nat. Rev. Nephrol.* **2017**, *13*, 368–380.
8. Boullier, A.; Bird, D.A.; Chang, M.-K.; Dennis, E.A.; Friedman, P.; Gillotte-Taylor, K.; Hörkkö, S.; Palinski, W.; Quehenberger, O.; Shaw, P.; et al. Scavenger receptors, oxidized LDL, and atherosclerosis. *Ann. N. Y. Acad. Sci.* **2001**, *947*, 214–222. [CrossRef]
9. Halliwell, B.; Cross, C.E. Oxygen-derived species: Their relation to human disease and environmental stress. *Environ. Health Perspect.* **1994**, *102* (Suppl. 10), 5–12.
10. Serafini, M.; Del Rio, D. Understanding the association between dietary antioxidants, redox status and disease: Is the total antioxidant capacity the right tool? *Redox Rep.* **2004**, *9*, 145–152. [CrossRef]
11. Hazell, L.J.; Stocker, R. Oxidation of low-density lipoprotein with hypochlorite causes transformation of the lipoprotein into a high-uptake form for macrophages. *Biochem. J.* **1993**, *290*, 165–172. [CrossRef]
12. Stocker, R.; Keaney, J.F., Jr. New insights on oxidative stress in the artery wall. *J. Thromb. Haemost.* **2005**, *3*, 1825–1834. [CrossRef] [PubMed]
13. Moore, K.J.; Tabas, I. Macrophages in the pathogenesis of atherosclerosis. *Cell* **2011**, *145*, 341–355. [CrossRef]
14. Wolf, D.; Ley, K. Immunity and inflammation in atherosclerosis. *Circ. Res.* **2019**, *124*, 315–327. [CrossRef] [PubMed]
15. Heinz, J.; Marinello, M.; Fredman, G. Pro-resolution therapeutics for cardiovascular diseases. *Prostaglandins Other Lipid Mediat.* **2017**, *132*, 12–16. [CrossRef] [PubMed]
16. Sharma, M.; Schlegel, M.P.; Afonso, M.S.; Brown, E.J.; Rahman, K.; Weinstock, A.; Sansbury, B.E.; Corr, E.M.; Van Solingen, C.; Koelwyn, G.J.; et al. Regulatory T cells license macrophage pro-resolving functions during atherosclerosis regression. *Circ. Res.* **2020**, *127*, 335–353. [CrossRef]
17. O'Donnell, V.B.; Rossjohn, J.; Wakelam, M.J. Phospholipid signaling in innate immune cells. *J. Clin. Investig.* **2018**, *128*, 2670–2679. [CrossRef]
18. Tjoelker, L.W.; Wilder, C.; Eberhardt, C.; Stafforinit, D.M.; Dietsch, G.; Schimpf, B.; Hooper, S.; Le Trong, H.; Cousens, L.S.; Zimmerman, G.A.; et al. Anti-inflammatory properties of a platelet-activating factor acetylhydrolase. *Nature* **1995**, *374*, 549–553. [CrossRef]
19. Min, J.H.; Jain, M.K.; Wilder, C.; Paul, L.; Apitz-Castro, R.; Aspleaf, D.C.; Gelb, M.H. Membrane-bound plasma platelet activating factor acetylhydrolase acts on substrate in the aqueous phase. *Biochemistry* **1999**, *38*, 12935–12942. [CrossRef]
20. Stafforini, D.M.; McIntyre, T.M.; E Carter, M.; Prescott, S.M. Human plasma platelet-activating factor acetylhydrolase. Association with lipoprotein particles and role in the degradation of platelet-activating factor. *J. Biol. Chem.* **1987**, *262*, 4215–4222.
21. Tselepis, A.D.; Chapman, M.D. Inflammation, bioactive lipids and atherosclerosis: Potential roles of a lipoprotein-associated phospholipase A2, platelet activating factor-acetylhydrolase. *Atheroscler. Suppl.* **2002**, *3*, 57–68. [CrossRef]
22. Khakpour, H.; Frishman, W.H. Lipoprotein-associated phospholipase A2: An independent predictor of cardiovascular risk and a novel target for immunomodulation therapy. *Cardiol. Rev.* **2009**, *17*, 222–229. [CrossRef] [PubMed]

23. Watson, A.D.; Navab, M.; Hama, S.Y.; Sevanian, A.; Prescott, S.M.; Stafforini, D.M.; McIntyre, T.M.; Du, B.N.; Fogelman, A.M.; A Berliner, J. Effect of platelet activating factor-acetylhydrolase on the formation and action of minimally oxidized low density lipoprotein. *J. Clin. Investig.* **1995**, *95*, 774–782. [CrossRef] [PubMed]
24. Stafforini, D.M.; McIntyre, T.M.; A Zimmerman, G.; Prescott, S.M. Platelet-activating factor acetylhydrolases. *J. Biol. Chem.* **1997**, *272*, 17895–17898. [CrossRef] [PubMed]
25. Marathe, G.K.; Pandit, C.; Lakshmikanth, C.L.; Chaithra, V.H.; Jacob, S.P.; D'Souza, C.J.M. To hydrolyze or not to hydrolyze: The dilemma of platelet-activating factor acetylhydrolase. *J. Lipid Res.* **2014**, *55*, 1847–1854. [CrossRef]
26. Maiolino, G.; Bisogni, V.; Rossitto, G.; Rossi, G.P. Lipoprotein-associated phospholipase A2 prognostic role in atherosclerotic complications. *World J. Cardiol.* **2015**, *7*, 609–620. [CrossRef]
27. Yamada, Y.; Yoshida, H.; Ichihara, S.; Imaizumi, T.; Satoh, K.; Yokota, M. Correlations between plasma platelet-activating factor acetylhydrolase (PAF-AH) activity and PAF-AH genotype, age, and atherosclerosis in a Japanese population. *Atherosclerosis* **2000**, *150*, 209–216. [CrossRef]
28. Unno, N.; Nakamura, T.; Kaneko, H.; Uchiyama, T.; Yamamoto, N.; Sugatani, J.; Miwa, M. Plasma platelet-activating factor acetylhydrolase deficiency is associated with atherosclerotic occlusive disease in japan. *J. Vasc. Surg.* **2000**, *32*, 263–267. [CrossRef]
29. Tsaoussis, V.; Vakirtzi-Lemonias, C. The mouse plasma PAF acetylhydrolase: II. It consists of two enzymes both associated with the HDL. *J. Lipid Mediat. Cell Signal.* **1994**, *9*, 317–331.
30. Singh, U.; Zhong, S.; Xiong, M.; Li, T.-B.; Sniderman, A.; Teng, B.-B. Increased plasma non-esterified fatty acids and platelet-activating factor acetylhydrolase are associated with susceptibility to atherosclerosis in mice. *Clin. Sci.* **2004**, *106*, 421–432. [CrossRef]
31. Chroni, A.; Mavri-Vavayanni, M. Characterization of a platelet activating factor acetylhydrolase from rat adipocyte. *Life Sci.* **2000**, *67*, 2807–2825. [CrossRef]
32. Elstad, M.R.; Stafforini, D.M.; McIntyre, T.M.; Prescott, S.M.; Zimmerman, G.A. Platelet-activating factor acetylhydrolase increases during macrophage differentiation. A novel mechanism that regulates accumulation of platelet-activating factor. *J. Biol. Chem.* **1989**, *264*, 8467–8470. [PubMed]
33. Stafforini, D.M.; Tjoelker, L.W.; McCormick, S.P.; Vaitkus, D.; McIntyre, T.M.; Gray, P.W.; Young, S.G.; Prescott, S.M. Molecular basis of the interaction between plasma platelet-activating factor acetylhydrolase and low density lipoprotein. *J. Biol. Chem.* **1999**, *274*, 7018–7024. [CrossRef]
34. Rosenson, R.S. Lp-PLA(2) and risk of atherosclerotic vascular disease. *Lancet* **2010**, *375*, 1498–1500. [CrossRef]
35. Macphee, C.H.; Moores, K.E.; Boyd, H.F.; Dhanak, D.; Ife, R.J.; Leach, C.A.; Leake, D.S.; Milliner, K.J.; Patterson, R.A.; Suckling, K.E.; et al. Lipoprotein-associated phospholipase A2, platelet-activating factor acetylhydrolase, generates two bioactive products during the oxidation of low-density lipoprotein: Use of a novel inhibitor. *Biochem. J.* **1999**, *338 Pt 2*, 479–487. [CrossRef]
36. Karakas, M.; Koenig, W. Lp-Pla2 Inhib. -Atheroscler. Panacea? *Pharmaceuticals* **2010**, *3*, 1360–1373. [CrossRef] [PubMed]
37. Carpenter, K.L.; Dennis, I.F.; Challis, I.R.; Osborn, D.P.; Macphee, C.H.; Leake, D.S.; Arends, M.J.; Mitchinson, M.J. Inhibition of lipoprotein-associated phospholipase A2 diminishes the death-inducing effects of oxidised LDL on human monocyte-macrophages. *FEBS Lett.* **2001**, *505*, 357–363. [CrossRef]
38. Quinn, M.T.; Parthasarathy, S.; Steinberg, D. Lysophosphatidylcholine: A chemotactic factor for human monocytes and its potential role in atherogenesis. *Proc. Natl. Acad. Sci. USA* **1988**, *85*, 2805–2809. [CrossRef] [PubMed]
39. Lavi, S.; McConnell, J.P.; Rihal, C.S.; Prasad, A.; Mathew, V.; Lerman, L.O.; Lerman, L.O. Local production of lipoprotein-associated phospholipase A2 and lysophosphatidylcholine in the coronary circulation: Association with early coronary atherosclerosis and endothelial dysfunction in humans. *Circulation* **2007**, *115*, 2715–2721. [CrossRef]
40. Hazen, S.L. Oxidized phospholipids as endogenous pattern recognition ligands in innate immunity. *J. Biol. Chem.* **2008**, *283*, 15527–15531. [CrossRef] [PubMed]
41. Wilensky, R.L.; Macphee, C.H. Lipoprotein-Assoc. Phospholipase A(2) Atherosclerosis. *Curr. Opin. Lipidol.* **2009**, *20*, 415–420. [CrossRef]
42. Zalewski, A.; Macphee, C. Role of lipoprotein-associated phospholipase A2 in atherosclerosis: Biology, epidemiology, and possible therapeutic target. *Arter. Thromb. Vasc. Biol.* **2005**, *25*, 923–931. [CrossRef] [PubMed]

43. Häkkinen, T.; Luoma, J.S.; Hiltunen, M.O.; Macphee, C.H.; Milliner, K.J.; Patel, L.; Rice, S.Q.; Tew, D.G.; Karkola, K.; Ylä-Herttuala, S. Lipoprotein-associated phospholipase A(2), platelet-activating factor acetylhydrolase, is expressed by macrophages in human and rabbit atherosclerotic lesions. *Arter. Thromb. Vasc. Biol.* **1999**, *19*, 2909–2917. [CrossRef] [PubMed]
44. Kolodgie, F.D.; Burke, A.P.; Taye, A.; Liu, W.H.; Sudhir, K.; Virmani, R. Lipoprotein-associated phospholipase A2 is highly expressed in macrophages of coronary lesions prone to rupture. *Circulation* **2004**, *110*, 246–247.
45. Stewart, R.A.; White, H.D. The role of lipoprotein-associated phospholipase a_2 as a marker and potential therapeutic target in atherosclerosis. *Curr. Atheroscler. Rep.* **2011**, *13*, 132–137. [CrossRef]
46. Kolodgie, F.D.; Burke, A.; Skorija, K.S.; Ladich, E.; Kutys, R.; Makuria, A.T.; Virmani, R. Lipoprotein-Associated Phospholipase A 2 Protein Expression in the Natural Progression of Human Coronary Atherosclerosis. *Arter. Thromb. Vasc. Biol.* **2006**, *26*, 2523–2529. [CrossRef] [PubMed]
47. Sairam, S.G.; Sola, S.; Barooah, A.; Javvaji, S.K.; Jaipuria, J.; Venkateshan, V.; Chelli, J.; Sanjeevi, C.B. The role of Lp-PLA2 and biochemistry parameters as potential biomarkers of coronary artery disease in Asian South-Indians: A case-control study. *Cardiovasc. Diagn. Ther.* **2017**, *7*, 589–597. [CrossRef]
48. Sakka, S.; Siahanidou, T.; Voyatzis, C.; Pervanidou, P.; Kaminioti, C.; Lazopoulou, N.; Kanaka-Gantenbein, C.; Chrousos, G.P.; Papassotiriou, I. Elevated circulating levels of lipoprotein-associated phospholipase A2 in obese children. *Clin. Chem. Lab. Med.* **2015**, *53*, 1119–1125. [CrossRef]
49. Donato, L.J.; Meeusen, J.W.; Callanan, H.; Saenger, A.K.; Jaffe, A.S. Advantages of the lipoprotein-associated phospholipase A2 activity assay. *Clin. Biochem.* **2016**, *49*, 172–175. [CrossRef]
50. Dennis, E.A. Introduction to Thematic Review Series: Phospholipases: Central Role in Lipid Signaling and Disease. *J. Lipid Res.* **2015**, *56*, 1245–1247. [CrossRef]
51. Tang, X.; Benesch, M.G.K.; Brindley, D.N. Lipid phosphate phosphatases and their roles in mammalian physiology and pathology. *J. Lipid Res.* **2015**, *56*, 2048–2060. [CrossRef]
52. Brindley, D.N.; Pilquil, C. Lipid phosphate phosphatases and signaling. *J. Lipid Res.* **2008**, *50*, S225–S230. [CrossRef] [PubMed]
53. McIntyre, T.M.; Pontsler, A.V.; Silva, A.R.; St Hilaire, A. Identification of an intracellular receptor for lysophosphatidic acid (LPA): LPA is a transcellular PPARgamma agonist. *Proc. Natl. Acad. Sci. USA* **2003**, *100*, 131–136. [CrossRef] [PubMed]
54. Tabbai, S.; Moreno-Fernández, R.D.; Zambrana-Infantes, E.; Nieto-Quero, A.; Chun, J.; García-Fernández, M.; Estivill-Torrús, G.; De Fonseca, F.R.; Santín, L.J.; Gil Oliveira, T.; et al. Effects of the LPA1 Receptor Deficiency and Stress on the Hippocampal LPA Species in Mice. *Front. Mol. Neurosci.* **2019**, *12*, 146. [CrossRef] [PubMed]
55. Choi, J.W.; Herr, D.R.; Noguchi, K.; Yung, Y.C.; Lee, C.-W.; Mutoh, T.; Lin, M.-E.; Teo, S.T.; Park, K.E.; Mosley, A.N.; et al. LPA Receptors: Subtypes and Biological Actions. *Annu. Rev. Pharmacol. Toxicol.* **2010**, *50*, 157–186. [CrossRef]
56. Loveridge, C.; Tonelli, F.; Leclercq, T.; Lim, K.G.; Long, J.S.; Berdyshev, E.; Tate, R.J.; Natarajan, V.; Pitson, S.M.; Pyne, N.J.; et al. The Sphingosine Kinase 1 Inhibitor 2-(p-Hydroxyanilino)-4-(p-chlorophenyl)thiazole Induces Proteasomal Degradation of Sphingosine Kinase 1 in Mammalian Cells. *J. Biol. Chem.* **2010**, *285*, 38841–38852. [CrossRef]
57. Sobel, K.; Menyhart, K.; Killer, N.; Renault, B.; Bauer, Y.; Studer, R.; Steiner, B.; Bolli, M.H.; Nayler, O.; Gatfield, J. Sphingosine 1-Phosphate (S1P) Receptor Agonists Mediate Pro-fibrotic Responses in Normal Human Lung Fibroblasts via S1P2 and S1P3 Receptors and Smad-independent Signaling. *J. Biol. Chem.* **2013**, *288*, 14839–14851. [CrossRef]
58. Wang, P.; Yuan, Y.; Lin, W.; Zhong, H.; Xu, K.; Qi, X. Roles of sphingosine-1-phosphate signaling in cancer. *Cancer Cell Int.* **2019**, *19*, 295. [CrossRef]
59. Zhang, Q.X.; Carlos, S.P.; Jay, D.; Luc, G.B.; David, N.B. Identification of structurally important domains of lipid phosphate phosphatase-1: Implications for its sites of action. *Biochem. J.* **2000**, *345 Pt 2*, 181–184. [CrossRef]
60. Humtsoe, J.O.; Bowling, R.A.; Feng, S.; Wary, K.K. Murine lipid phosphate phosphohydrolase-3 acts as a cell-associated integrin ligand. *Biochem. Biophys. Res. Commun.* **2005**, *335*, 906–919. [CrossRef]
61. Humtsoe, J.O.; Feng, S.; Thakker, G.D.; Yang, J.; Hong, J.; Wary, K.K. Regulation of cell–cell interactions by phosphatidic acid phosphatase 2b/VCIP. *EMBO J.* **2003**, *22*, 1539–1554. [CrossRef]

62. Tanyi, J.L.; Hasegawa, Y.; LaPushin, R.; Morris, A.J.; Wolf, J.K.; Berchuck, A.; Lu, K.; Smith, D.I.; Kalli, K.; Hartmann, L.C.; et al. Role of decreased levels of lipid phosphate phosphatase-1 in accumulation of lysophosphatidic acid in ovarian cancer. *Clin. Cancer Res.* **2003**, *9*, 3534–3545. [PubMed]
63. Smyth, S.S.; Sciorra, V.A.; Sigal, Y.J.; Pamuklar, Z.; Wang, Z.; Xu, Y.; Prestwich, G.D.; Morris, A.J. Lipid Phosphate Phosphatases Regulate Lysophosphatidic Acid Production and Signaling in Platelets. *J. Biol. Chem.* **2003**, *278*, 43214–43223. [CrossRef]
64. Panchatcharam, M.; Salous, A.K.; Brandon, J.; Miriyala, S.; Wheeler, J.; Patil, P.; Sunkara, M.; Morris, A.J.; Escalante-Alcalde, D.; Smyth, S.S. Mice with Targeted Inactivation of Ppap2b in Endothelial and Hematopoietic Cells Display Enhanced Vascular Inflammation and Permeability. *Arter. Thromb. Vasc. Biol.* **2014**, *34*, 837–845. [CrossRef]
65. Gutiérrez-Martínez, E.; Fernández-Ulibarri, I.; Lázaro-Diéguez, F.; Johannes, L.; Pyne, S.; Sarri, E.; Egea, G. Lipid phosphate phosphatase 3 participates in transport carrier formation and protein trafficking in the early secretory pathway. *J. Cell Sci.* **2013**, *126*, 2641–2655. [CrossRef]
66. Sciorra, V.A.; Morris, A.J. Sequential Actions of Phospholipase D and Phosphatidic Acid Phosphohydrolase 2b Generate Diglyceride in Mammalian Cells. *Mol. Biol. Cell* **1999**, *10*, 3863–3876. [CrossRef]
67. Schunkert, H.; Cardiogenics, J.F.C.; König, I.R.; Kathiresan, S.; Reilly, M.P.; Assimes, T.L.; Holm, H.; Preuss, M.; Stewart, A.F.R.; Barbalic, M.; et al. Large-scale association analysis identifies 13 new susceptibility loci for coronary artery disease. *Nat. Genet.* **2011**, *43*, 333–338. [CrossRef] [PubMed]
68. Mueller, P.A.; Yang, L.; Ubele, M.; Mao, G.; Brandon, J.; Vandra, J.; Nichols, T.C.; Escalante-Alcalde, D.; Morris, A.J.; Smyth, S.S. Coronary Artery Disease Risk-Associated Plpp3 gene and its product lipid phosphate phosphatase 3 regulate experimental atheroscerosis. *Arter. Thromb. Vasc. Biol.* **2019**, *39*, 2261–2272. [CrossRef] [PubMed]
69. Wirtwein, M.; Olle, M.; Marketa, S.; Michal, H.; Krzysztof, N.; Marcin, G.; Wojciech, S. Relationship between selected DNA polymorphisms and coronary artery disease complications. *Int. J. Cardiol.* **2017**, *228*, 814–820. [CrossRef]
70. Reschen, M.E.; Gaulton, K.J.; Lin, D.; Soilleux, E.J.; Morris, A.J.; Smyth, S.S.; O'Callaghan, C.A. Lipid-Induced Epigenomic Changes in Human Macrophages Identify a Coronary Artery Disease-Associated Variant that Regulates PPAP2B Expression through Altered C/EBP-Beta Binding. *PLoS Genet.* **2015**, *11*, e1005061. [CrossRef]
71. Siess, W.; Zangl, K.J.; Essler, M.; Bauer, M.; Brandl, R.; Corrinth, C.; Bittman, R.; Tigyi, G.; Aepfelbacher, M. Lysophosphatidic acid mediates the rapid activation of platelets and endothelial cells by mildly oxidized low density lipoprotein and accumulates in human atherosclerotic lesions. *Proc. Natl. Acad. Sci. USA* **1999**, *96*, 6931–6936. [CrossRef]
72. Escalante-Alcalde, D.; Hernandez, L.; Le Stunff, H.; Maeda, R.; Lee, H.S.; Sciorra, V.A.; Daar, I.; Spiegel, S.; Morris, A.J.; Stewart, C.L. The lipid phosphatase LPP3 regulates extra-embryonic vasculogenesis and axis patterning. *Development* **2003**, *130*, 4623–4637. [CrossRef] [PubMed]
73. Busnelli, M.; Manzini, S.; Hilvo, M.; Parolini, C.; Ganzetti, G.S.; Dellera, F.; Ekroos, K.; Jänis, M.; Escalante-Alcalde, D.; Sirtori, C.R.; et al. Liver-specific deletion of the Plpp3 gene alters plasma lipid composition and worsens atherosclerosis in apoE−/− mice. *Sci. Rep.* **2017**, *7*, 44503. [CrossRef]
74. Gresset, A.; Sondek, J.; Harden, T.K. The Phospholipase C Isozymes and Their Regulation. *Subcell. Biochem.* **2012**, *58*, 61–94. [CrossRef] [PubMed]
75. Nakamura, Y.; Fukami, K. Regulation and physiological functions of mammalian phospholipase C. *J. Biochem.* **2017**, *89*, 189–198. [CrossRef] [PubMed]
76. Cocco, L.; Follo, M.Y.; Manzoli, L.; Suh, P.-G. Phosphoinositide-specific phospholipase C in health and disease. *J. Lipid Res.* **2015**, *56*, 1853–1860. [CrossRef] [PubMed]
77. Rebecchi, M.J.; Pentyala, S.N. Structure, Function, and Control of Phosphoinositide-Specific Phospholipase C. *Physiol. Rev.* **2000**, *80*, 1291–1335. [CrossRef] [PubMed]
78. Cecchetti, S.; Spadaro, F.; Gessani, S.; Podo, F.; Fantuzzi, L. Phospholipases: At the crossroads of the immune system and the pathogenesis of HIV-1 infection. *J. Leukoc. Biol.* **2016**, *101*, 53–75. [CrossRef]
79. Bi, K.; Tanaka, Y.; Coudronniere, N.; Sugie, K.; Hong, S.; Van Stipdonk, M.J.B.; Altman, A. Antigen-induced translocation of PKC-θ to membrane rafts is required for T cell activation. *Nat. Immunol.* **2001**, *2*, 556–563. [CrossRef]

80. Schulze-Luehrmann, J.; Ghosh, S. Antigen-receptor signaling to nuclear factor kappa B. *Immunity* **2006**, *25*, 701–715. [CrossRef]
81. Lee, J.-U.; Kim, L.-K.; Choi, J.-M. Revisiting the Concept of Targeting NFAT to Control T Cell Immunity and Autoimmune Diseases. *Front. Immunol.* **2018**, *9*, 2747. [CrossRef]
82. Chuck, M.I.; Zhu, M.; Shen, S.; Zhang, W. The role of the LAT-PLC-gamma1 interaction in T regulatory cell function. *J. Immunol.* **2010**, *184*, 2476–2486. [CrossRef] [PubMed]
83. Barbosa, C.M.; Claudia, B.; Carlos, C.B.; Alice, T.F.; Edgar, J.P. PLCγ2 and PKC are important to myeloid lineage commitment triggered by M-SCF and G-CSF. *J. Cell. Biochem.* **2014**, *115*, 42–51. [CrossRef] [PubMed]
84. Luft, T.; Rodionova, E.; Maraskovsky, E.; Kirsch, M.; Hess, M.; Buchholtz, C.; Goerner, M.; Schnurr, M.; Skoda, R.; Ho, A. Adaptive functional differentiation of dendritic cells: Integrating the network of extra- and intracellular signals. *Blood* **2006**, *107*, 4763–4769. [CrossRef] [PubMed]
85. Bohdanowicz, M.; Schlam, D.; Hermansson, M.; Rizzuti, D.; Fairn, G.D.; Ueyama, T.; Somerharju, P.; Du, G.; Grinstein, S. Phosphatidic acid is required for the constitutive ruffling and macropinocytosis of phagocytes. *Mol. Biol. Cell* **2013**, *24*, 1700–1712. [CrossRef]
86. Yurdagul, A.J.; Doran, A.C.; Cai, B.; Fredman, G.; Tabas, I.A. Mechanisms and Consequences of Defective Efferocytosis in Atherosclerosis. *Front. Cardiovasc. Med.* **2018**, *4*, 86. [CrossRef]
87. Wang, Z.; Liu, B.; Wang, P.; Dong, X.; Fernandez-Hernando, C.; Li, Z.; Hla, T.; Li, Z.; Claffey, K.; Smith, J.D.; et al. Phospholipase C β3 deficiency leads to macrophage hypersensitivity to apoptotic induction and reduction of atherosclerosis in mice. *J. Clin. Investig.* **2008**, *118*, 195–204. [CrossRef]
88. Osto, E.; Alexey, K.; Pavani, M.; Alexander, A.; Christian, B.; Lucia, R.; Arnold, V.E.; Iliceto, S.; Volpe, M.; Lüscher, T.F.; et al. Inhibition of protein kinase Cbeta prevents foam cell formation by reducing scavenger receptor A expression in human macrophages. *Circulation* **2008**, *118*, 2174–2182. [CrossRef]
89. Harja, E.; Chang, J.S.; Lu, Y.; Leitges, M.; Zou, Y.S.; Schmidt, A.M.; Yan, S. Mice deficient in PKCbeta and apolipoprotein E display decreased atherosclerosis. *Faseb. J.* **2009**, *23*, 1081–1091. [CrossRef]
90. Tse, K.; Tse, H.; Sidney, J.; Sette, A.; Ley, K. T cells in atherosclerosis. *Int. Immunol.* **2013**, *25*, 615–622. [CrossRef] [PubMed]
91. Sage, A.P.; Tsiantoulas, D.; Binder, C.J.; Mallat, Z. The role of B cells in atherosclerosis. *Nat. Rev. Cardiol.* **2018**, *16*, 180–196. [CrossRef]
92. McDermott, M.I.; Wang, Y.; Wakelam, M.; Bankaitis, V. Mammalian phospholipase D: Function, and therapeutics. *Prog. Lipid Res.* **2020**, *78*, 101018. [CrossRef] [PubMed]
93. Divecha, N.; Mieke, R.; Jonathan, R.H.; Sabine, D.; Mar, F.-B.; Lauran, O.; Kahlid, M.S.; Michael, W.; Clive, D. Interaction of the type Ialpha PIPkinase with phospholipase D: A role for the local generation of phosphatidylinositol 4, 5-bisphosphate in the regulation of PLD2 activity. *EMBO J.* **2000**, *19*, 5440–5449. [CrossRef] [PubMed]
94. Bruntz, R.C.; Lindsley, C.W.; Brown, H.A. Phospholipase D Signaling Pathways and Phosphatidic Acid as Therapeutic Targets in Cancer. *Pharmacol. Rev.* **2014**, *66*, 1033–1079. [CrossRef] [PubMed]
95. Kim, J.H.; Kim, Y.; Lee, S.D.; Lopez, I.; Arnold, R.S.; Lambeth, J.; Suh, P.-G.; Ryu, S.H. Selective activation of phospholipase D2 by unsaturated fatty acid. *FEBS Lett.* **1999**, *454*, 42–46. [CrossRef]
96. Gomez-Cambronero, J. New Concepts in Phospholipase D Signaling in Inflammation and Cancer. *Sci. World J.* **2010**, *10*, 1356–1369. [CrossRef] [PubMed]
97. Malcolm, K.C.; Ross, A.H.; Qiu, R.G.; Symons, M.; Exton, J.H. Activation of rat liver phospholipase D by the small GTP-binding protein RhoA. *J. Biol. Chem.* **1994**, *269*, 25951–25954.
98. Hammond, S.M.; Jenco, J.M.; Nakashima, S.; Cadwallader, K.; Gu, Q.; Cook, S.; Nozawa, Y.; Prestwich, G.D.; Frohman, M.A.; Morris, A.J. Characterization of two alternately spliced forms of phospholipase D1. Activation of the purified enzymes by phosphatidylinositol 4,5-bisphosphate, ADP-ribosylation factor, and Rho family monomeric GTP-binding proteins and protein kinase C-alpha. *J. Biol. Chem.* **1997**, *272*, 3860–3868. [CrossRef]
99. Du, G.; Huang, P.; Liang, B.T.; Frohman, M.A. Phospholipase D2 Localizes to the Plasma Membrane and Regulates Angiotensin II Receptor Endocytosis. *Mol. Biol. Cell* **2004**, *15*, 1024–1030. [CrossRef]
100. Manifava, M.; Sugars, J.; Ktistakis, N.T. Modification of Catalytically Active Phospholipase D1 with Fatty Acidin Vivo. *J. Biol. Chem.* **1999**, *274*, 1072–1077. [CrossRef]
101. O'Brien, K.D.; Pineda, C.; Chiu, W.S.; Bowen, R.; Deeg, M.A. Glycosylphosphatidylinositol-Specific Phospholipase D Is Expressed by Macrophages in Human Atherosclerosis and Colocalizes With Oxidation Epitopes. *Circulation* **1999**, *99*, 2876–2882. [CrossRef]

102. Corrotte, M.; Chasserot-Golaz, S.; Huang, P.; Du, G.; Ktistakis, N.T.; Frohman, M.A.; Vitale, N.; Bader, M.-F.; Grant, N.J. Dynamics and Function of Phospholipase D and Phosphatidic Acid During Phagocytosis. *Traffic* **2006**, *7*, 365–377. [CrossRef] [PubMed]
103. Ganesan, R.; Henkels, K.M.; Wrenshall, L.E.; Kanaho, Y.; Di Paolo, G.; Frohman, M.A.; Gomez-Cambronero, J. Oxidized LDL phagocytosis during foam cell formation in atherosclerotic plaques relies on a PLD2-CD36 functional interdependence. *J. Leukoc. Biol.* **2018**, *103*, 867–883. [CrossRef] [PubMed]
104. Melendez, A.J.; Bruetschy, L.; Floto, R.A.; Harnett, M.M.; Allen, J.M. Functional coupling of FcγRI to nicotinamide adenine dinucleotide phosphate (reduced form) oxidative burst and immune complex trafficking requires the activation of phospholipase D1. *Blood* **2001**, *98*, 3421–3428. [CrossRef] [PubMed]
105. Watson, F.; Gordon, M.L.; Robinson, J.J.; Galvani, D.W.; Edwards, S.W. Phospholipase D-dependent and-independent activation of the neutrophil NADPH oxidase. *Biosci. Rep.* **1994**, *14*, 91–102. [CrossRef]
106. Bréchard, S.; Plançon, S.; Tschirhart, E.J. New Insights into the Regulation of Neutrophil NADPH Oxidase Activity in the Phagosome: A Focus on the Role of Lipid and Ca2+Signaling. *Antioxid. Redox Signal.* **2013**, *18*, 661–676. [CrossRef]
107. Nakamura, H.; Wakita, S.; Suganami, A.; Tamura, Y.; Hanada, K.; Murayama, T. Modulation of the activity of cytosolic phospholipase A2α (cPLA2α) by cellular sphingolipids and inhibition of cPLA2α by sphingomyelin[S]. *J. Lipid Res.* **2009**, *51*, 720–728. [CrossRef]
108. Kramer, R.M.; Sharp, J.D. Structure, function and regulation of Ca2+-sensitive cytosolic phospholipase A2 (cPLA2). *FEBS Lett.* **1997**, *410*, 49–53. [CrossRef]
109. Kudo, I.; Murakami, M. Phospholipase A2 enzymes. *Prostaglandins Other Lipid Mediat.* **2002**, *68*, 3–58. [CrossRef]
110. Choukroun, G.J.; Marshansky, V.; Gustafson, C.E.; McKee, M.; Hajjar, R.J.; Rosenzweig, A.; Brown, D.; Bonventre, J.V. Cytosolic phospholipase A2 regulates Golgi structure and modulates intracellular trafficking of membrane proteins. *J. Clin. Investig.* **2000**, *106*, 983–993. [CrossRef]
111. Lin, L.-L.; Wartmann, M.; Lin, A.Y.; Knopf, J.L.; Seth, A.; Davis, R.J. cPLA2 is phosphorylated and activated by MAP kinase. *Cell* **1993**, *72*, 269–278. [CrossRef]
112. Trelle, S.; Reichenbach, S.; Wandel, S.; Hildebrand, P.; Tschannen, B.; Villiger, P.M.; Egger, M.; Jüni, P. Cardiovascular safety of non-steroidal anti-inflammatory drugs: Network meta-analysis. *BMJ* **2011**, *342*, c7086. [CrossRef]
113. Schanstra, J.P.; Luong, T.T.; Makridakis, M.; Van Linthout, S.; Lygirou, V.; Latosisnska, A.; Alesutan, I.; Boehme, B.; Schelski, N.; Von Lewinski, D.; et al. Systems biology identifies cytosolic PLA2 as a target in vascular calcification treatment. *JCI Insight* **2019**, *4*, 4. [CrossRef]
114. Oestvang, J.; Bonnefont-Rousselot, D.; Ninio, E.; Hakala, J.K.; Johansen, B.; Anthonsen, M.W. Modification of LDL with human secretory phospholipase A2or sphingomyelinase promotes its arachidonic acid-releasing propensity. *J. Lipid Res.* **2004**, *45*, 831–838. [CrossRef]
115. Csaki, L.S.; Reue, K. Lipins: Multifunctional Lipid Metabolism Proteins. *Annu. Rev. Nutr.* **2010**, *30*, 257–272. [CrossRef] [PubMed]
116. Donkor, J.; Sariahmetoglu, M.; Dewald, J.; Brindley, D.N.; Reue, K. Three Mammalian Lipins Act as Phosphatidate Phosphatases with Distinct Tissue Expression Patterns. *J. Biol. Chem.* **2006**, *282*, 3450–3457. [CrossRef] [PubMed]
117. Han, G.-S.; Wu, W.-I.; Carman, G.M. TheSaccharomyces cerevisiaeLipin Homolog Is a Mg2+-dependent Phosphatidate Phosphatase Enzyme. *J. Biol. Chem.* **2006**, *281*, 9210–9218. [CrossRef]
118. Chen, Y.; Rui, B.-B.; Tang, L.-Y.; Hu, C.-M. Lipin Family Proteins—Key Regulators in Lipid Metabolism. *Ann. Nutr. Metab.* **2014**, *66*, 10–18. [CrossRef] [PubMed]
119. Finck, B.N.; Gropler, M.C.; Chen, Z.; Leone, T.C.; Croce, M.A.; Harris, T.E.; Lawrence, J.C., Jr.; Kelly, D.P. Lipin 1 is an inducible amplifier of the hepatic PGC-1alpha/PPARalpha regulatory pathway. *Cell Metab.* **2006**, *4*, 199–210. [CrossRef]
120. Harris, T.E.; Huffman, T.A.; Chi, A.; Shabanowitz, J.; Hunt, N.F.; Kumar, A.; Lawrence, J.C. Insulin Controls Subcellular Localization and Multisite Phosphorylation of the Phosphatidic Acid Phosphatase, Lipin 1. *J. Biol. Chem.* **2006**, *282*, 277–286. [CrossRef]
121. Donkor, J.; Zhang, P.; Wong, S.; O'Loughlin, L.; Dewald, J.; Kok, B.P.; Brindley, D.N.; Reue, K. A Conserved Serine Residue Is Required for the Phosphatidate Phosphatase Activity but Not the Transcriptional Coactivator Functions of Lipin-1 and Lipin-2. *J. Biol. Chem.* **2009**, *284*, 29968–29978. [CrossRef]

122. Péterfy, M.; Phan, J.; Xu, P.; Reue, K. Lipodystrophy in the fld mouse results from mutation of a new gene encoding a nuclear protein, lipin. *Nat. Genet.* **2001**, *27*, 121–124. [CrossRef] [PubMed]
123. Han, G.-S.; Carman, G.M. Characterization of the HumanLPIN1-encoded Phosphatidate Phosphatase Isoforms. *J. Biol. Chem.* **2010**, *285*, 14628–14638. [CrossRef] [PubMed]
124. Bartz, R.; Li, W.-H.; Venables, B.; Zehmer, J.K.; Roth, M.R.; Welti, R.; Anderson, R.G.W.; Liu, P.; Chapman, K.D. Lipidomics reveals that adiposomes store ether lipids and mediate phospholipid traffic. *J. Lipid Res.* **2007**, *48*, 837–847. [CrossRef] [PubMed]
125. Navratil, A.R.; Vozenilek, A.; Cardelli, J.A.; Green, J.M.; Thomas, M.J.; Sorci-Thomas, M.; Orr, A.W.; Woolard, M.D. Lipin-1 contributes to modified low-density lipoprotein-elicited macrophage pro-inflammatory responses. *Atherosclerosis* **2015**, *242*, 424–432. [CrossRef] [PubMed]
126. Coleman, R.; Coleman, R.A.; Lee, D.P. Enzymes of triacylglycerol synthesis and their regulation. *Prog. Lipid Res.* **2004**, *43*, 134–176. [CrossRef]
127. Zhang, P.; Reue, K. Lipin proteins and glycerolipid metabolism: Roles at the ER membrane and beyond. *Biochim. Biophys. Acta (BBA)-Biomembr.* **2017**, *1859*, 1583–1595. [CrossRef] [PubMed]
128. Kwiatek, J.M.; Carman, G.M. Yeast phosphatidic acid phosphatase Pah1 hops and scoots along the membrane phospholipid bilayer. *J. Lipid Res.* **2020**, *61*, 1232–1243. [CrossRef]
129. Murphy, D.J. The biogenesis and functions of lipid bodies in animals, plants and microorganisms. *Prog. Lipid Res.* **2001**, *40*, 325–438. [CrossRef]
130. Meana, C.; Pena, L.; Lordén, G.; Esquinas, E.; Guijas, C.; Valdearcos, M.; Balsinde, J.; Balboa, M.A. Lipin-1 Integrates Lipid Synthesis with Proinflammatory Responses during TLR Activation in Macrophages. *J. Immunol.* **2014**, *193*, 4614–4622. [CrossRef]
131. Valdearcos, M.; Esquinas, E.; Meana, C.; Gil-De-Gómez, L.; Guijas, C.; Balsinde, J.; Balboa, M.A. Subcellular Localization and Role of Lipin-1 in Human Macrophages. *J. Immunol.* **2011**, *186*, 6004–6013. [CrossRef]
132. Jung, Y.; Kwon, S.; Ham, S.; Lee, D.; Park, H.H.; Yamaoka, Y.; Jeong, D.; Artan, M.; Altintas, O.; Park, S.; et al. Caenorhabditis elegans Lipin 1 moderates the lifespan-shortening effects of dietary glucose by maintaining ω-6 polyunsaturated fatty acids. *Aging Cell* **2020**, *19*. [CrossRef] [PubMed]
133. Zhang, P.; Verity, M.A.; Reue, K. Lipin-1 Regulates Autophagy Clearance and Intersects with Statin Drug Effects in Skeletal Muscle. *Cell Metab.* **2014**, *20*, 267–279. [CrossRef]
134. Alshudukhi, A.A.; Zhu, J.; Huang, D.; Jama, A.; Smith, J.D.; Wang, Q.J.; Esser, K.A.; Ren, H. Lipin-1 regulates Bnip3–mediated mitophagy in glycolytic muscle. *FASEB J.* **2018**, *32*, 6796–6807. [CrossRef] [PubMed]
135. Peterson, T.R.; Sengupta, S.S.; Harris, T.E.; Carmack, A.E.; Kang, S.A.; Balderas, E.; Guertin, D.A.; Madden, K.L.; Carpenter, A.E.; Finck, B.N.; et al. mTOR Complex 1 Regulates Lipin 1 Localization to Control the SREBP Pathway. *Cell* **2011**, *146*, 408–420. [CrossRef]
136. McCarthy, C.; Lieggi, N.T.; Barry, D.; Mooney, D.; De Gaetano, M.; James, W.G.; McClelland, S.; Barry, M.C.; Escoubet-Lozach, L.; Li, A.C.; et al. Macrophage PPAR gamma Co-activator-1 alpha participates in repressing foam cell formation and atherosclerosis in response to conjugated linoleic acid. *EMBO Mol. Med.* **2013**, *5*, 1443–1457. [CrossRef] [PubMed]
137. Kim, H.E.; Bae, E.; Jeong, D.-Y.; Kim, M.-J.; Jin, W.-J.; Park, S.-W.; Han, G.-S.; Carman, G.M.; Koh, E.; Kim, K.-S. Lipin1 regulates PPARγ transcriptional activity. *Biochem. J.* **2013**, *453*, 49–60. [CrossRef]
138. Soskic, S.; Dobutović, B.D.; Sudar, E.; Obradovic, M.; Nikolic, D.; Zaric, B.; Stojanović, S.Đ.; Stokić, E.J.; Mikhailidis, D.P.; Isenovic, E.R. Peroxisome Proliferator-Activated Receptors and Atherosclerosis. *Angiology* **2011**, *62*, 523–534. [CrossRef]
139. Chen, H.; Shi, R.; Luo, B.; Yang, X.; Qiu, L.; Xiong, J.; Jiang, M.; Liu, Y.; Zhang, Z.-R.; Wu, Y. Macrophage peroxisome proliferator-activated receptor γ deficiency delays skin wound healing through impairing apoptotic cell clearance in mice. *Cell Death Dis.* **2015**, *6*, e1597. [CrossRef]
140. Chandran, S.; Schilke, R.M.; Blackburn, C.M.R.; Yurochko, A.; Mirza, R.; Scott, R.S.; Finck, B.N.; Woolard, M.D. Lipin-1 Contributes to IL-4 Mediated Macrophage Polarization. *Front. Immunol.* **2020**, *11*, 787. [CrossRef]
141. Kim, H.B.; Kumar, A.; Wang, L.; Liu, G.-H.; Keller, S.R.; Lawrence, J.C.; Finck, B.N.; Harris, T.E. Lipin 1 Represses NFATc4 Transcriptional Activity in Adipocytes to Inhibit Secretion of Inflammatory Factors. *Mol. Cell. Biol.* **2010**, *30*, 3126–3139. [CrossRef]
142. Grkovich, A.; Johnson, C.A.; Buczynski, M.W.; Dennis, E.A. Lipopolysaccharide-induced Cyclooxygenase-2 Expression in Human U937 Macrophages Is Phosphatidic Acid Phosphohydrolase-1-dependent. *J. Biol. Chem.* **2006**, *281*, 32978–32987. [CrossRef] [PubMed]

143. Grkovich, A.; Armando, A.; Quehenberger, O.; Dennis, E.A. TLR-4 mediated group IVA phospholipase A2 activation is phosphatidic acid phosphohydrolase 1 and protein kinase C dependent. *Biochim. Biophys. Acta (BBA)-Mol. Cell Biol. Lipids* **2009**, *1791*, 975–982. [CrossRef] [PubMed]
144. Vozenilek, A.E.; Navratil, A.R.; Green, J.M.; Coleman, D.T.; Blackburn, C.M.; Finney, A.C.; Pearson, B.H.; Chrast, R.; Finck, B.N.; Klein, R.L.; et al. Macrophage-Associated Lipin-1 Enzymatic Activity Contributes to Modified Low-Density Lipoprotein-Induced Proinflammatory Signaling and Atherosclerosis. *Arter. Thromb. Vasc. Biol.* **2017**, *38*, 324–334. [CrossRef] [PubMed]
145. Meana, C.; García-Rostán, G.; Peña, L.; Lordén, G.; Cubero, Á.; Orduña, A.; Győrffy, B.; Balsinde, J.; Balboa, M.A. The phosphatidic acid phosphatase lipin-1 facilitates inflammation-driven colon carcinogenesis. *JCI Insight* **2018**, *3*, 3. [CrossRef] [PubMed]
146. Schweitzer, G.G.; Collier, S.L.; Chen, Z.; Eaton, J.M.; Connolly, A.M.; Bucelli, R.C.; Pestronk, A.; Harris, T.E.; Finck, B.N.; Zschocke, J. Rhabdomyolysis-Associated Mutations in Human LPIN1 Lead to Loss of Phosphatidic Acid Phosphohydrolase Activity. *JIMD Rep.* **2015**, *23*, 113–122. [CrossRef]
147. Schweitzer, G.G.; Collier, S.L.; Chen, Z.; McCommis, K.S.; Pittman, S.K.; Yoshino, J.; Matkovich, S.; Hsu, F.-F.; Chrast, R.; Eaton, J.M.; et al. Loss of lipin 1-mediated phosphatidic acid phosphohydrolase activity in muscle leads to skeletal myopathy in mice. *FASEB J.* **2018**, *33*, 652–667. [CrossRef]
148. Reue, K.; Wang, H. Mammalian lipin phosphatidic acid phosphatases in lipid synthesis and beyond: Metabolic and inflammatory disorders. *J. Lipid Res.* **2019**, *60*, 728–733. [CrossRef]
149. Reue, K.; Doolittle, M.H. Naturally occurring mutations in mice affecting lipid transport and metabolism. *J. Lipid Res.* **1996**, *37*, 1387–1405.
150. Li, Y.Y.; Zhou, J.Y. Role of lipin-1 in the pathogenesis of alcoholic fatty liver disease. *Zhonghua gan zang bing za zhi = Zhonghua ganzangbing zazhi = Chin. J. Hepatol.* **2016**, *24*, 237–240.
151. Barrett, T.J. Macrophages in Atherosclerosis Regression. *Arter. Thromb. Vasc. Biol.* **2020**, *40*, 20–33. [CrossRef]
152. Reiss, A.B.; E Vagell, M. PPARgamma activity in the vessel wall: Anti-atherogenic properties. *Curr. Med. Chem.* **2006**, *13*, 3227–3238. [CrossRef] [PubMed]
153. Staels, B. PPARgamma and atherosclerosis. *Curr. Med. Res. Opin.* **2005**, *21*, 13–20. [CrossRef] [PubMed]
154. Taketa, K.; Matsumura, T.; Yano, M.; Ishii, N.; Senokuchi, T.; Motoshima, H.; Murata, Y.; Kim-Mitsuyama, S.; Kawada, T.; Itabe, H.; et al. Oxidized Low Density Lipoprotein Activates Peroxisome Proliferator-activated Receptor-α (PPARα) and PPARγ through MAPK-dependent COX-2 Expression in Macrophages. *J. Biol. Chem.* **2008**, *283*, 9852–9862. [CrossRef] [PubMed]
155. Kornmueller, K.; Vidakovic, I.; Prassl, R. Artificial High Density Lipoprotein Nanoparticles in Cardiovascular Research. *Molecules* **2019**, *24*, 2829. [CrossRef]

Publisher's Note: MDPI stays neutral with regard to jurisdictional claims in published maps and institutional affiliations.

© 2020 by the authors. Licensee MDPI, Basel, Switzerland. This article is an open access article distributed under the terms and conditions of the Creative Commons Attribution (CC BY) license (http://creativecommons.org/licenses/by/4.0/).

 *biomolecules* 

eview

Majeed Syndrome: A Review of the Clinical, Genetic and Immunologic Features

olly J. Ferguson * and Hatem El-Shanti 

Stead Family Department of Pediatrics, University of Iowa Carver College of Medicine, 200 Hawkins Drive, Iowa City, IA 52242, USA; hatem-el-shanti@uiowa.edu

* Correspondence: polly-ferguson@uiowa.edu

Abstract: Majeed syndrome is a multi-system inflammatory disorder affecting humans that presents with chronic multifocal osteomyelitis, congenital dyserythropoietic anemia, with or without a neutrophilic dermatosis. The disease is an autosomal recessive disorder caused by mutations in *LPIN2*, the gene encoding the phosphatidic acid phosphatase LIPIN2. It is exceedingly rare. There are only 24 individuals from 10 families with genetically confirmed Majeed syndrome reported in the literature. The early descriptions of Majeed syndrome reported severely affected children with recurrent fevers, severe multifocal osteomyelitis, failure to thrive, and marked elevations of blood inflammatory markers. As more affected families have been identified, it has become clear that there is significant phenotypic variability. Data supports that disruption of the phosphatidic acid phosphatase activity in LIPIN2 results in immune dysregulation due to aberrant activation of the NLRP3 inflammasome and overproduction of proinflammatory cytokines including IL-1β, however, these findings did not explain the bone phenotype. Recent studies demonstrate that *LPIN2* deficiency drives pro-inflammatory M2-macrophages and enhances osteoclastogenesis which suggest a critical role of lipin-2 in controlling homeostasis at the growth plate in an inflammasome-independent manner. While there are no approved medications for Majeed syndrome, pharmacologic blockade of the interleukin-1 pathway has been associated with rapid clinical improvement.

Keywords: majeed syndrome; LPIN2; LIPIN2; chronic non-bacterial osteomyelitis; chronic recurrent multifocal osteomyelitis; autoinflammatory; inflammasome; macrophage; osteoclast

itation: Ferguson, P.J.; El-Shanti, H. lajeed Syndrome: A Review of the linical, Genetic and Immunologic eatures. *Biomolecules* **2021**, *11*, 367. ttps://doi.org/10.3390/biom 030367

cademic Editor: María A. Balboa

eceived: 19 January 2021
ccepted: 23 February 2021
ublished: 28 February 2021

ublisher's Note: MDPI stays neutral ith regard to jurisdictional claims in ublished maps and institutional affil- tions.

opyright: © 2021 by the authors. icensee MDPI, Basel, Switzerland. his article is an open access article istributed under the terms and onditions of the Creative Commons ttribution (CC BY) license (https:// reativecommons.org/licenses/by/ .0/).

1. Introduction

1.1. Autoinflammatory Disorders

Autoinflammatory diseases present with recurrent or persistent inflammation that occurs in the absence of self-reactive T or B cells which distinguishes them from autoimmune disorders [1,2]. Instead, it is the innate immune system that is dysregulated and that leads to persistent or recurrent bouts of inflammation. The major breakthroughs in the understanding of autoinflammatory diseases came around the turn of the century with the identification of the genetic cause for several periodic fever syndromes, including familial Mediterranean fever (FMF) due to pathogenic variants in *MEFV* and cryopyrin associated periodic syndrome (CAPS) due to pathogenic variants in *NLRP3*, affecting the pyrin and NLRP3 inflammasome function, respectively, and leading to dysregulation of IL-1 production [3–5] Since that time, there has been a flurry of discovery in the autoinflammatory disorder field [6]. These disorders can affect many organ systems including a subgroup of autoinflammatory disorders that target the bone [7–9]. Their unifying features are sterile osteomyelitis or osteitis of one or more bones, often accompanied by inflammatory disorders of the skin and gastrointestinal tract.

1.2. Chronic Recurrent Multifocal Osteomyelitis (CRMO)

The most common autoinflammatory bone disorder is chronic recurrent multifocal osteomyelitis (CRMO) which predominantly affects children with an average age of onset around 9 to 10 years and presents with bone pain with or without associate swelling [10–14]. It was first described by Giedion et al. in 1972 as a symmetric, sterile, mu tifocal osteomyelitis [15]. An alternative name is chronic non-bacterial osteomyelitis (CNC which has been proposed as an umbrella term as some cases remain unifocal and not a patients have recurrent disease [8,16,17]. When adults present with a similar disease ph notype, the term synovitis, acne, pustulosis, hyperostosis and osteitis (SAPHO) syndrom is used [8,18–21]. Individuals with CRMO and their close relatives are more likely to hav or develop inflammatory bowel disease (most often Crohn disease), psoriasis or inflamm tory arthritis suggesting a shared pathogenetic process [10,22–27]. Most cases of CRMC are sporadic but there are cases of affected siblings, concordant monozygotic twins an parent-child duos supporting a genetic contribution to the etiology of the disease [10,28–30 A putative susceptibility locus was reported on chromosome 18q21.3–18q22 but this wa not reproduced in a larger cohort when sequencing rather than microsatellite marker ana ysis was used to assess for linkage [30,31]. CRMO can also occur as part of a Mendelia syndrome in which a single gene defect leads to sterile multifocal osteomyelitis furthe supporting a genetic contribution to the etiology of the disease [32–37].

The molecular mechanisms of non-syndromic CRMO remains unclear but the existin data suggests that it is a genetically complex disorder leading to an imbalance between pr and anti-inflammatory cytokines produced by innate immune cells, ultimately leading t osteoclast activation with osteolytic destruction of the bone [9]. Despite the phenotypi similarities to bacterial osteomyelitis, culture of the bone is typically negative, antibiotic do not produce sustained improvement and bacterial DNA is not detected by molecula methods [10,16]. Proinflammatory cytokines including IL-1β, IL-6 and TNF are elevated i the blood of CRMO patients whereas 'anti-inflammatory' or immunoregulatory cytokine including IL-10 are reduced [9,38–42]. Peripheral blood monocytes from patients wit CRMO produce lower levels of IL-10 and IL-19 when stimulated with lipopolysaccharid versus control monocytes [42]. Evaluation of human CRMO bone specimens for cytokine suggests dysregulation of IL-1 pathway but this data is from a few small studies [9,43 The proinflammatory cytokines found in the blood of these patients are known to activat osteoclasts via receptor activator of nuclear factor kappa-B (RANK) and RANK ligan signaling which could lead to bone resorption producing the osteolytic lesions. Epigeneti alterations have also been found in CRMO including reduced phosphorylation of histon H3 at position 10 (H3S10P) which would result in 'closure' of the IL-10 promotor leadin to reduced IL-10 transcription [44].

There are multiple animal models of the disease including humans, mice, rats, dog and non-human primates [45–51] several of which have been shown to be geneticall driven [31,33,36,38,49,52–59]. Utilizing murine models of CRMO, including the proline serine-threonine phosphatase interacting protein 2 (PSTPIP2) deficient and Ali18 model it has been shown that the disease is a hematopoietically-driven innate immune syster disorder, occurring independent of the adaptive immune system [37,38,55,56]. Furthe more, work on the PSTPIP2 deficient cmo mouse has shown that it is an IL-1β driver disorder with granulocytes, particularly neutrophils playing a key role in the inflammator process [49,60–63]. Due to these features, CRMO has been classified as an autoinflammator disorder [2,22].

1.3. Autoinflammatory Bone Disease Syndromes

There are two syndromic forms of CRMO in humans that have also been shown to b IL-1 mediated disorders by in vitro investigations and by favorable responses to IL-1 block ing drugs [34,35,64–66]. One is Majeed syndrome which is the focus of this review. Th other is the deficiency of the interleukin-1 receptor antagonist (DIRA) in which, sterile multi focal osteomyelitis/osteitis and skin pustulosis are the dominant phenotypes [34,35,67–78 Patients with DIRA present in infancy with systemic inflammation, severe multifocal os teomyelitis and pustulosis of the skin. It is caused by either deficiency or loss of functior mutations in the gene that encodes the IL-1 receptor antagonist leading to unfettered IL-1 signaling and results in a systemic inflammatory disorder that if left untreated is often fatal [34,35,67–78]. While a direct connection to IL-1 signaling is evident in DIRA, that connection had been less clear in the Majeed syndrome which is caused by pathogenic variants in *LPIN2*, in which the encoded protein plays a central role in lipid metabolism. However, clinical observations and basic research has confirmed that Majeed syndrome is also an IL-1 mediated disease. We review the clinical, genetic and immunologic features of Majeed syndrome and the recent data that links the derangements in lipid metabolism with the innate immune system dysfunction.

2. Majeed Syndrome

Majeed syndrome was first described in 1989, in three children from a consanguineous two-related-sibship family who presented with CRMO and a congenital dyserythropoietic anemia (CDA) that was characteristically microcytic [33]. This was followed with the addition of a fourth child [54] and the description of two siblings from another unrelated consanguineous family [54]. Two brothers from the initial family also had a neutrophilic dermatosis, Sweet syndrome in addition to CRMO and CDA [33]. The children were severely affected and had significant failure to thrive and growth delay [53]. Each child presented in the first 2 years of life with recurrent fevers and bone pain, sometimes with subsequent joint contractures due to severe multifocal sterile osteomyelitis, and often with hepatosplenomegaly [53]. All affected individuals had CDA that required multiple red blood cell transfusions and partially resolved in one patient after a splenectomy [33,53,54].

Inflammatory markers, including the erythrocyte sedimentation rate, were elevated in all 6 children [53]. Non-steroidal anti-inflammatory medications improved but did not control symptoms. Since the disease segregated in an autosomal recessive pattern, *LPIN2* was identified as the responsible gene using homozygosity mapping and positional cloning [52]. To date, only 24 individuals from 10 families with molecularly confirmed Majeed syndrome have been reported in the literature (Table 1) [33,52–54,64,79–85].

Clinical Features of Majeed Syndrome

Based on the earliest reports, Majeed syndrome became recognizable by the clinical triad of early onset CRMO, severe CDA and a neutrophilic dermatosis [53]. However, as more cases have been reported it has become evident that < 10% of those affected have all 3 features (Tables 1 and 2). Most individuals with Majeed syndrome present in the first 2 years of life and 91% have both CRMO and CDA, and the neutrophilic dermatosis, if present, may be transient. The median age at onset is 12 months old (average age of onset = 20.4 months), with CRMO being the prominent phenotype in Majeed syndrome with 91 percent having radiologic or clinical evidence of CRMO (only 2 individuals had no musculoskeletal symptoms but were not imaged).

Table 1. Clinical features in Majeed syndrome.

	A1	A2	A3	A4	B1	B2	C1	D1	D2	E1	E2	F1	F2	G1	H1	H2	I1	I2	I3	I4	I5	I6	J1	K1
Recurrent fever	+	+	+	+	+++	+++	+	–	+	–	–	NR	NR	–	+	–	+	NR	NR	–	–	–	+	–
Failure to thrive	+	+	+	+	+	+	–	NR	NR	+	–	NR	NR	+	–	–	+	NR	NR	NR	NR	NR	NR	–
CNO	+	+	+	+	+	+	+	+	+	+	+	+	+	+	+	+	+	+	+	Joint and bone pain	Joint and bone pain	Knee pains	–	+
Age at onset (months) $^{\Delta}$	12	19	9	1	0.75	9	15	6	3	24	96	6	48	72	13	15	infant	?	?	?	?	?	6	12
LE long bones	+	+	+	+	+	+	+	+	+	+	+	+	+	+	+	+	+							+
UE long bones	+	+	+	+	+	+	+	+	+	+	+	–	–	NR	NR	ND	NR							NR
Feet	–	–	–	–	–	+	+	–	–	+	–	+	–	+	–	ND	NR							NR
Spine #	–	–	–	–	–	–	–	–	–	–	–	–	–	–	ND	ND	NR							NR
Other			hands			hands		ribs	hands											Limb pain	Limb pain	Knee pain		
Bone biopsy	osteo	osteo	osteo	NR	NR	NR	ND	Osteo Φ	ND	ND	ND	ND	necrosis	osteo	c/w osteo	ND	ND	ND	ND	ND	ND	ND	ND	ND
Objective joint swelling	+	+	+	NR	+	+	+	–	+	+	+	+	+	+	+	–	NR	NR	NR	NR	NR	NR	–	–
Hepatosplenomegaly	+	+	+	+	+	NR	+	–	–	+	–	NR	NR	NR	–	–	NR	NR	NR	NR	NR	NR	–	NR
↑ ESR/CRP	+++	+++	+++	+++	++	+++	+++	+++	+++	+++	++	+++	++	+++	++	++	+++	+++	++	NR	NR	NR	+++	+++
Microcytic anemia	+	+	+	+	+ 8.8	+ 4.0	+	+°	+°	+	+*	+	+	NR	9.3 MCV	7.7@	6.9	+	+	anemia	anemia	–	8.5	10.1
Transfusion Rx	++	+++	++	++	++	+++	–	–	–	–	–	–	–	NR	–	–	–	–	–	–	–	–	–	–
Dyserythropoiesis on BM	+	+	+	+	+	+	+	+	+	+	ND	ND	+	NR	ND	+	+	ND	ND	ND	ND	ND	–	+
Neutropenia (ANC)							+ (750)			leuko	–				+ (588)	NR	NR	NR	NR	NR	NR	NR	+ (400)	–
Neutrophilic dermatosis	+	+	–	–	–	–	–	–	–	–	–	–	–	NR	–	–	–	–	–	–	–	–	–	–
Treatments	Pred, NSAID	Failed colchicine. Pred, NSAID	Failed colchicine. Pred, NSAID	Pred, NSAID	NSAID	NSAID	NSAID + prednisolone	Steroids, Failed etanercept, improved with anakinra, canakin	Steroids, Failed etanercept, improved with anakinra, canakin	NSAId, MTX, partial pam	MTX	NSAID	NSAID, canak	Anakinra	NSAID, Etanercept (failed), did well on anakinra	Failed etanercept, improved with anakinra	NSAID, steroid, partial resonse to bis-phos, anakinra	Anakinra	Anakinra	NSAID	NSAID	none	none	NSAID, canakin
Mutation	S734L	S734L	S734L	S734L	C181*	C181*	R776Sfs*66	L439fs*15	L439fs*15	Y747*	Y747*	S734L	S734L	R776Sfs*66	R776Sfs*66	R776Sfs*66	R736H	R736H	R736H	R736H	R736H	R736H	R776Sfs*66 And R564Kfs*3	R517H and 17.8 kb del

Osteo = osteomyelitis; ND = not done; NR = not reported; # whole body MRI not done in most; Φ granulomatous inflammation; ° anemic but MCV not reported; * mild anemia MCV 86.8. – = not present; + = present or mildly elevated; ++ = moderately elevated; +++ = markedly elevated. Reference for family A and B (1); family C (4), family D (5), family E (6), two patients reported by Moussa F (7), patient G (8). $^{\Delta}$ months to year conversion: 24 mo = 2 yrs; 48 mo = 4 yrs; 72 mo = 6 yrs; 96 mo = 8 yrs. Modified from Ferguson PJ; Elsevier, Edited by Cimaz and Lehman, Pediatric in Systemic Autoimmune Diseases, Volume 11, Chapter 15, Pages 315-339, 2016. @ dimorphic red cells on peripheral smear (A) Majeed 1989 (B) Majeed 1989 (C) Al-Mosawi 2007 (D) Herlin (E) Rao (F) Moussa (G) Pinto-Fernandez (H) Al-Mosai 2019 (I) Roy 2019 (J) Liu (K) Bhuyan 2020.

Table 2. Majeed Clinical Features.

Clinical Feature	%
Recurrent fever	46
Failure to thrive	38
Hepatosplenomegaly	30
Objective limb swelling	54
Neutrophilic dermatosis	8
↑inflammatory marker	88
Microcytic anemia/CDA	92
Neutropenia	13
Radiographic CNO++	83

Percentage of cases reporting this feature. Total $n = 24$. ++ No imaging was done in 4 but 3/4 had bone pain.

The CRMO of Majeed syndrome presents with recurrent episodes of bone pain affecting the long bones, often close to joints, with a predilection for the lower extremities. Unlike classic CRMO, involvement of the mandible, clavicle, anterior chest, spine and pelvis have not been reported, however, whole body magnetic resonance imaging has only been performed in a few Majeed syndrome patients. Seven individuals have had bone biopsies with six demonstrating culture negative osteomyelitis and the seventh demonstrating osteonecrosis [33,64,79,83,86].

The diagnosis of Majeed syndrome requires a high index of suspicion as about half of patients had objective changes on physical examination such as swelling or warmth overlying the involved bone or nearby joint.

The anemia of Majeed syndrome is hypochromic and microcytic and ranges in severity from very mild to quite severe with about a quarter of patients requiring red blood cell transfusions [33,80,84,86]. Bone marrow aspiration or biopsy demonstrates morphological abnormalities, such as erythroid hyperplasia with binuclearity or multinuclearity, which is typical of CDA [53,54,80,84,86]. Based on bone marrow morphology, associated anomalies and molecular etiology, the CDAs have been recently classified into five categories: CDAI, CDAII, CDAIII, transcription-factor-related CDA and CDA variants, with the CDA of Majeed syndrome falling in the last category [87]. While CRMO and CDA are commonly reported in Majeed syndrome, only 2 of 24 affected individuals had skin disease, both with the neutrophilic dermatosis Sweet syndrome [53]. There are two other reports of a patient with CRMO and Sweet syndrome in the absence of CDA, as a bone marrow biopsy was normal in one [87,88]. Thus, it remains to be determined if Sweet syndrome is part of Majeed syndrome.

Elevated blood inflammatory markers were present in 21 of 21 patients tested; this likely contributes to the failure to thrive, growth delay and organomegaly that are present in ~ 30% of affected individuals. Nearly half of the patients have recurrent fevers and for many this is one of the first manifestations of the disease, therefore, Majeed syndrome should be in the differential diagnosis of the periodic fever syndromes. While the initial reports of Majeed syndrome were of early onset severe CRMO and severe CDA, it has become evident that there is considerable phenotypic variability in the presentation. Moussa et al., Pinto-Fernandez et al. and Rao et al., reported children where the age of onset of disease was 4, 6 and 8 years, respectively [83,84,86]. Even more pronounced phenotypic variation was described by Roy et al. in a family of 6 who were all homozygous for a novel mutation in *LPIN2* for which 3 had a typical severe phenotype while 2 had mild symptoms and one adult had only mild knee pain throughout his childhood and adult life [85].

3. Genetics of Majeed Syndrome

The observations that males and females are equally affected with unaffected consanguineous parents and the presence of affected individuals from related sibships highly suggest that Majeed syndrome is a distinct entity with an autosomal recessive mode o inheritance [53]. Homozygosity mapping utilizing DNA from 2 unrelated affected fam lies demonstrated linkage with markers in a 1.8 Mb region on chromosome 18p. Sange sequencing of genes in the candidate interval revealed unique mutations in *LPIN2* in eac family [52]. Affected individuals from one family were homozygous for a missense muta tion replacing the highly conserved serine at amino acid 734 with a leucine (p.S734L), whil affected children from the other family had a frameshift mutation caused by a consecutiv 2 base pair deletion resulting in a premature stop codon in the first quarter of the codin sequence (p.C181*) [52]. To date, 9 unique mutations have been reported in families of Ara bic, Turkish, Chinese, Indian (East Asia), Pakistan, Spanish and mixed-race background (Figure 1). These come from 10 unrelated families (the reports by Al-Mosawi et al. in 200 and 2019 are from the same extended family). All patients reported to date have bee homozygous with only two patients having a compound heterozygous genotype [81,82 Most of the mutations are deleterious mutations predicted to lead to a truncated protei and loss of function. However, two families harbor missense mutations changing the highl conserved amino acids; one altering the serine at amino acid 734 to a leucine (p.S734L) an the other altering proline at amino acid 736 to a histidine (p.P736H) [33,52,85]. To dat there is no clear genotype-phenotype correlation pattern providing an explanation to th clinical heterogeneity of Majeed syndrome.

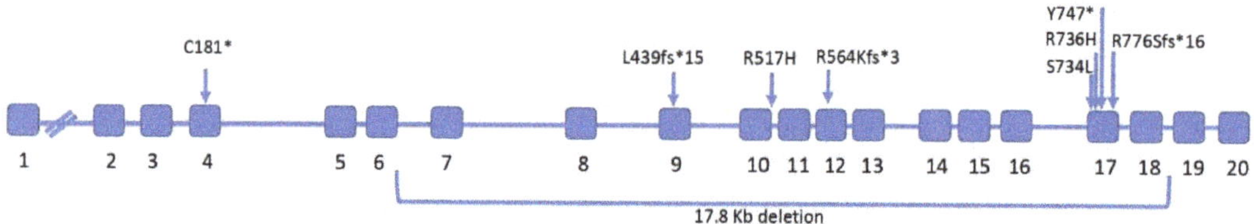

Figure 1. Majeed syndrome associated mutations. *LPIN2* is composed of 20 exons. Pathogenic mutations are located throughout the gene. Most disease-causing mutations reported to date are predicted to results in early truncation or in exon deletion. Two missense mutation. The in vitro functional work by Donkor et al. showed that the mutant protein is expressed but lacked PAP activity.

It is of note that a recent study found five out of 182 Turkish FMF patients are het erozygous for four variants in LPIN2, three variants are previously described and are likel benign, and the fourth is a novel nonsense variant, p.Y732X [89]. Further, another repor details a child diagnosed clinically with FMF and CDA who underwent successful bon marrow transplant for severe anemia. In addition to the anemia and fevers, that child als had bone pain and limb swelling. The child was heterozygous for pathogenic mutatior p.M680I but a second disease associated mutation in *MEFV* was not found. This case repor predates the discovery of LPIN2 as the disease causing mutation in Majeed syndrome an while the authors speculate that FMF and CDA II were co-segregating in this patient, w posit that the child likely had Majeed syndrome given the limb pain, limb swelling, marke inflammatory markers and CDA [90].

3.1. Pathogenesis of Majeed Syndrome

The LIPIN family (LIPIN1, LIPIN2, and LIPIN3) is a trio of cytosolic intracellular pro teins with phosphatidic acid phosphohydrolase (PAP) activity that converts phosphatidi acid to diacylglycerol (DAG) [89–91]. This is the penultimate step in triacylglycerol (TAG synthesis and a regulatory nodal point that can impact synthesis of phosphatidylcholine phosphatidylethanolamine, and other membrane lipids. Beyond this central role in lipid metabolism, the LIPINs can regulate lipid intermediates in cellular signaling pathways an have transcriptional co-regulatory capabilities [90,92,93]. These proteins have been show to be involved in a wide range of cellular processes including autophagy, inflammation and as a regulator of gene expression [52,65,66,90,92,94–97]. Mutations in LPIN1 and LPIN2 cause disease in mouse and humans with desperate phenotypes which is likely influenced by their differences in tissue expression and as well as by compensatory mechanisms. While tissue expression is broad in all of the LIPINs, LIPIN1 is most highly expressed in skeletal muscle, adipose tissue, peripheral nerve and testis, whereas LIPIN2 is the most abundant LIPIN in the liver, small intestine, macrophages and CNS [98]. The distribution of LIPIN3 overlaps with that of the other LIPINs but isn't the dominant protein in any tissue [98].

Deficiency of LPIN1 has significant consequences in adipose tissue, liver, skeletal muscle and nervous system [90,99,100]. *Lpin1* mutations where identified in a spontaneous mutant mouse called the fatty liver dystrophy or fld mouse [99,101,102]. The fld mouse has a phenotype consisting of lipodystrophy, a transient neonatal fatty liver and progressive non-inflammatory peripheral demyelinating neuropathy [102,103]. LPIN1 is highly expressed in skeletal muscle, yet these mice do not have an overt muscle phenotype under normal colony conditions but have abnormalities in skeletal muscle when stressed and have subtle skeletal muscle fiber changes seen on histopathologic exam [98,104]. This is in contrast to the phenotype seen in humans with homozygous or compound heterozygous loss of function mutations in LPIN1 which cause severe episodes of recurrent myoglobinuria [100]. Affected individuals have severe episodes of muscle necrosis (rhabdomyolysis) that causes profound muscle weakness accompanied by myoglobinuria which has led to renal failure and is potentially fatal [105–108]. Despite the severe phenotype during attacks, affected individuals are well in between episodes. The precise triggers have not been identified but the episodes are most often associated with febrile illnesses which implicates infections as a possible precipitating factor. Surprisingly, the episodes of myoglobinuria are rarely related to exercise. Unlike the murine models of Lpin1 deficiency, affected humans do not have lipodystrophy as part of the phenotype. The lack of lipodystrophy has been postulated to be from redundancy in the roles the LIPIN proteins have in triacylglycerol synthesis in human adipocytes [109]. Genetic dissection demonstrated that it is the lack of Lipin1 PAP activity, rather than co-activator activity, that drives the altered lipid metabolism that results in lipodystrophy in mice [110–112]. Thus, PAP activity is needed for normal adipocyte differentiation and triacylglcerol synthesis.

There are also key differences in the phenotypes seen in Majeed syndrome and that seen in *Lpin2* knockout mice. Mice deficient in Lipin2 do not develop sterile osteomyelitis by gross inspection and have normal bone histopathology and no osseous lesions by radiography; thus, they are missing a key feature of Majeed syndrome [113]. Yet there is phenotypic similarity. Similar to humans, *Lpin2* knockout mice develop anemia with features consistent with a congenital dyserythropoietic anemia [113]. The anemia in mice is mild, whereas, humans with Majeed syndrome the anemia is of variable severity from mild and asymptomatic to severe anemia that requires recurrent red blood cell transfusions [54]. Another difference is that there are neurologic abnormalities in Lipin2 deficient mice which have not been reported in Majeed syndrome including the development of tremor, ataxia and difficulties with maintaining their balance in the mice beginning around age 5 to 6 months [113]. Both Lipin1 and Lipin2 proteins are present in the cerebellum in young mice but with age, Lipin1 expression falls to undetectable levels in wildtype and Lipin2 knockout mice. This suggests that Lipin1 is able to compensate for the lack of Lipin2 in the cerebellum in young mice but that as the mice age, the lack of Lipin 1 and 2 leads to cerebellar dysfunction [113]. There is additional evidence for redundancy in the system as Lipin2 deficient mice do not have abnormal lipid homeostasis on a chow diet which is associated by a compensatory increase in hepatic Lipin1. However, stressing the mice by feeding them a high fat diet does result in lipid dysregulation. No abnormalities in fat metabolism have been identified in Majeed syndrome but investigations have been limited to analysis of blood lipids in a few children [80]. It is likely that a compensatory mechanism for lipid homeostasis is also occurring in humans with Majeed syndrome but this has not been experimentally proven. There is no information about long term effects of LIPIN2 deficiency on the central nervous system function in humans as the original cohorts described by Majeed et al. have been lost to follow up and the subjects i other reported cases are still too young to determine if cerebellar dysfunction will occu with age [33,53,54]. The importance of compensatory mechanisms in the Lipin family c proteins is demonstrated by embryonic lethality in mice that are deficient in both *Lpin1* an *Lpin2* [113]. To date, mutations in *LPIN3* have not been linked to human or murine diseas

Five of the 9 *LPIN2* mutations identified in affected individuals with Majeed syndrom are predicted to cause an early termination codon and a truncated protein, if the mRN escapes the nonsense-mediated RNA decay and one is a whole gene deletion. It is unclea if any LIPIN2 protein is produced in patients with these pathogenic variants. Three c the 9 pathogenic variants are single nucleotide substitutions resulting in non-synonymou variants (p.S734L, p.P736H and p.R517H) [52,85]. The first two amino acid substitution are separated by one amino acid, which implies that the disruption of protein function i this region of the LIPIN2 is important in the pathogenesis of Majeed syndrome. Donko et al. analyzed the functional consequences of the human p.S734L in a mouse model [114 The in vitro functional work showed that the mutant protein is expressed but lacked PA activity [114]. They also demonstrated that the mutation did not disrupt Lipin associatio with microsomal membranes. Further they demonstrated that, similar to Lipin1, Lipin can act as a transcriptional coactivator for peroxisome proliferator-activated recepto response elements and that the co-activator function is not disrupted by the p.S731 (equivalent to human p.S734L) mutation in their murine model [114]. This study strongl implicates LIPIN2 PAP activity in the pathogenesis of Majeed syndrome. Valdearcos et a demonstrated that LIPIN2 is important in controlling inflammation triggered by exces saturated fatty acids in vitro [66]. They under-expressed *LPIN2* in murine and huma monocytes and found that the monocytes produce excess proinflammatory cytokine including TNF and IL-6, when exposed to excessive amounts of the saturated fatty aci palmitic acid. In contrast, they found that when *LPIN2* is over-expressed in the sam experimental setting that the inflammatory response to palmitic acid was blunted [66 However, the connection between diminished PAP activity and inflammatory bone diseas remained unclear.

We and others have shown that the pro-inflammatory cytokine IL-1 drives the inflam matory bone disease and systemic inflammation in Majeed syndrome [64]. Therapeuticall blocking the IL-1RI or IL-1β (n = 10), but not TNF (n = 4), results in prompt resolution c systemic inflammation and healing of the sterile osteomyelitis seen in Majeed syndrom patients (Table 2) [64,79,81,83,85,86]. There is additional evidence that dysregulated IL- mediated signaling is central to the pathogenesis of sterile osteomyelitis. Pstpip2 deficien cmo mice develop kinked tails, paw deformities and inflammation of the skin and sof tissues of the ears [10,11]. Immune dysregulation is present with the over production o inflammatory cytokines including IL-6 and macrophage inflammatory protein 1- alph (MIP-1α) accompanied by increase production of IL-1 in the bones accompanied by in creased osteoclast function [12,13]. Bone marrow transfers in this model demonstrate tha the disease is hematopoietically driven [14]. Genetic crosses demonstrate that it is innat immune system driven as the adaptive immune system is dispensable [14]. The disease i completely blocked in cmo mice that lack an IL-1 receptor [13,15] and the effect of IL-1 i mediated through IL-1β rather than IL-1α [13,15]. While our group demonstrated that th disease can occur independent of the Nlrp3 inflammasome, others propose that the Nlrp inflammasome via caspase-1 and caspase-8 redundancy may play a role in disease. The Sr family kinases are also involved in aberrant signaling that can cause sterile osteomyeliti in both the cmo mouse which is dependent on function SYK [16] and in the Ali18 mous which is due to gain of function mutations in FGR [17].

3.2. Majeed Syndrome as an Inflammasomopathy

There are several IL-1 mediated human autoinflammatory disorders that are caused b genetic mutations in inflammasome components [115–118]. Given that Majeed syndrom is also IL-1 driven, investigators set out to determine if Majeed syndrome is an inflam masomopathy. Inflammasomes are intracellular macromolecular protein complexes that assembles in response to numerous stimuli (including various pathogens, uric acid crystals, certain lipids, and signal cellular stress molecules such as ATP) ultimately leading to IL-1 production and release. The NLRP3 inflammasome in macrophages requires 2 signals for activation and assembly including a priming step such as LPS binding to Toll-like receptors and a second signal such as ATP binding to the purinergic P2X7 receptor leading to efflux of intracellular K+ and subsequent inflammasome assembly. Given the role of IL-1 in Majeed syndrome and its phenotypic characteristics as an autoinflammatory disorder. Lorden et al. performed a series of experiments that established the important role of the NLRP3 inflammasome in the excess IL-1β production that occurs in the absence of LIPIN2. Using both human and murine in vitro systems, they demonstrated that there is enhanced production of IL-1β by Lipin2 deficient macrophages [65]. Further, that Lipin2 regulates MAPK activation, inhibits activation and sensitization of the purinergic P2X7 receptor, inhibits inflammasome assembly, and controls caspase-1 activation [65]. They showed that intracellular cholesterol levels were low in bone marrow derived macrophages from Lipin2 deficient mice and when Lipin2 is silenced in RAW264.7 cells, and further, that low intracellular cholesterol levels affected P2X7R function. Normalization of intracellular cholesterol in Lipin2 deficient cells normalized P2X7R function and reversed inflammasome overactivation. Lastly, they confirmed that LIPIN2 restrains inflammasome assembly and activation in vivo using *Lpin2* deficient mice [65]. Collectively, this provides support for Majeed syndrome being an NLRP3 inflammasomopathy. Further studies are needed to fully understand the interplay of cellular lipid alterations and immune function.

Most recently macrophage polarization has been implicated in the sterile osteomyelitis that is a prominent feature of Majeed syndrome. Bhuyan et al. report extensive investigations on cells obtained from the first Majeed syndrome patient who resides in the US who was found to be heterozygous for 2 novel *LPIN2* mutations (Table 1, K1) [81]. As expected, monocytes and monocyte-derived M1-like macrophages from patients with Majeed syndrome as well as monocytes from individuals with another genetically driven NLRP3 inflammasomopathy Neonatal Onset Multisystem Inflammatory Disorder [NOMID] had elevated caspase-1 activity and their cells secreted more IL-1β levels when compared to healthy controls [81]. Yet, only the cells from Majeed syndrome patients (versus NOMID or healthy controls) showed increased expression of osteoclastogenic mediators including IL-8, IL-6, TNF, CCL2, MIP1α, MIP-1β, CXCL8/IL-8 and CXCL1 in M2-like macrophages stimulated with LPS [81]. In addition, Majeed cells showed increased osteoclastogenesis in response to RANKL and M-CSF, associated with higher NFATc1 levels, showed enhanced JNK/MAP kinase activation and reduced Src kinase activation. The data on this single patient suggests that Lipin-2 modulates bone homeostasis through alteration of phosphorylation levels of JNK and Src kinases driving differentiation of macrophages to a pro -inflammatory M2- phenotype and by driving osteoclastogenesis. This could help explain why the bone is triggered in this IL-1 mediated disease but not other IL-1 mediated disorders (Figure 2). Further, these proinflammatory cellular changes could be attenuated by IL-1 inhibitors and JNK inhibitors suggesting that the JNK/MAP kinase pathways may be a novel target for treatment [81].

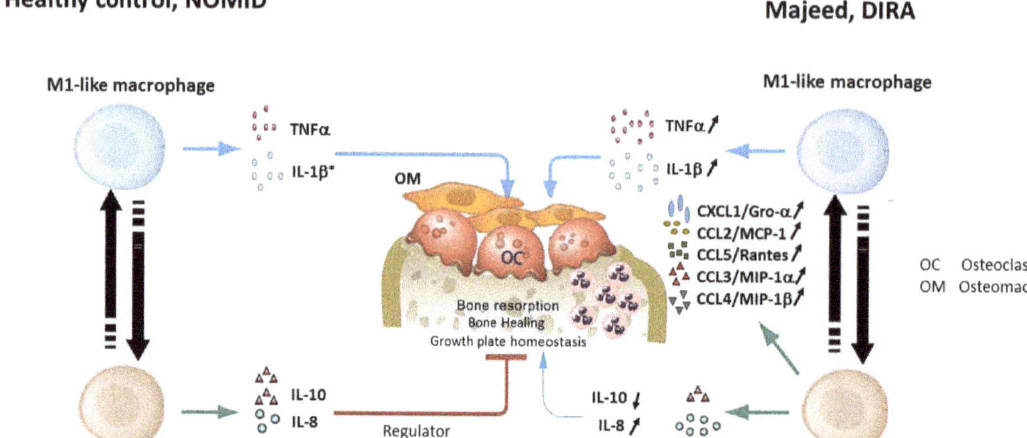

Figure 2. Impaired bone homeostasis as a disease model for the osteomyelitis phenotype in Majeed syndrome. In Majeed syndrome proinflammatory M2-like macrophages produce increased amounts of IL-8 and osteoclastogenic chemokines and lower IL-10 levels thus shifting towards a pro-inflammatory "environment". The increased production of chemokines such as CXCL1 with IL-8 lead to recruitment of neutrophils while MCP-1 and MIP-1α/β recruit monocytes and affect macrophage differentiation. The chemokines released by the inflammatory M2-like macrophages further propagate osteoclastogenesis leading to the bone destruction seen in Majeed syndrome and DIRA. In contrast, the M1-like macrophages in NOMID (an NLRP3 inflammasomopathy) also produce increased IL-1β, yet the M2-like macrophages are not inflammatory and osteoclastogenesis is not increased thus osteomyelitis is not part of their phenotype. (Reprinted from Bhuyan et al. Arthritis and Rheumatology, 2021—Reference [81]).

4. Conclusions

In summary, Majeed syndrome is a rare autosomal recessive disorder due to los of function mutations in *LPIN2*. Low intra cellular cholesterol leads to altered functio of the P2X7R and subsequent K+ efflux and NLRP3 inflammasome activation leading tc enhance production of proinflammatory cytokines including IL-1. There are no clinica trials in Majeed syndrome, so treatment is empiric. Therapeutically blocking the IL-1R or IL-1β has been utilized in 10 patients with Majeed syndrome and all have reporte significant benefit with resolution of the inflammatory bone disease and normalizatio of inflammatory markers. Several patients treated with IL-1 blocking agents have hac improvement in their anemia but none have had repeat bone marrow biopsies to determin if the dyserythropoeisis which is a classic part of the disease is reversed with IL-1 blockade Given that chronic inflammation can result in anemia of chronic disease, it remains unclea if the CDA is improved with IL-1 blockade or if simply that a component of the anemia was due to chronic inflammation which resolves with treatment. Further study is needec to determine best treatment for Majeed syndrome.

Author Contributions: P.J.F. and H.E.-S. performed literature searches. Writing: P.J.F. –original draf preparation; P.J.F. and H.E.-S. contributed to writing, review and editing. All authors have read anc agreed to the published version of the manuscript.

Funding: P.J.F. is funded by the University of Iowa Carver College of Medicine, the Marjorie K Lamb Professorship, funding from the National Institutes of Health/National Institute of Arthritis Musculoskeletal and Skin Disease grant number R01AR05970. HE is funded by the University o Iowa Carver College of Medicine and the Stead Family Department of Pediatrics.

Conflicts of Interest: P.J.F. has served as a consultant for Novartis.

References

1. Masters, S.L.; Simon, A.; Aksentijevich, I.; Kastner, D.L. Horror autoinflammaticus: The molecular pathophysiology of autoinflammatory disease (*). *Annu. Rev. Immunol.* **2009**, *27*, 621–668. [CrossRef]
2. McGonagle, D.; McDermott, M.F. A proposed classification of the immunological diseases. *PLoS Med.* **2006**, *3*, e297. [CrossRef]
3. A candidate gene for familial Mediterranean fever. The French FMF Consortium. *Nat. Genet.* **1997**, *17*, 25–31. [CrossRef] [PubMed]
4. Ancient missense mutations in a new member of the RoRet gene family are likely to cause familial Mediterranean fever. The International FMF Consortium. *Cell* **1997**, *90*, 797–807. [CrossRef]
5. Hoffman, H.M.; Mueller, J.L.; Broide, D.H.; Wanderer, A.A.; Kolodner, R.D. Mutation of a new gene encoding a putative pyrin-like protein causes familial cold autoinflammatory syndrome and Muckle-Wells syndrome. *Nat. Genet.* **2001**, *29*, 301–305. [CrossRef]
6. Oda, H.; Kastner, D.L. Genomics, Biology, and Human Illness: Advances in the Monogenic Autoinflammatory Diseases. *Rheum. Dis. Clin. N. Am.* **2017**, *43*, 327–345. [CrossRef]
7. Cox, A.J.; Zhao, Y.; Ferguson, P.J. Chronic Recurrent Multifocal Osteomyelitis and Related Diseases-Update on Pathogenesis. *Curr. Rheumatol. Rep.* **2017**, *19*, 18. [CrossRef]
8. Lenert, A.; Ferguson, P.J. Comparing children and adults with chronic nonbacterial osteomyelitis. *Curr. Opin. Rheumatol.* **2020**, *32*, 421–426. [CrossRef] [PubMed]
9. Hofmann, S.R.; Kapplusch, F.; Mabert, K.; Hedrich, C.M. The molecular pathophysiology of chronic non-bacterial osteomyelitis (CNO)-a systematic review. *Mol. Cell. Pediatr.* **2017**, *4*, 7. [CrossRef]
10. Jansson, A.; Renner, E.D.; Ramser, J.; Mayer, A.; Haban, M.; Meindl, A.; Grote, V.; Diebold, J.; Jansson, V.; Schneider, K.; et al. Classification of non-bacterial osteitis: Retrospective study of clinical, immunological and genetic aspects in 89 patients. *Rheumatology* **2007**, *46*, 154–160. [CrossRef]
11. Wipff, J.; Costantino, F.; Lemelle, I.; Pajot, C.; Duquesne, A.; Lorrot, M.; Faye, A.; Bader-Meunier, B.; Brochard, K.; Despert, V.; et al. A large national cohort of French patients with chronic recurrent multifocal osteitis. *Arthritis Rheumatol.* **2015**, *67*, 1128–1137. [CrossRef]
12. Ferguson, P.J.; Sandu, M. Current understanding of the pathogenesis and management of chronic recurrent multifocal osteomyelitis. *Curr. Rheumatol. Rep.* **2012**, *14*, 130–141. [CrossRef]
13. Hofmann, S.R.; Kapplusch, F.; Girschick, H.J.; Morbach, H.; Pablik, J.; Ferguson, P.J.; Hedrich, C.M. Chronic Recurrent Multifocal Osteomyelitis (CRMO): Presentation, Pathogenesis, and Treatment. *Curr. Osteoporos. Rep.* **2017**, *15*, 542–554. [CrossRef]
14. King, S.M.; Laxer, R.M.; Manson, D.; Gold, R. Chronic recurrent multifocal osteomyelitis: A noninfectious inflammatory process. *Pediatr. Infect. Dis. J.* **1987**, *6*, 907–911. [CrossRef]
15. Giedion, A.; Holthusen, W.; Masel, L.F.; Vischer, D. Subacute and chronic "symmetrical" osteomyelitis. *Ann. Radiol.* **1972**, *15*, 329–342.
16. Girschick, H.J.; Raab, P.; Surbaum, S.; Trusen, A.; Kirschner, S.; Schneider, P.; Papadopoulos, T.; Muller-Hermelink, H.K.; Lipsky, P.E. Chronic non-bacterial osteomyelitis in children. *Ann. Rheum. Dis.* **2005**, *64*, 279–285. [CrossRef]
17. Zhao, Y.; Ferguson, P.J. Chronic non-bacterial osteomyelitis and autoinflammatory bone diseases. *Clin. Immunol.* **2020**, *216*, 108458. [CrossRef]
18. Chamot, A.M.; Benhamou, C.L.; Kahn, M.F.; Beraneck, L.; Kaplan, G.; Prost, A. Acne-pustulosis-hyperostosis-osteitis syndrome. Results of a national survey. 85 cases. *Rev. Rhum. Mal. Osteoartic* **1987**, *54*, 187–196.
19. Cianci, F.; Zoli, A.; Gremese, E.; Ferraccioli, G. Clinical heterogeneity of SAPHO syndrome: Challenging diagnose and treatment. *Clin. Rheumatol.* **2017**, *36*, 2151–2158. [CrossRef]
20. Wu, N.; Shao, Y.; Huo, J.; Zhang, Y.; Cao, Y.; Jing, H.; Zhang, F.; Chenyang, Y.; Yanying, Y.; Li, C.; et al. Clinical characteristics of pediatric synovitis, acne, pustulosis, hyperostosis, and osteitis (SAPHO) syndrome: The first Chinese case series from a single center. *Clin. Rheumatol.* **2020**. [CrossRef]
21. Hayem, G.; Bouchaud-Chabot, A.; Benali, K.; Roux, S.; Palazzo, E.; Silbermann-Hoffman, O.; Kahn, M.F.; Meyer, O. SAPHO syndrome: A long-term follow-up study of 120 cases. *Semin Arthritis Rheum.* **1999**, *29*, 159–171. [CrossRef]
22. Ferguson, P.J.; El-Shanti, H.I. Autoinflammatory bone disorders. *Curr. Opin. Rheumatol.* **2007**, *19*, 492–498. [CrossRef] [PubMed]
23. Ferguson, P.J.; Brown, D.R.; Sundel, R.; Killian, J.; El-Shanti, H.I. Chronic inflammatory disorders in first and second degree relatives of children with chronic recurrent multifocal osteomyelitis. *Arthritis Rheum.* **2005**, *52*, S301–S302.
24. Paller, A.S.; Pachman, L.; Rich, K.; Esterly, N.B.; Gonzalez-Crussi, F. Pustulosis palmaris et plantaris: Its association with chronic recurrent multifocal osteomyelitis. *J. Am. Acad. Dermatol.* **1985**, *12*, 927–930. [CrossRef]
25. Laxer, R.M.; Shore, A.D.; Manson, D.; King, S.; Silverman, E.D.; Wilmot, D.M. Chronic recurrent multifocal osteomyelitis and psoriasis–a report of a new association and review of related disorders. *Semin Arthritis Rheum.* **1988**, *17*, 260–270. [CrossRef]
26. Bergdahl, K.; Bjorksten, B.; Gustavson, K.H.; Liden, S.; Probst, F. Pustulosis palmoplantaris and its relation to chronic recurrent multifocal osteomyelitis. *Dermatologica* **1979**, *159*, 37–45. [CrossRef]
27. Bognar, M.; Blake, W.; Agudelo, C. Chronic recurrent multifocal osteomyelitis associated with Crohn's disease. *Am. J. Med. Sci.* **1998**, *315*, 133–135.
28. Ben Becher, S.; Essaddam, H.; Nahali, N.; Ben Hamadi, F.; Mouelhi, M.H.; Hammou, A.; Hadj Romdhane, L.; Ben Cheikh, M.; Boudhina, T.; Dargouth, M. Recurrent multifocal periostosis in children. Report of a familial form. *Ann. Pediatr.* **1991**, *38*, 345–349.

29. Festen, J.J.; Kuipers, F.C.; Schaars, A.H. Multifocal recurrent periostitis responsive to colchicine. *Scand. J. Rheumatol.* **1985**, 1 8–14. [CrossRef] [PubMed]
30. Golla, A.; Jansson, A.; Ramser, J.; Hellebrand, H.; Zahn, R.; Meitinger, T.; Belohradsky, B.H.; Meindl, A. Chronic recurrer multifocal osteomyelitis (CRMO): Evidence for a susceptibility gene located on chromosome 18q21.3–18q22. *Eur. J. Hum. Gene* **2002**, *10*, 217–221. [CrossRef]
31. Cox, A.J.; Ferguson, P.J. Update on the genetics of nonbacterial osteomyelitis in humans. *Curr. Opin. Rheumatol.* **2018**, *30*, 521–52 [CrossRef]
32. El-Shanti, H.I.; Ferguson, P.J. Chronic recurrent multifocal osteomyelitis: A concise review and genetic update. *Clin. Orthop. Rela Res.* **2007**, *462*, 11–19. [CrossRef]
33. Majeed, H.A.; Kalaawi, M.; Mohanty, D.; Teebi, A.S.; Tunjekar, M.F.; al-Gharbawy, F.; Majeed, S.A.; al-Gazzar, A.H. Congenit dyserythropoietic anemia and chronic recurrent multifocal osteomyelitis in three related children and the association with Swee syndrome in two siblings. *J. Pediatr.* **1989**, *115*, 730–734. [CrossRef]
34. Aksentijevich, I.; Masters, S.L.; Ferguson, P.J.; Dancey, P.; Frenkel, J.; van Royen-Kerkhoff, A.; Laxer, R.; Tedgard, U.; Cowen, E.W Pham, T.-H.; et al. An autoinflammatory disease with deficiency of the interleukin-1-receptor antagonist. *N. Engl. J. Med.* **200** *360*, 2426–2437. [CrossRef]
35. Reddy, S.; Jia, S.; Geoffrey, R.; Lorier, R.; Suchi, M.; Broeckel, U.; Hessner, M.J.; Verbsky, J. An autoinflammatory disease due t homozygous deletion of the IL1RN locus. *N. Engl. J. Med.* **2009**, *360*, 2438–2444. [CrossRef]
36. Reiff, A.; Bassuk, A.G.; Church, J.A.; Campbell, E.; Bing, X.; Ferguson, P.J. Exome sequencing reveals RAG1 mutations in child with autoimmunity and sterile chronic multifocal osteomyelitis evolving into disseminated granulomatous disease. *J. Cli Immunol.* **2013**, *33*, 1289–1292. [CrossRef] [PubMed]
37. Abe, K.; Cox, A.; Takamatsu, N.; Velez, G.; Laxer, R.M.; Tse, S.M.L.; Mahajan, V.B.; Bassuk, A.G.; Fuchs, H.; Ferguson, P.J.; et a Gain-of-function mutations in a member of the Src family kinases cause autoinflammatory bone disease in mice and human *Proc. Natl. Acad. Sci. USA* **2019**, *116*, 11872–11877. [CrossRef]
38. Abe, K.; Fuchs, H.; Lisse, T.; Hans, W.; Hrabe de Angelis, M. New ENU-induced semidominant mutation, Ali18, cause inflammatory arthritis, dermatitis, and osteoporosis in the mouse. *Mamm. Genome Off. J. Int. Mamm. Genome Soc.* **2006**, 1 915–926. [CrossRef]
39. Hofmann, S.R.; Bottger, F.; Range, U.; Luck, C.; Morbach, H.; Girschick, H.J.; Suttorp, M.; Hedrich, C.M. Serum Interleukin-6 an CCL11/Eotaxin May Be Suitable Biomarkers for the Diagnosis of Chronic Nonbacterial Osteomyelitis. *Front. Pediatr.* **2017**, 256–266. [CrossRef]
40. Hofmann, S.R.; Kubasch, A.S.; Ioannidis, C.; Rosen-Wolff, A.; Girschick, H.J.; Morbach, H.; Hedrich, C. Altered expression o IL-10 family cytokines in monocytes from CRMO patients result in enhanced IL-1beta expression and release. *Clin. Immuno* **2015**, *161*, 300–307. [CrossRef]
41. Hofmann, S.R.; Kubasch, A.S.; Range, U.; Laass, M.W.; Morbach, H.; Girschick, H.J.; Hedrich, C.M. Serum biomarkers for th diagnosis and monitoring of chronic recurrent multifocal osteomyelitis (CRMO). *Rheumatol. Int.* **2016**, *36*, 769–779. [CrossRef]
42. Hofmann, S.R.; Morbach, H.; Schwarz, T.; Rosen-Wolff, A.; Girschick, H.J.; Hedrich, C.M. Attenuated TLR4/MAPK signaling i monocytes from patients with CRMO results in impaired IL-10 expression. *Clin. Immunol.* **2012**, *145*, 69–76. [CrossRef]
43. Scianaro, R.; Insalaco, A.; Bracci Laudiero, L.; De Vito, R.; Pezzullo, M.; Teti, A.; De Benedetti, F.; Prencipe, G. Deregulation of th IL-1beta axis in chronic recurrent multifocal osteomyelitis. *Pediatr. Rheumatol. Online J.* **2014**, *12*, 30. [CrossRef]
44. Hofmann, S.R.; Schwarz, T.; Moller, J.C.; Morbach, H.; Schnabel, A.; Rosen-Wolff, A.; Girschick, H.; Hedrich, C.M. Chroni non-bacterial osteomyelitis is associated with impaired Sp1 signaling, reduced IL10 promoter phosphorylation, and reduce myeloid IL-10 expression. *Clin. Immunol.* **2011**, *141*, 317–327. [CrossRef]
45. Harrus, S.; Waner, T.; Aizenberg Safra, N.; Mosenco, A.; Radoshitsky, M.; Bark, H. Development of hypertrophic osteodystroph and antibody response in a litter of vaccinated Weimaraner puppies. *J. Small Anim. Pract.* **2002**, *43*, 27–31. [CrossRef] [PubMed
46. Safra, N.; Hitchens, P.L.; Maverakis, E.; Mitra, A.; Korff, C.; Johnson, E.; Kol, A.; Bannasch, M.J.; Pedersen, N.C.; Bannasch, D. Serum levels of innate immunity cytokines are elevated in dogs with metaphyseal osteopathy (hypertrophic osteodytroph during active disease and remission. *Vet. Immunol. Immunopathol.* **2016**, *179*, 32–35. [CrossRef]
47. Safra, N.; Johnson, E.G.; Lit, L.; Foreman, O.; Wolf, Z.T.; Aguilar, M.; Karmi, N.; Finno, C.J.; Bannasch, D.L. Clinical manifestation response to treatment, and clinical outcome for Weimaraners with hypertrophic osteodystrophy: 53 cases (2009-2011). *J. Am. Ve Med. Assoc.* **2013**, *242*, 1260–1266. [CrossRef]
48. Backues, K.A.; Hoover, J.P.; Bahr, R.J.; Confer, A.W.; Chalman, J.A.; Larry, M.L. Multifocal pyogranulomatous osteomyeliti resembling chronic recurrent multifocal osteomyelitis in a lemur. *J. Am. Vet. Med. Assoc.* **2001**, *218*, 250–253. [CrossRef] [PubMed
49. Liao, H.J.; Chyuan, I.T.; Wu, C.S.; Lin, S.W.; Chen, K.H.; Tsai, H.F.; Hsu, P.-N. Increased neutrophil infiltration, IL-1 productio and a SAPHO syndrome-like phenotype in PSTPIP2-deficient mice. *Rheumatology* **2015**, *54*, 1317–1326. [CrossRef]
50. Byrd, L.; Grossmann, M.; Potter, M.; Shen-Ong, G.L. Chronic multifocal osteomyelitis, a new recessive mutation on chromosom 18 of the mouse. *Genomics* **1991**, *11*, 794–798. [CrossRef]
51. Yao, Y.; Cai, X.; Yu, H.; Xu, Q.; Li, X.; Yang, Y.; Meng, X.; Huang, C.; Li, J. PSTPIP2 attenuates joint damage and suppresse inflammation in adjuvant-induced arthritis. *Eur. J. Pharmacol.* **2019**, *859*, 172558. [CrossRef]

2. Ferguson, P.J.; Chen, S.; Tayeh, M.K.; Ochoa, L.; Leal, S.M.; Pelet, A.; Munnich, A.; Lyonnet, S.; Majeed, H.A.; El-Shanti, E. Homozygous mutations in LPIN2 are responsible for the syndrome of chronic recurrent multifocal osteomyelitis and congenital dyserythropoietic anaemia (Majeed syndrome). *J. Med. Genet.* **2005**, *42*, 551–557. [CrossRef]
3. Majeed, H.A.; Al-Tarawna, M.; El-Shanti, H.; Kamel, B.; Al-Khalaileh, F. The syndrome of chronic recurrent multifocal osteomyelitis and congenital dyserythropoietic anaemia. Report of a new family and a review. *Eur. J. Pediatr.* **2001**, *160*, 705–710. [CrossRef]
4. Majeed, H.A.; El-Shanti, H.; Al-Rimawi, H.; Al-Masri, N. On mice and men: An autosomal recessive syndrome of chronic recurrent multifocal osteomyelitis and congenital dyserythropoietic anemia. *J. Pediatr.* **2000**, *137*, 441–442. [CrossRef]
5. Ferguson, P.J.; Bing, X.; Vasef, M.A.; Ochoa, L.A.; Mahgoub, A.; Waldschmidt, T.J.; Tygrett, L.T.; Schlueter, A.J.; El-shanti, E. A missense mutation in pstpip2 is associated with the murine autoinflammatory disorder chronic multifocal osteomyelitis. *Bone* **2006**, *38*, 41–47. [CrossRef]
6. Grosse, J.; Chitu, V.; Marquardt, A.; Hanke, P.; Schmittwolf, C.; Zeitlmann, L.; Schropp, P.; Barth, B.; Yu, P.; Paffenholz, R.; et al. Mutation of mouse Mayp/Pstpip2 causes a macrophage autoinflammatory disease. *Blood* **2006**, *107*, 3350–3358. [CrossRef]
7. Abe, K.; Klaften, M.; Narita, A.; Kimura, T.; Imai, K.; Kimura, M.; Rubio-Aliaga, I.; Wagner, S.; Jakob, T.; Hrabe de Angelis, M. Genome-wide search for genes that modulate inflammatory arthritis caused by Ali18 mutation in mice. *Mamm. Genome Off. J. Int. Mamm. Genome Soc.* **2009**, *20*, 152–161. [CrossRef]
8. Abe, K.; Wechs, S.; Kalaydjiev, S.; Franz, T.J.; Busch, D.H.; Fuchs, H.; Soewarto, D.; Behrendt, H.; Wagner, S.; Jakob, T.; et al. Novel lymphocyte-independent mechanisms to initiate inflammatory arthritis via bone marrow-derived cells of Ali18 mutant mice. *Rheumatology* **2008**, *47*, 292–300. [CrossRef]
9. Cox, A.J.; Darbro, B.W.; Laxer, R.M.; Velez, G.; Bing, X.; Finer, A.L.; Erives, A.; Mahajan, V.B.; Bassuk, A.G.; Ferguson, P.J. Recessive coding and regulatory mutations in FBLIM1 underlie the pathogenesis of chronic recurrent multifocal osteomyelitis (CRMO). *PLoS ONE* **2017**, *12*, e0169687. [CrossRef]
0. Cassel, S.L.; Janczy, J.R.; Bing, X.; Wilson, S.P.; Olivier, A.K.; Otero, J.E.; Iwakura, Y.; Shayakhmetov, D.M.; Bassuk, A.G.; Abu-Amer, Y.; et al. Inflammasome-independent IL-1beta mediates autoinflammatory disease in Pstpip2-deficient mice. *Proc. Natl. Acad. Sci. USA* **2014**, *111*, 1072–1077. [CrossRef] [PubMed]
1. Lukens, J.R.; Gross, J.M.; Calabrese, C.; Iwakura, Y.; Lamkanfi, M.; Vogel, P.; Kanneganti, T.-D. Critical role for inflammasome-independent IL-1beta production in osteomyelitis. *Proc. Natl. Acad. Sci. USA* **2014**, *111*, 1066–1071. [CrossRef]
2. Young, S.; Sharma, N.; Lee, J.H.; Chitu, V.; Neumeister, V.; Sohr, E.; Stanley, E.R.; Hedrich, C.M.; Craig, A.W.B. Mast cells enhance sterile inflammation in chronic nonbacterial osteomyelitis. *Dis. Models Mech.* **2019**, *12*. [CrossRef] [PubMed]
3. Drobek, A.; Kralova, J.; Skopcova, T.; Kucova, M.; Novak, P.; Angelisova, P.; Otahal, P.; Alberich-Jorda, M.; Brdicka, T. PSTPIP2, a Protein Associated with Autoinflammatory Disease, Interacts with Inhibitory Enzymes SHIP1 and Csk. *J. Immunol.* **2015**, *195*, 3416–3426. [CrossRef]
4. Herlin, T.; Fiirgaard, B.; Bjerre, M.; Kerndrup, G.; Hasle, H.; Bing, X.; Ferguson, P.J. Efficacy of anti-IL-1 treatment in Majeed syndrome. *Ann. Rheum. Dis.* **2013**, *72*, 410–413. [CrossRef]
5. Lorden, G.; Sanjuan-Garcia, I.; de Pablo, N.; Meana, C.; Alvarez-Miguel, I.; Perez-Garcia, M.T.; Pelegrin, P.; Balsinde, J.; Balboa, M.A. Lipin-2 regulates NLRP3 inflammasome by affecting P2X7 receptor activation. *J. Exp. Med.* **2017**, *214*, 511–528. [CrossRef]
6. Valdearcos, M.; Esquinas, E.; Meana, C.; Peña, L.; Gil-de-Gómez, L.; Balsinde, J.; Balboa, M.A. Lipin-2 reduces proinflammatory signaling induced by saturated fatty acids in macrophages. *J. Biol. Chem.* **2012**, *287*, 10894–10904. [CrossRef]
7. Jesus, A.A.; Osman, M.; Silva, C.A.; Kim, P.W.; Pham, T.H.; Gadina, M.; Yang, B.; Bertola, D.R.; Carneiro-Sampaio, M.; Ferguson, P.J.; et al. A novel mutation of IL1RN in the deficiency of interleukin-1 receptor antagonist syndrome: Description of two unrelated cases from Brazil. *Arthritis Rheum.* **2011**, *63*, 4007–4017. [CrossRef] [PubMed]
8. Kutukculer, N.; Puel, A.; Eren Akarcan, S.; Moriya, K.; Edeer Karaca, N.; Migaud, M.; Cassanova, J.-L.; Aksu, G. Deficiency of Interleukin-1 Receptor Antagonist: A Case with Late Onset Severe Inflammatory Arthritis, Nail Psoriasis with Onychomycosis and Well Responsive to Adalimumab Therapy. *Case Rep. Immunol.* **2019**, *2019*, 1902817. [CrossRef] [PubMed]
9. Mendonca, L.O.; Grossi, A.; Caroli, F.; de Oliveira, R.A.; Kalil, J.; Castro, F.F.M.; Pontillo, A.; Ceccherini, I.; Barros, M.A.M.T.; Gattorno, M. A case report of a novel compound heterozygous mutation in a Brazilian patient with deficiency of Interleukin-1 receptor antagonist (DIRA). *Pediatr. Rheumatol. Online J.* **2020**, *18*, 67. [CrossRef]
0. Mendonca, L.O.; Malle, L.; Donovan, F.X.; Chandrasekharappa, S.C.; Montealegre Sanchez, G.A.; Garg, M.; Tedgard, U.; Castells, M.; Saini, S.S.; Dutta, S.; et al. Deficiency of Interleukin-1 Receptor Antagonist (DIRA): Report of the First Indian Patient and a Novel Deletion Affecting IL1RN. *J. Clin. Immunol.* **2017**, *37*, 445–451. [CrossRef] [PubMed]
1. Minkis, K.; Aksentijevich, I.; Goldbach-Mansky, R.; Magro, C.; Scott, R.; Davis, J.G.; Sardana, N.; Herzog, R. Interleukin 1 receptor antagonist deficiency presenting as infantile pustulosis mimicking infantile pustular psoriasis. *Arch. Dermatol.* **2012**, *148*, 747–752. [CrossRef] [PubMed]
2. Sakran, W.; Shalev, S.A.; Sakran, W.; Shalev, S.A.; El-Shanti, H.; Uziel, Y.; Uziel, Y. Chronic recurrent multifocal osteomyelitis and deficiency of interleukin-1-receptor antagonist. *Pediatr. Infect. Dis. J.* **2013**, *32*, 94. [CrossRef]
3. Schnellbacher, C.; Ciocca, G.; Menendez, R.; Aksentijevich, I.; Goldbach-Mansky, R.; Duarte, A.M.; Rivas-Chacon, R. Deficiency of interleukin-1 receptor antagonist responsive to anakinra. *Pediatr. Dermatol.* **2013**, *30*, 758–760. [CrossRef] [PubMed]
4. Stenerson, M.; Dufendach, K.; Aksentijevich, I.; Brady, J.; Austin, J.; Reed, A.M. The first reported case of compound heterozygous IL1RN mutations causing deficiency of the interleukin-1 receptor antagonist. *Arthritis Rheum.* **2011**, *63*, 4018–4022. [CrossRef]

75. Ulusoy, E.; Karaca, N.E.; El-Shanti, H.; Kilicoglu, E.; Aksu, G.; Kutukculer, N. Interleukin-1 receptor antagonist deficiency with novel mutation; late onset and successful treatment with canakinumab: A case report. *J. Med. Case Rep.* **2015**, *9*, 145. [CrossRef] [PubMed]
76. Ziaee, V.; Youssefian, L.; Faghankhani, M.; Jazayeri, A.; Saeidian, A.H.; Vahidnezhad, H.; Uitto, J. Homozygous IL1RN Mutation in Siblings with Deficiency of Interleukin-1 Receptor Antagonist (DIRA). *J. Clin. Immunol.* **2020**, *40*, 637–642. [CrossRef]
77. Kuemmerle-Deschner, J.B.; Welzel, T.; Hoertnagel, K.; Tsiflikas, I.; Hospach, A.; Liu, X.; Schlipf, S.; HAnsmann, S.; Samba, S.D. Griesinger, A. New variant in the IL1RN-gene (DIRA) associated with late-onset, CRMO-like presentation. *Rheumatology* **2020**, *5* 3259–3263. [CrossRef]
78. Sozeri, B.; Gerceker-Turk, B.; Yildiz-Atikan, B.; Mir, S.; Berdeli, A. A novel mutation of interleukin-1 receptor antagonist (IL1RN in a DIRA patient from Turkey: Diagnosis and treatment. *Turk. J. Pediatr.* **2018**, *60*, 588–592. [CrossRef] [PubMed]
79. Al Mosawi, Z.; Madan, W.; Al Moosawi, B.; Al-Wadaei, S.; Naser, H.; Ali, F. Dramatic Response of Familial Majeed Syndrome t Interleukin-1 Antagonist Therapy: Case report. *Arch. Rheumatol.* **2019**, *34*, 352–356. [CrossRef]
80. Al-Mosawi, Z.S.; Al-Saad, K.K.; Ijadi-Maghsoodi, R.; El-Shanti, H.I.; Ferguson, P.J. A splice site mutation confirms the role c LPIN2 in Majeed syndrome. *Arthritis Rheum.* **2007**, *56*, 960–964. [CrossRef]
81. Bhuyan, F.; de Jesus, A.A.; Mitchell, J.; Leikina, E.; VanTries, R.; Herzog, R.; Onel, K.B.; Oler, A.; Montealegre Sanchez, G.A. Johnson, K.A.; et al. Novel Majeed syndrome causing LPIN2 mutations link bone inflammation to inflammatory M2 macrophage and accelerated osteoclastogenesis. *Arthritis Rheumatol.* **2020**. [CrossRef]
82. Liu, J.; Hu, X.Y.; Zhao, Z.P.; Guo, R.L.; Guo, J.; Li, W.; Hao, C.-J.; Xu, B.-P. Compound heterozygous LPIN2 pathogenic variant in a patient with Majeed syndrome with recurrent fever and severe neutropenia: Case report. *BMC Med. Genet.* **2019**, *20*, 18 [CrossRef]
83. Pinto-Fernandez, C.; Seoane-Reula, M.E. Efficacy of treatment with IL-1RA in Majeed syndrome. *Allergol. Immunopathol.* **2017**, *4* 99–101. [CrossRef]
84. Rao, A.P.; Gopalakrishna, D.B.; Bing, X.; Ferguson, P.J. Phenotypic Variability in Majeed Syndrome. *J. Rheumatol.* **2016**, *4* 1258–1259. [CrossRef]
85. Roy, N.B.A.; Zaal, A.I.; Hall, G.; Wilkinson, N.; Proven, M.; McGowan, S.; Hipkiss, R.; Buckle, V.; Kavirayani, A.; Babbs, C. Majee syndrome: Description of a novel mutation and therapeutic response to bisphosphonates and IL-1 blockade with anakinr *Rheumatology* **2019**, *59*, 448–451. [CrossRef]
86. Moussa, T.; Bhat, V.; Kini, V.; Fathalla, B.M. Clinical and genetic association, radiological findings and response to biologica therapy in seven children from Qatar with non-bacterial osteomyelitis. *Int. J. Rheum. Dis.* **2016**, *20*, 1286–1296. [CrossRef]
87. Iolascon, A.; Andolfo, I.; Russo, R. Congenital dyserythropoietic anemias. *Blood* **2020**, *136*, 1274–1283. [CrossRef] [PubMed]
88. Edwards, T.C.; Stapleton, F.B.; Bond, M.J.; Barrett, F.F. Sweet's syndrome with multifocal sterile osteomyelitis. *Am. J. Dis. Chil* **1986**, *140*, 817–818. [CrossRef] [PubMed]
89. Bozgeyik, E.; Mercan, R.; Arslan, A.; Tozkir, H. Next-generation screening of a panel of genes associated with periodic feve syndromes in patients with Familial Mediterranean Fever and their clinical characteristics. *Genomics* **2020**, *112*, 2755–276 [CrossRef] [PubMed]
90. Milledge, J.; Shaw, P.J.; Mansour, A.; Williamson, S.; Bennetts, B.; Roscioli, T.; Curtin, J.; Christodoulou, J. Allogeneic bone marrov transplantation: Cure for familial Mediterranean fever. *Blood* **2002**, *100*, 774–777. [CrossRef] [PubMed]
91. Nurre, L.D.; Rabalais, G.P.; Callen, J.P. Neutrophilic dermatosis-associated sterile chronic multifocal osteomyelitis in pediatri patients: Case report and review. *Pediatr. Dermatol.* **1999**, *16*, 214–216. [CrossRef] [PubMed]
92. Zhang, P.; Reue, K. Lipin proteins and glycerolipid metabolism: Roles at the ER membrane and beyond. *Biochim. Biophys. Act Biomembr.* **2017**, *1859*, 1583–1595. [CrossRef] [PubMed]
93. Reue, K. The lipin family: Mutations and metabolism. *Curr. Opin. Lipidol.* **2009**, *20*, 165–170. [CrossRef]
94. Reue, K.; Wang, H. Mammalian lipin phosphatidic acid phosphatases in lipid synthesis and beyond: Metabolic and inflammator disorders. *J. Lipid. Res.* **2019**, *60*, 728–733. [CrossRef] [PubMed]
95. Finck, B.N.; Gropler, M.C.; Chen, Z.; Leone, T.C.; Croce, M.A.; Harris, T.E.; Lawrence, C., Jr.; Kelly, D.P. Lipin 1 is an inducibl amplifier of the hepatic PGC-1alpha/PPARalpha regulatory pathway. *Cell Metab.* **2006**, *4*, 199–210. [CrossRef]
96. Csaki, L.S.; Dwyer, J.R.; Fong, L.G.; Tontonoz, P.; Young, S.G.; Reue, K. Lipins, lipinopathies, and the modulation of cellular lipic storage and signaling. *Prog. Lipid. Res.* **2013**, *52*, 305–316. [CrossRef] [PubMed]
97. Zhang, P.; Verity, M.A.; Reue, K. Lipin-1 regulates autophagy clearance and intersects with statin drug effects in skeletal muscl *Cell Metab.* **2014**, *20*, 267–279. [CrossRef] [PubMed]
98. Zhang, P.; Reue, K. Lipin-1 flexes its muscle in autophagy. *Cell Cycle* **2014**, *13*, 3789–3790. [CrossRef] [PubMed]
99. Chen, Z.; Gropler, M.C.; Mitra, M.S.; Finck, B.N. Complex interplay between the lipin 1 and the hepatocyte nuclear factor 4 alph (HNF4alpha) pathways to regulate liver lipid metabolism. *PLoS ONE* **2012**, *7*, e51320.
100. Chen, Z.; Gropler, M.C.; Norris, J.; Lawrence, J.C.; Harris, T.E., Jr.; Finck, B.N. Alterations in hepatic metabolism in fld mice revea a role for lipin 1 in regulating VLDL-triacylglyceride secretion. *Arter. Thromb. Vasc. Biol.* **2008**, *28*, 1738–1744. [CrossRef]
101. Donkor, J.; Sariahmetoglu, M.; Dewald, J.; Brindley, D.N.; Reue, K. Three mammalian lipins act as phosphatidate phosphatase with distinct tissue expression patterns. *J. Biol. Chem.* **2007**, *282*, 3450–3457. [CrossRef]
102. Peterfy, M.; Phan, J.; Xu, P.; Reue, K. Lipodystrophy in the fld mouse results from mutation of a new gene encoding a nuclea protein, lipin. *Nat. Genet.* **2001**, *27*, 121–124. [CrossRef] [PubMed]

103. Zeharia, A.; Shaag, A.; Houtkooper, R.H.; Hindi, T.; de Lonlay, P.; Erez, G.; Hubert, L.; Saada, A.; de Keyzer, Y.; Eshel, G.; et al. Mutations in LPIN1 cause recurrent acute myoglobinuria in childhood. *Am. J. Hum. Genet.* **2008**, *83*, 489–494. [CrossRef] [PubMed]
104. Rowe, L.B.; Sweet, H.O.; Gordon, J.I.; Birkenmeier, E.H. The fld mutation maps near to but distinct from the Apob locus on mouse chromosome 12. *Mamm. Genome Off. J. Int. Mamm. Genome Soc.* **1996**, *7*, 555–557. [CrossRef] [PubMed]
105. Langner, C.A.; Birkenmeier, E.H.; Ben-Zeev, O.; Schotz, M.C.; Sweet, H.O.; Davisson, M.T.; Gordon, J.I. The fatty liver dystrophy (fld) mutation. A new mutant mouse with a developmental abnormality in triglyceride metabolism and associated tissue-specific defects in lipoprotein lipase and hepatic lipase activities. *J. Biol. Chem.* **1989**, *264*, 7994–8003. [CrossRef]
106. Nadra, K.; de Preux Charles, A.S.; Medard, J.J.; Hendriks, W.T.; Han, G.S.; Gres, S.; Carman, G.M.; Saulnier-Blache, J.-S.; Verheijen, M.H.G.; Chrast, R. Phosphatidic acid mediates demyelination in Lpin1 mutant mice. *Genes Dev.* **2008**, *22*, 1647–1661. [CrossRef] [PubMed]
107. Jiang, W.; Zhu, J.; Zhuang, X.; Zhang, X.; Luo, T.; Esser, K.A.; Hongmei, R. Lipin1 Regulates Skeletal Muscle Differentiation through Extracellular Signal-regulated Kinase (ERK) Activation and Cyclin D Complex-regulated Cell Cycle Withdrawal. *J. Biol. Chem.* **2015**, *290*, 23646–23655. [CrossRef]
108. Farmer, T.A., Jr.; Hammack, W.J.; Frommeyer, W.B., Jr. Idiopathic recurrent rhabdomyolysis associated with myoglobinuria: Report of a case. *N. Engl. J. Med.* **1961**, *264*, 60–66. [CrossRef]
109. Bowden, D.H.; Fraser, D.; Jackson, S.H.; Walker, N.F. Acute recurrent rhabdomyolysis (paroxysmal myohaemoglobinuria); a report of three cases and a review of the literature. *Medicine* **1956**, *35*, 335–353. [CrossRef] [PubMed]
110. Christensen, T.E.; Saxtrup, O.; Hansen, T.I.; Kristensen, B.H.; Beck, B.L.; Plesner, T.; Krogh, I.M.; Andersen, V.; Standgaard, S. amilial myoglobinuria. A study of muscle and kidney pathophysiology in three brothers. *Dan. Med. Bull.* **1983**, *30*, 112–115. [PubMed]
111. Ramesh, V.; Gardner-Medwin, D. Familial paroxysmal rhabdomyolysis: Management of two cases of the non-exertional type. *Dev. Med. Child Neurol.* **1992**, *34*, 73–79. [CrossRef]
112. Temprano, A.; Sembongi, H.; Han, G.S.; Sebastian, D.; Capellades, J.; Moreno, C.; Guardiola, J.; Wabitsch, M.; Richart, C.; Yanes, O.; et al. edundant roles of the phosphatidate phosphatase family in triacylglycerol synthesis in human adipocytes. *Diabetologia* **2016**, *59*, 1985–1994. [CrossRef] [PubMed]
113. Phan, J.; Peterfy, M.; Reue, K. Lipin expression preceding peroxisome proliferator-activated receptor-gamma is critical for adipogenesis in vivo and in vitro. *J. Biol. Chem.* **2004**, *279*, 29558–29564. [CrossRef]
114. Mitra, M.S.; Chen, Z.; Ren, H.; Harris, T.E.; Chambers, K.T.; Hall, A.M.; Nadra, K.; Klein, S.; Chrast, R.; Su, X.; et al. Mice with an adipocyte-specific lipin 1 separation-of-function allele reveal unexpected roles for phosphatidic acid in metabolic regulation. *Proc. Natl. Acad. Sci. USA* **2013**, *110*, 642–647. [CrossRef]
115. Schweitzer, G.G.; Collier, S.L.; Chen, Z.; Eaton, J.M.; Connolly, A.M.; Bucelli, R.C.; Pestronk, A.; Harris, T.E.; Finck, B.N. Rhabdomyolysis-Associated Mutations in Human LPIN1 Lead to Loss of Phosphatidic Acid Phosphohydrolase Activity. *JIMD Rep.* **2015**, *23*, 113–122.
116. Dwyer, J.R.; Donkor, J.; Zhang, P.; Csaki, L.S.; Vergnes, L.; Lee, J.M.; Dewald, J.; Brindley, D.N.; Atti, E.; Tetradis, S.; et al. Mouse lipin-1 and lipin-2 cooperate to maintain glycerolipid homeostasis in liver and aging cerebellum. *Proc. Natl. Acad. Sci. USA* **2012**, *109*, E2486–E2495. [CrossRef]
117. Donkor, J.; Zhang, P.; Wong, S.; O'Loughlin, L.; Dewald, J.; Kok, B.P.; Brindley, D.N.; Reue, K. A conserved serine residue is required for the phosphatidate phosphatase activity but not the transcriptional coactivator functions of lipin-1 and lipin-2. *J. Biol. Chem.* **2009**, *284*, 29968–29978. [CrossRef] [PubMed]
118. Jesus, A.A.; Goldbach-Mansky, R. IL-1 blockade in autoinflammatory syndromes. *Ann. Rev. Med.* **2014**, *65*, 223–244. [CrossRef] [PubMed]

Review

Sphingolipid Metabolism in Glioblastoma and Metastatic Brain Tumors: A Review of Sphingomyelinases and Sphingosine-1-Phosphate

Cyntanna C. Hawkins [1], Tomader Ali [2], Sasanka Ramanadham [1,3] and Anita B. Hjelmeland [1,*]

[1] Department of Cell, Developmental, and Integrative Biology, University of Birmingham at Alabama, Birmingham, AL 35233, USA; cyntanna@uab.edu (C.C.H.); sramvem@uab.edu (S.R.)
[2] Research Department, Imperial College London Diabetes Centre, Abu Dhabi P.O. Box 48338, UAE; tfali@icldc.ae
[3] Comprehensive Diabetes Center, University of Birmingham at Alabama, Birmingham, AL 35294, USA
* Correspondence: hjelmea@uab.edu

Received: 31 August 2020; Accepted: 20 September 2020; Published: 23 September 2020

Abstract: Glioblastoma (GBM) is a primary malignant brain tumor with a dismal prognosis, partially due to our inability to completely remove and kill all GBM cells. Rapid tumor recurrence contributes to a median survival of only 15 months with the current standard of care which includes maximal surgical resection, radiation, and temozolomide (TMZ), a blood–brain barrier (BBB) penetrant chemotherapy. Radiation and TMZ cause sphingomyelinases (SMase) to hydrolyze sphingomyelins to generate ceramides, which induce apoptosis. However, cells can evade apoptosis by converting ceramides to sphingosine-1-phosphate (S1P). S1P has been implicated in a wide range of cancers including GBM. Upregulation of S1P has been linked to the proliferation and invasion of GBM and other cancers that display a propensity for brain metastasis. To mediate their biological effects, SMases and S1P modulate signaling via phospholipase C (PLC) and phospholipase D (PLD). In addition, both SMase and S1P may alter the integrity of the BBB leading to infiltration of tumor-promoting immune populations. SMase activity has been associated with tumor evasion of the immune system, while S1P creates a gradient for trafficking of innate and adaptive immune cells. This review will explore the role of sphingolipid metabolism and pharmacological interventions in GBM and metastatic brain tumors with a focus on SMase and S1P.

Keywords: glioblastoma; sphingolipid; sphingosine-1-phosphate; sphingomyelinase; sphingomyelin; metastasis

---

1. Introduction

In recent years, studies of the role of sphingolipid metabolism have become an integral part of cancer research. Sphingomyelins (SMs), predominant sphingophospholipids in the outer leaflet of cell membranes, and their hydrolysis by sphingomyelinases (SMase) are essential to the efficacy of chemo- and radiotherapy [1–4]. SMases are distinguished according to their subcellular location and optimal pH for activity: SMases are named based on the pH at which they are active, with acid SMase in the lysosome, neutral SMase at the plasma membrane, and alkaline SMase in the endoplasmic reticulum [5,6]. Activation of SMase results in the production of phosphorylcholine and a ceramide, the central lipid in sphingolipid metabolism [7]. Ceramide can also be produced by the salvage pathway (Figure 1). The salvage pathway and de novo synthesis involve ceramide synthases and serine palmitoyl transferase (SPT), respectively (See Gault et al. for a more detailed review of de novo synthesis) [8]. Ceramide has been linked to decreased cell motility and angiogenesis but is most well-characterized as a pro-apoptotic signal [9–11]. However, cells can escape apoptosis if ceramide is hydrolyzed by ceramidases (CDases) to sphingosine [7]. Like the SMases, the CDases are also distinguished by their subcellular location and optimum pH for activity: acid CDase, neutral CDase, and alkaline CDase [12–14]. The CDases catalyze cleavage of the fatty acid from ceramide to produce sphingosine, which can subsequently be phosphorylated by sphingosine kinases (SK1 and SK2) to generate sphingosine-1-phosphate (S1P) [8,15]. S1P is linked to increased cellular proliferation, angiogenesis, and motility [10,16–18]. The levels of ceramides and S1P can be modulated based on cellular stress through pathways described as a series of "drains" and "faucets" [19]. This has led to the concept of the sphingolipid rheostat, which illustrates the consequence of shifting the balance between ceramide (pro-apoptotic) and S1P (pro-proliferative) on cell survival [20,21].

In cancer, the sphingolipid rheostat tilts toward S1P, promoting cell signaling that increases survival, proliferation, and migration [20,22]. S1P signals through five G-protein coupled receptors designated S1P receptor 1-5 (S1PR1-5) by autocrine and paracrine mechanisms [23–25]. Initially referred to as endothelial differentiation genes (EDG), recognition of their ability to bind S1P prompted a name change to S1PRs (S1PR1/Edg-1, S1PR2/Edg-5, S1PR3/Edg-3, S1PR4/Edg-6, S1PR5/Edg-8) [26–29]. Each receptor can couple to different G-proteins based on their motifs with primary functions through G_i, G_q, and G_{12}. Both G_i and G_{12} promote downstream effects through phospholipase C (PLC) and phospholipase D (PLD) [30–32]. PLC cleaves the proximal phosphodiester bond of glycerophospholipids to produce diacylglycerols and a phosphorylated headgroup, while PLD cleaves the distal phosphodiester bond to produce the headgroup and phosphatidic acid [33]. PLC can signal through protein kinase C to cause the intracellular release of Ca^{2+} and promotion of cell proliferation [34]. Additionally, phosphatidic acid produced by PLD can attract and bind SK1 at the plasma membrane where S1P is produced to fuel cell growth and survival [35].

The sphingolipid rheostat plays an important role in the progression of glioblastoma (GBM)—the most common primary malignant brain tumor. Patients with GBM have a median survival of only 15 months with standard of care, which includes maximum surgical resection, radiation, and chemotherapy with temozolomide (TMZ), and adjuvant TMZ [39,40]. Unfortunately, GBM recurs in almost all cases due to both the inability to remove tumor cells invading normal brain and development of therapeutic resistance. Contributing to the latter are a number of factors, including multiple facets of intra-tumoral heterogeneity, some of which involve a subset of cells called brain tumor-initiating cells (BTICs) [41,42]. The BTICs are a less-differentiated, neural stem cell-like population that preferentially survive chemo- and radiotherapy, propagate tumors in animal models, and are suggested to repopulate the tumor after therapeutic intervention [41–44]. Alterations in sphingolipid metabolism have recently been implicated in both differentiated GBM cells and the less differentiated BTICs [10,45]. Samples from GBM patients have shown an increase in S1P concurrent with a decrease in ceramides, as compared with normal brain, indicating tilting of the sphingolipid rheostat toward a pro-tumor phenotype [10]. Further, GBM cells grown in cell culture under conditions, which enrich for the BTIC population had even higher S1P compared to their differentiated GBM counterparts [45].

While brain tumors can arise from cells within the brain as in GBM, metastases of other cancers to the brain (including breast, colon, lung, and skin cancers) are more common causes of tumor development in the brain [46–49]. Altered sphingolipid metabolism is also evident in these cancers with a propensity for metastasis to the brain [50–52]. Metastasis to distant organs is characteristic of advanced stages of disease, with 10–30% of all cancer patients exhibiting brain metastasis [53]. Once brain metastasis is established, the disease becomes much more difficult to treat [51,54–56]. Although brain metastases vary in mutational load and immune system alterations from GBM, any advances that benefit GBM may also benefit metastatic cancers. This warrants further research of sphingolipid metabolism in the context of both GBM and metastatic brain tumors [57].

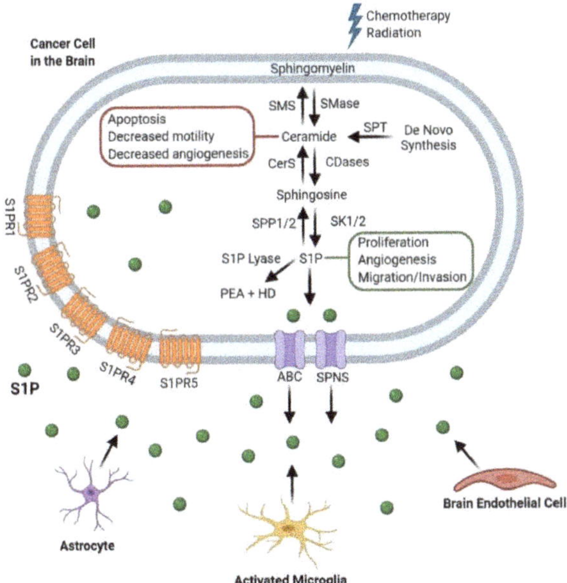

Figure 1. Sphingolipid Metabolism and its role in Cancer Progression. After chemotherapy and radiation, sphingomyelin is broken down into ceramide which has roles in blocking cancer progression. Cancer cells can convert ceramide to sphingosine-1-phosphate (S1P), which is transported out of the cell by either ATP-binding cassette (ABC) or spinster (SPNS) transporters [36,37]. S1P then exerts its pro-tumor effects through both intracellular and extracellular mechanisms. Alternatively, S1P can be degraded by S1P lyase to produce Phosphatidylethanolamine (PEA) and Hexadecenal (HD) [6,38]. These processes also occur in the other cell populations within the brain tumor microenvironment including astrocytes, microglia, and endothelial cells.Sphingomyelin Synthase (SMS); Ceramide synthase (CerS); Sphingosine phosphate phosphatase 1/2 (SPP1/2); Sphingomyelinase (SMase); Ceramidase (CDase); Sphingosine kinase 1/2 (SK1/2); Serine palmitoyltransferase (SPT); Sphingosine-1-phosphate (S1P); Phosphatidylethanolamine (PEA); Hexadecenal (HD); ATP-binding cassette (ABC); Spinster (SPNS).

2. Sphingomyelinases (SMases)

Since sphingomyelinases are imperative to the efficacy of both radio- and chemotherapy, the study of their roles has provided novel information concerning treatment regiments and established new directions for research. Of particular interest is the modulation of cell death in patients resistant to radio- and chemotherapy which is common in both GBM and metastatic cancers, as discussed in earlier.

2.1. Glioblastoma (GBM)

Studies analyzing lung tissue from mice deficient for acid SMase show resistance to apoptosis following radiation and that the p53- and ceramide-induced apoptosis were distinct [58]. Additionally, others have shown that while p53 induction by irradiation can increase ceramides, induction of ceramide generation by other means does not always induce p53 expression in leukemia or fibroblast cell lines [59]. Initial studies showed that acid SMase overexpression sensitized glioma cells to chemotherapies, gemcitabine and doxorubicin [60]. In contrast, subsequent studies found that overexpression of acid SMase increased ceramide levels but failed to sensitize GBM cells to radiation or chemotherapy with TMZ, the current standard of care [61]. To control for differences in p53, GBM cell lines with mutant p53 expression or without p53 expression were assessed [62,63]. While acid SMase overexpression increased ceramide levels, the acid SMase overexpression may not have been high enough to sensitize cells to TMZ. The same study demonstrated that other methods of increasing ceramides, specifically a glucosylceramide synthase inhibitor or direct addition of C2- or C6-ceramide, decreased survival of TMZ-resistant glioma cell lines [61]. Acid SMase, but not neutral SMase, caused the hydrolysis of sphingomyelin to ceramides to induce apoptosis in p53-deficient GBM cells. The same study showed that wildtype p53 expression was capable of blocking the ceramide response by upregulating acid CDase to shift towards increased production of S1P from ceramides and allow the cells to evade apoptosis [64]. Conversely, the presence of p53 was able to generate ceramide through formation of reactive oxygen species and subsequent activation of neutral SMase [65]. Other studies suggest that p53 alters ceramide production by increasing alkaline CDase expression indicating a potential cell type-dependent mechanism [66]. The exact mechanism for the p53/ceramide interaction is not fully understood and continues to be an active area of investigation. Activation of neutral SMase in C6 glioma cells has been suggested to an increase in mitogen-activated protein kinase (MAPK) activation through upregulation of ceramides, leading to apoptosis [67]. Together, these data suggest that SMases can regulate ceramide levels and apoptosis in glioma cells, with differential responses, in part, due to p53.

2.2. Metastatic Cancers

Studies directly correlating SMases, expression or activity, to brain metastasis are limited. However, various studies suggest that SMase alterations can affect metastasis of cancers with higher propensities to metastasize to the brain. Human studies on non-small cell lung cancer demonstrated that these patients have increased acid SMase expression, which was suggested to be pro-tumorigenic via immunosuppression [68]. In a mouse model of melanoma, acid SMase deficiency showed prevention of lung metastasis by inhibition of secretory acid SMase in platelets. The metastatic phenotype was re-established when wildtype platelets were returned, again suggesting a pro-metastatic role for secreted acid SMase [69]. Conversely, high levels of sphingomyelins, which could suggest lower levels or activity of SMases, have been associated with a highly metastatic subset of prostate cancer cells [70]. Additional data in melanoma cells suggest that low acid SMase could be pro-tumorigenic via promotion of therapeutic resistance. Cells with low acid SMase expression display higher resistance to cisplatin, possibly due to the inability to produce ceramide [71]. Overexpression of acid SMase or addition of recombinant acid SMase to melanoma-bearing mice sensitized the tumors to irradiation. Additionally, in vitro culture of the B16 melanoma cells at a pH of 6.5 increased the activity of acid SMase at the cell membrane, suggesting that the pH of many solid tumors may increase the activity of acid SMase and increase radiosensitivity [72]. These studies suggest a role for acid SMase in metastasis and therapeutic resistance that may be cell- or level-dependent, making it an important enzyme to consider when looking at cancers that readily metastasize to the brain.

3. Sphingosine-1-Phosphate (S1P)

S1P has long been evaluated for its ability to promote tumor progression, but research in this context has been restrained by the availability of rigorous methods for quantification. However, as mass spectrometry approaches have improved, there has been an explosion of research into this ubiquitous sphingolipid species. This research advancement has created more questions than answers regarding its function, particularly concerning its signaling receptors. Both GBM and metastatic cancers have alterations in sphingolipid metabolism pushing them towards this pro-tumor species.

3.1. Glioblastoma (GBM)

S1P has been implicated in many of the aggressive phenotypes that arise in GBM. In patient samples of GBM, the sphingolipid rheostat is shifted toward S1P with a concurrent decrease in ceramides [10]. C18 ceramide showed the most dramatic decrease compared to C24:1 ceramide, C16 ceramide [10]. Abuhusain et al. showed that S1P concentration increased with tumor grade and led to an increase in angiogenesis with reported levels of 1 pmole/mg of tissue in GBM brain samples, compared to 0.2 pmole/mg of tissue in the normal brain samples. S1P has also been implicated in GBM invasion [73], with S1P upregulation of urokinase plasminogen activator as a possible mechanism [74]. The upregulation of S1P in GBM patients may be partially due to an upregulation of the enzymes that produce S1P in cell surviving radiotherapy [75,76]. Acid CDase and sphingosine kinase 1 (SK1), which shunt ceramide to S1P, were shown to be higher in GBM tissue compared to normal brain [10]. The acid CDase enzyme has also been associated with markers of the neural stem cell-like BTIC fraction [77]. Furthermore, extracellular rather than intracellular S1P, has been shown to promote survival of the BTIC population; with BTICs exporting more S1P than their differentiated GBM cell counterparts [78]. Marfia et al. subsequently reported that S1P could induce the proliferative effects of BTICs via autocrine signaling [45]. The extracellular S1P was also shown by Abdel Hadi et al. to be produced by brain endothelial cells in a co-culture model [79]. When considering the opposite side of the sphingolipid rheostat, the direct addition of C6-ceramide to GBM cells in culture induced apoptosis, further implicating the importance of ceramide and S1P ratio in cell fate decisions between apoptosis and survival [80].

Mechanistically, to initiate a cell signaling cascade, S1P binds to one of five different receptors (S1PR1-5). The S1PR1, S1PR2, and S1PR3 were shown to be elevated in patient brain tumor samples compared to normal brain, while S1PR4 was not expressed, and S1PR5 remained unchanged [76]. S1PR1-5 are critical for mediating different functions of S1P, but the direction in which they alter cellular phenotypes is not entirely clear (Table 1). For instance, inhibiting S1PR1 using siRNA had been reported to increase GBM proliferation, but conflicting studies suggested that signaling through S1PR1-3 all increase GBM proliferation with S1PR1 having the greatest effects [81,82]. For S1PR2, this receptor was reported to attenuate migration of GBM cells through the Rho kinase pathway, but was also involved in increasing invasion [83,84]. Studies from other patient tumor samples have implicated S1PR5 as an independent prognostic factor in GBM, which aligns with data that S1PR5 increased proliferation [82,85]. The discrepancies between Bien-Möller et al. and Quint et al. regarding S1PR5 expression is likely due to small samples sizes and the vast inter- and intratumoral heterogeneity exhibited in GBM [86]. Recent studies have investigated how S1P receptor levels and signaling may also be affected by the brain tumor microenvironment. Upon co-culture of GBM cells with brain endothelial cells, expression of S1PR1 and S1PR3 was elevated in GBM cells [79]. In the normal brain, S1P promoted the survival of mature oligodendrocytes through a protein kinase B (AKT)-dependent pathway, and S1PR5 was required for process retraction in immature oligodendrocytes, which is necessary during development [87]. Pharmacological alteration of S1PR expression by fingolimod, a sphingosine analog which leads to the internalization of S1PR1, also known as FTY720, decreased human astrocyte activation and altered C-X-C motif chemokine 5 (CXCL5) release from both astrocytes and microglia [88,89]. CXCL5 is known to increase proliferation and invasion in GBM cells, emphasizing the importance of this interaction with the tumor microenvironment [90]. These studies suggest the importance of S1PR signaling in brain tumor cells and the brain microenvironment, but additional understanding of biological consequences is needed to more fully predict the benefits and potential risks of S1PR modulation.

3.2. Metastatic Cancers

Metastatic cancers have been shown to produce and secrete more S1P, as compared to primary tumors [91]. Such secreted S1P has been reported to be capable of establishing pre-metastatic niches in distant organs, such as the brain, through mechanisms involving S1PR1 [91,92]. Inhibition of S1P signaling using fingolimod in multiple myeloma revealed that metastasis to the bone marrow was due to the C-X-C chemokine receptor 4 (CXCR4)/C-X-C motif chemokine 12 (CXCL12) pathway [92]. While metastasis specific to the brain has not been studied with respect to S1P, the expression of CXCL12 was positively correlated with brain metastasis in solid tumors [93]. This links CXCL12 and its downstream signaling pathways to brain metastasis.

For the remainder of this section, we will consider the broader role of S1P in metastasis with a focus on cancers with the highest propensity of metastasizing to the brain. However, we recognize that metastatic niches and molecular mediators of metastasis can vary by organ, and not all of the signals discussed here may be relevant for brain-specific metastasis. Of the breast cancer subtypes, triple-negative breast cancer (TNBC) has the highest and earliest likelihood of metastasizing to the brain [94]. Studies of TNBC report high expression of SK1 with a concurrent increase in S1P, which promotes growth through the S1P/S1PR3/Notch signaling pathway [95,96]. A study concerning obesity-related progression of breast cancer found S1P elevated in obese patients and mice [97]. S1P contributed to the establishment of a pre-metastatic niche, and targeting S1PR1 with fingolimod decreased metastasis to the lungs [97]. In non-small-cell lung carcinoma (NSCLC), elevated SK1 and SK2 mRNA expression was associated with a worse prognosis in patients, likely due to an increase in S1P produced by the cancer cells [98]. Zhao et al. further considered the role of S1P in NSCLC metastasis through S1PR3. When S1PR3 was either genetically or pharmacologically targeted, a decrease in metastasis was observed due to attenuation of the Transforming Growth Factor-β (TGF-β)/Mothers against decapentaplegic homolog 3 (SMAD3) signaling axis [99]. Studies in metastatic breast cancer showed a similar decrease in migration when inhibiting S1PR3, potentially through a S1P/S1PR3/Cyclooxygenase-2 (COX-2) pathway [100]. Paradoxically, chemotherapy itself may induce metastasis—one of the more serious complications of chemotherapy in solid tumors [101]. Inhibition of S1P signaling through S1PR1 was able to mitigate this serious side effect [102]. In contrast, JTE013, an antagonist of S1PR2, increased migration and invasion in melanoma, showing opposing signaling roles related to S1PR1 and S1PR3 [103]. Albeit less studied, S1PR4 was associated with a decrease in overall survival in estrogen receptor (ER)-negative breast cancer [104]. Beyond the S1PRs, overexpression of acid CDase in melanoma cells enhanced resistance to dacarbazine, the DNA-alkylating agent often used in patients [105]. Consistent with this finding for a pro-tumorigenic role of acid CDase, knockdown of acid CDase in melanoma cells decreased both growth and invasion [106]. Concerning the tumor microenvironment, S1P produced by higher SK1 expression was found to increase the differentiation of surrounding fibroblasts, further promoting metastasis of melanoma cells [107]. Lastly, a screen of over 800 mutant mice revealed that mice deficient in S1P transporter spinster homologue 2, the protein responsible for transporting S1P from the cell, had the least amount of pulmonary metastases [108]. Together, these data demonstrate critical roles for S1P/S1PR signaling in cancers that have a propensity to metastasize to the brain.

Table 1. Summary of the current research on S1PR effects. S1PRs have multiple effects on cancer cells, illustrating a lack of consensus on the predominant effects in all cancer types. Glioblastoma (GBM); triple-negative breast cancer (TNBC); non-small cell lung cancer (NSCLC).

S1PR	Cancer Type	Alteration	Phenotype	Study
S1PR1	GBM	Absence	Increased proliferation	Yoshida et al., 2010 [81]
	GBM	Presence	Increased proliferation	Young et al., 2007 [82]
	Breast cancer	Decrease	Decreased metastasis	Nagahashi et al., 2018 [97]
	Solid tumors	Decrease	Decreased migration	Liu et al., 2015 [102]
S1PR2	GBM	Absence	Increased migration	Lepley et al., 2005 [83]; Malchinkkuu et al., 2008 [84]
	Melanoma	Decrease	Increased migration/invasion	Arikawa et al., 2003 [103]
S1PR3	GBM	Presence	Increased proliferation	Young et al., 2007 [82]
	TNBC	Presence	Increased metastasis	Wang et al., 2018 [96]
	Breast cancer	Decrease	Decreased migration	Filipenko et al., 2016 [100]
	NSCLC	Decrease	Decreased metastasis	Zhao et al., 2016 [99]
S1PR4	Breast cancer	Presence	Decreased survival	Ohotski et al., 2012 [104]
S1PR5	GBM	Presence	Increased proliferation	Young et al., 2007 [82]

4. Phospholipase-Mediated Signaling

One of the predominant ways that S1P can signal within the cell is through phospholipase-mediated pathways. Phospholipases C (PLC) and D (PLD) have emerged as major contributors to the aggressive phenotypes seen in GBM including invasion and chemoresistance, as well as promotion of metastasis in other cancer types [109–112]. The intersection between sphingolipid metabolism and phospholipase signaling provides a greater understanding of how these pathways synergize to promote cancer progression in both GBM and metastatic cancers.

4.1. Glioblastoma (GBM)

Signaling through the S1P receptors can alter phospholipase signaling, particularly that of PLC (Figure 2). Addition of S1P induced activation of matrix metalloproteinase-9 (MMP-9), which is known to increase invasion in breast cancer cells. S1P exerted its effects through Gαq and S1PR3, and this further induced the expression of PLC-β_4 (PLC-β_4) via Rac1 [113]. In other studies, MMP-9 expression was reported to be increased at an extracellular acidic pH (5.4–6.5), which is common in the GBM tumor microenvironment [114]. Later studies from the same group reported induction of SMase by extracellular acidic pH mediated MMP-9 activity, through nuclear factor kappa-light-chain-enhancer of activated B cells (NF-κB) activation [115]. In C6 rat glioma cells, activation of S1PR2 led to downstream signaling via the PLC-Ca^{2+} system, as well as PLD. This further activated extracellular-signal-regulated kinase (ERK), which can stimulate proliferation, migration, and angiogenesis [116]. Notably, only S1PR1 and S1PR2 were expressed on C6 glioma cells, which is not consistent with the reported expression in glioma patient samples [76]. Overall, these findings suggest the importance of understanding S1P-mediated phospholipase signaling in order to elucidate the mechanisms behind GBM progression.

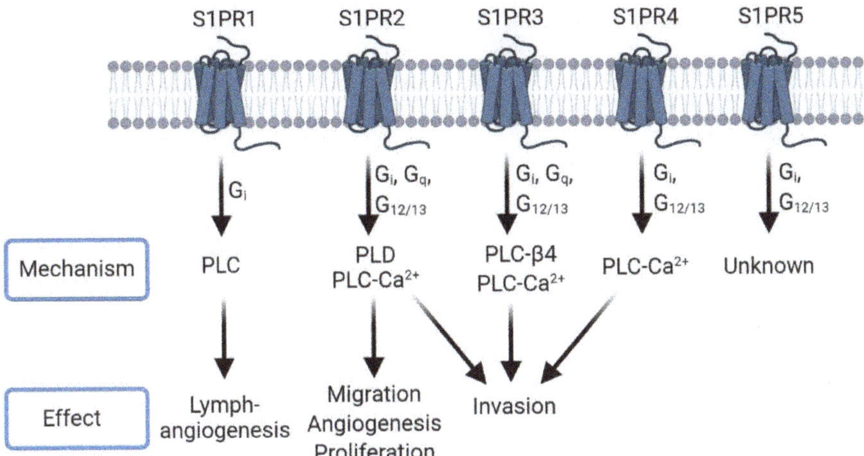

Figure 2. S1PRs can signal through phospholipase mechanisms. Each S1PR can couple to one or more G Protein-Coupled Receptors (GPCRs) to signal through different phospholipases and induce phenotypes such as angiogenesis, proliferation, and invasion. Many of these mechanisms overlap between receptors. Conversely, PLD signaling can lead to further production of S1P through the interaction of phosphatidic acid with sphingosine kinase 1 (SK1) [35].

4.2. Metastatic Cancers

The interaction between sphingolipid metabolism and phospholipase signaling in metastatic cancers extends our mechanistic insight into GBM pathology. Recent investigations show S1P can trigger the release of Ca^{2+} to increase phosphatidic acid through activation of PLD [117]. Phenotypically,

S1P can promote the formation of lymphatic vessels [118]. This process, called lymphangiogenesis, is thought to be similar to angiogenesis, occurs in cancer, and is associated with lymph node metastasis [119,120]. Using human lymphatic endothelial cells (HLECs) and in vivo models, Yoon et al. found that S1P promoted lymphangiogenesis through S1PR1 and phospholipase C [118]. In lung adenocarcinoma cells, S1P added in vitro increased the activity of phospholipase D, leading to a dramatic increase in RhoA [121]. The induction of RhoA as part of signaling through PLCε and G Protein-Coupled Receptors (GPCRs) with S1P treatment has also been shown in astrocytes as part of a neuroinflammation model [122]. To date, there is very limited knowledge on the exact convergence of sphingolipid metabolism and phospholipase signaling in metastatic cancers, particularly to the brain, but investigations in other cancers and of the pathways independently suggest that this will be a very active area of exploration in upcoming years.

5. Blood–Brain Barrier Integrity

The blood–brain barrier (BBB) consists of endothelial cells connected by tight junctions and protects the brain by preventing invasion of deleterious molecules present in circulation [123]. The BBB also prevents the passage of many cancer therapeutics, precluding a large number of drugs from effectively targeting GBM or metastatic brain tumors. While there are leaky areas of the BBB in a tumor-setting, they are not evenly dispersed throughout, so not all portions of the brain tumor will receive the drug uniformly (Figure 3). As GBM has been suggested to be a "whole brain disease," failure to target even a small section of the tumor or the invading tumor cells will lead to recurrence of the disease [124]. During aging, the BBB begins to deteriorate and correlates with elevated acid SMase levels in the plasma of humans. Park et al. recapitulated this finding in an aged mouse model and determined that increases in acid SMase led to endothelial cell death and disruptions in the BBB via altering caveolae internalization [125]. Apoptosis of endothelial cells post-irradiation in a rat model also caused breakdown of the BBB through an acid SMase-dependent mechanism [126].

Furthermore, the role of S1P signaling in BBB integrity has become an interesting avenue of research [127]. A study of ischemia-reperfusion injury demonstrated that S1P increased STAT3 activation, leading to BBB dysfunction [128]; these data suggested that inhibition of S1P could be a strategy to prevent BBB breakdown. However, knockdown of S1P lyase (an enzyme that breaks down S1P) in endothelial cells of the BBB increased expression of adherens junction molecules, leading to increased BBB integrity in vitro [129]. Thus, inhibition of S1P lyase may prevent the breakdown of the BBB caused by inflammatory factors [129]. One explanation of the discrepancies in S1P effects on the BBB was provided by Li et al. using human umbilical vein endothelial cells (HUVECs) in culture: they showed that physiologic concentrations of S1P promoted assembly of tight junctions via S1PR1 and Rac1 activation, but higher concentrations actually led to the disassembly of tight junctions via S1PR2 and the RhoA/ROCK pathway [130]. Additional work by van Doorn et al. found that expression of S1PR5 on brain endothelial cells was crucial for maintaining BBB integrity [131]. Additionally, a study utilizing co-cultured endothelial cells and astrocytes to mimic the BBB has observed that fingolimod increased endothelial cell survival when exposed to inflammatory cytokines by inducing the release of Granulocyte-macrophage Colony-Stimulating Factor (GM-CSF) from astrocytes [132].

Emerging therapies for GBM are exploring ways to open the BBB in order to allow anti-cancer therapies to cross the BBB. For instance, a clinical trial (NCT03712293) where patients received standard of care and magnetic resonance-guided focused ultrasound to disrupt the BBB proved to be safe and accurate and such trials are continuing [133]. Other in vivo studies demonstrated that pharmacologic inhibition of S1PR1 allowed for transient opening of the BBB by altering tight junction protein localization [134]. These reports open the door for many new treatments in GBM and metastatic brain tumors, but the risks of disrupting the BBB must be considered, particularly if it cannot be reliably restored post-treatment.

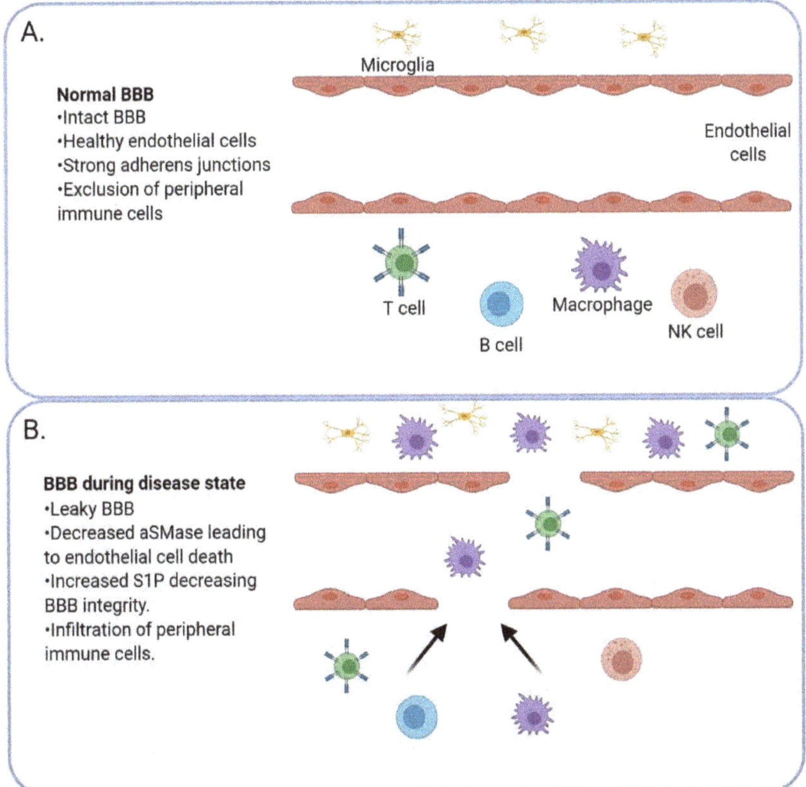

Figure 3. The blood–brain barrier (BBB) is altered in the context of a brain tumor. A healthy BBB (**A**) with strong adherens junctions prevents peripheral immune cells from entering the brain. When the BBB is compromised (**B**) by a decrease in acid SMase and increase in S1P, peripheral immune cells can enter the brain.

6. Immune Trafficking

In recent years, research into cancer immunology has focused on activating the patients' immune system for tumor elimination, including checkpoint inhibitors to alter the adaptive immune response. While checkpoint inhibitors have largely failed in GBM, they have shown promising results in cancers with higher mutational loads such as melanoma [135]. Much of the research in GBM immunotherapy has focused on altering the innate immune system [136]. Sphingolipid metabolism plays a role in the trafficking of both adaptive and innate immune cells.

6.1. Adaptive Immunity

Multiple components of sphingolipid metabolism have been shown to regulate adaptive immunity, which is characterized by T and B cell responses days after infection [137]. Under normal conditions, there are very few, if any, T-cells that cross the BBB. During neurological disease, the breakdown of the BBB allows T-cells, among others, to enter the brain [123,138]. Compared to GBM, metastatic brain tumors have a substantially higher number of infiltrating lymphocytes [139]. While this process could lead to the elimination of a tumor, it often allows tumor-promoting populations to enter the brain such as regulatory T-cells (Tregs). which are $CD4^{+}CD25^{+}FoxP3^{+}$ [140]. Tregs serve to protect the host from autoimmune disorder, but in cancer, they can promote tumor growth through their immunosuppressive functions [141]. Interestingly, acid SMase-mediated activation has been linked to CD4$^+$ T-cells proliferation [142]. Further, mice deficient in acid SMase demonstrate an increase in Tregs globally, indicating a shift toward the immunosuppressive phenotype [143]. However, an increase in activated CD4$^+$ and CD8$^+$ T-cells with no change in Tregs was seen when NSCLC cells were injected into acid SMase deficient mice [68]. These discrepancies highlight the importance of the tumor microenvironment and the differential effects of acid SMase expression by the tumor cells, as compared to the immune cells.

S1P has long been appreciated for its ability to create a gradient for T-cell trafficking. Recent studies demonstrated that the presence of a brain tumor in mice, whether primary or metastatic, causes downregulation of S1PR1 on the surface of T cells, and leads to the homing and sequestration of T-cells in bone marrow [144]. This downregulation of S1PR1 has been exploited for the treatment of multiple sclerosis using fingolimod. Fingolimod is a sphingosine analog; it acts as a functional antagonist of S1PR1 by the phosphorylation of FTY720 to FTY720-Pi, alleviating the symptoms of autoimmunity [145]. This double-edged sword of S1P makes targeting the pro-proliferative and migratory lipid challenging. In the context of multiple sclerosis, blocking immune trafficking leads to a decrease in disease severity [145,146]. However, the same effect is not beneficial to brain tumor patients who already exhibit lymphopenia [144]. Conversely, van der Weyden et al. showed that mice deficient in the S1P transporter spinster homologue 2, the protein responsible for transporting S1P from the cell, had an increase in T cells and natural killer cells in the lung preventing metastasis [108]. This research suggests site-specific roles of S1P and the regulation of immune trafficking, revealing an exciting new area of investigation in GBM and metastatic cancers.

6.2. Innate Immunity

Sphingolipid metabolism contributes to the innate immune system, which provides rapid defense against foreign bodies and involves multiple cell types, including macrophages, dendritic cells, mast cells, and granulocytes. In GBM, macrophages constitute up to 50% of the bulk tumor, making them the primary innate immune cell in the tumor microenvironment [147,148]. In simplest terms, macrophages can be described as classically (M1) or alternatively activated (M2) with M1 being proinflammatory and M2 being the immunosuppressive and pro-tumorigenic tumor-associated macrophages (TAMs) [149]. Immunosuppressive TAMs express inducible nitric oxide synthase (iNOS), which can produce nitric oxide (NO) and lead to resistance of cancer cells to cisplatin—a commonly used chemotherapeutic agent. The NO induction also decreased acid SMase in glioma cells, allowing them to escape apoptosis [150]. Studies using melanoma cell lines demonstrated that low aSMase expression contributes to a pro-tumor immune response by allowing myeloid-derived suppressor cell accumulation [151].

As a result of apoptosis, expression of SK1 leading to S1P production has been shown to be a chemoattractant for macrophages [152]. In physiologically normal cells, this process is beneficial as migrating macrophages arrive to clear the waste of dying cells. Unfortunately, when SK1 is increased as a response to chemo- or radiotherapy, macrophages that are recruited to the tumor microenvironment can shift towards immunosuppressive phenotypes and promote tumor growth [10,153]. To this point, melanomas that express high levels of SK1 have a greater infiltration of macrophages and polarization to the immunosuppressive phenotype [154]. In fact, the expression of S1PRs on the surface of macrophages can be altered under different polarizing conditions in vitro. For example, M1 polarized macrophages had decreased expression of S1PR4 compared to unpolarized bone marrow-derived macrophages. In contrast, S1P1R was decreased in both M1 and M2 polarized macrophages in comparison to unpolarized bone marrow-derived macrophages. Differential downregulation of S1PRs with polarization could suggest that S1P activity alters macrophage biology. While S1P did not alter the phagocytic activity [155], deletion of S1PR1 on macrophages prevented pulmonary metastasis and lymphangiogenesis. This effect was mediated through attenuation of Nucleotide-Binding Oligomerization Domain, Leucine-Rich Repeat and Pyrin Domain Containing 3 (NLRP3) expression and interleukin 1-β (IL-1β) production in a breast cancer mouse model [156]. Interleukin 22 Receptor 1 (IL-22R1) signaling through S1PR1 also leads to the recruitment of macrophages to the tumor microenvironment in breast cancer [157]. As discussed above, fingolimod can also have an effect on the myeloid-derived suppressor cells (MDSCs). The MDSCs are a pro-tumorigenic immune population that accumulate in tissues when mice were treated with fingolimod. In long-term usage, this was shown to increase the risk of cancer development [158]. Subsequent investigations revealed that fingolimod decreased recruitment of macrophages to the brain tumor microenvironment, as well as pushed them toward a proinflammatory (M1) phenotype via C-X-C Motif Chemokine Receptor 4 (CXCR4) internalization [159]. Outside of the tumor setting, mice without S1P lyase expression had greater microglial activation in the brain. The accumulation of S1P signaled through S1PR2 to mediate this inflammation [160]. Figure 4 shows an overview of the alterations in both the adaptive and innate immune cells in GBM and metastatic cancers.

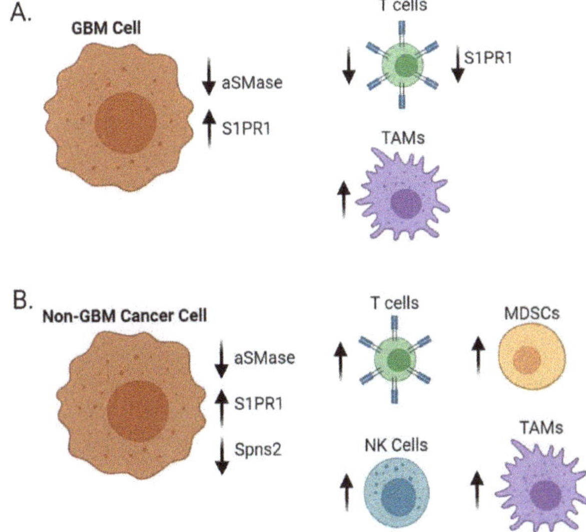

Figure 4. Alterations of sphingolipid metabolism affect surrounding immune populations. In GBM cells (**A**), low acid SMase and high S1PR1 on GBM cells but low S1PR1 on T cells decrease T cell recruitment while increasing macrophage recruitment and polarization towards an immunosuppressive (M2) phenotype. In metastatic brain cancers (**B**), such as melanoma, breast, and lung, decreases in acid SMase and Spns2 with increases in S1PR1 can increase recruitment and activation of both adaptive and innate immune cells.

7. Altering Sphingolipid Metabolism for Therapeutic Intervention

As research continues into the role of sphingolipid metabolism, many new and repurposed therapeutics have emerged to modulate this pathway and improve the survival of patients with GBM as well as metastatic brain cancers [161]. Many of these emerging therapies are already in clinical trials, but there are still many unanswered questions regarding the efficacy of these treatments.

7.1. Glioblastoma

While the BBB serves as a barrier to many GBM therapies, other research efforts have focused on restoring the BBB to prevent the recruitment of immune suppressive cell populations. Drugs that target acid SMase, such as amitriptyline hydrochloride, have the potential to block the progression of both GBM and metastatic cancers by restoring the BBB [133]. However, that same inhibition of acid SMase can potentially decrease the efficacy of chemo- and radiotherapy. Thus, when considering combinatorial therapies, timing can prove to be very important. The small molecule inhibitor ARC39 has shown specificity of acid SMase in vitro, as well [162]. However, this would likely blunt the effects of chemo- and radiotherapy as they partially rely on the breakdown of sphingomyelin by SMases to induce apoptosis. The majority of therapeutic focus in this area has been on enzymes involved in S1P production, although some have suggested that vitamin D metabolites could activate the sphingomyelin pathway [163]. Inhibitors of SK1 have shown mixed results in GBM with one study showing that selective SK1 inhibitors, SKI-1a and SKI-1b, did not affect cell death but instead blocked angiogenesis, while others have shown that SK1 inhibition using SKI-II could be more effective than TMZ due to the induction of reactive oxygen species [10,164]. This difference in findings could be due to the selectivity of the inhibitors as SKI-II targets both SK1 and SK2 making SKI-II more efficacious at decreasing S1P production [165]. Additionally, the study that showed no difference in cell death used concentrations up to 1 μM, while the study that showed cell death used up to 20 μM, a 20-fold difference in concentrations. Others have shown that inhibitors of SK1 have efficacy in GBM cells which are TMZ-resistant [166]. Within the conflicting reports, some have suggested SK inhibitors be added as a maintenance therapy to prevent S1P formation and sustain ceramide induction [167]. In GBM, epidermal growth factor receptor (EGFR) is often mutated or constitutively active. EGFR inhibitors have also been shown to decrease activation of SK1, consequentially leading to a decrease in S1P and invasion [168]. Growth of GBM cells in vitro can be inhibited by acid CDase inhibitors such as carmofur [77,169]. Additionally, tamoxifen, a treatment for ER-positive breast cancer, has been shown to inhibit acid CDase and readily crosses the BBB [170]. As an already approved therapy, this provides potential application for the treatment of GBM. Other inhibitors of acid CDase such as B13 and the LCL series of compounds have yet to be tested in clinic despite showing promise in mouse models of prostate cancer [171,172]. One reason they have not been investigated further in GBM is that their structures make them unlikely to cross the BBB. Lastly, fenretinide (4-HPR), a synthetic retinoid with an early FDA approval for T-cell lymphoma, has been investigated for its ability to increase ceramide production by p53-independent mechanisms in cancer cells [173–176]. While 4-HPR has shown promise in vitro, its poor solubility and bioavailability have limited its use clinically, with both of the glioma clinical trials showing no improvement in progression-free survival at the doses administered (Table 2) [175,177].

Table 2. Current and previous clinical trials targeting sphingolipid metabolism. Multiple clinical trials have attempted to target sphingolipid metabolism by altering enzymes, receptor expression, and ceramide accumulation in the sphingolipid pathway.

Cancer Type	NCT Number	Dates	Drug Name	Primary Target	Phase	Results	FDA Approval?
GBM	NCT02490930	July 2015–September 2017	Fingolimod	S1PR1 antagonist	Early Phase 1	Not Published	Yes
Recurrent Glioma	NCT00006080	September 2000–September 2004	Fenretinide (Single Agent)	Ceramide	Phase 2	Ineffective at the doses given	Yes
Recurrent GBM	NCT00075491	December 2003–March 2005	Fenretinide (Combination treatment)	Ceramide	Phase 2	Not Published due to termination	Yes
Advanced solid tumors	NCT01488513	August 2014–August 2015	ABC294640	SK inhibitor	Phase 1	Tolerated up to 500mg bid	No
Advanced solid tumors	NCT02834611	March 2017–August 2019	Ceramide NanoLiposome	Ceramide	Phase 1	Not Published	No

Intriguingly, there have been reports of multiple sclerosis patients treated with fingolimod that later developed high grade gliomas [178]. The mechanism behind this etiology is unclear, but possibly linked to the immunosuppression or activation of additional S1P receptors. These alternative effects have generated interest in developing drugs that target specific S1P receptors in order to mitigate the effect on T-cell trafficking or tumor cell migration. Studies in multiple sclerosis using fingolimod as a functional antagonist of S1PR1 have indicated that it does not alter BBB integrity nor change MMP-9 expression [179]. These data indicate that fingolimod may not function through the same mechanisms as endogenous S1P [179]. Additionally, phosphorylated fingolimod was reported to bind to S1PR3, S1PR4, and S1PR5 as well as S1PR1, the receptor primarily responsible for immune trafficking [180]. Since the drug is phosphorylated by SK2 only, it can act as a SK1 inhibitor as demonstrated in prostate cancer [181,182]. Nonetheless, due to favorable toxicity profile of fingolimod, it has been clinically evaluated for safety in combination with radiation and TMZ in glioma patients (NCT02490930). Interestingly, fingolimod was given one week prior to radio- and chemotherapy with the hope of sequestering lymphocytes away from systemic circulation, thus protecting them from the immunosuppressive effects of radiation. The final results of that study are not yet available. As a whole, there are many promising avenues for new therapeutics in GBM patients, particularly as more specific inhibitors to enzymes in this pathway are discovered. Subsequently, a more thorough understanding of sphingolipid metabolism will provide more information for therapeutic design.

7.2. Metastatic Cancers

Developments in GBM treatments may translate to metastatic brain cancers and, thereby, open the door to more effective treatment in advanced stages of other cancers. A study in TNBC found that a nanoparticle containing docetaxel and fingolimod could abrogate lymphopenia while still blocking progression [183]. Carmofur, a derivative of 5′ fluorouracil and an inhibitor of acid CDase, has been used as adjuvant therapy in colorectal cancer patients to prevent metastasis [184]. The drug, which is known to cross the BBB, has been used for years in Japan as maintenance therapy following standard of care but has not been subjected to FDA approval in the United States [185]. While relatively rare, carmofur has been linked to cases of leukoencephalopathy in patients [186]. This side effect is likely one of the reasons it has not been approved for use in the United States. It is unclear whether this is a side effect of the sphingolipid metabolism alterations or an off-target effect of carmofur. A study of hispidulin, a polyphenolic flavonoid, found that the treatment shifted the sphingolipid rheostat to ceramide by inhibiting SK1 [187]. ABC294640 is the first-in-class inhibitor of SK2 and has been clinically tested in a variety of cancers (Table 2) [188]. ABC294640 has proven effective in both in vitro and in vivo models of TNBC [189]. While efficacy has yet to be determined in clinic, the treatment had minimal side effects in a phase I clinical trial (NCT01488513). The ability to repurpose already known modulators of sphingolipid metabolism may serve as exciting new treatments in metastatic cancers. Additionally, advancements in GBM therapeutics are likely to translate to metastatic brain cancers. Unfortunately, it is still unknown if many of these novel modulators of sphingolipid metabolism are able to cross the BBB, reinforcing the need for further investigation of these treatments for GBM and metastatic brain cancers.

8. Conclusions

Research in the field of sphingolipid metabolism has grown exponentially recently. This is particularly so with regards to the most aggressive cancers including GBM and cancers that have metastasized to the brain such as breast and lung cancers. As experimental methodologies and sphingolipid analysis techniques continue to advance, more therapeutically targeted approaches with more specific and efficacious treatments become feasible. Continuing to elucidate the evidently complex mechanistic interplay of the various molecules involved, and the resultant physiological implications, is imperative for the enhancement of cancer research and for working towards overall positive clinical outcomes.

Author Contributions: Original draft preparation, C.C.H.; reviewing and editing, T.A., S.R., A.B.H. All authors have read and agreed to the published version of the manuscript.

Funding: This research was funded by NIH/NIDDK (R01DK110292) and NIH/NIAID (R21AI146743)—S.R.; UAB Start Up Funds, NS104339, and CDIB Pilot Grant—A.B.H.

Acknowledgments: We would like to extend our deepest gratitude to the members of the Hjelmeland laboratory for their review of this manuscript—Sajina G.C., Amber Jones, Kaysaw Tuy, and Sarah Williford. All images were created using BioRender.com.

Conflicts of Interest: The authors declare no conflict of interest.

References

1. Aureli, M.; Murdica, V.; Loberto, N.; Samarani, M.; Prinetti, A.; Bassi, R.; Sonnino, S. Exploring the link between ceramide and ionizing radiation. *Glycoconj. J.* **2014**, *31*, 449–459. [CrossRef] [PubMed]
2. Brady, R.O.; Kanfer, J.N.; Mock, M.B.; Fredrickson, D.S. The metabolism of sphingomyelin. II. Evidence of an enzymatic deficiency in Niemann-Pick diseae. *Proc. Natl. Acad. Sci. USA* **1966**, *55*, 366. [CrossRef]
3. Haimovitz-Friedman, A.; Kan, C.C.; Ehleiter, D.; Persaud, R.S.; McLoughlin, M.; Fuks, Z.; Kolesnick, R.N. Ionizing radiation acts on cellular membranes to generate ceramide and initiate apoptosis. *J. Exp. Med.* **1994**, *180*, 525–535. [CrossRef]
4. Senchenkov, A.; Litvak, D.A.; Cabot, M.C. Targeting Ceramide Metabolism—A Strategy for Overcoming Drug Resistance. *J. Natl. Cancer Inst.* **2001**, *93*, 347–357. [CrossRef]
5. Hannun, Y.A.; Luberto, C.; Argraves, K.M. Enzymes of sphingolipid metabolism: From modular to integrative signaling. *Biochemistry* **2001**, *40*, 4893–4903. [CrossRef]
6. Wegner, M.S.; Schiffmann, S.; Parnham, M.J.; Geisslinger, G.; Grösch, S. The enigma of ceramide synthase regulation in mammalian cells. *Prog. Lipid Res.* **2016**, *63*, 93–119. [CrossRef]
7. Nganga, R.; Oleinik, N.; Ogretmen, B. Mechanisms of Ceramide-Dependent Cancer Cell Death. In *Advances in Cancer Research*; Academic Press: Cambridge, MA, USA, 2018; Volume 140, pp. 1–25.
8. Gault, C.R.; Obeid, L.M.; Hannun, Y.A. An overview of sphingolipid metabolism: From synthesis to breakdown. *Adv. Exp. Med. Biol.* **2010**, *688*, 1–23. [CrossRef]
9. Mizushima, N.; Koike, R.; Kohsaka, H.; Kushi, Y.; Handa, S.; Yagita, H.; Miyasaka, N. Ceramide induces apoptosis via CPP32 activation. *FEBS Lett.* **1996**, *395*, 267–271. [CrossRef]
10. Abuhusain, H.J.; Matin, A.; Qiao, Q.; Shen, H.; Kain, N.; Day, B.W.; Stringer, B.W.; Daniels, B.; Laaksonen, M.A.; Teo, C.; et al. A metabolic shift favoring sphingosine 1-phosphate at the expense of ceramide controls glioblastoma angiogenesis. *J. Biol. Chem.* **2013**, *288*, 37355–37364. [CrossRef]
11. Jung, J.S.; Ahn, Y.H.; Moon, B.I.; Kim, H.S. Exogenous C2 Ceramide Suppresses Matrix Metalloproteinase Gene Expression by Inhibiting ROS Production and MAPK Signaling Pathways in PMA-Stimulated Human Astroglioma Cells. *Int. J. Mol. Sci.* **2016**, *17*, 477. [CrossRef]
12. Gatt, S. Enzymic hydrolysis of sphingolipids: Hydrolysis of ceramide glucoside by an enzyme from ox brain. *Biochem. J.* **1966**, *101*, 687–691. [PubMed]
13. Momoi, T.; Ben-Yoseph, Y.; Nadler, H.L. Substrate-specificities of acid and alkaline ceramidases in fibroblasts from patients with Farber disease and controls. *Biochem. J.* **1982**, *205*, 419–425. [CrossRef] [PubMed]
14. Mao, C.; Obeid, L.M. Ceramidases: Regulators of cellular responses mediated by ceramide, sphingosine, and sphingosine-1-phosphate. *Biochim. Biophys. Acta* **2008**, *1781*, 424–434. [CrossRef] [PubMed]
15. Buehrer, B.M.; Bell, R.M. Inhibition of sphingosine kinase in vitro and in platelets. Implications for signal transduction pathways. *J. Biol. Chem.* **1992**, *267*, 3154–3159. [PubMed]
16. Goetzl, E.J.; Kong, Y.; Mei, B. Lysophosphatidic acid and sphingosine 1-phosphate protection of T cells from apoptosis in association with suppression of Bax. *J. Immunol.* **1999**, *162*, 2049–2056. [PubMed]
17. Augé, N.; Nikolova-Karakashian, M.; Carpentier, S.; Parthasarathy, S.; Nègre-Salvayre, A.; Salvayre, R.; Merrill, A.H., Jr.; Levade, T. Role of sphingosine 1-phosphate in the mitogenesis induced by oxidized low density lipoprotein in smooth muscle cells via activation of sphingomyelinase, ceramidase, and sphingosine kinase. *J. Biol. Chem.* **1999**, *274*, 21533–21538. [CrossRef]
18. Clair, T.; Aoki, J.; Koh, E.; Bandle, R.W.; Nam, S.W.; Ptaszynska, M.M.; Mills, G.B.; Schiffmann, E.; Liotta, L.A.; Stracke, M.L. Autotaxin hydrolyzes sphingosylphosphorylcholine to produce the regulator of migration, sphingosine-1-phosphate. *Cancer Res.* **2003**, *63*, 5446–5453.
19. Shaw, J.; Costa-Pinheiro, P.; Patterson, L.; Drews, K.; Spiegel, S.; Kester, M. Novel Sphingolipid-Based Cancer Therapeutics in the Personalized Medicine Era. *Adv. Cancer Res.* **2018**, *140*, 327–366. [CrossRef]

20. Cuvillier, O.; Pirianov, G.; Kleuser, B.; Vanek, P.G.; Coso, O.A.; Gutkind, J.S.; Spiegel, S. Suppression of ceramide-mediated programmed cell death by sphingosine-1-phosphate. *Nature* **1996**, *381*, 800–803. [CrossRef]
21. Newton, J.; Lima, S.; Maceyka, M.; Spiegel, S. Revisiting the sphingolipid rheostat: Evolving concepts in cancer therapy. *Exp. Cell Res.* **2015**, *333*, 195–200. [CrossRef]
22. Nagahashi, M.; Abe, M.; Sakimura, K.; Takabe, K.; Wakai, T. The role of sphingosine-1-phosphate in inflammation and cancer progression. *Cancer Sci.* **2018**, *109*, 3671–3678. [CrossRef] [PubMed]
23. Alvarez, S.E.; Milstien, S.; Spiegel, S. Autocrine and paracrine roles of sphingosine-1-phosphate. *Trends Endocrinol. Metab.* **2007**, *18*, 300–307. [CrossRef] [PubMed]
24. Johnson, K.R.; Becker, K.P.; Facchinetti, M.M.; Hannun, Y.A.; Obeid, L.M. PKC-dependent activation of sphingosine kinase 1 and translocation to the plasma membrane. Extracellular release of sphingosine-1-phosphate induced by phorbol 12-myristate 13-acetate (PMA). *J. Biol. Chem.* **2002**, *277*, 35257–35262. [CrossRef]
25. Anelli, V.; Bassi, R.; Tettamanti, G.; Viani, P.; Riboni, L. Extracellular release of newly synthesized sphingosine-1-phosphate by cerebellar granule cells and astrocytes. *J. Neurochem.* **2005**, *92*, 1204–1215. [CrossRef] [PubMed]
26. Im, D.S.; Clemens, J.; Macdonald, T.L.; Lynch, K.R. Characterization of the human and mouse sphingosine 1-phosphate receptor, S1P5 (Edg-8): Structure-activity relationship of sphingosine1-phosphate receptors. *Biochemistry* **2001**, *40*, 14053–14060. [CrossRef] [PubMed]
27. Hla, T.; Maciag, T. An abundant transcript induced in differentiating human endothelial cells encodes a polypeptide with structural similarities to G-protein-coupled receptors. *J. Biol. Chem.* **1990**, *265*, 9308–9313. [PubMed]
28. Masana, M.I.; Brown, R.C.; Pu, H.; Gurney, M.E.; Dubocovich, M.L. Cloning and characterization of a new member of the G-protein coupled receptor EDG family. *Recept. Channels* **1995**, *3*, 255–262. [PubMed]
29. Zondag, G.C.; Postma, F.R.; Etten, I.V.; Verlaan, I.; Moolenaar, W.H. Sphingosine 1-phosphate signalling through the G-protein-coupled receptor Edg-1. *Biochem. J.* **1998**, *330 Pt 2*, 605–609. [CrossRef]
30. Siehler, S.; Manning, D.R. Pathways of transduction engaged by sphingosine 1-phosphate through G protein-coupled receptors. *Biochim. Biophys. Acta Mol. Cell Biol. Lipids* **2002**, *1582*, 94–99. [CrossRef]
31. Sanchez, T.; Hla, T. Structural and functional characteristics of S1P receptors. *J. Cell. Biochem.* **2004**, *92*, 913–922. [CrossRef]
32. Windh, R.T.; Lee, M.J.; Hla, T.; An, S.; Barr, A.J.; Manning, D.R. Differential coupling of the sphingosine 1-phosphate receptors Edg-1, Edg-3, and H218/Edg-5 to the G(i), G(q), and G(12) families of heterotrimeric G proteins. *J. Biol. Chem.* **1999**, *274*, 27351–27358. [CrossRef] [PubMed]
33. Aloulou, A.; Rahier, R.; Arhab, Y.; Noiriel, A.; Abousalham, A. Phospholipases: An Overview. In *Lipases and Phospholipases: Methods and Protocols*; Sandoval, G., Ed.; Springer: New York, NY, USA, 2018.
34. Nishizuka, Y. Intracellular signaling by hydrolysis of phospholipids and activation of protein kinase C. *Science* **1992**, *258*, 607–614. [CrossRef] [PubMed]
35. Delon, C.; Manifava, M.; Wood, E.; Thompson, D.; Krugmann, S.; Pyne, S.; Ktistakis, N.T. Sphingosine kinase 1 is an intracellular effector of phosphatidic acid. *J. Biol. Chem.* **2004**, *279*, 44763–44774. [CrossRef] [PubMed]
36. Mitra, P.; Oskeritzian, C.A.; Payne, S.G.; Beaven, M.A.; Milstien, S.; Spiegel, S. Role of ABCC1 in export of sphingosine-1-phosphate from mast cells. *Proc. Natl. Acad. Sci. USA* **2006**, *103*, 16394–16399. [CrossRef] [PubMed]
37. Fukuhara, S.; Simmons, S.; Kawamura, S.; Inoue, A.; Orba, Y.; Tokudome, T.; Sunden, Y.; Arai, Y.; Moriwaki, K.; Ishida, J.; et al. The sphingosine-1-phosphate transporter Spns2 expressed on endothelial cells regulates lymphocyte trafficking in mice. *J. Clin. Investig.* **2012**, *122*, 1416–1426. [CrossRef]
38. Zhou, J.; Saba, J.D. Identification of the first mammalian sphingosine phosphate lyase gene and its functional expression in yeast. *Biochem. Biophys. Res. Commun.* **1998**, *242*, 502–507. [CrossRef] [PubMed]
39. Miller, C.R.; Perry, A. Glioblastoma. *Arch. Pathol. Lab. Med.* **2007**, *131*, 397–406. [CrossRef]
40. Stupp, R.; Mason, W.P.; van den Bent, M.J.; Weller, M.; Fisher, B.; Taphoorn, M.J.; Belanger, K.; Brandes, A.A.; Marosi, C.; Bogdahn, U.; et al. Radiotherapy plus concomitant and adjuvant temozolomide for glioblastoma. *N. Engl. J. Med.* **2005**, *352*, 987–996. [CrossRef]

41. Bao, S.; Wu, Q.; McLendon, R.E.; Hao, Y.; Shi, Q.; Hjelmeland, A.B.; Dewhirst, M.W.; Bigner, D.D.; Rich, J.N. Glioma stem cells promote radioresistance by preferential activation of the DNA damage response. *Nature* **2006**, *444*, 756–760. [CrossRef]
42. Bao, S.; Wu, Q.; Sathornsumetee, S.; Hao, Y.; Li, Z.; Hjelmeland, A.B.; Shi, Q.; McLendon, R.E.; Bigner, D.D.; Rich, J.N. Stem cell-like glioma cells promote tumor angiogenesis through vascular endothelial growth factor. *Cancer Res.* **2006**, *66*, 7843–7848. [CrossRef]
43. Libby, C.J.; Tran, A.N.; Scott, S.E.; Griguer, C.; Hjelmeland, A.B. The pro-tumorigenic effects of metabolic alterations in glioblastoma including brain tumor initiating cells. *Biochim. Biophys. Acta Rev. Cancer* **2018**, *1869*, 175–188. [CrossRef] [PubMed]
44. Lathia, J.D.; Gallagher, J.; Myers, J.T.; Li, M.; Vasanji, A.; McLendon, R.E.; Hjelmeland, A.B.; Huang, A.Y.; Rich, J.N. Direct in vivo evidence for tumor propagation by glioblastoma cancer stem cells. *PLoS ONE* **2011**, *6*, e24807. [CrossRef] [PubMed]
45. Marfia, G.; Campanella, R.; Navone, S.E.; Di Vito, C.; Riccitelli, E.; Hadi, L.A.; Bornati, A.; de Rezende, G.; Giussani, P.; Tringali, C.; et al. Autocrine/paracrine sphingosine-1-phosphate fuels proliferative and stemness qualities of glioblastoma stem cells. *Glia* **2014**, *62*, 1968–1981. [CrossRef]
46. Liebner, S.; Dijkhuizen, R.M.; Reiss, Y.; Plate, K.H.; Agalliu, D.; Constantin, G. Functional morphology of the blood-brain barrier in health and disease. *Acta Neuropathol.* **2018**, *135*, 311–336. [CrossRef] [PubMed]
47. Johnson, J.D.; Young, B. Demographics of brain metastasis. *Neurosurg. Clin. N. Am.* **1996**, *7*, 337–344. [CrossRef]
48. Nussbaum, E.S.; Djalilian, H.R.; Cho, K.H.; Hall, W.A. Brain metastases. Histology, multiplicity, surgery, and survival. *Cancer* **1996**, *78*, 1781–1788. [CrossRef]
49. Graf, A.H.; Buchberger, W.; Langmayr, H.; Schmid, K.W. Site preference of metastatic tumours of the brain. *Virchows Arch. A* **1988**, *412*, 493–498. [CrossRef]
50. Nayak, L.; Lee, E.Q.; Wen, P.Y. Epidemiology of Brain Metastases. *Curr. Oncol. Rep.* **2012**, *14*, 48–54. [CrossRef]
51. Lowery, F.J.; Yu, D. Brain metastasis: Unique challenges and open opportunities. *Biochim. Biophys. Acta Rev. Cancer* **2017**, *1867*, 49–57. [CrossRef]
52. Ogretmen, B. Sphingolipid metabolism in cancer signalling and therapy. *Nat. Rev. Cancer* **2018**, *18*, 33–50. [CrossRef]
53. Kotecha, R.; Gondi, V.; Ahluwalia, M.S.; Brastianos, P.K.; Mehta, M.P. Recent advances in managing brain metastasis. *F1000Research* **2018**, *7*. [CrossRef]
54. Hubbs, J.L.; Boyd, J.A.; Hollis, D.; Chino, J.P.; Saynak, M.; Kelsey, C.R. Factors associated with the development of brain metastases: Analysis of 975 patients with early stage nonsmall cell lung cancer. *Cancer* **2010**, *116*, 5038–5046. [CrossRef]
55. Dawood, S.; Broglio, K.; Esteva, F.J.; Yang, W.; Kau, S.W.; Islam, R.; Albarracin, C.; Yu, T.K.; Green, M.; Hortobagyi, G.N.; et al. Survival among women with triple receptor-negative breast cancer and brain metastases. *Ann. Oncol.* **2009**, *20*, 621–627. [CrossRef] [PubMed]
56. Zakrzewski, J.; Geraghty, L.N.; Rose, A.E.; Christos, P.J.; Mazumdar, M.; Polsky, D.; Shapiro, R.; Berman, R.; Darvishian, F.; Hernando, E.; et al. Clinical variables and primary tumor characteristics predictive of the development of melanoma brain metastases and post-brain metastases survival. *Cancer* **2011**, *117*, 1711–1720. [CrossRef] [PubMed]
57. Winkler, F. The brain metastatic niche. *J. Mol. Med.* **2015**, *93*, 1213–1220. [CrossRef] [PubMed]
58. Santana, P.; Peña, L.A.; Haimovitz-Friedman, A.; Martin, S.; Green, D.; McLoughlin, M.; Cordon-Cardo, C.; Schuchman, E.H.; Fuks, Z.; Kolesnick, R. Acid sphingomyelinase-deficient human lymphoblasts and mice are defective in radiation-induced apoptosis. *Cell* **1996**, *86*, 189–199. [CrossRef]
59. Dbaibo, G.S.; Pushkareva, M.Y.; Rachid, R.A.; Alter, N.; Smyth, M.J.; Obeid, L.M.; Hannun, Y.A. p53-dependent ceramide response to genotoxic stress. *J. Clin. Investig.* **1998**, *102*, 329–339. [CrossRef] [PubMed]
60. Grammatikos, G.; Teichgräber, V.; Carpinteiro, A.; Trarbach, T.; Weller, M.; Hengge, U.R.; Gulbins, E. Overexpression of acid sphingomyelinase sensitizes glioma cells to chemotherapy. *Antioxid. Redox Signal.* **2007**, *9*, 1449–1456. [CrossRef]
61. Gramatzki, D.; Herrmann, C.; Happold, C.; Becker, K.A.; Gulbins, E.; Weller, M.; Tabatabai, G. Glioma cell death induced by irradiation or alkylating agent chemotherapy is independent of the intrinsic ceramide pathway. *PLoS ONE* **2013**, *8*, e63527. [CrossRef] [PubMed]

62. Van Meir, E.G.; Kikuchi, T.; Tada, M.; Li, H.; Diserens, A.C.; Wojcik, B.E.; Huang, H.J.; Friedmann, T.; de Tribolet, N.; Cavenee, W.K. Analysis of the p53 gene and its expression in human glioblastoma cells. *Cancer Res.* **1994**, *54*, 649–652. [PubMed]
63. Wang, X.; Chen, J.X.; Liu, Y.H.; You, C.; Mao, Q. Mutant TP53 enhances the resistance of glioblastoma cells to temozolomide by up-regulating O(6)-methylguanine DNA-methyltransferase. *Neurol. Sci.* **2013**, *34*, 1421–1428. [CrossRef]
64. Hara, S.; Nakashima, S.; Kiyono, T.; Sawada, M.; Yoshimura, S.; Iwama, T.; Banno, Y.; Shinoda, J.; Sakai, N. p53-Independent ceramide formation in human glioma cells during gamma-radiation-induced apoptosis. *Cell Death Differ.* **2004**, *11*, 853–861. [CrossRef] [PubMed]
65. Sawada, M.; Nakashima, S.; Kiyono, T.; Nakagawa, M.; Yamada, J.; Yamakawa, H.; Banno, Y.; Shinoda, J.; Nishimura, Y.; Nozawa, Y.; et al. p53 regulates ceramide formation by neutral sphingomyelinase through reactive oxygen species in human glioma cells. *Oncogene* **2001**, *20*, 1368–1378. [CrossRef]
66. Xu, R.; Garcia-Barros, M.; Wen, S.; Li, F.; Lin, C.-L.; Hannun, Y.A.; Obeid, L.M.; Mao, C. Tumor suppressor p53 links ceramide metabolism to DNA damage response through alkaline ceramidase 2. *Cell Death Differ.* **2017**, 1–16. [CrossRef] [PubMed]
67. Tseng, T.H.; Shen, C.H.; Huang, W.S.; Chen, C.N.; Liang, W.H.; Lin, T.H.; Kuo, H.C. Activation of neutral-sphingomyelinase, MAPKs, and p75 NTR-mediating caffeic acid phenethyl ester-induced apoptosis in C6 glioma cells. *J. Biomed. Sci.* **2014**, *21*, 61. [CrossRef] [PubMed]
68. Kachler, K.; Bailer, M.; Heim, L.; Schumacher, F.; Reichel, M.; Holzinger, C.D.; Trump, S.; Mittler, S.; Monti, J.; Trufa, D.I.; et al. Enhanced Acid Sphingomyelinase Activity Drives Immune Evasion and Tumor Growth in Non-Small Cell Lung Carcinoma. *Cancer Res.* **2017**, *77*, 5963–5976. [CrossRef] [PubMed]
69. Carpinteiro, A.; Beckmann, N.; Seitz, A.; Hessler, G.; Wilker, B.; Soddemann, M.; Helfrich, I.; Edelmann, B.; Gulbins, E.; Becker, K.A. Role of Acid Sphingomyelinase-Induced Signaling in Melanoma Cells for Hematogenous Tumor Metastasis. *Cell Physiol. Biochem.* **2016**, *38*, 1–14. [CrossRef]
70. Dahiya, R.; Boyle, B.; Goldberg, B.C.; Yoon, W.H.; Konety, B.; Chen, K.; Yen, T.S.; Blumenfeld, W.; Narayan, P. Metastasis-associated alterations in phospholipids and fatty acids of human prostatic adenocarcinoma cell lines. *Biochem. Cell Biol.* **1992**, *70*, 548–554. [CrossRef]
71. Cervia, D.; Assi, E.; De Palma, C.; Giovarelli, M.; Bizzozero, L.; Pambianco, S.; Di Renzo, I.; Zecchini, S.; Moscheni, C.; Vantaggiato, C.; et al. Essential role for acid sphingomyelinase-inhibited autophagy in melanoma response to cisplatin. *Oncotarget* **2016**, *7*, 24995–25009. [CrossRef]
72. Smith, E.L.; Schuchman, E.H. Acid Sphingomyelinase Overexpression Enhances the Antineoplastic Effects of Irradiation In Vitro and In Vivo. *Mol. Ther.* **2008**, *16*, 1565–1571. [CrossRef]
73. Van Brocklyn, J.R.; Young, N.; Roof, R. Sphingosine-1-phosphate stimulates motility and invasiveness of human glioblastoma multiforme cells. *Cancer Lett.* **2003**, *199*, 53–60. [CrossRef]
74. Young, N.; Pearl, D.K.; Van Brocklyn, J.R. Sphingosine-1-phosphate regulates glioblastoma cell invasiveness through the urokinase plasminogen activator system and CCN1/Cyr61. *Mol. Cancer Res.* **2009**, *7*, 23–32. [CrossRef] [PubMed]
75. Doan, N.B.; Nguyen, H.S.; Al-Gizawiy, M.M.; Mueller, W.M.; Sabbadini, R.A.; Rand, S.D.; Connelly, J.M.; Chitambar, C.R.; Schmainda, K.M.; Mirza, S.P. Acid ceramidase confers radioresistance to glioblastoma cells. *Oncol. Rep.* **2017**, *38*, 1932–1940. [CrossRef] [PubMed]
76. Bien-Möller, S.; Lange, S.; Holm, T.; Böhm, A.; Paland, H.; Küpper, J.; Herzog, S.; Weitmann, K.; Havemann, C.; Vogelgesang, S.; et al. Expression of S1P metabolizing enzymes and receptors correlate with survival time and regulate cell migration in glioblastoma multiforme. *Oncotarget* **2016**, *7*, 13031–13046. [CrossRef] [PubMed]
77. Doan, N.B.; Alhajala, H.; Al-Gizawiy, M.M.; Mueller, W.M.; Rand, S.D.; Connelly, J.M.; Cochran, E.J.; Chitambar, C.R.; Clark, P.; Kuo, J.; et al. Acid ceramidase and its inhibitors: A de novo drug target and a new class of drugs for killing glioblastoma cancer stem cells with high efficiency. *Oncotarget* **2017**, *8*, 112662–112674. [CrossRef] [PubMed]
78. Riccitelli, E.; Giussani, P.; Di Vito, C.; Condomitti, G.; Tringali, C.; Caroli, M.; Galli, R.; Viani, P.; Riboni, L. Extracellular sphingosine-1-phosphate: A novel actor in human glioblastoma stem cell survival. *PLoS ONE* **2013**, *8*, e68229. [CrossRef] [PubMed]
79. Abdel Hadi, L.; Anelli, V.; Guarnaccia, L.; Navone, S.; Beretta, M.; Moccia, F.; Tringali, C.; Urechie, V.; Campanella, R.; Marfia, G.; et al. A bidirectional crosstalk between glioblastoma and brain endothelial

cells potentiates the angiogenic and proliferative signaling of sphingosine-1-phosphate in the glioblastoma microenvironment. *Biochim. Biophys. Acta Mol. Cell Biol. Lipids* **2018**, *1863*, 1179–1192. [CrossRef] [PubMed]

80. Qin, L.S.; Yu, Z.Q.; Zhang, S.M.; Sun, G.; Zhu, J.; Xu, J.; Guo, J.; Fu, L.S. The short chain cell-permeable ceramide (C6) restores cell apoptosis and perifosine sensitivity in cultured glioblastoma cells. *Mol. Biol. Rep.* **2013**, *40*, 5645–5655. [CrossRef]
81. Yoshida, Y.; Nakada, M.; Sugimoto, N.; Harada, T.; Hayashi, Y.; Kita, D.; Uchiyama, N.; Hayashi, Y.; Yachie, A.; Takuwa, Y.; et al. Sphingosine-1-phosphate receptor type 1 regulates glioma cell proliferation and correlates with patient survival. *Int. J. Cancer* **2010**, *126*, 2341–2352. [CrossRef]
82. Young, N.; Van Brocklyn, J.R. Roles of sphingosine-1-phosphate (S1P) receptors in malignant behavior of glioma cells. Differential effects of S1P2 on cell migration and invasiveness. *Exp. Cell Res.* **2007**, *313*, 1615–1627. [CrossRef]
83. Lepley, D.; Paik, J.H.; Hla, T.; Ferrer, F. The G protein-coupled receptor S1P2 regulates Rho/Rho kinase pathway to inhibit tumor cell migration. *Cancer Res.* **2005**, *65*, 3788–3795. [CrossRef]
84. Malchinkhuu, E.; Sato, K.; Maehama, T.; Mogi, C.; Tomura, H.; Ishiuchi, S.; Yoshimoto, Y.; Kurose, H.; Okajima, F. S1P(2) receptors mediate inhibition of glioma cell migration through Rho signaling pathways independent of PTEN. *Biochem. Biophys. Res. Commun.* **2008**, *366*, 963–968. [CrossRef]
85. Quint, K.; Stiel, N.; Neureiter, D.; Schlicker, H.U.; Nimsky, C.; Ocker, M.; Strik, H.; Kolodziej, M.A. The role of sphingosine kinase isoforms and receptors S1P1, S1P2, S1P3, and S1P5 in primary, secondary, and recurrent glioblastomas. *Tumour Biol.* **2014**, *35*, 8979–8989. [CrossRef] [PubMed]
86. Patel, A.P.; Tirosh, I.; Trombetta, J.J.; Shalek, A.K.; Gillespie, S.M.; Wakimoto, H.; Cahill, D.P.; Nahed, B.V.; Curry, W.T.; Martuza, R.L.; et al. Single-cell RNA-seq highlights intratumoral heterogeneity in primary glioblastoma. *Science* **2014**, *344*, 1396–1401. [CrossRef]
87. Jaillard, C.; Harrison, S.; Stankoff, B.; Aigrot, M.S.; Calver, A.R.; Duddy, G.; Walsh, F.S.; Pangalos, M.N.; Arimura, N.; Kaibuchi, K.; et al. Edg8/S1P5: An oligodendroglial receptor with dual function on process retraction and cell survival. *J. Neurosci.* **2005**, *25*, 1459–1469. [CrossRef]
88. Rothhammer, V.; Kenison, J.E.; Tjon, E.; Takenaka, M.C.; de Lima, K.A.; Borucki, D.M.; Chao, C.C.; Wilz, A.; Blain, M.; Healy, L.; et al. Sphingosine 1-phosphate receptor modulation suppresses pathogenic astrocyte activation and chronic progressive CNS inflammation. *Proc. Natl. Acad. Sci. USA* **2017**, *114*, 2012–2017. [CrossRef] [PubMed]
89. O'sullivan, S.A.; O'Sullivan, C.; Healy, L.M.; Dev, K.K.; Sheridan, G.K. Sphingosine 1-phosphate receptors regulate TLR4-induced CXCL5 release from astrocytes and microglia. *J. Neurochem.* **2018**, *144*, 736–747. [CrossRef]
90. Dai, Z.; Wu, J.; Chen, F.; Cheng, Q.; Zhang, M.; Wang, Y.; Guo, Y.; Song, T. CXCL5 promotes the proliferation and migration of glioma cells in autocrine- and paracrine-dependent manners. *Oncol. Rep.* **2016**, *36*, 3303–3310. [CrossRef]
91. Rostami, N.; Nikkhoo, A.; Ajjoolabady, A.; Azizi, G.; Hojjat-Farsangi, M.; Ghalamfarsa, G.; Yousefi, B.; Yousefi, M.; Jadidi-Niaragh, F. S1PR1 as a Novel Promising Therapeutic Target in Cancer Therapy. *Mol. Diagn. Ther.* **2019**, *23*, 467–487. [CrossRef] [PubMed]
92. Beider, K.; Rosenberg, E.; Bitner, H.; Shimoni, A.; Leiba, M.; Koren-Michowitz, M.; Ribakovsky, E.; Klein, S.; Olam, D.; Weiss, L.; et al. The Sphingosine-1-Phosphate Modulator FTY720 Targets Multiple Myeloma via the CXCR4/CXCL12 Pathway. *Clin. Cancer Res.* **2017**, *23*, 1733–1747. [CrossRef] [PubMed]
93. Azizidoost, S.; Asnafi, A.A.; Saki, N. Signaling-chemokine axis network in brain as a sanctuary site for metastasis. *J. Cell Physiol.* **2019**, *234*, 3376–3382. [CrossRef]
94. Foulkes, W.D.; Smith, I.E.; Reis-Filho, J.S. Triple-negative breast cancer. *N. Engl. J. Med.* **2010**, *363*, 1938–1948. [CrossRef] [PubMed]
95. Maiti, A.; Takabe, K.; Hait, N.C. Metastatic triple-negative breast cancer is dependent on SphKs/S1P signaling for growth and survival. *Cell Signal.* **2017**, *32*, 85–92. [CrossRef]
96. Wang, S.; Liang, Y.; Chang, W.; Hu, B.; Zhang, Y. Triple Negative Breast Cancer Depends on Sphingosine Kinase 1 (SphK1)/Sphingosine-1-Phosphate (S1P)/Sphingosine 1-Phosphate Receptor 3 (S1PR3)/Notch Signaling for Metastasis. *Med. Sci. Monit.* **2018**, *24*, 1912–1923. [CrossRef] [PubMed]
97. Nagahashi, M.; Yamada, A.; Katsuta, E.; Aoyagi, T.; Huang, W.C.; Terracina, K.P.; Hait, N.C.; Allegood, J.C.; Tsuchida, J.; Yuza, K.; et al. Targeting the SphK1/S1P/S1PR1 Axis That Links Obesity, Chronic Inflammation, and Breast Cancer Metastasis. *Cancer Res.* **2018**, *78*, 1713–1725. [CrossRef] [PubMed]

98. Wang, Y.; Shen, Y.; Sun, X.; Hong, T.L.; Huang, L.S.; Zhong, M. Prognostic roles of the expression of sphingosine-1-phosphate metabolism enzymes in non-small cell lung cancer. *Transl. Lung Cancer Res.* **2019**, *8*, 674–681. [CrossRef] [PubMed]
99. Zhao, J.; Liu, J.; Lee, J.F.; Zhang, W.; Kandouz, M.; VanHecke, G.C.; Chen, S.; Ahn, Y.H.; Lonardo, F.; Lee, M.J. TGF-β/SMAD3 Pathway Stimulates Sphingosine-1 Phosphate Receptor 3 Expression: IMPLICATION OF SPHINGOSINE-1 PHOSPHATE RECEPTOR 3 IN LUNG ADENOCARCINOMA PROGRESSION. *J. Biol. Chem.* **2016**, *291*, 27343–27353. [CrossRef]
100. Filipenko, I.; Schwalm, S.; Reali, L.; Pfeilschifter, J.; Fabbro, D.; Huwiler, A.; Zangemeister-Wittke, U. Upregulation of the S1P(3) receptor in metastatic breast cancer cells increases migration and invasion by induction of PGE(2) and EP(2)/EP(4) activation. *Biochim. Biophys. Acta* **2016**, *1861*, 1840–1851. [CrossRef]
101. Karagiannis, G.S.; Condeelis, J.S.; Oktay, M.H. Chemotherapy-induced metastasis: Mechanisms and translational opportunities. *Clin. Exp. Metastasis* **2018**, *35*, 269–284. [CrossRef]
102. Liu, G.; Chen, Y.; Qi, F.; Jia, L.; Lu, X.A.; He, T.; Fu, Y.; Li, L.; Luo, Y. Specific chemotherapeutic agents induce metastatic behaviour through stromal- and tumour-derived cytokine and angiogenic factor signalling. *J. Pathol.* **2015**, *237*, 190–202. [CrossRef]
103. Arikawa, K.; Takuwa, N.; Yamaguchi, H.; Sugimoto, N.; Kitayama, J.; Nagawa, H.; Takehara, K.; Takuwa, Y. Ligand-dependent inhibition of B16 melanoma cell migration and invasion via endogenous S1P2 G protein-coupled receptor. Requirement of inhibition of cellular RAC activity. *J. Biol. Chem.* **2003**, *278*, 32841–32851. [CrossRef] [PubMed]
104. Ohotski, J.; Long, J.S.; Orange, C.; Elsberger, B.; Mallon, E.; Doughty, J.; Pyne, S.; Pyne, N.J.; Edwards, J. Expression of sphingosine 1-phosphate receptor 4 and sphingosine kinase 1 is associated with outcome in oestrogen receptor-negative breast cancer. *Br. J. Cancer* **2012**, *106*, 1453–1459. [CrossRef] [PubMed]
105. Bedia, C.; Casas, J.; Andrieu-Abadie, N.; Fabriàs, G.; Levade, T. Acid ceramidase expression modulates the sensitivity of A375 melanoma cells to dacarbazine. *J. Biol. Chem.* **2011**, *286*, 28200–28209. [CrossRef] [PubMed]
106. Lai, M.; Realini, N.; La Ferla, M.; Passalacqua, I.; Matteoli, G.; Ganesan, A.; Pistello, M.; Mazzanti, C.M.; Piomelli, D. Complete Acid Ceramidase ablation prevents cancer-initiating cell formation in melanoma cells. *Sci. Rep.* **2017**, *7*, 7411. [CrossRef]
107. Pyne, N.J.; Pyne, S. Sphingosine kinase 1 enables communication between melanoma cells and fibroblasts that provides a new link to metastasis. *Oncogene* **2014**, *33*, 3361–3363. [CrossRef]
108. Van der Weyden, L.; Arends, M.J.; Campbell, A.D.; Bald, T.; Wardle-Jones, H.; Griggs, N.; Velasco-Herrera, M.D.; Tüting, T.; Sansom, O.J.; Karp, N.A.; et al. Genome-wide in vivo screen identifies novel host regulators of metastatic colonization. *Nature* **2017**, *541*, 233–236. [CrossRef]
109. Kang, D.W.; Hwang, W.C.; Noh, Y.N.; Park, K.S.; Min, D.S. Phospholipase D1 inhibition sensitizes glioblastoma to temozolomide and suppresses its tumorigenicity. *J. Pathol.* **2020**. [CrossRef]
110. Walker, K.; Boyd, N.H.; Anderson, J.C.; Willey, C.D.; Hjelmeland, A.B. Kinomic profiling of glioblastoma cells reveals PLCG1 as a target in restricted glucose. *Biomark. Res.* **2018**, *6*, 22. [CrossRef]
111. Mercurio, L.; Cecchetti, S.; Ricci, A.; Pacella, A.; Cigliana, G.; Bozzuto, G.; Podo, F.; Iorio, E.; Carpinelli, G. Phosphatidylcholine-specific phospholipase C inhibition down- regulates CXCR4 expression and interferes with proliferation, invasion and glycolysis in glioma cells. *PLoS ONE* **2017**, *12*, e0176108. [CrossRef]
112. Yao, Y.; Wang, X.; Li, H.; Fan, J.; Qian, X.; Li, H.; Xu, Y. Phospholipase D as a key modulator of cancer progression. *Biol. Rev. Camb. Philos. Soc.* **2020**, *95*, 911–935. [CrossRef]
113. Kim, E.S.; Kim, J.S.; Kim, S.G.; Hwang, S.; Lee, C.H.; Moon, A. Sphingosine 1-phosphate regulates matrix metalloproteinase-9 expression and breast cell invasion through S1P3-Gαq coupling. *J. Cell Sci.* **2011**, *124*, 2220–2230. [CrossRef] [PubMed]
114. Kato, Y.; Nakayama, Y.; Umeda, M.; Miyazaki, K. Induction of 103-kDa gelatinase/type IV collagenase by acidic culture conditions in mouse metastatic melanoma cell lines. *J. Biol. Chem.* **1992**, *267*, 11424–11430. [PubMed]
115. Kato, Y.; Ozawa, S.; Tsukuda, M.; Kubota, E.; Miyazaki, K.; St-Pierre, Y.; Hata, R. Acidic extracellular pH increases calcium influx-triggered phospholipase D activity along with acidic sphingomyelinase activation to induce matrix metalloproteinase-9 expression in mouse metastatic melanoma. *FEBS J.* **2007**, *274*, 3171–3183. [CrossRef] [PubMed]

116. Sato, K.; Ui, M.; Okajima, F. Differential roles of Edg-1 and Edg-5, sphingosine 1-phosphate receptors, in the signaling pathways in C6 glioma cells. *Mol. Brain Res.* **2000**, *85*, 151–160. [CrossRef]
117. Desai, N.N.; Zhang, H.; Olivera, A.; Mattie, M.E.; Spiegel, S. Sphingosine-1-phosphate, a metabolite of sphingosine, increases phosphatidic acid levels by phospholipase D activation. *J. Biol. Chem.* **1992**, *267*, 23122–23128.
118. Yoon, C.M.; Hong, B.S.; Moon, H.G.; Lim, S.; Suh, P.G.; Kim, Y.K.; Chae, C.B.; Gho, Y.S. Sphingosine-1-phosphate promotes lymphangiogenesis by stimulating S1P1/Gi/PLC/Ca2+ signaling pathways. *Blood* **2008**, *112*, 1129–1138. [CrossRef]
119. Stacker, S.A.; Achen, M.G.; Jussila, L.; Baldwin, M.E.; Alitalo, K. Lymphangiogenesis and cancer metastasis. *Nature Rev. Cancer* **2002**, *2*, 573–583. [CrossRef]
120. Paduch, R. The role of lymphangiogenesis and angiogenesis in tumor metastasis. *Cell Oncol.* **2016**, *39*, 397–410. [CrossRef]
121. Meacci, E.; Nuti, F.; Catarzi, S.; Vasta, V.; Donati, C.; Bourgoin, S.; Bruni, P.; Moss, J.; Vaughan, M. Activation of Phospholipase D by Bradykinin and Sphingosine 1-Phosphate in A549 Human Lung Adenocarcinoma Cells via Different GTP-Binding Proteins and Protein Kinase C Delta Signaling Pathways. *Biochemistry* **2003**, *42*, 284–292. [CrossRef]
122. Dusaban, S.S.; Purcell, N.H.; Rockenstein, E.; Masliah, E.; Cho, M.K.; Smrcka, A.V.; Brown, J.H. Phospholipase C epsilon links G protein-coupled receptor activation to inflammatory astrocytic responses. *Proc. Natl. Acad. Sci. USA* **2013**, *110*, 3609–3614. [CrossRef]
123. Obermeier, B.; Verma, A.; Ransohoff, R.M. The blood-brain barrier. *Handb. Clin. Neurol.* **2016**, *133*, 39–59. [CrossRef] [PubMed]
124. Sarkaria, J.N.; Hu, L.S.; Parney, I.F.; Pafundi, D.H.; Brinkmann, D.H.; Laack, N.N.; Giannini, C.; Burns, T.C.; Kizilbash, S.H.; Laramy, J.K.; et al. Is the blood-brain barrier really disrupted in all glioblastomas? A critical assessment of existing clinical data. *Neuro Oncol.* **2018**, *20*, 184–191. [CrossRef]
125. Park, M.H.; Lee, J.Y.; Park, K.H.; Jung, I.K.; Kim, K.T.; Lee, Y.S.; Ryu, H.H.; Jeong, Y.; Kang, M.; Schwaninger, M.; et al. Vascular and Neurogenic Rejuvenation in Aging Mice by Modulation of ASM. *Neuron* **2018**, *100*, 167–182.e169. [CrossRef]
126. Li, Y.Q.; Chen, P.; Haimovitz-Friedman, A.; Reilly, R.M.; Wong, C.S. Endothelial apoptosis initiates acute blood-brain barrier disruption after ionizing radiation. *Cancer Res.* **2003**, *63*, 5950–5956.
127. Prager, B.; Spampinato, S.F.; Ransohoff, R.M. Sphingosine 1-phosphate signaling at the blood-brain barrier. *Trends Mol. Med.* **2015**, *21*, 354–363. [CrossRef] [PubMed]
128. Nakagawa, S.; Aruga, J. Sphingosine 1-Phosphate Signaling Is Involved in Impaired Blood-Brain Barrier Function in Ischemia-Reperfusion Injury. *Mol. Neurobiol.* **2020**, *57*, 1594–1606. [CrossRef] [PubMed]
129. Stepanovska, B.; Lange, A.I.; Schwalm, S.; Pfeilschifter, J.; Coldewey, S.M.; Huwiler, A. Downregulation of S1P Lyase Improves Barrier Function in Human Cerebral Microvascular Endothelial Cells Following an Inflammatory Challenge. *Int. J. Mol. Sci.* **2020**, *21*, 1240. [CrossRef]
130. Li, Q.; Chen, B.; Zeng, C.; Fan, A.; Yuan, Y.; Guo, X.; Huang, X.; Huang, Q. Differential activation of receptors and signal pathways upon stimulation by different doses of sphingosine-1-phosphate in endothelial cells. *Exp. Physiol.* **2015**, *100*, 95–107. [CrossRef]
131. Van Doorn, R.; Lopes Pinheiro, M.A.; Kooij, G.; Lakeman, K.; van het Hof, B.; van der Pol, S.M.; Geerts, D.; van Horssen, J.; van der Valk, P.; van der Kam, E.; et al. Sphingosine 1-phosphate receptor 5 mediates the immune quiescence of the human brain endothelial barrier. *J. Neuroinflammation* **2012**, *9*, 133. [CrossRef]
132. Spampinato, S.F.; Obermeier, B.; Cotleur, A.; Love, A.; Takeshita, Y.; Sano, Y.; Kanda, T.; Ransohoff, R.M. Sphingosine 1 Phosphate at the Blood Brain Barrier: Can the Modulation of S1P Receptor 1 Influence the Response of Endothelial Cells and Astrocytes to Inflammatory Stimuli? *PLoS ONE* **2015**, *10*, e0133392. [CrossRef]
133. Park, S.H.; Kim, M.J.; Jung, H.H.; Chang, W.S.; Choi, H.S.; Rachmilevitch, I.; Zadicario, E.; Chang, J.W. Safety and feasibility of multiple blood-brain barrier disruptions for the treatment of glioblastoma in patients undergoing standard adjuvant chemotherapy. *J. Neurosurg.* **2020**, 1–9. [CrossRef]
134. Yanagida, K.; Liu, C.H.; Faraco, G.; Galvani, S.; Smith, H.K.; Burg, N.; Anrather, J.; Sanchez, T.; Iadecola, C.; Hla, T. Size-selective opening of the blood-brain barrier by targeting endothelial sphingosine 1-phosphate receptor 1. *Proc. Natl. Acad. Sci. USA* **2017**, *114*, 4531–4536. [CrossRef]

135. Fecci, P.E.; Heimberger, A.B.; Sampson, J.H. Immunotherapy for Primary Brain Tumors: No Longer a Matter of Privilege. *Clin. Cancer Res.* **2014**, *20*, 5620. [CrossRef] [PubMed]
136. Prionisti, I.; Bühler, L.H.; Walker, P.R.; Jolivet, R.B. Harnessing Microglia and Macrophages for the Treatment of Glioblastoma. *Front. Pharmacol.* **2019**, *10*, 506. [CrossRef] [PubMed]
137. Maceyka, M.; Spiegel, S. Sphingolipid metabolites in inflammatory disease. *Nature* **2014**, *510*, 58–67. [CrossRef] [PubMed]
138. Sweeney, M.D.; Zhao, Z.; Montagne, A.; Nelson, A.R.; Zlokovic, B.V. Blood-Brain Barrier: From Physiology to Disease and Back. *Physiol. Rev.* **2019**, *99*, 21–78. [CrossRef] [PubMed]
139. Klemm, F.; Maas, R.R.; Bowman, R.L.; Kornete, M.; Soukup, K.; Nassiri, S.; Brouland, J.-P.; Iacobuzio-Donahue, C.A.; Brennan, C.; Tabar, V.; et al. Interrogation of the Microenvironmental Landscape in Brain Tumors Reveals Disease-Specific Alterations of Immune Cells. *Cell* **2020**, *181*, 1643–1660.e1617. [CrossRef]
140. Fecci, P.E.; Mitchell, D.A.; Whitesides, J.F.; Xie, W.; Friedman, A.H.; Archer, G.E.; Herndon, J.E., 2nd; Bigner, D.D.; Dranoff, G.; Sampson, J.H. Increased regulatory T-cell fraction amidst a diminished CD4 compartment explains cellular immune defects in patients with malignant glioma. *Cancer Res.* **2006**, *66*, 3294–3302. [CrossRef] [PubMed]
141. Wainwright, D.A.; Dey, M.; Chang, A.; Lesniak, M.S. Targeting Tregs in Malignant Brain Cancer: Overcoming IDO. *Front. Immunol.* **2013**, *4*, 116. [CrossRef]
142. Bai, A.; Guo, Y. Acid sphingomyelinase mediates human CD4(+) T-cell signaling: Potential roles in T-cell responses and diseases. *Cell Death Dis.* **2017**, *8*, e2963. [CrossRef]
143. Zhou, Y.; Salker, M.S.; Walker, B.; Münzer, P.; Borst, O.; Gawaz, M.; Gulbins, E.; Singh, Y.; Lang, F. Acid Sphingomyelinase (ASM) is a Negative Regulator of Regulatory T Cell (Treg) Development. *Cell Physiol. Biochem.* **2016**, *39*, 985–995. [CrossRef] [PubMed]
144. Chongsathidkiet, P.; Jackson, C.; Koyama, S.; Loebel, F.; Cui, X.; Farber, S.H.; Woroniecka, K.; Elsamadicy, A.A.; Dechant, C.A.; Kemeny, H.R.; et al. Sequestration of T cells in bone marrow in the setting of glioblastoma and other intracranial tumors. *Nat. Med.* **2018**, *24*, 1459–1468. [CrossRef]
145. Brinkmann, V.; Billich, A.; Baumruker, T.; Heining, P.; Schmouder, R.; Francis, G.; Aradhye, S.; Burtin, P. Fingolimod (FTY720): Discovery and development of an oral drug to treat multiple sclerosis. *Nat. Rev. Drug Discov.* **2010**, *9*, 883–897. [CrossRef]
146. Volpi, C.; Orabona, C.; Macchiarulo, A.; Bianchi, R.; Puccetti, P.; Grohmann, U. Preclinical discovery and development of fingolimod for the treatment of multiple sclerosis. *Expert Opin. Drug Discov.* **2019**, *14*, 1199–1212. [CrossRef] [PubMed]
147. Chen, Z.; Feng, X.; Herting, C.J.; Garcia, V.A.; Nie, K.; Pong, W.W.; Rasmussen, R.; Dwivedi, B.; Seby, S.; Wolf, S.A.; et al. Cellular and Molecular Identity of Tumor-Associated Macrophages in Glioblastoma. *Cancer Res.* **2017**, *77*, 2266–2278. [CrossRef] [PubMed]
148. Hambardzumyan, D.; Gutmann, D.H.; Kettenmann, H. The role of microglia and macrophages in glioma maintenance and progression. *Nat. Neurosci.* **2016**, *19*, 20–27. [CrossRef]
149. Murray, P.J. Macrophage Polarization. *Annu. Rev. Physiol.* **2017**, *79*, 541–566. [CrossRef]
150. Perrotta, C.; Cervia, D.; Di Renzo, I.; Moscheni, C.; Bassi, M.T.; Campana, L.; Martelli, C.; Catalani, E.; Giovarelli, M.; Zecchini, S.; et al. Nitric Oxide Generated by Tumor-Associated Macrophages Is Responsible for Cancer Resistance to Cisplatin and Correlated With Syntaxin 4 and Acid Sphingomyelinase Inhibition. *Front. Immunol.* **2018**, *9*, 1186. [CrossRef]
151. Assi, E.; Cervia, D.; Bizzozero, L.; Capobianco, A.; Pambianco, S.; Morisi, F.; De Palma, C.; Moscheni, C.; Pellegrino, P.; Clementi, E.; et al. Modulation of Acid Sphingomyelinase in Melanoma Reprogrammes the Tumour Immune Microenvironment. *Mediat. Inflamm.* **2015**, *2015*, 370482. [CrossRef]
152. Gude, D.R.; Alvarez, S.E.; Paugh, S.W.; Mitra, P.; Yu, J.; Griffiths, R.; Barbour, S.E.; Milstien, S.; Spiegel, S. Apoptosis induces expression of sphingosine kinase 1 to release sphingosine-1-phosphate as a "come-and-get-me" signal. *FASEB J.* **2008**, *22*, 2629–2638. [CrossRef]
153. Weigert, A.; Tzieply, N.; von Knethen, A.; Johann, A.M.; Schmidt, H.; Geisslinger, G.; Brüne, B. Tumor cell apoptosis polarizes macrophages role of sphingosine-1-phosphate. *Mol. Biol. Cell* **2007**, *18*, 3810–3819. [CrossRef]

154. Mrad, M.; Imbert, C.; Garcia, V.; Rambow, F.; Therville, N.; Carpentier, S.; Ségui, B.; Levade, T.; Azar, R.; Marine, J.C.; et al. Downregulation of sphingosine kinase-1 induces protective tumor immunity by promoting M1 macrophage response in melanoma. *Oncotarget* **2016**, *7*, 71873–71886. [CrossRef] [PubMed]
155. Müller, J.; von Bernstorff, W.; Heidecke, C.D.; Schulze, T. Differential S1P Receptor Profiles on M1- and M2-Polarized Macrophages Affect Macrophage Cytokine Production and Migration. *Biomed. Res. Int.* **2017**, *2017*, 7584621. [CrossRef] [PubMed]
156. Weichand, B.; Popp, R.; Dziumbla, S.; Mora, J.; Strack, E.; Elwakeel, E.; Frank, A.C.; Scholich, K.; Pierre, S.; Syed, S.N.; et al. S1PR1 on tumor-associated macrophages promotes lymphangiogenesis and metastasis via NLRP3/IL-1β. *J. Exp. Med.* **2017**, *214*, 2695–2713. [CrossRef]
157. Kim, E.Y.; Choi, B.; Kim, J.E.; Park, S.O.; Kim, S.M.; Chang, E.J. Interleukin-22 Mediates the Chemotactic Migration of Breast Cancer Cells and Macrophage Infiltration of the Bone Microenvironment by Potentiating S1P/SIPR Signaling. *Cells* **2020**, *9*, 131. [CrossRef] [PubMed]
158. Li, Y.; Zhou, T.; Wang, Y.; Ning, C.; Lv, Z.; Han, G.; Morris, J.C.; Taylor, E.N.; Wang, R.; Xiao, H.; et al. The protumorigenic potential of FTY720 by promoting extramedullary hematopoiesis and MDSC accumulation. *Oncogene* **2017**, *36*, 3760–3771. [CrossRef]
159. Guo, X.D.; Ji, J.; Xue, T.F.; Sun, Y.Q.; Guo, R.B.; Cheng, H.; Sun, X.L. FTY720 Exerts Anti-Glioma Effects by Regulating the Glioma Microenvironment Through Increased CXCR4 Internalization by Glioma-Associated Microglia. *Front. Immunol.* **2020**, *11*, 178. [CrossRef]
160. Karunakaran, I.; Alam, S.; Jayagopi, S.; Frohberger, S.J.; Hansen, J.N.; Kuehlwein, J.; Hölbling, B.V.; Schumak, B.; Hübner, M.P.; Gräler, M.H.; et al. Neural sphingosine 1-phosphate accumulation activates microglia and links impaired autophagy and inflammation. *Glia* **2019**, *67*, 1859–1872. [CrossRef]
161. Tea, M.N.; Poonnoose, S.I.; Pitson, S.M. Targeting the Sphingolipid System as a Therapeutic Direction for Glioblastoma. *Cancers* **2020**, *12*, 111. [CrossRef]
162. Naser, E.; Kadow, S.; Schumacher, F.; Mohamed, Z.H.; Kappe, C.; Hessler, G.; Pollmeier, B.; Kleuser, B.; Arenz, C.; Becker, K.A.; et al. Characterization of the small molecule ARC39, a direct and specific inhibitor of acid sphingomyelinase in vitro. *J. Lipid Res.* **2020**, *61*, 896–910. [CrossRef]
163. Magrassi, L.; Adorni, L.; Montorfano, G.; Rapelli, S.; Butti, G.; Berra, B.; Milanesi, G. Vitamin D metabolites activate the sphingomyelin pathway and induce death of glioblastoma cells. *Acta Neurochir.* **1998**, *140*, 707–714. [CrossRef] [PubMed]
164. Oancea-Castillo, L.R.; Klein, C.; Abdollahi, A.; Weber, K.J.; Régnier-Vigouroux, A.; Dokic, I. Comparative analysis of the effects of a sphingosine kinase inhibitor to temozolomide and radiation treatment on glioblastoma cell lines. *Cancer Biol. Ther.* **2017**, *18*, 400–406. [CrossRef] [PubMed]
165. French, K.J.; Schrecengost, R.S.; Lee, B.D.; Zhuang, Y.; Smith, S.N.; Eberly, J.L.; Yun, J.K.; Smith, C.D. Discovery and Evaluation of Inhibitors of Human Sphingosine Kinase. *Cancer Res.* **2003**, *63*, 5962. [PubMed]
166. Bektas, M.; Johnson, S.P.; Poe, W.E.; Bigner, D.D.; Friedman, H.S. A sphingosine kinase inhibitor induces cell death in temozolomide resistant glioblastoma cells. *Cancer Chemother. Pharmacol.* **2009**, *64*, 1053–1058. [CrossRef]
167. Sordillo, L.A.; Sordillo, P.P.; Helson, L. Sphingosine Kinase Inhibitors as Maintenance Therapy of Glioblastoma After Ceramide-Induced Response. *Anticancer Res.* **2016**, *36*, 2085–2095.
168. Cattaneo, M.G.; Vanetti, C.; Samarani, M.; Aureli, M.; Bassi, R.; Sonnino, S.; Giussani, P. Cross-talk between sphingosine-1-phosphate and EGFR signaling pathways enhances human glioblastoma cell invasiveness. *FEBS Lett.* **2018**, *592*, 949–961. [CrossRef]
169. Dementiev, A.; Joachimiak, A.; Nguyen, H.; Gorelik, A.; Illes, K.; Shabani, S.; Gelsomino, M.; Ahn, E.-Y.E.; Nagar, B.; Doan, N. Molecular Mechanism of Inhibition of Acid Ceramidase by Carmofur. *J. Med. Chem.* **2019**, *62*, 987–992. [CrossRef]
170. White-Gilbertson, S.; Lu, P.; Jones, C.M.; Chiodini, S.; Hurley, D.; Das, A.; Delaney, J.R.; Norris, J.S.; Voelkel-Johnson, C. Tamoxifen is a candidate first-in-class inhibitor of acid ceramidase that reduces amitotic division in polyploid giant cancer cells—Unrecognized players in tumorigenesis. *Cancer Med.* **2020**, *9*, 3142–3152. [CrossRef]
171. Mahdy, A.E.; Cheng, J.C.; Li, J.; Elojeimy, S.; Meacham, W.D.; Turner, L.S.; Bai, A.; Gault, C.R.; McPherson, A.S.; Garcia, N.; et al. Acid ceramidase upregulation in prostate cancer cells confers resistance to radiation: AC inhibition, a potential radiosensitizer. *Mol. Ther.* **2009**, *17*, 430–438. [CrossRef]

172. Proksch, D.; Klein, J.J.; Arenz, C. Potent inhibition of Acid ceramidase by novel B-13 analogues. *J. Lipids* **2011**, *2011*, 971618. [CrossRef]
173. Reynolds, C.P.; Maurer, B.J.; Kolesnick, R.N. Ceramide synthesis and metabolism as a target for cancer therapy. *Cancer Lett.* **2004**, *206*, 169–180. [CrossRef] [PubMed]
174. Corazzari, M.; Lovat, P.E.; Oliverio, S.; Di Sano, F.; Donnorso, R.P.; Redfern, C.P.; Piacentini, M. Fenretinide: A p53-independent way to kill cancer cells. *Biochem. Biophys. Res. Commun.* **2005**, *331*, 810–815. [CrossRef] [PubMed]
175. Orienti, I.; Salvati, V.; Sette, G.; Zucchetti, M.; Bongiorno-Borbone, L.; Peschiaroli, A.; Zolla, L.; Francescangeli, F.; Ferrari, M.; Matteo, C.; et al. A novel oral micellar fenretinide formulation with enhanced bioavailability and antitumour activity against multiple tumours from cancer stem cells. *J. Exp. Clin. Cancer Res.* **2019**, *38*, 373. [CrossRef] [PubMed]
176. Song, M.M.; Makena, M.R.; Hindle, A.; Koneru, B.; Nguyen, T.H.; Verlekar, D.U.; Cho, H.; Maurer, B.J.; Kang, M.H.; Reynolds, C.P. Cytotoxicity and molecular activity of fenretinide and metabolites in T-cell lymphoid malignancy, neuroblastoma, and ovarian cancer cell lines in physiological hypoxia. *Anti Cancer Drugs* **2019**, *30*, 117–127. [CrossRef] [PubMed]
177. Puduvalli, V.K.; Yung, W.K.; Hess, K.R.; Kuhn, J.G.; Groves, M.D.; Levin, V.A.; Zwiebel, J.; Chang, S.M.; Cloughesy, T.F.; Junck, L.; et al. Phase II study of fenretinide (NSC 374551) in adults with recurrent malignant gliomas: A North American Brain Tumor Consortium study. *J. Clin. Oncol.* **2004**, *22*, 4282–4289. [CrossRef]
178. Erbay, M.F.; Kamisli, O. A case of high grade glioma following treatment of relapsing-remitting multiple sclerosis with fingolimod. *Neurol. India* **2020**, *68*, 478–480. [CrossRef]
179. Schuhmann, M.K.; Bittner, S.; Meuth, S.G.; Kleinschnitz, C.; Fluri, F. Fingolimod (FTY720-P) Does Not Stabilize the Blood-Brain Barrier under Inflammatory Conditions in an in Vitro Model. *Int. J. Mol. Sci.* **2015**, *16*, 29454–29466. [CrossRef]
180. Cartier, A.; Leigh, T.; Liu, C.H.; Hla, T. Endothelial sphingosine 1-phosphate receptors promote vascular normalization and antitumor therapy. *Proc. Natl. Acad. Sci. USA* **2020**, *117*, 3157–3166. [CrossRef]
181. Pchejetski, D.; Bohler, T.; Brizuela, L.; Sauer, L.; Doumerc, N.; Golzio, M.; Salunkhe, V.; Teissié, J.; Malavaud, B.; Waxman, J.; et al. FTY720 (fingolimod) sensitizes prostate cancer cells to radiotherapy by inhibition of sphingosine kinase-1. *Cancer Res.* **2010**, *70*, 8651–8661. [CrossRef]
182. Paugh, S.W.; Payne, S.G.; Barbour, S.E.; Milstien, S.; Spiegel, S. The immunosuppressant FTY720 is phosphorylated by sphingosine kinase type 2. *FEBS Lett.* **2003**, *554*, 189–193. [CrossRef]
183. Alshaker, H.; Wang, Q.; Srivats, S.; Chao, Y.; Cooper, C.; Pchejetski, D. New FTY720-docetaxel nanoparticle therapy overcomes FTY720-induced lymphopenia and inhibits metastatic breast tumour growth. *Breast Cancer Res. Treat.* **2017**, *165*, 531–543. [CrossRef]
184. Ito, K.; Yamaguchi, A.; Miura, K.; Kato, T.; Baba, S.; Matsumoto, S.; Ishii, M.; Takagi, H. Effect of oral adjuvant therapy with Carmofur (HCFU) for distant metastasis of colorectal cancer. *Int. J. Clin. Oncol.* **2000**, *5*, 29–35. [CrossRef]
185. Realini, N.; Solorzano, C.; Pagliuca, C.; Pizzirani, D.; Armirotti, A.; Luciani, R.; Costi, M.P.; Bandiera, T.; Piomelli, D. Discovery of highly potent acid ceramidase inhibitors with in vitro tumor chemosensitizing activity. *Sci. Rep.* **2013**, *3*, 1035. [CrossRef]
186. Matsumoto, S.; Nishizawa, S.; Murakami, M.; Noma, S.; Sano, A.; Kuroda, Y. Carmofur-induced leukoencephalopathy: MRI. *Neuroradiology* **1995**, *37*, 649–652. [CrossRef] [PubMed]
187. Gao, M.Q.; Gao, H.; Han, M.; Liu, K.L.; Peng, J.J.; Han, Y.T. Hispidulin suppresses tumor growth and metastasis in renal cell carcinoma by modulating ceramide-sphingosine 1-phosphate rheostat. *Am. J. Cancer Res.* **2017**, *7*, 1501–1514. [PubMed]
188. Lewis, C.S.; Voelkel-Johnson, C.; Smith, C.D. Targeting Sphingosine Kinases for the Treatment of Cancer. *Adv. Cancer Res.* **2018**, *140*, 295–325. [CrossRef] [PubMed]
189. Antoon, J.W.; White, M.D.; Driver, J.L.; Burow, M.E.; Beckman, B.S. Sphingosine kinase isoforms as a therapeutic target in endocrine therapy resistant luminal and basal-A breast cancer. *Exp. Biol. Med.* **2012**, *237*, 832–844. [CrossRef] [PubMed]

© 2020 by the authors. Licensee MDPI, Basel, Switzerland. This article is an open access article distributed under the terms and conditions of the Creative Commons Attribution (CC BY) license (http://creativecommons.org/licenses/by/4.0/).

Review

Sphingomyelinases and Liver Diseases

Naroa Insausti-Urkia [1,2,3,4,†], Estel Solsona-Vilarrasa [1,2,3,4,†], Carmen Garcia-Ruiz [1,2,3,5,*] and Jose C. Fernandez-Checa [1,2,3,5,*]

1 Department of Cell Death and Proliferation, Instituto de Investigaciones Biomédicas de Barcelona, Consejo Superior de Investigaciones Científicas, 08036 Barcelona, Spain; naroaiu@hotmail.com (N.I.-U.); estelsolsona@gmail.com (E.S.-V.)

2 Liver Unit Hospital Clínic i Provincial, IDIBAPS, 08036 Barcelona, Spain

3 Centro de Investigación Biomédica en Red (CIBERehd), 28029 Madrid, Spain

4 Department of Biomedical Sciences, School of Medicine, University of Barcelona, 08036 Barcelona, Spain

5 Southern California Research Center for ALPD and Cirrhosis, Keck School of Medicine, USC, Los Angeles, CA 90033, USA

* Correspondence: cgrbam@iibb.csic.es (C.G.-R.); checa229@yahoo.com (J.C.F.-C.); Tel.: +34-933638310 (C.G.-R.); +34-932275400 (ext. 2905) (J.C.F.-C.)

† These authors contributed equally to the work.

Received: 2 October 2020; Accepted: 28 October 2020; Published: 30 October 2020

Abstract: Sphingolipids (SLs) are critical components of membrane bilayers that play a crucial role in their physico-chemical properties. Ceramide is the prototype and most studied SL due to its role as a second messenger in the regulation of multiple signaling pathways and cellular processes. Ceramide is a heterogeneous lipid entity determined by the length of the fatty acyl chain linked to its carbon backbone sphingosine, which can be generated either by de novo synthesis from serine and palmitoyl-CoA in the endoplasmic reticulum or via sphingomyelin (SM) hydrolysis by sphingomyelinases (SMases). Unlike de novo synthesis, SMase-induced SM hydrolysis represents a rapid and transient mechanism of ceramide generation in specific intracellular sites that accounts for the diverse biological effects of ceramide. Several SMases have been described at the molecular level, which exhibit different pH requirements for activity: neutral, acid or alkaline. Among the SMases, the neutral (NSMase) and acid (ASMase) are the best characterized for their contribution to signaling pathways and role in diverse pathologies, including liver diseases. As part of a Special Issue (Phospholipases: From Structure to Biological Function), the present invited review summarizes the physiological functions of NSMase and ASMase and their role in chronic and metabolic liver diseases, of which the most relevant is nonalcoholic steatohepatitis and its progression to hepatocellular carcinoma, due to the association with the obesity and type 2 diabetes epidemic. A better understanding of the regulation and role of SMases in liver pathology may offer the opportunity for novel treatments of liver diseases.

Keywords: ceramide; sphingomyelin; acidic sphingomyelinase; neutral sphingomyelinase; hepatocellular carcinoma; alcoholic and nonalcoholic steatohepatitis

---

1. Introduction: Metabolism and Regulation of Sphingolipids

Sphingolipids (SLs) are a family of lipids ubiquitously present in all cells that contain a long-chain base called sphingosine. Among the best understood and characterized SLs are ceramides, a heterogeneous group of lipids featuring an acyl chain linked to the sphingosine via an amide bond. The molecular identity of each ceramide molecule is determined by the specific fatty acyl moiety, which encompasses short to very long fatty acids (C2–C34). Complex SLs consist of ceramide and various head groups covalently attached to the hydroxyl group of the sphingosine through an ester or a glycosidic bond. For instance, a phosphocholine molecule is found in sphingomyelin (SM), while either a simple sugar residue or more complex carbohydrates determine the head group of glycosphingolipids (GSLs), a complex family of SLs that include cerebrosides, globosides and gangliosides [1,2].

Following their discovery from brain extracts in the 19th century, SLs have been merely considered as structural components of biological membranes. Their distinctive association with cholesterol defines specific domains of membrane bilayers that exhibit unique physical properties, and are used as scaffolds for key signaling platforms involved in diverse cellular processes [2]. Recent evidence over the last decade indicated that SLs are involved in a plethora of cellular functions, including cell growth, cell death, inflammation, immune responses, cell adhesion and migration, angiogenesis, nutrient uptake and responses to stress stimuli and autophagy. The wide range of biological effects of SLs is related not only to their structural diversity but also to their subcellular distribution and mechanism of generation [3,4]. Ceramide is the prototype SL that has been the most intensively characterized due to its role as a second messenger in the regulation of metabolism and cell death pathways in response to stress, apoptotic triggers and chemotherapy [5–8]. Ceramide can arise from the endoplasmic reticulum (ER), where the molecular machinery for its de novo synthesis resides [9]. In addition, ceramide can also be generated from the hydrolysis of sphingomyelin (SM) either at the plasma membrane or in intracellular acidic compartments (endosomes/lysosomes) by the activation of sphingomyelinases (SMases) [10]. Once generated, ceramide can be converted into a variety of metabolites. The deacylation of ceramide by ceramidases (CDases) yields sphingosine, which can be phosphorylated by sphingosine kinase (SK) to sphingosine-1-phosphate (S1P), another SL with an important role in cell signaling (Figure 1), or reacylated back to ceramide in the so-called salvage pathway [11,12]. Besides, the trafficking of ceramides to the Golgi mediated by the ceramide transfer protein (CERT) fuels the synthesis of complex GSLs and SMs (Figure 1). While the basal levels of ceramides in healthy cells are low, in response to many deleterious stimuli causing stress, apoptosis and cell death, cells trigger a rapid and transient mechanism of ceramide generation in specific sites due to SMase activation in distinct intracellular compartments, predominantly in lysosomes and the plasma membrane, that activate particular signaling pathways. Given the role of ceramide in hepatocellular apoptosis and fibrosis (see below), and since SMases represent the predominant pathway for the sudden generation of ceramide, in the present review, we summarize the role of SMases in liver pathology, including predominant chronic liver diseases, such as alcoholic and nonalcoholic steatohepatitis (ASH/NASH) and their progression to hepatocellular carcinoma (HCC) [13–15].

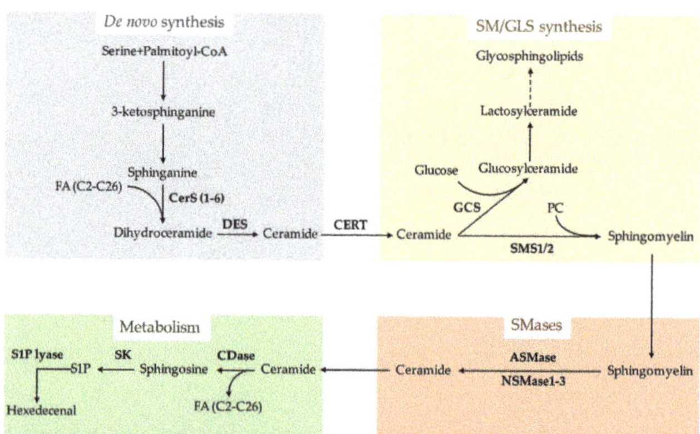

Figure 1. Synthesis and metabolism of sphingolipids (SLs). Ceramide is the prototype SL, which is synthesized de novo in the ER from serine and palmitoyl-CoA (upper left panel). The molecular identity of ceramide is determined by the length of the acyl chain linked to the carbon backbone. Six different

ceramide synthases (CerS 1–6) exhibit differential affinity towards fatty acids of different length (C2–C34). Once synthesized in the ER, ceramide is transported to the Golgi (upper right panel) and serves as the substrate for glucosylceramide synthase (GCS) to generate glucosylceramide or sphingomyelin synthases (SMS1/2) to yield sphingomyelin from phosphatidylcholine (PC). The distribution and subsequent hydrolysis of sphingomyelin in different membrane bilayers by SMases (lower right panel) represents a fast mechanism of almost instant ceramide generation. Ceramide can be catabolized by ceramidase (CDase) (lower left panel) to generate sphingosine. Sphingosine can be phosphorylated by sphingosine kinase (SK) into sphingosine 1-phosphate (S1P), a bioactive lipid, which can be further degraded by S1P lyase into hexadecenal. In addition, the pool of sphingosine generated by CDase can be reacylated by CerS back into ceramide in the so-called salvage pathway.

2. Ceramide Generation: De Novo Synthesis and Sphingomyelin Hydrolysis by Sphingomyelinases

2.1. De Novo Synthesis

The de novo pathway of ceramide generation occurs in the ER (Figure 1). In this pathway, the amino acid serine is conjugated with palmitoyl-CoA in a step catalyzed by the rate-limiting enzyme serine palmitoyl transferase (SPT). The product of the reaction sphinganine is acylated by ceramide synthases (CerSs) to dihydroceramide. Subsequent dehydrogenation catalyzed by dihydroceramide desaturase (DES) generates ceramide. In addition, CerSs also catalyze the reacylation of sphingosine to ceramide in the salvage pathway. Six different CerSs have been identified [16,17], which exhibit tissue-specific expression and substrate selectivity, thereby providing the basis for the generation of singular ceramide species of variable acyl chains in particular tissues. For instance, the ceramide synthase CerC2 is widely expressed and of major importance in the liver and preferentially incorporates long-chain C20–C24 acyl residues to generate C20–C24 ceramides. CerS3 is preferentially expressed in the skin and catalyzes the acylation of very long acyl chains up to C34:0 to sphinganine. The ceramide synthase CerC5 specifically catalyzes the generation of C16 ceramide, while the ceramide synthase CerC6 shows a wide substrate selectivity, and it is involved in the generation of C14, C16 and C18 ceramides [16,18]. Importantly, ceramides with different acyl chain lengths are generated in specific physiological and pathophysiological contexts in a tissue- and cell-dependent fashion. Despite this defined specific profile of ceramide synthesized by the different CerSs, there are compensatory mechanisms that offset the absence of specific ceramide species. In this regard, an increase in a particular CerS may regulate a specific ceramide pool that may affect the integrity and function of individual cell compartments, such as lysosomes, the ER or mitochondria. For instance, CerS2 knockout mice exhibit a compensatory increase in the levels of C16 in the liver, which triggers hepatocyte apoptosis and proliferation, leading to hepatocellular hyperplasia [19]. These changes in ceramide homeostasis translate into increased rates of hepatocyte apoptosis, mitochondrial dysfunction and mitochondrial ROS generation, as well as proliferation, that progress to the widespread formation of nodules of regenerative hepatocellular hyperplasia in aged mice. Progressive hepatomegaly and noninvasive liver tumors are observed in 10-month-old CerS2$^{-/-}$ mice [20]. An important factor that controls the de novo ceramide synthesis involves the availability of the substrate palmitoyl-CoA, which is required for sphinganine synthesis and whose level increases in obesity, metabolic syndrome and related disorders (e.g., NASH) [21,22]. Thus, the obesity-related increase in palmitoyl-CoA is expected to enhance ceramide synthesis.

The de novo synthesized ceramide can then be distributed to distinct intracellular compartments, such as the Golgi, where it acts as a source of SM or GSLs. Besides, SM trafficking to lysosomes or the plasma membrane can generate discrete ceramide species due to the local activation of SMases, which initiate specific signaling pathways that account for the diverse biological actions of ceramide.

2.2. Sphingomyelinases: Types and Function

The SMases comprise a family of enzymes that catalyze SM hydrolysis with different biochemical characteristics, yielding ceramide and phosphocholine. The SMases encompass three subclasses based on their optimal pH and subcellular localization: acid (ASMase), neutral (NSMase) and alkaline (Alk-SMase). While Alk-SMase is mainly localized in the gastrointestinal tract and to some extent in the liver, NSMase and ASMase are ubiquitous and account for the generation of ceramide in specific intracellular compartments, predominantly in the plasma membrane and lysosomes, respectively (Figure 2).

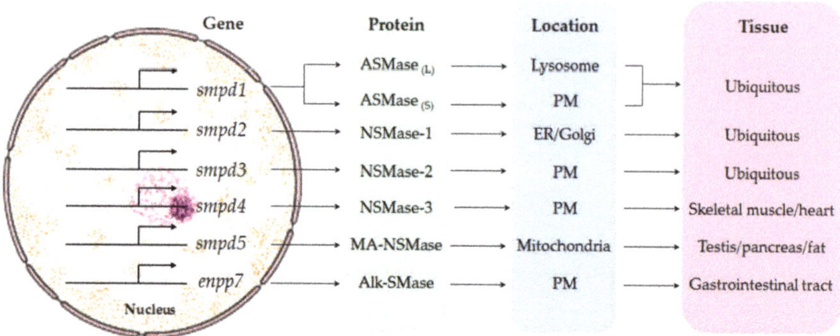

Figure 2. Types and characteristics of mammalian sphingomyelinases (SMases). SMases are encoded by different genes (smp1–5; enpp7), which results in 7 different proteins: two acid sphingomyelinases (ASMases), four neutral sphingomyelinases (NSMases), including the mitochondrial-associated NSMase (MA-NSMase) and alkaline sphingomyelinase (Alk-SMase). Please note that the lysosomal ASMase (ASMase $_L$) and the secretory ASMase (ASMase $_S$) are encoded by smpd1 and localized in different membrane bilayers, namely, lysosomes and the plasma membrane (PM), respectively. ASMase and NSMase differ in their optimal pH for maximal activity and requirement for specific cations for activation and exhibit differential distribution within the cell and in specific tissues.

Two forms of ASMases are encoded by the gene Smpd1. The ASMase associated with the endosomal/lysosomal compartment hydrolyses lysosomal SM delivered by lipoproteins or through the endocytic pathways. On the other hand, secretory ASMase is found in the plasma and in the conditioned medium of stimulated cells, and has a complex pattern of glycosylation as well as a longer in vivo half-life [23–25]. Although the secretory ASMase form has been reported to be dependent on Zn^{2+} [24], recent findings describing the crystal structure of mammalian ASMase revealed an N-terminal saposin domain and a catalytic domain, which adopts a calcineurin-like fold with two Zn^{2+} ions [26]. Whether this accounts for the selective dependence of the secretory ASMase on Zn^{2+} remains to be established. The mechanisms involved in the generation of the secretory ASMase are not fully understood. While the trafficking of lysosomal ASMase to the Golgi and processing by S1P are thought to generate secretory ASMase, recent findings in the protozoan parasite *Trypanosoma cruzi* suggested that conventional lysosomes fuse with the plasma membrane in response to an increase in intracellular Ca^{2+}, releasing their contents extracellularly, where the resultant exocytosed ASMase from lysosomes remodels the outer leaflet of the plasma membrane [27]. Further work would be required to examine if this mechanism contributes to the genesis of the mammalian secretory ASMase. The Smpd2 and Smpd3 genes encode NSMase-1 and NSMase-2, respectively, both of which are Mg^{2+}-dependent but differ in their subcellular localization and role in signaling pathways. In mammalian cells, NSMase-1 is found in the ER and the Golgi apparatus, while NSMase-2 promotes SM hydrolysis on the cytosolic face of the plasma membrane as well as in multilamellar bodies and the nuclear envelope (Figure 2) [28–31]. NSMase-2 is regulated by the TNFα and IL-1β cytokines and mediates cellular responses to stress and inflammation [32–35]. By contrast, due to the subcellular location of NSMase-1 in the ER and Golgi, it is unlikely that this isoform plays a role in signaling pathways, in line with data suggesting a lack of influence of NSMase-1 overexpression in TNFα-induced signaling pathways [36] or Fas-induced apoptosis [37]. The Smpd4 gene encodes a novel form of

NSMase, NSMase-3, which is mostly found in skeletal muscle and the heart but not in the liver. Lastly, a mitochondrion-specific Smpd5 has been recently identified, with the highest expression in the testis, pancreas and fat tissue (MA-NSMase) [38]. Moreover, Alk-SMase does not share any structural similarity with NSMases or ASMases, belongs to the ectonucleotide pyrophosphatase/phosphodiesterase (NPP) family, and therefore is also known as NPP7 and is encoded by the ENPP7 gene. Following transcription, Alk-SMase is anchored on the surface of cell membranes, and in addition, it can also be released from this location by bile salts and pancreatic trypsin. Moreover, Alk-SMase accumulates in the gallbladder, and its activity depends on bile salts [39,40] and is thought to be mainly involved in hydrolyzing dietary SM and stimulating cholesterol absorption [41].

3. Physiological and Signaling Function of Sphingomyelinases

As mentioned, SMase's enzymatic activity leads to a rapid and transient release of ceramide in specific discrete intracellular sites depending on the type of SMase activated. In the following section, we will briefly describe the mechanisms and signaling pathways underlying the effects of NSMase and ASMase. Consistent with their sites of activation and requirements for optimal activity, the resultant ceramide species generated target different pathways, such as PKCδ or KSR for NSMase or MAT1A or cathepsins for ASMase (Figure 3).

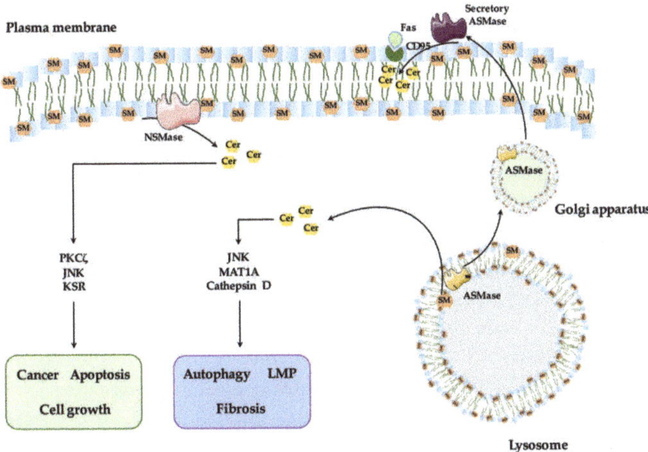

Figure 3. Functional role of NSMase and ASMase in signaling pathways. NSMase and ASMase are localized in different membrane bilayers, where they hydrolyze specific sphingomyelin (SM) pools, mostly in the plasma membrane and lysosomes, respectively, consistent with their pH optima for activity. NSMase-induced ceramide generation in the vicinity of the plasma membrane activates specific targets, e.g., PKCd, KSR or JNK, and is mainly involved in cancer, apoptosis and cell growth. ASMase, on the other hand, hydrolyzes lysosomal SM, and its deficiency causes Niemann–Pick type A (NPA) disease, a lysosomal storage disorder characterized by the accumulation of SM in lysosomes. Ceramide generated by ASMase activation has been shown to target MAT1A and cathepsin D as well as JNK and is involved in the regulation of autophagy, hepatic fibrosis and lysosomal membrane permeabilization (LMP). A subset of ASMase traffics to the Golgi and is secreted to the plasma membrane. The secretory ASMase hydrolyzes SM at the outer leaflet of the plasma membrane, and the resulting ceramide causes the death activation receptors, i.e., CD95, to bind Fas ligand. This pool of ceramide mediates Fas-induced liver injury and failure.

3.1. NSMase

From the different mammalian NSMases characterized to date, NSMase-2 appears to be the predominant isoform involved in cell physiology and in the activation of different signaling pathways. NSMase-2 specifically hydrolyzes the phosphocholine headgroup from SM at the plasma membrane and does not exhibit any phospholipase C-type activity against phospholipids, such as phosphatidylcholine (PC) or lysophosphatidylcholine. As mentioned, the principal biochemical features of NSMase-2 are its requirement for a neutral pH and divalent cations, such as Mg^{2+}, for optimal activity. In addition, phosphatidylserine (PS), as well as other anionic phospholipids, including phosphatidic acid or phosphatidylinositol, stimulates enzymatic activity, while unsaturated fatty acids have been shown to mimic this behavior only in vitro [35,42]. Besides anionic phospholipids, NSMase-2 is regulated by phosphorylation in conserved serine residues, consistent with its identification as a phosphoprotein. In this regard, protein phosphatase 2B was found to bind a PQIKIY motif between the N-terminus and the C-terminus to dephosphorylate NSMase-2 [43]. However, it is still unknown whether protein kinases, such as p38 or PKCδ, directly phosphorylate NSMase-2.

NSMase-2 has been characterized as a mediator of TNF-α signaling, which involves the formation of a pentacomplex containing TNFR-1, NSMase-2, EED, RACK1 and FAN that results in the regulation of neurological (synaptic plasticity and neuronal cytotoxicity), vascular (vasodilation and adhesion) and inflammatory effects. As a mediator of TNF-α's biological effects, NSMase-2 has been shown to induce apoptosis that can be prevented by the overexpression of Bcl-xL [44]. Recent studies indicated that NSMase-2 cooperates with Bax and Bcl-2 to activate the mitochondrial-dependent apoptotic machinery [45]. Interestingly, emerging data implicate NSMase-2 as a component of exosomes whose release takes place through a non-canonical pathway independent of endosomal sorting complexes required for transport (ESCRT) proteins. This new role of NSMase-2 has been involved in cancer development in breast xenografts, as the blockade of the NSMase-2-mediated exosomal release from 4T1 xenografts reduced tumor growth and lung metastasis through alterations in endothelial function [46]. Additionally, involving the action of NSMase-2 via exosomal release, it has been shown that NSMase-2 plays an emerging role in Alzheimer´s disease. Primary astrocytes from murine cortices treated with Aβ25–35 or Aβ1–42 died in parallel with the production of ceramide and caspase-3 activation. The treatment of wild-type mice and a 5XFAD Alzheimer's disease mouse model with GW4869, an NSMase-2 inhibitor, resulted in lower exosome accumulation in the brains of 5XFAD mice, resulting in lower concentrations of Aβ1–42 and reducing Alzheimer´s pathology [47]. Linked to its role in cancer development, NSMase-2 has been shown to regulate cell differentiation and growth arrest. MCF7 breast cancer cells overexpressing NSMase-2 have a similar growth phenotype to control cells, and upon serum starvation, NSMase-2 expression prevents the progression of the cell cycle, which is retained in the G0/G1 phase, compared to control cells overexpressing empty plasmid. Moreover, it has been shown that cell confluence upregulates NSMase-2 to arrest cells in G0/G1 with the hypophosphorylation of the retinoblastoma protein and induction of p21, while NSMase-2 downregulation prevents this phenotype [32]. Thus, these findings identify NSMase-2 as a potential target for the modulation of inflammation, cell growth and apoptosis, emerging as novel target in cancer development and neurodegeneration.

3.2. *ASMase*

The generation of ceramide via ASMase regulates multiple signaling pathways, which are central to metabolism, Ca^{2+} regulation, autophagy and lysosomal homeostasis, and hence, ASMase emerges as an important signaling molecule regulating diverse cellular processes.

3.2.1. ASMase and ER Stress

ER stress is a condition in which there is an accumulation of misfolded proteins in the ER. The unfolded protein response (UPR) is a complex signaling network, which is designed to restore protein homeostasis by reducing protein synthesis and increasing protein folding. Three signaling proteins are initially activated in the UPR—inositol requiring 1 alpha (IRE1α), PKR-like ER kinase (PERK) and activating transcription factor 6a (ATF6)—and this signaling is followed by the activation of several downstream targets [48–50]. The master regulator of UPR activation is the glucose-regulated protein 78 (GRP78, also known as BiP), which in physiological conditions, binds to IRE1α, PERK and ATF6 and prevents their activation. Upon an accumulation of misfolded proteins, GRP78 is released, enabling IRE1α, PERK and ATF6 to trigger the UPR.

ASMase per se has been shown to trigger ER stress through ceramide production and a subsequent impact on Ca^{2+} signaling. The disruption of ER Ca^{2+} homeostasis induced by exogenous ASMase treatment in hepatocytes caused ER stress due to Ca^{2+} release to the cytosol triggered by ceramide, which affected the ability of the chaperone BiP to bind UPR transducers [51]. Although the detailed mechanism whereby ASMase-induced ceramide generation perturbs Ca^{2+} regulation in the ER is not fully understood, the role of ASMase in this effect is consistent with reports showing that an aberrant lipid composition in the ER triggers Ca^{2+} release through sarco-endoplasmic reticulum Ca^{2+}-ATPase (SERCA) regulation [52–54]. In addition, whether an ASMase-mediated increase in ceramide impacts the ER lipid composition remains to be investigated.

3.2.2. ASMase and Autophagy

Autophagy is a highly regulated and complex catabolic process that is involved in the degradation of damaged or dysfunctional cell components, such as organelles such as mitochondria or peroxysomes, protein aggregates, lipid droplets or inflammasomes, through the fusion of autophagosomes with lysosomes [55]. During this process, cytoplasmic materials are recruited into a double membrane structure (the phagophore), which distends to form an autophagosome, a spherical structure with double layer membranes. This structure further fuses with lysosomes, creating an autolysosome, where all the contained materials are degraded by lysosomal hydrolases, while monomers such as free fatty acids (FFA) and amino acids are recycled. A failure to "digest" altered organelles can contribute to sustained alterations in metabolism and homeostasis, leading to cell dysfunction. This is best illustrated in the case of the defective elimination of disrupted mitochondria, as their accumulation can cause the release of stimulated ROS, disruption of lipid intermediate signaling and stimulation of proinflammatory cytokines.

Autophagy can be divided into several subtypes [56,57]. Macroautophagy is a protective non-selective mechanism that is activated during scarce nutrient availability with the aim of degrading cellular components in order to enhance the energy supply. By contrast, selective autophagy is a well-orchestrated process involving the recruitment of the autophagic molecular machinery to digest specific targets through specific protein signaling pathways [58]. This specificity makes autophagy a very important degradation process for preventing the accumulation of dysfunctional organelles and cell waste. Lipophagy, one of the several subtypes of autophagy, performs the selective degradation of intracellular lipids, and it has been described to have an important role in lipid metabolism and hepatic steatosis [59]. Besides regulating lipid storage, lipophagy also controls cellular energy homeostasis by providing FA to mitochondria to fuel β-oxidation and ATP synthesis. Cytosolic lipases have been known to catabolize triglycerides (TG) and lipid droplets in hepatocytes.

Autophagy disruption mediates the progression of many liver diseases, such as ASH/NASH, in which defects in autophagy promote steatotic and fibrogenic mechanisms [55]. Recent evidence suggests a role for ASMase in autophagy-mediated liver injury. For instance, hepatocytes from ASMase$^{-/-}$ mice exhibit impaired autophagic flux, reflected specifically in the accumulation of dysfunctional mitochondria and resistance to high-fat diet (HFD)-induced hepatic steatosis [60]. Whether the role of ASMase in autophagy is mediated via the regulation of lysosomal cholesterol homeostasis and its impact on the fusion of lysosomes with autophagosomes remains to be fully elucidated [61]. Overall, while some data suggest a role for ASMase in autophagy-mediated steatosis in ASH/NASH, the potential involvement of ASMase in fibrosis via autophagy is still unknown and requires further investigation.

3.2.3. ASMase and Lysosomal Membrane Permeabilization

Lysosomes are specialized membrane-bound organelles that contain a variety of hydrolytic enzymes responsible for the digestion and removal of macromolecules and organelles. Furthermore, lysosomes also play a crucial role in cell death regulation, in which lysosomal membrane permeabilization (LMP) followed by the leakage of lysosomal content is enough for cell death initiation [62,63]. Complete LMP causes a massive release of lysosomal content into the cytosol, which increases cytosolic acidification and enhances hydrolytic damage in different cellular components [62,64]. For instance, cathepsins (Cts), the major class of lysosomal proteases, are targeted to other organelles such as mitochondria, where they induce the release of proapoptotic factors. In addition, the acidification of mitochondria triggers mitochondrial membrane depolarization and Ca^{2+} handling impairment, which can trigger the recruitment of Bax in mitochondria to also initiate apoptosis [65]. Moreover, the acidification of the cytosol allows for some of the lysosomal proteases, such as Cts, to maintain their enzymatic activity and induce the proteolytic degradation of key cellular proteins during apoptosis [66,67].

LMP is induced by several different stimuli, such as ROS, lipids such as saturated FA or sphingosine, as well as cell death mediators, such as Bax. As lipids can trigger LMP, this event has been reported as one of the molecular mechanisms involved in NASH progression. For instance, palmitic acid (PA), one of the most abundant saturated FA in Western diets, induces LMP in hepatocytes, whic triggers CtsB leakage into the cytosol, mitochondrial dysfunction followed by cytochrome c release and, ultimately, cell death [68,69]. As with ER stress and autophagy, ASMase is also related to LMP in NASH. ASMase deficiency causes resistance to PA-induced lipotoxicity in primary mouse hepatocytes. Consistent with the resistance to PA-mediated lipotoxicity, hepatocytes from $ASMase^{-/-}$ mice are also resistant to amphiphilic lysosomotropic detergent-induced cell death. The increased levels of cholesterol in lysosomes upon ASMase deficiency are involved in such protection, as decreasing the lysosomal cholesterol content reverses the resistance of $ASMase^{-/-}$ hepatocyes to amphiphilic lysosomotropic detergent and PA-induced cell death [60]. Thus, a deficiency of ASMase raises lysosomal cholesterol, which results in reduced LMP and a resistance to PA-induced lipotoxicity.

4. SMases and Liver Diseases

The liver is an important organ for lipid metabolism since it is involved in fatty acid β-oxidation, ketone body generation, cholesterol metabolism, lipoprotein synthesis and phospholipid metabolism. Hepatocytes secrete up to 5% of the newly synthesized SLs in the form of VLDL, constituting an important source of SLs. Specifically, the hepatic SM content is 7–8 fold higher than in subcutaneous and intra-abdominal adipose tissues [70]. Such high levels of SM in the liver are due to the fact that the liver effectively absorbs choline-containing compounds from phospholipid digestion in the intestinal tract, which are subsequently used for PC and SM synthesis [71,72]. Apart from synthesizing SM, the liver is also involved in SM hydrolysis, as the hepatic ASMase activity is higher than that in most organs. SMases are important for liver physiology and SM and ceramide homeostasis, and defects in SMases, particularly ASMase, result in profound alterations in liver function and the accumulation of SM, as illustrated in Niemann–Pick type A/B diseases (NPA/NPB) (see below) [73]. A high-fat diet (HFD), endotoxins or hepatitis B virus infections and liver cancer also influence hepatic SM homeostasis [74–77].

Alterations of SM levels subsequent to changes in the activity of SMases have been associated with the onset and progression of chronic liver diseases, of which NASH is of particular prevalence in the world. Several studies employing animal models as well as human specimens have provided strong evidence for ASMase and NSMase (e.g., NSMase-2) in liver diseases [1]. While ASMase is abundant in the healthy liver, hepatic NSMase-2 activity is very low in physiological conditions. Nevertheless, NSMase-2 is regulated by antioxidants, and hence, it can be activated during oxidative stress and hepatic GSH depletion [1]. Both NSMase and ASMase can be activated by cytokines and proinflammatory conditions, contributing to the progression of chronic liver diseases [35,78].

In the following section, we will briefly describe the contribution of both SMases to major chronic liver diseases (Table 1).

Table 1. Major highlights of the role of NSMase and ASMase in liver diseases.

Disease	Protein	Function
Alcoholic and non-alcoholic steatohepatitis (ASH/NASH)	ASMase	Triggers hepatocellular apoptosis in response to TNF and Fas-induced fulminant liver injury. Is required in alcohol or HFD-induced lipogenesis and macrosteatosis. Is required for ER stress (either alcohol or HFD-induced or autophagy suppression-mediated). Is activated during HSC activation and required for their transdifferentiation to myofibroblast-like cells that promote fibrogenesis. Is a crucial link in the regulation of methionine metabolism and PC homeostasis mediating NASH progression.
	NSMase	Less characterized in ASH/NASH. Controversial function of TNF-induced hepatocellular apoptosis.
Hepatocellular carcinoma (HCC)	NSMase-1	Is downregulated in HCC tissues
	NSMase-2	Its deficiency promotes liver tumor development by regulating the survival and proliferation of cancer stem-like cells
	ASMase	Promotes cell death by increasing ER stress and autophagy
Niemann–Pick A/B (NPA/B)	ASMase	Its deficiency affects lysosomal sphingolipid accumulation, resulting in lipid-loaded foam cells in a wide variety of organs having a severe impact in their correct functioning. Its deficiency impairs cholesterol trafficking causing oxidative stress and affects vesicle trafficking pathways mediated by Rab proteins as well as fusion of the late endosomal/lysosomal compartments Its deficiency alters lysosomal–mitochondrial interactions, involving impaired mitophagy, resulting in mitochondrial dysfunction and overall contributing to disease progression.
Ischemia–reperfusion (I/R) liver injury	ASMase	Its inhibition prevents ceramide increase after hepatic I/R injury, attenuating serum ALT levels, hepatocellular necrosis, cytochrome c release and caspase 3 activation.
	NSMase	Its inhibition decreases enhanced levels of nitrosative and oxidative stress in I/R injury. Its inhibition downregulates apoptotic stimuli during I/R injury
Drug-induced liver injury (DILI)	ASMase	Its deficiency alters lysosomal–mitochondrial interactions, involving impaired mitophagy, resulting in mitochondrial dysfunction and sensitization to APAP hepatotoxicity.
Viral hepatitis B (HBV)	ASMase	Is required for the production of HBV-DNA carrying extracellular vesicles (EV), essential for hepatocyte infection.
Hepatobiliary diseases	Alk-SMase	Its activity is reduced in the bile and liver from primary sclerosing cholangitis (PSC) patients Bile salt diversion strongly reduces Alk-SMase activity in the small intestinal content, which may also affect intestinal SM digestion
Wilson disease	ASMase	Cu^{2+} triggers hepatocyte apoptosis through activation of ASMase and the release of ceramide

4.1. Alcoholic and Non-Alcoholic Steatohepatitis

Fatty liver disease is a spectrum of disorders that begin with hepatic fat accumulation (steatosis) that can progress to cirrhosis and HCC. Steatohepatitis is an advanced stage of fatty liver disease characterized by liver steatosis, inflammation, fibrosis and hepatocellular death [79,80] and encompasses both ASH and NASH, which share common histological and biochemical features [81]. The prevalence of NASH is of significant relevance due to its association with obesity and the type 2 diabetes epidemic. Unfortunately, the current available therapy for NASH is limited due to our incomplete understanding of the underlying mechanisms. In this regard, recent data have disclosed a pivotal role for ASMase in both ASH and NASH. In the case of ASH, ASMase activity and ASMase mRNA levels have been shown to be upregulated in liver biopsies from patients with acute alcoholic hepatitis [51,82–84]. In line with these observations, ASMase activity has been recently linked to biomarkers and alcohol use and abuse phenotypes in patients and healthy controls [85]. On the other hand, the expression and activity of ASMase have been shown to be increased in liver and serum samples from patients with NASH [86,87]. In line with the reported role of ASMase-induced ceramide generation in mediating hepatocellular apoptosis and fibrosis, ASMase has emerged as a potential new target for treatment for both ASH and NASH [88].

ASMase mediates the stress response and triggers hepatocellular apoptosis in response to TNF and Fas-induced fulminant liver injury. ASMase is required in TNF-induced hepatocellular apoptosis, TNF plus D-(+)-galactosamine (TNF/Gal)-mediated fulminant liver failure and Fas-induced lethal hepatitis [89–91], in which the recruitment of mitochondria is involved through ganglioside GD3 generation and further apoptosis via reactive oxygen species (ROS) generation and nuclear factor kappa-light-chain-enhancer of activated B cells (NFκB) inhibition [92,93]. In addition to cell death regulation, ASMase$^{-/-}$ mice are resistant to alcohol- or HFD-induced lipogenesis and macrosteatosis, and hence, the pharmacological inhibition of ASMase using amitriptyline or imipramine (tricyclic antidepressants) in wild-type mice (WT) blocked alcohol- and HFD-induced steatosis [30,51,60,84]. As described above, there is a strong link between lipogenesis and ER stress, and ASMase has been shown to be required for either alcohol- or HFD-induced ER stress by regulating ER Ca^{2+} homeostasis. Besides the proapoptotic and prosteatotic role of ASMase, ASMase promotes fibrosis. Fibrosis is a wound-healing response to diverse types of insults, in which chronic injury drives the activation of hepatic stellate cells (HSCs) involved in the remodeling of the extracellular matrix and deposition of collagen, resulting in tissue scarring. Recent data have shown that ASMase is activated during HSC activation and required for HSCs' transdifferentiation to myofibroblast-like cells that promote fibrogenesis [87]. In this event, ASMase regulates CtsB and CtsD, which are necessary for HSC activation and fibrosis initiation. Consistent with these findings, the pharmacological inhibition of ASMase using amitriptyline reduces hepatic fibrosis in a well-established CCl_4-mediated-fibrosis model [94].

Autophagy and ER stress are mutually regulated, as defective autophagy induces ER stress [95]. In fact, ASMase is required for autophagy suppression-mediated ER stress in primary mouse hepatocytes [60]. Therefore, the induction of lipogenic genes through impaired-autophagy-mediated ER stress may be a significant pathway in the regulation of hepatic steatosis, in which ASMase has a central regulatory role. On the other hand, ER stress can also induce autophagy and further HSC activation and fibrosis [96]. The most characteristic feature of HSCs in physiological conditions is the presence of perinuclear membrane-bound droplets filled with retinyl esters. Thus, an increased autophagic flux may contribute to the loss of these lipid droplets during HSC activation [97,98]. Moreover, the genetic or pharmacologic inhibition of autophagy prevented HSC activation and fibrogenesis. Interestingly, the specific deletion of Atg7 in HSCs decreased fibrosis following sustained liver injury [99]. Overall, autophagy enhances energy production through the release of FFA from retinyl esters, which are used by HSsC as an energy supply for their activation and following fibrogenesis. In addition to its role in promoting apoptosis, steatosis and fibrosis, ASMase has also emerged as a crucial link in the regulation of methionine metabolism and PC homeostasis, which have been shown to mediate NASH progression. In particular, methionine adenosyltransferases I/III (encoded by MAT1A), the key enzymes involved in methionine metabolism, and ASMase-induced ceramide generation engage in a reciprocal inhibitory process in which increased ceramide levels generated by ASMase repress MAT1A expression, leading to increased ASMase expression, causing a self-sustained vicious cycle of relevance to ASH/NASH development [100,101]. The lower expression of MAT1A from increased ASMase expression results in decreased SAM levels, which are key for the synthesis of PC from phosphatidylethanolamine (PE) via PE methyltransferase (PEMT). Thus, targeting ASMase may boost MAT1A activity and SAM levels, which in turn, may impact the maintenance of an adequate PC to PE ratio to improve NASH. In contrast to ASMase, the role of NSMase in ASH/NASH has been less characterized, and consistent with its controversial function as a mediator of TNF-induced hepatocellular apoptosis, the contribution of NSMase to NASH remains to be fully established [102,103].

4.2. Hepatocellular Carcinoma (HCC)

HCC is the end stage of chronic liver disease and the culmination of metabolic diseases, particularly NASH. HCC is the most prevalent form of liver cancer and one of the leading causes of cancer-related deaths in the world. Unfortunately, current therapy is limited and inefficient, and advanced HCCs develop resistance to chemotherapy, making early diagnosis essential for survival. As some lipids such as SLs have been known to mediate cell death pathways, several therapeutic approaches target the regulation of these molecules in order to halt HCC progression. Ceramide levels are markedly reduced in HCC tissues due to the increased expression of strategies that stimulate its degradation, and hence, raising ceramide levels specifically within the tumor may be a relevant therapeutic approach [104]. In this regard, the celecoxib-mediated activation of ER-stress has been shown to induce de novo ceramide biosynthesis, resulting in enhanced apoptosis in hepatoma HepG2 cells [105]. Targeting ACDase, an enzyme that hydrolyzes ceramide (Figure 1), with LCL521 or carmofur has been shown to be of potential relevance in cancer [106,107]. Furthermore, NSMase-1 was reported to be downregulated in HCC tissues [108], and NSMase-2 deficiency promotes liver tumor development by regulating the survival and proliferation of cancer stem-like cells [109]. Despite these findings, many solid tumors, including HCC, develop strategies contributing to the development of chemoresistance. For instance, ceramide-modifying enzymes, particularly glucosylceramide synthase (GCS), are upregulated during sorafenib treatment in hepatoma cells (HepG2 and Hep3B), decreasing ceramide-induced cell death activation and therefore conferring resistance to the sorafenib treatment. In line with these findings, GCS silencing or pharmacological GCS inhibition sensitized hepatoma cells to sorafenib exposure [110].

The contribution of SLs and SMases to HCC has been further suggested to be mediated through their interactions with the mTOR pathway and autophagy, both of which are involved in HCC pathogenesis. The upregulation of the mTORC2 pathway has been shown to promote hepatocarcinogenesis, in part, through the stimulation of de novo fatty acid and lipid synthesis, which leads to steatosis and tumor development [111]. Indeed, increased lipogenesis correlated with elevated mTORC2 activity and human HCC incidence, and the inhibition of fatty acid or SL synthesis prevented tumor development, indicating a link between steatosis and tumor generation and therefore further suggesting that hepatosteatosis acts as a driving force for HCC progression [112]. Although autophagy acts as a mechanism to provide nutrients for energy generation to maintain key cellular functions, autophagy plays a paradoxical role in HCC. For instance, autophagy has been shown to protect cancer cells from the accumulation of damaged organelles and protein aggregates, preventing cell death and the toxicity of cancer therapies [113]. Consequently, the pharmacological or siRNA-mediated inhibition of autophagy has been reported to sensitize HCC cells to the multikinase inhibitor linifanib [114]. However, decreased autophagy markers have been associated with more aggressive HCC phenotypes [115], and several tumor suppressors (e.g., XPD, PTPRO, TAK1 and Klotho) have also been reported to activate autophagy in HCC cells [116–119]. Among drugs that affect autophagy, vorinostat and sorafenib have been reported to promote carcinoma cell death by increasing ER stress and autophagy via a ASMase/ceramide-dependent pathway [120]. In addition, recombinant human ASMase has emerged as a potential adjuvant treatment with sorafenib in HCC [121]. Moreover, exosomal NSMase-1 has been reported to suppress HCC via decreasing the ratio of

SM/ceramide [108]. Overall, although there are emerging data regarding the role of SMases in HCC, further understanding about this link is still needed to optimize the targeting of SLs/SMases as potential therapeutics.

4.3. Niemann–Pick Disease Type A/B

Niemann–Pick diseases are a group of autosomal recessive disorders mainly characterized by an excessive accumulation of SLs and cholesterol, primarily in lysosomes. There are two distinct genetic Niemann–Pick disorders. NPA and NPB are caused by a deficient activity of ASMase mostly affecting lysosomal SM homeostasis. Niemann–Pick type C disease (NPC) is characterized by defects in the lysosomal resident proteins NPC1 and NPC2, which are responsible for cholesterol efflux from lysosomes [122,123]. Due to the close association between SLs and cholesterol, both types of lipids accumulate in lysosomes in NPA/B or NPC, regardless of whether the cause is ASMase or NPC1/NPC2 deficiency. Besides SM and cholesterol, other SLs such as gangliosides and sphingosine also accumulate in the lysosomes in NPA/B [73,124,125]. The primary organs affected in NPA patients are the spleen, liver and lung. Consequently, lipid-loaded foam cells are found in a wide variety of organs such as the liver, spleen, lymph nodes, adrenal cortex, lungs and bone marrow, having a severe impact on their correct functioning [126]. NPA and NPB differ in the degree of residual ASMase activity, which determines the impact on clinical features, being more severe in NPA than in type B, and the differential occurrence of neurological symptoms allows diagnosis and prognosis [127]. NPA patients, besides exhibiting hepatosplenomegaly, present a rapidly progressive neurodegeneration with a deep hypotonia, leading to the patient's death beyond the third year of life [125,128]. NPB patients have no signs of central nervous system involvement, although they may present severe hepatosplenomegaly and liver failure [129,130]. Many NPB patients die before or in early adulthood, often from respiratory or liver failure. An early diagnosis and appropriate handling are thus crucial to lower complication risks, improving quality of life and avoiding extreme procedures such as splenectomy [131–133]. Liver biopsy, the diffusion capacity measured by spirometry, the spleen volume and several plasma markers of lipid-laden cells, fibrosis or inflammation are among the biomarkers that are currently used to diagnose NPA and NPB to ensure an appropriate management of the disease [134,135]. Currently, an efficient treatment for NPD type A/B is lacking, and symptomatic therapy has been the only available option for these patients. Enzyme replacement therapy with olipudase alfa, a recombinant form of human ASMase, aims to reduce the non-neurological manifestations of NPD type A/B, showing promising effects in improving liver function and lipid profiles [134]. Although the subsequent increase in ASMase activity following olipudase administration increases ceramide levels, this outcome seems to be transient, and ceramide contents are stabilized after 3 months of treatment [131,132,134,136,137].

ASMase-deficient mice were generated as a mouse model for NPA to study the effects of ASMase deficiency and unravel the molecular pathways involved in this devastating disease [138–141]. Like NPA patients, $ASMase^{-/-}$ mice present profoundly impaired lipid metabolism and trafficking. The accumulation of lipids in lysosomes is the primary consequence of ASMase deficiency, although this outcome is also observed in other subcellular organelles, including mitochondria, resulting in an overall disruption of cellular homeostasis. Lysosomal SM accumulation has been reported to inhibit the lysosomal transient receptor potential Ca^{2+} channel, impairing endolysosomal trafficking, protein degradation and macroautophagy [142]. In addition, the impaired cholesterol trafficking observed in NPA causes oxidative stress and affects the vesicle trafficking pathways mediated by Rab proteins as well as the fusion of the late endosomal/lysosomal compartments [143–145]. Interestingly, despite ASMase deficiency, $ASMase^{-/-}$ mice present elevated ceramide levels in the affected organs, which is possibly due to a breakdown of the accumulated SM in non-lysosomal compartments by other functional SMases. Whether the increase in ceramide levels could contribute to the pathogenesis of NPA remains to be fully investigated [123]. In addition to the ASMase function in lysosomes, some studies have also suggested a role for ASMase in response to stress at the plasma membrane, where it seems to participate in different signaling pathways [146,147]. Thus, the consequences of ASMase deficiency may extend beyond lysosomes and affect other subcellular compartments. In line with this possibility, emerging evidence indicates that lysosomal–mitochondrial interactions are also altered in NPA disease, involving impaired mitophagy due to increased lysosomal cholesterol-mediated impairment of the fusion of autophagosomes containing mitochondria with lysosomes, resulting in mitochondrial dysfunction and overall contributing to disease progression [148,149].

5. Role of SMases in Liver Injury and Metabolic Liver Diseases

Besides the aforementioned prevalent chronic liver diseases, SMases have also been related to a wide variety of other liver disorders.

5.1. Ischemia–Reperfusion (I/R) Liver Injury

Hepatic ischemia/reperfusion (I/R) injury is a serious complication that compromises liver function because of extensive hepatocellular loss, which impacts diverse clinical settings such as liver surgery and liver transplantation. I/R is caused by the restoration of blood circulation after a period of ischemia in which the supply of oxygen and nutrients is curtailed. I/R results in severe cellular injury, with inflammation and oxidative stress as the main culprits. In the liver, SLs and ceramides, in particular, have been described as signaling lipid intermediates playing a significant role in the stress response and cell death [15,150–152]. SMases, as mediators of ceramide production, have also been reported to participate in cell death events and thus play a crucial role in I/R liver injury. ASMase inhibition, either pharmacologically using imipramine or by siRNA-mediated silencing, prevented ceramide increase after hepatic I/R injury and attenuated serum ALT levels, hepatocellular necrosis, cytochrome c release and caspase-3 activation, indicating the relevance of ASMase in this type of liver injury [153]. Apart from ASMase, NSMase has also been linked to I/R, as it has been reported that the inhibition of NSMase decreases the enhanced levels of nitrosative and oxidative stress in I/R injury [154]. Moreover, although NSMase inhibition does not alleviate ER stress, this event downregulates apoptotic stimuli during I/R injury, arguing in favor of a direct role of NSMase in I/R liver injury and the significant protective effect of selective NSMase inhibition for future therapies [155].

5.2. Drug-Induced Liver Injury (DILI)

DILI is a major cause of liver failure due to hepatocellular demise upon exposure to a toxic dose of drugs or xenobiotics. Acetaminophen (APAP) hepatotoxicity is the prototype DILI paradigm since APAP is one of the most used pain killers worldwide. Although relatively safe, APAP is a dose-dependent hepatotoxin and a major cause of acute liver failure requiring liver transplantation [156,157]. APAP metabolism generates N-acetyl-p-benzo-quinoenimine (NAPQI), a toxic electrophile, which is detoxified by conjugation with GSH. Excess APAP consumption or a limited hepatic GSH pool favors the binding of NAPQI to mitochondrial protein thiols, leading to the disruption of mitochondrial function and release of generated ROS and oxidative stress. Excess mitochondrial ROS generation potentiates mitochondrial JNK translocation, which amplifies the induction of mitochondrial permeability transition pore opening, leading to further ROS generation, ATP depletion and subsequent hepatocellular death [158–163]. As APAP-induced injury mainly impacts mitochondria, the elimination of APAP-induced mitochondrial damage by mitophagy protects against APAP hepatotoxicity.

As indicated above, ASMase deficiency induces the accumulation of SM and other lipids within lysosomal membranes, affecting membrane structure and dynamics. Lysosomal cholesterol accumulation induced by ASMase deficiency decreases mitophagy due to the defective fusion of mitochondrion-containing autophagosomes with lysosomes, resulting in sensitization to APAP hepatotoxicity. Thus, a protective role for ASMase in APAP emerges due to the maintenance of SM/cholesterol homeostasis, which in turn, impacts the turnover of mitochondria and the clearance of defective organelles that sustain APAP-mediated liver injury [148].

5.3. Viral Hepatitis B (HBV)

Along with viral hepatitis C, viral hepatitis B (HBV) is a major form of chronic liver disease characterized by severe inflammation and liver injury that can further lead to complications such as cirrhosis, liver failure and liver cancer. The link between SMases and viral hepatitis is still not completely clear, but it is related to the capacity of SLs to control extracellular vesicle (EV) formation. EVs are bilayered particles that carry diverse types of molecules such as proteins, lipids and nucleic acids. Their role as signaling complexes in order to perform intercellular communication has been reported to impact several physiological processes [164–166]. SLs, and particularly ceramides, can control the formation of EVs due to the effects on structural and physical properties exerted in lipid membranes [167–169]. Ceramides can induce lateral phase separation and domain formation in membrane bilayers and can promote negative curvature in the membrane as well as membrane invagination [170]. Furthermore, lipid phases have been shown to be dependent on ASMase activity. ASMase regulates the structural domains of scaffold molecules, which in lysosomes, have been reported to have a severe impact in death-receptor related liver diseases [171]. EVs play a particular role in viral hepatitis, as these vesicles can transport viral particles, overall enabling the expansion of the viral infection to surrounding cells [172,173]. Specifically, EVs are important for HBV-infected hepatocytes [174]. The production of EVs carrying HBV DNA from HBV-infected hepatocytes has been reported to depend on the ASMase/ceramide system to mediate exosome formation [175], indicating that the regulation of this pathway could be an important therapeutic approach to preventing EV-mediated HBV infection.

5.4. Hepatobiliary Diseases

Hepatobiliary diseases comprise a large and heterogeneous group of diseases that affect the hepatic and biliary system. These disorders can be developmental or congenital and can arise at different stages during life. Alk-SMases have been widely studied in the digestion context, as intestinal Alk-SMases have been reported to play an important role in SM digestion, colon cancer prevention and cholesterol absorption [176–178]. However, little is known about the role of Alk-SMases in liver diseases. A reduction of Alk-SMase activity was detected in liver specimens from patients with primary sclerosing cholangitis (PSC) [179]. Moreover, a reduction of Alk-SMase activity was also reported in the bile from PSC patients and in patients with cholangiocarcinoma [179,180]. Overall, current evidence suggests that there may be an association between reduced Alk-SMase activity and the progression of hepatobiliary disease. In addition, as Alk-SMases are regulated by bile salts, bile salt diversion was found to strongly reduce Alk-SMase activity in the small intestinal content and feces in rats, indicating that hepatobiliary diseases may also affect intestinal SM digestion [181,182].

5.5. Wilson Disease

Wilson disease is an autosomal recessive disorder caused by inactivating mutations in ATP7B, an enzyme involved in the secretion of Cu^{2+} from the liver. The defect results in the accumulation of Cu^{2+} in hepatocytes and other tissues including neuronal, blood or muscle cells [183,184]. An excess of Cu^{2+} ions in cells and tissues induces severe disorders including progressive hepatic cirrhosis, chronic active hepatitis or even progressive hepatic failure, Fanconi syndrome, neurological and psychiatric symptoms, cardiomyopathy, osteomalacia and, in some individuals, anemia. Although Cu^{2+} is an essential trace element in the human diet and is required as a cofactor for the function of diverse proteins, Cu^{2+} triggers the release of ROS that seem to be crucially involved in the induction of cell death by Cu^{2+} [185–187]. The release of ROS after cellular treatment with Cu^{2+} has been described as occurring predominantly in lysosomes and mitochondria, and lysosomal membrane damage precedes Cu^{2+} cytotoxicity. Furthermore, Cu^{2+} triggers, at least in erythrocytes, the formation of lipid peroxides and inhibits the activity of antioxidant enzymes, finally resulting in oxidative cell damage, the denaturation of hemoglobin and hemolytic anemia.

Recent studies have shown that Cu^{2+} triggers hepatocyte apoptosis through the activation of ASMase and the release of ceramide [188]. A genetic deficiency of ASMase prevented Cu^{2+}-induced hepatocyte apoptosis, while ASMase inhibition with desipramine in rats with a mutation in the Atp7b gene, a genetic model of Wilson's disease, protects against Cu^{2+} -induced hepatocyte death and liver failure. In line with these findings, individuals with Wilson disease showed elevated plasma levels of ASMase and displayed a constitutive increase in ceramide and phosphatidylserine-positive erythrocytes. The concentrations of free Cu^{2+} (1–3 mM) required to elicit the activation of the ASMase, ceramide release and the apoptosis of hepatocytes are in the range of the concentrations encountered in the plasma of individuals with Wilson disease, thus supporting the clinical significance of the activation of ASMase by Cu^{2+} and its involvement in Wilson´s disease.

6. Future Perspectives

Ceramide has been the preferential target of biomedical research deciphering the biological role of SLs. Although considered as critical players in membrane bilayers due to their crucial role in determining their physical properties, SLs and ceramide, in particular, have been recognized as second messengers acting as intermediates of a number of stimuli that regulate multiple cellular functions. While de novo ceramide synthesis is slow but sustained, its generation via NSMase or ASMase represents a fast mechanism for ceramide generation, and hence, both SMases stand as efficient molecular devices for releasing ceramide almost instantly in response to different stimuli, which lends further support to positioning both enzymes as key intermediates in signaling pathways. A common feature of NSMase and ASMase is that both hydrolyze SM embedded in membrane bilayers. However, their pH dependence marks an important difference between the enzymes, and accounts for the generation of discrete pools of ceramide in specific cellular sites. In this regard, NSMase requires a neutral pH for optimal activity and generates ceramide in the vicinity of the inner leaflet of the plasma membrane, with NSMase-2 being of major relevance in pathophysiology. On the other hand, ASMase, which requires an acid pH for activity, is located in acidic compartments, predominantly in lysosomes, where it hydrolyzes lysosomal SM pools. Quite interestingly, lysosomal ASMase processing at the Golgi or following its exocytosis generates a secretory form of ASMase, which acts at the plasma membrane to generate ceramide from local SM hydrolysis. Since the molecular identity of ceramide is determined by the length of the fatty-acyl chain linked to the carbon backbone, the hydrolysis of SM at different membrane bilayers by individual SMases contributes to the generation of unique molecular species of ceramide. Thus, the activation of either NSMase or ASMase represents a specific mechanism for the generation of different ceramide species that likely accounts for their diverse biological effects. Although in physiological settings, the generation of ceramide occurs predominantly by de novo synthesis, the activation of SMases in response to different triggers, such as inflammatory cytokines, stress or chemotherapy, results in the rapid generation of ceramide in specific, discrete cellular sites that targets particular pathways involved in chronic liver diseases. Of particular relevance is the role of SMases in NASH and its progression to HCC, which has escalated to become the most important chronic liver disease worldwide and is expected to increase in the near future due to its association with obesity and the type 2 diabetes epidemic. Thus, elucidating the mechanisms of activation and identification of intermediates involved in SMase signaling may be of relevance for the treatment of prevalent liver diseases. As a proof of concept, targeting ASMase with tricyclic antidepressants, such as amitriptyline, has been initially reported in preclinical models of both ASH and NASH. Since the complete inhibition of ASMase may result in serious complications as disclosed by the deletion of ASMase in NPA, the exposure time and dose for ASMase inhibitors need to be carefully chosen to allow residual ASMase activity to avoid an undesirable NPA phenotype, a further complication that may not be relevant for NSMase. Further research would be required to develop specific and reversible SMase inhibitors as potential treatments for liver diseases.

Author Contributions: N.I.-U. and E.S.-V. searched PubMed for the literature and drafted the manuscript. C.G.-R. and J.C.F.-C. discussed the outline, supervised the writing and generated the final version. N.I.-U. and J.C.F.-C. crafted the figures. All authors have read and agreed to the published version of the manuscript.

Funding: We acknowledge the support from grants SAF2017-85877R and PID2019-111669RB from Plan Nacional de I+D funded by the Agencia Estatal de Investigación (AEI) and the Fondo Europeo de Desarrollo Regional (FEDER) and from the CIBEREHD; the center grant P50AA011999 Southern California Research Center for ALPD and Cirrhosis funded by NIAAA/NIH; and support from AGAUR of the Generalitat de Catalunya SGR-2017-1112, European Cooperation in Science & Technology (COST) ACTION CA17112 Prospective European Drug-Induced Liver Injury Network, "ER stress-mitochondrial cholesterol axis in obesity-associated insulin resistance and comorbidities"—Ayudas FUNDACION BBVA and the Red Nacional 2018-102799-T de Enfermedades Metabólicas y Cáncer. Project 201916/31 Contribution of mitochondrial oxysterol and bile acid metabolism to liver carcinogenesis 2019 by Fundació Marato TV3.

Acknowledgments: Some images in the figures were adapted from Smart Servier Medical Art under the Creative Commons Attribution 3.0 Unported License.

Conflicts of Interest: The authors declare no conflict of interest.

References

1. Nikolova-Karakashian, M. Alcoholic and non-alcoholic fatty liver disease: Focus on ceramide. *Adv. Biol. Regul.* **2018**, *70*, 40–50. [CrossRef] [PubMed]
2. Colacios, C.; Sabourdy, F.; Andrieu-Abadie, N.; Ségui, B.; Levade, T. Basics of Sphingolipid Metabolism and Signalling. In *Bioactive Sphingolipids in Cancer Biology and Therapy*; Hannun, Y.A., Luberto, C., Mao, C., Obeid, L.M., Eds.; Springer International Publishing: Cham, Switzerland, 2015; pp. 1–20, ISBN 978-3-319-20750-6.
3. Hannun, Y.A.; Luberto, C.; Mao, C.; Obeid, L.M. *Bioactive Sphingolipids in Cancer Biology and Therapy*; Springer International Publishing: Cham, Switzerland, 2015.
4. Hannun, Y.A.; Obeid, L.M. Sphingolipids and their metabolism in physiology and disease. *Nat. Rev. Mol. Cell Biol.* **2018**, *19*, 175–191. [CrossRef] [PubMed]
5. Ryland, L.K.; Fox, T.E.; Liu, X.; Loughran, T.P.; Kester, M. Dysregulation of sphingolipid metabolism in cancer. *Cancer Biol. Ther.* **2011**, *11*, 138–149. [CrossRef] [PubMed]
6. Ogretmen, B.; Hannun, Y.A. Biologically active sphingolipids in cancer pathogenesis and treatment. *Nat. Rev. Cancer* **2004**, *4*, 604–616. [CrossRef] [PubMed]
7. Ponnusamy, S.; Meyers-Needham, M.; Senkal, C.E.; Saddoughi, S.A.; Sentelle, D.; Selvam, S.P.; Salas, A.; Ogretmen, B. Sphingolipids and cancer: Ceramide and sphingosine-1-phosphate in the regulation of cell death and drug resistance. *Future Oncol. Lond. Engl.* **2010**, *6*, 1603–1624. [CrossRef] [PubMed]
8. Chavez, J.A.; Summers, S.A. A ceramide-centric view of insulin resistance. *Cell Metab.* **2012**, *15*, 585–594. [CrossRef] [PubMed]
9. Hirschberg, K.; Rodger, J.; Futerman, A.H. The long-chain sphingoid base of sphingolipids is acylated at the cytosolic surface of the endoplasmic reticulum in rat liver. *Biochem. J.* **1993**, *290*, 751–757. [CrossRef]
10. Jaffrézou, J.P.; Levade, T.; Bettaïeb, A.; Andrieu, N.; Bezombes, C.; Maestre, N.; Vermeersch, S.; Rousse, A.; Laurent, G. Daunorubicin-induced apoptosis: Triggering of ceramide generation through sphingomyelin hydrolysis. *EMBO J.* **1996**, *15*, 2417–2424. [CrossRef]
11. Kitatani, K.; Idkowiak-Baldys, J.; Hannun, Y.A. The sphingolipid salvage pathway in ceramide metabolism and signaling. *Cell. Signal.* **2008**, *20*, 1010–1018. [CrossRef]
12. Spiegel, S.; Merrill, A.H. Sphingolipid metabolism and cell growth regulation. *FASEB J. Off. Publ. Fed. Am. Soc. Exp. Biol.* **1996**, *10*, 1388–1397. [CrossRef]
13. Hannun, Y.A. The sphingomyelin cycle and the second messenger function of ceramide. *J. Biol. Chem.* **1994**, *269*, 3125–3128.
14. Kolesnick, R.; Golde, D.W. The sphingomyelin pathway in tumor necrosis factor and interleukin-1 signaling. *Cell* **1994**, *77*, 325–328. [CrossRef]
15. Kolesnick, R.N.; Krönke, M. Regulation of ceramide production and apoptosis. *Annu. Rev. Physiol.* **1998**, *60*, 643–665. [CrossRef]
16. Hannun, Y.A.; Obeid, L.M. Principles of bioactive lipid signalling: Lessons from sphingolipids. *Nat. Rev. Mol. Cell Biol.* **2008**, *9*, 139–150. [CrossRef]

17. Mizutani, Y.; Kihara, A.; Chiba, H.; Tojo, H.; Igarashi, Y. 2-Hydroxy-ceramide synthesis by ceramide synthase family: Enzymatic basis for the preference of FA chain length. *J. Lipid Res.* **2008**, *49*, 2356–2364. [CrossRef]
18. Grösch, S.; Schiffmann, S.; Geisslinger, G. Chain length-specific properties of ceramides. *Prog. Lipid Res.* **2012**, *51*, 50–62. [CrossRef]
19. Pewzner-Jung, Y.; Park, H.; Laviad, E.L.; Silva, L.C.; Lahiri, S.; Stiban, J.; Erez-Roman, R.; Brügger, B.; Sachsenheimer, T.; Wieland, F.; et al. A Critical Role for Ceramide Synthase 2 in Liver Homeostasis. *J. Biol. Chem.* **2010**, *285*, 10902–10910. [CrossRef] [PubMed]
20. Pewzner-Jung, Y.; Brenner, O.; Braun, S.; Laviad, E.L.; Ben-Dor, S.; Feldmesser, E.; Horn-Saban, S.; Amann-Zalcenstein, D.; Raanan, C.; Berkutzki, T.; et al. A critical role for ceramide synthase 2 in liver homeostasis: II. Insights into molecular changes leading to hepatopathy. *J. Biol. Chem.* **2010**, *285*, 10911–10923. [CrossRef] [PubMed]
21. Bartke, N.; Hannun, Y.A. Bioactive sphingolipids: Metabolism and function. *J. Lipid Res.* **2009**, *50*, S91–S96. [CrossRef] [PubMed]
22. Breslow, D.K.; Weissman, J.S. Membranes in balance: Mechanisms of sphingolipid homeostasis. *Mol. Cell* **2010**, *40*, 267–279. [CrossRef] [PubMed]
23. Schissel, S.L.; Schuchman, E.H.; Williams, K.J.; Tabas, I. Zn^{2+}-stimulated sphingomyelinase is secreted by many cell types and is a product of the acid sphingomyelinase gene. *J. Biol. Chem.* **1996**, *271*, 18431–18436. [CrossRef] [PubMed]
24. Schissel, S.L.; Keesler, G.A.; Schuchman, E.H.; Williams, K.J.; Tabas, I. The cellular trafficking and zinc dependence of secretory and lysosomal sphingomyelinase, two products of the acid sphingomyelinase gene. *J. Biol. Chem.* **1998**, *273*, 18250–18259. [CrossRef] [PubMed]
25. Jenkins, R.W.; Canals, D.; Idkowiak-Baldys, J.; Simbari, F.; Roddy, P.; Perry, D.M.; Kitatani, K.; Luberto, C.; Hannun, Y.A. Regulated secretion of acid sphingomyelinase: Implications for selectivity of ceramide formation. *J. Biol. Chem.* **2010**, *285*, 35706–35718. [CrossRef]
26. Gorelik, A.; Illes, K.; Heinz, L.X.; Superti-Furga, G.; Nagar, B. Crystal structure of mammalian acid sphingomyelinase. *Nat. Commun.* **2016**, *7*, 12196. [CrossRef]
27. Andrews, N.W. Solving the secretory acid sphingomyelinase puzzle: Insights from lysosome-mediated parasite invasion and plasma membrane repair. *Cell. Microbiol.* **2019**, *21*, e13065. [CrossRef]
28. Airola, M.V.; Hannun, Y.A. Sphingolipid metabolism and neutral sphingomyelinases. *Handb. Exp. Pharmacol.* **2013**, 57–76. [CrossRef]
29. Tomiuk, S.; Zumbansen, M.; Stoffel, W. Characterization and subcellular localization of murine and human magnesium-dependent neutral sphingomyelinase. *J. Biol. Chem.* **2000**, *275*, 5710–5717. [CrossRef]
30. Neuberger, Y.; Shogomori, H.; Levy, Z.; Fainzilber, M.; Futerman, A.H. A lyso-platelet activating factor phospholipase C, originally suggested to be a neutral-sphingomyelinase, is located in the endoplasmic reticulum. *FEBS Lett.* **2000**, *469*, 44–46. [CrossRef]
31. Rodrigues-Lima, F.; Fensome, A.C.; Josephs, M.; Evans, J.; Veldman, R.J.; Katan, M. Structural requirements for catalysis and membrane targeting of mammalian enzymes with neutral sphingomyelinase and lysophospholipid phospholipase C activities. Analysis by chemical modification and site-directed mutagenesis. *J. Biol. Chem.* **2000**, *275*, 28316–28325. [CrossRef]
32. Marchesini, N.; Osta, W.; Bielawski, J.; Luberto, C.; Obeid, L.M.; Hannun, Y.A. Role for mammalian neutral sphingomyelinase 2 in confluence-induced growth arrest of MCF7 cells. *J. Biol. Chem.* **2004**, *279*, 25101–25111. [CrossRef]
33. Karakashian, A.A.; Giltiay, N.V.; Smith, G.M.; Nikolova-Karakashian, M.N. Expression of neutral sphingomyelinase-2 (NSMase-2) in primary rat hepatocytes modulates IL-beta-induced JNK activation. *FASEB J. Off. Publ. Fed. Am. Soc. Exp. Biol.* **2004**, *18*, 968–970. [CrossRef]
34. De Palma, C.; Meacci, E.; Perrotta, C.; Bruni, P.; Clementi, E. Endothelial nitric oxide synthase activation by tumor necrosis factor alpha through neutral sphingomyelinase 2, sphingosine kinase 1, and sphingosine 1 phosphate receptors: A novel pathway relevant to the pathophysiology of endothelium. *Arterioscler. Thromb. Vasc. Biol.* **2006**, *26*, 99–105. [CrossRef] [PubMed]
35. Shamseddine, A.A.; Airola, M.V.; Hannun, Y.A. Roles and regulation of neutral sphingomyelinase-2 in cellular and pathological processes. *Adv. Biol. Regul.* **2015**, *57*, 24–41. [CrossRef] [PubMed]
36. Tomiuk, S.; Hofmann, K.; Nix, M.; Zumbansen, M.; Stoffel, W. Cloned mammalian neutral sphingomyelinase: Functions in sphingolipid signaling? *Proc. Natl. Acad. Sci. USA* **1998**, *95*, 3638–3643. [CrossRef] [PubMed]

37. Tepper, A.D.; Ruurs, P.; Borst, J.; van Blitterswijk, W.J. Effect of overexpression of a neutral sphingomyelinase on CD95-induced ceramide production and apoptosis. *Biochem. Biophys. Res. Commun.* **2001**, *280*, 634–639. [CrossRef] [PubMed]
38. Wu, B.X.; Rajagopalan, V.; Roddy, P.L.; Clarke, C.J.; Hannun, Y.A. Identification and characterization of murine mitochondria-associated neutral sphingomyelinase (MA-nSMase), the mammalian sphingomyelin phosphodiesterase 5. *J. Biol. Chem.* **2010**, *285*, 17993–18002. [CrossRef]
39. Cheng, Y.; Nilsson, A.; Tömquist, E.; Duan, R.D. Purification, characterization, and expression of rat intestinal alkaline sphingomyelinase. *J. Lipid Res.* **2002**, *43*, 316–324.
40. Duan, R.-D.; Cheng, Y.; Hansen, G.; Hertervig, E.; Liu, J.-J.; Syk, I.; Sjostrom, H.; Nilsson, A. Purification, localization, and expression of human intestinal alkaline sphingomyelinase. *J. Lipid Res.* **2003**, *44*, 1241–1250. [CrossRef]
41. Duan, R.-D. Alkaline sphingomyelinase (NPP7) in hepatobiliary diseases: A field that needs to be closely studied. *World J. Hepatol.* **2018**, *10*, 246–253. [CrossRef]
42. Hofmann, K.; Tomiuk, S.; Wolff, G.; Stoffel, W. Cloning and characterization of the mammalian brain-specific, Mg^{2+}-dependent neutral sphingomyelinase. *Proc. Natl. Acad. Sci. USA* **2000**, *97*, 5895–5900. [CrossRef]
43. Filosto, S.; Fry, W.; Knowlton, A.A.; Goldkorn, T. Neutral sphingomyelinase 2 (nSMase2) is a phosphoprotein regulated by calcineurin (PP2B). *J. Biol. Chem.* **2010**, *285*, 10213–10222. [CrossRef] [PubMed]
44. Wiesner, D.A.; Kilkus, J.P.; Gottschalk, A.R.; Quintáns, J.; Dawson, G. Anti-immunoglobulin-induced apoptosis in WEHI 231 cells involves the slow formation of ceramide from sphingomyelin and is blocked by bcl-XL. *J. Biol. Chem.* **1997**, *272*, 9868–9876. [CrossRef] [PubMed]
45. Chipuk, J.E.; Green, D.R. How do BCL-2 proteins induce mitochondrial outer membrane permeabilization? *Trends Cell Biol.* **2008**, *18*, 157–164. [CrossRef] [PubMed]
46. Kosaka, N.; Iguchi, H.; Hagiwara, K.; Yoshioka, Y.; Takeshita, F.; Ochiya, T. Neutral sphingomyelinase 2 (nSMase2)-dependent exosomal transfer of angiogenic microRNAs regulate cancer cell metastasis. *J. Biol. Chem.* **2013**, *288*, 10849–10859. [CrossRef] [PubMed]
47. Dinkins, M.B.; Dasgupta, S.; Wang, G.; Zhu, G.; Bieberich, E. Exosome reduction in vivo is associated with lower amyloid plaque load in the 5XFAD mouse model of Alzheimer's disease. *Neurobiol. Aging* **2014**, *35*, 1792–1800. [CrossRef]
48. Malhi, H.; Kaufman, R.J. Endoplasmic reticulum stress in liver disease. *J. Hepatol.* **2011**, *54*, 795–809. [CrossRef]
49. Kaplowitz, N.; Than, T.A.; Shinohara, M.; Ji, C. Endoplasmic reticulum stress and liver injury. *Semin. Liver Dis.* **2007**, *27*, 367–377. [CrossRef]
50. Henkel, A.; Green, R.M. The unfolded protein response in fatty liver disease. *Semin. Liver Dis.* **2013**, *33*, 321–329. [CrossRef]
51. Fernandez, A.; Matias, N.; Fucho, R.; Ribas, V.; Von Montfort, C.; Nuño, N.; Baulies, A.; Martinez, L.; Tarrats, N.; Mari, M.; et al. ASMase is required for chronic alcohol induced hepatic endoplasmic reticulum stress and mitochondrial cholesterol loading. *J. Hepatol.* **2013**, *59*, 805–813. [CrossRef]
52. Fu, S.; Watkins, S.M.; Hotamisligil, G.S. The role of endoplasmic reticulum in hepatic lipid homeostasis and stress signaling. *Cell Metab.* **2012**, *15*, 623–634. [CrossRef]
53. Fu, S.; Yang, L.; Li, P.; Hofmann, O.; Dicker, L.; Hide, W.; Lin, X.; Watkins, S.M.; Ivanov, A.R.; Hotamisligil, G.S. Aberrant lipid metabolism disrupts calcium homeostasis causing liver endoplasmic reticulum stress in obesity. *Nature* **2011**, *473*, 528–531. [CrossRef] [PubMed]
54. Pinton, P.; Ferrari, D.; Rapizzi, E.; Di Virgilio, F.; Pozzan, T.; Rizzuto, R. The Ca^{2+} concentration of the endoplasmic reticulum is a key determinant of ceramide-induced apoptosis: Significance for the molecular mechanism of Bcl-2 action. *EMBO J.* **2001**, *20*, 2690–2701. [CrossRef]
55. Czaja, M.J.; Ding, W.-X.; Donohue, T.M.; Friedman, S.L.; Kim, J.-S.; Komatsu, M.; Lemasters, J.J.; Lemoine, A.; Lin, J.D.; Ou, J.J.; et al. Functions of autophagy in normal and diseased liver. *Autophagy* **2013**, *9*, 1131–1158. [CrossRef] [PubMed]
56. Parzych, K.R.; Klionsky, D.J. An overview of autophagy: Morphology, mechanism, and regulation. *Antioxid. Redox Signal.* **2014**, *20*, 460–473. [CrossRef] [PubMed]
57. Mizushima, N.; Komatsu, M. Autophagy: Renovation of cells and tissues. *Cell* **2011**, *147*, 728–741. [CrossRef]
58. Kraft, C.; Peter, M.; Hofmann, K. Selective autophagy: Ubiquitin-mediated recognition and beyond. *Nat. Cell Biol.* **2010**, *12*, 836–841. [CrossRef]

59. Singh, R.; Kaushik, S.; Wang, Y.; Xiang, Y.; Novak, I.; Komatsu, M.; Tanaka, K.; Cuervo, A.M.; Czaja, M.J. Autophagy regulates lipid metabolism. *Nature* **2009**, *458*, 1131–1135. [CrossRef]
60. Fucho, R.; Martínez, L.; Baulies, A.; Torres, S.; Tarrats, N.; Fernandez, A.; Ribas, V.; Astudillo, A.M.; Balsinde, J.; Garcia-Rovés, P.; et al. Asmase Regulates Autophagy and Lysosomal Membrane Permeabilization and its Inhibition Prevents Early Stage Nonalcoholic Steatohepatitis. *J. Hepatol.* **2014**. [CrossRef]
61. Moles, A.; Tarrats, N.; Fernández-Checa, J.C.; Marí, M. Cathepsin B overexpression due to Acid sphingomyelinase ablation promotes liver fibrosis in niemann-pick disease. *J. Biol. Chem.* **2012**, *287*, 1178–1188. [CrossRef]
62. Boya, P.; Kroemer, G. Lysosomal membrane permeabilization in cell death. *Oncogene* **2008**, *27*, 6434–6451. [CrossRef]
63. Marí, M.; Fernandez-Checa, J.C. Damage Mediated by Dysfunction of Organelles and Cellular Systems: Lysosomes. In *Pathobiology of Human Disease*; McManus, L.M., Mitchell, R.N., Eds.; Academic Press: San Diego, CA, USA, 2014; pp. 97–107. ISBN 978-0-12-386457-4.
64. Brunk, U.T.; Dalen, H.; Roberg, K.; Hellquist, H.B. Photo-oxidative disruption of lysosomal membranes causes apoptosis of cultured human fibroblasts. *Free Radic. Biol. Med.* **1997**, *23*, 616–626. [CrossRef]
65. Gursahani, H.I.; Schaefer, S. Acidification reduces mitochondrial calcium uptake in rat cardiac mitochondria. *Am. J. Physiol. Heart Circ. Physiol.* **2004**, *287*, H2659–H2665. [CrossRef]
66. Nilsson, C.; Johansson, U.; Johansson, A.-C.; Kågedal, K.; Ollinger, K. Cytosolic acidification and lysosomal alkalinization during TNF-alpha induced apoptosis in U937 cells. *Apoptosis* **2006**, *11*, 1149–1159. [CrossRef] [PubMed]
67. Mrschtik, M.; Ryan, K.M. Lysosomal proteins in cell death and autophagy. *FEBS J.* **2015**, *282*, 1858–1870. [CrossRef]
68. Li, Z.; Berk, M.; McIntyre, T.M.; Gores, G.J.; Feldstein, A.E. The lysosomal-mitochondrial axis in free fatty acid-induced hepatic lipotoxicity. *Hepatology* **2008**, *47*, 1495–1503. [CrossRef] [PubMed]
69. Egnatchik, R.A.; Leamy, A.K.; Jacobson, D.A.; Shiota, M.; Young, J.D. ER calcium release promotes mitochondrial dysfunction and hepatic cell lipotoxicity in response to palmitate overload. *Mol. Metab.* **2014**, *3*, 544–553. [CrossRef]
70. Kotronen, A.; Seppänen-Laakso, T.; Westerbacka, J.; Kiviluoto, T.; Arola, J.; Ruskeepää, A.-L.; Yki-Järvinen, H.; Oresic, M. Comparison of lipid and fatty acid composition of the liver, subcutaneous and intra-abdominal adipose tissue, and serum. *Obesity* **2010**, *18*, 937–944. [CrossRef] [PubMed]
71. Le Kim, D.; Betzing, H. Intestinal absorption of polyunsaturated phosphatidylcholine in the rat. *Hoppe Seylers Z. Physiol. Chem.* **1976**, *357*, 1321–1331. [CrossRef]
72. Nilsson, Å. Metabolism of sphingomyelin in the intestinal tract of the rat. *Biochim. Biophys. Acta BBA Lipids Lipid Metab.* **1968**, *164*, 575–584. [CrossRef]
73. Vanier, M.T. Niemann-Pick diseases. *Handb. Clin. Neurol.* **2013**, *113*, 1717–1721. [CrossRef]
74. Choi, S.; Snider, A.J. Sphingolipids in High Fat Diet and Obesity-Related Diseases. *Mediat. Inflamm.* **2015**, *2015*, 520618. [CrossRef]
75. Memon, R.A.; Holleran, W.M.; Moser, A.H.; Seki, T.; Uchida, Y.; Fuller, J.; Shigenaga, J.K.; Grunfeld, C.; Feingold, K.R. Endotoxin and cytokines increase hepatic sphingolipid biosynthesis and produce lipoproteins enriched in ceramides and sphingomyelin. *Arterioscler. Thromb. Vasc. Biol.* **1998**, *18*, 1257–1265. [CrossRef] [PubMed]
76. Li, J.-F.; Qu, F.; Zheng, S.-J.; Wu, H.-L.; Liu, M.; Liu, S.; Ren, Y.; Ren, F.; Chen, Y.; Duan, Z.-P.; et al. Elevated plasma sphingomyelin (d18:1/22:0) is closely related to hepatic steatosis in patients with chronic hepatitis C virus infection. *Eur. J. Clin. Microbiol. Infect. Dis. Off. Publ. Eur. Soc. Clin. Microbiol.* **2014**, *33*, 1725–1732. [CrossRef] [PubMed]
77. Koumanov, K.; Infante, R. Phospholipid-transfer proteins in human liver and primary liver carcinoma. *Biochim. Biophys. Acta* **1986**, *876*, 526–532. [CrossRef]
78. Zeidan, Y.H.; Hannun, Y.A. The acid sphingomyelinase/ceramide pathway: Biomedical significance and mechanisms of regulation. *Curr. Mol. Med.* **2010**, *10*, 454–466. [CrossRef]
79. Tilg, H.; Diehl, A.M. Cytokines in alcoholic and nonalcoholic steatohepatitis. *N. Engl. J. Med.* **2000**, *343*, 1467–1476. [CrossRef]
80. Ratziu, V.; Bellentani, S.; Cortez-Pinto, H.; Day, C.; Marchesini, G. A position statement on NAFLD/NASH based on the EASL 2009 special conference. *J. Hepatol.* **2010**, *53*, 372–384. [CrossRef]

81. Guicciardi, M.E.; Malhi, H.; Mott, J.L.; Gores, G.J. Apoptosis and necrosis in the liver. *Compr. Physiol.* **2013**, *3*, 977–1010. [CrossRef]
82. Leake, I. ASMase implicated in alcoholic liver disease. *Nat. Rev. Gastroenterol. Hepatol.* **2013**, *10*, 384–385. [CrossRef]
83. Deaciuc, I.V.; Nikolova-Karakashian, M.; Fortunato, F.; Lee, E.Y.; Hill, D.B.; McClain, C.J. Apoptosis and dysregulated ceramide metabolism in a murine model of alcohol-enhanced lipopolysaccharide hepatotoxicity. *Alcohol. Clin. Exp. Res.* **2000**, *24*, 1557–1565. [CrossRef]
84. Liangpunsakul, S.; Rahmini, Y.; Ross, R.A.; Zhao, Z.; Xu, Y.; Crabb, D.W. Imipramine blocks ethanol-induced ASMase activation, ceramide generation, and PP2A activation, and ameliorates hepatic steatosis in ethanol-fed mice. *Am. J. Physiol. Gastrointest. Liver Physiol.* **2012**, *302*, G515–G523. [CrossRef] [PubMed]
85. Mühle, C.; Weinland, C.; Gulbins, E.; Lenz, B.; Kornhuber, J. Peripheral Acid Sphingomyelinase Activity Is Associated with Biomarkers and Phenotypes of Alcohol Use and Dependence in Patients and Healthy Controls. *Int. J. Mol. Sci.* **2018**, *19*, 4028. [CrossRef]
86. Grammatikos, G.; Mühle, C.; Ferreiros, N.; Schroeter, S.; Bogdanou, D.; Schwalm, S.; Hintereder, G.; Kornhuber, J.; Zeuzem, S.; Sarrazin, C.; et al. Serum acid sphingomyelinase is upregulated in chronic hepatitis C infection and non alcoholic fatty liver disease. *Biochim. Biophys. Acta* **2014**, *1841*, 1012–1020. [CrossRef] [PubMed]
87. Moles, A.; Tarrats, N.; Morales, A.; Domínguez, M.; Bataller, R.; Caballería, J.; García-Ruiz, C.; Fernández-Checa, J.C.; Marí, M. Acidic sphingomyelinase controls hepatic stellate cell activation and in vivo liver fibrogenesis. *Am. J. Pathol.* **2010**, *177*, 1214–1224. [CrossRef] [PubMed]
88. Caballero, F.; Fernández, A.; Matías, N.; Martínez, L.; Fucho, R.; Elena, M.; Caballeria, J.; Morales, A.; Fernández-Checa, J.C.; García-Ruiz, C. Specific contribution of methionine and choline in nutritional nonalcoholic steatohepatitis: Impact on mitochondrial S-adenosyl-L-methionine and glutathione. *J. Biol. Chem.* **2010**, *285*, 18528–18536. [CrossRef]
89. García-Ruiz, C.; Colell, A.; Marí, M.; Morales, A.; Calvo, M.; Enrich, C.; Fernández-Checa, J.C. Defective TNF-alpha-mediated hepatocellular apoptosis and liver damage in acidic sphingomyelinase knockout mice. *J. Clin. Investig.* **2003**, *111*, 197–208. [CrossRef] [PubMed]
90. Marí, M.; Colell, A.; Morales, A.; Pañeda, C.; Varela-Nieto, I.; García-Ruiz, C.; Fernández-Checa, J.C. Acidic sphingomyelinase downregulates the liver-specific methionine adenosyltransferase 1A, contributing to tumor necrosis factor-induced lethal hepatitis. *J. Clin. Investig.* **2004**, *113*, 895–904. [CrossRef] [PubMed]
91. Lin, T.; Genestier, L.; Pinkoski, M.J.; Castro, A.; Nicholas, S.; Mogil, R.; Paris, F.; Fuks, Z.; Schuchman, E.H.; Kolesnick, R.N.; et al. Role of acidic sphingomyelinase in Fas/CD95-mediated cell death. *J. Biol. Chem.* **2000**, *275*, 8657–8663. [CrossRef]
92. Colell, A.; García-Ruiz, C.; Roman, J.; Ballesta, A.; FernándezCheca, J.C. Ganglioside GD3 enhances apoptosis by suppressing the nuclear factor-κB-dependent survival pathway. *FASEB J.* **2001**, *15*, 1068–1070. [CrossRef]
93. García-Ruiz, C.; Colell, A.; Morales, A.; Calvo, M.; Enrich, C.; Fernández-Checa, J.C. Trafficking of ganglioside GD3 to mitochondria by tumor necrosis factor-alpha. *J. Biol. Chem.* **2002**, *277*, 36443–36448. [CrossRef] [PubMed]
94. Quillin, R.C.; Wilson, G.C.; Nojima, H.; Freeman, C.M.; Wang, J.; Schuster, R.M.; Blanchard, J.A.; Edwards, M.J.; Gandhi, C.R.; Gulbins, E.; et al. Inhibition of acidic sphingomyelinase reduces established hepatic fibrosis in mice. *Hepatol. Res.* **2015**, *45*, 305–314. [CrossRef] [PubMed]
95. Yang, L.; Li, P.; Fu, S.; Calay, E.S.; Hotamisligil, G.S. Defective hepatic autophagy in obesity promotes ER stress and causes insulin resistance. *Cell Metab.* **2010**, *11*, 467–478. [CrossRef] [PubMed]
96. Hernández-Gea, V.; Hilscher, M.; Rozenfeld, R.; Lim, M.P.; Nieto, N.; Werner, S.; Devi, L.A.; Friedman, S.L. Endoplasmic Reticulum Stress Induces Fibrogenic Activity in Hepatic Stellate Cells through Autophagy. *J. Hepatol.* **2013**, *59*, 98–104. [CrossRef] [PubMed]
97. Hernández-Gea, V.; Ghiassi-Nejad, Z.; Rozenfeld, R.; Gordon, R.; Fiel, M.I.; Yue, Z.; Czaja, M.J.; Friedman, S.L. Autophagy releases lipid that promotes fibrogenesis by activated hepatic stellate cells in mice and in human tissues. *Gastroenterology* **2012**, *142*, 938–946. [CrossRef]
98. Thoen, L.F.R.; Guimarães, E.L.M.; Dollé, L.; Mannaerts, I.; Najimi, M.; Sokal, E.; van Grunsven, L.A. A role for autophagy during hepatic stellate cell activation. *J. Hepatol.* **2011**, *55*, 1353–1360. [CrossRef] [PubMed]
99. Hernández-Gea, V.; Friedman, S.L. Autophagy fuels tissue fibrogenesis. *Autophagy* **2012**, *8*, 849–850. [CrossRef]

100. Mato, J.M.; Martínez-Chantar, M.L.; Lu, S.C. Methionine metabolism and liver disease. *Annu. Rev. Nutr.* **2008**, *28*, 273–293. [CrossRef]
101. Mato, J.M.; Lu, S.C. Role of S-adenosyl-L-methionine in liver health and injury. *Hepatology* **2007**, *45*, 1306–1312. [CrossRef] [PubMed]
102. Montfort, A.; Martin, P.G.P.; Levade, T.; Benoist, H.; Ségui, B. FAN (factor associated with neutral sphingomyelinase activation), a moonlighting protein in TNF-R1 signaling. *J. Leukoc. Biol.* **2010**, *88*, 897–903. [CrossRef]
103. Ségui, B.; Cuvillier, O.; Adam-Klages, S.; Garcia, V.; Malagarie-Cazenave, S.; Lévêque, S.; Caspar-Bauguil, S.; Coudert, J.; Salvayre, R.; Krönke, M.; et al. Involvement of FAN in TNF-induced apoptosis. *J. Clin. Investig.* **2001**, *108*, 143–151. [CrossRef]
104. Krautbauer, S.; Meier, E.M.; Rein-Fischboeck, L.; Pohl, R.; Weiss, T.S.; Sigruener, A.; Aslanidis, C.; Liebisch, G.; Buechler, C. Ceramide and polyunsaturated phospholipids are strongly reduced in human hepatocellular carcinoma. *Biochim. Biophys. Acta BBA Mol. Cell Biol. Lipids* **2016**, *1861*, 1767–1774. [CrossRef]
105. Maeng, H.J.; Song, J.-H.; Kim, G.-T.; Song, Y.-J.; Lee, K.; Kim, J.-Y.; Park, T.-S. Celecoxib-mediated activation of endoplasmic reticulum stress induces de novo ceramide biosynthesis and apoptosis in hepatoma HepG2 cells. *BMB Rep.* **2017**, *50*, 144–149. [CrossRef] [PubMed]
106. Bai, A.; Mao, C.; Jenkins, R.W.; Szulc, Z.M.; Bielawska, A.; Hannun, Y.A. Anticancer actions of lysosomally targeted inhibitor, LCL521, of acid ceramidase. *PLoS ONE* **2017**, *12*, e0177805. [CrossRef] [PubMed]
107. Dementiev, A.; Joachimiak, A.; Nguyen, H.; Gorelik, A.; Illes, K.; Shabani, S.; Gelsomino, M.; Ahn, E.-Y.E.; Nagar, B.; Doan, N. Molecular Mechanism of Inhibition of Acid Ceramidase by Carmofur. *J. Med. Chem.* **2019**, *62*, 987–992. [CrossRef] [PubMed]
108. Lin, M.; Liao, W.; Dong, M.; Zhu, R.; Xiao, J.; Sun, T.; Chen, Z.; Wu, B.; Jin, J. Exosomal neutral sphingomyelinase 1 suppresses hepatocellular carcinoma via decreasing the ratio of sphingomyelin/ceramide. *FEBS J.* **2018**, *285*, 3835–3848. [CrossRef]
109. Zhong, L.; Kong, J.N.; Dinkins, M.B.; Leanhart, S.; Zhu, Z.; Spassieva, S.D.; Qin, H.; Lin, H.-P.; Elsherbini, A.; Wang, R.; et al. Increased liver tumor formation in neutral sphingomyelinase-2-deficient mice. *J. Lipid Res.* **2018**, *59*, 795–804. [CrossRef] [PubMed]
110. Stefanovic, M.; Tutusaus, A.; Martinez-Nieto, G.A.; Bárcena, C.; de Gregorio, E.; Moutinho, C.; Barbero-Camps, E.; Villanueva, A.; Colell, A.; Marí, M.; et al. Targeting glucosylceramide synthase upregulation reverts sorafenib resistance in experimental hepatocellular carcinoma. *Oncotarget* **2016**, *7*, 8253. [CrossRef]
111. Xu, Z.; Xu, M.; Liu, P.; Zhang, S.; Shang, R.; Qiao, Y.; Che, L.; Ribback, S.; Cigliano, A.; Evert, K.; et al. The mTORC2-Akt1 Cascade Is Crucial for c-Myc to Promote Hepatocarcinogenesis in Mice and Humans. *Hepatology* **2019**, *70*, 1600–1613. [CrossRef]
112. Guri, Y.; Colombi, M.; Dazert, E.; Hindupur, S.K.; Roszik, J.; Moes, S.; Jenoe, P.; Heim, M.H.; Riezman, I.; Riezman, H.; et al. mTORC2 Promotes Tumorigenesis via Lipid Synthesis. *Cancer Cell* **2017**, *32*, 807–823. [CrossRef]
113. Wu, W.K.K.; Coffelt, S.B.; Cho, C.H.; Wang, X.J.; Lee, C.W.; Chan, F.K.L.; Yu, J.; Sung, J.J.Y. The autophagic paradox in cancer therapy. *Oncogene* **2012**, *31*, 939–953. [CrossRef]
114. Pan, H.; Wang, Z.; Jiang, L.; Sui, X.; You, L.; Shou, J.; Jing, Z.; Xie, J.; Ge, W.; Cai, X.; et al. Autophagy inhibition sensitizes hepatocellular carcinoma to the multikinase inhibitor linifanib. *Sci. Rep.* **2014**, *4*, 6683. [CrossRef] [PubMed]
115. Qiu, D.-M.; Wang, G.-L.; Chen, L.; Xu, Y.-Y.; He, S.; Cao, X.-L.; Qin, J.; Zhou, J.-M.; Zhang, Y.-X.; Qun, E. The expression of beclin-1, an autophagic gene, in hepatocellular carcinoma associated with clinical pathological and prognostic significance. *BMC Cancer* **2014**, *14*, 327. [CrossRef] [PubMed]
116. Zheng, J.-F.; Li, L.-L.; Lu, J.; Yan, K.; Guo, W.-H.; Zhang, J.-X. XPD Functions as a Tumor Suppressor and Dysregulates Autophagy in Cultured HepG2 Cells. *Med. Sci. Monit. Int. Med. J. Exp. Clin. Res.* **2015**, *21*, 1562–1568. [CrossRef] [PubMed]
117. Zhang, W.; Hou, J.; Wang, X.; Jiang, R.; Yin, Y.; Ji, J.; Deng, L.; Huang, X.; Wang, K.; Sun, B. PTPRO-mediated autophagy prevents hepatosteatosis and tumorigenesis. *Oncotarget* **2015**, *6*, 9420–9433. [CrossRef]
118. Inokuchi-Shimizu, S.; Park, E.J.; Roh, Y.S.; Yang, L.; Zhang, B.; Song, J.; Liang, S.; Pimienta, M.; Taniguchi, K.; Wu, X.; et al. TAK1-mediated autophagy and fatty acid oxidation prevent hepatosteatosis and tumorigenesis. *J. Clin. Investig.* **2014**, *124*, 3566–3578. [CrossRef]

119. Shu, G.; Xie, B.; Ren, F.; Liu, D.; Zhou, J.; Li, Q.; Chen, J.; Yuan, L.; Zhou, J. Restoration of klotho expression induces apoptosis and autophagy in hepatocellular carcinoma cells. *Cell. Oncol. Dordr.* **2013**, *36*, 121–129. [CrossRef]
120. Park, M.A.; Zhang, G.; Martin, A.P.; Hamed, H.; Mitchell, C.; Hylemon, P.B.; Graf, M.; Rahmani, M.; Ryan, K.; Liu, X.; et al. Vorinostat and sorafenib increase ER stress, autophagy and apoptosis via ceramide-dependent CD95 and PERK activation. *Cancer Biol. Ther.* **2008**, *7*, 1648–1662. [CrossRef]
121. Savić, R.; He, X.; Fiel, I.; Schuchman, E.H. Recombinant human acid sphingomyelinase as an adjuvant to sorafenib treatment of experimental liver cancer. *PLoS ONE* **2013**, *8*, e65620. [CrossRef]
122. Patterson, M.C.; Walkley, S.U. Niemann-Pick disease, type C and Roscoe Brady. *Mol. Genet. Metab.* **2017**, *120*, 34–37. [CrossRef]
123. Schuchman, E.H.; Desnick, R.J. Types A and B Niemann-Pick disease. *Mol. Genet. Metab.* **2017**, *120*, 27–33. [CrossRef]
124. Rodriguez-Lafrasse, C.; Rousson, R.; Pentchev, P.G.; Louisot, P.; Vanier, M.T. Free sphingoid bases in tissues from patients with type C Niemann-Pick disease and other lysosomal storage disorders. *Biochim. Biophys. Acta* **1994**, *1226*, 138–144. [CrossRef]
125. Valle, D.; Beaudet, A.L.; Vogelstein, B.; Kinzler, K.W.; Antonarakis, S.E.; Ballabio, A.; Gibson, K.M.; Mitchell, G.A.; FACMG. Lysosomal Disorders. In *The Online Metabolic and Molecular Bases of Inherited Disease;* David Valle, M., Arthur, L., Beaudet, M., Bert Vogeslstein, M., Kenneth, W., Kinzler, P.D., Stylianos, E., Antonarakis, M.D., Andrea Ballabio, M.K., Michael Gibson, P.D., et al., Eds.; McGraw-Hill Education: New York, NY, USA, 2018.
126. Pick, L., II. Niemann-Pick's disease and other forms of so-called xanthomatosis. *Am. J. Med. Sci.* **1933**, *185*, 615–616. [CrossRef]
127. Pavlů-Pereira, H.; Asfaw, B.; Poupctová, H.; Ledvinová, J.; Sikora, J.; Vanier, M.T.; Sandhoff, K.; Zeman, J.; Novotná, Z.; Chudoba, D.; et al. Acid sphingomyelinase deficiency. Phenotype variability with prevalence of intermediate phenotype in a series of twenty-five Czech and Slovak patients. A multi-approach study. *J. Inherit. Metab. Dis.* **2005**, *28*, 203–227. [CrossRef] [PubMed]
128. McGovern, M.M.; Aron, A.; Brodie, S.E.; Desnick, R.J.; Wasserstein, M.P. Natural history of Type A Niemann-Pick disease: Possible endpoints for therapeutic trials. *Neurology* **2006**, *66*, 228–232. [CrossRef]
129. Wasserstein, M.P.; Desnick, R.J.; Schuchman, E.H.; Hossain, S.; Wallenstein, S.; Lamm, C.; McGovern, M.M. The natural history of type B Niemann-Pick disease: Results from a 10-year longitudinal study. *Pediatrics* **2004**, *114*, e672–e677. [CrossRef] [PubMed]
130. McGovern, M.M.; Wasserstein, M.P.; Giugliani, R.; Bembi, B.; Vanier, M.T.; Mengel, E.; Brodie, S.E.; Mendelson, D.; Skloot, G.; Desnick, R.J.; et al. A prospective, cross-sectional survey study of the natural history of Niemann-Pick disease type B. *Pediatrics* **2008**, *122*, e341–e349. [CrossRef]
131. McGovern, M.M.; Avetisyan, R.; Sanson, B.-J.; Lidove, O. Disease manifestations and burden of illness in patients with acid sphingomyelinase deficiency (ASMD). *Orphanet J. Rare Dis.* **2017**, *12*, 41. [CrossRef]
132. Lidove, O.; Belmatoug, N.; Froissart, R.; Lavigne, C.; Durieu, I.; Mazodier, K.; Serratrice, C.; Douillard, C.; Goizet, C.; Cathebras, P.; et al. Déficit en sphingomyélinase acide (maladie de Niemann-Pick B): Une étude rétrospective multicentrique de 28 patients adultes. *Rev. Med. Interne* **2017**, *38*, 291–299. [CrossRef]
133. Cassiman, D.; Packman, S.; Bembi, B.; Turkia, H.B.; Al-Sayed, M.; Schiff, M.; Imrie, J.; Mabe, P.; Takahashi, T.; Mengel, K.E.; et al. Cause of death in patients with chronic visceral and chronic neurovisceral acid sphingomyelinase deficiency (Niemann-Pick disease type B and B variant): Literature review and report of new cases. *Mol. Genet. Metab.* **2016**, *118*, 206–213. [CrossRef]
134. Thurberg, B.L.; Wasserstein, M.P.; Jones, S.A.; Schiano, T.D.; Cox, G.F.; Puga, A.C. Clearance of Hepatic Sphingomyelin by Olipudase Alfa Is Associated with Improvement in Lipid Profiles in Acid Sphingomyelinase Deficiency. *Am. J. Surg. Pathol.* **2016**, *40*, 1232–1242. [CrossRef]
135. Eskes, E.C.B.; Sjouke, B.; Vaz, F.M.; Goorden, S.M.I.; van Kuilenburg, A.B.P.; Aerts, J.M.F.G.; Hollak, C.E.M. Biochemical and imaging parameters in acid sphingomyelinase deficiency: Potential utility as biomarkers. *Mol. Genet. Metab.* **2020**, *130*, 16–26. [CrossRef] [PubMed]
136. Wasserstein, M.P.; Jones, S.A.; Soran, H.; Diaz, G.A.; Lippa, N.; Thurberg, B.L.; Culm-Merdek, K.; Shamiyeh, E.; Inguilizian, H.; Cox, G.F.; et al. Successful within-Patient Dose Escalation of Olipudase Alfa in Acid Sphingomyelinase Deficiency. *Mol. Genet. Metab.* **2015**, *116*, 88–97. [CrossRef] [PubMed]

137. Wasserstein, M.P.; Diaz, G.A.; Lachmann, R.H.; Jouvin, M.-H.; Nandy, I.; Ji, A.J.; Puga, A.C. Olipudase alfa for treatment of acid sphingomyelinase deficiency (ASMD): Safety and efficacy in adults treated for 30 months. *J. Inherit. Metab. Dis.* **2018**, *41*, 829–838. [CrossRef] [PubMed]
138. Gulbins, E.; Palmada, M.; Reichel, M.; Lüth, A.; Böhmer, C.; Amato, D.; Müller, C.P.; Tischbirek, C.H.; Groemer, T.W.; Tabatabai, G.; et al. Acid sphingomyelinase-ceramide system mediates effects of antidepressant drugs. *Nat. Med.* **2013**, *19*, 934–938. [CrossRef]
139. Hua, G.; Kolesnick, R. Using ASMase Knockout Mice to Model Human Diseases. *Sphingolipids Dis.* **2012**, *216*, 29–54. [CrossRef]
140. Jenkins, R.W.; Canals, D.; Hannun, Y.A. Roles and regulation of secretory and lysosomal acid sphingomyelinase. *Cell. Signal.* **2009**, *21*, 836–846. [CrossRef]
141. Horinouchi, K.; Erlich, S.; Perl, D.P.; Ferlinz, K.; Bisgaier, C.L.; Sandhoff, K.; Desnick, R.J.; Stewart, C.L.; Schuchman, E.H. Acid sphingomyelinase deficient mice: A model of types A and B Niemann-Pick disease. *Nat. Genet.* **1995**, *10*, 288–293. [CrossRef]
142. Shen, D.; Wang, X.; Li, X.; Zhang, X.; Yao, Z.; Dibble, S.; Dong, X.; Yu, T.; Lieberman, A.P.; Showalter, H.D.; et al. Lipid storage disorders block lysosomal trafficking by inhibiting a TRP channel and lysosomal calcium release. *Nat. Commun.* **2012**, *3*, 1–11. [CrossRef]
143. Vázquez, M.C.; Balboa, E.; Alvarez, A.R.; Zanlungo, S. Oxidative Stress: A Pathogenic Mechanism for Niemann-Pick Type C Disease. *Oxid. Med. Cell. Longev.* **2012**, *2012*. [CrossRef]
144. Choudhury, A.; Sharma, D.K.; Marks, D.L.; Pagano, R.E. Elevated Endosomal Cholesterol Levels in Niemann-Pick Cells Inhibit Rab4 and Perturb Membrane Recycling. *Mol. Biol. Cell* **2004**, *15*, 4500–4511. [CrossRef]
145. Vance, J.E.; Karten, B. Niemann-Pick C disease and mobilization of lysosomal cholesterol by cyclodextrin. *J. Lipid Res.* **2014**, *55*, 1609–1621. [CrossRef]
146. Grassme, H.; Jekle, A.; Riehle, A.; Schwarz, H.; Berger, J.; Sandhoff, K.; Kolesnick, R.; Gulbins, E. CD95 signaling via ceramide-rich membrane rafts. *J. Biol. Chem.* **2001**, *276*, 20589–20596. [CrossRef] [PubMed]
147. Charruyer, A.; Grazide, S.; Bezombes, C.; Müller, S.; Laurent, G.; Jaffrézou, J.-P. UV-C light induces raft-associated acid sphingomyelinase and JNK activation and translocation independently on a nuclear signal. *J. Biol. Chem.* **2005**, *280*, 19196–19204. [CrossRef]
148. Baulies, A.; Ribas, V.; Núñez, S.; Torres, S.; Alarcón-Vila, C.; Martínez, L.; Suda, J.; Ybanez, M.D.; Kaplowitz, N.; García-Ruiz, C.; et al. Lysosomal Cholesterol Accumulation Sensitizes to Acetaminophen Hepatotoxicity by Impairing Mitophagy. *Sci. Rep.* **2016**, *5*, 18017. [CrossRef]
149. Torres, S.; Balboa, E.; Zanlungo, S.; Enrich, C.; Garcia-Ruiz, C.; Fernandez-Checa, J.C. Lysosomal and Mitochondrial Liaisons in Niemann-Pick Disease. *Front. Physiol.* **2017**, *8*, 982. [CrossRef] [PubMed]
150. Hannun, Y.A.; Luberto, C. Ceramide in the eukaryotic stress response. *Trends Cell Biol.* **2000**, *10*, 73–80. [CrossRef]
151. García-Ruiz, C.; Colell, A.; Marí, M.; Morales, A.; Fernández-Checa, J.C. Direct effect of ceramide on the mitochondrial electron transport chain leads to generation of reactive oxygen species. Role of mitochondrial glutathione. *J. Biol. Chem.* **1997**, *272*, 11369–11377. [CrossRef]
152. Scarlatti, F.; Bauvy, C.; Ventruti, A.; Sala, G.; Cluzeaud, F.; Vandewalle, A.; Ghidoni, R.; Codogno, P. Ceramide-mediated macroautophagy involves inhibition of protein kinase B and up-regulation of beclin 1. *J. Biol. Chem.* **2004**, *279*, 18384–18391. [CrossRef] [PubMed]
153. Llacuna, L.; Marí, M.; Garcia-Ruiz, C.; Fernandez-Checa, J.C.; Morales, A. Critical role of acidic sphingomyelinase in murine hepatic ischemia-reperfusion injury. *Hepatology* **2006**, *44*, 561–572. [CrossRef]
154. Unal, B.; Ozcan, F.; Tuzcu, H.; Kırac, E.; Elpek, G.O.; Aslan, M. Inhibition of neutral sphingomyelinase decreases elevated levels of nitrative and oxidative stress markers in liver ischemia-reperfusion injury. *Redox Rep. Commun. Free Radic. Res.* **2017**, *22*, 147–159. [CrossRef]
155. Tuzcu, H.; Unal, B.; Kırac, E.; Konuk, E.; Ozcan, F.; Elpek, G.O.; Demir, N.; Aslan, M. Neutral sphingomyelinase inhibition alleviates apoptosis, but not ER stress, in liver ischemia–reperfusion injury. *Free Radic. Res.* **2017**, *51*, 253–268. [CrossRef]
156. Lee, W.M. Acetaminophen and the U.S. Acute Liver Failure Study Group: Lowering the risks of hepatic failure. *Hepatology* **2004**, *40*, 6–9. [CrossRef] [PubMed]
157. Bernal, W.; Auzinger, G.; Dhawan, A.; Wendon, J. Acute liver failure. *Lancet* **2010**, *376*, 190–201. [CrossRef]

158. Hanawa, N.; Shinohara, M.; Saberi, B.; Gaarde, W.A.; Han, D.; Kaplowitz, N. Role of JNK translocation to mitochondria leading to inhibition of mitochondria bioenergetics in acetaminophen-induced liver injury. *J. Biol. Chem.* **2008**, *283*, 13565–13577. [CrossRef] [PubMed]
159. Matsumaru, K.; Ji, C.; Kaplowitz, N. Mechanisms for sensitization to TNF-induced apoptosis by acute glutathione depletion in murine hepatocytes. *Hepatology* **2003**, *37*, 1425–1434. [CrossRef] [PubMed]
160. Meyers, L.L.; Beierschmitt, W.P.; Khairallah, E.A.; Cohen, S.D. Acetaminophen-induced inhibition of hepatic mitochondrial respiration in mice. *Toxicol. Appl. Pharmacol.* **1988**, *93*, 378–387. [CrossRef]
161. Jaeschke, H.; Bajt, M.L. Intracellular signaling mechanisms of acetaminophen-induced liver cell death. *Toxicol. Sci. Off. J. Soc. Toxicol.* **2006**, *89*, 31–41. [CrossRef] [PubMed]
162. Jaeschke, H. Glutathione disulfide formation and oxidant stress during acetaminophen-induced hepatotoxicity in mice in vivo: The protective effect of allopurinol. *J. Pharmacol. Exp. Ther.* **1990**, *255*, 935–941. [PubMed]
163. Kon, K.; Kim, J.-S.; Jaeschke, H.; Lemasters, J.J. Mitochondrial permeability transition in acetaminophen-induced necrosis and apoptosis of cultured mouse hepatocytes. *Hepatology* **2004**, *40*, 1170–1179. [CrossRef]
164. De Toro, J.; Herschlik, L.; Waldner, C.; Mongini, C. Emerging roles of exosomes in normal and pathological conditions: New insights for diagnosis and therapeutic applications. *Front. Immunol.* **2015**, *6*, 203. [CrossRef]
165. Robbins, P.D.; Morelli, A.E. Regulation of immune responses by extracellular vesicles. *Nat. Rev. Immunol.* **2014**, *14*, 195–208. [CrossRef]
166. Masyuk, A.I.; Masyuk, T.V.; Larusso, N.F. Exosomes in the pathogenesis, diagnostics and therapeutics of liver diseases. *J. Hepatol.* **2013**, *59*, 621–625. [CrossRef] [PubMed]
167. Trajkovic, K.; Hsu, C.; Chiantia, S.; Rajendran, L.; Wenzel, D.; Wieland, F.; Schwille, P.; Brügger, B.; Simons, M. Ceramide triggers budding of exosome vesicles into multivesicular endosomes. *Science* **2008**, *319*, 1244–1247. [CrossRef]
168. Zhang, F.; Sun, S.; Feng, D.; Zhao, W.-L.; Sui, S.-F. A novel strategy for the invasive toxin: Hijacking exosome-mediated intercellular trafficking. *Traffic* **2009**, *10*, 411–424. [CrossRef]
169. Bianco, F.; Perrotta, C.; Novellino, L.; Francolini, M.; Riganti, L.; Menna, E.; Saglietti, L.; Schuchman, E.H.; Furlan, R.; Clementi, E.; et al. Acid sphingomyelinase activity triggers microparticle release from glial cells. *EMBO J.* **2009**, *28*, 1043–1054. [CrossRef] [PubMed]
170. Huang, H.W.; Goldberg, E.M.; Zidovetzki, R. Ceramides perturb the structure of phosphatidylcholine bilayers and modulate the activity of phospholipase A2. *Eur. Biophys. J. EBJ* **1998**, *27*, 361–366. [CrossRef]
171. Gilbert, S.; Loranger, A.; Omary, M.B.; Marceau, N. Keratin impact on PKCδ- and ASMase-mediated regulation of hepatocyte lipid raft size—implication for FasR-associated apoptosis. *J. Cell Sci.* **2016**, *129*, 3262–3273. [CrossRef]
172. Wurdinger, T.; Gatson, N.N.; Balaj, L.; Kaur, B.; Breakefield, X.O.; Pegtel, D.M. Extracellular vesicles and their convergence with viral pathways. *Adv. Virol.* **2012**, *2012*, 767694. [CrossRef] [PubMed]
173. Urbanelli, L.; Buratta, S.; Tancini, B.; Sagini, K.; Delo, F.; Porcellati, S.; Emiliani, C. The Role of Extracellular Vesicles in Viral Infection and Transmission. *Vaccines* **2019**, *7*, 102. [CrossRef]
174. Kakizaki, M.; Yamamoto, Y.; Yabuta, S.; Kurosaki, N.; Kagawa, T.; Kotani, A. The immunological function of extracellular vesicles in hepatitis B virus-infected hepatocytes. *PLoS ONE* **2018**, *13*, e0205886. [CrossRef]
175. Sanada, T.; Hirata, Y.; Naito, Y.; Yamamoto, N.; Kikkawa, Y.; Ishida, Y.; Yamasaki, C.; Tateno, C.; Ochiya, T.; Kohara, M. Transmission of HBV DNA Mediated by Ceramide-Triggered Extracellular Vesicles. *Cell. Mol. Gastroenterol. Hepatol.* **2017**, *3*, 272–283. [CrossRef]
176. Duan, R.-D. Alkaline sphingomyelinase: An old enzyme with novel implications. *Biochim. Biophys. Acta* **2006**, *1761*, 281–291. [CrossRef] [PubMed]
177. Duan, R.-D.; Nilsson, A. Metabolism of sphingolipids in the gut and its relation to inflammation and cancer development. *Prog. Lipid Res.* **2009**, *48*, 62–72. [CrossRef] [PubMed]
178. Chen, Y.; Zhang, P.; Xu, S.-C.; Yang, L.; Voss, U.; Ekblad, E.; Wu, Y.; Min, Y.; Hertervig, E.; Nilsson, Å.; et al. Enhanced colonic tumorigenesis in alkaline sphingomyelinase (NPP7) knockout mice. *Mol. Cancer Ther.* **2015**, *14*, 259–267. [CrossRef] [PubMed]
179. Cheng, Y.; Wu, J.; Hertervig, E.; Lindgren, S.; Duan, D.; Nilsson, A.; Duan, R.-D. Identification of aberrant forms of alkaline sphingomyelinase (NPP7) associated with human liver tumorigenesis. *Br. J. Cancer* **2007**, *97*, 1441–1448. [CrossRef]

180. Duan, R.-D.; Hindorf, U.; Cheng, Y.; Bergenzaun, P.; Hall, M.; Hertervig, E.; Nilsson, Å. Changes of activity and isoforms of alkaline sphingomyelinase (nucleotide pyrophosphatase phosphodiesterase 7) in bile from patients undergoing endoscopic retrograde cholangiopancreatography. *BMC Gastroenterol.* **2014**, *14*, 138. [CrossRef]
181. Duan, R.D.; Verkade, H.J.; Cheng, Y.; Havinga, R.; Nilsson, A. Effects of bile diversion in rats on intestinal sphingomyelinases and ceramidase. *Biochim. Biophys. Acta* **2007**, *1771*, 196–201. [CrossRef] [PubMed]
182. Duan, R.D.; Cheng, Y.; Tauschel, H.D.; Nilsson, A. Effects of ursodeoxycholate and other bile salts on levels of rat intestinal alkaline sphingomyelinase: A potential implication in tumorigenesis. *Dig. Dis. Sci.* **1998**, *43*, 26–32. [CrossRef]
183. Cumings, J.N. The copper and iron content of brain and liver in the normal and in hepato-lenticular degeneration. *Brain J. Neurol.* **1948**, *71*, 410–415. [CrossRef]
184. Gitlin, J.D. Wilson disease. *Gastroenterology* **2003**, *125*, 1868–1877. [CrossRef]
185. Krumschnabel, G.; Manzl, C.; Berger, C.; Hofer, B. Oxidative stress, mitochondrial permeability transition, and cell death in Cu-exposed trout hepatocytes. *Toxicol. Appl. Pharmacol.* **2005**, *209*, 62–73. [CrossRef] [PubMed]
186. Pourahmad, J.; Ross, S.; O'Brien, P.J. Lysosomal involvement in hepatocyte cytotoxicity induced by Cu(2+) but not Cd(2+). *Free Radic. Biol. Med.* **2001**, *30*, 89–97. [CrossRef]
187. Sokol, R.J.; Twedt, D.; McKim, J.M.; Devereaux, M.W.; Karrer, F.M.; Kam, I.; von Steigman, G.; Narkewicz, M.R.; Bacon, B.R.; Britton, R.S. Oxidant injury to hepatic mitochondria in patients with Wilson's disease and Bedlington terriers with copper toxicosis. *Gastroenterology* **1994**, *107*, 1788–1798. [CrossRef]
188. Lang, P.A.; Schenck, M.; Nicolay, J.P.; Becker, J.U.; Kempe, D.S.; Lupescu, A.; Koka, S.; Eisele, K.; Klarl, B.A.; Rübben, H.; et al. Liver cell death and anemia in Wilson disease involve acid sphingomyelinase and ceramide. *Nat. Med.* **2007**, *13*, 164–170. [CrossRef] [PubMed]

Publisher's Note: MDPI stays neutral with regard to jurisdictional claims in published maps and institutional affiliations.

© 2020 by the authors. Licensee MDPI, Basel, Switzerland. This article is an open access article distributed under the terms and conditions of the Creative Commons Attribution (CC BY) license (http://creativecommons.org/licenses/by/4.0/).

MDPI
St. Alban-Anlage 66
4052 Basel
Switzerland
Tel. +41 61 683 77 34
Fax +41 61 302 89 18
www.mdpi.com

Biomolecules Editorial Office
E-mail: biomolecules@mdpi.com
www.mdpi.com/journal/biomolecules

www.ingramcontent.com/pod-product-compliance
Lightning Source LLC
LaVergne TN
LVHW071548170726
843515LV00008B/2242

9783036540641